W9-CLE-335

Geology and Paleontology of the Miocene Sinap Formation, Turkey

Geology and Paleontology of the Miocene Sinap Formation, Turkey

Edited by

Mikael Fortelius

John Kappelman

Sevket Sen

Raymond L. Bernor

Columbia University Press
New York

QE
881
.G38
2003

This volume is dedicated to Professor Fikret Ozansoy,
teacher and pioneering spirit of Turkish vertebrate paleontology

Columbia University Press
Publishers since 1893
New York Chichester, West Sussex

© 2003 Columbia University Press
All rights reserved.

Library of Congress Cataloging-in-Publication Data

Geology and paleontology of the Miocene Sinap Formation / edited by Mikael Fortelius . . . [et al.]
p. cm.
Includes bibliographical references and index.
ISBN 0-231-11358-7 (alk. paper)
1. Mammals, Fossil—Turkey—Ankara Region. 2. Geology, Stratigraphic—Miocene. 3. Geology—Turkey—Ankara Region. 4. Paleontology—Miocene. 5. Paleontology—Turkey—Ankara Region. I. Fortelius, Mikael.
QE881.G38 2003
566'.09563—dc21 2002041515
CIP

Columbia University Press books are printed on durable and acid-free paper.

Printed in the United States of America

10 9 8 7 6 5 4 3 2 1

LONGWOOD UNIVERSITY LIBRARY
REDFORD AND RACE STREET
FARMVILLE, VA 23909

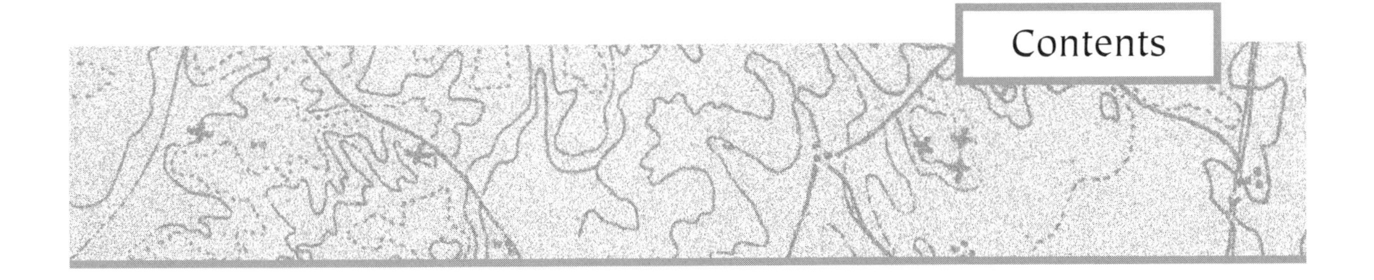

Contents

LONGWOOD UNIVERSITY LIBRARY
REDFORD AND RACE STREET
FARMVILLE, VA 23909
LONGWOOD LIBRARY
1000451249

Contributors

Miranda Armour-Chelu
Virginia Museum of Natural History
1001 Douglas Avenue
Martinsville, VA 24112
USA

Fehmi Aslan
Maden Tetkik ve Arama Enstitüsu
Orta Anadolu 3 Gölge Müdürlügü
Kızılcahamam
Turkey

Raymond L. Bernor
College of Medicine
Department of Anatomy,
Laboratory of Evolutionary Biology
Howard University
515 W Street NW
Washington, DC 20059
USA

Jeff Crabaugh
Department of Geology and Geophysics
University of Wyoming
PO Box 3006
Laramie, WY 82071
USA

Alexander Duncan
Department of Anthropology
University of Texas
1 University Station C3200
Austin, TX 78712-0303
USA

Douglas Ekart
Department of Geology and Geophysics
University of Utah
Salt Lake City, UT 84112
USA

Mulugeta Feseha
Department of Geological Sciences
University of Texas
1 University Station C1100
Austin, TX 78712-0254
USA

Mikael Fortelius
Department of Geology
PO Box 64
University of Helsinki
Helsinki 00014
Finland

Alan W. Gentry
c/o Department of Palaeontology
The Natural History Museum
Cromwell Road
London SW7 SBD
UK

Philip L. Gibbard
Godwin Institute of Quaternary Research
Department of Geography
University of Cambridge
Downing Place
Cambridge CB2 3EN
UK

Kurt Heissig
Institut für Palaeontologie und Historische Geologie
Richard Wagner Strasse 10
Munich 80333
Germany

Kati Huttunen
Naturhistorisches Museum Wien
Geologisch-Paläontologische Abteilung
Burgring 7
Vienna 1014
Austria

John Kappelman
Department of Anthropology
University of Texas
1 University Station C3200
Austin, TX 78712-0303
USA

Juha Pekka Lunkka
Department of Geology
University of Helsinki
PO Box 64
Helsinki 00014
Finland

A. Murat Maga
Department of Anthropology
University of Texas
1 University Station C3200
Austin, TX 78712-0303
USA

Fred McDowell
Department of Geological Sciences
University of Texas
1 University Station C1100
Austin, TX 78712-0254
USA

Jorge Morales
Departamento Paleobiologia
Museo Nacional de Ciencias Naturales (CSIC)
José Gutiérrez Abascal 2
Madrid 28006
Spain

Sirpa Nummela
Department of Ecology and Systematics
University of Helsinki
PO Box 65
Helsinki 00014
Finland

Brian G. Richmond
Department of Anthropology
George Washington University
2110 G Street NW
Washington, DC 20052
USA

Timothy M. Ryan
Department of Biological Anthropology and Anatomy
Duke University
3705 Erwin Road
Durham, NC 27705
USA

William J. Sanders
Museum of Paleontology
University of Michigan
1109 Geddes Avenue
Ann Arbor, MI 48109-1079
USA

Gerçek Saraç
Maden Tetkik ve Arama Enstitüsu
Genel Müdürlügü
Ankara
Turkey

Nuran Sarica
Laboratoire de Paleontologie
UMR8569 du CNRS
Museum National d'Histoire Naturelle
8 Rue Buffon
Paris 75005
France

Robert S. Scott
Department of Anthropology
University of Texas
1 University Station C3200
Austin, TX 78712-0303
USA

Eric R. Seiffert
Department of Biological Anthropology and Anatomy
Duke University
3705 Erwin Road
Durham, NC 27705
USA

Lena Selänne
Department of Geology
University of Helsinki
PO Box 64
Helsinki 00014
Finland

Sevket Sen
Laboratoire de Paléontologie
UMR8569 du CNRS
Museum National d'Histoire Naturelle
8 Rue Buffon
Paris 75005
France

Carl C. Swisher III
Department of Geological Sciences
Wright Geological Laboratory
610 Taylor Road
Piscataway, NJ 08854-8066
USA

Jan van der Made
Consejo Superior de Investigaciones Cientificas
Museo Nacional de Ciencias Naturales
José Gutiérrez Abascal 2
Madrid 28006
Spain

Suvi Viranta
Department of Geology
University of Helsinki
PO Box 64
Helsinki 00014
Finland

Lars Werdelin
Department of Palaeozoology
Swedish Museum of Natural History
PO Box 50007
Stockholm 104 05
Sweden

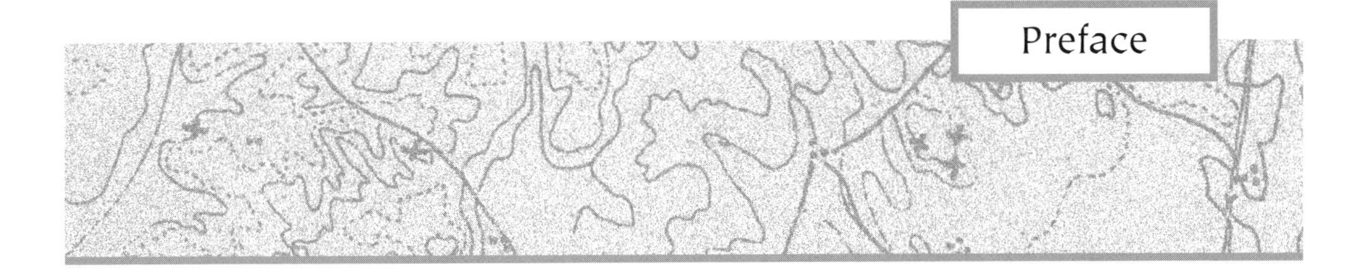

Preface

The Museum of Anatolian Civilizations in Ankara was founded over three-quarters of a century ago, and through most of its history has focused its efforts on the recovery, documentation, and preservation of the numerous archaeological sites found all across central Anatolia. This region was then and remains today at the crossroads of the civilizations of Europe, Africa, and Asia, and for many millennia has witnessed the intersection of these various cultures. The Museum, along with its extensive collections, is widely regarded as one of the most important and popular centers of its kind for research on these civilizations.

Beginning in 1989, the Museum, in partnership with Ankara University, the University of Texas at Austin, and Helsinki University, embarked on a new direction of international collaboration that had as its focus the study of the paleontology and natural history of central Anatolia. Many of the factors that serve to make the archaeology of the region so fascinating were also found to be at work in the more distant past; for example, central Anatolia of ten million years ago was also at the crossroads of the Old World and witnessed the exchange of faunas from Africa, Europe, and Asia. The collaboration that came to be called the "Sinap Project" documented the evolution, immigration, and extinction of the many species that once lived in this region, and broke new ground in carefully integrating this research with detailed studies of the geology and geochronology of the region.

One of the high points of the Sinap Project was the recovery in 1995 of the remains of *Ankarapithecus meteai,* a fossil ape known from previous excavations by an earlier generation of Turkish paleontologists. This new discovery provided portions of the skull and skeleton that were not represented by the earlier discoveries and have shed new light on the anatomy of this species.

The final success of the Sinap Project can be measured by the results of the studies documented in the volume that is now before you. It was my great pleasure to be associated with the numerous scholars from all around the world who came to call the village of Kazan their home during the summers of 1989–1995, and I thank all of them for their commitment to their studies and the collegiality that pervaded those times. Their efforts have not only provided new answers to long-standing questions about the faunas and geology of central Anatolia—they have also raised many new and important questions that await answers from the next generation of scholars.

İlhan Temizsoy
Director (retired)
Museum of Anatolian Civilizations

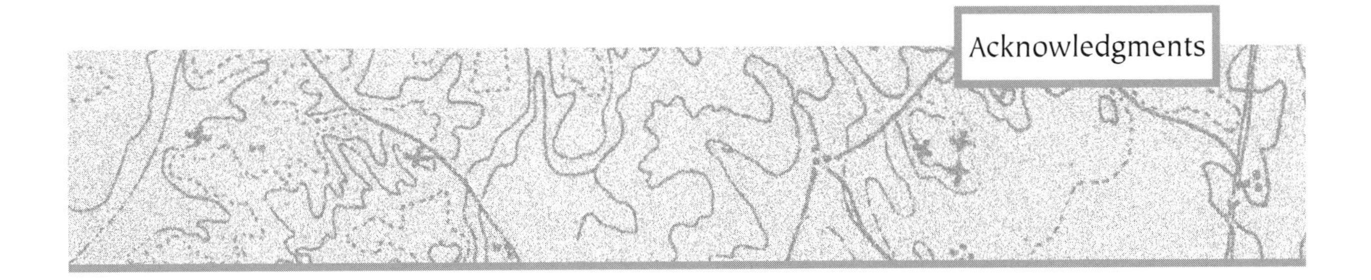

Acknowledgments

We extend our sincere thanks and gratitude to the following agencies and foundations for their support of the Sinap Project: National Science Foundation (grant EAR 9304302), USA; Academy of Finland (grants 1011579, 1588 and 33558); L.S.B. Leakey Foundation, San Francisco, CA, USA; Ankara University; The University of Helsinki; The University of Texas; Autodesk, Inc.; Finnair; Monsanto Company; and Motorola, Inc.

We are also very grateful to the General Directorate of Antiquities, T. C. Ministry of Culture and Tourism, Ankara, for its support of the project, and the Anadolu Medenyetleri Müzesi, Ankara, Turkey, and the Maden Tetkik ve Arama, Ankara, Turkey, for their support and assistance in the field and laboratory.

A generous grant from the University Cooperative Society, Austin, Texas, assisted with the production of this volume.

Geology and Paleontology of the Miocene Sinap Formation, Turkey

History of Paleontologic Research in Neogene Deposits of the Sinap Formation, Ankara, Turkey

S. Sen

The first fossil mammals were discovered at Sinap in 1951, and the first paper on the occurrence of Neogene mammals in the area was published in 1955. Indeed, the discoveries made in the 1950s led to a series of research projects, excavations, and studies that made the Sinap Tepe area one of the most interesting sites for mammal paleontology in Turkey. In addition, the occurrence of the hominoid primate *Ankarapithecus meteai* in two sites contributed to the celebrity of this area. After 30 years of field training for hunting fossil mammals in Turkey, I am in a good position to confirm that Sinap is paleontologically extraordinary because of its abundance of mammal localities and the taxonomic and chronologic diversity of its faunas. The Sinap area faunas are important for understanding the evolution of mammal communities in Central Anatolia between 16 and 2 Ma, for building a mammal biochronology in Turkey, and for achieving a better understanding of the environmental and climatic evolution of Central Anatolia during the Neogene. The purpose of this chapter is not to discuss these aspects in great depth, but to present the historical background of paleontologic investigations in this area.

Over the past 50 years, a series of national and international teams prospected, excavated, and collected mammal remains in the Sinap area. Large collections have been assembled, which are stored in museums and universities in Turkey, Germany, and France. I assure the next generation of paleontologists that the Sinap fossil deposits are not completely worked out: there are many new localities and taxa to be discovered.

The number of papers describing the Sinap Formation mammals reflects its mammal diversity and the quantity of material currently available in various collections. The Natural History Museum and Museum of Anatolian Civilizations in Ankara display in their exhibitions and keep in their collections thousands of specimens still undescribed and poorly studied; in particular, those collected in the 1960s and the early 1990s. Conflicts between persons and/or institutions have complicated access to these collections when studies were undertaken on Sinap Formation mammals to be published in the present volume.

To write this chapter, I visited those who were involved in the first discoveries of fossil bones in the Sinap Formation, Professors Fikret Ozansoy and Oguz Erol, in their retirement at Çardak near the Dardanelles and in Istanbul, respectively. I also collected information from old books, papers, and unpublished reports that are not readily available to the interested scientist or layperson. I interviewed Janine Brondel, Leonard Ginsburg, Emile Heintz, Robert Hoffstetter, Nizamettin Kazanci, Germaine Petter, Jacques Richir, and Gerçek Saraç, asking them to dredge their memories and describe to me what they remembered from past field campaigns in the Sinap area and from their contact with those who were the first to undertake geologic and paleontologic investigations there. They opened their memories as well as their archives to me. I am grateful to all of them.

I have been personally involved in the paleontologic exploration of the Sinap Formation since May 1967. Being a student of professors Ozansoy and Erol in the Faculty of Languages, History, and Geography (hereafter DTCF) at Ankara University and, at the same time, a novice paleontologist in the Mineral Research and Exploration Institute (currently known as the MTA), and in the MTA's future Natural History Museum, I have participated since 1967 in almost all Sinap field campaigns and studied some fossil mammals from several localities. Thus, my interest in Sinap Formation paleontology for almost 35 years and participation in joint research projects with many of the Sinap workers allowed me to assemble documents and data on the work and workers at Sinap over the past half century.

Here I present the history of paleontologic research in the Sinap Formation; specifically, in the area situated north and northwest of the town of Kazan, some 50–60 km northwest of Ankara, between the villages of Virancik (now Örencik) and Çalta, where the fossiliferous Sinap Formation

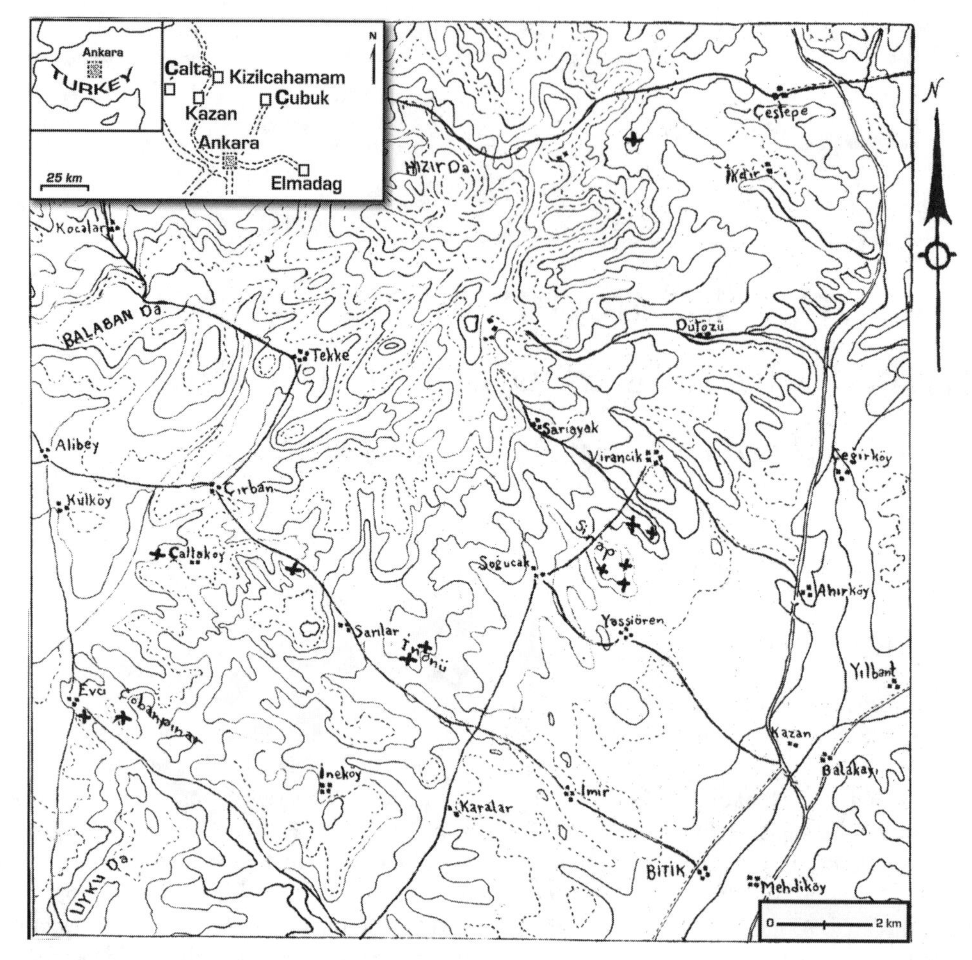

Figure I.1. Location map of the Sinap Formation mammal localities excavated by Ozansoy between 1951 and 1957. Note that the localities are indicated by X, and the names (village, mountains, locality, etc.) are handwritten by Ozansoy himself. Modified from Ozansoy (1958: fig. 4).

deposits are exposed along a line directed southwest to northeast (fig. I.1).

First Light on Mammal Paleontology in Turkey

Turkey is not the cradle of geology, much less of vertebrate paleontology—the earliest descriptions of vertebrate faunas from this territory date from the 1880s. However, Turkey's position at the crossroads between the Old World continents has made it important paleontologically and biogeographically. Interest in Turkish paleontology has grown throughout the twentieth century, and since the 1960s, Turkey has hosted several national and international research teams.

In the eastern Aegean region, fossil mammals have been known from Samos Island since Major (1888), an English officer, discovered their abundant remains at several sites. To the east, Kittl (1887) first described late Miocene mammals from Maragheh sites in Iran. In the late nineteenth century, the Ottoman Empire still extended from the Balkans to Yemen and Egypt. On its territories, geologic investigations were carried out by foreigners working for mining companies active in the empire, amateur paleontologists posted to embassies and consulates, or travelers in the country of the Sublime Porte. Thus the oldest discoveries of mammals are those of Calvert and Neumayr (1880), who described a few remains of giraffids and proboscideans from the late Miocene deposits along the Dardanelles. Later the same area would attract other geoscientists, such as Toula (1891, 1896), English (1904), Newton (1904), and Andrews (1918), who announced the discovery of some late Miocene to Pleistocene large mammals in sites in northwestern Turkey. The study of the Küçük Çekmece (Istanbul) mammal fauna by Nafiz and Malik (1933) is the first undertaken by a Turkish team.

It is difficult to explain why vertebrate (particularly mammal) fossil sites were not well explored in Turkey before the 1950s. This land, encompassing some 777,000 km^2, is more than 50% covered by terrestrial Neogene deposits that are particularly well exposed because of the sparse vegetation and steppic environment found in most parts of the country. Paleontologic investigations undertaken in the past 50 years have revealed that in almost every region of the country, these deposits are richly fossiliferous. A recent inventory by G. Saraç (MTA, Ankara; unpubl. ms.) lists more than 400 localities containing the remains of large and small mammals cited in the literature between 1950 and 2000. The only explanation I have for the dearth of early explorations is the circumstances that prevailed in the country: political and social problems during the nineteenth century and early years of twentieth century and the poor economic situation in the emerging Turkish Republic impeded research. Paleontologic investigations (and many other scientific endeavors) were not among the priorities of the period because there was no immediate economic return on such studies.

Before describing the beginning of paleontologic investigations by Turkish scientists, I note some examples that highlight the spirit of the young Turkish Republic in its efforts to promote intellectual life and modernize its institutions.

After the foundation of the Republic of Turkey (1923), Mustafa Kemal Atatürk undertook a series of reforms to make a clean break with policies of the former Ottoman Empire—to change its administration and model it after those of western European countries. Among these reforms, the foundation of Türk Tarih Kurumu (Turkish Society of History) in 1927 is of interest here. Atatürk knew that for Turkey to be a modern nation, there was a need for historical awareness among its citizens. He personally organized the Turkish Society of History and was its honorary president. The intention of the young Turkish Republic in the late 1920s and 1930s was that historical research (including prehistory and paleoanthropology) should be conducted by this society to reveal the history of the country and its populations (Pittard 1939). Under the leadership of this society, archaeological research was undertaken in late 1920s in some important sites of Anatolia, such as Alaca Hüyük, Çatal Hüyük, Bogazköy, and Pazarli. Besides these archaeological investigations, some prehistoric sites were also recognized. These research projects revealed that Anatolia was a country occupied by Paleolithic humans, as shown by discoveries of prehistoric tools in many sites. Atatürk also named a young doctor and anthropologist, Professor Sevket Aziz Kansu, to organize a large project to evaluate the diversity of the Turkish population and seek evidence of their ancestry in Anatolia. In early 1930 Kansu formed a large team, mainly composed of doctors, to collect biometric and morphometric data on living populations in Turkey that he, his students, and collaborators later analyzed. Kansu also took an interest in paleoanthropology and hoped to find evidence of prehistoric man in Turkey. He visited and excavated some caves and terrasse deposits around Ankara and Adiyaman. His research projects uncovered many Quaternary faunas and prehistoric tools, but found no human remains. These archeological and prehistoric discoveries fill the museums in Ankara and, to a lesser degree, museums in the provinces. In 1937, the Second Congress of Turkish History in Istanbul was the occasion for a large exhibition on the archaeology and prehistory of Turkey. It was held in the Dolmabahçe Palace, where the last Ottoman emperors resided, and was at that time the residence of the president of the Republic in Istanbul. This exhibition presented new discoveries in archaeology and prehistory to demonstrate the ancestry of history and humans in Anatolia and also the results of studies on the diversity of past and present civilizations in Turkey.

In the Ankara area in central Anatolia, archaeological and prehistoric investigations started in the late 1920s and resulted in the discovery of paleolithic tools in many terrasse deposits. Nothing was accomplished in mammal paleontology until 1942. This is because this area is far from the capital of the Ottoman Empire (Istanbul, until 1923) and it had no economic or tourist attractions until the 1940s. B. C. Tschachtli (1942) was the first to announce the discovery of mammal remains in Neogene deposits near Küçükyozgat (renamed Elmadag in late 1950s), some 30 km east of Ankara. The fossiliferous deposits are situated between the towns of Elmadag and Karacahasan. Between 1951 and 1959, Muzaffer Şenyürek undertook extensive excavations in this region and also at Gökdere (about 30 km southeast of Ankara) and collected abundant late Miocene faunas that he described in several papers (see Şenyürek 1960a,b and references therein). This area being outside the scope of this volume, I will not pursue the history of its exploration any further here. However, I must mention that during the course of the Sinap Project (1989–1995) these areas were visited several times by a small team under the leadership of M. Fortelius and J. Kappelman; surface collecting and some small excavations allowed the discovery of several hundred new specimens. These are for the most part stored in the Museum of Anatolian Civilizations in Ankara, under the supervision of Professor B. Alpagut. M. Şenyürek also excavated in the Sinap Formation during late 1950s (see the next section).

After these beginnings in mammal paleontology of the Ankara area, the investigations mainly concerned the Sinap Formation, northwest of the town of Kazan. To understand the historical background of scientific studies at Sinap, it is important to explain how and why the area was initially investigated in the 1950s, and the context in which the first mammal fossils were recovered.

Main Actors in the Mammal Paleontologic Investigations in the Sinap Formation

Four individuals—Suat Erk, Oguz Erol, H. Fikret Ozansoy, and Muzaffer Şenyürek—initiated geologic research in the Sinap Formation. They were the first to explore the Neogene deposits, discover mammal localities, and study the mammal faunas of these deposits. Their role was crucial in developing paleontologic interest in the Ankara area for the study of mammal evolution and biochronology in Central Anatolia.

Suat Erk

In a book honoring him (Kazanci 1993), there is a short biography of Erk (fig. I.2) that I expand here with additional

Figure I.2. From left to right, Professors Suat Erk, Oguz Erol, and Fikret Ozansoy, who were the pioneers of paleontologic research in the Sinap Formation in the early 1950s.

information obtained from his colleagues and students from the Faculty of Sciences. Erk was born in 1912 in Istanbul and he died in this same city on August 29, 1993. He studied at Istanbul University between 1932 and 1936 and graduated from the Natural Sciences Department of the Faculty of Sciences. In 1936 he was accepted as an assistant in this department under Professors E. Chaput, H. N. Pamir, and E. Parejas. In Turkey before the 1950s, geology and related sciences were taught only in the universities in Ankara and Istanbul, where the faculties of sciences had a Department of Natural Sciences to teach geology together with zoology, botany, and the other natural sciences. Subsequently new universities were established in Turkey, and the departments were divided to allow greater specialization.

In September 1938, Erk obtained a grant from MTA and went to Switzerland to prepare a Ph.D. thesis. He stayed first in Geneva to work with Professors L. Collet and M. Gysin and then in Zurich with Professors R. Staub, A. Jannet, and P. Niggli. He subsequently moved to Basel to complete his doctoral dissertation under Professor Reitschel. The title of his Ph.D. thesis was "The Geology of the Bursa-Gemlik Area" (in northwestern Turkey). He defended his thesis in 1942 at the University of Basel. In 1942 he began to work in MTA as a geologist in the Department of Geology and he was the chief of the Paleontology Division in this department from 1950 to 1955. He left MTA in 1955 to open the Erk Laboratory, a private company for consulting and sample analyses. He worked primarily for oil companies, which became very active at that time in Turkey. However, a law established in 1962 regulated oil exploration in Turkey, ending the activities of private oil companies. As a consequence, the activity of the Erk Laboratory rapidly declined and the company closed at the end of the same year.

In addition, Erk began to teach stratigraphy and paleontology as lecturer in the Faculty of Sciences at Ankara University in 1953, and later, in the Middle East Technical University's Geological Engineering Department (1958–1968) (fig. I.3). From 1962 until his retirement in 1982, he was engaged as a permanent teacher in the Faculty of Sciences, where he was promoted to the position of docent in 1971 and professor in 1976. His studies were mainly concerned with the stratigraphy and invertebrate paleontology of late Paleozoic and Mesozoic marine deposits in Turkey.

Figure I.3. Professor Erk (fifth from the right) together with his students, circa 1979. Courtesy of Professor Dr. N. Kazanci, Ankara.

Oguz Erol

Erol was born in Bursa on February 25, 1926. He studied in the Department of Geography at Ankara University (DTCF) between 1942 and 1946. He was a student of Professor Dr. W. J. McCallien from Scotland, whom the Turkish Ministry of Education had invited to teach geology in the Physical Geography and Geomorphology Division of DTCF. McCallien spent his free time investigating the geology around Ankara, and Erol often participated in these excursions with his professor. After graduating from the DTCF, he was invited to prepare a Ph.D. thesis with a grant of 50 Turkish liras (almost $50) awarded by the Ministry of Education.

Erol prepared his Ph.D. thesis under the supervision of McCallien on the Ankara Melange, and he defended it in 1949. At that time, the term "melange" had not yet been coined, and the volcanic-sedimentary formations that cover large areas around Ankara were interpreted as being composed of a mixture of volcanic and sedimentary deposits disturbed by active tectonics. McCallien (1946) called them "inverted graded beds." Erol spent several summers in the field investigating these formations. He had no car, so he walked every day from one village to another, and he slept in village guest houses along the roadside, accepting the hospitality provided by local people that he still remembers with considerable nostalgia. To make his geologic map of the Ankara area, he obtained the army's newly published 1/25,000 topographic maps of the Ankara area (provided by a friend). Such maps were top secret at that time (curiously, they still are today). Erol was probably the first Turkish geologist to use such maps for geologic mapping (Sengör 2001).

Erol carefully studied the volcanic-sedimentary deposits and observed that they are formed of a marine muddy matrix that includes volcanic blocks. He recorded all lithologic variation of the matrix in various places; the nature, shape, and lithology of blocks; the sedimentary-volcanic relationships, and so on. He termed these formations "*karmasik*" [mixture] and explained their origin as deep-sea volcanic explosions mixed with the bottom sediments, or sea-bottom debris flows mixed with volcanic rocks. (We now know that these kinds of rocks are formed in subduction zones.)

After defending his thesis in 1949, Erol was drafted into army service. Meanwhile, Professor McCallien invited his friend Sir Edward Bailey, a famous Scottish geologist, to visit Ankara. These two gentlemen visited the areas studied by Erol, and Erol guided them in the field when he was on leave from the army. Bailey immediately recognized the origin of these deposits and their significance for the geodynamic evolution of the area. McCallien left Ankara at the end of 1949, and in Scotland he prepared three papers together with Bailey to be published in the *MTA Bulletin* and *Nature* in 1950, and in the *Bulletin of the Royal Society of Edinburg* in 1953. They termed these geologic beds the "Ankara Melange." In these papers, Erol was acknowledged but not credited with the coinage of this term. After his army service, Erol tried to publish his Ph.D. thesis, but did not complete this work until 1956. Thus, the fruits of his work and priority for the term "Ankara Melange" were scooped by McCallien and Bailey.

Returning from the army in 1951, Erol was engaged as an assistant doctor in the Physical Geography and Geomorphology Division in the DTCF. He was promoted to the rank of docent in 1955 and professor in 1960. He taught geology and geomorphology in the DTCF and in the Faculty of Sciences of Ankara University until 1988. In 1988 he moved to Istanbul University, Department of Geography; he retired from this position in 1995. Professor Erol lives in the district of Yenibosna in Istanbul and continues his research activities with his old collaborators and geologists from the Technical University of Istanbul.

In 1951 MTA invited him to participate in a geologic mapping project of the area between Ayas and Kazan. He worked on this project each year for several months from 1951 until 1954. He discovered the first Sinap Formation mammal localities at Sinap Tepe and Çobanpinar. His geologic observations are recorded in several successive unpublished reports that are today preserved in the MTA archive. They describe in detail the geologic structures he had studied, their stratigraphic and geometric relationships, and his interpretation of the geodynamic evolution of Central Anatolia. These reports include detailed geologic maps of the Ankara area and several crosscut and vertical sections along the formations studied. The last and most complete of these reports (Erol 1954) was used for many years by generations of geologists as the principal document on the geology of this large area. This document also served to fill in the 1:500,000 geologic map of Turkey. His studies during the 1950s and 1960s mainly concerned the geology of several Neogene basins and surrounding formations in Anatolia (Erol 1956). In the late 1960s, he became more interested in geomorphology, studying the relationships between landscape morphology and tectonic evolution during the late Neogene and Quaternary. He published more than 100 books and papers providing detailed field observations, geologic and geomorphologic maps, stratigraphic logs, and sedimentologic analyses.

H. Fikret Ozansoy

I know H. Fikret Ozansoy personally better than I do the other scientists whom I review here. He was my professor at DTCF between 1966 and 1972, and I learned paleontology under his instruction, both in the field and the laboratory. In addition, I possess several documents concerning his studies in Ankara and Paris, such as the manuscript of his Ph.D. thesis, his Ph.D. diploma (fig. I.4), and many letters he exchanged with his colleagues from Paris and with me. I also went to his summer house in July 1998 to interview him.

Ozansoy was born in Istanbul on October 8, 1913. His father, Ali Hilmi, was a pasha of the Ottoman army, so he received an upper-class education as a child and teenager, studying in the best schools and having a family tutor. He received his diploma from the *Isik Lisesi* (high school), from the literature section, in 1935. His life between 1935 and 1944 is somewhat of a mystery—he never cared to discuss this chapter of his life with anyone. It is certain that he served in the army for two years. Little is known about how he spent the remainder of this time. It seems that Ozansoy attempted studies in several fields, including art, music, and medicine, but finally was admitted to the Department of

RÉPUBLIQUE FRANÇAISE

Diplôme de Docteur ès Sciences naturelles

Le Ministre de l'Education Nationale,

Vu le Certificat d'aptitude au grade de Docteur ès Sciences naturelles accordé le 4 Décembre 1958 par les Professeurs de la Faculté des Sciences de l'Université de Paris

Académie de Paris à Monsieur Ozansoy Hüseyin Fikret

né à Istanbul ~~département~~ Turquie le 8 Octobre 1913

Vu la dispense de la Licence accordée en application de la Décision Ministérielle du 10 Mars 1956.

~~Vu l'approbation donnée à ce Certificat par le Recteur de ladite Académie~~

Ratifiant le susdit Certificat,

Donne, par les présentes, à Monsieur Ozansoy Hüseyin le Diplôme de Docteur ès Sciences naturelles, pour en jouir avec les droits et prérogatives qui y sont attachés par les lois, décrets et règlements.

Fait à Paris, sous le Sceau du Ministère de l'Education Nationale, le 25 JUIL 1959

Le Ministre de l'Education Nationale, Signé: André BOULLOCHE

Pour expédition conforme:

Le Directeur de l'Enseignement supérieur, Pour le Directeur: Le Chef du 2e Bureau HENRI RACHOU

Signature de l'Impétrant:

Délivré par le Recteur de l'Académie de Paris le 18 AOUT 1959

Par le RECTEUR Le Doyen Délégué

No 310

Figure I.4. Ph.D. diploma (Docteur ès Sciences) of F. Ozansoy, obtained on December 4, 1958, from the Sorbonne, Paris.

Geography at the University of Ankara (DTCF) in 1944. Professor Dr. W. J. McCallien and Professor Dr. Sevket Aziz Kansu were among his principal professors. Ozansoy studied geology with McCallien and anthropology with Kansu. Professor Kansu was scientifically active between 1930 and 1950 and had important responsibilities within the Historical Society of Turkey. He was a medical doctor, but taught anthropology at several institutions (mainly in Ankara but also in Istanbul) and he founded the Department of Anthropology at Ankara University (DTCF). In 1944 he was named the rector of Ankara University. As he listened to Kansu's lectures, Ozansoy became increasingly interested in anthropology and eventually decided to change his studies to that field.

In the 1944–1945 academic year, Ozansoy discovered fossil mammal bones in the excavations made for the Ankara Hospital that were just behind the DTCF building. Visiting the foundations of this hospital day after day, he collected a number of specimens and showed them to McCallien. Ozansoy studied books from MTA (where he was working as an office employee) to identify his fossils. McCallien suggested to Ozansoy that he show the finds to Kansu. Kansu is reported to have told Ozansoy, "You have to leave what you are doing and begin work on these fossils." His request to transfer to the Department of Anthropology was a matter of formality only. In this department, Kansu and Şenyürek were his principal professors. Ozansoy studied in this department between 1945 and 1948. In his spare time he was allowed to use the *Kemikhane* (the departmental laboratory where the bones used for teaching and study are stored), where he prepared and studied the bones he had collected. At Kansu's request, Ozansoy also classified the anthropological and paleontologic material held by the department. Professor Kansu and subsequently Professor Şenyürek proved to be mentors who greatly stimulated Ozansoy's interest in anthropology and mammal paleontology.

Ozansoy graduated from the DTCF Department of Anthropology in June 1948. In Turkey, the bachelor degree exams included a short thesis. Under Professor Şenyürek's supervision, Ozansoy produced a thesis titled "Foramen mentale in ancient and living populations of Anatolia." With his diploma in hand, Ozansoy secured a position as vertebrate paleontologist in the newly formed paleontology section of the MTA. In the 1950s, Suat Erk was the chief of this section. Ozansoy quickly secured his participation in field expeditions as a vertebrate paleontologist, launching his career in paleontology. He excavated the Sinap Formation between 1951 and 1953, and in 1956 and 1957. In the meantime, the MTA offered him a grant to go to Paris and undertake a Ph.D. thesis. A certificate signed by Professor Şenyürek (then the head of the Paleoanthropology section in the DTCF) notes that on February 26, 1953, the Council of Professors approved the subject of F. Ozansoy's thesis as being "The Çobanpinar (Ayas) mammal fauna and its relationships with similar faunas in Turkey," and allowed him to "pursue his studies further in France, where he will go in the autumn."

Before Ozansoy left for Paris in the autumn of 1953, he spent time learning French. At his arrival, he was received at the Sorbonne by the powerful Professor Jean Piveteau, but his official registration at this university as a Ph.D. student was delayed: it was first necessary to acquire various certificates from the University of Ankara to prove that he had earned the claimed degrees, and then to ask the Turkish consulate office in Paris for their translation from Turkish to French. He was registered at the Sorbonne as student in mid-1954, and initially attended some lectures in the Department of Geology of this university to complete his undergraduate work. His registration as a Ph.D. student at the Sorbonne dates from March 10, 1956.

Ozansoy brought to Paris the most complete specimens he had recovered in the Sinap Formation localities (but nothing from Çobanpinar); he left the others in the Paleontology Section of MTA in Ankara. In the Natural History Museum of Paris, he had access to all of its mammal collections—those collected by Gaudry, Mecquenem, Arambourg, and Teilhard de Chardin from almost every corner of the world, and most significantly, those from Pikermi and Salonica in Greece, Maragheh in Iran, and from several sites in China, France, and Spain that were particularly useful for comparison with Turkish faunas. He told me that "the Paris Natural History Museum was like a paradise" for him. He haunted the museum's rich collection rooms and the large paleontology exhibition, where he learned the meaning of the evolutionary process. In this museum, he was helped in his paleontologic studies mainly by Professors C. Arambourg and R. Hoffstetter, and by the staff of the Paleontology Laboratory as well. He learned methods for documentation, preparation of specimens, and photography, among other topics. Ozansoy worked in this laboratory during the years 1954 and 1955 and from October 1957 to January 1959. From 1954 until the end of 1958, Ozansoy had weekly meetings with Professor Piveteau at the university. Early in his relationship with Piveteau, he offered the professor the opportunity to study and publish the *Ankarapithecus* mandible. Piveteau declined the offer indignantly, informing Ozansoy that this mandible was "a pivotal point in his [Ozansoy's] thesis."

Ozansoy defended his thesis on December 4, 1958 (fig. I.4) and then left Paris to reclaim his position at the MTA. He was unpleasantly surprised at his return. During the years he was absent, some important changes had taken place in the Paleontology Section of the MTA: Erk had left, and the new chief of the section, Yunus Pekmen, was not very interested in vertebrate paleontology. The fossils Ozansoy had left in Ankara were "stepped on and crushed" or else lost (probably thrown away), and nothing much remained of the collections he made in the 1950s. Professors Kansu and Şenyürek were still active in the DTCF. So, while working in MTA, in 1960 he began to teach vertebrate paleontology and biostratigraphy as chargé de cours in the Department of Anthropology. In 1964, he moved to this department, where he assumed the position of docent. During these years, his field activities were mainly concentrated on the exploration of the Dardanelles area, where Neogene Paratethys marine deposits interfinger with terrestrial deposits. He studied the stratigraphy of these deposits and recorded remains of mainly marine and some land mammals.

In 1966, the general director of MTA, S. Alpan, decided to open a natural history museum on the new campus of MTA in Ankara. He formed a committee to prepare a plan for this museum. Ozansoy was invited to sit on this committee, as head of its vertebrate paleontology section. While continuing to teach at the university, Ozansoy prepared plans for the vertebrate exhibition in this museum. To collect new mammal fossils to be exhibited in this museum, he organized excavations at Sinap Tepe in the summer of 1967. The MTA Natural History Museum was inaugurated in February 1968, but his work on its executive committee continued until 1971. The vertebrate paleontology exhibition still shown on the ground level of this museum is mainly due to Ozansoy's efforts. Establishing relationships with foreign museums, in particular with the Natural History Museum of Paris, he obtained casts of specimens for exhibition, such as the skeleton of *Gomphotherium angustidens* from Sansan and skulls of Mesozoic reptiles from the United States.

Ozansoy moved to Izmir in 1973 as professor at the Aegean University (Fig. I.5). This was the beginning of another adventure for him. Ozansoy was widely known in Turkey for his belief that paleontology should be taught in a faculty of sciences, not a faculty of literature as in Ankara, and that every major city of Turkey should have a natural history museum. He started teaching paleontology in the Department of Engineering Geology at the Aegean University. He was not satisfied with the term "engineering," because in Turkish universities, engineering was too often treated as an unscientific subject. For months Ozansoy waged battles through various committees and the general university heirarchy to found an Institute of Paleontology related to a natural history museum. He succeeded in founding this institute from scratch, and built a natural history museum on the campus of the Aegean University at Izmir. He organized this museum in less than two years, opening it in early 1975 with the layout it still has today. Unfortunately, after his retirement in 1982, the museum was allowed to go into decline, and since late 1980s it and the related Institute of Paleontology have not flourished.

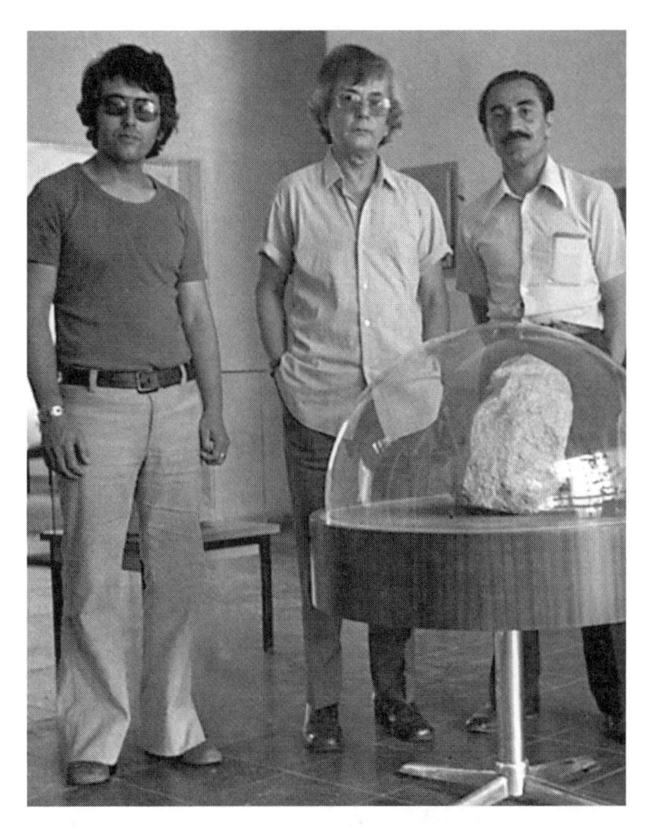

Figure I.5. Professor Ozansoy flanked by his student Dr. Gerçek Saraç (left) and MTA photographer Ismail Koçer (right) in the Natural History Museum of Izmir, in 1975. Courtesy of Dr. G. Saraç.

On September 21, 1981, the Turkish army was vested with the power to stop all internal conflicts within the state. A swift and ruthless manhunt was initiated to identify and arrest people, mainly intellectuals, that opposed the state's authority. Universities were at the center of such police activities. Ozansoy was classified by these groups as being a communist, and although he has never been, he was threatened with arrest if he did not retire from the university. Ozansoy's precipitous retirement in January 1982 effectively ended his academic career. Soon after he retired, Ozansoy moved to his home in Istanbul, where he lives today in the District of Moda. In the summertime, he moves to a modest summer house in Çardak, which is on the Anatolian side of the Dardanelles. He now spends time writing poetry, which has always been his hobby.

Muzaffer Şenyürek

At first glance, the contribution of Şenyürek to Sinap Formation paleontology may appear minor, because his papers mainly concern anthropology, paleoanthropology of

Figure I.6. Professor Dr. Muzaffer Şenyürek a few months before his death in 1961. Şenyürek explored the late Miocene mammal localities of Elmadag (in the Ankara area) and excavated in the Sinap Formation. From Ozansoy (1961).

prehistoric humans from various sites, and Turolian mammals from the Elmadag area, some 30 km east of Ankara. However, he is an important figure for the development of Turkish mammal paleontology, and he investigated the Sinap Formation between 1957 and 1959. The biography presented here mainly follows Ozansoy (1961) (fig. I.6).

Şenyürek was born in Izmir in 1915. In 1932 he was a student in the Faculty of Medicine at Istanbul University. At the beginning of the 1930s, the Turkish Ministry of Education decided to send students to Western countries to be trained in preparation for joining the country's elite. Şenyürek was the brightest student in his class; as a result, the ministry offered him a grant, and he left in 1934 for the United States. As a student at Harvard University, Şenyürek pursued his interest in anthropology. After defending his thesis on the cranial morphology of primates in 1939, he returned to Turkey and assumed a position as lecturer in the DTCF. He was promoted to docent in 1940 and professor in 1950 in the same faculty, and was its dean from December 1958 until his death. He died in an airplane crash near Ankara, together with two university colleagues, on June 1, 1961. He visited the United States and Europe on several occasions, attending congresses and conducting museum research. Şenyürek studied many remains of prehistoric humans from Turkey, Iraq, and Algeria and published valuable contributions on the morphologic characteristics of primate and human populations.

Şenyürek undertook excavations of Neogene mammals at several late Miocene and Pliocene localities in the departments of Nevsehir, Kayseri, and (most extensively) Ankara (1951–1959, between Elmadag and Karacahasan). He published 15 papers describing the fossils he unearthed. In the Sinap Formation, he apparently excavated some sites northeast of Sinap Tepe, at Üst Yoncalik and Alt Yoncalik, among other places, according to the labels on the material he collected. I could not obtain any information about exactly when and where he excavated fossils at these localities. A few inquiries to local villagers proved unreliable, except that Ibrahim Abaci from Yassiören revealed a number of localities investigated by Şenyürek. Şenyürek (1961: 340) includes a passing note, which states that Erol notified him of fossil localities at Akdogan village, which is situated on the highway between Kazan and Kizilcahamam. The DTCF provided grants enabling him to conduct excavations in this region in 1957, 1958, and 1959. Şenyürek's fine material from this site and from Sinap is still stored in the DTCF collections.

Most of his papers on Neogene mammals are in English, but they appeared in Turkish journals published in Ankara (*Belleten* and *DTCF Bulletin*). The papers are remarkable because of their concise descriptions of the material and precise morphologic and biometric comparisons. The taxonomic determinations are well documented and clearly demonstrated. His papers are still very useful for studies on late Neogene Turkish mammals.

Chronology of Paleontologic Research in the Sinap Formation

1951–1960

In the early 1950s, the MTA decided that a 1:500,000 geologic map of Turkey should be completed by compiling the 1:100,000 geologic maps that MTA geologists had been drafting for many years. In the Ankara area, some sedimentary deposits were not yet mapped; among them, the Kazan area, just northwest of Ankara. MTA assigned Dr. Suat Erk the task of undertaking the first study of this area. In mid-July 1951, Erk established the MTA camp at Orhaniye village, about 30 km northwest of Ankara and quite central to the Mesozoic and Paleogene deposits of interest. He asked the geologist Erol to map for MTA the Neogene deposits and the volcanics which crop out northwest of the Ovaçayi river, between Kazan and Ayas. Erol accepted this offer because of his great interest in the geology of the Ankara area and to gain access to MTA's facilities: a jeep, a driver, and a guide were placed at his disposal to investigate this region's geology. When Erol had been studying the Ankara Melange for his Ph.D. thesis, his only means of transport was his own feet, or, on occasion, local horse-drawn carts. During these later field seasons, Erol stayed in a water mill that his wife's father owned at Köprübasi (today included in the town of Kazan) along the Ovaçayi River. He started mapping the area around the villages of Yassiören, Soğucak, Sarılar, Ineköy, and Evciköy.

On August 7, 1951, the Yassiören watchman Şevket Ünal showed Erol an isolated tusk from the western slopes of

Figure I.7. A proboscidean tusk discovered at Sinap Tepe on September 20, 1951, by Sevket Sirel from Yassiören village (second from the left) and Satilmis (sitting), who was the field guide for Erol and Ozansoy. Courtesy of Professor Dr. Erol.

Sinap Tepe (fig. I.7). This was the first mammal fossil ever recorded in the Sinap Formation. However, because this specimen was of unknown provenance, Erol could only note the find in his diary. However, in the days that followed, Erol was frequently informed of the occurrence of bones in these deposits, and he discovered many bones and teeth at the locality of Çobanpinar. In his diary, Erol noted: "August 21, 1951; I ran from Ineköy upward. . . . On the way to Evciköy, 500 m before Çobanpinar, I found mammalian bones and hipparion teeth. The horizon seems to be below chert levels." He informed Erk of his discoveries. Erk was aware of the importance of these mammal remains for understanding the area's geology. In the following week, he went to Ankara to invite F. Ozansoy to join at the Orhaniye camp. According to Erol, Ozansoy hesitated, but was obliged to accept the order of his chief. One week after the discovery of the Çobanpinar locality, he joined the team. The team had only two vehicles (jeeps), one for Erk and the other for Erol. Consequently, Erk decided that Ozansoy should stay on the site where two tents were used for sleeping, one for Ozansoy and the other for two local workmen, Satilmis and Sakalli, from the village of Orhaniye. These workmen helped Ozansoy excavate the site until the end of the field season.

The excavation at Çobanpinar started on August 30 and ended October 10, 1951. This site produced a great number of specimens. However, it is situated in a dry area, far from any amenities. Consequently, almost every day, Erol ferried food and drink for the team. Ozansoy often became angry with Erol, complaining about the harsh conditions under which he had to work in this hot desert environment. In September, the sun is strong in the Ankara area, and there are no trees to provide shade around the vicinity of Çobanpinar. Ozansoy often told me that it was hard to excavate this locality, and in the following years he excavated many other sites in the Sinap Formation, but never again Çobanpinar. Erol told me that, conscious of Ozansoy's discomfort with the heat and precarious living conditions, he often brought bottles of wine along with with food for Ozansoy. Although during the day it is blistering hot for most of September in this region, later in the month the temperature frequently drops precipitously at night. Ozansoy would then complain of the cold sleeping conditions. He finally resolved his problems by bringing a bicycle from Ankara, which he used to ride to the village of Evciköy, where he was allowed to take refuge in the local primary school building.

At the end of the excavation season, Ozansoy returned to Ankara with a truck full of fossils. He spent the rest of the year preparing them and making preliminary determinations. Enthusiastic about the discoveries at Çobanpinar, Erk decided to pursue mammal paleontologic research in the area and encouraged Ozansoy to prospect and excavate new localities. As I mentioned previously, during the 1951 field season, the villagers from Yassiören and Sogucak showed Erol (and later, Ozansoy) several bone-bearing horizons in the Sinap Tepe area. For the 1952 and 1953 field seasons, it was decided to settle the camp north of Yassiören, just below the southwestern slopes of Sinap Tepe. Ozansoy told me that each field season he stayed several months in the field. Erk decided to camp together with Ozansoy. They had only one car between them. During the summer of 1952, a few days of prospecting were enough to find several localities. During 1952 and 1953, they excavated localities along the eastern, southern, and western slopes of Sinap Tepe. Ozansoy's monographs of 1958 and 1965 do not precisely indicate the localities they excavated. The locality that yielded the *Ankarapithecus* mandible is unique, and certainly corresponds to Loc. I of Ozansoy (1965) in the Igdelik gully.

During these field seasons, Erk stayed at the Yassiören camp together with his wife, who was a violin virtuoso. Ozansoy had learned a lot of opera arias during his boyhood years. After a good day, the Sinap Tepe slopes echoed with song and laughter washed down with liberal quantities of wine late into the night.

Ozansoy went to Paris in the autumn of 1953 and remained there until early 1956. He came back to Ankara to pursue excavations in the Sinap Formation. I did not find detailed information about the field and laboratory activities of Ozansoy for the years 1956–1957. In his thesis (1958), he notes that he prospected the Lower Sinap series and found a locality containing *"Hyaena minor."* He also prospected the Kavakdere and Çalta areas and made some small excavations in both localities. The excavation at Inönü probably dates from these years. In his thesis, there are photos of excavations at Inönü showing elephant tusks and limb bones. In addition, he excavated a few spots in the Middle Sinap Series around Sinap Tepe (1958: 66).

In his Ph.D. thesis, Ozansoy describes part of the material he collected from the Sinap Formation: primates, carnivores, an *Orycteropus,* some proboscideans, hipparions, suids, giraffids, and bovids. The other taxa, mainly those from Çobanpinar, Kavakdere, and Inönü localities, are not mentioned in his thesis because these taxa were already known from the classical Pikermi and Maragheh faunas. Most of the species he describes he considered to be new to the literature, and when naming them, he dedicated them to MTA, Ankara, or his professors and collaborators:

Anatolia:	*Choerolophodon anatolicus*
Ankara:	*Hipparion ankyranum, Ankarapithecus meteai*
C. Arambourg:	*Hyaena arambourgi, Ictitherium arambourgi,* and *Schizochoerus arambourgi*
S. Erk:	*Synconolophus erki*
R. Hoffstetter:	*Palaeotragus hoffstetteri* and *Megantereon hoffstetteri*
MTA:	*Ankarapithecus meteai, Synconolophus meteai,* and *Dicoryphochoerus meteai*
H. N. Pamir:	*Felis pamiri* and *Samotherium pamiri*
J. Piveteau:	*Eomellivora piveteaui, Hyaenictis piveteaui, Megantereon piveteaui, Synconolophus piveteaui,* and *Listriodon piveteaui*
M. Şenyürek:	*Hyaena senyureki* and *Qurliqnoria senyureki*

In all, Ozansoy named about 30 new species from Sinap, but unfortunately did not adequately describe all of them. As a consequence, many of them are *nomen nuda*.

In late 1950s, Şenyürek also undertook excavations in the Sinap Formation. At his request, Erol informed Şenyürek of his June 1957 discovery of some mammal bones near Akdogan village, which is situated on the highway between Kazan and Kizilcahamam. Şenyürek went to the site and recovered a few specimens from two horizons. One of them yielded only a proboscidean molar that Şenyürek (1960b, 1961; see also Erol 1966) described as *Archidiscodon meridionalis*. The second locality, stratigraphically lower, yielded various remains of mammals from a small bone pocket. Since this locality is not far from the Sinap Tepe area, Şenyürek decided to visit Ozansoy's localities and the related formations. Finding several bone occurrences, he undertook new excavations along the slopes of Sinap Tepe during the summer of 1958. With the help of local workmen, he collected abundant fossil vertebrate material from several localities. When Ozansoy came back from Paris to Ankara, Şenyürek had a terse conversation with him about his new excavations at Sinap:

Şenyürek:	I excavated at Sinap Tepe and collected faunas very similar to yours.
Ozansoy:	From which horizons?
Şenyürek:	From what you called Middle Sinap.
Ozansoy:	You don't intended to publish them, I hope?
Şenyürek:	Yes, I intend to publish them soon.
Ozansoy:	This means that you will do what Wallace did with Darwin's work!
Şenyürek:	I am not Wallace, but I will publish the material I collected.
Ozansoy:	If you publish before me, I will inform the international societies of geology and paleontology of what a bad man you are!
Şenyürek:	I am not a bad man, so I will wait for your monograph to be published.

Şenyürek did not publish one word on his discoveries from the Sinap Formation. He died in June 1961, long before Ozansoy's monograph was published (1965). In fact, Ozansoy's thesis manuscript was not submitted immediately for publication. He was busy in Ankara, working at MTA and preparing lectures at the university. Professor J. P. Lehman, director of the Institute of Paleontology at the Paris Museum of Natural History, asked Dr. Germaine Petter, at that time a young mammal paleontologist in this laboratory, to improve the French in Ozansoy's manuscript prior to submitting it to the *Mémoires de la Société Géologique de France*. This was not an easy task for Petter, but she succeeded in improving the text, reducing the number of illustrations and plates, and preparing the entire monograph in the proper format for the *Mémoires*.

1967–1969

In 1967 Ozansoy undertook new excavations in the Sinap Tepe area to collect well-preserved mammal fossils to be displayed at the future Museum of Natural History in Ankara. For this work MTA provided all the facilities needed for a large excavation campaign. Ibrahim Tekkaya and I were chosen to be Ozansoy's collaborators. Just before this campaign began, Tekkaya, newly graduated from the DTCF, had obtained a position at MTA to work in the future Museum of Natural History as a vertebrate paleontologist. During the 1966–1967 academic year, I had been awarded an internship at MTA to pursue fossil prospecting at the Sinap Formation as part of my first year of education at the DTCF under Ozansoy's direction (Ozansoy had requested my participation at Sinap from General Director S. Alpan). MTA added a cook to this team, Emin; a technical staff member, Cafer; and the drivers Hasan and Halit, along with their jeeps. We were given all the equipment we needed from the MTA stores—tents, tables and chairs, beds, covers and sheets, jerricans for water and gas, pots and plates, picks and shovels, and the like—and we loaded all in an MTA truck. The team started on a sunny day in late May 1967 to set up camp below the slopes of Sinap Tepe, exactly where Ozansoy had established his camps in 1952 and 1953. We spent the week setting up this camp on the almost flat meadowland just north of Yassiören village. Ozansoy traveled daily to Ankara in Halit's jeep. On several days Ozansoy had to stay in Ankara, leaving Tekkaya and me—both with little practical experience—to undertake the field work at Sinap. Setting up tents was easy enough, but installing a shower without any special equipment proved to be a challenge. Near a small stream running down from the mountain, we used the natural slope to make a 2-m-deep trench and we placed a large water tank on wooden beams close to the steepest border of the trench. We then covered

Figure I.8. The 1967 Sinap team. Standing, from left to right, a local workman, Ibrahim Tekkaya, Hasan the driver, Fileret Ozansoy, Sevket Sen, and Cafer Türk. Photograph by S. Sen.

the bottom of the trench with gravel from a nearby stream bed. The tank robinet served as the shower. When Ozansoy discovered our ingenious shower, he found it to be too primitive for his own use. Therefore, he never stayed at the camp and returned to Ankara every day after work (fig. I.8).

After the camp was settled, Ozansoy showed Tekkaya and me the Sinap localities that he had excavated in the 1950s. I remember the first time we climbed the slopes of Sinap Tepe from its western side: Ozansoy was in the center, and Tekkaya and I on either side of him. Toward the middle of the slope, Ozansoy suddenly stopped. He picked up a nearby wood stick and partly drove it into the loose sediment. He said that a fossil bone would appear like the stick on the ground, and we had to be careful when unearthing it. With his fingers, he cleaned first the upper surface of the stick, then its sides; finally he pulled the stick out of the ground and threw it. For Tekkaya and me, this was our first lesson on paleontological excavation.

From late May until early August, our team excavated four localities. The excavation started at Loc. I, where Ozansoy unearthed the *Ankarapithecus* lower jaw in 1953. This locality is situated on quite a steep slope in the Igdelik gully southeast of Sinap Tepe. As a result, it was necessary to dig deep to reach the fossiliferous horizon. We engaged a dozen local workmen from the villages of Yassiören and Sogucak, who made successive platforms. They worked at the site for a week. When the bone horizon was reached, Ozansoy personally taught us how to excavate it carefully. Our discoveries consisted of a few limb bones of horses and ruminants of no particular interest. In fact, the bone pocket was virtually empty, having been extensively excavated in previous years. While this excavation was going on, Ozansoy prospected the adjacent geologic beds and found new bone occurrences at Kayincak Tepe, about 1 km east of Sinap Tepe (Loc. II). We started a new excavation at this site, where several bone pockets were rapidly uncovered. However, a new problem arose: the fossiliferous horizon was covered by a thick and hard boulder conglomerate that prevented us from opening a large excavation at this site. Ozansoy went to Ankara to request a dynamite specialist from MTA and he came back a day later with a dynamite expert, Ramiz, and a truck full of explosives and other tools. The operation lasted one week. During the day Ramiz made holes in the conglomerate and filled them with dynamite to blow it up. In the evenings he showed equal gusto downing bottles of *Raki* (a local brandy). Local workmen cleaned the slopes, and it was then possible to open the fossiliferous horizon over a large area along the saddle. During this operation the general director of MTA, Professor S. Alpan, visited the excavation and proposed the use of a bulldozer to clean up the entire hill. Fortunately, Ozansoy refused this offer, arguing that the removal of the conglomerate in toto would expose the fossiliferous horizon to excessive weathering that would damage fossils.

I excavated in this locality, together with a dozen workmen for more than a month. Thousands of well-preserved specimens were recovered. In the meantime, another locality was found along the southern slopes of Sinap Tepe, which Ozansoy named Loc. Ia. Because this new locality was quite rich in bovid remains, it was decided that Tekkaya and a few of the workmen would excavate it. Tekkaya was pleased to accept this because he was preparing a Ph.D. on bovids, and because Loc. Ia was much closer to the camp than Loc. II. I supervised the excavation myself at Loc. II. The silty sediment of this locality is quite soft, which allowed relatively rapid excavation of a large number of bones and jaws, including a large turtle, a bear upper jaw, several skulls and jaws of other carnivores, suids, and hipparions. On some days so many bones were spread out on the ground that we needed a watchman to keep them safe from the ubiquitous bystanders, who often visited the excavation at night. Finally, in late June or early July 1967, I discovered a primate palate and partial face in a square that I was excavating.

In early August 1967, one of us discovered another locality, just 300 m northeast of the camp along the western slopes of Sinap Tepe. We quickly abandoned all other localities and concentrated our effort on this new site that we named Loc. Ib. It yielded perfectly preserved specimens speckled black with manganese (incidentally, this is commonly found on fossils recovered from Maragheh, Iran). I still remember a pig skull, some complete *Choerolophodon* jaws, and an *Ictitherium* skull with its atlas and axis still articulated to the base of the skull. The season ended around August 20. We carried our entire collection to Ankara in two trucks fully loaded with boxes of fossils.

Parallel to the excavations in the Sinap Tepe area, Gerçek Saraç, a graduate student of Ozansoy from the DTCF undertook excavations at Çalta, together with his friend Bülent Can, to prepare his master's thesis on the Çalta fauna. They stayed in a small tent at that site for two weeks in July 1967. Saraç (1968) described the specimens that they found at Çalta and then gave them to the MTA Natural History Museum.

During the summer of 1969, a new campaign of bone collecting was organized by Ozansoy, this time near Sarilar village. We stayed at the nearby village of Kazan, where we rented a house. In addition to the 1967 team, Ergun

Kaptan and Zeki Atalay from MTA participated in this excavation. We started our excavations at Inönü, where we observed a large concentration of proboscidean remains. We called these beds the "elephant cemetery," because the remains of that animal were so abundant. I still remember a huge skull and several tusks that we spent several weeks excavating. It was possible to withdraw the tusks without much damage, but the huge skull in its plaster jacket was too heavy to move, and the plaster jacket was not strong enough to protect it. As a result, the specimen shattered into a thousand fragments when we attempted to move it to a nearby truck.

In the second part of this campaign, we excavated at Kavakdere, first in the pinkish marls of the upper horizon, and later in the gray-beige marls of the lower horizon. These two sites also yielded well-preserved specimens mainly concentrated in bone pockets in both horizons. I remember a complete rhinoceros skull that we found in the lower horizon, and several complete skulls and jaws of hipparions from the upper horizon. The excavations at Inönü and Kavakdere lasted throughout June and July 1969.

Only a few specimens have ever been studied of the huge amount of material we recovered at Sinap in 1967 and 1969 (from the localites of Sinap, Kayincak, Inönü, and Kavakdere). Tekkaya began preparing his Ph.D. thesis on the Sinap bovids in 1967 and defended it in 1970. He published some of his results (Tekkaya 1970, 1973, 1974a, 1975). Some specimens were given to foreign researchers to publish, including P. Andrews (*Ankarapithecus* palate), D. Begun and E. Güleç (*Ankarapithecus* again), M. Pickford (suids), and D. Geraads (some carnivores and bovids). In addition there are several internal reports in the MTA Museum archives with preliminary determinations, but most of the specimens remain unpublished. In the years following these excavations, many people handled the collections, intending to study some of the mammal groups. Unfortunately, labels are not always correct, and some nice specimens have been badly damaged. Extreme care is needed to study any of these specimens today.

1972–1974

A new Turkish-French expedition was organized by Ozansoy and E. Heintz in October 1971. After a brief survey in 1971, Heintz and L. Ginsburg (both from the Natural History Museum of Paris) traveled to Turkey in July 1972 to reengage field work in the Sinap Tepe area (Sen 1998). The MTA director assigned M. Gürbüz and me to the fieldwork at Sinap. We established our camp at Kazan by renting rooms at the home of Mehmet and Yilmaz. Because Heintz was principally interested in Villafranchian mammals, we commenced our excavation at Sarikol Tepe, a small hill about 2 km west of Sinap Tepe, where Ozansoy had collected some material that included the remains of *"Hyaena" arambourgi* and *Equus stenonis*. Three local workmen were hired to help with the digging. The excavation lasted one week—we soon found that fossil specimens were scarce and badly preserved and had been damaged by transport and roots.

Heintz asked me if I knew of another Villafranchian locality in the area. I mentioned Çalta, because in his monograph, Ozansoy mentioned a record of *Equus stenonis* in this locality. We all moved to Çalta, where the fossil bones were abundant and well-preserved. However, we soon realized that Çalta was not a Villafranchian fauna, but older. We excavated Çalta for 20 days and made plaster jackets for a number of nice specimens. In the meantime, we observed the occurrence of small mammals in the same bed, so we collected sacks of dirt to screenwash in the river at Kazan. The rich discoveries were gratifying to the entire team, but everybody suffered greatly from heat and sunburn.

At the end of the season, Ginsburg requested that the team visit all of the Sinap Tepe localities previously excavated by Ozansoy's team. I guided the team to some of them. At Loc. I we saw some bones cropping out from a new bone pocket. We then excavated two more days to recover some of them; we also made a large plaster jacket for a block of mixed bone. I observed the presence of small mammals in the same geologic horizon. This field season ended around August 15.

In July 1974, I returned to Çalta together with Fehmi Aslan (MTA) to collect more small mammal material. We screen washed about half a ton of dirt in the spring nearest the locality. The preliminary results of the Çalta excavations were presented in two papers (Ginsburg et al. 1974; Sen et al. 1974). Between 1974 and 1976, I studied the Çalta rodents for my Ph.D. thesis (Sen 1977); I subsequently edited a small monograph on the whole Çalta fauna (Sen 1998).

1975–1976

MTA undertook new excavations in the Sinap Formation in 1975 and 1976. The objectives included completion of Sinap Tepe's stratigraphic framework, and the start of reconstructing the basin's structural geology. I. Tekkaya (team leader), F. Aslan, Ç. Ertürk, and G. Saraç collaborated in this campaign. Each year the field season lasted from late May until October. In 1975 the field team stayed in a hotel at Kazan, but in 1976 they used the MTA hydrogeologic research station near Kazan. Logging studies were undertaken in several locations and some fossil localities were excavated. In 1975 the field team discovered the middle Miocene vertebrate locality of Inönü-I. In prior years no localites of such antiquity had been found in the Sinap Formation, and the Inönü area itself had been known for yielding late Miocene (Turolian) mammals only. The occurrence of a middle Miocene locality 200–300 m from the previously explored Turolian localities was interpreted as being due to the presence of an angular unconformity between the deposits of both localities.

The Inönü-I locality was intensively excavated during two field seasons and yielded quite abundant remains of large mammals. When excavating Inönü-I, the sites of the classical Inönü localities were also explored. During the 1976 field season, the team had at its disposal about 15 MTA workmen. This enabled several sites to be excavated, the most significant being Sarikol Tepe (Upper Sinap), Çalta, Çobanpinar, Sehlek (or Pinaryaka), and Basbereket. G. Saraç informed me in 1976 that the expedition found a complete skull of a large carnivore at Inönü-I on a Friday afternoon. Saraç prepared a plaster jacket for this specimen. Tekkaya decided that the team would take a weekend of rest, and in the rush to leave for Ankara, he removed the skull, although the plaster was not yet dry. Only the mandible survived the trip intact. The mandible referred to by Gürbüz (1981) as *Hemicyon sansaniensis* is what remained from this skull. Some suids from the same locality were studied by Pickford and Ertürk (1979) and later by Fortelius et al. (1996). Saraç (1994) produced a Ph.D. thesis on the rhinos. Geraads et al. (1995) studied some of the bovids from Inönü-I. Other material retrieved by this team in 1975 and 1976 from various localities remains, again, unpublished.

1981

I returned to Sinap Tepe in May 1981 to sample a section across Loc. I for magnetostratigraphic correlation. I also sampled some dirt from Loc. I to wash for small mammals. This trip lasted from May 18 to May 27, 1981. The results of this work were published in Sen (1991).

1989–1995

During 1980s, Berna Alpagut (DTCF, University of Ankara) and Peter Andrews (British Museum of Natural History, London) collaborated on an excavation of the middle Miocene locality of Paşalar in northwestern Turkey. Lawrence Martin (Stony Brook, New York) was involved in this joint work as a student of Andrews. As a continuation of the Paşalar collaboration, Alpagut and Martin decided to start a new project to discover more hominoids from Sinap. This project commenced in June 1989 with a survey permit from the Division of Antiquities and Museums, Ministry of Culture, Turkey. June was considered the best period for investigations in the Sinap area, and every year, between 1989 and 1995, the field campaign started in early June and ended at the beginning of July. In 1989, a paleontologic team prospected the areas previously developed by Ozansoy (1965): Sinap Tepe, Inönü, Kavakdere, Çalta, and so on. In a field season report, Alpagut and Fortelius (1991, p. 334) enumerated the goals of the Sinap Project:

- excavation of the richest localities;
- mapping of the geology to establish internal correlations between local exposures;
- geologic sampling for radioisotopic and magnetostratigraphic dating;
- collecting additional faunas to develop a better regional biostratigraphy for the later Miocene of Central Anatolia; and
- prospecting both within and outside the area covered in 1989 (i.e., around the villages of Yassiören, Sogucak, Sarilar, and Çalta).

In fact, the aim of the Sinap Project was to discover new hominoid fossils in the Sinap Formation. In the late 1980s, only three Turkish localities were known to yield hominoid primates: Paşalar, Çandir, and Sinap Tepe. Paşalar had been excavated by Alpagut and Andrews since 1983. The site of Çandir had been undergoing research under the leadership of Erksin Güleç since the mid-1980s. The only available site known to contain hominoid fossils in 1989 was Sinap Tepe. Martin, whose only field season at Sinap was in 1989, was replaced by Mikael Fortelius (Helsinki) and John Kappelman (Austin), who served as co-leaders with B. Alpagut. The project was funded by grants awarded to Fortelius (Academy of Finland) and Kappelman (National Science Foundation and L.S.B. Leakey Foundation).

During the first two years, the team tried to locate old localities excavated by Ozansoy, Şenyürek, and other researchers from MTA and collected surface findings over a large area. Beginning in 1991, the team obtained a permit to survey selected localities. In the following years, the permit was extended to include excavation. The project grew substantially and included myself (in charge of micromammal collection and analysis) and more than 25 other researchers and students from several countries. In six years of field work, more than 150 mammal localities were discovered in the Sinap Formation and other areas, including Çubuk (Igbek, or Loc. 49) and Elmadag (Locs. 76, 77, and 78). Several sites in the Sinap Formation were excavated. Altogether, surface prospecting, trenching, and excavations produced a large collection of fossil mammals, as well as other nonmammal vertebrate material, from localities ranging in age from early middle Miocene to the latest Pliocene. This project also undertook intensive geologic, stratigraphic, sedimentologic, and taphonomic studies in the fossiliferous deposits. John Kappelman and his collaborators sampled sections >1000 m in thickness for magnetostratigraphic correlation. The Sinap Project also trained many Turkish, American, and Finnish students, who learned prospecting, excavation, faunal identification, cataloging, and other field paleontology and geology techniques.

The fossil material accumulated daily and was stored in cabinets in the underground flats that the project rented at Kazan. Several students stayed at the flats to study various groups of mammals. A general spirit of great enthusiasm infused scientific collaborators and students alike. However, everything ended abrubtly a few days after the discovery of the Loc. 12 hominoid face and mandible in late June 1995. Although this locality had been prepared by her collaborators for excavation, Berna Alpagut continued the excavation at Loc. 12 alone with her students, claiming that she was the sole scientist responsible for the study of these specimens. Despite night-long discussions behind closed doors at the Kazan dormatory, Alpagut refused to collaborate with her colleagues on the new hominoid material. In the end, she kept the hominoid remains for herself, and allowed the other co-leaders to have only a brief look at the material and serve as junior authors on a letter to *Nature* (Alpagut et al. 1996).

Alpagut terminated the Sinap collaboration at the end of 1995. Not a single member of the Sinap research group has been allowed to continue their studies of the fauna. Not one of her collaborators in this Sinap project group actually knows where the fossil material she recovered is stored, or if they will ever be able to study it.

Hominoid Primates in the Sinap Formation

For mammal paleontologists, and especially paleoanthropologists, *Ankarapithecus meteai* Ozansoy, 1965 is now a familiar species amongst the Hominoidea. The stories of the discoveries of the first mandible in 1953 and of other, more complete specimens in later years reflect the spirit of research in the 1950s, 1960s, and 1990s. The type mandible of *Ankarapithecus meteai* was discovered during the summer excavations of 1953 in Ozansoy's original Locality I, situated in the Igdelik ravine along the middle of the eastern slopes of Sinap Tepe. Ozansoy told me that he was not on the site at the time of the mandible's discovery, but out prospecting the area. It was found by a local workman, trained for excavation, who did not know what he had discovered in the bone pocket under his care. During a break, Yunus Pekmen, a young invertebrate paleontologist from MTA, and occasionally a member of the team, accidentally walked on bones still in situ, thus breaking the mandible and some other specimens. When Ozansoy discovered what had happened, it was too late to save the complete mandible. Consequently, only the symphysis bearing one incisor, two canines and a p3, and isolated left p4 and molars survived this disaster. Ozansoy never forgave Pekmen and nicknamed him "Yunus the Crab" because of his somewhat oblique walk, insinuating that he is not fit to work on an excavation.

Ozansoy described this mandible in detail in his Ph.D. thesis (1958). He illustrated this specimen in no fewer than 11 pictures. A new genus and species, *Ankarapithecus meteai,* was established, paying tribute to the city of Ankara and to MTA. However, the nomen *Ankarapithecus meteai* n. g. n. sp. was already published without any description but with an illustration in a paper of 1957, in which Ozansoy drew up a faunal list of 31 Turkish Neogene mammal localities, including that of the Sinap Tepe area. The new genus and species names became valid only in 1965, when Ozansoy's monograph (based on his Ph.D. thesis) was published. In this monograph, Ozansoy compared this species with other fossil and living apes and some fossil hominids. He revealed its resemblances to *Sugrivapithecus* (presence of a chin process), *Pongo* (canine index), *Gigantopithecus* (canine situated on the incisor plane), and *Sivapithecus* and *Sugrivapithecus* (lack of diastema). He found that *Ankarapithecus meteai* enjoyed a constellation of characters that warranted generic distinction. The main difference he noted (and subsequently generally retained by other primate students) is that m3 is larger than m2 in *Ankarapithecus,* whereas the opposite is always true in the other anthropoids used for comparison.

In discovering and describing this hominoid mandible, Ozansoy became increasingly interested in primate evolution, which led him to write a book in Turkish (1966) on the ancestry of Turkish hominoids and their biostratigraphic bases. In this book, *Ankarapithecus* occupies the central position; the book also tells the history of primates before and after the rise of *Ankarapithecus.* Ozansoy further discusses influence of paleobiogeographic and paleoenvironmental changes on higher primate evolution.

The discovery of another hominoid fossil, an isolated p4, was mentioned by Ozansoy (1958, p. 16) as "Dryopithécinés?". Ozansoy noted that it originated from the Lower Sinap Series and was found in 1956 in a "couche de marnes située juste au-dessous des blocs anguleux," which forms "une couche de 1 m d'épaisseur." The specific locality remains obscure, but Ozansoy reported that this locality also yielded *"Hyaena minor* n. sp., *Listriodon piveteaui* n. sp." and some ruminants and turtles. *H. minor* was described by Ozansoy (1958, 1965) and later on referred to the genus *Allohyaena* (Howell and Petter 1985). But *"Listriodon piveteaui* n. sp." has never been described. It consists of a palate with both tooth rows and is still preserved, as is the mandible of *H. minor,* in the collections of Museum National d'Histoire Naturelle in Paris. Ozansoy (1966, p. 72) later determined that the "Dryopithécinés ?" specimen was an upper right M1 and described its morphology: three roots below the protocone, paracone, and metacone; four cusps plus one accessory cusp (protoconulid); presence of the fovea posterior and crista obliqua; small hypocone; small but clear parastyle; and parastyle fossette on the buccal wall. A very poor picture (fig. 9 in Ozansoy 1966) illustrates the occlusal surface. The description and illustration are insufficient to determine this specimen accurately. I did not find this specimen in the collection of the Paris Museum, and no one knows where it currently rests.

I found the next hominoid specimen in 1967. As noted above, in late May 1967, MTA organized a new campaign of excavation in the Sinap Tepe area under the leadership of Ozansoy, and I was a member of the team. After several weeks of excavation at Locality II, where we collected several hipparion metapodials and bovid jaws, an ape palate with a part of the face was excavated from a bone pocket in late June or early July. I remember that Ozansoy came to the locality after spending most of the day prospecting, and I showed him my discovery with some hesitation. He was so happy that he made an exception of staying with us for dinner in the camp and offered the team bottles of a local champagne to celebrate the event.

The following day I prepared the specimen. Ozansoy then took it to his office in Ankara to study. We did not see him in Sinap Tepe for several days following his departure. When he returned to the excavation, he told us many things about the hominid features and other particularities

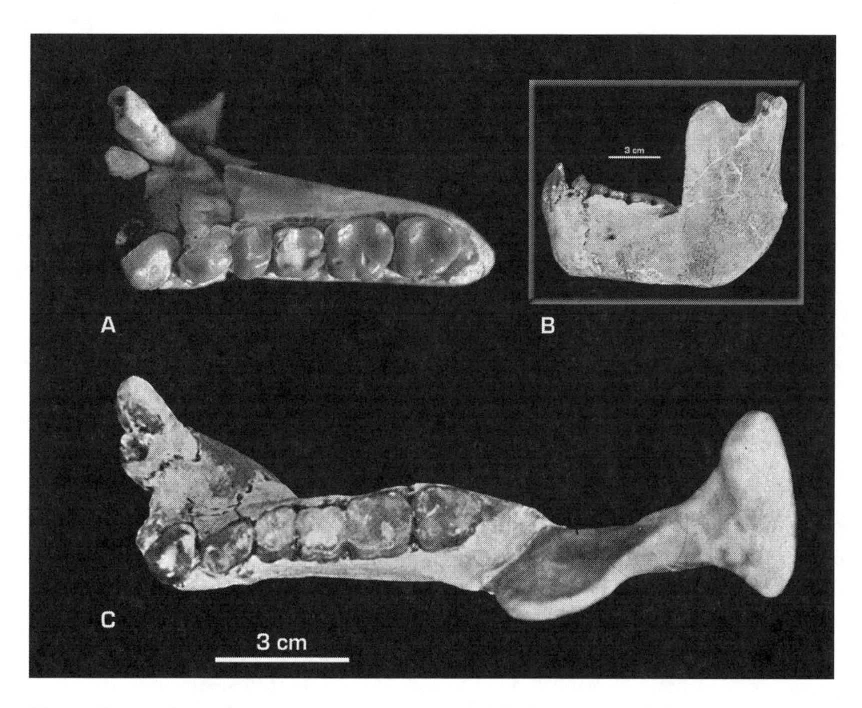

Figure I.9. *Ankarapithecus meteai* Ozansoy, 1965. (**A**) the type mandible fragments (symphysis with incisors, canines and left p3, and isolated p4–m3 on a plaster pedestal) assembled together in their first illustration. From Ozansoy (1957a). (**B**) The type mandible fragments mounted on a gorilla mandible by S. Sen in late 1967, in lateral view, and (**C**) the same reconstruction seen in occlusal view. From Ozansoy (1970a).

of the dentition and face and made many other observations that confirmed his opinion about the importance of *Ankarapithecus* to human ancestry. This specimen (numbered MTA-2125) and the previous one from Loc. I are kept in the collections of the Natural History Museum of Ankara. I remember Ozansoy, during the months following the 1967 summer Sinap excavations, studying this specimen and the type mandible in his office at the museum. He chain-smoked while poring over all the available books and reprints piled high on his office desk and floor. It is remarkable that Ozansoy never published a single word on this new specimen. Even in the papers he published in later years on hominoids (Ozansoy 1970a, b) he neglected to make any reference to this specimen. I have never been told why he did not publish something about it.

Ozansoy left the Museum in 1971, and in 1973, he moved to Izmir to assume a professorship at the Aegean University. In so doing his relations with MTA and its Natural Natural History Museum became distant. Ibrahim Tekkaya became the scientist in charge of vertebrate fossil material at MTA, and soon offered Andrews (London) an opportunity to publish on the MTA-2125 specimen. Andrews went to Ankara to study this specimen, and their paper was published in the journal *Palaeontology* (Andrews and Tekkaya 1980). This paper describes in detail MTA-2125 and compares the *Ankarapithecus* remains mainly with those of *Sivapithecus darwini, S. sivalensis,* and *Ouranopithecus macedoniensis.* They "concluded that the new Turkish material confirms the distinction at the species level between *A. meteai* and *S. indicus,* contrary to the synonymy proposed by Simons and Pilbeam (1965), but there appears to be no good reason for separating them generically" (1980, p. 93). In addition, they observed that "In the description of the maxilla and dentition of *S. meteai,* the closest comparisons in most cases were with the orangutan, and this could indicate some degree of relationship between it and the orangutan." The principal shared characters are deep and widely flaring zygomatic process, marked alveolar prognatism, short upper face, narrow interorbital distance, large I1 relative to I2, and large squared molars. Consequently, the Sinap ape was referred to a distinct species of the genus *Sivapithecus, S. meteai.*

Soon after publication of Ozansoy's 1965 monograph and with renewed vigor when the palate MTA-2125 was published, several fossil hominoid specialists took up discussion of the Sinap ape. Space constraints here do not permit an extensive review of the literature (but see Kappelman et al., this volume). As far as I know, the first paper discussing the characteristics of this species and its systematics was that of Simons and Pilbeam (1965) who as mentioned, proposed the synonymy of *Ankarapithecus meteai* and the Siwaliks ape *Sivapithecus indicus* Pilgrim, 1910. Many hominoid specialists believe that *Ankarapithecus* is best left as a distinct genus, a nomenclature followed in this volume. Among the more radical papers proposing otherwise is that of Kay and Simons (1983), in which the authors reconsider the systematic status and relationships of hominoids. Concerning the Sinap ape, they state that "the published dimensions of the lower molars and the palatal teeth are all within the range of Siwalik *Sivapithecus sivalensis.* The proportions of the palate and face closely resemble comparable parts of *S. indicus* specimens GSP 9977 and GSP 15000 recently recovered from Pakistan" (p. 598). For these authors, the recognition of *S. indicus* in Turkey serves as a major argument to include the recently described *Ouranopithecus macedoniensis* in the same species: "The presence of *S. sivalensis* in Turkey further supports the possibility that the smaller Macedonian *Sivapithecus* also belongs to this species."

I add here some background on the preservation of the *Ankarapithecus meteai* type mandible. In December 1967 or January 1968, Ozansoy asked me to reconstruct the fragments of the type mandible in such a way that it could be exibited in the future MTA Natural History Museum, together with the then-recently recovered palate from Sinap. To satisfy Ozansoy's request, I made a plaster cast of a living gorilla mandible in which I withdrew the symphysis and teeth and then fixed in their place the casts of the fragment of symphysis and isolated teeth of *A. meteai.* Coloring the plaster as the original bone, I obtained something that was reasonably complete and convincing to be exhibited in a museum cabinet. Later this "mandible" was exhibited together with the 1967 palate in the cabinet devoted to the primates in the Natural History Musem. In a newspaper report on *Ankarapithecus,* Ozansoy (1970a) illustrated this reconstruction of the "mandible" with two pictures taken in occlusal and labial views (fig. I.9).

I was told that Professor Elwyn Simons (Durham, North Carolina) visited the Natural History Museum of Ankara in 1977. At that time I was no longer working in Ankara, but living in Paris (since 1972). He was received in this Museum with all the honors that a scientist of his distinction deserves and given the opportunity to study all of the available *Ankarapithecus* remains. The resultant paper (Kay and Simons 1983) presents an extensive discussion on the characters and affinities of *Ankarapithecus,* together with those from the Siwaliks of Pakistan, China, Greece, Central Europe, and other localities. It is concluded that *Ankarapithecus* from Sinap Tepe as well as *Ouranopithecus* from northern Greece are best assigned to the genus *Sivapithecus.* Among many other arguments, their remarks on the characters of the mandibular corpus in *Ankarapithecus meteai* are very interesting: "There is an evident failure by Andrews and Tekkaya (1980) to consider any characters with distributions that do not support the conclusion that *Sivapithecus meteai* is related to orangutans. For example, the mandibular corpus of the type specimen of this animal is shallow and broad in the molar region, corresponding to Siwalik *Sivapithecus,* but not to be seen even as an extreme variant in any living pongid species. Robust mandibular corpora occur elsewhere only in later hominids and appear to represent a shared derived character linking the two groups. Failure to consider this character and others mentioned below leads to an unbalanced view of the probable affinities of *Sivapithecus*" (Kay and Simons 1983, p. 601). As correctly argued by Bonis and Melentis (1984, p. 17), "on the type mandible of *S. meteai,* the mandibular corpus is not preserved. It seems that these authors were a victim of a plaster reconstruction based on a cast of a gorilla mandible. For Kay and Simons, this 'specimen' apparently has some unexpected hominoid features. This needs no more comment." I apologize to Kay and Simons for their assumption that the mandible was reconstructed without their knowing it. However, I am also astonished that a well trained primate paleontologist did not distinguish the original fragments from the plaster reconstruction. It is still more incredible for Prof. Simons who has discussed the phylogenetic position of *Ankarapithecus meteai* for many years now based on this very same material (Simons and Pilbeam 1965).

The third discovery of *Ankarapithecus* remains occurred in mid-June 1995, at Loc. 12, where palate MTA-2125 was excavated in 1967. When the 1989–1995 Sinap Project began, the trench made for the 1967 excavations was still obvious. In 1991 a preliminary excavation of this locality was undertaken and revealed bone concentrations still in situ. However, the presence of the loose boulders on the saddle prevented secure excavation of the site. At the end of the 1993 field season, a bulldozer was used to remove part of the saddle and to open a large surface for excavation. Trenches and preliminary excavations made in the 1993 and 1994 field seasons did not provide a substantial record. In 1994, Peter Andrews was invited to direct excavations at Loc. 12 with a group of students. When I visited the excavation, I told him that he was digging in the wrong horizon, and was actually substantially above the fossiliferous level. I do not know why my advice was not taken by Dr. Andrews that year. The year after in June 1995, B. Alpagut decided that she would direct the Loc. 12 excavations, which she did with her Turkish students. Trenching deep in various parts of the excavation, bone pockets were soon identified. A partial skull and mandible of *Ankarapithecus* was discovered by Zeynep Bostan on June 20, 1995. This inexperienced team damaged parts of the skull and mandible before removing them from the quarry. Further damage was incurred during their preparation, and most of the fragments were lost. In the descriptions and illustrations given from these specimens (Alpagut et al. 1996; Andrews and Alpagut 2001; Kappelman et al., this volume), this damage is only briefly mentioned. The face and mandible as well as some limb bones have been extensively described in these publications.

Peter Andrews has been involved as an author in most of the papers dealing with Turkish Miocene hominoids. I consider his opinion on the systematic status of *Ankarapithecus meteai* remarkable. In papers describing the palate MTA-2125 (Andrews and Tekkaya 1980; Andrews 1982), Andrews attributed the Sinap ape to the genus *Sivapithecus,* following in part the opinion of Simons and Pilbeam (1965). However, in later papers he co-authored (Alpagut, Andrews et al. 1996; Andrews and Alpagut 2001), Andrews recognizes the generic identity of the Sinap ape following a small discovery by Begun and Güleç (1995, 1998). On reading Andrews et al. (1996), I was not satisfied by this new assessment.

During a bus excursion in the Valles-Penedes Basin near Barcelona, on October 26, 1996, I sat next to Andrews and questioned him about this important discovery that changed the generic status of the Sinap ape. He referred me to what was written in Alpagut et al. (1996, p. 350): "The naso-alveolar region is preserved in MTA-2125, but unfortunately the alveolar margin of the premaxillary region and anterior palate are heavily eroded in AS95-500. . . . Recent cleaning of MTA 2125 reveals a mildly stepped morphology with a large incisive canal and foramen in contrast to earlier descriptions." Apparently, a mild stepping in nasal morphology and some size variation in the incisive canal and foramen are significant enough to change the generic status of a taxon. My paleontologic experience is insufficient to comprehend the systematic importance of such subtleties.

Concerning hominoid field research in Turkey, there is currently an active competition between Berna Alpagut and Erksin Güleç, both professors in the Anthropology Department of the DTCF in Ankara. This competition has been ongoing since the 1980s, and has at several times reached such acrimonious proportions that it has led them to seek relief in the Turkish courts. Alpagut has excavated at Paşalar since 1983, mainly together with Andrews. Güleç has excavated at Çandir (another hominoid locality that contained the "*Sivapithecus alpani*" lower jaw described in Tekkaya [1974b]) during several years in the late 1980s and early 1990s, and she continues to excavate other localities elsewhere in Turkey for fossil hominoids. Neither Alpagut or Güleç have a background in vertebrate paleontology, so they lack the requisite training to excavate mammal fossil localities. The primate remains collected by these paleoanthropologists were in fact studied by either Andrews or D. Begun (Toronto) and co-authored with Alpagut or with Güleç.

Peter Andrews is a noted celebrity in Turkish hominoid paleontology. He introduced himself in Turkey when Tekkaya discovered and described the Çandir hominoid mandible. He proposed to Tekkaya to restudy the Çandir material (Andrews and Tekkaya 1976). This collaboration has apparently been mutually beneficial to these scientists, since their collaboration was pursued on the Sinap Loc. II palate, MTA-2125 (Andrews and Tekkaya 1980; Andrews 1982), discovered by the 1967 MTA team. In the 1980s, Andrews collaborated with Alpagut on the Paşalar locality excavations. Paşalar was initially discovered by the German Lignite Exploration program conducted in the 1960s. Paşalar has yielded large collections of hominoid remains that were described in several papers by Alpagut and Andrews. Andrews was not included as a co-leader in the 1989–1995 Sinap Project, but Alpagut invited him for a short field season in 1994. When the 1995 face and mandible were discovered, Andrews was invited by Alpagut to contribute to the *Nature* letter (Alpagut et al. 1996). I find it remarkable that Turkish paleontologists and paleoanthropologists are allowing their discoveries to be published by foreign nationals. Is it possible that my Turkish colleagues are too busy with their internal squabbling to study their own material?

A Natural History Museum in Ankara

I sometimes wonder whether the Turkish people—particularly the bureaucrats—are really interested in the preservation and development of their country's natural treasures. Before 1968, there did not exist a single institution committed to housing, preserving, and displaying natural objects such as rocks, minerals, and living or fossil animals and plants. There was not one exhibition, permanent or temporary, displaying the natural diversity of Turkey. This can be explained by the political and economic conditions in the declining Ottoman Empire and the emerging Turkish Republic from the mid-1800s to the 1960s. Toward the end of the Ottoman Empire, political intrigue and a series of wars destroyed the economy. These factors contributed to the low educational level of the public and the lack of development of the sciences. After a long war between 1914 and 1923, the new Turkish Republic had many other and more urgent priorities than the development of natural sciences. Although teaching geology began in 1850 at the University of Istanbul, nineteenth-century Turkish geologists were employed solely as mining engineers. The first academic studies by Turkish geologists started in the late 1920s. However, the young republic needed an institution to organize research on ore deposits and the geology of Turkey. Consequently, in 1935 the Mineral Research and Exploration Institute of Turkey (MTA) was founded. Since that time, MTA has organized all kinds of geologic research and the exploration of ore deposits in Turkey. This institution provides basic training to geologists who subsequently move on to universities to pursue a professional career in either geology or paleontology.

In 1960s, the general director of MTA was Sadrettin Alpan, a professor of mining techniques. Alpan was immensely interested in the natural sciences. He was aware that in its 30 years, MTA had accumulated knowledge in all branches of earth sciences, and that its stores were full of rocks, minerals, and fossils collected by geologists during their field studies. To add value to these MTA treasures and to bring a cultural aspect to MTA, he decided in 1967 to form an executive committee to establish a natural history museum. Ozansoy, at that time associate professor at the Ankara University, was a member of this committee, together with four other paleontologists, one mineralogist, two geologists, and one administrator (fig. I.10).

In 1967, the executive committee first identified specimens to be displayed in the museum. The available collection of vertebrate fossils was insufficient to fill the cabinets and rooms devoted to paleontology. It was decided to buy or to borrow some skeletons and skulls of large animals—such as those of dinosaurs and mastodons—from European and American museums. At the same time Ozansoy was tapped to organize large excavations in the Sinap Tepe area, already known to be rich in mammal fossils. Between May and August 1967, extensive excavations along the Sinap Tepe slopes and at Kayincak Tepe provided thousands of skulls and bones, among them an anthropoid palate, that

Figure I.10. The executive committee of the MTA Natural History Museum together with the general director of the institute in 1968. From left to right: Mükerrem Türkünal, Dr. Cahide Kiragli, Sehavet Mersinoglu, Professor Hamit Nafiz Pamir, Professor Sadrettin Alpan (general director of MTA), Cemal Öztemür, Dr. Lütfiye Erentöz, Docent Dr. Fikret Ozansoy, and Yüksel Sezginman. From Natural History Museum (1968).

Figure I.11. The Natural History Museum of MTA in Ankara in late 1967, before its opening. Note that the trees are newly planted on the grounds and there is still building rubble in the foreground. From Natural History Museum (1968).

were prepared for display in the museum. The remaining material was stored in the collections. This museum project was the primary reason for renewing the paleontologic research in the Sinap Formation, and incidentally influenced my entry into paleontology.

In the presentation booklet of the museum (Natural History Museum, 1968), Alpan writes "there has never been an institution in our country which concerns itself with such a long period of time, enveloping the nature and history of this world from the beginning up to today, and representing the process, progress and evolution of material and living beings during this period." Indeed, under his supervision, 10 months of continuous work resulted in a Natural History Museum that contains three floors of exhibits, including rocks and minerals, fossils, and zoologic samples in natural environmental reconstructions (fig. I.11). The museum was inaugurated on February 7, 1968, by the prime minister and other high authorities of the country. It was the first natural history museum in Turkey. A new and much larger building was built in the late 1990s and furnished in 2000 to be the new Natural History Museum in Ankara. The collections were moved to the new building in 2002–2003, and the inauguration of the new exhibition is planned for June 2003. The old building was at the center of the MTA campus and consequently less accessible to the public, whereas the new one is on the campus but along a main road, and thus more convenient for the casual visitor.

The initial aim of the museum was to present the natural world to the public and to develop those natural sciences of primary interest to MTA, namely, the earth sciences. The efforts of the years following the opening of the museum were insufficient to reach these goals. The museum directors, appointed by the general director of MTA, was unable to develop research and museum activities; the staff employed there, mainly archeologists and anthropologists, were not trained enough to develop this museum; and relationships with other departments and sections of MTA, such as geology, paleontology, or minerology have too often suffered from lack of cooperation. Recent discussion with MTA authorities gave me the impression that they do intend to open the new museum building as soon as possible with a modern exhibition and to adopt a new regimen of research, museology, and conservation.

Stratigraphy of Neogene Deposits in the Kazan Area

In the foreword to his monographs, Ozansoy (1958, 1965) summarized the results of his studies:

> Nos principaux résultats sont la mise en évidence, en Turquie, d'une succession pratiquement continue, s'échelonnant depuis le Miocène moyen jusque dans le Pléistocène, et caractérisée par une série de faunes de Mammifères, dont certains éléments étaient déjà connus en Turquie et ailleurs, tandis que beaucoup d'autres sont nouveaux.

Indeed, Ozansoy did not approach paleontologic study devoid of a biostratigraphic dimension. Before describing mammal remains from the Sinap Formation, his papers discuss first the biostratigraphic context of the localities and faunas. Reading his succesive papers on Neogene mammals, I wonder whether his principal aim was to enlighten the systematics and phylogeny of taxa and evolution of mammal communities or to establish the stratigraphic succession in which they are included. I believe the latter objective dominated his studies, even influencing the systematic assignments and phylogenetic interpretations of the groups he studied.

Concerning the stratigraphy of the Sinap Formation, from Kömürlük Dere to Çalta, Ozansoy (1965: 8) presented the succession of deposits and related mammal localities. In light of later studies, we now know that such a stratigraphic chart is erroneous; current data suggest the correlation of these deposits as given in Kappelman et al. (this volume). In particular, the lower and middle Sinap horizons are older than all other deposits, and their faunas are therefore referred to the late Astaracian and Vallesian land mammal ages, respectively, whereas the other horizons yielded mammal faunas varying in age between the early Turolian and Villanyian ages. What is the origin of such a spectacular error? Several elements taken from his papers and some explanations given to me by Ozansoy himself help to clarify these stratigraphic misinterpretations.

Ozansoy went to the Sinap area because O. Erol had discovered some bone occurrences there. Erol (1954) studied the geology and stratigraphy of bone bearing deposits actually included in the Sinap Formation. He observed that in the Sarilar area (which includes the mammal localities of Inönü, Kavakdere, and Çalta), the deposits are often strongly folded and faulted, whereas in the Sinap Tepe area, the deposits are horizontal or gently dipping to the southwest. These tectonic observations convinced Erol that the deposits of the Sinap Tepe area should overlie those of the Sarilar area. Ozansoy uncritically accepted Erol's opinion, although he never mentioned this in his papers and monograph. On the contrary, he stated that the "Série de Sinap . . . vient en concordance avec la formation de Kavakdere près de Çamlibayir et aux environs de Çalta" (Ozansoy 1958, p. 15), believing thus to demonstrate that the mammal-bearing deposits around Sinap Tepe are younger than those of the Sarilar area (Inönü, Kavakdere, and also Çobanpinar). Consequently, the systematics and the evolutionary stage of the mammal taxa from the various localities of the Sinap Formation were interpreted within the context of this erroneous stratigraphy. When I interviewed him in 1998, he confessed to me that "he had failed under the influence of the geologists."

Another explanation is related to the changing definitions of the Miocene-Pliocene boundary, which was particularly heavily disputed by international scientists until the late 1970s. Until the mid-1970s, the definition of the Miocene-Pliocene boundary in the continental realm was debated by stratigraphers and paleontologists, dividing them into two schools: the French one placing the boundary above the Pontian (i.e., long after the appearence of *Hipparion*), and the American school, which started the Pliocene with the first appearence of *Hipparion*. Ozansoy prepared his Ph.D. thesis in Paris under professors Piveteau and Arambourg. Thus it would have been logical for him to adopt the opinion of the French school. In fact, he did the opposite. In a paper published in 1957 and in the introduction to his thesis, Ozansoy discussed the question, but he preferred the American definition. Consequently, he placed all deposits with *Hipparion* faunas in the Pliocene, except the site of Kömürlük Dere, where the poor material he recorded did not contain *Hipparion* (and was therefore classified as older, i.e., Miocene), and the site of "Sinap supérieur," which yielded *Equus* (and was therefore younger, i.e., Pleistocene).

In his stratigraphic chart, the oldest horizon that Ozansoy recognized is that of Kömürlük Dere, which is situated east of the study area. It is quite far from the other localities in the Sinap Formation, and this prevents any clear stratigraphic correlation between this locality and the others. It was briefly excavated by Ozansoy in 1956, resulting in the discovery of "quelques rares Cavicornes associés à des Reptiles" (Ozansoy 1965, p. 11). Ozansoy did not describe any fossils from this locality. The Sinap Project team prospected the area in 1990 and found a lagomorph p3 that was determined to be *Ochotonoma* sp., suggesting a late Ruscinian age (Sen, this volume).

The deposits at Inönü yielded one middle Miocene age locality that is tentatively correlated to early Astaracian (Pickford and Ertürk 1979; Gürbüz 1981; Geraads et al. 1995; Fortelius et al. 1996). The localities excavated by Ozansoy in the 1950s yielded (perhaps early) Turolian faunas in several spots. The stratigraphy of the area is complicated because of intense faulting and folding of deposits due to compressive tectonics. As far as I know, nobody tried to tie these deposits with other outcrops of the Sinap Formation in the east (Sinap Tepe area) or the west (Kavakdere and Çalta area).

In the Kavakdere area, Ozansoy recorded a fauna from the upper horizon equivalent to Loc. 26 of the 1989–1995 Sinap Project. Ozansoy correlated this fauna to the "Epi-Pikermian," (i.e., a level above the classical Pikermian fauna). This horizon is currently magnetostratigraphically dated as being 8.12 Ma, which corresponds to the early Turolian (Kappelman et al., this volume). However, the faunal community compares more closely with middle Turolian faunas of Western Europe.

The lower Sinap deposits yielded sparse large mammal faunas without *Hipparion*. Ozansoy mentioned from this deposit *Hyaena minor* n. sp., *Listriodon piveteaui* n. sp. (undescribed), some antelopes, and a "Dryopithecine." The lack of *Hipparion* was particularly troubling. Thus, Ozansoy (1958, p. 17) noted that this deposit

> ne contient pas le genre *Hipparion*, bien qu'il se trouve entre les deux niveaux à *Hipparion*. Je ne peux pas m'expliquer pourqoui il n'existe pas dans cette faune de Sinap inférieur. Est-ce que l'absence de ce genre peut nous faire penser à une migration provisoire? ou bien est-ce que ce niveau était, pendant ces temps paléogéographiques, une grande île que les *Hipparion* ne pouvaient pas atteindre? Ou bien, tout simplement, les conditions n'étaient-elles pas favorables à la fossilisation?

In summary, Ozansoy believed this horizon was intermediate between the Kavakdere and middle Sinap localities, and tried to explain the lack of *Hipparion* by some hypotheses (e.g., migration, insularity, or fossilization process). Two new localities (Locs. 64 and 65) discovered by the 1989–1995 Sinap Project in the lower Sinap deposits yielded reasonably rich assemblages of small mammals that correlate with the late Astaracian.

The middle Sinap series conformably cover underlying strata. They are rich in small and large mammal fossils. Ozansoy mentioned the occurrence of bones in three horizons and correlated all with the late Pliocene. Remember that *Ankarapithecus meteai* was found only in these horizons. Recent magnetostratigraphic and biostratigraphic data bracket the related deposits within the Vallesian mammal age (Kappelman et al., this volume). The local Hipparion Datum corresponds approximately to the boundary between the lower and middle Sinap members.

The stratigraphic position of the upper Sinap deposits was another source of confusion. Ozansoy (1955) named it first as "Villafranchien des Monts Sinap" and recognized an erosional unconformity between this horizon and the underlying middle Sinap deposits. In his 1957 paper, Ozansoy named it "Sinap supérieur" and no longer recognized the presence of an unconformity, but defended the series as continuous from the lower Sinap to the upper Sinap member. He maintained the same opinion in his later papers. The main faunal elements that Ozansoy recovered from this member are *Equus stenonis* and *Hyaena arambourgi* n. sp. In 1972, I excavated the Ozansoy's upper Sinap locality (Sarikol Tepe) together with E. Heintz, L. Ginsburg, and M. Gürbüz, and recovered about a hundred specimens, reasonably well preserved, belonging to large and small mammals. Our study of this material (Kostopoulos and Sen 1999) shows that its age fits better with MN 17. Our present state of knowledge indicates that in the Sinap Tepe area, there is a gap of about seven million years between the last horizons of the middle Sinap member and the overlying upper Sinap member.

Conclusions

My review of the studies at Sinap Tepe and its immediate surroundings is meant to provide an historical context to this volume. My recounting of the involvement of Turkish and foreign scientists is based on my interviews with several colleagues, especially Turkish scientists, whose interest in Sinap Tepe preceded my own. It is my hope that this historical accounting will serve well future generations of scientists working at Sinap, and that it will help them build on the current record there as we understand it and report it in this volume.

Here I have highlighted some crucial periods of paleontologic investigations in the Sinap Tepe area. It appears that this is the most thoroughly investigated region in Turkey, and it has yielded the most complete fossil record of the Turkish Neogene. All Turkish mammal paleontologists, past and present, have been more or less involved in studies of the Sinap Formation faunas. Some of them have even undertaken their first paleontologic work in these deposits. I do not believe that the Sinap Formation has revealed all its secrets, and there is yet more to be discovered by future generations.

Acknowledgments

I am grateful to all Sinap workers and their collaborators who provided information on the history of investigations in the Sinap Formation. Thank you, John, Mikael, and Ray for encouraging me to write this chapter. The illustrations were scanned and improved by P. Loubry.

Literature Cited

Alpagut, B., and M. Fortelius, 1991, Survey results for the Sinap project, Kazan and Çubuk provinces, Ankara, Turkey, 1990. IX. Arastirma Sonuçlari Toplantisi, T. C. Kültür Bakanligi, Ankara, pp. 333–356.

Alpagut, B., P. Andrews, M. Fortelius, J. Kappelman, I. Temizsoy, H. Çelebi, and W. Lindsay, 1996, A new specimen of *Ankarapithecus meteai* from the Sinap Formation of central Anatolia: Nature, v. 382, pp. 349–351.

Andrews, C. W., 1918, Note on some fossil mammals from Salonica and Imbros: Geological Magazine, v. 5, pp. 540–543.

Andrews, P., 1982, The relationships of *Sivapithecus* and *Ramapithecus* and the evolution of the orang-utan: Nature, v. 297, pp. 541–546.

Andrews, P., and B. Alpagut, 2001, Functional morphology of *Ankarapithecus meteai, in* L. de Bonis, G. D. Koufos, and P. Andrews, eds., Phylogeny of the Neogene hominoid primates in Eurasia: Cambridge, Cambridge University Press, pp. 213–230.

Andrews, P., and I. Tekkaya, 1976, *Ramapithecus* from Kenya and Turkey, *in* P. V. Tobias, and Y. Coppens, eds., Les plus anciens hominidés, IX Congrès de l'Union Internationale des Sciences Préhistoriques et Protohistoriques, September 1976, Nice, pp. 7–25.

Andrews, P., and I. Tekkaya, 1980, A revision of the Turkish Miocene hominoid *Sivapithecus meteai:* Palaeontology, v. 23, pp. 85–95.

Bailey, E. B., and W. J. McCallien, 1950a, Ankara Melanji ve Anadolu sariaji (The Ankara Mélange and the Anatolian thrust): Maden Tetkik ve Arama Dergisi, v. 40, pp. 12–22.

Bailey, E. B., and W. J. McCallien, 1950b, The Ankara Mélange and the Anatolian thrust: Nature, v. 166, p. 4231.

Bailey, E. B., and W. J. McCallien, 1953, Serpentine lavas, the Ankara Melange and the Anatolian thrust: Transactions of the Royal Society of Edinburgh, v. 62, pp. 403–442.

Begun D., and E. Güleç, 1995, Restauration and reinterpretation of the facial specimen attributed to *Sivapithecus meteai* from Kayincak (Yassiören), central Anatolia, Turkey: American Journal of Physical Anthropology Supplement, v. 20, p. 26.

Begun D., and E. Güleç, 1998, Restauration of the type and palate of *Ankarapithecus meteai:* Taxonomic and phylogenetic implications: American Journal of Physical Anthropology, v. 105, pp. 279–314.

Bonis, L. de, and J. Melentis, 1984, La position phylétique d'*Ouranopithecus:* Courier Forschunginstitut Senckenberg, v. 69, pp. 13–23.

Calvert, F., and M. Neumayer, 1880, Die jungen Ablagerungen am Hellespont. Denkschriften der Kaiserlischen Akademie der Wissenschaften. Mathematisch-Naturwissenschaftliche Klasse, Wien, v. 40, pp. 357–378.

English, T., 1904, Eocene and later formations surrounding the Dardanelles: Quaterly Journal of the Geological Society of London, v. 60, pp. 243–275.

Erol, O., 1954, Ankara ve civarinin jeolojisi hakkinda rapor, unpublished report, no. 2491: MTA Enstitüsü, Archives, pp. 1–69.

Erol, O. 1956. Ankara güneydogusundaki Elmadag ve çevresinin jeoloji ve jeomorfolojisi üzerine bir arastirma. MTA Enstitüsü Yayinlari, v. 9, pp. 1–99.

Erol, O., 1966, The geomorphological importance of the remains of fossil mammals found between Üçbas and Akdogan villages in the northwest of Ankara: Ankara Üniversitesi Cografya Arastirmalari Dergisi, v. 1, pp. 109–120.

Fortelius, M., J. van der Made, and R. L. Bernor, 1996, A new listriodont suid, *Bunolistriodon meidamon* sp. nov., from the middle Miocene of Anatolia: Journal of Vertebrate Paleontology, v. 16, pp. 149–164.

Geraads, D., E. Güleç, and G. Saraç, 1995, Middle Miocene ruminants from Inönü, central Turkey: Neues Jahrbuch für Geologie und Paläontologie, v. 8, pp. 462–474.

Ginsburg, L., E. Heintz, and S. Sen,1974, Le gisement Pliocène à mammifères de Çalta (Ankara, Turquie): Comptes Rendus de l'Académie des Sciences de Paris, v. D278, pp. 2739–2742.

Gürbüz, M., 1981, Inönü (KB Ankara) Orta Miyosenindeki *Hemicyon sansaniensis* (Ursidae) türünün tanimlanmasi ve stratigrafik yayilimi: Türkiye Jeoloji Kurumu Bülteni, v. 24, pp. 85–90.

Howell, F. C., and G. Petter, 1985, Comparative observations on some Middle and Upper Miocene hyaenids. Genera : *Percrocuta* Kretzoi, *Allohyaena* Kretzoi, *Adcrocuta* Kretzoi (Mammalia, Carnivora, Hyaenidae): Géobios, v. 18, pp. 419–476.

Kay, R. F., and E. L. Simons, 1983, A reassessment of the relationship between later Miocene and subsequent Hominoidea, *in* R. L. Ciochon, and R. S. Corruccini, eds., New interpretations of ape and human ancestry: New York, Plenum Press, pp. 577–624.

Kazanci, N., 1993, A. Suat Erk Jeoloji Simpozyumu (2-5 Eylül 1991) bildirileri: Ankara, Turkey, Ankara Üniversitesi Basimevi, 450 pp.

Kittl, E., 1887, Beiträge zur Kenntniss der fossilen Säugetiere von Maragha in Persien. I. Carnivoren: Annalen des Kaiserlich-Koniglichen naturhistorischen Hofmuseums in Wien, v. 2, pp. 317–338.

Kostopoulos D. S., and S. Sen, 1999, Late Pliocene (Villafranchian) mammals from Sarikol Tepe, Ankara, Turkey: Mitteilungen der Bayerischen Staatssammlung für Paläontologie und Historiche Geologie, v. 39, pp. 165–202.

Major, C. J. F., 1888, Sur un gisement d'ossements fossiles dans l'île de Samos, contemporain de l'âge de Pikermi: Comptes Rendus de l'Académie des Sciences de Paris, v. 107, pp. 1178–1182.

McCallien, W. J., 1946, Inverted graded bedding at Köserelik, north of Ankara: Maden Tetkik ve Arama Enstitüsü Dergisi, v. 35, pp. 121–125.

Nafiz, H., and A. Malik, 1933, Vertébrés fossiles de Küçük Çekmece: Bulletin de la Faculté des Sciences de l'Université d'Istanbul, v. 8, pp. 99–120.

Natural History Museum, 1968, Natural History Museum MTA Editions, Ankara, Turkey, pp. 1–16.

Newton, R. B., 1904, Notes on the post-Tertiary and Tertiary fossils from the district surrounding the Dardanelles: Quaterly Journal of the Geological Society of London, v. 60, pp. 277–292.

Ozansoy, F., 1955, Sur les gisements continentaux et les mammifères du Néogène et du Villafranchien d'Ankara (Turquie): Comptes Rendus de l'Académie des Sciences de Paris, v. 240, pp. 992–994.

Ozansoy, F., 1957a, Faunes de mammifères du Tertiaire de Turquie et leurs révisions stratigraphiques: Bulletin of the Mineral Research and Exploration Institute of Turkey, v. 49, pp. 29–48.

Ozansoy, F., 1957b, Position stratigraphique des formations continentales du Tertiaire de l'Eurasie au point de vue de la chronologie Nord-Américaine: Bulletin of the Mineral Research and Exploration Institute of Turkey, v. 49, pp. 11–28.

Ozansoy, F., 1958, Etude des gisements continentaux et des mammifères du Cénozoïque de Turquie [Ph.D. thesis]: Paris, Faculté des Sciences de l'Université de Paris, 179 pp. + 30 plates.

Ozansoy, F., 1961, After Muzaffer Şenyürek, the great Turkish scientist: Bulletin of the Mineral Research and Exploration Institute of Turkey, v. 57, pp. 121–130.

Ozansoy, F., 1965, Etude des gisements continentaux et des mammifères du Cénozoïque de Turquie: Mémoires de la Société Géologique de France, v. 102, pp. 1–92.

Ozansoy, F., 1966, Türkiye Senozoik çaglarinda fosil insan formu ve biostratigrafik dayanaklari: Ankara Üniversitesi, Dil ve Tarih Cografya Fakültesi Yayinlari, v. 172, pp. 1–104.

Ozansoy, F., 1970a, *Ankarapithecus meteai,* Pongidé fossile aux traits humains du Pliocène de Turquie: Belleten, v. 34, pp. 1–15.

Ozansoy, F., 1970b, Türkiye Orta Pleistosen fosil insanlari ve Paleolitik öncesi: Ankara Üniversitesi Rektörlügü Yayinlari, v. 67, pp. 1–56.

Pickford, M., and Ç. Ertürk, 1979, Suidae and Tayassuidae from Turkey: Bulletin of the Geological Society of Turkey, v. 22, pp. 141–154.

Pittard, E., 1939, Un chef d'etat animateur de l'anthropologie et de la préhistoire: Kemal Atatürk: Revue Anthropologique, v. 49, pp. 1–12.

Saraç, G., 1968, Çaltaköy Villafransiyen tortullari [MS. thesis]: Ankara, Turkey, Ankara Üniversitesi, Dil ve Tarih Cografya Fakültesi, 60 pp.

Saraç, G., 1994, Ankara yöresindeki karasal Neojen çökellerinin Rhinocerotidae (Mammalia-Perissodactyla) biyostratigrafisi ve paleontolojisi [Ph.D. thesis]: Ankara, Turkey, Üniversitesi Fen Fakültesi, 214 pp.

Sen, S., 1977, La faune de rongeurs Pliocènes de Çalta (Ankara, Turquie): Bulletin du Muséum National d'Histoire Naturelle, v. 61, pp. 89–172.

Sen, S., 1991, Stratigraphie, faunes de mammifères et magnétostratigraphie du Néogène de Sinap Tepe, Province d'Ankara, Turquie: Bulletin du Muséum National d'Histoire Naturelle, v. 12, pp. 243–277.

Sen, S., 1998, Pliocene vertebrate locality of Çalta, Ankara, Turkey: Geodiversitas, v. 20, pp. 325–513.

Sen, S., E. Heintz, and L. Ginsburg, 1974, Premiers résultats des fouilles effectuées à Çalta, Ankara, Turquie: Bulletin of the Mineral Research and Exploration Institute of Turkey, v. 83, pp. 112–118.

Sengör, C., 2001, Oguz Erol'un hikayesi: Cumhuriyet Bilim Teknik, v. 20, no. 10, p. 5.

Şenyürek, M., 1960a, The Pontian ictitheres from the Elmadag district: Publications of the Faculty of Languages, History and Geography, University of Ankara, Turkey, Supplement 1, 223 pp. + 14 plates.

Şenyürek, M., 1960b, A note on the remains of fossil elephants preserved in the University of Ankara: Belleten, v. 24, pp. 693–698.

Şenyürek, M., 1961, The molar of *Archidiscodon* from Akdogan: Belleten, v. 25, pp. 339–350.

Simons, E. L., and D. R. Pilbeam, 1965, Preliminary revision of the Dryopithecinae (Pongidae, Anthropoidea): Folia Primatologia, v. 3, pp. 81–152.

Tekkaya, I., 1970, A horn-core of *Gazella deperdita* Gervais (n. var.) from Middle Sinap: Bulletin of the Mineral Research and Exploration Institute of Turkey, v. 74, pp. 59–60.

Tekkaya, I., 1973, Une nouvelle espèce de Gazelle de Sinap Moyen: Bulletin of the Mineral Research and Exploration Institute of Turkey, v. 80, pp. 118–143.

Tekkaya, I., 1974a, The Bovidae fauna of Middle Sinap of Turkey: Bulletin of the Geological Society of Turkey, v. 17, pp. 174–186.

Tekkaya, I., 1974b, A new species of Tortonian anthropoid (Primates, Mammalia) from Anatolia: Bulletin of the Mineral Research and Exploration Institute of Turkey, v. 83, pp. 148–165.

Tekkaya, I., 1975, Orta Sinap Bovinae faunasi: Bulletin of the Geological Society of Turkey, v. 18, pp. 27–32.

Toula, F., 1891, Saugetierreste gesammelt bei gelegenheit des baues der Eisenbahn von Scutari nach Ismit. Anzaigere der Akademie der Wissenschaftliche in Wien: Mathematisch-Naturwissenschaftliche Klasse, v. 27, pp. 112–114.

Toula, F., 1896, Saugetierreste von Eskihisar. Anzaigere der Akademie der Wissenschaftliche in Wien: Mathematisch-Naturwissenschaftliche Klasse, v. 32, pp. 92–97.

Tschachtli, B., 1942, Fossile Säugetiere aus der Gegend von Küçükyozgat (östlich Ankara): Bulletin of the Mineral Research and Exploration Institute of Turkey, v. 7, pp. 325–327.

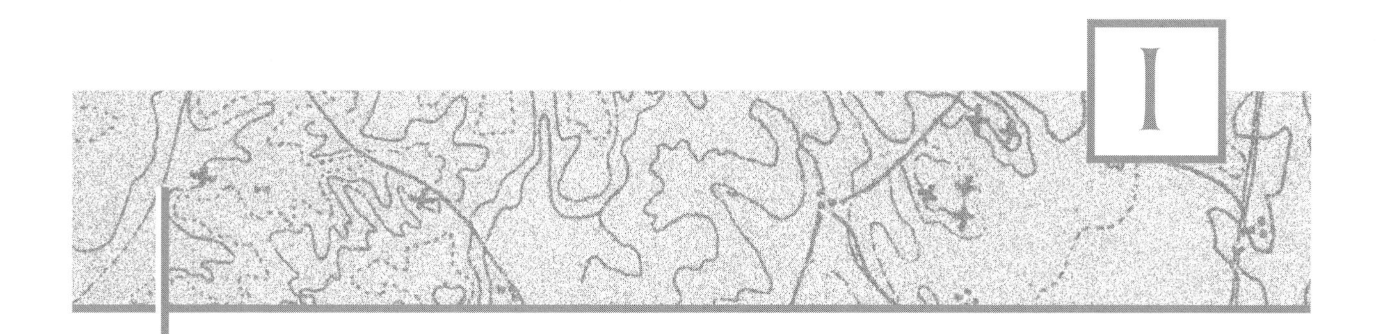

I

Geology and Chronology

1

Geology

J.-P. Lunkka, J. Kappelman, D. Ekart, J. Crabaugh, and P. Gibbard

The study area of the Sinap Project is situated about 37 km northwest of Ankara, Central Anatolia, Turkey. In this area, paleontologic prospecting revealed fossil-rich terrestrial sediments in the vicinity of the towns of Kazan and Çubuk (fig. 1.1). Before the 1993 field season, geologic research in the area was undertaken mainly in connection with paleomagnetic work, and it was not until that season that full-scale geologic work began, including observations on the mesoscale tectonic history of the area. The aims of the geologic research in the project were twofold. The first goal was to produce a geologic map of the area to establish a firm basis for an inter-area correlation of individual fossil localities. Although geologic mapping alone cannot provide time-dependent correlation between various fossil localities, it can provide lithostratigraphic control for the magneto- and biostratigraphic results. The second and principal aim of the geologic work was to apply conventional sedimentologic techniques to model the paleoenvironments that existed when the fossiliferous sediments were being deposited. Although the area under investigation is highly tectonized and complex, it was possible to reconstruct a lithostratigraphic scheme for the area that has shed light on the evolution of the Sinap paleoenvironments in space and time.

Geologic Background

The study area around towns of Kazan and Çubuk (see fig. 1.1) is located in the general region of Ankara and is part of the collisonal-tectonic system that included the Eurasian, Sakarya, and Gondwana continents (Sengör and Yilmaz 1981). Pre-Miocene rocks after Koçyiğit (1991; see fig. 1.1) comprise Karakaya Complex, Ankara Group, Anatolian Complex, and upper Cretaceous-middle Eocene Memlik Group. Karakaya Complex represents a tectono-sedimentary melange that consists of metamorphic, ultramafic, and mafic rocks together with crystallized and fossiliferous limestone, radiolarian chert, and clastic blocks in litharenitic shaley matrix. The Ankara Group includes six different lithofacies of sedimentary origin (marine sediments and sedimentary melange). The Anatolian Complex is a tectonic melange characterized by a chaotic tectonic mixture of ophiolithic sandstone, shale, and pelagic mudstone, whereas the Memlik Group consists of deep marine flysh that passes into terrestrial clastics (conglomerates, sandstones, and mudstones) and finally into reefal limestone representing forearc basin fill.

Miocene to Pliocene rocks rest on the Memlik Group with an angular unconformity. These mainly volcaniclastic sediments and subaerial basalts outcrop on the flanks of prominent hills and are often covered by recent alluvium in the general region around the towns of Kazan and Çubuk (cf. Lüttig and Steffens 1976; Lunkka et al. 1995; Kappelman et al. 1996; Lunkka et al. 1998; see fig. 1.2). Fossiliferous sediments of the Sinap Formation discussed in this chapter form the upper part of the Miocene strata and are relatively well exposed in the area northwest of Kazan as well as at fossil locality 49 at Igbek village (Loc. 49 in fig. 1.1). However, in a few areas such as at Copanpinar (Loc. 42 in fig. 1.1), fossil occurrences were not possible to place in a lithostratigraphic context due to lack of vertical exposures.

The structural geology of the study area is very complicated as a result of tectonic movements and volcanic activity during the Neogene. There is clear evidence of increased tectonic activity and volcanism since the middle Miocene that continued well into the Pleistocene (cf. Lüttig and Steffens 1976). Tectonic movements are related to the extensional neotectonics of Anatolia and the north Anatolian transform fault (Angelier et al. 1981; Inci 1991). It is typical for this part of central Anatolia that extensional tectonics created small fault-bounded basins (Graben and Horst features) or so-called intermontane basins of various sizes (cf. Lüttig and Steffens 1976; Erol 1981). In addition, folding events that were synchronous with, as well as pre- and postdating normal and strike-slip faulting events, have further complicated the structural geology.

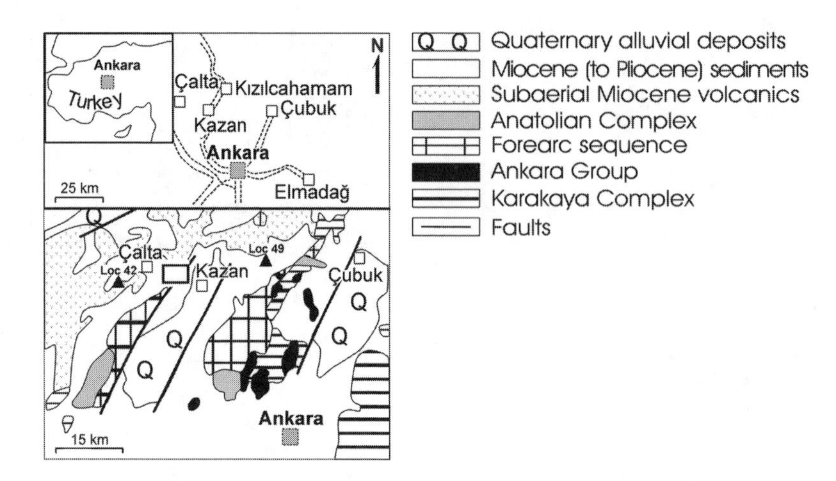

Figure 1.1. Location of the study area north of Ankara, Turkey, and a generalized geologic map of the area modified after Koçyiğit (1991). The small rectangle with a heavy black border indicates the area shown in figure 1.2, where most of the fossil localities were found. Fossil sites Loc. 42 (Copanpinar) and Loc. 49 (Igbek) are also indicated in the lower map.

Most of the fossil localities are concentrated in the area northwest of Kazan; only one fossil locality (Loc. 49 in fig. 1.1) was discovered at Igbek. In the former area, a geologic map covering ~40 km^2 was constructed and is shown in figure 1.2. Extensive mapping was not carried out in the area surrounding Igbek (Loc. 49) or that of Cobanpinar (Loc. 42), and the correlation between these two sites with those in the main section in the Sinap Tepe-Kavakdere area is based on magneto and biostratigraphy, respectively.

The main structural features in the Sinap Tepe-Beycedere-Kavakdere area, northwest of the town of Kazan (fig. 1.2) show two major fault zones that run approximately northeast–southwest across the study area. Fault zone 1 (FZ 1) represents a normal fault whereas fault zone 2 (FZ 2) represents a strike-slip fault. The two fault zones eventually join together in the southwestern part of the study area. Southeast of the Sinap Tepe-Kavakdere area, several minor faults and folds were also found. The southernmost fault zone (FZ 1) borders the northern flank of the Graben basin. Most of the fossil localities are situated south of FZ 1 (fig. 1.2). Northwest of FZ 2, sediments at the base of the Beycedere are the oldest and pass with a considerable hiatus into sediments in the Kavakdere syncline. There is no direct lithostratigraphic correlation between the sediments south of FZ 1 and those northwest of FZ 2, although bio- and magnetostratigraphic results together with structural observations provide a firm base for inter-area correlation (see the next section).

General Lithostratigraphy

Stratigraphic work previously carried out in the area is limited. Ozansoy (1957, 1965) and later Öngür (1976) established a lithostratigraphic scheme for the Miocene sequence in the study area. Ozansoy (1957, 1965) named three members of the Sinap Formation and recognized a conformable contact between the lower and middle members, and an erosional unconformity between the upper and middle members. Later work by Öngür (1976) confirmed Ozansoy's (1957, 1965) results but recognized an additional, older unit, the Pazar Formation, and stated that the Pazar and Sinap Formations are separated by a volcanic unit (basaltic lava flow). Thereafter work on the Miocene strata was carried out by the Sinap Project (Sen 1991; Kappelman et al. 1996; Lunkka et al. 1998; Lunkka et al. 1999).

The general lithostratigraphic scheme for the Miocene sequence in the study area presented in figure 1.3 indicates the relative chronostratigraphic relationships between individual members of the Sinap Formation. North of FZ 1, the oldest Miocene sediments that belong to the Pazar Formation are current bedded volcaniclastics (SB in fig. 1.2) that lie below the basaltic lava flow northeast of the Sinap Tepe area and dated at 15.2 ± 0.3 Ma. In the study area, there is only one fossil locality (Loc. 125 in fig. 1.2) that can be clearly placed in the Pazar Formation. In addition to the basaltic flow observed in the Sinap Tepe area, another intermediate volcanic rock (trachyte) was also found some 4 km northeast of Örencik village that most likely represents an older, more acidic lava flow event than that which occurs in the Sinap-Beycedere area.

The chronostratigraphic position of the Beycedere member is somewhat problematic because the basaltic lava flow at the base of the Beycedere sequence is mapped to be the same as that elsewhere north of FZ 1 (see fig. 1.2); that is, the lava flow that was laid down at 15.2 ± 0.3 Ma according to a whole rock K-Ar radiometric date (Kappelman et al. 1996). In contrast, $^{40}Ar/^{39}Ar$ radiometric dates performed using plagioclase separates from two ash beds from the upper part of the Beycedere section, located well

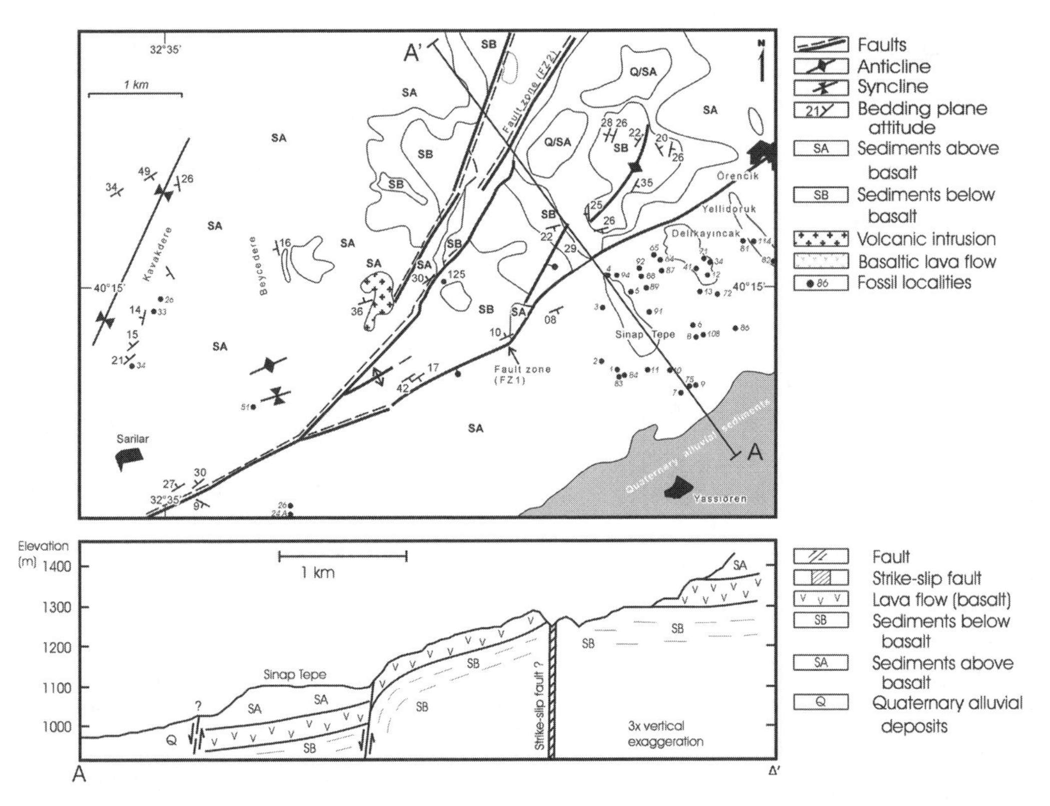

Figure 1.2. Geologic map of the Sinap Tepe-Kavakdere area and a cross section A′–A indicating the main structural and stratigraphic units in the Sinap Tepe area. The main structural elements consist of two main fault zones (FZ 1 and FZ 2) and associated folds. Note also the Kavakdere synform. Fossil localities in the Yellidoruk, Delikayincak Tepe, Sinap Tepe, and Beycedere-Kavakdere areas are also indicated.

above basaltic lava, yield ages of 15.88 ± 0.07 Ma and 15.96 ± 0.08 Ma (Kappelman et al., this volume). This obvious discrepancy between dating results and the results obtained from detailed geologic mapping makes it difficult to place the Beycedere member in a stratigraphic context. However, geologic mapping also demonstrates that there are no basaltic lava flows above dated ash beds in the study area correlative to the flow dated at 15.2 ± 0.3 Ma. Furthermore, sediments in the Beycedere area are conformably overlain by the Kavakdere member, although there appears to be a major hiatus between these two members with a duration of ~5 Ma as estimated by the $^{40}Ar/^{39}Ar$ dates and magnetostratigraphic results. Therefore it is assumed that sediments in the Beycedere section belong to the Sinap Formation. The data also suggest that the basalt at the base of the Beycedere section is correlative to that found elsewhere in the study area, and that it is possibly somewhat older than the date indicated by the whole rock K-Ar radiometric method.

As stated above, the basal sediments in the Kavakdere area rest conformably on silicified flat iron horizons that occur at the top of the Beycedere sequence. These sediments in the Kavakdere syncline and farther west are divided into Kavakdere and Çalta members of the Sinap Formation (see fig. 1.3), with the Çalta member dating to the Pliocene (Lunkka et al. 1998).

Volcaniclastic sediments southeast of FZ 1 occupy the southern flank of a fault-bounded basin with sedimentary units that dip gently (5–12°) toward the center of the basin (see fig. 1.2). The correlation of these sediments to those north of FZ 1 is based on magneto- and biostratigraphy and to some extent on the structural setting of FZ 1, as mentioned previously. These observations clearly indicate that the sediments on the flank of the basin margin postdate the emplacement of the basaltic lava flow. The sedimentary sequence at Igbek that contains fossil locality 49 is located ~22 km east-northeast of the Sinap Tepe area on the northern slope of the same fault-bounded basin that runs across the Sinap Tepe area to the northwest of Kazan. We consider the Igbek sequence as an individual member of the Sinap Formation, and it is correlated magnetostratigraphically to the rest of the members of this formation.

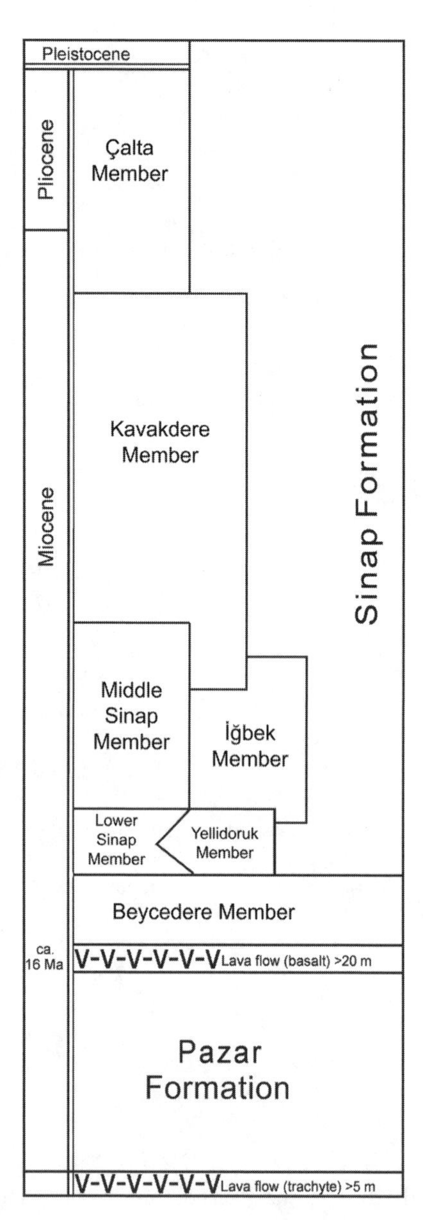

Figure 1.3. Stratigraphic division of the Sinap Formation and its relationship to the basaltic lava flow, the volcaniclastic Pazar Formation, and the more acidic (trachyte) lava flow in the study area.

Sedimentation and Lithostratigraphy of the Sinap Formation

Here we discuss the local lithostratigraphy, style of sedimentation, and position of individual fossil localities for three separate areas: Sinap Tepe (south of FZ 1), Beycedere-Kavakdere (northwest of FZ 2), and Igbek. Because most of the fossil localities are found in the Sinap Tepe area that lies to the southeast of FZ 1, we give a more detailed description of the sediments and style of deposition for this area. Only brief descriptions are given for the Beycedere-Kavakdere and Igbek areas.

Lithostratigraphy of the Sinap Tepe Area

There are three prominent hills—Sinap Tepe, Delikayincak, and Yellidoruk—along with several smaller buttes (see fig. 1.2), where sediment exposures are extensive enough to conduct reliable sedimentologic observations and measurements in the Sinap Tepe area. The correlation of sedimentary units between individual sediment exposures was constructed using marker horizons, bedding plane attitudes, and lateral tracing of sediment units across the area. Six sections were logged on Sinap Tepe, where the Sinap I section was located in the northwest and Sinap VI in the southeast (see the logs of Sinap I–VI in fig. 1.4). The correlation between the Sinap III and VI sections is primarily based on tracing a prominent paleosol horizon across the area. In addition, two paleosol horizons lower in the Sinap III–V sections tie these exposures together. The Sinap I and II sections located in the northwest part of Sinap Tepe were correlated using bedding plane attitude information and tracing one distinct paleosol horizon between the sections. It was not possible to trace any clear marker beds between Sinap II and III. However, we correlated these two sections by sighting using the bedding attitude with Abney level and Jacob staff. This correlation was compared and also confirmed with tachymetric leveling and the correlation based on photographs. We consider these techniques to be accurate for this correlation and those discussed later because there are no known faults in the Sinap Tepe area that would invalidate the tachymetric correlations.

We correlated the Sinap I–VI sections and the Delikayincak section (see fig. 1.2 for their locations) by tracing and measuring the thickness from Sinap II to the base of the Delikayincak section. Using tacheometer tracings, it also became evident that the well-developed paleosol horizons in the upper part of the Delikayincak section correspond well with the top marker paleosol in the Sinap III–VI sections (see fig. 1.4).

The correlation between the Yellidoruk sections was done by the lateral tracing of individual sediment beds along the cliff exposures. However, the tie between sediment beds in the Yellidoruk exposures and those of Delikayincak is based solely on tacheometric tracings, because there are no traceable marker horizons exposed across these two hills. According to the tacheometer results, the prominent paleosol marker horizons at the top of the logged sections at Yellidoruk correspond to those at the top of the Delikayincak section. The stratigraphic position of Yellidoruk is further confirmed by the petrographic results that are discussed below.

Note that the correlations given here are also independently supported by the results of the paleomagnetic reversal stratigraphy (see Kappelman et al., this volume).

We logged the individual geologic exposures using conventional sedimentologic techniques (cf. Tucker 1982) to shed light on the style of deposition and the lateral and vertical variation of sediment beds. We divided the vertical sections into lithofacies based on lithology and depositional features using facies codes and architectural element

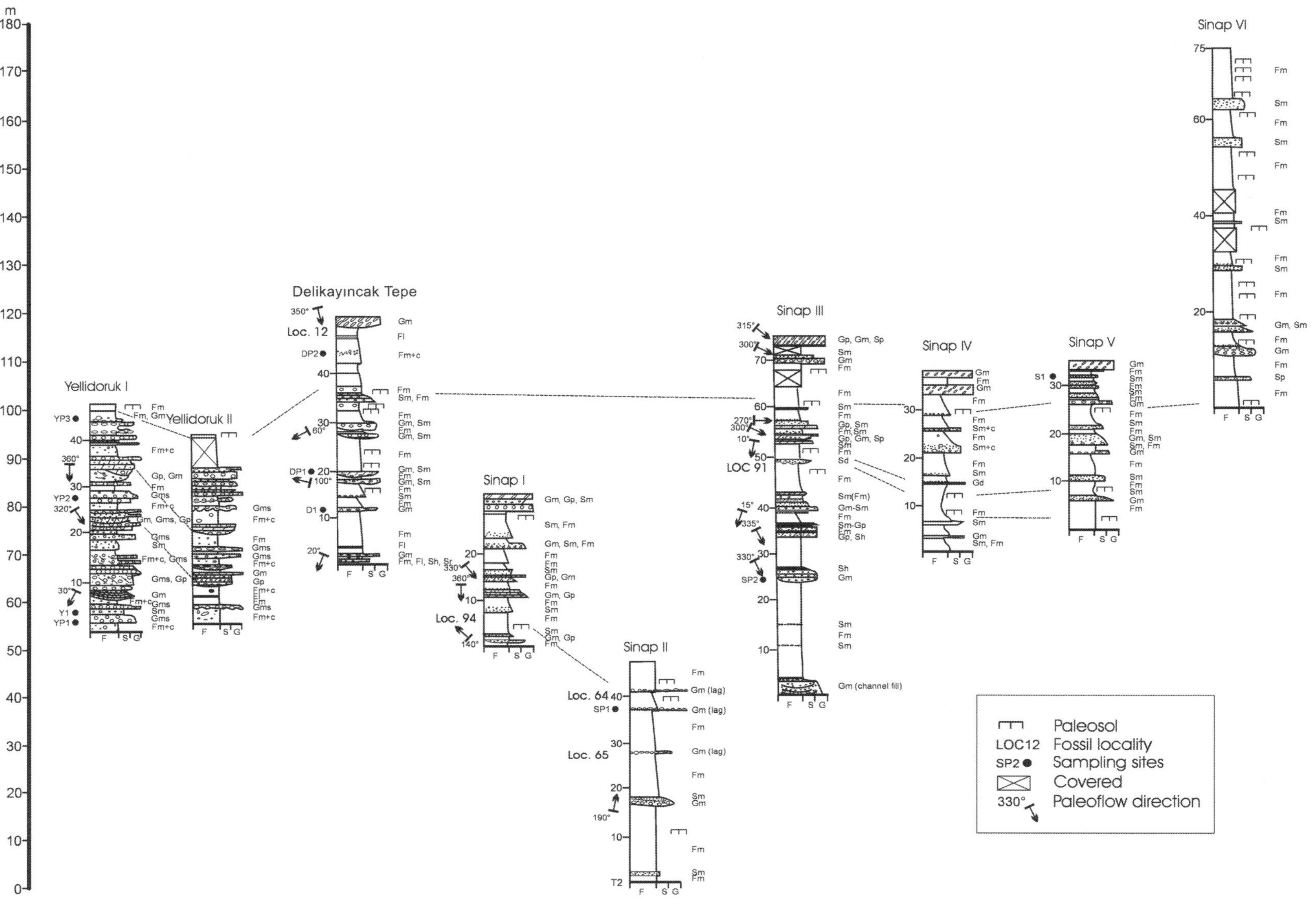

Figure 1.4. Logged sections in the hills of Yellidoruk-Delikayincak Tepe and Sinap Tepe areas. Correlation of individual sections is based on the tracing of different marker horizons (mainly paleosols) throughout the area. Main fossil localities in these sections are also indicated. Facies codes: Gms, matrix-supported and massive gravel; Gm, crudely bedded or stratified gravel; Gp, planar cross-bedded gravel; Gd, syndepositionally deformed gravel; Sm, massive sand; Sm + c, massive sand with outsized clasts; Sh, horizontally bedded sand; Sr, ripple-bedded sand; Sp, planar cross-bedded sand; St, trough cross-bedded sand; Sd, syndepositionally deformed sand; Fm, massive fines (fines defined as silt and clay); Fm + c, massive fines with clasts; Fl, laminated fines; Fd, deformed fines.

analysis modified from Miall (1978, 1985). Sedimentary units in the Sinap Tepe area are laterally continuous and dip 5–12° away from FZ 1 toward the center of the basin (fig. 1.2). The main gravel, sand, and silt-clay facies associations that occur in the Yellidoruk, Delikayincak, and Sinap Tepe areas with their genetic interpretations are described in the following sections.

Gravel Lithofacies

Sediment Gravity-Flow Lithofacies. Massive matrix-supported gravel and massive or crudely bedded gravel facies (Gms and Gm facies) occur mainly in the Yellidoruk sections (fig. 1.4). This facies consists of poorly sorted, matrix-supported, pebble-to-cobble gravel conglomerate. Clasts are typically subangular to subrounded and the matrix is composed of clay silt or fine sand. Stratification is lacking, but Gms units show normal grading and on one occasion, reverse grading was also observed. Gms beds are 0.2–2.2 m thick and typically have nonerosional bases; some of the beds have loaded contact between the underlying finer units. Gms units are wedge-shaped sheets or lobes, although some are lenticular-shaped, and their observed lateral extent is several tens of meters in cross section. Gms facies is interbedded with gravel bars and bedforms (i.e., Gm, Sm, and Gp facies and also with Fm facies; for facies codes, see fig. 1.4).

Viscous debris-flow events commonly generate features such as mudstone-matrix support, lack of internal structures, as well as very poor sorting and nonchannelized bases (cf. Hubert and Filipov 1989; De Celles et al. 1991). However, it is also possible that thin units of texturally bimodal, pebble-to-cobble matrix-supported conglomerates that occur in close association with crudely bedded or massive conglomerates (Gm facies) might have resulted from postdepositional floral or faunal disturbances, in which larger clasts might have been mixed with fine sediments (cf. DeCelles et al. 1987). The Gms facies is interpreted here as a debris-flow deposit, although only decimeter-thick Gms units could have originated from postdepositional disturbances.

Gravel Bars and Bedforms. Crudely bedded gravel with horizontal bedding and imbrication and stratified planar cross-bedded gravel facies (Gm, Gp, Gt facies) are present in the Yellidoruk section as well as in the Sinap Tepe sections (upward-fining cycles in sections III, IV, and V in fig. 1.4). Gm facies is typically very crudely stratified and pebble-to-cobble gravel conglomerates show some imbrication. Gm units have erosional lower contacts and they occur in very broad and shallow channel-forms or tabular gravel bodies in close association with planar cross-bedded gravel units (Gp facies). In Sinap section II, Gm facies occur as <10-cm thick lag horizons. Gp facies consists of up to a few-decimeter thick sets of moderately to well-sorted granule-to-pebble gravel. The thickness of these elements ranges between 0.8 and 3.4 m, and lateral extent is at least few hundred m in cross section; units become finer in the down-current direction. Structures in these gravels and associated sands indicate bedload transport in flows that were not channelized. Their upward-fining trend together with main features described above suggest that these sediments were deposited under sheet flow conditions. Most of the Gm and Gp facies are interpreted as representing gravel bars and bedforms.

Fluvial Channel Deposits. In the Delikayincak section as well as in the lower and upper parts of the Sinap sections (see the Delikayincak and Sinap I, II, III, and VI sections in fig. 1.4), granule-to-pebble gravel conglomerate occupies basal parts of shallow channel fill forms rarely wider than 10 m. This faintly stratified Gm facies occurs in close connection with planar and occasionally trough cross-bedded granule gravel units (Gp and Gt facies), individual cross-bedded sets being only a few cm thick. Channels have a concave-up erosional base, and channel fills often fine upward into sand and finally into silty clay.

Individual channel fill gravel units at Delikayincak are 0.2–0.3 m thick where individual bedsets are limited to a few cm. In the Delikayincak section, channel features themselves occur in rather continuous and laterally extensive channel fill trains. In these trains, even the deepest parts of the channels are very shallow (maximum depth 0.3–2.0 m), and laterally, the gravel may extend only a few cm between channel troughs.

Channel elements at the Delikayincak and Sinap sections are interpreted as representing bedload deposits of migrating small streams formed during flood events.

Sand Lithofacies

Sheet Flood Sands. Massive sand facies (Sm facies) consists of 0.2–1.5-m thick units of massive, structureless silty sand to coarse sand beds that include randomly distributed isolated pebbles (1–6 cm in diameter). Individual beds are wedge- and tabular-shaped. Lower contacts of Sm facies are normally sharp, conformably overlying pre-existing units. Sm facies is under- and overlain by Gms and Gm facies or sandwiched between massive, clast bearing silt beds. In the Delikayincak and Sinap sections, massive sands overlie gravel facies in channels and extend farther away from channel features. Massive sands, particularly in the Delikayincak and Sinap sections, are often mottled and sometimes contain calcite nodules and root burrows. Massive sands with floating pebbles are thought to have been deposited as sandy mudflows that spilled out of channels and spread over the pre-existing ground surface as sheet flood deposits; pedogenic processes subsequently modified these deposits.

Channel Sands. Horizontal to subhorizontal lamination (Sh facies) and planar cross-bedded, medium to coarse sand facies (Sp facies) occur either in channels as lenses that are a few m wide or form laterally extensive, wedge-shaped bodies in the upward-fining sequences that are especially common in the Sinap sections. These sand units are 0.2–1.8 m thick with individual sand sets 0.1–0.4 m thick and most commonly occur in association with Gm facies. Lower contacts of these sands are sharp.

Sh and Sp facies were usually laid down in channels. Sp facies represent dunes where overlying Sh facies were laid down during conditions of waning flow.

Flood Plain Fine Sands. Thin beds (7 cm on average) of parallel bedded and ripple-bedded fine-to-medium sand sets (Sr and Sh facies) occur and are interbedded with massive fines in the lower part of the Delikayincak section. These thin sand sets seem to be laterally extensive for at least tens of m. The horizontally and ripple-bedded sands that are associated with thick accumulations of massive and laminated fines are thought to have been laid down in the lower flow regime in the flood plain.

Silt and Clay Lithofacies

Debris Flow and Mudflow Deposits. Massive fine-grained facies with clasts (Fm + c facies) consists of 0.2–3.9-m thick units of red-brown and gray clay silt to sandy silt. Clay silt and sandy silt contain clasts 1–25 cm in diameter. The clasts are normally floating and occur throughout the massive mudbeds. In some clay silt units, clasts occur as isolated pockets and discontinuous stripes that have sunk into massive fines. Their occurrence decreases toward the top of some units. The lower contacts of the fine units are normally sharp and conformable, although gradational contacts fining upward from Gms and Sm facies are also present.

The Fm + c facies at the Yellidoruk sections is interpreted as mudflow or cohesive debris-flow deposits (Lowe 1982). The larger clusters of clasts and stripes could have been in contact with each other during downslope movement while some of the evenly distributed smaller clasts were being supported by buoyancy and cohesiveness of the silt/clay mixture (cf. Bagnold 1956; Lowe 1979).

Flood Plain Fines. Thick units of massive fines (silt and silt clay, Fm facies) including floating pebbles and occasional lag horizons are particularly common in the lower parts of the Delikayincak and Sinap sections. These fines display features indicative of paleosols such as calcite nodules, layers, and veins. Mottling and color changes of these fine units also suggest pedogenic alterations of the fines.

In the upper part of the Delikayincak and Sinap I, IV, and V sections, massive silts and clays occur in upward-fining sequences between distinct horizons of channelized and nonchannelized gravels and sands. These fines also show features typical of paleosol development, although somewhat less so than the fines that occur in the lower parts of the sections. Accumulations of silt and silty clay represent pedogenically modified overbank deposits.

Only a few laminated 0.1–0.2-m thick silt/clay units (Fl facies) were observed in the sections. Individual laminae in Fl facies are only few mm to a cm thick; these rest conformably upon coarser sediments. These thin units most likely represent nonbioturbated parts of the fines deposited during conditions of waning flow into small ponds on the flood plain during flood events.

In addition to these examples, at least one 0.7-m thick laminated bed was observed ~20 m west of the hominoid fossil site (Loc. 12). Although the section was poorly exposed, we concluded that this bed represents a pond deposit on the ancient flood plain.

Lithology of Pebbles and Sands and Paleocurrents in the Sinap Tepe Area

We collected samples from Yellidoruk, Delikayincak Tepe, and Sinap Tepe exposures to determine the lithology of the sands and gravels and thus to shed light on their provenance. Thin sections were prepared using three different grain-size fractions (0.5–1.0 mm, 0.250–0.500 mm, and 0.125–0.250 mm) from three different levels of the sedimentary sequence. Of the three fractions, the most representative (0.250–0.500 mm) was used for lithologic studies, with 500 grain counts per slide. In addition, stone counts on pebble gravel fraction were carried out from six different levels of the sequence.

The lithologic results from sand and pebble fraction are listed in table 1.1. The amount of intermediate volcanic rocks (trachyte) stays rather consistent (16.1%–24.4%) in the sand fraction in all of the sand samples studied. However, at the lowest level (58 m) in the Yellidoruk section, volcanic basic rock fragments (basalt) are not present and this rock-type is also very rare (2.0%) at 79-m level in the Delikayincak section, whereas its percentage increases to 67.8 in the Sinap V section at the 108-m level. Similarly, the pebble counts indicate that trachyte is a dominant rock type throughout the Yellidoruk sequence and in the lowermost part (36-m level) of the Sinap sequence (Sinap II section), but basalt is the dominant rock type in the Delikayincak section as well as in the middle part of the Sinap III section.

Paleocurrent indicators in sedimentary units are relatively difficult to recognize in the field. However, in well-exposed beds it was possible to carry out paleoflow measurements from planar foresets, trough axis azimuths, asymmetric small-ripple foresets, and crest trends. Occasionally clast imbrication in relation to channel trend could be also taken into account when paleocurrent direction was determined. We recorded four readings from each suitable sediment bed, corrected the readings for tilt, and averaged them. Results of the paleocurrent direction measurements are shown in figure 1.4. Paleoflow measurements indicate widely varied current directions. In general, the lower part of the sedimentary sequence (Sinap I and II sections in fig. 1.4) indicates current flows from a southerly direction. In the Yellidoruk area, current flow ranged mainly from the northwest/northeast, whereas in the Delikayincak section it

Table 1.1a. Petrological Composition of Sand Grains (0.250–0.500 mm Fraction) from Three Different Levels in the Yellidoruk-Delikayincak Tepe-Sinap Tepe Area

Sample	Rock Type	Percentage
Sample Y1, Yellidoruk Section I (58 m level)	A. Igneous rock fragments	
	1. Volcanic, intermediate rocks (trachyte)	20.40
	2. Crystal (fsp) fragments (derived from A1 rocks)	33.40
	B. Sedimentary rock fragments	
	1. Volcaniclastic rocks (volcanic sandstone)	14.40
	2. Chert	26.40
	3. Shale or slate	5.40
Sample D1, Delikayincak Tepe section (79-m level)	A. Igneous rock fragments	
	1. Volcanic, intermediate rocks (trachyte)	24.40
	2. Volcanic basic rocks (basalt)	2.00
	3. Crystal (fsp from A1 rocks, very rare cpx from A2 rocks) fragments	28.40
	B. Sedimentary rock fragments	
	1. Volcaniclastic rocks (volcanic sandstone)	19.00
	2. Chert	25.80
	3. Shale or slate	0.40
Sample S1, Sinap Tepe V Section (108 m level)	A. Igneous rock fragments	
	1. Volcanic, intermediate rocks (trachyte)	16.10
	2. Volcanic basic rocks (basalt)	67.80
	3. Crystal (phenocrysts: cpx, fsp) fragments derived from A1/A2 rocks	11.40
	B. Sedimentary rock fragments	
	1. Carbonate (sparite)	4.70

Notes: Number of grains counted is 500/sample. For exact sedimentary units where samples (Y1, D1, and S1) were taken, see fig. 1.4. Abbreviations: cpx, clinopyroxene (augite); fsp = feldspars (plagioclase, sanidine).

Table 1.1b. Stone Count Results from Seven Different Levels in the Yellidoruk-Delikayincak Tepe-Sinap Tepe Sections

Sample	Number of Pebbles	Level (m)	Trachyte (%)	Basalt (%)	Chert (%)	Others (%)
Yellidoruk section						
YP1	95	56	99	0	1	0
YP2	202	82	97	0	3	0
YP3	131	98	88	0	10	2
Delikayincak Tepe Section						
DP1	142	85	27	59	10	4
DP2	185	110	20	74	2	4
Sinap Tepe Section						
SP1	42	36	96	0	2	2
SP2	159	64	22	60	6	12

Notes: Number of pebbles is number counted in the 4–10 cm size. Level is measured as m above the base line. See fig. 1.4 for exact sections where samples were taken.

ranged from the north/east. In the upper part of the Sinap sequence, current flow was mainly from the northwest.

The results from petrographic studies as well as paleocurrent indicators strongly suggest that the provenance area must have changed during deposition in the Yellidoruk-Delikayincak Tepe-Sinap Tepe area. The lower part of the sequence (sections Sinap I and the lower part of the Sinap II and the entire Yellidoruk sequence) is rich in trachyte and shows a paleocurrent pattern that differs from that of the upper part of the Sinap sequence. It seems that the provenance area for trachyte was somewhat northeast of the Yellidoruk area, whereas that for the sediments in the basalt-rich upper part of the Sinap sequence was situated in the northwest of the area.

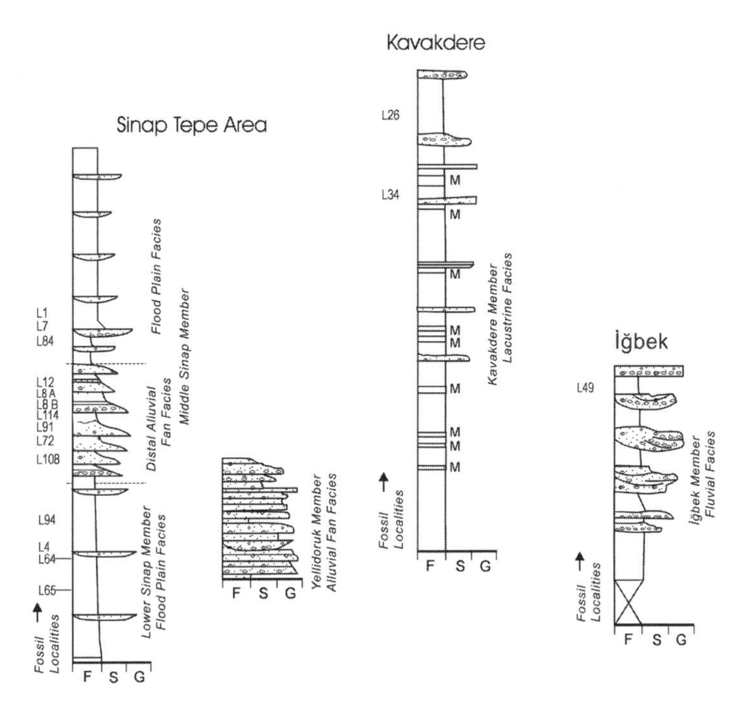

Figure 1.5. Paleoenvironmental interpretation of the Sinap Tepe area, Kavakdere, and Igbek sequences studied (i.e., Lower Sinap Member, Yellidoruk Member, Middle Sinap Member, Kavakdere Member and Igbek Member). The Sinap Tepe area includes a composite log of the Delikayincak section and Sinap sections I–VI (see fig 1.4.), whereas the Yellidoruk alluvial fan facies represents a separate, lateral facies to the former. The main fossil localities are also listed. M, lake marl.

Lithostratigraphy and Sedimentation in the Yellidoruk, Delikayincak Tepe, and Sinap Tepe Areas

In the Sinap Tepe area, sediments display a suite of facies that indicates deposition by channelized and nonchannelized alluvial systems. Paleontologic data from mudstones and conglomerates show that sediments were deposited in a terrestrial setting. However, the distribution and facies associations indicate considerable variations in sedimentation style and depositional environments in this terrestrial setting.

Based on the lithostratigraphic correlation of individual sections in the Sinap Tepe area (see fig. 1.4), the sedimentary sequence can be divided into four members based on lithofacies characteristics (fig. 1.5). Here we refer to these members as the Yellidoruk, lower Sinap, middle Sinap, and upper Sinap members.

Yellidoruk Member

Sediments in the Yellidoruk section are characterized by coarse-grained Gms-Gm facies and fine-grained facies with clasts (Fm+c facies). These facies are interpreted as debris flow and mudflow or cohesive debris-flow deposits, respectively, whereas the less frequent Gm-Gp-Gt facies represents migrating small stream deposits laid down during flood events and the Sm facies are interpreted as sheet flood deposits. These facies characteristics, together with the paleocurrent patterns, the rapid fining of conglomerates in a down-current direction, and the relative absence of fossil material suggest that sedimentation occurred in a proximal alluvial fan facies (fan head). The depocenter was located north or northeast of the Yellidoruk section. Sand petrography and pebble counts indicating a high percentage of trachyte show that most of the material was derived from a volcanic flow that is older than the basaltic lava dated at ~15–16 Ma. Note, however, that the stratigraphic position of the Yellidoruk sediments below the prominent paleosol marker horizons at the top of the Yellidoruk sections cannot be directly correlated to the rest of the Sinap or Delikayincak sections by stratigraphic means. Because the paleocurrent pattern and the high percentage of trachyte clasts are similar in the Yellidoruk and the lower parts of the Delikayincak and Sinap II and II sections, it is reasonable to assume that the Yellidoruk sediments correlate to these sections. This correlation is also independently supported by paleomagnetic reversal stratigraphy (see Kappelman et al., this volume).

Lower Sinap Member

The lower Sinap member is found in the Sinap II and the lower parts of the Sinap I, III, and Delikayincak sections.

This member is almost entirely composed of massive silty clay (Fm facies) with lag horizons (Gm) and rare intervals of massive, parallel, and ripple-bedded sand facies (Sm, Sh, Sr facies). This member displays features indicative of paleosols such as calcite nodules, layers, and veins, as well as mottling and color changes. The lower Sinap member represents pedogenically modified overbank deposits that were laid down in a flood plain.

Middle Sinap Member

The lower Sinap member passes gradually into the middle Sinap member, which is characterized in its lower part by numerous laterally extensive horizons of upward-fining channel fills (see the Sinap III, IV, and V sections in fig. 1.4). These channelized flows pass laterally and vertically into nonchannelized, upward-fining cycles of sheet flow deposits, which in turn pass into flood plain fines and channel deposits. The top part of this member (Sinap VI section in fig. 1.4) represents flood plain fines and small channel deposits laid down in the main river valley.

As a whole, the middle Sinap member most likely represents a facies change from small braided channels occurring on the lower fan segment to flood deposits occurring on a distal fan that was close to the adjacent flood plain of the trunk river. The distal fan sediments show increased interfingering with flood plain deposits in the down-current direction. Paleocurrent measurements suggest the depocenter for distal alluvial fan sediments was situated in the northwest, whereas fluvial sands in the trunk river valley were deposited from a northeasterly direction.

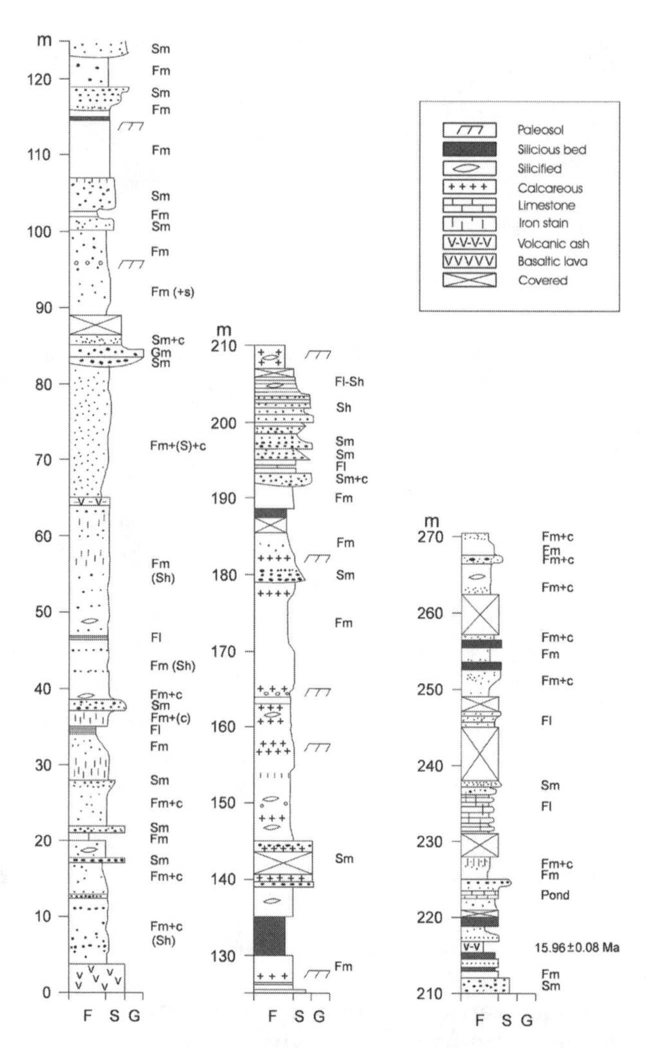

Figure 1.6. Sediment log of the Beycedere section. For facies codes see caption to figure 1.4.

Upper Sinap Member (post-Miocene)

The upper Sinap member overlies the middle Sinap member with an erosional and unconformable contact. The sediments of this member are coarse and represent mainly crudely stratified gravel facies with relatively thin intervals of planar cross-bedded gravel and massive mudstone. Although the whole of the upper Sinap member (estimated to be >20 m thick) is not well exposed, the poorly sorted, massive pebble-to-boulder conglomerates are thought to have been deposited from high-concentration floods, whereas the moderately sorted pebble-to-conglomerate and cross-bedded gravels represent longitudinal bars, side bars, and channel fill deposits. It is assumed that the upper Sinap member represents post-Miocene accumulation that took place considerably later than that of the older members; therefore we do not discuss it here.

Sedimentation and Lithostratigraphy of the Beycedere-Kavakdere Area

The Beycedere-Kavakdere area is situated northwest of FZ 2 (see fig. 1.2). In this area, a conformable sequence above the basalt flow at the base can be grouped into an older Beycedere member (280 m) and a younger Kavakdere member (180 m) (see figs 1.3 and 1.5). Although the sequence is conformable, there seems to be a considerable hiatus between these two members that is manifested by a flat iron surface of silicified sediments. Dates of 15.88 ± 0.07 and 15.96 ± 0.08 Ma obtained from two volcanic ash layers in the upper part of the Beycedere section (see fig. 1.6) and age estimates based on paleomagnetic stratigraphy (see Kappelman et al., this volume) suggest that this hiatus represents a time period of >5 Ma. In the Beycedere section, only a few indeterminable fossil bones have been discovered, whereas in the Kavakdere area, several very rich fossil localities occur in the upper part of the sequence.

Beycedere Member

Sedimentary units in the Beycedere section (fig. 1.6) are slightly silicified and primary sedimentary structures are virtually absent. Here we describe only the main sedimentary characteristics of a relatively poor exposure. The entire sedimentary sequence is dominated by three main facies above the basal basaltic lava flow: channel fill and bar deposits, overbank fines and sheet flood sands, and pond sediments.

Channel Fill and Bar Deposits

The coarse-grained facies is uncommon in the Beycedere section. Only one stratified gravel unit (Gm facies), 1.5 m thick at the 83-m level was observed in the entire sequence but there are numerous fluvial channel sand units (Sh and Sm facies) at the 116-m, 140-m, and 193-m levels. Although it was not possible to determine the internal structure of many of these sand units, their basal and upper contacts together with their shape and lateral extent suggest that these sands represent channel fills and longitudinal and transverse bars. We did not find facies characteristics typical for point bars (see Walker and Cant 1984; Miall 1995).

Overbank Fines and Sheet Flood Sands

The sediments in the Beycedere section are mainly composed of massive fines (Fm facies) that are often silicified to various degrees and display features typical of poorly developed paleosols such as carbonate nodules, veins, and mottling. Relatively thin beds of massive sand (Sm facies) also occur within these fines. Here we interpret thin tabular and wedge-shaped massive sands as sheet flood sands and pedogenically altered massive fines as overbank deposit.

Pond Sediments

There are a few laminated silt and clay units (0.8–5.0 m thick) and one distinct, 1-m thick marl bed that occur in the upper part of the section (Fl facies and pond in fig. 1.6). These laminated fine sediments extend at least several tens of m laterally. We suggest that these sediments were deposited in small lakes or ponds that existed in the flood plain similar to those described by Lunkka et al. (1998) from Çalta. Based on lithologic characteristics of the Beycedere section, we conclude that the sediments were laid down in a fluvial environment where ephemeral ponds or small lakes existed.

Kavakdere Member

The Kavakdere sections are located in the western part of the study area (see fig. 1.1). One 182-m thick section was logged in the northern part of the Kavakdere valley. The other site logged is situated in the southern part of the valley (fig. 1.7). Based on these logs and vertical tracing of individual marker horizons, it became clear that the overall facies succession in the sequence is twofold. The lower part of the sequence is dominated by silicified mudstone and marls, whereas in the upper part, mudstone beds are frequently interbedded with conglomerates and sandstones. Two main lithofacies dominate the sequence (see fig. 1.7).

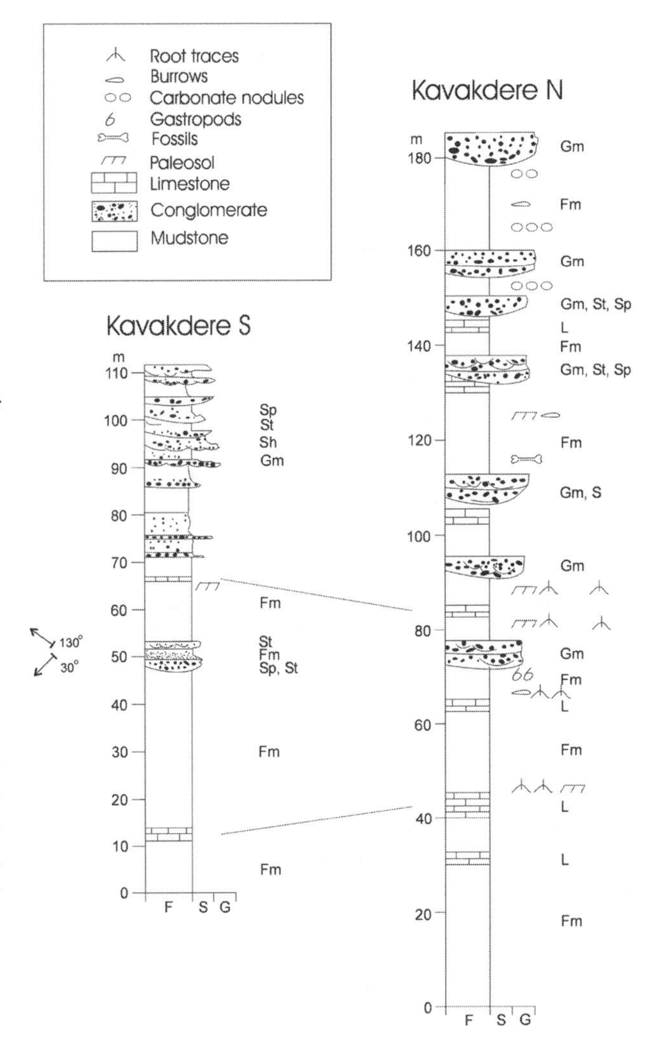

Figure 1.7. Sediment logs of the Kavakdere sections. Limestone unit is used as a marker horizon. For facies codes see caption to figure 1.4. L, lake or pond sediment.

Channel Fill and Bar Deposits

Stratified gravel and sand facies (Gm, St, Sp, and Sh facies) occur in the upper part of the sequence (fig. 1.7). Coarser material consists of pebble-to-cobble–sized gravel that normally fines upward into cross-bedded and parallel-bedded sands and finally into massive mudstones. Conglomerates are composed of clasts of basalt, opal-replaced smectite mudstone, and marls. We think that the coarse units have been deposited as channel fills and bars under fluvial activity.

Lake and Flood Basin Sediments

The lower part of the sequence is dominated almost solely by massive silicified mudstone and marl (limestone) units (Fm and L facies). In the upper part of the sections, mudstones also occur but are frequently interbedded with gravel and sand layers described above. Smectite-rich mudstone alternates in color between brown and white in 1–3-m intervals, and burrow molds and root traces are sparsely distributed throughout the mudstone units. Carbonate nodules and other features indicative of paleosols are particularly common in the upper part of the sections.

Fine-grained, internally bedded or laminated marl (limestone) averaging 1 m in thickness is well exposed as ledge-forming units in both sections (fig. 1.7). The weathered-surface textures of these carbonate units range from laminar to pitted and rich zones of gastropod molds and casts were frequently found between 70 and 144 m.

The lower part of the sequence is composed almost entirely of silt and clay with marl horizons. The presence of these units together with the lack of features typical of paleosols suggest that the fine sediments in the lower part of the sequence accumulated in a broad lake basin. However, in the upper part of the sequence (at ~70-m level or above in the Kavakdere north and south sections), massive mudstones with paleosol features such as root traces, burrows, and carbonate nodules suggest that mudstone units were laid down in a flood plain, whereas the marl beds indicate more restricted pond deposits in the flood basin.

As a whole, the sediments in the Kavakdere sections appear to have been laid down within a broad lake basin that was subsequently filled by fluviatile sediments. The basin was characterized by periods of low terrigenous sediment influx (which is when marl beds formed) that alternated with periods of increased terrigenous influx. The presence of paleosols and marl beds particularly in the upper part of the sequence (see the Kavakdere north section in fig. 1.7) indicates that lakes in the basin fluctuated in volume and extent. The occurrence of high-spired gastropods, although not diagnostic alone, also supports the idea of a lacustrine setting for the basal sediments, whereas lacustrine conditions alternated with fluvial ones in the upper part of the sequence.

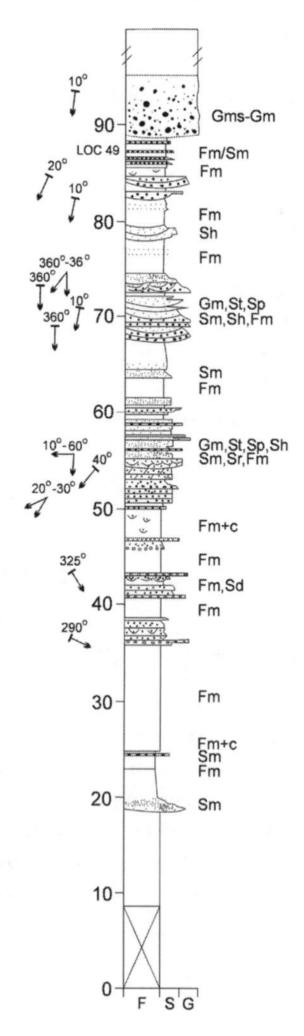

Figure 1.8. Sediment log of the Igbek section. Fossil locality 49 is at ~90-m level. For facies codes see the caption to figure 1.4.

Igbek Member

The exposures at Igbek are situated ~12 km northwest of Çubuk (see fig. 1.1), where we were able to record a ~94-m long section. Fossil locality 49 is situated in the upper part of the section at the ~90-m level (see fig 1.8). Because the top part of the sequence is poorly exposed and is mainly composed of cobble-to-boulder gravel, we do not discuss it here. Most of this unit seems to be post-Miocene.

At Igbek there are basically three major facies that characterize the sediment succession. The lower part of the logged section (<37 m) is dominated by mudstone with occasional coarser units. Farther up, beneath the top conglomerate, there are three distinct sand and gravel accumulations that are interbedded with finer sediments.

Lag Horizons in Channels

A crudely bedded, few-dm thick gravel horizon (Gm facies) occurs with an erosional lower contact at the basal parts of the channel features. These gravel units form the base of upward-fining sequences that represent traction deposits in channels and are related to sand facies described below.

Channel Fill and Bar Deposits

A typical succession in the Igbek section consists of gravel lag horizons overlain by cross-bedded sand as well as horizontal and ripple-bedded sand (Sp, St, Sh, and Sr facies). The lower contact of these units is erosional and their form is concave. Individual sediment successions fine upward and represent traction load in fluvial channels representing bars as well as channel fills.

Levee and/or Crevasse Splay Sands

Massive sands (Sm facies) that fine upward into mudstones are rather common in the Igbek sequence. These massive sands commonly occur at the top of channel and bar sequences but can also be unrelated to those units within vertical sequences. We think these sands were deposited either as levee or crevasse splay sands in which primary sedimentary structures were subsequently destroyed as a result of pedogenic alterations.

Massive structureless sand also occurs within the stratified sand bodies. These normally lenticular-shaped bodies were most likely laid down from suspension during flood events (cf. McCabe 1977).

Flood Plain Mudstones

Massive fines (Fm facies) with pedogenic features are most common in the lower part of the section but are also frequently present between the sediments that were deposited in channels. We think that these massive fines also represent flood plain deposits.

The sedimentary succession at Igbek was deposited in a fluvial environment and shows characteristic elements of channel, crevasse splay, levee, and flood plain deposits. Several paleocurrent measurements were recorded from current-induced structures and particularly from the channel forms and trough-trends of through cross-bedded units in channel deposits. These structures indicate that a paleocurrent flow from a northerly direction persisted throughout the time represented by the Igbek sediments.

Interpretation of the Paleoenvironments of the Sinap Formation in the Kazan Area

Terrestrial conditions existed in Central Anatolia, including the study area, for the duration of the Miocene (cf. Luttig and Steffens 1976 and references therein). The tectonic development of Anatolia and Asia Minor as a whole is related to movements associated with the North Anatolian transform fault. These movements, dating to at least the beginning of the middle Miocene, caused normal faulting events and the formation of intermontane basins that subsided as volcanic activity produced by crustal thinning intensified in Central Anatolia during the late Miocene. It was within this general tectonic setting that terrestrial Neogene sediments in the Kazan area were deposited.

All the evidence obtained from sediments in the study area clearly indicates that the landscape underwent drastic changes throughout the Neogene as a result of continuous tectonic activity and volcanism. Two separate lava flows and one volcanic vent have been observed in the area. The youngest basaltic lava flow that was laid down (~16 Ma) forms the boundary on which the sediments of the Sinap Formation were deposited, and another lava flow (trachyte) predates this basaltic flow (see fig. 1.3). Between these two lava flows there is a volcaniclastic Pazar Formation in the upper part of which a fossil site, Locality 125, was found. Sediments at Locality 125 include lacustrine pond sediments and current-bedded volcaniclastic and ash layers indicating a flood plain environment.

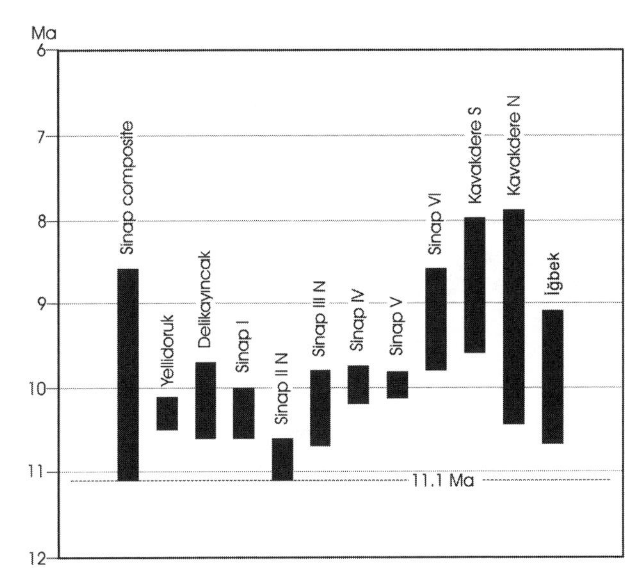

Figure 1.9. Correlation scheme of individual sediment sequences of the Sinap Formation based on magnetostratigraphic and lithostratigraphic results (excluding the Beycedere member, which was deposited ~16 Ma; see fig. 1.5).

The oldest sediments of the Sinap Formation that rest upon the youngest basaltic lava are those in the Beycedere area. This sequence was deposited from a northeasterly direction in a flood plain where shallow, ephemeral lakes existed immediately after the basaltic flow event. Beycedere sediments also include pyroclastic material related to volcanic eruptions that occurred ~15–16 Ma.

Following the chronology of Kappelman et al. (this volume), fossiliferous sediments investigated in the Kazan area that belong to the Sinap Formation (Sinap Tepe, Yellidoruk, Delikayincak Tepe, Igbek, and Kavakdere sections) cover the time span ~11.1–7.9 Ma (fig. 1.9); the interpolated ages of the fossil localities in this formation range from ~10.9–8.1 Ma. Although individual sedimentary units that were investigated represent only short time slices, it is possible to obtain a relatively clear picture of two major changes in the landscape evolution for the Kazan area during this time period.

The oldest sediments of the Sinap Formation that follow the deposition of the Beycedere member occur in the Sinap Tepe area (the Sinap II section and the lower parts of the Delikayincak and Sinap I and III sections) and the Igbek area (see fig. 1.10). The fine sediments in the oldest units of the Sinap Tepe and Igbek areas accumulated in a flood plain environment (fig. 1.10A). Rare sand horizons in the sedimentary sequence are likely to represent crevasse splay sands, with the active channel pattern being some distance to the south of the Sinap Tepe area. Well-developed paleosols and lag horizons characterize these flood plain deposits. Lag horizons (deflation surfaces) as well as some well-sorted silt beds may indicate increased aeolian activity at times that are probably related to semiarid climatic conditions; this conclusion is also suggested by some elements in small mammal fauna at fossil locality 64 (see Sen,

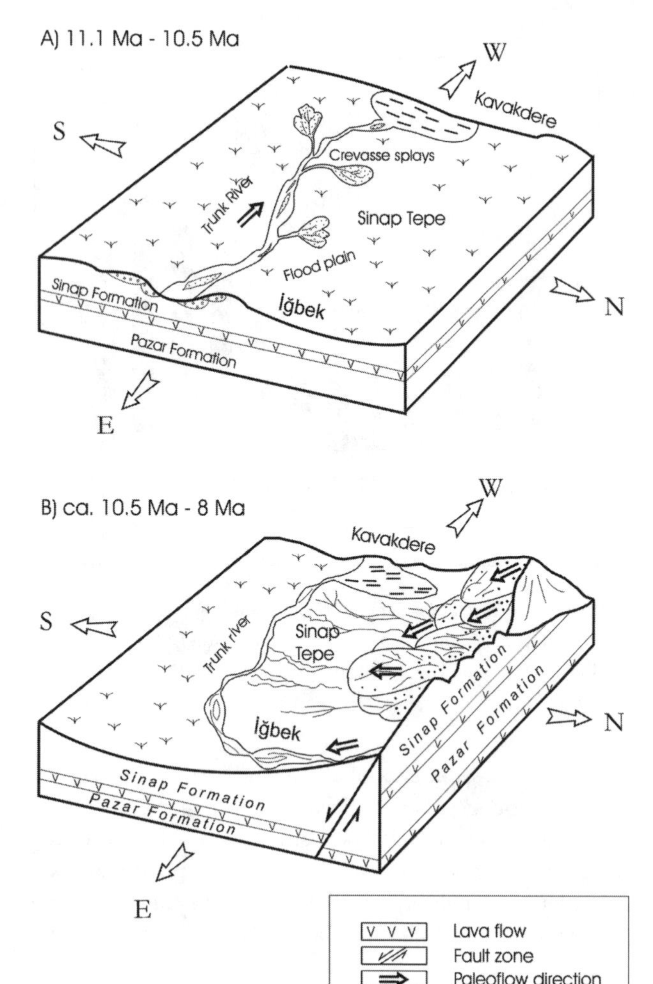

Figure 1.10. An idealized model of the landscape evolution in the study area. **(A)** Sediments investigated in the Igbek-Sinap Tepe-Kavakdere area indicate that the flood plain fines (with occasional crevasse splay sands and lake sediments) were deposited ~11.1–10.5 Ma. **(B)** The land surface profile changed as a result of tectonic activity ~10.5 Ma; subsequent deposition took place in alluvial fan, fluvial channel, flood plain, and lacustrine environments.

chapter 5, this volume). Fine sediments in the lower parts of Sinap sequence and Igbek section are roughly contemporaneous and may indicate that both of these areas belonged to the same drainage basin. It is also plausible that a lake basin already existed in the Kavakdere area and the lake marls in the upper part of the Beycedere section were also initially laid down into this basin.

A marked change in sedimentation and palaeogeography of the area must have taken place ~10.5 Ma. The sediments in the Yellidoruk sequence clearly indicate accumulation by gravity flow mechanisms from the northerly direction. As stated above, these gravity flow deposits accumulated in an alluvial fan perhaps close to the fan apex. Farther to the west (Delikayicak section and Sinap I, III, and IV sections), the gravel, sand, and silt successions represent sediments laid down in a lower fan segment of a separate alluvial fan with its depocenter situated in the northwest. At roughly the same time (10.5 Ma), coarse fluvial sediments started to accumulate in the Igbek area from the north/northeast (see figs. 1.6 and 1.10B). All of these characteristics of the style and pattern of sedimentation suggest that the land surface gradient became steeper as a result of faulting somewhere to the north of but close to the study area. Such tectonic movements (i.e., faulting events) are the primary reason for changes in sedimentation around this time. The Igbek, Sinap Tepe, and Kavakdere areas are thought to be parts of the same large sedimentary basin, because changes in sedimentation style that followed the faulting event at ~10.5 Ma are found in all of these areas.

Fluvial sedimentation continued in the Igbek area where channel deposits and related overbank mudstones were laid down. In the Sinap Tepe area, the Sinap VI section (see fig. 1.4) shows some interfingering with distal alluvial fan facies, but the upper part of the section displays normal flood plain sediments with tributary channel gravels, crevasse splay sands, and frequent paleosol horizons, suggesting that the active trunk river was situated farther south. In the Kavakdere area, the water level of a lacustrine basin fluctuated in space and time, and eventually the basin began to fill with fluviatile sediments. After this infilling, fluvial conditions were established in the Kavakdere area that lasted well into the Pliocene, as described in Lunkka et al. (1998) for the Çalta area adjacent to the Kavakdere sequence.

Most of the fossil localities occur in the Sinap Tepe and Delikayincak Tepe areas (see figs 1.2 and 1.5). In contrast, fossils are very sparse in the Beycedere and Yellidoruk sequences and the lower parts of the Igbek and Kavakdere sequences. The densest occurrence of fossils (e.g., fossil localities 108, 72, 91, 114, 8B, 8A, and 12) were found in the boundary between the distal alluvial fan and the proximal flood plain subenvironments. We assume that deposition in this setting was relatively continuous and thus more favorable for the burial and preservation of bone. In the flood plain environment, fossil accumulations (e.g., fossil localities 65, 64, 4, 94, 26, 34) are frequent but do not occur in as dense intervals as seen in the previous environmental setting. The virtual absence of fossil accumulations in the Yellidoruk sequence and the channel-gravel dominated parts of the Igbek section are most probably the result of the destructive nature of these high-energy environments.

Conclusion

The results from geologic mapping and sedimentologic investigations in the Sinap study area point to a complex history of depositional dynamics that was strongly influenced by regional tectonics. Numerous depositional settings are documented from across the study area and include alluvial fan, channel and stream deposits, lake, and flood plain deposits. The oldest sediments date from the early middle Miocene and document the presence of lacustrine as well as flood plain environments. After ~10.5 Ma, the landscape was dramatically altered in response to increased tectonic activity and reflects the effect of a steeper land surface gradient. The earlier preponderance of pedogenically altered flood plain sediments with only occasional crevasse splay deposits in the area of Sinap Tepe gives way to flood plain sediments that include a higher proportion of small scale tributary channel gravels, crevasse splay sands, and paleosol horizons. Other areas such as that at Kavakdere indicate the presence of a lake that fluctuated in extent and depth and eventually filled with fluviatile sediments. Fossil localities are most frequent in depositional environments at the interface between distal alluvial fan and proximal flood plain subenvironments. This low-energy setting probably received a nearly continuous influx of sediments and therefore provided an environment that was more favorable for the burial and preservation of bone.

Acknowledgments

The Sinap Project (1989–1995) enjoyed the efforts of a large number of colleagues and the results presented here reflect the collective efforts of the entire team. We thank Professor Dr. B. Alpagut and Dr. L. Martin, with whom we initiated this project, and İ. Temizsoy, director of the Museum of Anatolian Civilizations, who held the excavation permits and greatly facilitated our efforts in the field. The field and laboratory work was funded by the following grants: the Academy of Finland, the U.S. National Science Foundation (EAR 9304302), the L.S.B. Leakey Foundation, and assistance from Ankara University. Equipment loans from Motorola and software grants from Autodesk, Inc., are greatly appreciated. The research was supported by the General Directorate of Antiquities, T. C. Ministry of Culture and Tourism; we thank the Ministry for its seven years of support. The manuscript was greatly improved by insightful comments from J. C. Barry and M. O. Woodburne, and we sincerely thank them for their assistance.

Literature Cited

Angelier, R., Dumont, J. J. F., Karamandersei, H., Poisson, A., Sinsek, S., and Uys, S., 1981, Analyses of fault mechanisms and expansion of southwestern Anatolia since the Late Miocene: Tectonophysics, v. 75, pp. T1–T9.

Bagnold, R. A., 1956, The flow of cohesionless grains in fluids: Philosophical Transactions of the Royal Society of London, ser. A, v. 249, pp. 235–297.

De Celles, P. G., Tolson, R. G., Graham, S. A., Smith, G. A., Ingersol, R. V., White, J., Schmidt, C. J., Rice, R., Moxon, I., Lemke, L., Handschy, J. W., Follo, M. F., Edwards, D. P., Cavazza, W., Caldwell, M., and Bargar, E., 1987, Laramide thrust-generated alluvial-fan sedimentation, Sphinx Conglomerate, southwestern Montana: American Association of Petroleum Geologists Bulletin, v. 71, pp. 135–155.

De Celles, P. G., Grey, M. B., Ridgway, K. D., Cole, R. B., Pivnik, D. A., Pequera, N., and Srivastava, P., 1991, Controls on synorogenic alluvial-fan architecture, Beartooth Conglomerate (Paleocene), Wyoming and Montana: Sedimentology, v. 38, pp. 567–590.

Erol, O., 1981, Neotectonic and geomorphological evolution of Turkey: Zeitschrift für Geomorphologie, Neue Folge, Supplement Band: v. 40, pp. 193–211.

Hubert, J. F., and Filipov, A. J., 1989, Debris-flow deposits in alluvial fans on the western flank of the White Mountains, Owens Valley, California, USA: Sedimentary Geology, v. 61, pp. 177–205.

Inci, U., 1991, Miocene alluvial fan-alkaline playa lignite-trona bearing deposits from an inverted basin in Anatolia: Sedimentology and tectonic controls of deposition: Sedimentary Geology, v. 71, pp. 73–97.

Kappelman, J., Sen, S., Fortelius, M., Duncan, A., Alpagut, B., Crabaugh, J., Gentry, A., Lunkka, J. P., McDowell, F., Solounias, N., Viranta, S., and Werdelin, L., 1996, Chronology and biostratigraphy of the Miocene Sinap Formation of Central Turkey, *in* Bernor, R. L., Fahlbusch, V., and Mittman, H. W., eds., The evolution of western Eurasian Neogene mammal faunas: New York, Columbia University Press, pp. 78–95.

Koçyiğit, A., 1991, An example of an accretionary basin from northern Central Anatolia; its implications for the history of subduction of Neo-Tethys in Turkey: Geological Society of America Bulletin, v. 103, pp. 22–36.

Lowe, R. D., 1979, Sediment gravity flows: Their classification and some problems of application to natural flows and deposits, *in* Doyle, J. D., and Pilkey, O. H., eds., Geology of continental slopes: Tulsa, Oklahoma, Special Publication of the Society of the Economical Paleontologists and Mineralogists, v. 27, pp. 72–82.

Lowe, R. D., 1982, Sediment gravity flows: II. Depositional models with special reference to the deposits of high-density turbidity currents: Journal of Sedimentary Petrology, v. 52, pp. 279–297.

Lunkka, J.-P., Kappelman, J., Ekart, D., Fortelius, M., McDowell, F., Sen S., and Alpagut, B., 1995, Sedimentology and chronology of the vertebrate bearing Miocene Sinap Formation, Central Turkey: Geological Society of America Abstracts with Programs, v. 27, no. 6, p. A278.

Lunkka, J.-P., Kappelman, J., Ekart, D., and Sen, S., 1998, The Pliocene vertebrate locality of Çalta, Ankara, Turkey: 1. Sedimentation and lithostratigraphy: Geodiversitas, v. 20, pp. 329–338.

Lunkka, J.-P., Fortelius, M., Kappelman, J., and Sen, S., 1999, Chronology and mammal faunas of the Miocene Sinap Formation, Turkey, *in* Agusti, J., Rook, L., and Andrews, P., eds., Climate and environmental change in the Neogene of Europe: Cambridge, UK, Cambridge University Press, pp. 238–264.

Lüttig, G., and Steffens, P., 1976, Explanatory notes for the Paleogeographic atlas of Turkey from the Oligocene to the Pleistocene: Hannover, Germany, Bundesanstalt für Geowissenschaften und Rohstoffe, 64 pp.

McCabe, P. J., 1977, Deep distributary channels and giant bedforms in the Upper Carboniferous of the Central Pennines, northern England: Sedimentology, v. 24, pp. 271–290.

Miall, A. D., 1978, Lithofacies types and vertical profile models in braided river deposits: A summary, *in* Miall, A. D., ed., Fluvial sedimentology, Calgary, AB, Canada, Canadian Society of Petroleum Geology Memoir, pp. 597–604.

Miall, A. D., 1985, Architectural-element analysis: A new method of facies analysis applied to fluvial deposits: Review: Earth Science, v. 22, pp. 261–308.

Öngür, T., 1976, Kızılcahamam, Camlidere, Celtikci ve Kazan dolayinin jeoloji durumu ve jeotermal enerji olanakrari, Unpublished Report, MTA, Ankara.

Ozansoy, F., 1957, Faunes de mammifères du Tertiaire de Turquie et leurs revisions stratigraphiques: Bulletin of the Mineral Research and Exploration Institute of Turkey (Foreign Ed.), v. 49, pp. 29–48.

Ozansoy, F., 1965, Étude des gisements continentaux et de mammifères du Cénozoique de Turquie: Mémoires de la Société Géologique de France (nouvelle série), v. 44, pp. 1–92.

Sen, S., 1991, Stratigraphie, faunes de mammifères et magnétostratigraphie du Néogène de Sinap Tepe, province d'Ankara, Turquie: Bulletin du Museum National d'Histoire Naturelle, Paris, 4e séries, v. 12, pp. 243–277.

Sengör, A. M. C., and Yilmaz, Y., 1981, Tethyan evolution of Turkey: A plate tectonic approach: Tectonophysics v. 75, pp. 181–241.

Tucker, M. E., 1982, The field description of sedimentary rocks, The Geological Society of London, Handbook Series: Milton Keynes, UK, Open University Press. 112 pp.

Walker, R. G., and Cant, D. J., 1984, Sandy fluvial systems, *in* Walker, R. G., ed., Facies models: Waterloo, ON, Canada, Geological Association of Canada, pp. 71–89.

2

Chronology

J. Kappelman, A. Duncan, M. Feseha, J.-P. Lunkka, D. Ekart, F. McDowell, T. M. Ryan, and C. C. Swisher III

The Sinap Formation of central Turkey is well known for its record of Neogene mammals that date from the middle and late Miocene. These faunas have been studied for nearly 50 years (Ozansoy 1957, 1965), with more intensive study taking place during the early 1990s (Sen, Introduction, this volume; Temizsoy, Preface, this volume). A dated framework is essential for evaluating the changing composition of the Sinap faunas through time and for allowing the faunas to be compared with faunas from other fossil localities. Only recently has a program of paleomagnetic and radioisotopic dating been undertaken to provide a dated chronology for the formation (Sen 1991; Kappelman et al. 1996; Lunkka et al. 1999). The outcrops of Sinap Formation near the village of Yassiören, located ~37 km northwest of Ankara in central Anatolia, have been most intensively studied. Other outcrops found all along the Kazan valley (Lunkka et al. 1999, this volume) and as far east as the village of Igbek also contain very rich fossil occurrences (fig. 2.1). The aim of the dating program is to combine these various fossil localities into an integrated chronology.

Previous Studies in Paleomagnetic Reversal Stratigraphy

Ozansoy (1957, 1965) named three members of the Sinap Formation. He noted a conformable contact between the lower and the middle members and described an erosional unconformity that separates the upper and the middle members. Later work by Öngür (1976) recognized an additional older unit, the Pazar Formation, which is separated from the overlying Sinap Formation by a volcanic unit (basaltic flow).

The first study that attempted to provide a chronology for the Sinap Formation based on paleomagnetic reversal stratigraphy was that of Sen (1991). His study demonstrated that a stable primary magnetic signal is carried in the Sinap sediments and that these rocks are ideal for a program of paleomagnetic reversal stratigraphy. Sen studied a short section of ~41 m exposed along the southeastern flank of Sinap Tepe, a prominent butte that lies to the north of Yassiören and is nearly at the center of the fossil-bearing area (Sen 1991, fig. 1). Sen identified a sequence of normal (~30 m thick) and reversed (~11 m thick) polarity intervals that occurs near the upper part of the middle Sinap member. Sen tentatively suggested that this sequence preceded the long normal polarity event of Anomaly 5, which, under the geomagnetic polarity time scale (GPTS) of Harland et al. (1982), would place the section at >10.30 Ma (or at >10.949 Ma by correlation with the beginning of Chron 5 under the GPTS of Cande and Kent 1995).

Later work by the Sinap Project built on the results of Sen (1991) and recognized that longer stratigraphic sections that sampled more time would be required to establish secure correlations with the GPTS. A detailed program of geologic mapping and stratigraphic section logging was carried out in tandem with paleomagnetic sampling. This work documented the presence of geologic sections of sufficient thickness so that paleomagnetic reversal stratigraphy could be used to date the formation and produce correlations between isolated rock outcroppings. Preliminary results from these studies are given in Kappelman et al. (1996) and Lunkka et al. (1999). We provide the results from additional sampling efforts here. We use the GPTS of Cande and Kent (1995) (hereafter referred to as CK95) throughout this chapter. Age estimates are reported to three decimal places to facilitate detailed comparisons of the placements of fossil localities in the stratigraphic sections.

Sampling Methods

The sediments of the Sinap Formation were deposited in a variety of environments that included distal-to-proximal floodplains as well as lakes (Lunkka et al. 1999, this volume).

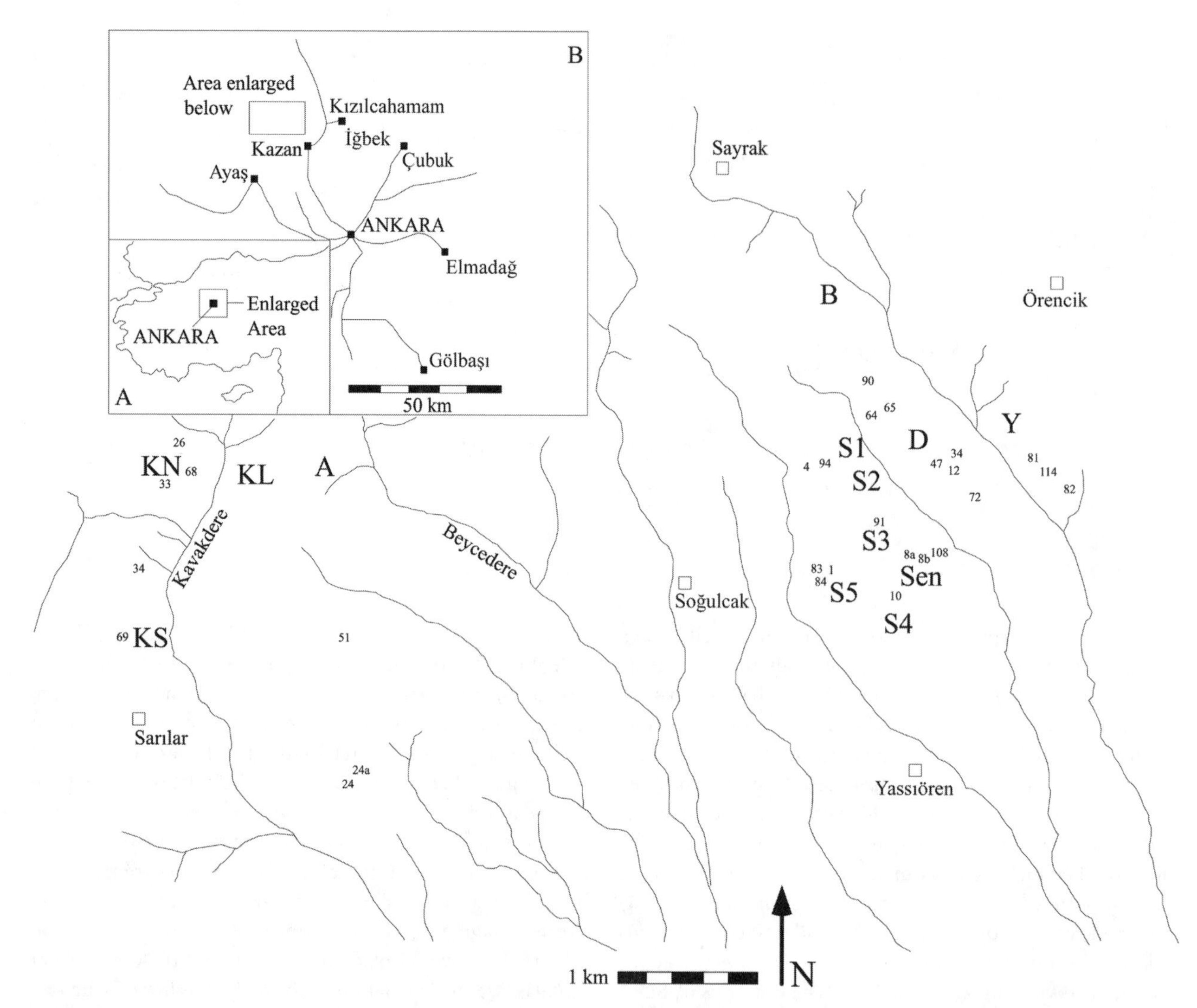

Figure 2.1. Map of the region containing the Sinap Formation in Turkey. The insets show the location of the study region within Turkey (**A**) and within the Ankara region (**B**). The detailed map shows the locations of the stratigraphic sections, select fossil localities, sites of volcanic rocks that have yielded radioisotopic dates, and major villages and stream drainages. A, location of ash beds B-50 and B-51; B, location of Karakaya basalt; D, the Delikayincak section; KL, Kavakdere lower section; KN, Kavakdere north section; KS, Kavakdere south section; S1, Sinap northeast 1 section; S2, Sinap northeast 2 section; S3, Sinap central east section; S4, Sinap south central section; S5, Sinap southwest section; Sen, section of Sen (1991); Y, Yellidoruk section. Small numbers indicate fossil localities.

Numerous measured sections that could be traced to one another using marker beds were combined to produce three composite stratigraphic sections: the Yellidoruk, Delikayincak, and Sinap sections located north and northeast of Yassiören; the Kavakdere and Beycedere sections located north and northeast of the village of Sarilar; and the Igbek section located ~20 km northeast of Kazan and ~12 km northwest of the village of Çubuk (fig. 2.1). Sections were measured with either a 1.6-m Jacob's staff and sliding mount Abney level or a Zeiss Total Station, with subsequent mapping completed in AutoCAD R.13 (Autodesk; Sausalito, California) on a Windows workstation. The sediments of Yellidoruk Tepe, Delikayincak Tepe, and Sinap Tepe are exposed across an area of ~10 km² and represent different portions of the same large-scale depositional system. We think it possible that the Kavakdere and Beycedere sediments that outcrop to the west represent a more distal portion of this same basin, but a major fault separates this area from that of Sinap Tepe (Lunkka et al. 1999, this volume) and precludes any direct stratigraphic correlation. The Igbek section probably documents a completely different depositional basin.

Rock samples for paleomagnetic analyses were collected from unweathered fine-grained sediments (claystones, siltstones, and sandy siltstones). In some cases extensive surface weathering necessitated digging into the outcrop about 1 m to expose unweathered sediments. Sampling avoided obviously bioturbated units and mature paleosols. We photographed the location of each sample site with a Polaroid camera to facilitate future resampling, mapping, or logging efforts. A minimum of three samples per site were collected and oriented on the outcrop using a hand rasp and Brunton compass. In many cases a Pomeroy water-cooled diamond drill (ASC Scientific; Cardiff, California) was used

Table 2.1. Paleomagnetic Site Spacings

Spacing (m)	Kavakdere Lower	Kavakdere South Upper	Kavakdere North	Igbek	Yellidoruk Tepe	Delikayincak Tepe	Sinap Tepe	All Sections[1]
Mean	7.71	1.67	2.64	1.65	3.51	2.20	1.73	2.01
S.D.	7.59	1.18	1.56	1.37	1.86	1.45	1.17	1.44
Minimum	0.4	0.1	0.3	0.2	0.6	0.1	0.1	0.1
Maximum	29.6	7.5	6.7	6.2	8.5	7.6	5.6	8.5

[1]Does not include the Kavakdere lower section.

to collect oriented samples from more indurated sediments. We collected 1367 samples from 467 sites. The mean site spacing is 2.01 m (standard deviation [SD], 1.44 m; range, 0.1–8.5 m); this value varied slightly among the different sections (table 2.1). The spacing is greater at the base of the Kavakdere section because many of these beds are silicified (see below).

We used a cut-off saw with a masonry blade and stationary vertical platter disc grinder to mill the hand-rasped samples into cubes 2.5 cm on each side, and a diamond cut-off saw was used to trim the 2.5-cm diameter drilled cores to 2.5-cm lengths. Measurements were made with an SCT superconducting magnetometer (2G Enterprises; Pacific Grove, California) housed in a magnetically shielded room at the University of Texas at Austin.

Rock Magnetic Studies and Demagnetization Protocol

Studies of the magnetic properties of the Sinap sediments including isothermal remanent magnetism (IRM) are reported in Sen (1991), Kappelman et al. (1996), and Feseha (1996). All of these studies agree in demonstrating a dominance of low-coercivity grains from 0–0.1 T that indicates the presence of magnetite or maghemite, with a minor contribution of higher coercivity grains.

The majority of samples collected during the project were subjected to stepwise thermal demagnetization using a Schonstedt Thermal Demagnetizer TSD-1 (Schonstedt Instrument; Reston, Virginia). This protocol was used in earlier studies (Sen 1991; Kappelman et al. 1996; Feseha 1996) and was shown to be effective in removing secondary magnetic components and isolating a stable magnetic vector. This demagnetization protocol generally included 6–12 steps from 0–650 °C. The results for two samples are shown in figure 2.2A, D. Given the IRM studies noted above and the results of the thermal demagnetization treatments (showing magnetite or maghemite to be the primary magnetic carrier), we also subjected some samples to alternating field (AF) demagnetization using a Schonstedt AF demagnetizer GSD-1. Samples were first measured for their natural remanent magnetism (NRM) and then subjected to a 150 °C or 200 °C blanket treatment to remove any secondary magnetic components. Stepwise AF demagnetization treatments were generally successful in isolating a stable component with a nearly linear decay to the origin (fig. 2.2B), but in some cases the origin was not reached (fig. 2.2C, E, F).

Site Statistics

We used a principal component analysis (PCA) (Kirschvink 1980) to calculate a best-fit line for the magnetic vector for each sample from every site by using the PCA routine in PaleoMagCalc© (Crow, Gose, and Kappelman; Austin, Texas) with data corrected for bedding dip. In many cases the low-temperature steps isolated a normal overprint (fig. 2.2D) that was not included in the analysis. Values were accepted as statistically significant if the maximum angular deviation (MAD) was <15°. In cases where the MAD value exceeded 15° but the polarity of the sample was not in question, a mean vector was calculated using the various demagnetization steps with Fisher's (1953) statistics, and statistical significance was evaluated with Watson's (1956) test for randomness. Site means and virtual geomagnetic pole latitudes were then calculated using Fisher statistics for those sites with three or more statistically significant samples and evaluated with the Watson criterion.

Of the 466 sites with three or more samples, 336 sites (72.0%) satisfied the Watson criterion and were classified as class A. Forty-nine sites (10.5%) did not meet this criterion. It was not possible to calculate a mean site vector for 44 sites (9.4%) because these sites had only two samples, but in most cases, both samples agreed closely on the mean polarity for the site. Both of the latter groups were classified as class B sites and are plotted in the figures presenting our results with different symbols. It was not possible to calculate a mean site vector for 37 sites (7.9%) because there was no agreement among the three samples; we excluded these sites from further analysis.

Stereoplots of the declinations and inclinations for the statistically significant class A sites after thermal or AF demagnetization treatments are shown in figure 2.3. These data illustrate the predominance of normal polarity sites and show a general division into normal and reversed populations (fig. 2.3) that becomes more apparent when all of the data are combined into a single stereoplot (fig. 2.4A). These two populations are roughly antipodal, but this

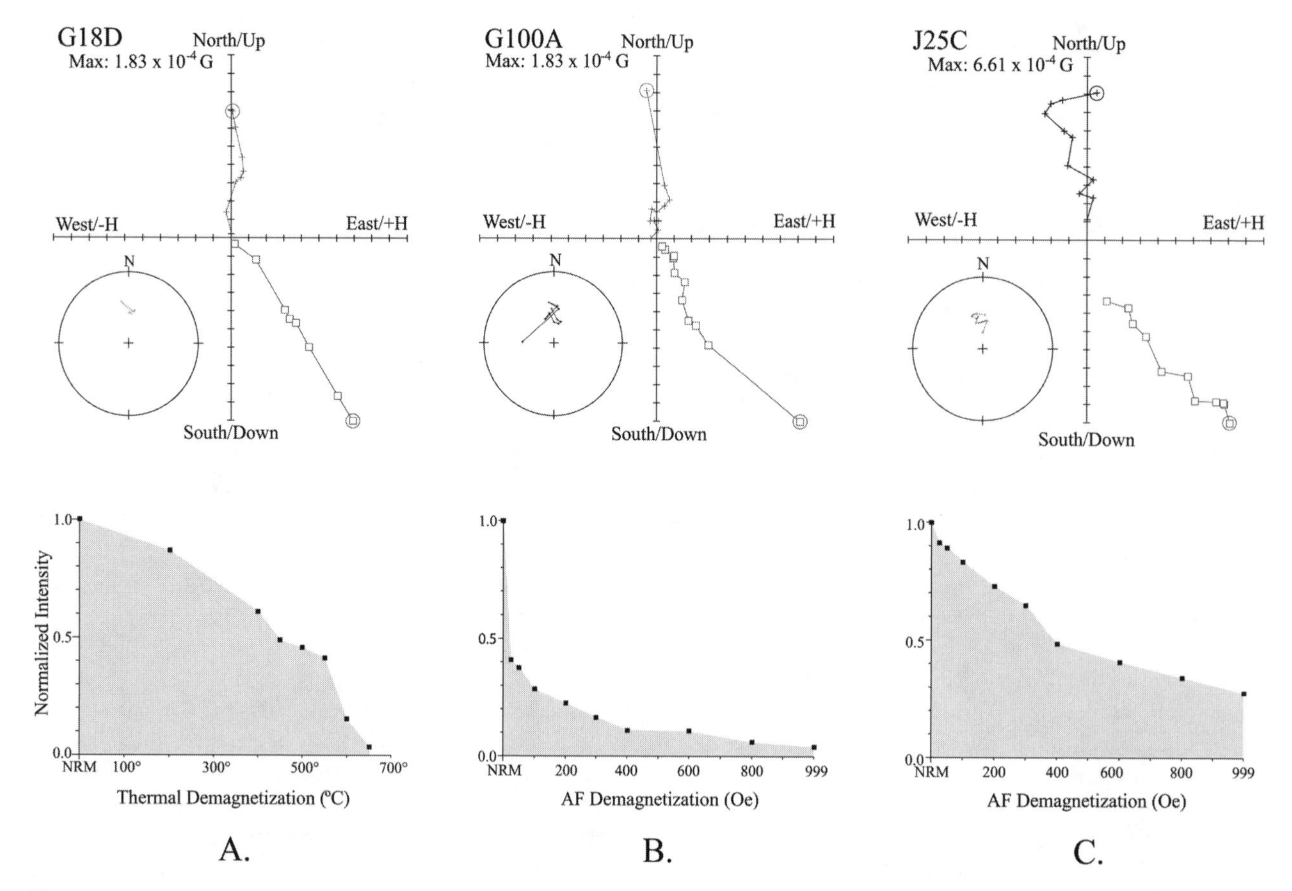

Figure 2.2. Vector end-point diagrams, stereoplots, and normalized intensity and step plots for representative samples after thermal (°C) and AF demagnetization (Oe). (**A**) Sample G18D is from the Delikayincak section. Thermal demagnetization was successful in isolating a stable normal component at temperatures above 200 °C. (**B**) Sample G100A is also from the Delikayincak section. AF demagnetization was successful in isolating a normal component. The vector endpoint projections indicate a nearly linear decay to the origin and both samples are normally magnetized. (**C**) Sample J25C is from the Kavakdere south section and also indicates a normally magnetized sample, but the AF demagnetization treatment does not produce a complete decay to the origin, suggesting the presence of some hematite. The sample is clearly normally magnetized. (**D–F**) Results from typical reversely magnetized samples from Sinap Tepe (**D**) and Kavakdere (**E–F**) show demagnetization behaviors that are similar to those of the normally magnetized samples. (**D**) In some cases an overprint is indicated, which is removed by 300 °C. Symbols for vector end-point diagrams: crosses, declination; hollow squares, inclination. Symbols for stereoplots: crosses, upper hemisphere; hollow circles, lower hemisphere.

conclusion is somewhat obscured by a large number of normal and reversed sites with intermediate directions. We conducted a reversal test using the class A site means. Sites exceeding an angular SD (ASD) of 1.5 ($n = 38$) were deleted from the test. The resulting sample of 298 site means is plotted in figure 2.4B and more clearly illustrates the antipodal nature of the data. The reversed polarity site means are inverted and combined with the normal polarity data in figure 2.4C, and figure 2.4D shows only the normal and inverted reversed polarity means along with their $\alpha 95\%$ circles and the present field. These data illustrate minor overlap at the $\alpha 95\%$ level. A test of the Fisher precision estimates shows the normal and reversed populations to be significantly different from each other at the 0.05 level (max k/min $k = X_{k,v}$; $24.03/7.95 = 3.023_{2,296}$); this result is apparent from the greater dispersion of the reversed-magnetized site means. A second test compares the mean vectors from each population and showed that there is no difference between the two populations at the 0.05 level:

$$F_{2(S-1),\,2(N-S)} = 2(N-S)/[2(S-1)] \times \Sigma R_i - R/N - \Sigma R_i$$

$$F_{2,296} = 1.957$$

The failure of the first test suggests that some caution should be used in the interpretation of the data. The results from the second test, however, along with the results from previous studies show that the sediments of the Sinap Formation preserve a record of stable magnetic vectors of normal and reversed polarity that are a signature of the ancient magnetic field.

Paleomagnetic Reversal Stratigraphy

Outcroppings of sedimentary rocks in the study area are generally of limited exposure and in only rare cases is it possible to trace a particular unit across long distances. To assess the relative relationships among the various fossil

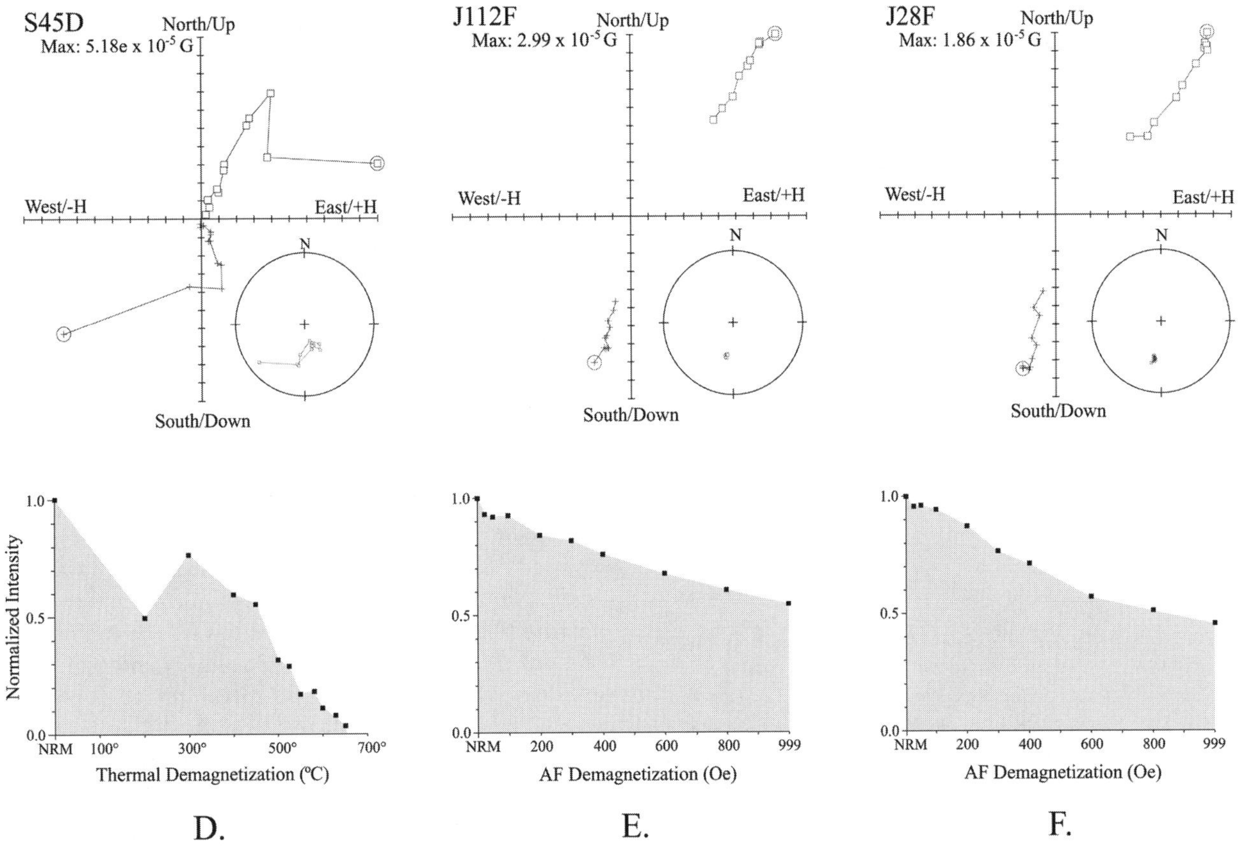

Figure 2.2. *(continued)*

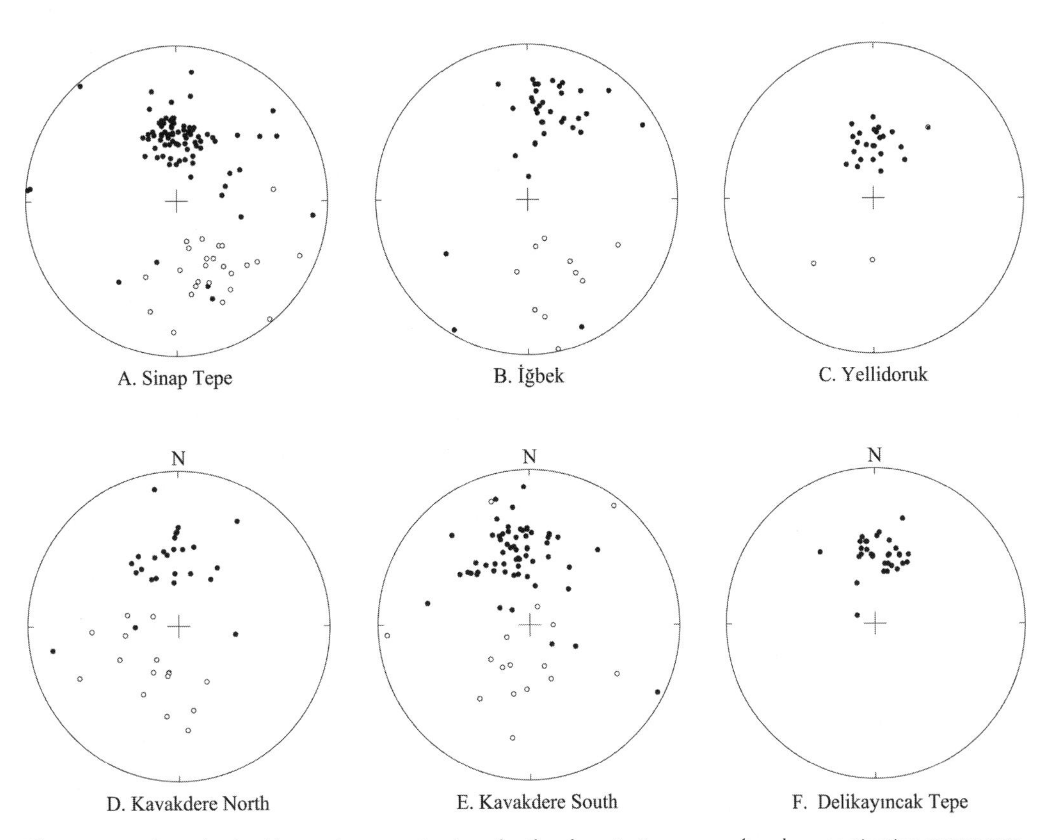

Figure 2.3. Plots of Schmidt equal area projections for the class A site means after demagnetization treatments, corrected for bedding dip. **(A)** Sinap Tepe (n = 112); **(B)** Igbek (n = 46); **(C)** Yellidoruk (n = 25); **(D)** Kavakdere north (n = 42); **(E)** Kavakdere south (n = 78); and **(F)** Delikayincak Tepe (n = 33). These data show a predominance of normal polarity sites and smaller populations of reversed polarity sites, with the two groups roughly antipodal to one another. Solid circles, upper hemisphere; hollow circles, lower hemisphere.

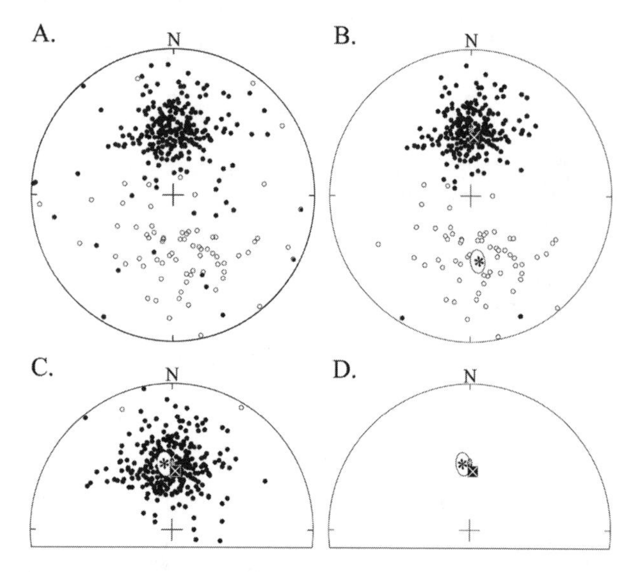

Figure 2.4. Plots of Schmidt equal area projections for **(A)** the class A sites (n = 336) of all data shown in figure 2.3 combined into a single plot. Normal population: declination = 0.9°, inclination = 51.4°, α95 = 2.8°, k = 10.80, n = 261. Reversed population: declination = 174.1°, inclination = −46.3°, α95 = 9.7°, k = 3.85, n = 75. **(B)** Same sample after the removal of sites having an ASD > 1.5. The antipodal nature of the two populations is more clearly illustrated with these data. Normal population: declination = 0.0°, inclination = 52.4°, α95 = 1.9°, k = 24.03, n = 234. Reversed population: declination = 173.2°, inclination = −52.5°, α95 = 6.7°, k = 7.95, n = 64. **(C)** When the reversed polarity sample is inverted and plotted with the normal polarity sample, the resulting plot shows the broad overlap of the two samples. Mean: declination = −1.4°, inclination = 52.4°, α95 = 2.1°, k = 16.74, n = 298. **(D)** Sample means and α95% of the normal (small white star, black ellipse) and reversed (large black star, white ellipse) populations show minor overlap. The present field (declination = 3.5°, inclination = 57.0°) is shown by the large X. Symbols as in figure 2.3.

localities and estimate their ages, it was necessary to sample and establish correlations among multiple sections. Preliminary results from the paleomagnetic reversal stratigraphy for sediments located around Sinap Tepe and Kavakdere are presented in Kappelman et al. (1996) and Lunkka et al. (1999); Feseha (1996) presented the results for Igbek. Here we combine these results for the first time and include additional sections and data. Detailed lithologic logs are provided by Lunkka et al. (this volume).

Kavakdere

The density of fossil occurrences in this region is generally low and restricted to the upper portion of the section; however, some isolated vertebrate and invertebrate remains are found on occasion in the lower part of the section. The original Kavakdere section is located in the northern portion of the valley and extends from the agricultural field cover in the valley base to the ridge top in west (see Kappelman et al. 1996: fig. 6.9). This section is shown in figure 2.1 as KN and in figure 2.5 as Kavakdere north. A new section was sampled in the southern portion of the valley (KS in fig. 2.1) and is shown as Kavakdere south in figure 2.5. Because field cover is absent in this part of the valley, it was possible to extend the base of this section through the top of a series of silicified beds that crop out in the valley bottom (see Lunkka et al. 1999; fig. 1.7); these are some of the beds that are hidden beneath field cover below the base of the section in the northern part of the valley (KN). The same beds crop out along the ridge top of Beycedere to the east as flat irons, and a third, short (65-m) section was sampled along this ridge top down to the point of field cover (KL in fig. 2.1 and Kavakdere lower in fig. 2.5); the KL section overlaps with the basal portion of the KS section but does not overlap with KN.

All of the Kavakdere sections are marked by a long interval of normal polarity (N1) at their base, with a single reversed polarity interval documented in the Kavakdere lower section. Numerous intervals of normal and reversed polarity (R1–N5 or N9) are found toward the top of both sections. More than two sites document most of these intervals; however, several polarity intervals are documented by single sites only (e.g., N6, R7–R8 in Kavakdere south), and their existence could be tested with additional sampling.

The correlation between the two sections shown here differs from that given in Lunkka et al. (1999) in which the Kavakdere north normal interval N1 was correlated with Kavakdere south normal interval N4–N6. The previous correlation was based on simple strike mapping, whereas the correlation presented here in figure 2.5 is based on marker bed tracings from the total station data and photographs (see fig. 1.7). The reversed polarity interval R1 in the northern section is correlated with intervals R1–R2 in the southern section; the short normal interval N2 in the southern section appears to have not been sampled in the northern section but is predicted to be located in the unsampled interval between 55 and 60 m. Note that interval N3 in the southern section and N2 in the northern section both demonstrate intermediate virtual geomagnetic poles (VGP) as polarity reversed to R3 and R2, respectively, with the reversal being an apparently protracted event.

Our early work in the Kavakdere area established that this valley is one limb of a plunging syncline whose axis lies only a km or so to the west. Other sampling at the Pliocene locality of Çalta (Lunkka et al. 1998) attempted to establish a reversal stratigraphy through this important vertebrate locality, but these samples suffered from a magnetic overprint that possibly dates to about or shortly after the time of folding. Other fossil localities are known from the western and southern portions of this syncline, and future work could easily tie these fossil localities into the overall framework and perhaps even add the important locality of Cobanpinar that is located 6 km to the southwest.

Sinap Tepe Area

The area around Sinap Tepe contains the richest concentration of fossil localities as well as the most variable

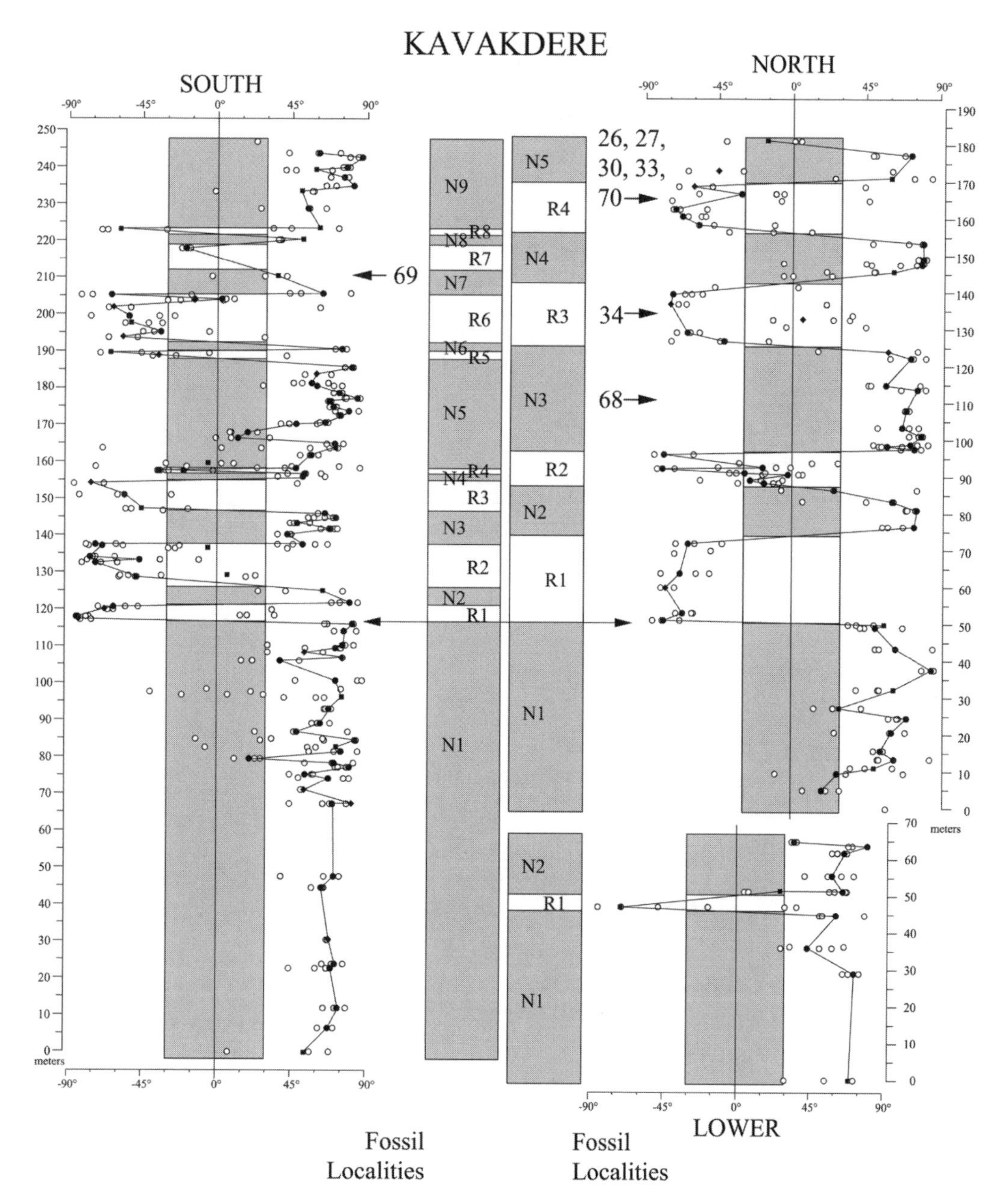

Figure 2.5. Paleomagnetic reversal stratigraphy of the Kavakdere south (left) and north (right) sections, with the lower Kavakdere section shown at bottom right. The two major sections are aligned at the top of the N1 interval in each section. Hollow circles, statistically significant samples; solid circles, class A site means; solid squares, class B site means of three samples; solid diamonds, class B site means of two samples only; arrows indicate level of numbered fossil localities. Class B sites with VGP values less than 30° are not connected by lines.

degree of sedimentary outcrop exposure in the region (see fig. 1.4). Three primary sections were sampled and are shown in figure 2.6: Sinap Tepe in the west, Delikayincak in the center, and Yellidoruk in the east. Because bedding dips in this region average <10° and the relief is generally low, each section contained one (Yellidoruk at 66 m; Delikayincak at 20 m) or several (Sinap Tepe, see fig. 2.6) section offsets. Exposures around Sinap Tepe are generally good with limited tree and shrub cover, and the Sinap sections illustrated in figure 2.6 trace the exposures of outcrop around the shoulders of the butte in a counterclockwise fashion (see also fig. 2.1). Fossil localities occur all around the perimeter of Sinap Tepe and are plotted in the nearest stratigraphic section, with the relative stratigraphic position of each locality mapped onto the composite section with regard to its stratigraphic relationship to the polarity intervals.

The Sinap composite section totals 204 m (fig. 2.6). The lower 125 m (intervals R1–N6 in the composite section) shows a predominance of normal polarity (N1–N6), a reversed polarity interval of moderate thickness at the base (R2), and several thin reversed intervals in the middle (R3–R6). Some of these reversed intervals are sampled in multiple sections (e.g., R3: thin reversed interval at the 2-m level in Delikayincak and the 22-m level in Sinap northeast 1),

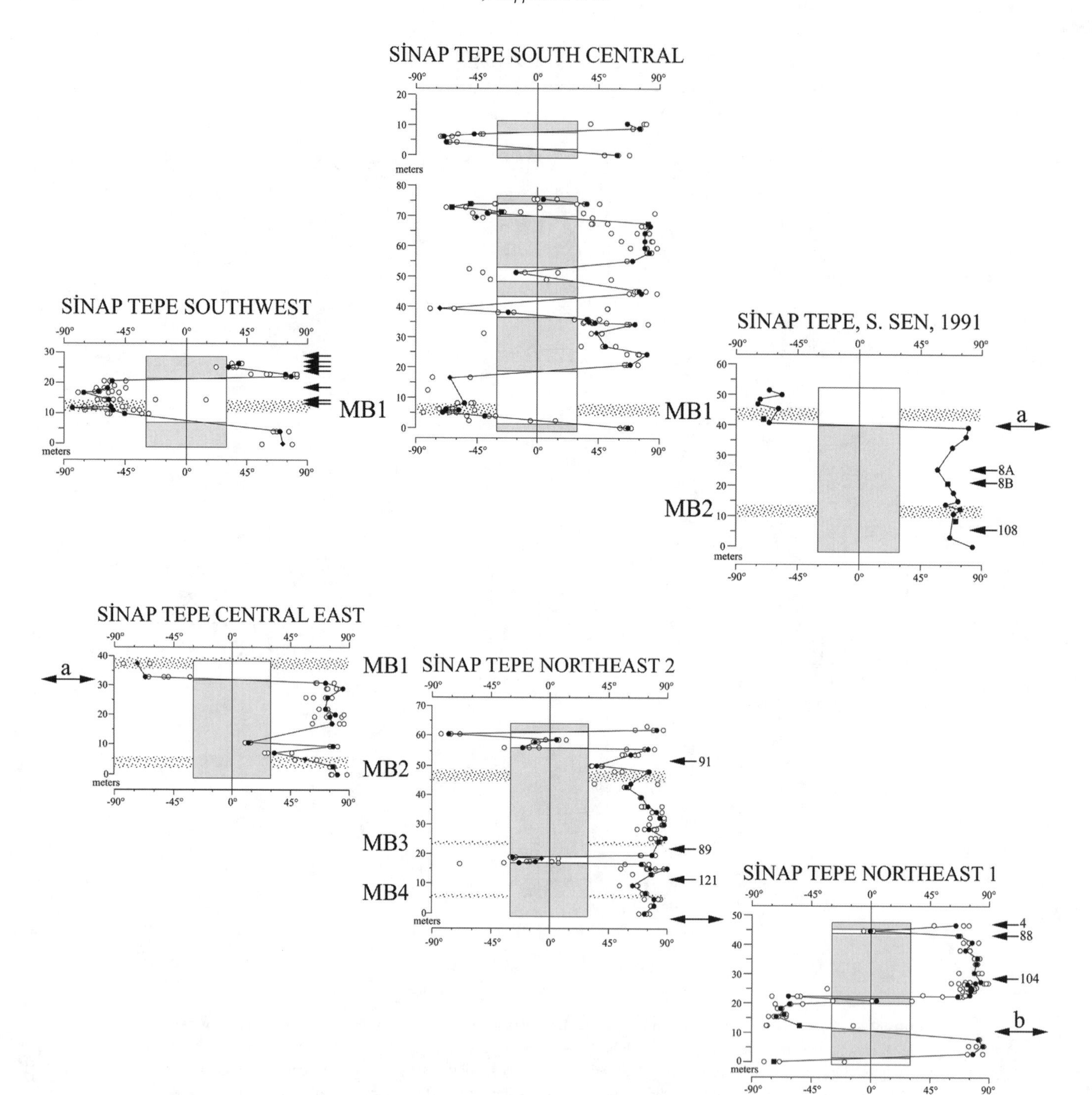

whereas others are documented as both reversals and VGP shallowings (e.g., R6: thin reversed interval at the 57-m level in the Sinap northeast 2 and the 83-m level in Yellidoruk, and shallow VGP at the 10-m level in the Sinap central east and the 25-m level in Sinap [Sen 1991]). The correlation among the sections is further strengthened by a few marker beds (stippled units labeled MB1–4 in fig. 2.6) that can be traced between sections (see fig. 1.4) or by three-dimensional mapping projections. For example, the paleosols that compose the double white marker beds (MB2 in fig. 2.6) prominently exposed on the northwestern ridge of Delikayincak Tepe (75-m level) can be projected in three dimensions and mapped to other white paleosols on Yellidoruk (73-m level) and Sinap Tepe (96-m level), even though any physical tracing of this bed is precluded by intervening tree and field or valley fill cover. This bed is preceded by an interval of shallow VGP values on Yellidoruk (70-m level) and followed by the thin reversed (or shallow VGP) interval of R6. This interval may be represented at Delikayincak Tepe by the single sample of reversed polarity at 89 m, and/or the unsampled 5-m interval below it. Other marker beds are shown in figure 2.6, and some of the distinctive VGP values and their trends can be correlated among the sections.

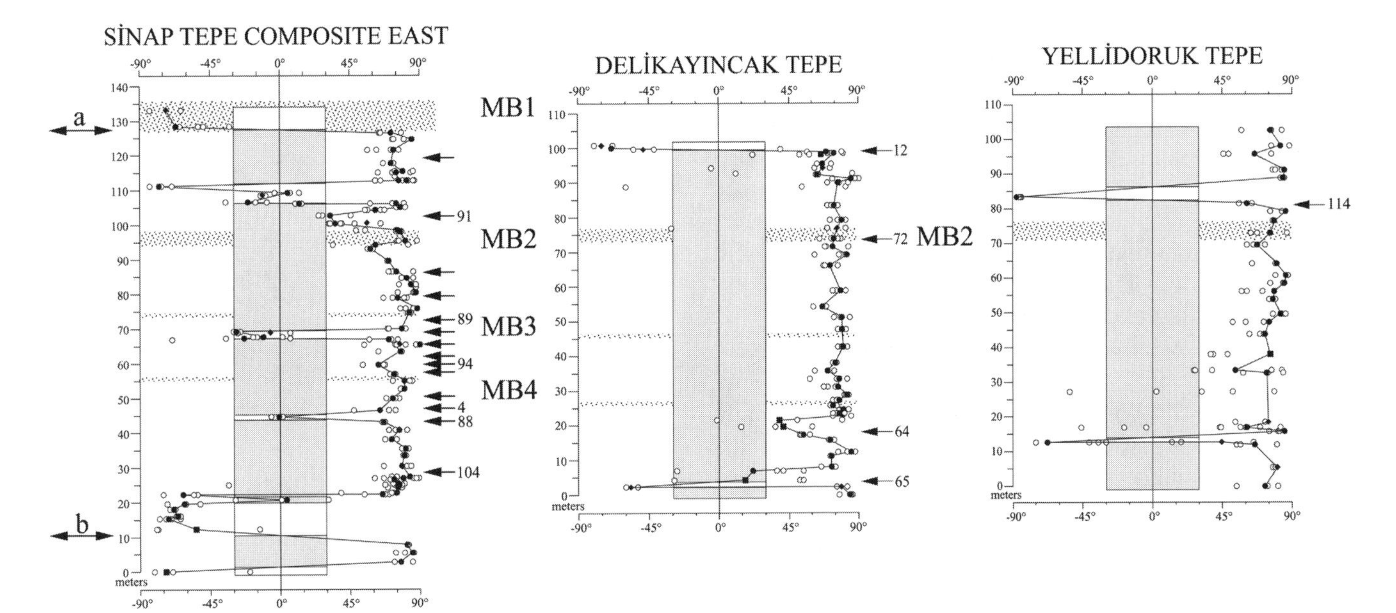

Figure 2.6. Paleomagnetic reversal stratigraphy in the Sinap area showing the individual sections sampled around the perimeter of Sinap Tepe (see fig. 2.1). Facing page: Correlations between sections are based on marker beds MB1–4. The upper panel of three sections ties into the lower panel of three sections at the double-headed arrow (a). This page: the Sinap Tepe composite east stratigraphy at top left ties into the sections on the facing page at the double-headed arrows (a and b) and is correlated with the Delikayncak Tepe and Yellidoruk Tepe sections by marker beds MB2–4. The composite section at bottom left ties into the other sections at the double-headed arrows (a and b). Stippled intervals, marker beds MB1–4. Scale in m. Symbols as in figure 2.5.

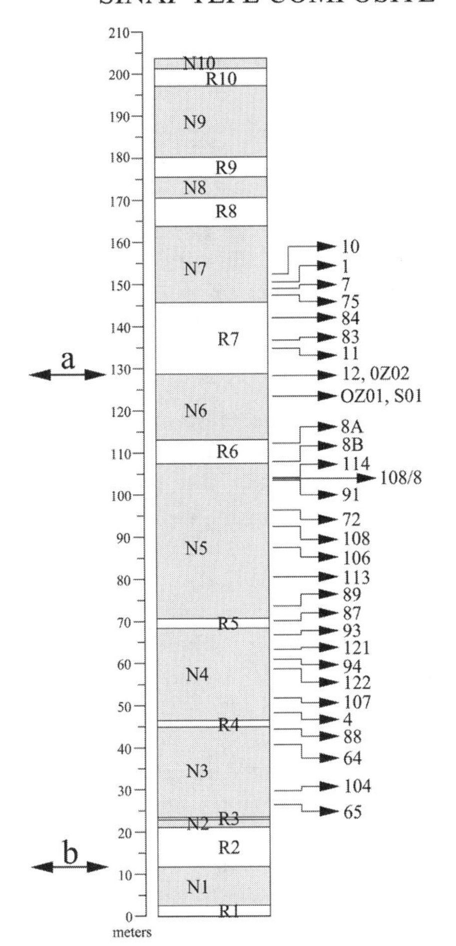

The upper part of Sinap Tepe is marked by intervals of normal and reversed polarity of nearly equal thicknesses. The first of these, R7, is sampled in multiple sections and contains a distinctive white caliche unit that, along with the R7 reversed interval, can be traced around the southern and southwestern outcrops of Sinap Tepe. The uppermost portion of the Sinap composite section from N7 upward is only exposed in the Sinap south central section. Three intervals within this section, R8–R9, suffer from a low sampling density and their definition could be improved by resampling. The top of normal interval N7 is distinctive in that the VGPs record intermediate directions as the field reversed direction to R8; these data suggest that this reversal was a somewhat protracted event. The uppermost part of Sinap south central includes a small exposure of sediments that appears to be unconformable to and above the main section (see fig. 2.6, top of Sinap south central). This 11-m section documents an interval of reversed polarity bracketed by normal polarity sediments but is not included in the composite section.

The polarity intervals documented in the various sections and illustrated in figure 2.6 are of nearly equal thickness. The Yellidoruk section differs from the other sections in that it consists of much coarser sediments, including gravels (see Lunkka et al. 1999, this volume). If the thin interval of reversed polarity at 12 m is the same interval as the 44-m reversed interval on Sinap northeast 1 and the VGP shallowing at 20 m on Delikayincak, then the Yellidoruk sedimentation rates would have been higher than those of the other sections. Lunkka et al. (1999, this volume) conclude that the Yellidoruk sediments represent the more proximal portion of the ancient alluvial fan. Our results provide additional support for their idea.

A second interesting difference is seen in the upper portion of the Delikayincak section. The marker bed at 75 m on Delikayincak Tepe (see above) is followed by ~23 m of normal polarity sediments, with reversed polarity sediments occurring at 98 m. The correlative section at Sinap Tepe central east contains a normal polarity interval ~25 m thick. A difference of 2 m is not striking, but samples from the top of Delikayincak Tepe suggest that its uppermost reversed interval may not be the same interval as that of R7. Sampling carried out below the gravel conglomerate that caps the ridge top of Delikayincak Tepe showed that the bone-bearing level of fossil locality 12 in the northernmost trench was of reversed polarity, as were the sediments immediately below the base of the overlying conglomerate. The capping conglomerate was removed in 1993 and a sample taken ~0.5 m above the bone level in a trench located ~70 m to the southeast was of normal polarity. The conglomerate unit that caps this section has an erosional base of variable relief that might have cut out a thin normal polarity interval at the northern part of the fossil locality that was preserved at its southern portion. This hypothesis could easily be checked with additional sampling. Its significance to the correlation with the GPTS is discussed later in this chapter.

Igbek

The Igbek section is a relatively short section ~94 m thick that is located ~20 km northeast of Kazan and ~12 km northwest of Çubuk. Feseha (1996) completed a study of this section, and his results are reproduced in figure 2.7. (This section was also reported in Lunkka et al. 1999, fig. 12.3 [p. 244] but the directions [colors] of the middle polarity intervals were inverted during the final production of the book and are incorrect.) The lithologic log is presented in figure 1.8. The basal 47 m of this section are of normal polarity (N1–N2) with one thin reversed interval at 45 m (R1), whereas the upper 47 m are of predominantly reversed polarity (R2–R4) with two intervals of normal polarity (N3 and N4) that are separated by a thin interval of reversed polarity documented by a single site with a shallow VGP (R3). The presence of erosional channels at the 70-m level leads us to suspect that a greater duration of reversed polarity was possibly once recorded here. These channels have erosional bases that cut into and probably eroded the underlying sediments. Sediments directly below the base of this channel are reversely magnetized (R3) while those in and above the channel are of normal polarity. Locality 49 is located near the top of this section in an interval of normal polarity (N5).

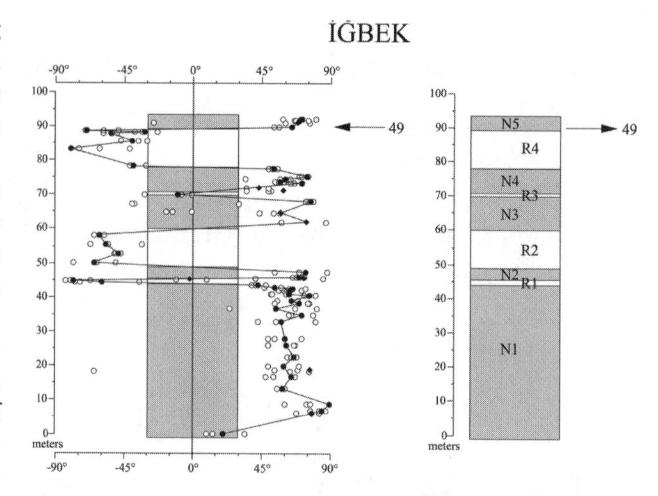

Figure 2.7. Paleomagnetic reversal stratigraphy of Igbek. Symbols as in figure 2.5.

Isotopic Dates of Volcanic Units

There are now several radioisotopically dated volcanic units that provide a maximum age for the fossil bearing sediments in the Sinap Tepe region. Earlier work in this region reported a whole rock K-Ar date of 15.2 ± 0.3 Ma for a basalt flow that separates the underlying Pazar Formation from the Sinap Formation (Kappelman et al. 1996), as mapped by Öngür (1976). This sample was collected in the area to the north-northeast of Sinap Tepe on Karakaya Tepe (see fig. 2.1), and even though this flow is separated from the Sinap Formation by a fault, it does provide a maximum age for the fossil bearing sediments in the area of Sinap Tepe. The basalt is of reversed polarity (Kappelman et al. 1996) and its age of 15.2 ± 0.3 Ma places it within Chron C5Br of CK95.

Other mapping to the west along the continuous Beycedere-Kavakdere section (fig. 2.1) identified two ash beds in the upper part of Beycedere at ~218 m (see fig. 1.6) and 216 m. Plagioclase feldspar separates were dated by the $^{40}Ar/^{39}Ar$ total fusion method and yielded dates (weighted means) of 15.88 ± 0.07 Ma and 15.96 ± 0.08 Ma for the upper and lower ashes, respectively (see table 2.2). These ages place the ashes within Chron C5Br of the GPTS of Cande and Kent (1995), a moderately long interval of reversed polarity (CK95: 15.155–16.014 Ma). However, preliminary analysis of the magnetic polarity of several sites from this section (Kappelman, in prep.) shows that the bed with the younger ash is of normal polarity, whereas a site 2 m below the older ash is reversed. If these polarities were acquired at or shortly after the time of deposition, the results suggest that these ashes may sample the slightly older transition boundary between Chron C5Cn.1r and C5Cn.1n and place it at about 15.92 Ma instead of 16.293

Table 2.2. $^{40}Ar/^{39}Ar$ Total Fusion Data for Beycedere, Turkey, Plagioclases

L	^{40}Ar moles (× 10^{-14})	$^{40}Ar/^{39}Ar$	$^{37}Ar/^{39}Ar$	$^{36}Ar/^{39}Ar$	$^{40}Ar^{*}/^{39}Ar$	$\%^{40}Ar^{*}$	Age (Ma ± 1σ)	
Sample B-50/7854								
–01N	13.9	2.717	6.124	0.0024	2.525	92.6	15.95	0.10
–02	11.8	2.500	4.671	0.0013	2.507	100.0	15.83	0.23
–03	9.91	2.749	7.197	0.0027	2.546	92.2	16.08	0.19
–04	7.46	2.529	7.300	0.0021	2.522	99.2	15.93	0.22
–05	5.84	2.856	6.407	0.0029	2.532	88.3	15.99	0.28
Arithmetic mean							15.96	0.09
Weighted mean							15.96	0.08
Sample B-51/7864								
–02	11.0	4.739	5.557	0.0091	2.513	52.8	15.87	0.19
–03	18.0	3.293	5.787	0.0042	2.525	76.4	15.95	0.13
–04	3.48	2.549	6.443	0.0020	2.501	97.7	15.79	0.42
–05	3.46	3.755	3.426	0.0051	2.517	66.9	15.90	0.36
–06	7.08	2.769	4.863	0.0022	2.519	90.7	15.91	0.15
–07	5.12	2.577	4.507	0.0015	2.518	97.4	15.90	0.18
–08	3.57	2.838	5.551	0.0027	2.505	88.0	15.82	0.27
–09	6.51	2.638	4.598	0.0018	2.488	94.1	15.72	0.19
Arithmetic mean							15.86	0.08
Weighted mean							15.88	0.07

Notes: For samples B-50/7854 and B-51/7864: J (Irrad. #102C) = 0.003516 ± 0.0000025. Js were determined from replicate single crystal analyses of the co-irradiated monitor mineral, Fish Canyon sanidine with an age of 27.84 Ma. Correction factors: $(^{39}Ar/^{37}Ar)_{Ca} = 6.85 \pm 0.24 \times 10^{-4}$; $(^{36}Ar/^{37}Ar)_{Ca} = 2.74 \pm 0.05 \times 10^{-4}$; $(^{40}Ar/^{39}Ar)_{K} = 9.0 \pm 3 \times 10^{-4}$; $^{40}Ar/^{36}Ar$ discrimination = 1.005 ± 0.0005. Decay constants are those recommended by Steiger and Jager (1977). The analytical procedures employed follow those described in Swisher et al. (1994) and references therein.

Total fusion of single crystals of the monitor mineral and the plagioclase unknowns was accomplished with a 6 W coherent Ar-ion laser. Released gases were purified by two SAES C-50 getters operated at approximately 400 °C. Argon was measured in an on-line Mass Analyzer Product 215 noble-gas mass spectrometer operated in the static mode, using automated data collection techniques. Laser heating, gas purification, and mass spectrometry were completely automated following computer programmed schedules.

The uncertainties associated with the individual incremental apparent ages are 2σ errors; those accompanying the arithmetic means are 1σ. The weighted means and standard errors are calculated following Taylor (1982).

Recently, the Berkeley Geochronology Center (Berkeley, CA) has adopted an age of 28.02 Ma for the Fish Canyon Sanidine Standard (Renne et al. 1998). However, this change is not reflected in currently available GPTS. We have therefore used the prior calibration age of 27.84 Ma for Fish Canyon Sanidine, following that used by Berggren et al. (1995) for the GPTS. The dates reported here would be about 0.64% older, or16.06 and 15.97 Ma, if the new FC calibration is adopted.

Ma. This interpretation receives support from the observation that sites above the B51 level are of reversed polarity and probably document Chron C5Br, whereas other sites below B50 document intervals of normal and reversed polarity and probably represent the brief normal and reversed events older than Chron C5Cn.1r but still within Chron C5Cn. If this interpretation is correct, it could necessitate some revisions to the age interpretations for this interval of the GPTS. We note, however, that calibration differences between the dates reported here and those estimated in CK95 will also need to be taken into consideration. Although much of CK95 used Ar dates calibrated by Fish Canyon Sanidine at 27.84 Ma, this particular interval was anchored by Plio-Pleistocene astronomical age calibrations and two Miocene K-Ar dates of uncertain standard calibration. Proposed changes in the Ar standard Fish Canyon Sanidine from 27.84 Ma to 28.02 Ma (Renne et al. 1998) would increase the ages of the two Beycedere ashes to 15.97 and 16.06 Ma, for a mean of 16.02 Ma instead of 15.92 Ma for the transition boundary between Chron C5Cn.1r and C5Cn.1n. We are now researching this topic for future publication.

The two ashes discussed here are ~100 m below the base of the Kavakdere sections and are separated from these sections by several thick silicified beds. These beds constitute the ridge-forming flat irons that separate the two valleys, and it has been proposed that these units represent a nondepositional unconformity of perhaps three to four million years (see Lunkka et al. 1999, this volume).

Fossil vertebrates, invertebrates, and plants are rare finds in the sedimentary rocks that underlie the silicified flat irons, but some occurrences are known. It appears likely that the rich concentration of fossil vertebrates found at the well-known site of Inönü I (fossil localities 24 and 24a; see fig. 2.1) is below the silicified flatirons. However, these localities occur in a small, isolated fault block that we have not been able to map with confidence into the main section. This fauna was assigned to Mammal Neogene (MN) 6 by Gübüz (1981). The base of MN 6 is placed between 16.5 and 15.2 Ma (Steininger et al. 1996), with its upper limit placed at 12.5 Ma. In an earlier paper we suggested that localities 24 and 24a might be in stratigraphic proximity to the basalt that we dated at 15.2 ± 0.3 Ma (Kappelman et al. 1996; Lunkka et al. 1999), but if they are in closer proximity to the ashes discussed above, their age could be closer to 16 Ma.

Correlations with the GPTS

The radioisotopic dates noted above provide a maximum age for the various fossil localities shown in figures 2.5–2.7. The GPTS shows that the interval of time from 15 Ma to ~12 Ma is marked by numerous nearly equal duration normal and reversed polarity events (C5Bn–C5An) that are followed by the relatively long duration reversed Chron C5r (11.935–10.949 Ma), normal Chron C5n (10.949–9.740 Ma), and reversed Chron C4Ar (9.740–9.025 Ma) intervals (CK95). Together these three intervals provide a unique signature for the latter portion of the Miocene. All of the sections from the Sinap Formation are marked by a long interval of normally magnetized sediments at their base, and we have concluded that this interval represents Chron C5n (Kappelman et al. 1996; Lunkka et al. 1999). This interpretation draws additional support from a critical biostratigraphic event, the first appearance datum of *"Hipparion,"* a tridactyl equid that entered the Old World during the late Miocene. This "datum" has been an issue of some controversy for the past several years (see Sen 1989), but recent reassessments suggest that this event occurred near the beginning of Chron C5n (Woodburne and Swisher 1995; Woodburne et al. 1996; Pilbeam et al. 1996) at ~10.7 Ma, or perhaps within the Chron C5r.1n at a somewhat earlier date of 11.1 Ma (Agustí et al. 1997; Garcés et al. 1997). As noted elsewhere (Bernor et al., this volume), the oldest fossil hipparionine horses occur in locality 4, 47 m above the base of the Sinap section; four well-sampled localities below this level are without this taxon. Based on these observations, we can conclude that it is very unlikely that the base of the Sinap section is older than 11 Ma.

The correlation between the individual sections and the GPTS presented below are expansions and revisions of our previous work, but each section is anchored to the distinctive long normal of Chron C5n.

Kavakdere

Both the north and south sections in Kavakdere are marked by a variable thickness of normal polarity sediments at their base (fig. 2.5). Figure 2.8A correlates the normal interval of N1 in the southern section with Chron C5n in the GPTS of CK95. Interestingly, none of the brief reversed intervals noted by Cande and Kent (1995) as cryptochrons are identified within N1. It is, however, important to note that several samples at the 97-m and 83-m levels are of reversed polarity, but the site means are not statistically significant (fig. 2.5); the upper level may represent the brief reversed polarity event of Chron C5n.1r whereas the lower level is C5n.2n-1. This prediction could be tested with additional sampling. The next interval in the southern section is of predominantly reversed polarity, and R1–R4 is correlated with C4Ar. Two normal polarity zones (N2 and N3) within this reversed interval at 123 m and 144 m are correlated with C4Ar.2n and C4Ar.1n, respectively. Cande and Kent (1995) do not include a brief normal event within C4Ar.1r, but studies of continental rocks (Tauxe 1979; Tauxe and Opdyke 1982; Meigs et al. 1995), Icelandic basalts (Saemundsson et al. 1980), and marine sediments (Schneider, 1995) provide clear support for the existence of this cryptochron, C4Ar.1r-1. The thin normal interval noted here as N4 is probably this event. The next interval, N5–N6, is a thick, predominantly normal polarity zone that has a thin reversed interval, R5, at its top. The thin normal interval, N6, is correlated with cryptochron C4r.2r-1. The next interval, R6–R8, is of predominantly reversed polarity but the upper four intervals (N7–R8) are supported by one or two sites only, the upper two of which are class B sites. This reversed interval is correlated with C4r, and the N7 interval is correlated with Chron C4Ar.1n. The upper normal of moderate thickness that caps the section, N9, is correlated with C4An.2n. There is no cyptochron noted in Chron C4n in the GPTS of CK95 that could in turn be correlated with R8. If only class A sites were used to construct figure 2.5, interval N8 and R8 would disappear and R7 would extend upward to ~224 m, thus more closely matching the pattern seen in the Kavakdere north section. The existence of these short upper intervals could be tested with additional sampling.

The reversal stratigraphy of the Kavakdere north largely mirrors that of the southern section but does show some interesting differences (fig. 2.8B). First, a thick interval of normal polarity, N1, at the base of the section is correlated with the upper portion of Chron C5n. The basal extension KL (marked as "Lower" in figure 2.5) is also primarily normal but does document a brief reversed interval at 47 m, which appears to represent one of the cryptochrons of Chron C5n. Because cover obscures the exact relationship between the KL and Kavakdere north sections, it is not known exactly how many meters separate these two sections, but this thin reversed interval probably represents C5n.2n-2 or C5n.2n-3. The corresponding portion of Kavakdere south demonstrates variable site densities through this part of the section and dense resampling here could test for the existence of the KL and other cryptochrons. The relatively thick interval of reversed polarity, R1, in Kavakdere north is correlated with C4Ar in figure 2.8B. It appears that C4Ar.2n was missed and if it was, this reversed interval is predicted to be at the 54–58 m level of the section. An alternative correlation (not shown) limits R1 to Chron C4Ar.2r only and argues that the reversed interval of Chron C4Ar.3r was missed and occurs within the N1 interval. The next intervals, N2 and R2, are of approximately the same thickness as the southern section; the transition between the two appears to be a protracted event. The thick interval of N3 is correlated with C4An but the thin normal events

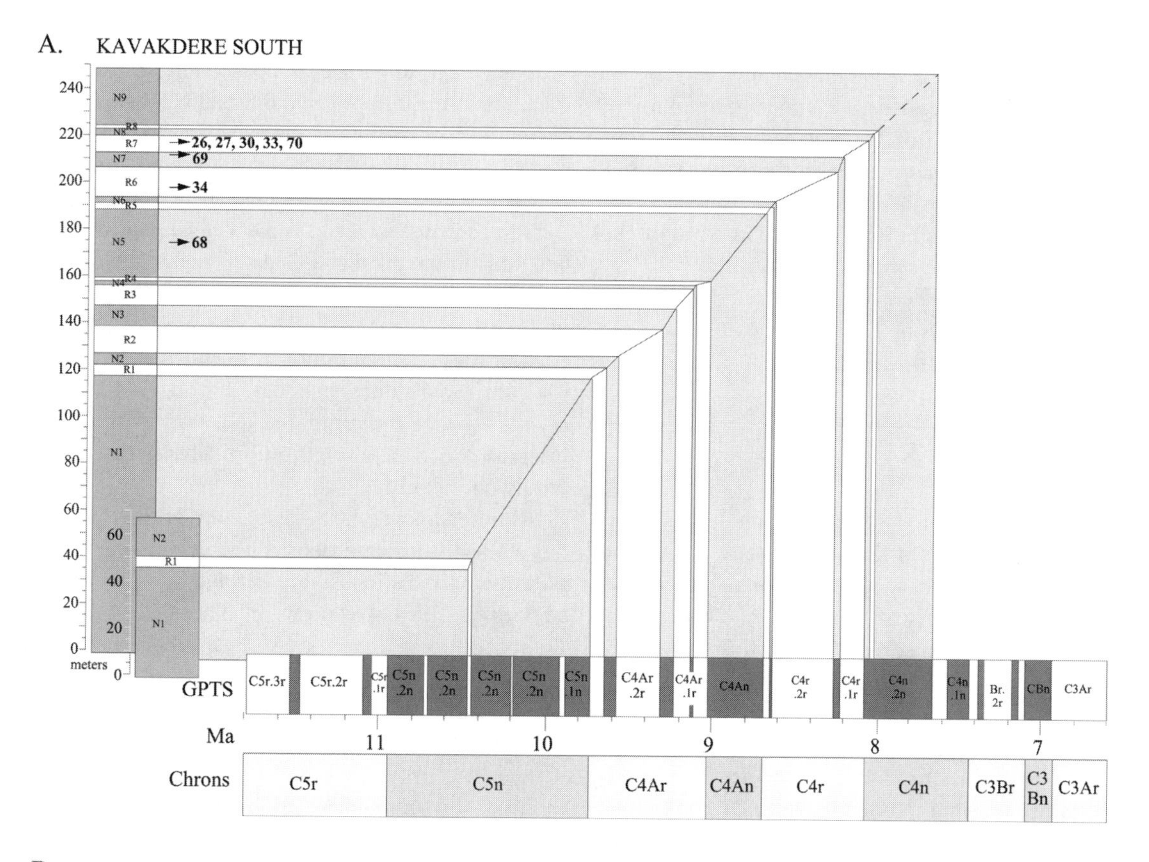

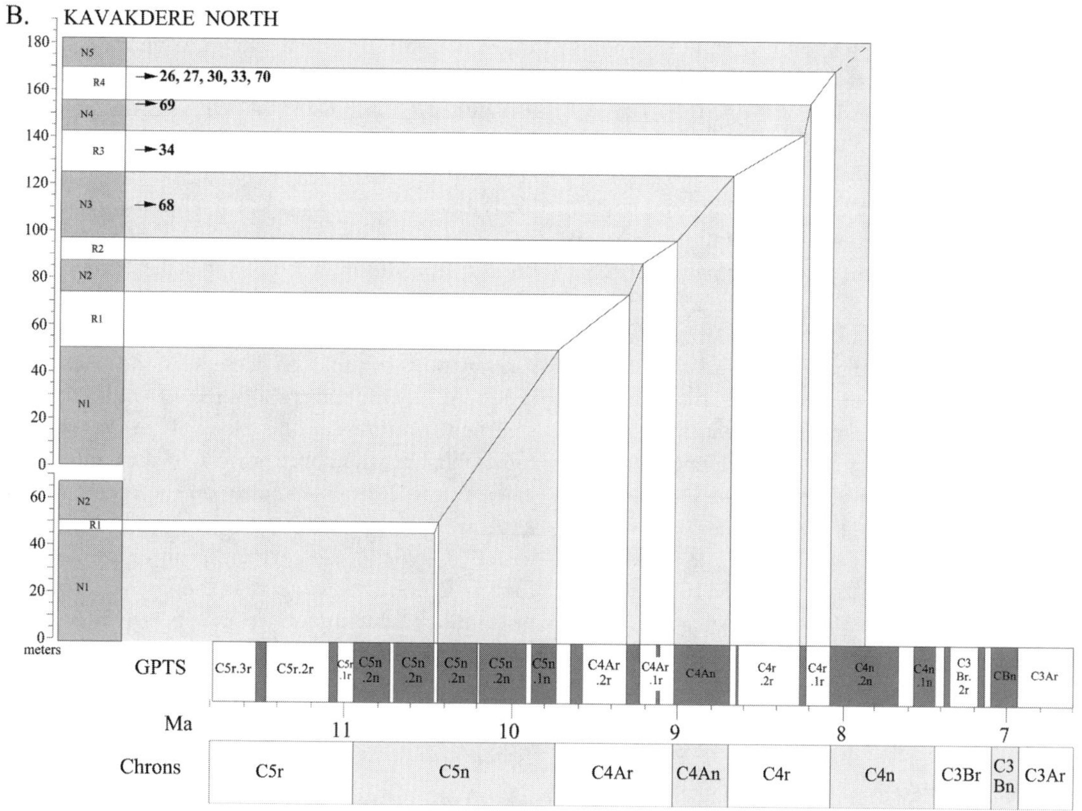

Figure 2.8. Correlation of the Kavakdere (**A**) south and (**B**) north from figure 2.5 with the GPTS of CK95. The lower Kavakdere section is shown in both (**A**) and (**B**). Fossil localities are bracketed between 8 and 9 Ma in this section.

seen in the southern section were not documented at the top and bottom of N3; the normal but shallow VGP site at 133 m (fig. 2.5) might represent the upper parts of these zones. The upper portion of the northern section from N4–N5 largely matches that of the southern section, but the intervals differ somewhat in thickness. The shallow reversed VGP at 166 m might represent the interval of the thin normal N8 seen in the southern section.

Fossil localities in Kavakdere are mostly restricted to the upper portion of the section. Their interpolated ages are given in table 2.3.

Sinap Tepe Area

Previous interpretations of the Sinap Tepe paleomagnetic reversal stratigraphy given in Kappelman et al. (1996) and Lunkka et al. (1999) correlated the long normal interval at the bottom of the section with Chron C5n. This correlation remains the preferred correlation for the majority of the section, but more detailed analysis (see fig. 2.6) suggests that the upper part of the section is younger than previously thought. The very base of the section appears to document evidence for Chron C5r, and figure 2.9A correlates the thin normal interval of N1 with C5r.1n. At least three cryptochrons appear to be preserved with the thick normal interval of N2–N5; the oldest of these, R3, does not appear in CK95, but is present in marine sediments (Schneider 1995) and at least one continental section (Agustí et al., 1997; Garcés et al. 1997). The next two thin reversed intervals, R4 and R5, appear to represent cryptochrons C5n.2n-3 and C5n.2n-2; cryptochron C5n.2n-1 is not seen but is predicted to be at the 87-m level; perhaps the VGP shallowing at the 92-m level found in several of the Sinap sections (see fig. 2.6) represents an incomplete record of this event. The next intervals, R6 and N6, can be correlated with some combination of C5n.1r and C4Ar.2n. The preferred correlation is given in figure 2.9A and matches N6 with both C5n.1n and C4Ar.2n. This correlation assumes that the brief reversed interval of C4Ar.3r was missed, but, as discussed (see fig. 2.6), there is some evidence that this event might be present at the very top of the Delikayincak section. Future sampling could test this idea. Alternative correlations would match R6 with C5n.1r and N6 with C5n.1n and assume that C4Ar.2n was missed (fig. 2.9B), or correlate R6 with C4Ar.3r and N6 with C4Ar.2n (fig. 2.9C). Both of these correlations are considered to be somewhat less likely than the correlation given in figure 2.9A because both would argue for more variable sedimentation rates (see below), but the correlation in figure 2.9B seems more likely than that in figure 2.9C because C4Ar.2n as seen in the Kavakdere south section is a very brief event that appears not to have been sampled in the Kavakdere north section.

The moderately thick reversed interval R7 is correlated with the long reversed interval C4Ar.2r in figures 2.9A–C. This interpretation seems to be quite secure because Chron C4Ar is the first major reversed interval that follows the distinctive long normal of C5n. Interval R7 is the thickest reversed interval in the Sinap section and it follows the thick predominantly normal interval N2–N6. The preferred correlation of N7 illustrated in figures 2.9A–C is with C4Ar.1n. This correlation does point to an inferred jump in sedimentation rates that may not in fact be as high as what is inferred from figures 2.9A–C. The CK95 gives C4Ar.1n as 0.078 Ma in duration and C4Ar.2n as 0.062 Ma in duration, with the younger interval being about 1.3 times the length of the older. Recent results from study of a marine core suggest that C4Ar.1n might be as much as twice as long as C4Ar.2n (Schneider 1995: figs. 8 and 9). If future work revises upward the estimated duration of this Chron, the inferred jump in sedimentation rates across interval N7 would be reduced.

As noted above, the Sinap section above the 170-m level has a lower sampling density than the lower section, and polarity intervals R8–R9 are defined by one or two class A sites only. The correlations in figures 2.9A–C match N8 with new cryptochron C4Ar.1r-1, which also appears to be present in the Kavakdere south section (see discussion above), and N9 with C4An. The uppermost normal interval N10 is incompletely sampled but may represent C4r.2r-1. The absence of a thick interval of reversed polarity sediments at the top of the section strongly suggests that these sediments do not continue far into Chron C4r.

A fourth correlation is given in figure 2.9D. This correlation is identical to that shown in figure 2.9B in that the lower part of the section is matched with Chron C5n, but it matches R7 with C4Ar, N7 with C4An, the next interval of predominantly reversed polarity (R8–R9) with C4r, and N9 with C4An.2n. This correlation is considered to be less likely than the other three correlations because many of the polarity events sampled in the Kavakdere sections are missed. In addition, because the sedimentary rocks in this upper part of the sequence continue to be dominated by siltstones and thin coarser-grained channel deposits, there is no support for a marked decrease in the rate of compacted sediment accumulation above the 129-m level.

Fossil localities in the Sinap Tepe region have been found in all but the upper portion of the composite section. Their interpolated ages based on the correlation in figure 2.9A are given in table 2.3. If future work provides evidence in favor of the correlation in figure 2.9B or C over that in figure 2.9A, note that the change would result in only minor adjustments to the interpolated ages of the fossil localities.

Igbek

As noted earlier, the Igbek section is located ~20 km northeast of Kazan (fig. 2.1) and preserves a 94-m thick section (see fig. 2.7). There are no volcanic units in this section and so radioisotopic age control for the section is lacking. *Hipparion* is found at fossil locality 49 near the top of the section. This biostratigraphic marker provides evidence that at least the upper part of the section postdates the beginning

Table 2.3. Interpolated Ages of Fossil Localities

Stratigraphic Section and Fossil Locality	Polarity Interval	Correlative Chron from GPTS		Interpolated Age (Ma)
Kavakdere north				
26, 27, 30, 33, 70	R4		C4r.1r	8.121
Kavakdere south				
69	N7	C4r.1n		8.230
Kavakdere north				
34	R3		C4r.2r	8.440
68	N3	C4An		8.866
Igbek				
49	N5		C4r.1r	9.130
Sinap Tepe composite				
10	N7	C4Ar.1n		9.279
1	N7	C4Ar.1n		9.288
7	N7	C4Ar.1n		9.295
75	N7	C4Ar.1n		9.301
84	R7		C4Ar.2r	9.367
83	R7		C4Ar.2r	9.452
11	R7		C4Ar.2r	9.483
12, OZ02	N6	C4Ar.2n-C5n.1n		9.590
OZ01, S01	N6	C4Ar.2n-C5n.1n		9.683
8A	R6		C5n.1r	9.886
8B	R6		C5n.1r	9.918
114	N5	C5n.2n-1n, −2n		9.967
108/8	N5	C5n.2n-1n, −2n		9.970
91	N5	C5n.2n-1n, −2n		9.977
72	N5	C5n.2n-1n, −2n		10.080
108	N5	C5n.2n-1n, −2n		10.135
106	N5	C5n.2n-1n, −2n		10.206
113	N5	C5n.2n-1n, −2n		10.306
89	N5	C5n.2n-1n, −2n		10.406
87	R5		C5n.2n-3r	10.452
93	N4	C5n.2n-3n		10.488
121	N4	C5n.2n-3n		10.526
94	N4	C5n.2n-3n		10.551
122	N4	C5n.2n-3n		10.577
107	N4	C5n.2n-3n		10.653
4	N4	C5n.2n-3n		10.692
88	N3	C5n.2n-4n		10.730
64	N3	C5n.2n-4n		10.765
104	N3	C5n.2n-4n		10.868
65	N3	C5n.2n-4n		10.899
Inönü				
24, 24A	—[1]	—		~15–16

Note: Chron ages are from GPTS of Cande and Kent (1995).

[1]No data available.

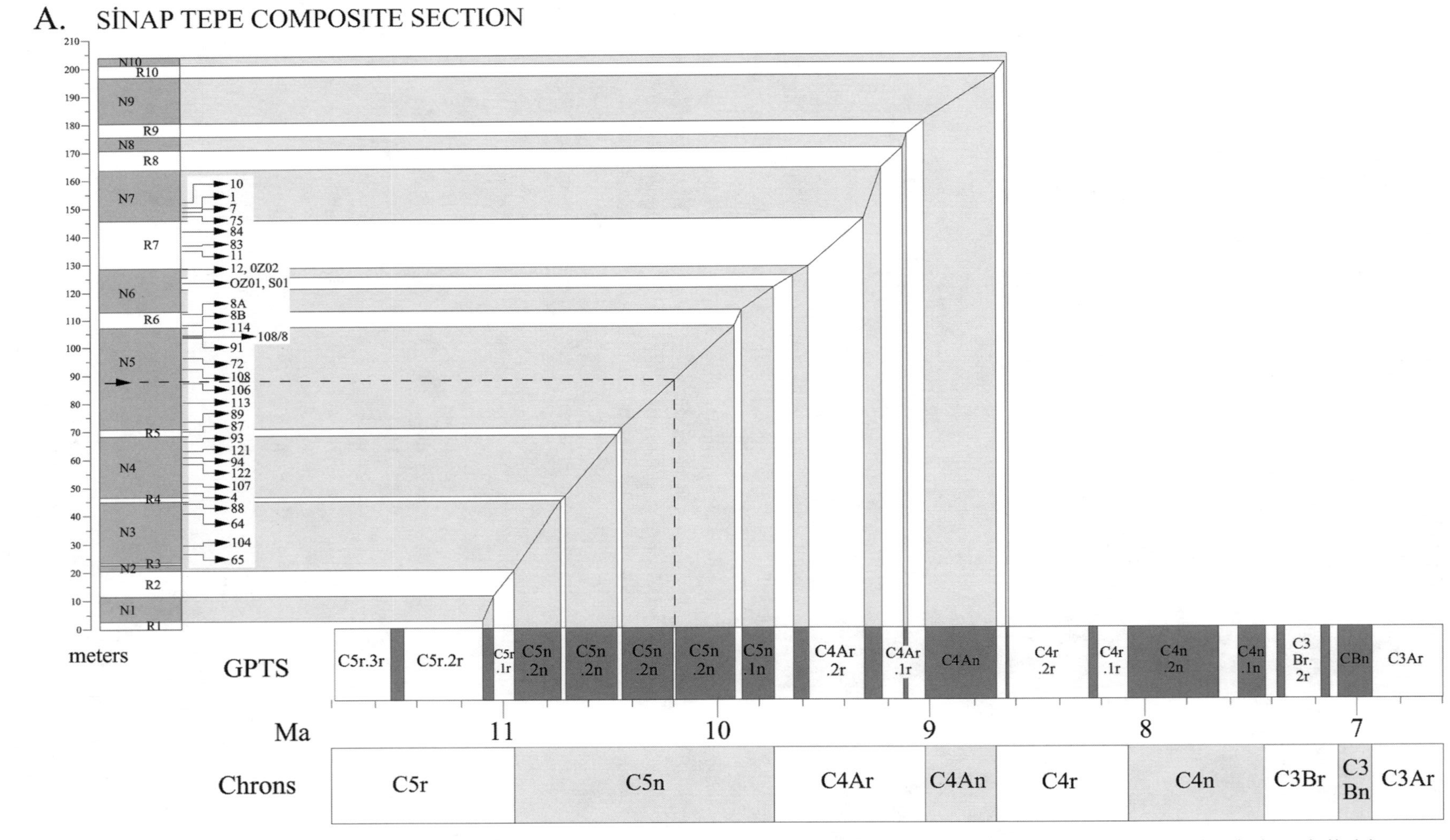

Figure 2.9. Correlation of the Sinap composite section (fig. 2.6) with the GPTS of CK95. **(A)** The preferred correlation matches the long normal interval in the lower half of the section with C5n and the reversed interval of R7 with C4Ar.2r, and assumes that the brief reversed interval of C4Ar.3r is within N6. The N7 interval is matched with C4Ar.1n. **(B)** This correlation matches N6 with C5n.1n and assumes that C4Ar.2n was missed. **(C)** This correlation matches N6 with C4Ar.2n and the lower normal intervals with C5n. **(D)** This correlation assumes that the normal events in C4Ar were missed, and that N7 is correlated with the C4An. Whereas correlations **(A)**–**(C)** have only minor impacts on the estimated ages of the fossil localities, correlation **(D)** estimates much younger ages for the uppermost localities. Dashed lines indicate predicted level of cryptochrons in C5n.

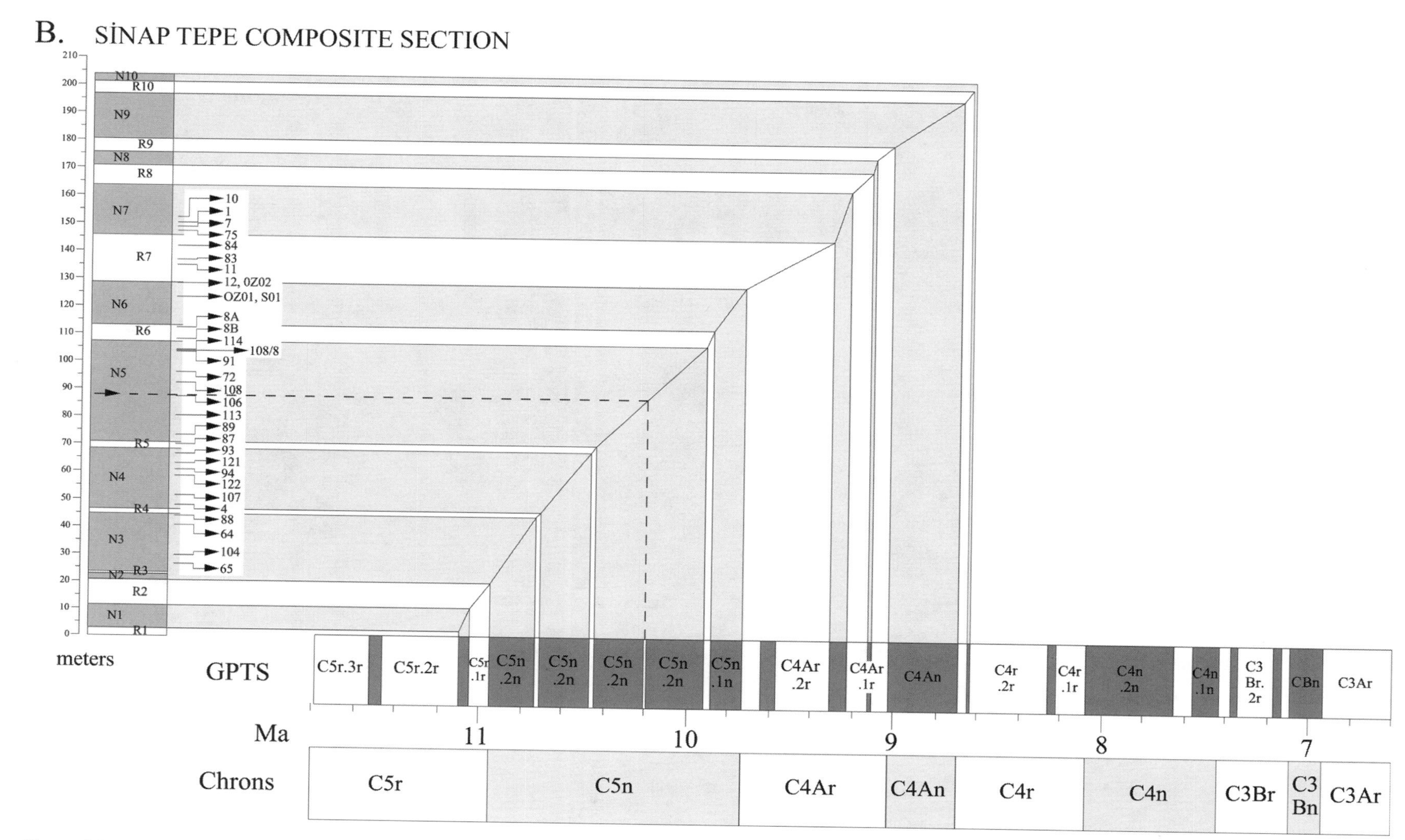

Figure 2.9. *(continued)*

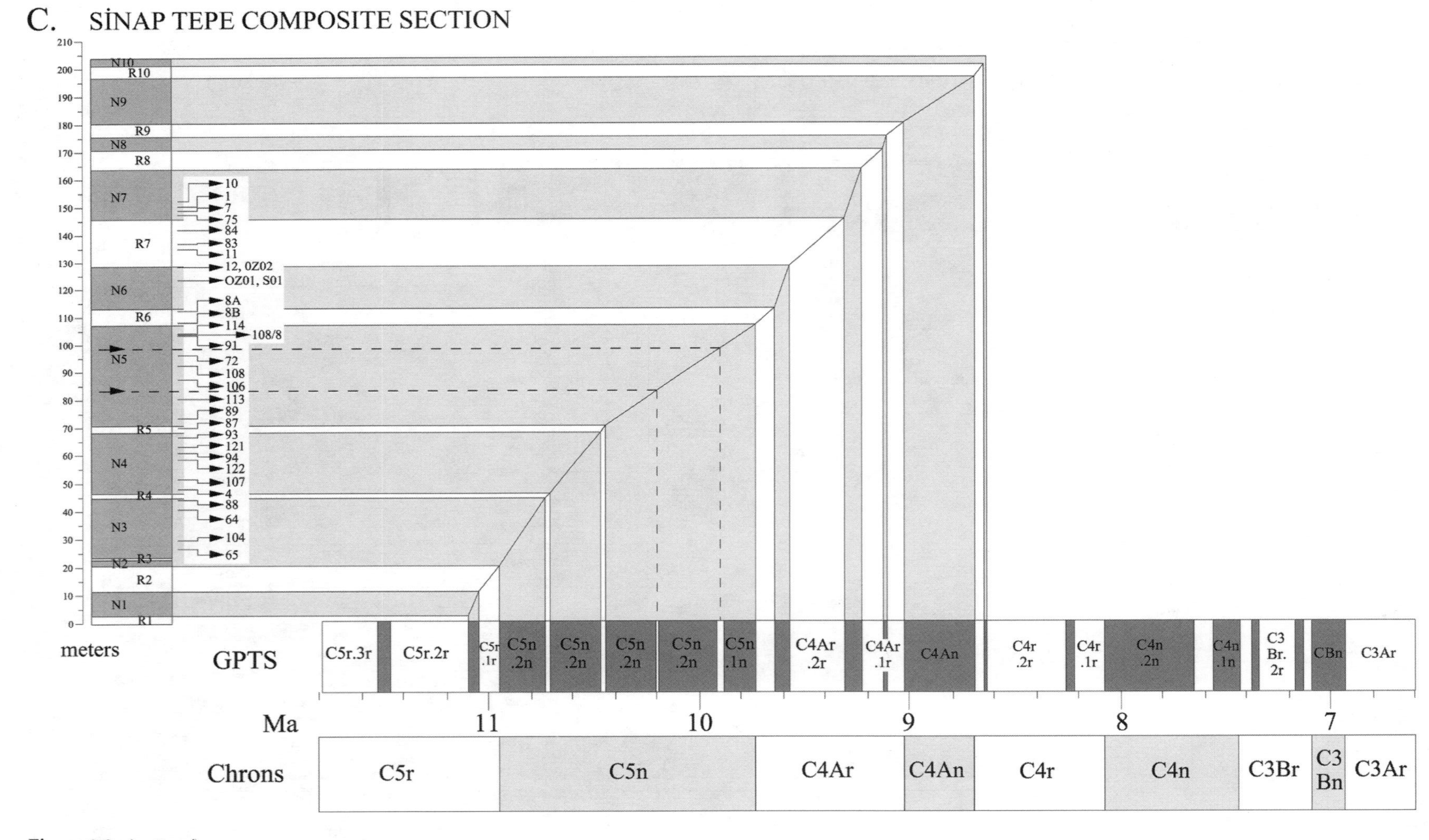

Figure 2.9. *(continued)*

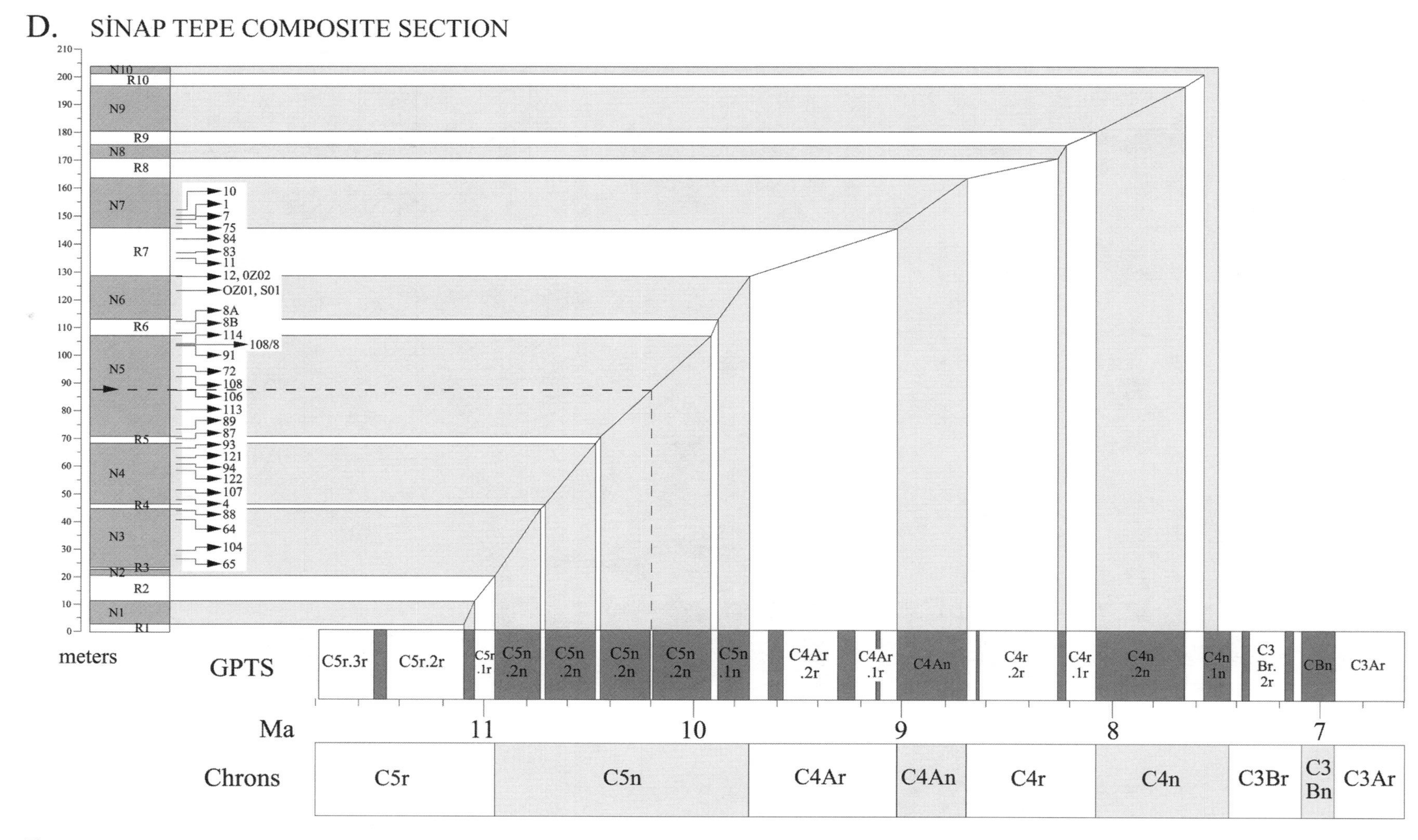

Figure 2.9. *(continued)*

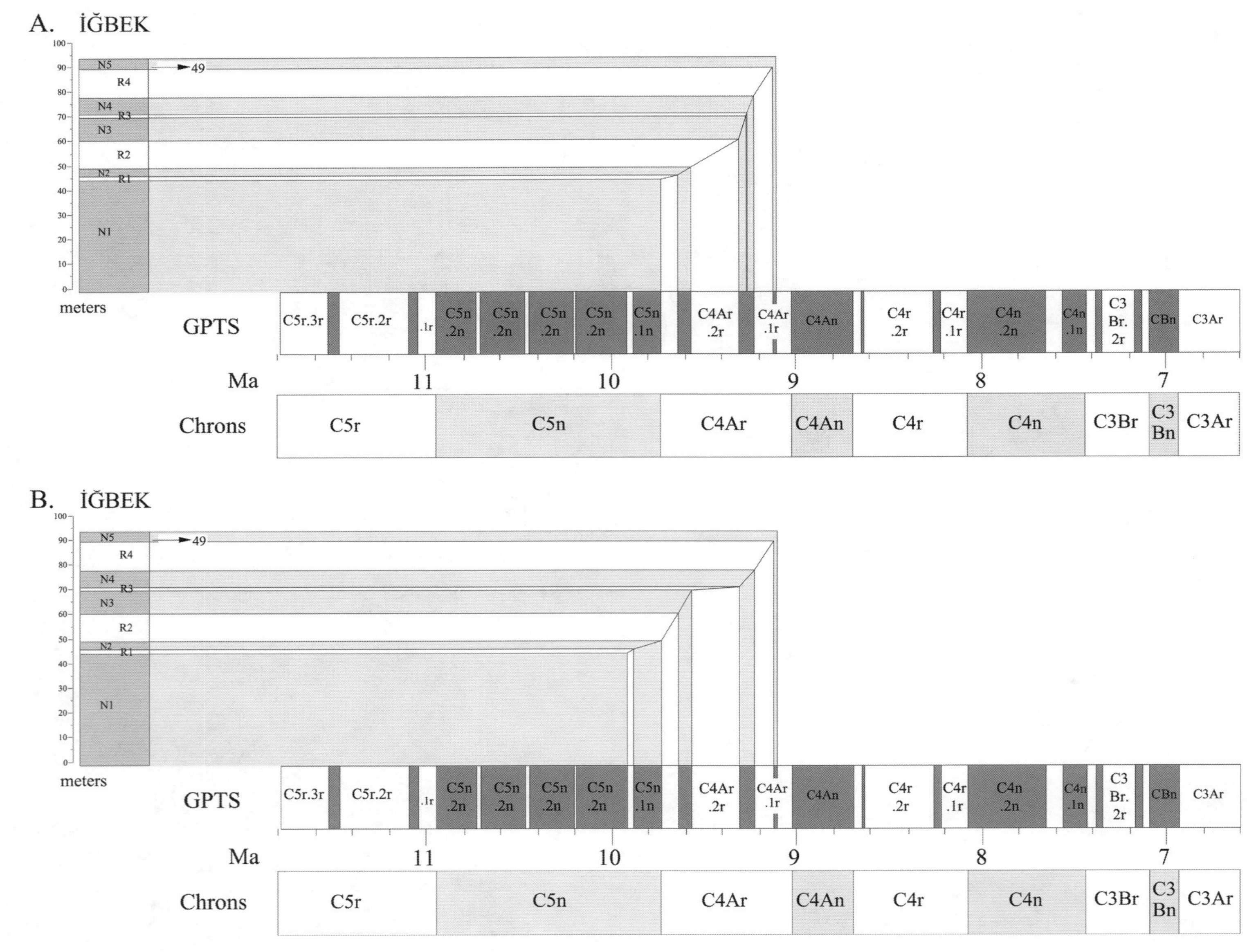

Figure 2.10. Correlation of Igbek (fig. 2.7) with the GPTS of CK95. (**A**) The preferred correlation matches the normal interval at N3–N4 with C4Ar.1n and implies a jump in sedimentation rates at this level in the section, which is correlated with an increased percentage of channel units. (**B**) This correlation assumes that the thin reversed zone at R3 represents C4Ar.2r and that the bulk of this interval was removed by erosion. This correlation implies highly variable sedimentation rates through the section.

of Chron C5n. The correlation with the GPTS shown in figure 2.10A follows Feseha (1996) and matches the thick normal polarity interval N1 with Chron C5n and the thin reversed (R1) and normal (N2) intervals with C4Ar.3r and C4Ar.2n, respectively. It is also possible that R1 and N2 represent C5n.1r and C5n.1n, respectively, and that C4Ar.3r and C4Ar.2n are both contained but not separately identified within R2. The former correlation (shown in figure 2.10A) is considered more likely than the latter (not shown) because the reversed polarity event of C4Ar.3r is estimated to be nearly 2.5 times the duration of C5n.1r and should be more readily identifiable. None of the cryptochrons known for C5n are found in N1 at Igbek, but one clearly reversed sample at 18 m and a shallow class A VGP site at the base of the section might represent some of these cryptochrons. Resampling could test for the existence of these brief reversed intervals.

The next interval in the Igbek section is the moderately thick zone of reversed polarity R2, which is correlated with C4Ar. This seems to be a fairly secure correlation because this is the first thick interval of reversed polarity that follows a thick normal interval (but see below). There are two possible matches that follow from this correlation. The first correlates R2 with C4Ar.2r, and the next predominantly normal interval N3–N4 with C4Ar.1n. This correlation in turn argues for a large jump in sedimentation rates across the latter interval (fig. 2.10A). A second correlation focuses on the thin normal interval R3 (70 m) that is documented by a single class A site with a low VGP value and correlates N3 with C4Ar.2n, the short interval R3 with C4Ar.2r, and N4 with C4Ar.1n (fig. 2.10B). The second correlation finds some support in the observation that a series of channels comes into the section at ~70 m. These channels have an erosional base, and samples taken directly below the base of the channels are reversely magnetized with a very shallow VGP (R3), whereas samples in and above the channels are of normal polarity (N4). The presence of these erosional channels makes it possible that a longer duration reversed polarity interval was once preserved here but was subsequently eroded by the down-cutting channels. However, given the long duration of C4Ar.2r, this correlation requires an extended period of no deposition and/or a significant loss of section by erosion. It can, however, also be argued that the increased incidence of channels above the 60-m level (see Lunkka et al. 1999; fig. 1.8) and the general coarsening-upward aspect of this section represent an increase in sedimentation rate, thus supporting the first correlation. In either event, the next reversed interval of R4 is correlated with C4Ar.1r, and the capping normal interval of N5 with the newly identified cryptochron C4Ar.1r-1. It is, however, also possible that N5 represents the very base of C4An.

The only fossils found at Igbek occur in fossil locality 49 at the 90-m level. The correlation presented here gives an age estimate of 9.130 Ma (revised estimate of the age of C4Ar.1r-1 from Schneider [1995] using CK95; see table 2.3) or 9.025 (base of C4An, CK95). Other correlations that match the base of the section with even younger normal events such as C4An are considered very unlikely because the fauna from locality 49 would then be estimated to be ~7.5 Ma in age, and the biostratigraphy does not support such a conclusion (Fortelius et al., chapter 12; Sen, chapter 5; Gentry, chapter 15; Sanders, chapter 10; all in this volume).

Discussion

Comparison of Sedimentation Rates

The correlations with the GPTS given in figures 2.8–2.10 permit a comparative assessment of the compacted sediment accumulation rates estimated for the various sections. These data are shown as individual sections in figure 2.11 and are layered upon one another in the upper part of the figure using the preferred correlations for Sinap Tepe (fig. 2.9A) and Igbek (fig. 2.10A). This figure demonstrates the great variability that is often seen with brief Chrons that is probably an artifact of sampling densities and depositional hiatuses, as well as the uncertainties that surround the interpolated ages of Chron boundaries, which are an inherent feature of the GPTS. It is obvious that there are only minor differences among the Sinap correlations shown in figure 2.9A–C, and the difference between these curves and the inferred decrease in sediment accumulation rates in the upper portion of the Sinap correlation in figure 2.9D is apparent. The overlaid sections in the upper part of figure 2.11 point to the general similarity in accumulation rates in the four areas. The mean rates are Kavakdere north, 0.061 mm/yr; Kavakdere south, 0.072 mm/yr; Igbek, 0.075 mm/yr; and Sinap Tepe, 0.081 mm/yr.

Unique Signatures in the GPTS

The portion of the GPTS recorded in each of the sections discussed above has several distinctive signatures that are well established. For example, the cryptochrons within the distinctive long normal of Chron C5n have been sampled in numerous continental sections, and the documentation of these events as reversals of the magnetic field shows that these are not simply fluctuations in the intensity of the magnetic field. These brief polarity reversals appear to be present in many of the sections reported here. There are several other interesting features in the data we collected that point to the existence of perhaps other unique aspects of field behavior during this time period. If these features can be identified, they could in turn prove to be useful for establishing more confident correlations with the GPTS. Here we discuss two of these features.

One interesting feature seen in nearly all of the sections concerns the first sizable interval of reversed polarity that follows the basal long duration interval of normal polarity. Note that there is a brief shallowing of the VGP values near the middle portion of this interval (fig. 2.12, arrow 1). This feature is found in four of the five sections in this study that sample the thick reversed interval correlated with C4Ar.2r; the Kavakdere north section is the only one

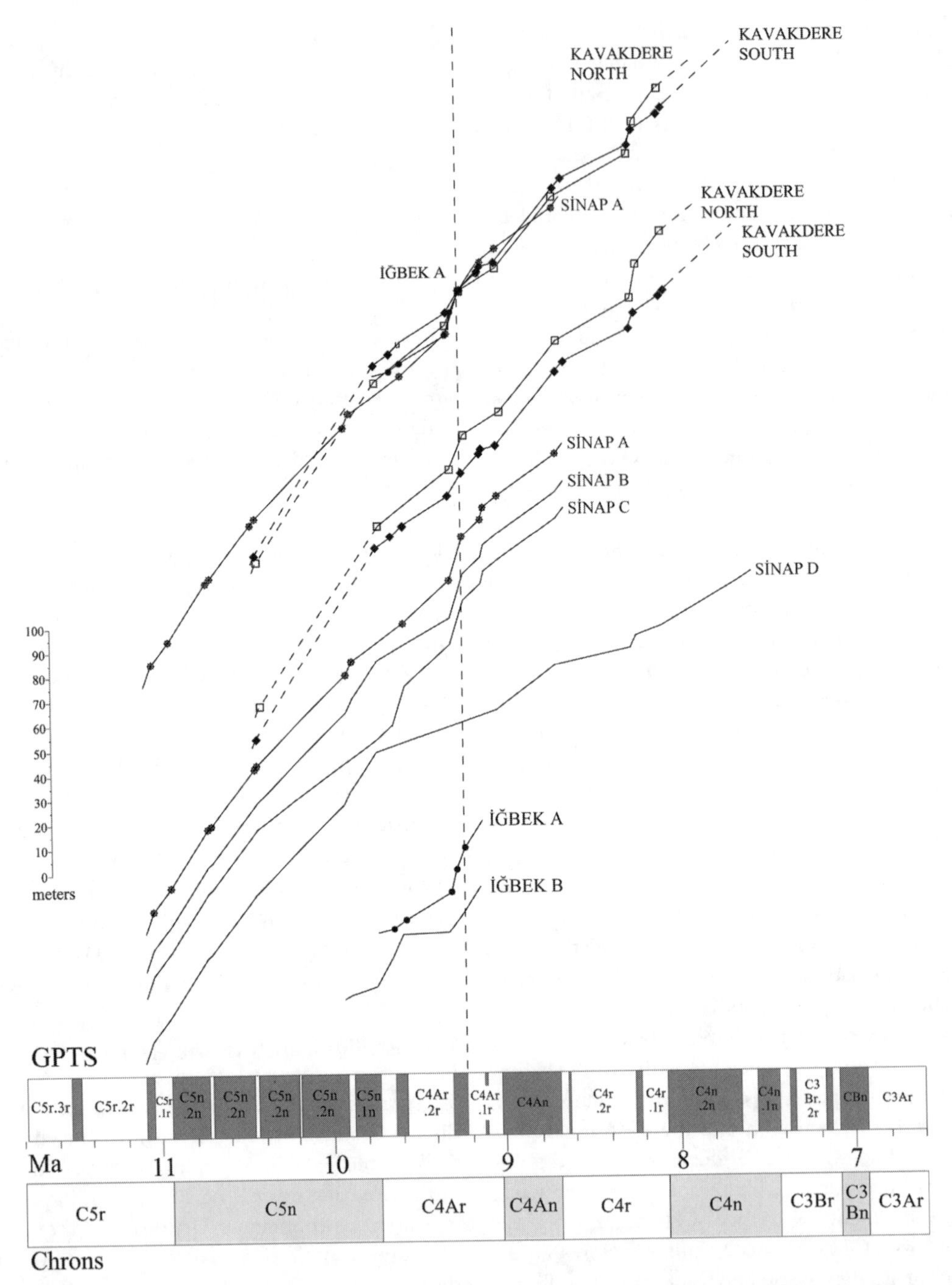

Figure 2.11. Curves of compacted sediment accumulation rates shown separately for each of the correlations in figures 2.8–2.10, with the preferred correlations for Sinap Tepe (fig. 2.9A) and Igbek (fig. 2.10A) shown in the overlaid curves at the top of the figure. Overlaid curves are centered on the inferred endpoint of C4Ar.1n. The stacked curves reveal similar compacted sediment accumulation rates for all sections.

that does not report this shallow VGP, and it is likely that it was missed because of the low sampling density through this part of the section. Other sections with secure correlations to the GPTS that also record this shallowing are found in the study by Tauxe (1979) of Ratha Kas and the newly reported Kangra and Nalad Khad sections by Brozovic and Burbank (2000), all in the Siwaliks of Pakistan. Four additional Siwaliks sections that do not record the shallowing are those by Johnson et al. (1983), Meigs et al. (1995), Khan et al. (1997), and Ojha et al. (2000), but these studies all have somewhat lower sampling densities through this interval. The high resolution log of Schneider (1995) on marine core 845A-14H shows that there is a flip from negative to positive and then back to negative inclination that is coupled with a drop in intensity and a slightly more positive declination in the middle of the reversed interval correlated with C4Ar.2r (Schneider 1995, fig. 3, 131 mbsf level). Because this interval of the GPTS is so distinctive, it would be fairly easy to study a densely sampled continental section to test for the existence of this suggested field behavior.

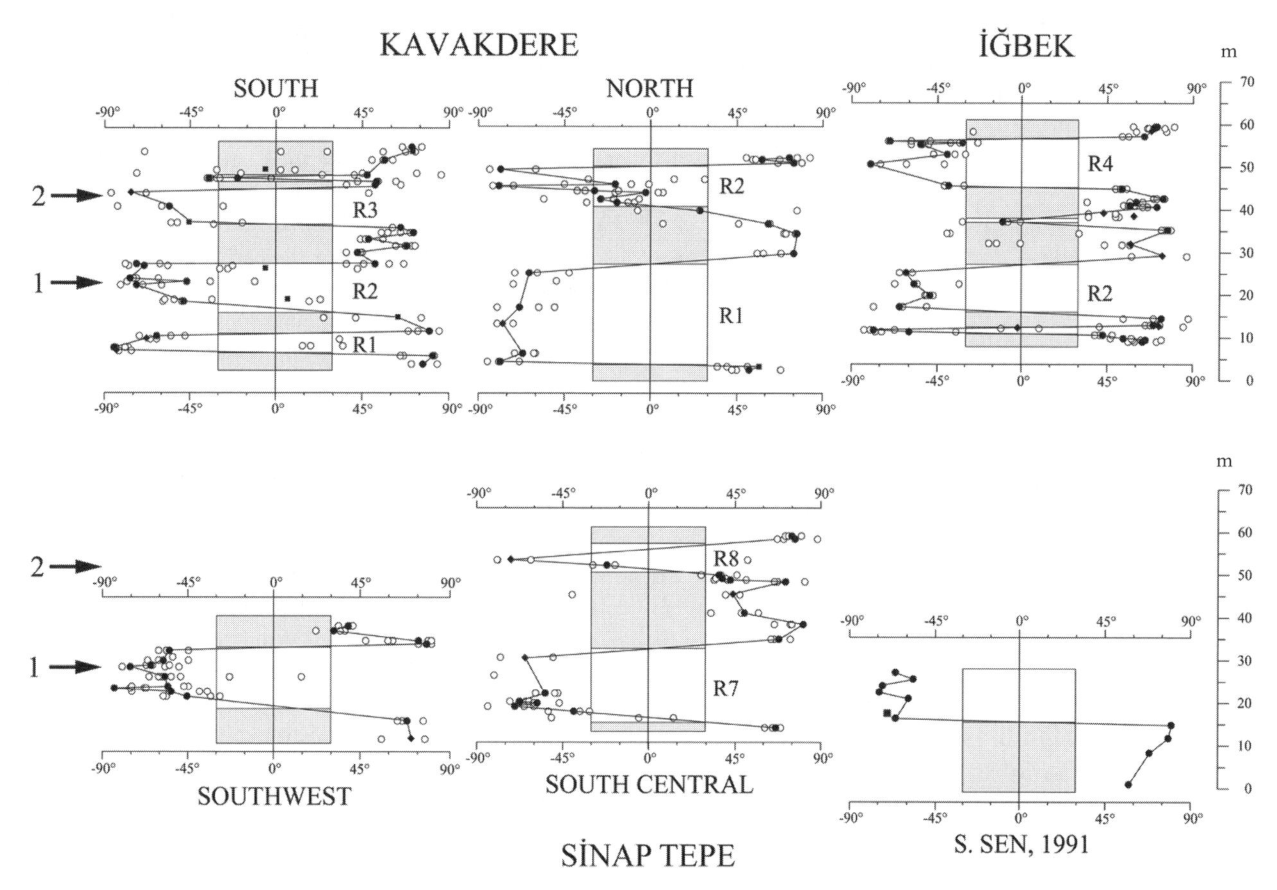

Figure 2.12. Detail of the reversal stratigraphies from the various sections shown in figures 2.5–2.7. Arrow 1 compares the reversed polarity interval correlated with C4Ar.2r and reveals a VGP shallowing in about the middle of this interval. Arrow 2 compares the protracted nature of the transition from normal to reversed polarity across C4Ar.1n to C4Ar.1r. Both of these signatures are seen in other sections from around the world. Symbols as in figure 2.5.

A second interesting feature we noted in our study concerns the transition from normal to reversed polarity that is recorded in the interval correlated with the C4Ar.1n-C4Ar.1r boundary. As noted above, this interval is distinctive in the sections reported here because of what appears to be the protracted manner of the transition. Each of the sections that preserves this interval contains sites with intermediate VGPs that record the transition to reversed polarity (fig. 2.12, arrow 2). A stereoplot of the individual samples and sites from the Sinap south central section (fig. 2.13) illustrates the polarity transition. Although evidence for polarity transitions in magnetostratigraphic sections is sometimes found, note that none of the other boundaries across intervals of opposite polarity from the sections presented here preserve any indication of transitional sites, even though the sampling densities across these boundaries are roughly equivalent to those of this particular event. These data suggest that either (1) the transition event across the C4Ar.1n-C4Ar.1r boundary is typical of other transitions thought to occur on a scale of ~4×10^3 years (Opdyke et al., 1973; Butler 1992), and the sampling in the Sinap Tepe area just happens to record evidence of a typically rapid transition event in multiple sections; or (2) the transition event across this boundary represents an atypical event characterized by a protracted reversal transition that is more readily sampled than are more typical, rapid transitions.

It is again possible to examine other well-calibrated geologic sections to search for the pattern displayed by the transition from C4Ar.1n to C4Ar.1r. Perhaps the best-known reversal transition from the entire Siwaliks sequence is the one reported by Tauxe and Opdyke (1982) in the Ganda Kas section that was subsequently studied in multiple parallel sections at the same stratigraphic level in Pakistan (Behrensmeyer and Tauxe 1982; Tauxe and Badgley 1984, 1988; Badgley and Tauxe 1990). Interestingly, other studies that focused on longer stratigraphic sections in other parts of the Siwaliks have also documented intermediate sites that record this reversal transition (Meigs et al. 1995; Khan et al. 1997) or at least the overall pattern of the VGP shallowing (see Johnson et al. 1983, but note that these authors correlate the normal interval at ~860 m with C4An; Meigs et al. [1995] resampled the Haritalyangar section and now correlate this interval with C4Ar.1n). It seems very unlikely that this reversal transition is typical because the latter two studies had lower overall sampling densities than the studies by Tauxe and her colleagues. In fact, Tauxe and Opdyke (1982) first identified this polarity transition in a study that used a low rather than high sampling density. On balance,

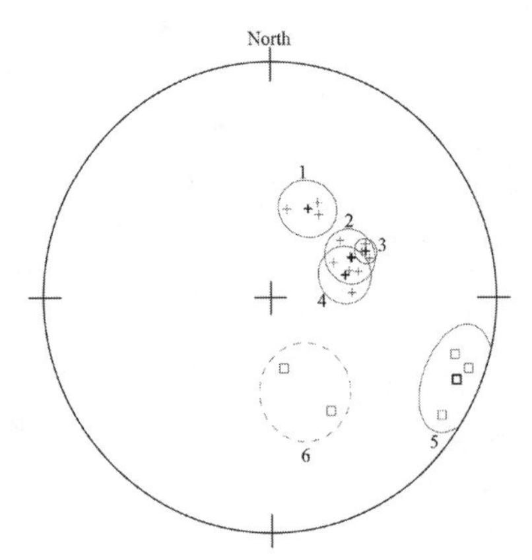

Figure 2.13. Equal area projection of the transition from normal to reversed polarity recorded by six sites across the N7–R8 boundary in the Sinap south central section in figure 2.6. Crosses, upper hemisphere; hollow squares, lower hemisphere; bold symbols, means; circles of confidence at the 95% level. The dashed circle for the two samples from site 6 simply encloses these samples.

the evidence suggests that this polarity transition was unusual in that the magnetic field took much longer on average to complete its reversal transition.

The nature of the C4Ar.1n event is also interesting and may point to why this polarity transition appears to be protracted. In the course of tracing the reversal boundary, Tauxe and Badgley (1984, 1988) densely sampled several short sections along strike and showed that this normal event actually consists of one or two brief intervals of reversed polarity that occur before the full transition to the reversed interval that is correlated with C4Ar.1r (Tauxe and Badgley 1984: fig. 3A at 6 m and 11 m; Tauxe and Badgley 1988: figs. 6–11). The detailed pattern that they document for this interval is very similar to the detailed record seen in N3–N4 at Igbek with its intervening thin R3 reversal, and N7 at Sinap Tepe (although the VGPs in this section only shallow and do not show a full reversal). A similar pattern but again without the reversal is seen at Haritalyangar (Johnson et al. 1983). Given the nature of what appears to be a very complex polarity transition, it may be that typical sampling densities will only reveal a pattern that appears to represent a protracted transition because only fragments of the pattern will be sampled. Whatever the case may be, future work should aim to better characterize the pattern of the magnetic field behavior in C4Ar.2r, C4Ar.1n, and the polarity transition to C4Ar.1r because a high-resolution approach would offer the potential for more confident correlations to the GPTS.

Age Determinations and Faunal Correlations

Earlier work on the Sinap faunas at both local and regional scales suggested that the lower and middle members of the Sinap Formation documented the transition from the Astaracian to the Vallesian, whereas sites in Kavakdere dated to the Turolian (see Kappelman et al. 1996 for a review). Our work places fossil localities 65 and 64 in MN 8 or the late Astaracian on the basis of their small mammal assemblages (see Sen, chapter 5, this volume; Sarica and Sen, chapter 6, this volume) and the absence of hipparionines. We place locality 4 in MN 9 or the early Vallesian with the local first appearance of hipparionines (see Bernor et al., chapter 11, this volume). The date for the MN 8-9 transition as documented at Sinap Tepe is estimated to be midway between localities 64 and 4 at 10.728 Ma; this date is younger than that of 11.2 Ma estimated in Steininger et al. (1996). The first local appearance of hipparionine fossils at locality 4 is estimated to be 10.692 Ma, which is indistinguishable from the age of 10.7 Ma for its appearance in the Siwaliks (Pilbeam et al. 1996). The dates from both of these well-calibrated sections are significantly younger than the date of 11.1 Ma estimated for this FAD in Spain (Agustí et al. 1997; Garcés, et al. 1997). Locality 8A documents the co-occurrence of *Progonomys* cf. *cathalai* and *Cricetulodon hartenbergeri* (see Sen, chapter 5, this volume) at an estimated age of 9.886 Ma. The latter species is known only from the MN 9 localities of Can Llobateres and Pedregueras IIc in Spain; its earliest occurrence is estimated at 10.4 Ma (Agustí et al. 1997). The first occurrence of *Progonomys* in Spain is estimated to be in Chron C4Ar.3r at 9.642–9.740 Ma (Agustí et al. 1997), whereas in Pakistan this genus first occurs at 12.3 Ma (Pilbeam et al. 1996); taken together, these data suggest that the movement of this taxon from Asia to Europe occurred over several million years. The estimated ages of *Ankarapithecus* of 9.886 Ma at locality 8A and 9.590 Ma at locality 12 (see Kappelman et al., this volume) are somewhat younger than our original estimate (Alpagut et al. 1996). These ages show broad overlap with *Dryopithecus* in Spain (Moyà-Solà and Köhler 1993; Agustí et al. 1996), *Ouranopithecus* in Greece (Bonis et al. 1990; Bonis and Koufos 1999), and *Sivapithecus* in the Nagri Formation of the Siwaliks (Barry 1986), but all of these hominoids are nearly three million years younger than the earliest occurrence of large-bodied hominoids in Pakistan (Kappelman et al. 1991). The upper Kavakdere localities are placed in MN 10–11 and the dates of 8.866–8.121 Ma for these localities place them squarely within the age range of the Turolian as estimated by Steininger et al. (1996).

Conclusion

Studies of rock magnetism of samples collected from the Sinap Formation of central Turkey reveal that these sedimentary rocks retain a reliable signature of the ancient magnetic field and support the idea that these samples can be used to construct a paleomagnetic reversal stratigraphy. Several radioisotopic dates from volcanic units that underlie the Sinap Formation, along with an estimate for the first appearance datum of *"Hipparion,"* provide critical anchors to the time scale. Correlations between the paleomagnetic reversal stratigraphies of several densely sampled sections in the Sinap Formation and the GPTS of Cande and Kent (1995) demonstrate that the majority of fossil localities are dated to between 8.1 and 10.9 Ma, with the most densely sampled interval in the area around Sinap Tepe dating to between 9.3 and 10.9 Ma. The older fossil locality known as Inönü I may date to ~15–16 Ma. The first local occurrence of hipparionines in the Sinap Formation is at locality 4 and is estimated to be 10.692 Ma. The hominoid *Ankarapithecus* is known from two localities, 8A and 12, with age estimates of 9.886 Ma and 9.590 Ma, respectively.

Acknowledgments

The Sinap Project (1989–1995) enjoyed the efforts of a large number of colleagues and the results presented here reflect the collective efforts of the entire team. Paleomagnetic lab work was assisted by A. Bean, P. Hake Shehan, D. Johnson, J. Stence, and P. Stubblefield; we thank them for their many hours of sample preparation and measurement. Dr. W. Gose offered important advice and lab assistance. We also express our gratitude to J. Crabaugh and P. Gibbard, who participated in the geologic fieldwork. The efforts of Mr. Ali Vzcan were indispensable in sample collection, and we thank him for his many weeks on the outcrop and his perseverance under stormy skies. We thank Professor Dr. B. Alpagut, with whom we initiated this project, and İ. Temizsoy, director, Museum of Anatolian Civilizations, who held the excavation permits and greatly facilitated our efforts in the field. The field and laboratory work was funded by grants from U.S. National Science Foundation (EAR 9304302), the L.S.B. Leakey Foundation, and the Academy of Finland, and assistance from Ankara University. Equipment loans from Motorola and software grants from Autodesk, Inc., are greatly appreciated. The research was supported by the general Directorate of Antiquities, T. C. Ministry of Culture and Tourism; we thank the Ministry for its seven years of support. The manuscript was greatly improved by the insightful comments from J. C. Barry, L. Flynn, and M. O. Woodburne, and we sincerely thank them for their assistance.

Literature Cited

Agustí, J., M. Köhler, S. Moyà-Solà, L. Cabrera, M. Garcès, and J. M. Parés, 1996, Can llobateres: The pattern and timing of the Vallesian hominoid radiation reconsidered: Journal of Human Evolution, v. 31, pp. 143–155.

Agustí, J., L. Cabrera, M. Garcés, and J. M. Parés, 1997, The Vallesian mammal succession in the Valles-Pendes basin, (NE Spain): A paleomagnetic calibration and correlation with global events: Palaeogeography, Palaeoclimatology, Palaeoecology, v. 133, pp. 149–180.

Alpagut, B., P. Andrews, M. Fortelius, J. Kappelman, I. Temizsoy, H. Çelebi, and W. Lindsay, 1996, A new specimen of *Ankarapithecus meteai* from the Sinap Formation of Central Anatolia: Nature, v. 382, pp. 349–351.

Badgley, C., and L. Tauxe, 1990, Paleomagnetic stratigraphy and time in sediments: Studies in alluvial Siwalik rocks of Pakistan: Journal of Geology, v. 98, pp. 457–477.

Barry, J., 1986, A review of the chronology of Siwalik hominoids, *in* J. C. Else, and P. C. Lee, eds., Primate evolution: Cambridge, Cambridge University Press, pp. 93–106.

Behrensmeyer, A. K., and L. Tauxe, 1982, Isochronous fluvial systems in Miocene deposits of northern Pakistan: Sedimentology, v. 29, pp. 331–352.

Berggren, W. A., D. V. Kent, C. C. Swisher III, and M.-P. Aubry, 1995, A revised Cenozoic geochronology and chronostratigraphy, *in* W. A. Berggren, D. V. Kent, M.-P. Aubry, and J. Hardenbol, eds., Geochronology, time scales and global stratigraphic correlation: Society for Sedimentary Geology Special Publication 54, pp. 129–212.

Bonis, L. De, and G. Koufos, 1999, The Miocene large mammal succession in Greece, *in* J. Augusti, L. Rook, and P. Andrews, eds., Hominid evolution and climatic change in Europe: Cambridge, Cambridge University Press, pp. 205–237.

Bonis, L. De, G. Bouvrain, D. Geraads, and G. Koufos, 1990, New hominoid skull material from the late Miocene of Macedonia in Northern Greece: Nature, v. 345, pp. 712–714.

Brozovic, N., and D. W. Burbank, 2000, Dynamic fluvial systems and gravel progradation in the Himalayan foreland: Geological Society of America Bulletin, v. 112, pp. 394–412.

Butler, R. F., 1992, Paleomagnetism: Magnetic domains to geologic terranes: Boston, Blackwell Scientific Publications, 319 pp.

Cande, S. C., and D. V. Kent, 1995, Revised calibration of the geomagnetic polarity timescale for the Late Cretaceous and Cenozoic: Journal of Geophysical Research, v. 100, no. B4, pp. 6093–6095.

Feseha, M., 1996, Magnetostratigraphy of the fossil-bearing Igbek Section of the Miocene Sinap Formation [M.A. thesis]: Austin, Texas, University of Texas, 66 pp.

Fisher, R. A., 1953, Dispersion on a sphere: Proceedings of the Royal Society, v. 217, pp. 295–305.

Garcés, M., L. Cabrera, J. Agustí, and J. M. Parés, 1997, Old World's first appearance datum of 'Hipparion' horses: Late Miocene large-mammal dispersal and global events: Geology, v. 25, pp. 19–22.

Gürbüz, M., 1981, Inönü (KB Ankara) Orta Miyosenindeki *Hemicyon sansaniensis* (Ursidae) turunum tanimlanmasi ve stratigrafik yayilimi: Türkiye Jeoloji Kurumu Bülteni C, v. 24, pp. 85–90.

Harland, W. B., A. V. Cox, P. G. LLewellyn, C. A. G. Picton, A. G. Smith, and R. Walters, 1982, A geologic time scale: Cambridge, Cambridge University Press, 131 pp.

Johnson, G. D., N. D. Opdyke, S. K. Tandon, and A. C. Nanda, 1983, The magnetic polarity stratigraphy of the Siwalik Group at Haritalyangar (India) and a new last appearance datum for *Ramapithecus* and *Sivapithecus* in Asia: Palaeogeography, Palaeoclimatology, Palaeoecology, v. 44, pp. 223–249.

Kappelman, J., J. Kelley, D. Pilbeam, K. A. Sheikh, S. Ward, M. Anwar, J. C. Barry, B. Brown, P. Hake, N. M. Johnson, S. M. Raza, and S. M. I. Shah, 1991, The earliest occurrence of *Sivapithecus* from the middle Miocene Chinji Formation of Pakistan: Journal of Human Evolution, v. 21 pp. 61–73.

Kappelman, J., S. Sen, M. Fortelius, A. Duncan, B. Alpagut, J. Crabaugh, A. Gentry, J.-P. Lunkka, F. McDowell, N. Solounias, S. Viranta, and L. Werdelin, 1996, Chronology and biostratigraphy of the Miocene Sinap Formation of Central Turkey, *in* R. L. Bernor, V. Fahlbusch, and H.-W. Mittman, eds., The evolution of western Eurasian Neogene mammal faunas: New York, Columbia University Press, pp. 78–95.

Khan, I. A., J. S. Bridge, J. Kappelman, and R. Wilson, 1997, Evolution of Miocene fluvial environments, eastern Potwar plateau, northern Pakistan: Sedimentology, v. 44, pp. 221–251.

Kirschvink, J. L., 1980, The least-squares line and plane and the analysis of palaeomagnetic data: Geophysical Journal of the Royal Astronomical Society, v. 62, pp. 699–718.

Lunkka, J.-P., J. Kappelman, D. Ekart, and S. Sen, 1998, Pliocene vertebrate locality of Calta, Ankara, Turkey: 1. Sedimentation and lithostratigraphy: Geodiversitas, v. 20, no. 3, pp. 329–338.

Lunkka, J.-P., M. Fortelius, J. Kappelman, and S. Sen, 1999, Chronology and mammal faunas of the Miocene Sinap Formation, Turkey, *in* J. Augusti, L. Rook, and P. Andrews, eds., Hominid evolution and climatic change in Europe: Cambridge, Cambridge University Press, pp. 238–264.

Meigs, A. J., D. W. Burbank, and R. A. Beck, 1995, Middle-late Miocene (>10 Ma) formation of the Main Boundary thrust in the western Himalaya: Geology, v. 23, pp. 423–426.

Moyà-Solà, S., and M. Köhler, 1996, A *Dryopithecus* skeleton and the origins of great-ape locomotion: Nature, v. 379, pp. 156–159.

Ojha, T. P., R. F. Butler, J. Quade, P. G. DeCelles, D. Richards, and B. N. Upreti, 2000, Magnetic polarity stratigraphy of the Neogene Siwalik Group at Khutia Khola, far western Nepal: Geological Society of America Bulletin, v. 112, pp. 424–434.

Öngür, T., 1976, Kizilcahamam, Camlidere, Celtikci ve Kazan dolayinin jeoloji durumu ve jeotermal enerji olanakrari, Unpublished Report, MTA, Ankara.

Opdyke, N. D., D. V. Kent, and W. Lowrie, 1973, Details of magnetic-polarity transitions in high deposition rate deep sea cores: Earth and Planetary Science Letters, v. 20, pp. 315–324.

Ozansoy, F., 1957, Faunes de mammifères du Tertiaire de Turquie et leurs revisions stratigraphiques: Bulletin of the Mineral Research and Exploration Institute of Turkey (Foreign Ed.), v. 49, pp. 29–48.

Ozansoy, F., 1965, Étude des gisements continentaux et de mammifères du Cènozoique de Turquie: Mémoires de la Société Géologique de France (nouvelle série), v. 44, pp. 1–92.

Pilbeam D., M. Morgan, J. C. Barry, and L. Flynn, 1996, European MN units and the Siwalik faunal sequence of Pakistan, *in* R. L. Bernor, V. Fahlbusch, and H.-W. Mittman, eds., The evolution of western Eurasian Neogene mammal faunas: New York, Columbia University Press, pp. 96–105.

Renne, P. R., C. C. Swisher III, A. L. Deino, D. B. Karner, T. L. Owens, and D. J. DePaolo, 1998, Intercalibration of standards, absolute ages and uncertainties in $^{40}Ar/^{39}Ar$ dating: Chemical Geology, v. 145, pp. 117–152.

Saemundsson, K., L. Kristjansson, I. McDougall, and N. D. Watkins, 1980, K-Ar dating, geological and paleomagnetic study of a 5-km lava succession in northern Iceland: Journal of Geophysical Research, v. 85, pp. 3628–3646.

Schneider, D. A., 1995, Paleomagnetism of some leg 138 sediments: Detailing Miocene magnetostratigraphy, *in* N. G. Pisias, L. A. Mayer, T. R. Janecke, A. Palmer-Julson, and T. H. van Andel, eds., Proceedings of the Ocean Drilling Program, Scientific Results, Volume 138: College Station, Texas, Ocean Drilling Program, pp. 59–72.

Sen, S., 1989, *Hipparion* datum and its chronological evidence in the Mediterranean area, *in* E. H. Lindsay, V. Fahlbusch, and P. Mein, eds., European Neogene mammal chronology: New York, Plenum Press, pp. 495–505.

Sen, S., 1991, Stratigraphie, faunes de mammifères et magnétostratigraphie du Néogène de Sinap Tepe, Province d'Ankara, Turquie: Bulletin du Museum National d'Histoire Naturelle, ser. 4e, v. 12, pp. 243–277.

Steiger, R. H., and Jager, E., 1977, Subcommission on geochronology: Convention on the use of decay constants in geo- and cosmochronology: Earth and Planetary Science Letters, v. 36, pp. 359–361.

Steininger, F. F., W. A. Berggren, D. V. Kent, R. L. Bernor, S. Sen, and J. Agustí, 1996, Circum-Mediterranean Neogene (Miocene-Pliocene) marine-continental chronologic correlations of European mammal units, *in* R. L. Bernor, V. Fahlbusch, and H.-W. Mittman, eds., The evolution of western Eurasian Neogene mammal faunas: New York, Columbia University Press, pp. 7–46.

Swisher, C. C., G. H. Curtis, T. Jacob, A. G. Getty, A. Surojo, and Widiasmoro, 1994, Age of the earliest known hominids in Java, Indonesia: Science, v. 263, pp. 1118–1121.

Tauxe, L., 1979, A new date for *Ramapithecus:* Nature, v. 282, pp. 399–401.

Tauxe, L., and N. D. Opdyke, 1982, A time framework based on magnetostratigraphy for the Siwalik sediments of the Khaur area, northern Pakistan: Palaeogeography, Palaeoclimatology, Palaeoecology, v. 37, pp. 43–61.

Tauxe, L., and C. Badgley, 1984, Transition stratigraphy and the problem of remanence lock-in time in the Siwalik red beds: Geophysical Research Letters, v. 11, pp. 611–613.

Tauxe, L., and C. Badgley, 1988, Stratigraphy and remanence acquisition of a palaeomagnetic reversal in alluvial Siwalik rocks of Pakistan: Sedimentology, v. 35, pp. 697–715.

Taylor, J. R., 1982, An introduction to error analysis: New York, Oxford University Press, 270 pp.

Watson, G. S., 1956, A test for randomness: Monthly Notices of the Royal Astronomical Society, Geophysics Supplement, v. 7, pp. 160–161.

Woodburne, M. O., and C. C. Swisher III, 1995, Land mammal high resolution geochronology, intercontinental overland dispersals, sea-level, climate, and vicariance, *in* W. A. Berggren, D. V. Kent, and J. Hardenbol, eds., Geochronology, time scales, and global stratigraphic correlations: A unified temporal framework for an historical geology: Society of Economic Mineralogists and Paleontologists Special Publication No. 54, pp. 338–364.

Woodburne, M. O., R. L. Bernor, C. C. Swisher III, 1996, An appraisal of the stratigraphic and phylogenetic bases for the *"Hipparion"* datum in the Old World, *in* R. L. Bernor, V. Fahlbusch, and H.-W. Mittman, eds., The evolution of western Eurasian Neogene mammal faunas: New York, Columbia University Press, pp. 124–136.

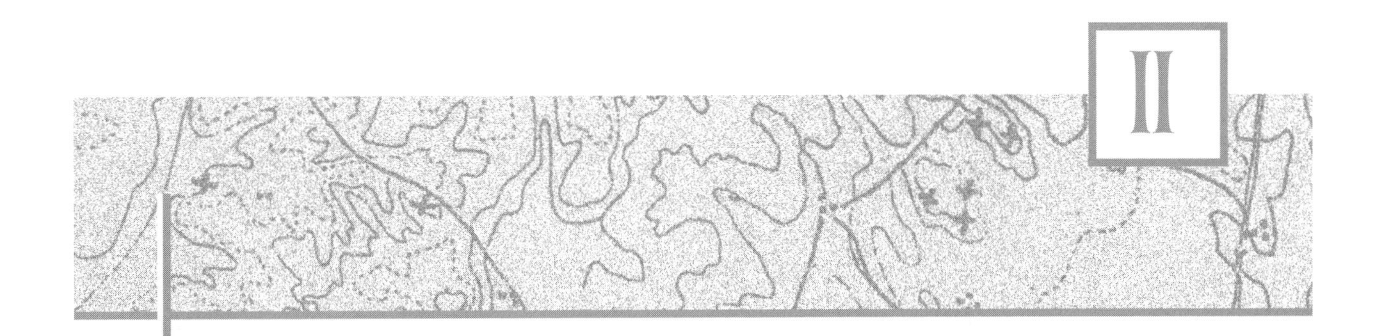

II

Mammal Paleontology

3

Genus *Schizogalerix* (Insectivora)

L. Selänne

Among small mammals in the Sinap Formation, the genus *Schizogalerix* Engesser, 1980 is well represented. Over 400 dental specimens have been recovered from ten localities in the lower and middle Sinap members. These localities range from ~8 to 11 Ma, spanning the Mammal Neogene (MN) faunal units MN 7/8–12.

Schizogalerix is an extinct genus of moonrats (subfamily Echinosoricinae) that lived during the middle and late Miocene. During this time the genus underwent a distinct evolution. Engesser (1980) separated *Schizogalerix* from the genus *Galerix* as an independent evolutionary lineage. He interpreted the most primitive species known, *S. pasalarensis,* as basal to the radiation of all other *Schizogalerix* species. The advanced species *S. sinapensis* from the Sinap Formation has already been described by Sen (1990).

The *Schizogalerix* material in the Sinap Formation brought new information about the morphology and evolution of the genus in Turkey. The lower anterior teeth of *Schizogalerix* could now be identified and the lower dentition of *S. sinapensis* described. The evolution of the genus involves a mosaic of continuously changing morphologic features in a stratigraphically dense record. The different morphologic forms are connected by intermediate forms that document a gradual evolution. This led to the problem of reconciling neontologic species concepts and morphotypic classification within the accepted nomenclature. The *Schizogalerix* populations in the Sinap localities were thus separated using the relative abundance of the different morphotypes (particularly in m_1, m_2 and M^1) into three polytypic species.

The gradual change of the genus *Schizogalerix* reaches over geologic time and across geographic space. *Schizogalerix* is divided into four evolutionary (and phylogenetic) lineages, of which the Turkish lineage is best known (Engesser 1980, Bi et al. 1999). There are no species common to the Turkish species elsewhere, but the resemblance is close. The European, African, and Chinese species seem to belong to separate evolutionary lineages (see page 82).

Methodology

The samples were produced by screen washing during field work from 1989 to 1995. The washing operation causes the isolated and fragmentary character of the samples. However, some fairly complete jaws of *Schizogalerix* have also been recovered.

To describe the different parts of the teeth, I follow the common dental terminology used by Engesser (1980). The greatest length and width were measured when possible with a precision of 0.05 mm as shown in figure 3.1. The height of the mandible was measured under the posterior root of p_4.

A statistical analysis of the measurements was made of p_3–m_3 and P^3–M^3. Due to the uncertainty of normal distribution in the sparse samples, a nonparametric Kruskal-Wallis one-way analysis of variance was chosen. When $p < 0.01$, the difference was considered significant.

The minimum number of individuals (MNI) was counted among the most common elements in a sample, where fragments were also considered to form whole teeth. NISP is an abbreviation for the number of identifiable specimens.

In the figures, the scale of magnification is either 10× or 15×, and all the teeth are drawn as sinistral (dextral tooth samples are denoted by *invert*). The figures of the teeth from the Sinap Formation are drawn after the originals; others are taken from the literature mentioned. All figures are drawn by the author.

All samples from the Sinap Formation are stored in the Museum of Anatolian Civilizations in Ankara, Turkey.

Specimens of the Genus Schizogalerix in the Sinap Formation

At least 42 individuals (based on MNI) of the genus *Schizogalerix* have been found in the Sinap Tepe area. The *Schizogalerix* material in the Sinap Formation is in some

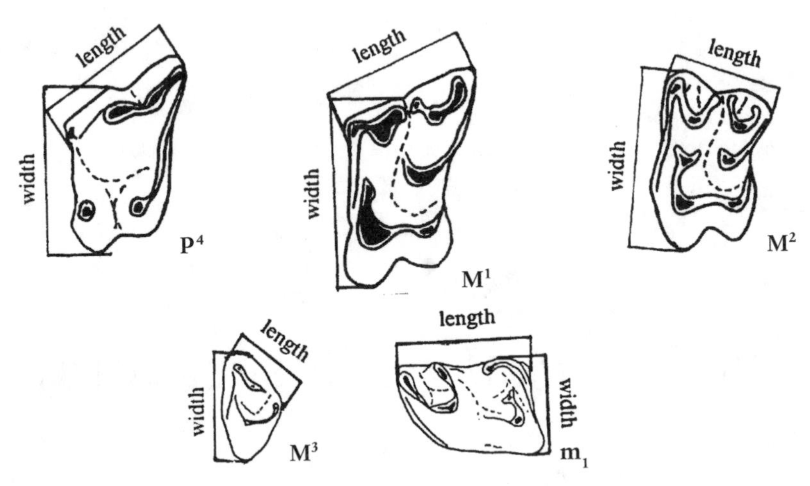

Figure 3.1. The definition of measurement points for P^4, M^1, M^2, M^3, and m_1.

aspects very homogeneous (e.g., dimensions, basic morphology), but distinct differences are also seen. The differences in the morphology, however, occur across the localities and can thus be grouped as morphotypes. Wear and individual variation may, to a certain extent, bridge the morphologic features between the forms, but usually the different forms may be recognized without difficulty. I first describe the basic morphology of the teeth and the morphotypes from all localities, after which I list the distribution of the different morphotypes and some other special features in the samples by locality. The lower incisors, the lower canine, and the lower first premolar of the genus *Schizogalerix* are described here for the first time.

Systematic Position

Order Insectivora Bowdich 1821
Family Erinaceidae Bonaparte 1838
Subfamily Galericinae Pomel 1848
Tribe Galericini Pomel 1848
Genus *Schizogalerix* Engesser 1980

Description

Mandible. The mandible is relatively pronounced, the average height of the mandibles being 2.82 mm (Appendix 3.1). The foramen mentale lies under either the posterior root of p_3 or the anterior root of p_4. The ramus ascendens makes an angle of 90° with the ramus horizontalis. The anterior part of the mandible is short (from the posterior end of the i_1 alveolus to the posterior root of p_4: 6.00 mm) (fig. 3.2A).

i_1. The first lower incisor has a long and labially concave root in an anteriorly slanting alveolus. The labial side is broad and round. There is a small cuspule on the posterior side and no cingulum (fig. 3.2A,B).

i_2. The second incisor is very similar to i_1, but it is narrower and more slender (fig. 3.2B).

i_3. The third incisor is a tiny single-rooted tooth with a round crown situated labially between the canine and i_2 (fig. 3.2A,B).

$c_{inf.}$. The lower canine is a robust single-rooted tooth in a large, anteriorly slanting alveolus. The posterior side of the tip rises to a higher point. Analogous to the two first incisors, there is a small cuspule on the posterior side and no cingulum (fig. 3.2A,B).

p_1. p_1 is a single-rooted tooth in a moderately large, anteriorly slanting round alveolus. There is no tooth in situ, but some isolated specimens can be regarded as the first premolar. The labial view of the tooth has some similarities to that in incisors and canines: the tip of the tooth is blunt and oblique, there is a small cuspule on the posterior side, and no cingulum. p_1 is short and one clear cusp is seen on the occlusal view (fig. 3.2A; see also fig. 3.13D,E).

p_2. p_2 has the shape of an ellipse. It consists of one main cusp in the middle and two smaller accessory cusps on both ends, of which the anterior cusp is much smaller. There are two roots (fig. 3.5A,D).

p_3. Its shape is similar to p_2, but p_3 is bigger. The posterior end is broader and has a pronounced triangular cingulum, which in some specimens forms a transverse ridge in the middle. There are two roots (fig. 3.2C).

p_4. The last lower premolar has a clear trigonid (fig. 3.2C, D). The broad posterior end has a pronounced posterior cingulum, which in a few specimens forms a transverse ridge in the middle. There are two roots.

m_1. m_1 has a very narrow trigonid and a projecting entostylid. The anterior cingulum ends at the base of the pro-

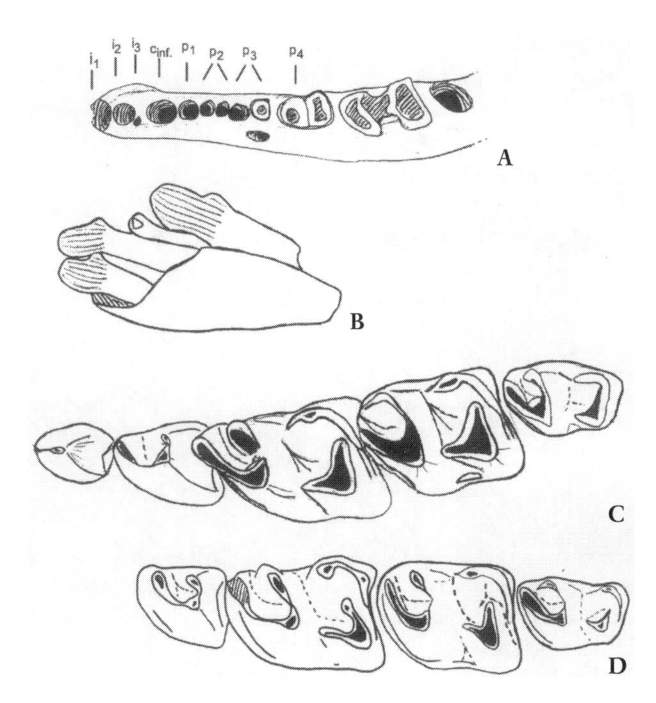

Figure 3.2. The genus *Schizogalerix*. **(A)** Locality 120, mandible fragment sin. (magnification 10×); **(B)** locality 4, mandible fragment sin. with i_1–$c_{inf.}$; **(C)** locality 94, mandible fragment sin. with p_3–m_2, locality 65 m_3 sin.; **(D)** locality 84, mandible fragment sin. (invert) with p_4–m_1, m_2 sin., and m_3 sin. Magnification 15×; reproduced at 50%.

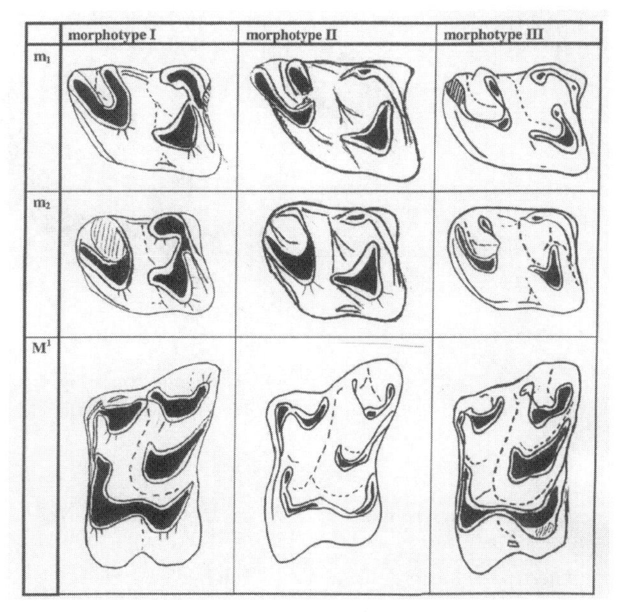

Figure 3.3. The morphotypes in m_1, m_2, and M^1 of the genus *Schizogalerix* in the Sinap Formation.

toconid. The construction of the talonid differs in three morphotypes (figs. 3.2C,D, 3.3):

1. The entoconid is continuous with the posterior cingulum. It widens toward the posterior arm of the hypoconid and connects with it.
2. The entoconid is continuous with the posterior cingulum without any connection to the hypoconid. The posterior arm of the hypoconid points obliquely toward the entoconid/posterior cingulum, making the hypoconid quite open. The valley between the hypocone and entoconid/posterior cingulum does not pass the posterior arm of the hypocone.
3. The posterior cingulum is continuous with the entoconid, which is very pronounced and high. The hypoconid is usually narrow and the posterior arm is vertically directed toward the middle of the entoconid. There is a distinct groove between the hypoconid and the posterior cingulum. In some specimens, the posterior arm of the hypoconid forms a cusp. There are two roots.

m_2. Compared with m_1, m_2 is rounder, has a narrower trigonid, and the anterior cingulum continues further under the protoconid. The entostylid is not as projecting. The construction of the talonid shows the same three morphotypes as m_1. There are two roots (figs. 3.2C,D, 3.3).

m_3. The trigonid is narrow; the anterior labial cingulum ends at the base of the protoconid. The hypoconid is either continuous with the entoconid (no posterior cingulum) or is separated from the entoconid (fig. 3.2C,D). There are two roots.

The upper dentition anterior to P^2 is unknown.

P^2. P^2 has one main cusp in the middle and a broader posterior end with a cingulum. There are two roots (fig. 3.4).

P^3. The labial cusps are high and joined. On the lingual side there are two lower cusps, of which the protocone is higher. A cingulum extends from the hypocone to the posterolabial corner. There is either a very small parastyle or none. P^3 is only posterolabially widened. There are three roots (fig. 3.4).

P^4. P^4 has almost the same morphology as P^3. The differences are that P^4 is bigger and anterolingual/posterolabially widened, the parastyle is more pronounced, and the groove between the paracone and the metacone is deeper. There are three roots (fig. 3.4).

M^1. M^1 is a large anterolingual/posterolabially widened tooth. The posterior arm of the metaconule and the hypocone are long and oblique. The protoconule is D-shaped. The parastyle is less pronounced and the metastyle is a distinct cusp. There is only a trace or no labial cingulum next to the paracone. The anterior cingulum starts from the parastyle and continues to the base of the protocone. The shape of the paracone and the metacone is expressed in three morphotypes (fig. 3.4):

1. The paracone and the metacone are broad. Mesostyles are slightly connected and the trigon basin does not continue to the labial end.

Figure 3.4. The genus *Schizogalerix*. Locality 8A, maxilla fragment sin. with P^2–M^2 and locality 4, M^3 sin. Magnification 15×; reproduced at 50%.

2. The paracone is narrow and the metacone narrower. Mesostyles are clearly split and form two small cusps. The trigon basin continues to the labial end.
3. The paracone and the metacone have a narrow V-shape. The mesostyles are split and the posterior mesostyle is branched. There are three roots.

M^2. M^2 has the same basic morphology as M^1, but it is less angular and not diagonally widened. Mesostyles are clearly split. The paracone and the metacone are narrow. The protoconule (Y-shaped) and the parastyle are more pronounced and the metastyle less so than those in M^1. There are three roots (fig. 3.4).

M^3. The mesostyles may be completely or partly divided. In some specimens, the anterior arm of the protocone ends in a forked protoconule. The anterior cingulum is long and the posterior cingulum varies in size or may not be present at all. There are three roots (fig. 3.4).

Schizogalerix Material in the Sinap Tepe Localities

The localities are in order of age, starting with the oldest.

Locality 65

Locality 65 lies in the lower Sinap member, stratigraphically below locality 64, and is dated to the late middle Miocene (MN 7/8) (Kappelman et al. 1996). The absolute age is ~10.9 Ma (Kappelman et al., this volume)

Material

Two mandible fragments with p_3–m_2 sin., one mandible fragment with p_3–p_4 dext., one mandible fragment with m_2–m_3 dext., two p_2 sin., two p_3 dext., one p_4 dext., two m_1 sin., two m_1 dext., one m_2 sin., three m_2 dext., one m_3 sin., one m_3 dext., one P^3 sin., two P^3 dext., four M^1 sin., three M^1 dext., one M^2 sin., one M^2 dext., one M^3 sin. MNI 4, NISP 46. Shown in figures 3.2C, 3.5.

Description

The material from locality 65 is very fragmentary: it contains isolated teeth and a few mandibles. There are only a few diagnostic specimens.

m_1. Three specimens belong to morphotype 1 (fig. 3.5F,G) and one specimen to morphotype 2 (fig. 3.5C).

m_2. Six specimens represent morphotype 1 (fig. 3.5F,G).

m_3. In two specimens, the hypoconid connects directly with the entoconid (fig. 3.2C).

M^1. One specimen shows morphotype 1 (fig. 3.5I).

Locality 64

Locality 64 belongs to the lower Sinap member, close to the middle/late Miocene boundary. The fauna assemblage suggests an age of Late Astracian (MN 7/8) (Kappelman et al. 1996). Paleomagnetic studies have given an age of ~10.8 Ma (Kappelman et al., this volume)

Material

Two $c_{inf.}$ sin., one $c_{inf.}$ dext., one p_1 sin., one p_2 sin., one p_2 dext., one p_3 sin., one p_3 dext., one p_4 sin., one p_4 dext., one m_1 sin., one m_1 dext., one m_2 sin., one m_2 dext., one P^3 sin., one P^3 dext., three P^4 sin., two P^4 dext., two M^2 sin., five M^2 dext., one M^3 sin., three M^3 dext. MNI 3, NISP 40. Shown in figure 3.6

Description

The material from locality 64 is quite poor, consisting only of isolated teeth.

m_1. One specimen exhibits the characteristics of morphotype 1 (fig. 3.6H).

m_2. One specimen represents morphotype 1 (fig. 3.6I) and one morphotype 2 (fig. 3.6J).

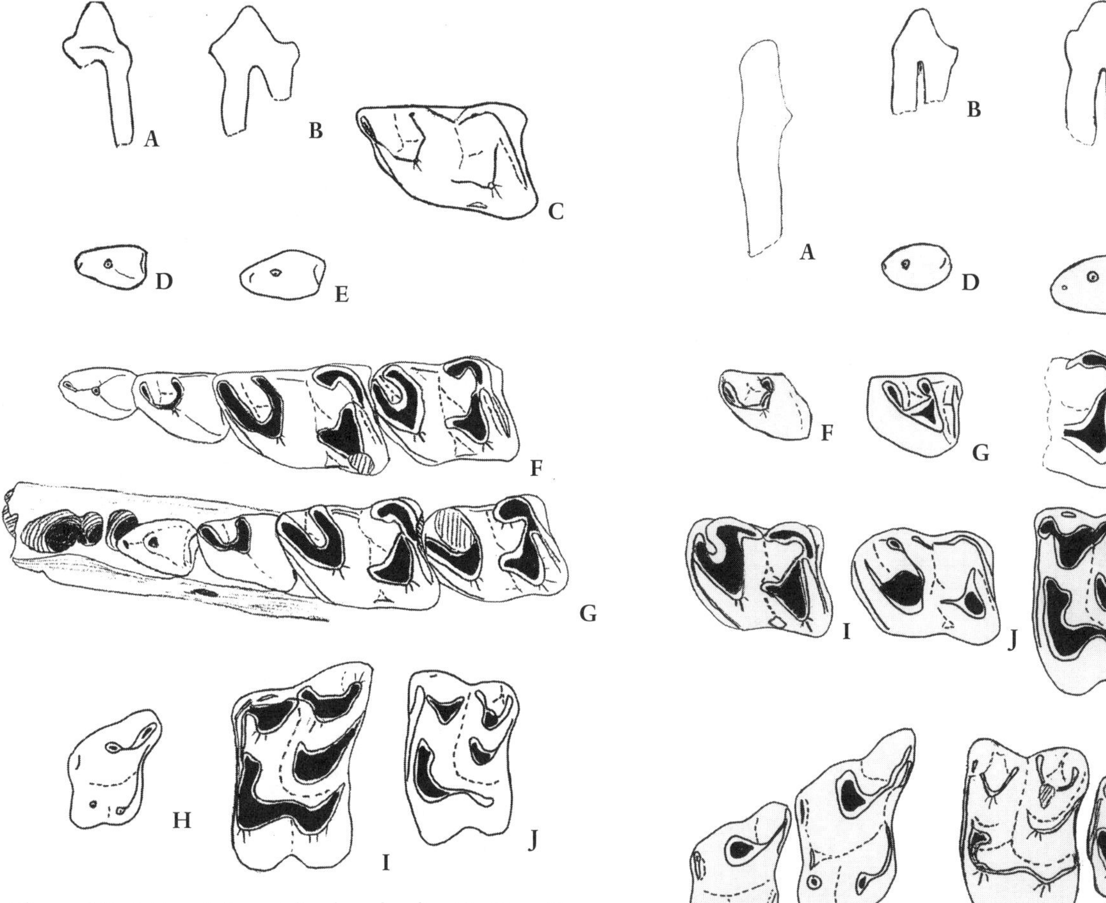

Figure 3.5. *Schizogalerix anatolica* from locality 65, Sinap Formation. **(A)** p_2 sin. buccal view; **(B)** p_3 sin. (invert) buccal view; **(C)** m_1 sin.; **(D)** p_2 sin. occlusal view; **(E)** p_3 sin. (invert) occlusal view; **(F)** mandible fragment with p_3–m_2 sin.; **(G)** mandible fragment with p_3–m_2 sin.; **(H)** P^3 sin. (invert); **(I)** M^1 sin.; **(J)** M^2 sin. (invert). Magnification 15×; reproduced at 50%.

Figure 3.6. *Schizogalerix anatolica* from locality 64, Sinap Formation. **(A)** $c_{inf.}$ sin.; **(B)** p_2 sin. buccal view; **(C)** p_3 sin. buccal view; **(D)** p_2 sin. occlusal view; **(E)** p_3 sin. occlusal view; **(F)** p_4 sin.; **(G)** p_4 sin. (invert); **(H)** m_1 fragment sin. (invert); **(I)** m_2 sin.; **(J)** m_2 sin. (invert); **(K)** M^2 sin. (invert); **(L)** P^3 sin. (invert); **(M)** P^4 sin. (invert); **(N)** M^2 sin.; **(O)** M^3 sin. Magnification 15×; reproduced at 50%.

M^2. In one very worn M^2, the posterior arm of the metacone is branched toward the center and connects with the paracone (fig. 3.6K).

M^3. The mesostyles are not distinct and the protoconule is less pronounced (fig. 3.6O).

Locality 4

Locality 4 is situated in the middle Sinap member a few m below locality 94, near the MN 7/8–9 boundary (Kappelman et al. 1996). The absolute age is ~10.7 Ma (Kappelman et al., this volume).

Material

One mandible fragment with i_1–$c_{inf.}$ sin., one mandible fragment with p_4 sin., one mandible fragment with p_4 dext., one i_1 sin., one $c_{inf.}$ sin., one p_2 sin., three p_2 dext., one p_3 dext., one p_4 sin., two p_4 dext., nine m_1 sin., ten m_1 dext., three m_2 sin., six m_2 dext., four m_3 sin., two m_3 dext., two P^2 dext., five P^3 sin., six P^3 dext., four P^4 sin., ten M^1 sin., three M^1 dext., five M^2 sin., eight M^2 dext., two M^3 sin., four M^3 dext. MNI 10, NISP 130. Shown in figures 3.2B, 3.4, 3.7.

Description

Locality 4 has a peculiar, rich pocket of small mammals that can also be seen in the abundance of the *Schizogalerix* material. The sample consists of a few mandible fragments and many isolated teeth.

m_1. Eight specimens resemble morphotype 2 (fig. 3.7J) and nine morphotype 3 (fig. 3.7E,N).

m_2. Morphotype 1 is represented by one (fig. 3.7F), morphotype 2 by four (fig. 3.7K), and morphotype 3 by three specimens (fig. 3.7O).

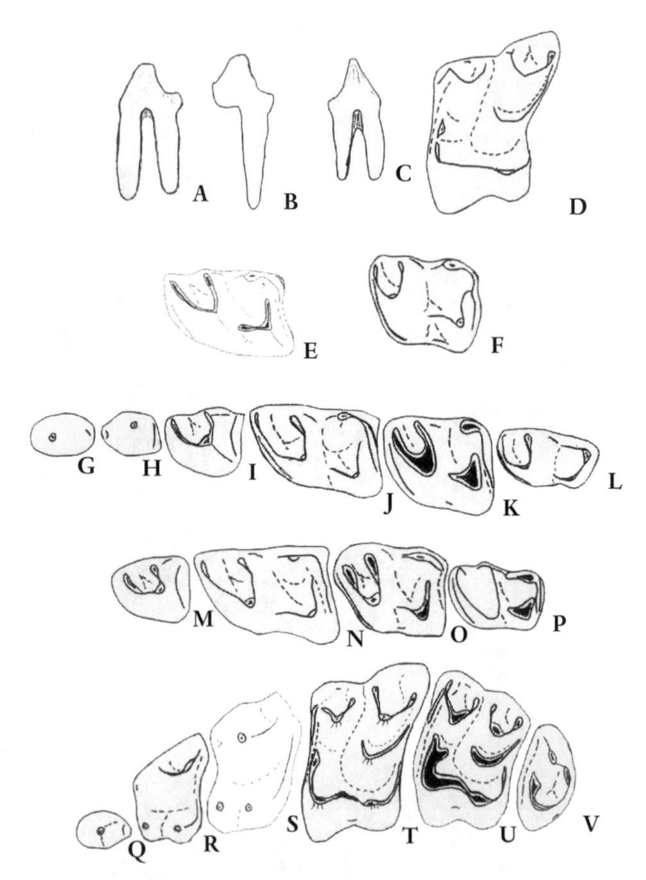

Figure 3.7. *Schizogalerix intermedia* from locality 4, Sinap Formation. **(A)** p_2 sin. buccal view; **(B)** p_3 sin. (invert) buccal view; **(C)** P^2 sin. (invert) buccal view; **(D)** M^1 sin. (invert); **(E)** m_1 sin. (invert); **(F)** m_2 sin.; **(G)** p_2 sin.; **(H)** p_3 sin. (invert); **(I)** p_4 sin. (invert); **(J)** m_1 sin.; **(K)** m_2 sin. (invert); **(L)** m_3 sin.; **(M)** p_4 sin. (invert); **(N)** m_1 sin. (invert); **(O)** m_2 sin.; **(P)** m_3 sin. (invert); **(Q)** P^2 sin. (invert) occlusal view; **(R)** P^3 sin.; **(S)** P^4 sin.; **(T)** M^1 sin.; **(U)** M^2 sin.; **(V)** M^3 sin. (invert). Magnification 15×; reproduced at 50%.

m_3. In four specimens the hypoconid is continuous with the entoconid (fig.3.7L), in one they are distinct (fig. 3.7P).

M^1. In four specimens the mesostyles are split (morphotype 2) (fig. 3.7T), in one the posterior mesostyle is branched (morphotype 3) (fig. 3.7D).

M^3. All the specimens have a forked protoconule and a weak connection between the mesostyles (figs. 3.4, 3.7V).

Locality 94

Locality 94 lies stratigraphically near the middle/late Miocene boundary (MN 9) (Kappelman et al. 1996). The absolute date is ~10.6 Ma (Kappelman et al., this volume)

Material

One mandible fragment with p_3 dext., one mandible fragment with p_3–p_4 sin., one mandible fragment with p_3–m_2 sin., one mandible fragment with p_4–m_3 sin., one mandible fragment with m_1–m_3 dext., one i_1 sin., one i_1 dext., one $c_{inf.}$ sin., one p_1 sin., one p_2 sin., one p_4 sin., two p_4 dext., two m_1 sin., one m_2 dext., three m_3 sin., one m_3 dext., two P^3 sin., two P^3 dext., one P^4 sin., two P^4 dext., three M^1 sin., three M^1 dext., three M^2 sin., two M^2 dext., two M^3 sin., one M^3 dext. MNI 4, NISP 56. Shown in figures 3.2C, 3.8, 3.9.

Description

The material from locality 94 consists of isolated teeth and some mandible fragments.

m_1. Two specimens belong to morphotype 2 (figs. 3.2C, 3.9A) and three to morphotype 3 (fig. 3.8G,H).

m_2. Three specimens represent morphotype 2 (figs. 3.2C, 3.9A,B) and two morphotype 3 (fig. 3.8G,H).

m_3. In four specimens the hypoconid is continuous with the entoconid (figs. 3.8G,H, 3.9A,B), in one they are distinct (fig. 3.9C).

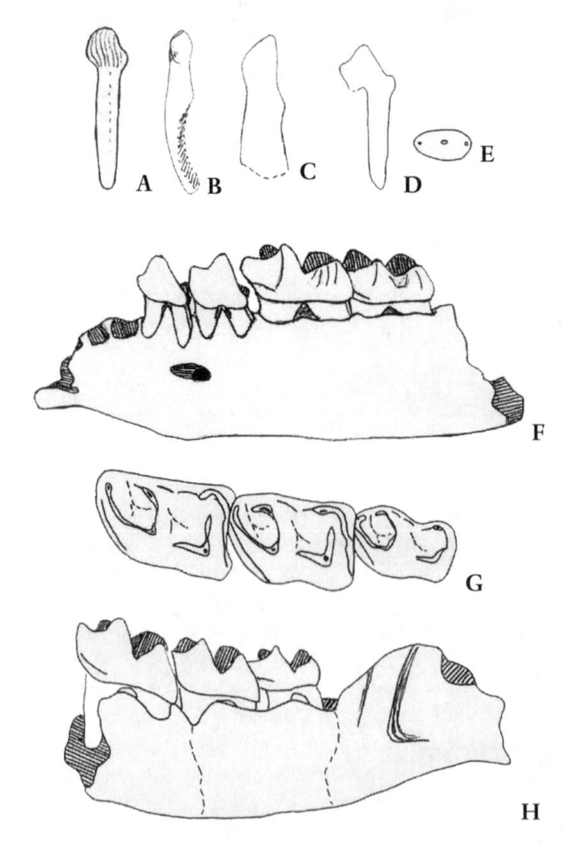

Figure 3.8. *Schizogalerix intermedia* from locality 94, Sinap Formation. **(A)** i_1 sin. buccal view; **(B)** i_1 sin. distal view; **(C)** $c_{inf.}$ sin.; **(D)** p_2 sin. (invert) buccal view; **(E)** p_2 sin. (invert) occlusal view; **(F)** mandible fragment with p_3–m_2 sin. buccal view (magnification 10×); **(G)** mandible fragment with m_1–m_3 sin. (invert) occlusal view; **(H)** mandible fragment with m_1–m_3 sin. (invert) buccal view (magnification 10×). Magnification 15× unless otherwise stated; reproduced at 50%.

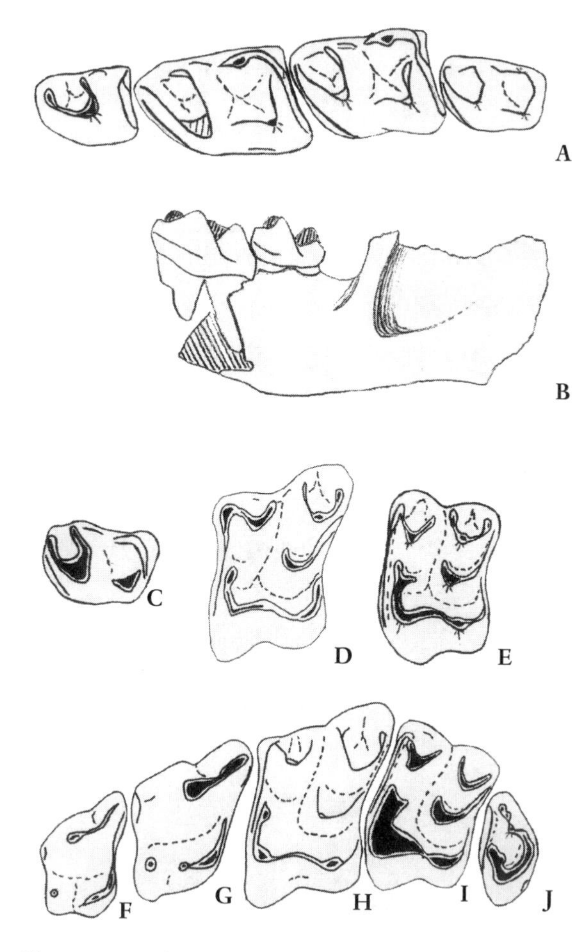

Figure 3.9. *Schizogalerix intermedia* from locality 94, Sinap Formation. **(A)** Mandible fragment with p_4–m_3 sin. occlusal view; **(B)** mandible fragment with m_2–m_3 sin. buccal view (magnification 10×); **(C)** m_3 sin.; **(D)** M^1 sin. (invert) (holotype of *S. intermedia*); **(E)** M^2 sin.; **(F)** P^3 sin.; **(G)** P^4 sin. (invert); **(H)** M^1 sin. (invert); **(I)** M^2 sin.; **(J)** M^3 sin. Magnification 15× unless otherwise stated; reproduced at 50%.

M^1. Three specimens belong to morphotype 2 (fig. 3.9D) and three to morphotype 3 (fig. 3.9H).

M^3. Three specimens have a protoconule and unsplit mesostyles (fig. 3.9J).

3.5 Locality 108

Locality 108 lies stratigraphically between localities 94 and 8A and thus belongs to MN 9 (S. Sen, pers. comm.). The paleomagnetic studies have given an age of ~10.1 Ma (Kappelman et al., this volume).

Material

One $c_{inf.}$ dext., one p_2 sin., two P^3 dext., one M^2 sin., one M^1 dext. MNI 2, NISP 16. Shown in figure 3.10A–C.

Description

There are very few remains of *Schizogalerix* in locality 108: only some fragmentary isolated teeth.

Locality 8A

Locality 8A has yielded a small mammal fauna that is referable to the early Vallesian and MN 9 (Kappelman et al. 1996). The absolute age is ~9.9 Ma (Kappelman et al., this volume).

Material

One mandible fragment with p_4 sin., one $c_{inf.}$ dext., one p_4 sin., two m_1 sin., two m_1 dext., two m_2 dext., two m_3 dext., one maxilla fragment with P^3–M^2 sin. (holotype of *S. sinapensis*), one P^2 sin., four P^3 sin., one P^3 dext., one P^4 sin., three P^4 dext., three M^2 sin., one M^2 dext., one M^3 sin., one M^3 dext. MNI 4, NISP 24. Shown in figures 3.4, 3.10D–G.

Description

The *Schizogalerix* material from locality 8A (locality I in Sen 1990) is rather poor, but there is one good maxilla fragment with P^3–M^2, which Sen (1990) has described as *S. sinapensis.*

m_1. Two specimens have the features of morphotype 3 (fig. 3.10E).

m_2. One m_2 represents morphotype 3.

m_3. In m_3 the hypoconid and the entoconid are joined (fig. 3.10F).

M^1. M^1 of the holotype shows the characteristics of morphotype 3 (fig. 3.4).

M^2. In one M^2 the posterior mesostyle is branched (fig. 3.10G).

Locality 120

Locality 120 is stratigraphically situated between localities 12 and 84, thus belonging to MN 9 (S. Sen, pers. comm.).

Material

One mandible fragment sin., one mandible fragment with $c_{inf.}$ and p_2 dext., one mandible fragment with p_4 sin., one m_1 sin., one m_2 sin., one m_3 dext., one M^1 sin. MNI 2, NISP 7. Shown in figures 3.2A, 3.10H–L.

Description

The material consists of some mandible fragments and a few isolated teeth.

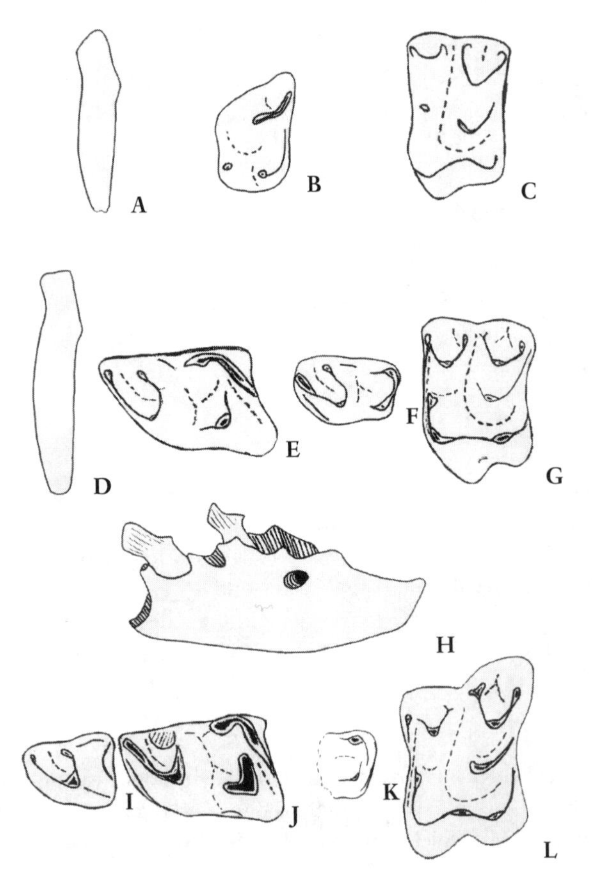

Figure 3.10. (A–C) *Schizogalerix* indet. from locality 108, Sinap Formation. (A) $c_{inf.}$ sin. (invert); (B) P^3 sin. (invert); (C) M^2 sin. (D–G) *Schizogalerix sinapensis* from locality 8A, Sinap Formation (D) $c_{inf.}$ sin. (invert); (E) m_1 sin.; (F) m_3 sin. (invert); (G) M^2 sin. (H–L) *Schizogalerix sinapensis* from locality 120, Sinap Formation. (H) mandible fragment with $c_{inf.}$ and p_2 sin. (invert) (magnification 10×); (I) p_4 sin.; (J) m_1 sin.; (K) m_3 fragment sin. (invert); (L) M^1 sin. Magnification 15× unless otherwise stated; reproduced at 50%.

m_1. m_1 has the features of morphotype 3 (fig. 3.10J).

m_3. There is no connection between the entoconid and the hypoconid (fig. 3.10K).

M^1. M^1 belongs to morphotype 3 and even the anterior mesostyle is branched. The parastyle is distinct from the paracone (fig. 3.10L).

Locality 84

Locality 84 is situated near the top of the middle Sinap member (Kappelman et al. 1996). The absolute age is ~9.4 Ma (Kappelman et al., this volume) and thus locality 84 belongs to MN 9.

Material

One mandible fragment dext., one mandible fragment with i_1 and $c_{inf.}$ sin., one mandible fragment with $c_{inf.}$ dext., one mandible fragment with p_3–p_4 sin., one mandible fragment with p_4 sin., one mandible fragment with p_4–m_1 sin., one mandible fragment with p_3–m_1 dext., one mandible fragment with p_4–m_1 dext., one mandible fragment with m_1 sin., two mandible fragments with m_2 dext., one i_1 sin., one $c_{inf.}$ sin., one p_2 sin., three p_4 sin., two p_4 dext., six m_1 sin., nine m_1 dext., five m_2 sin., one m_2 dext., four m_3 sin., one m_3 dext., one maxilla fragment with P^4–M^1 sin., one P^2 sin., one P^2 dext., three P^3 sin., three P^3 dext., two P^4 sin., three P^4 dext., six M^1 sin., four M^1 dext., two M^2 sin., two M^2 dext., one M^3 sin., one M^3 dext. MNI 11, NISP 85. Shown in figures 3.2D, 3.11, 3.12A–G.

Description

Locality 84 is the second richest locality of *Schizogalerix*. The sample consists of mandible and maxilla fragments and isolated teeth.

m_1. All specimens exhibit the characteristics of morphotype 3 (figs. 3.2D, 3.11D,E). In one m_1 and one m_2 the posterior arm of the hypoconid has formed a branched cusp (fig. 3.11E).

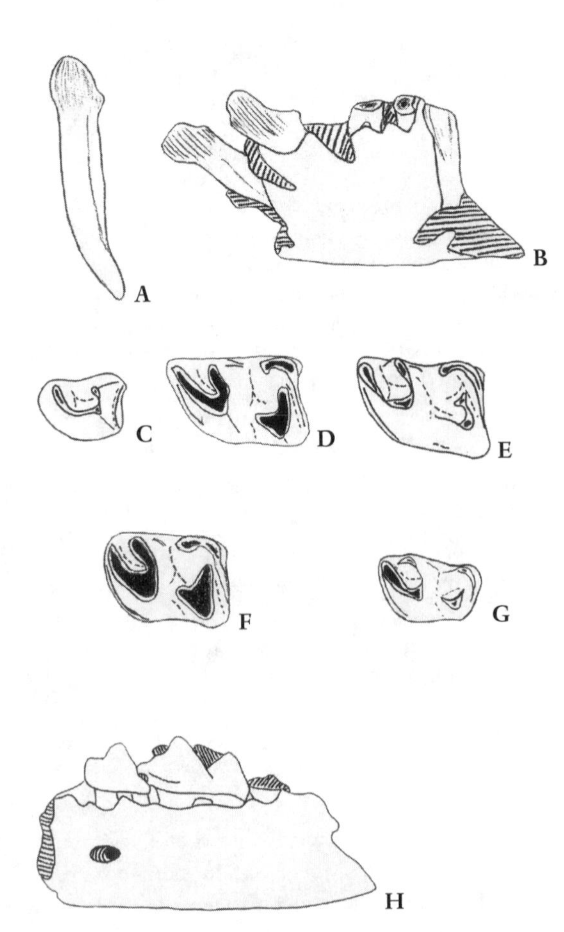

Figure 3.11. *Schizogalerix sinapensis* from locality 84, Sinap Formation. (A) i_1 sin.; (B) mandible fragment with i_1 and $c_{inf.}$ sin.; (C) p_4 sin. (invert); (D) m_1 sin.; (E) m_1 sin. (invert); (F) m_2 sin.; (G) m_3 sin.; (H) mandible fragment with p_4–m_1 sin. (invert) (magnification 10×). Magnification 15× unless otherwise stated; reproduced at 50%.

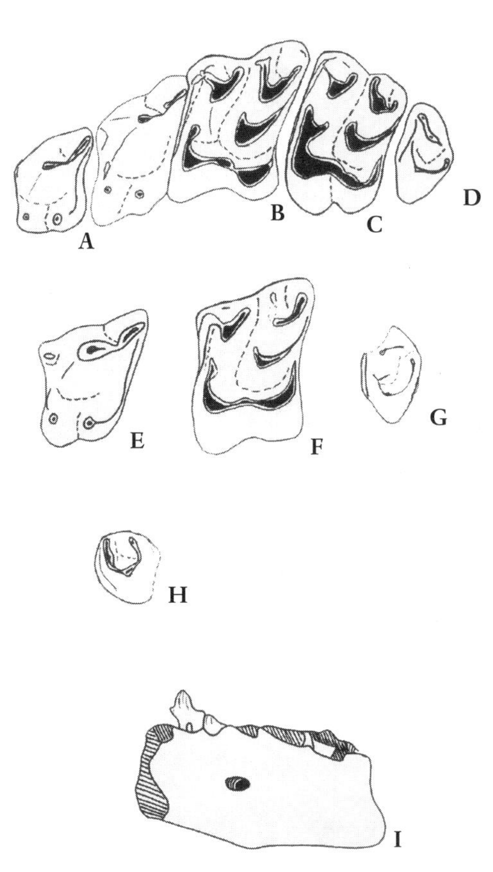

Figure 3.12. (A–G) *Schizogalerix sinapensis* from locality 84, Sinap Formation. (A) P^3 sin.; (B) maxilla fragment with P^4–M^1 sin.; (C) M^2 sin.; (D) M^3 sin. (invert); (E) P^4 sin. (invert); (F) M^1 sin.; (G) M^3 frag. sin. (H) *Schizogalerix* indet. from locality 42, Sinap Formation. m_3 fragment sin. (I) *Schizogalerix* cf. *sinapensis* from locality Inönü, Sinap Formation. Mandible fragment with p_2 sin. (invert) (magnification 10×). Magnification 15× unless otherwise stated; reproduced at 50%.

m_2. One m_2 has the pattern of morphotype 2 (fig. 3.11F), the rest belong to morphotype 3 (fig. 3.2D).

m_3. In one m_3 the hypoconid is connected with the entoconid. In five specimens it is clearly separated from the entoconid (figs. 3.2D, 3.11G).

P^4. In one specimen there is a small anterior cusp between the paracone and the protocone (fig. 3.12B). One specimen has a very pronounced parastyle (fig. 3.12E).

M^1. Six specimens belong to morphotype 3 (fig. 3.12B); only one belongs to morphotype 2 (fig. 3.12F).

M^3. Mesostyles are split and the protoconule is forked in one specimen (fig. 3.12D,G).

Locality 42 (Çobanpinar)

Locality 42 is situated ~10 km southwest of the Sinap Tepe and is commonly placed in the middle Turolian (MN 12) (Kappelman et al. 1996).

Material

One m_3 sin. MNI 1, NISP 2. Shown in figure 3.12H.

Description

Only three small fragments of Insectivora have been found from Çobanpinar. One is a trigonid of a lower third molar of the genus *Schizogalerix*.

Locality Inönü

The stratigraphy in the Inönü area (~4 km southwest of Sinap Tepe) is very complex and contains localities of different ages. The exact stratigraphic position of this unnumbered locality (~300 m south-southeast of Loc. 24) is unknown, but an Ochotonic specimen collected from this locality indicates that it belongs to MN 11 or MN 12 (S. Sen, pers. comm.).

Material

One mandible fragment with p_2 dext., one mandible fragment with p_3–m_3 sin., one i_1 sin., one i_1 dext., one i_3 dext., one $c_{inf.}$ sin., one $c_{inf.}$ dext., one p_1 sin., one p_1 dext., one p_2 sin., one P^2 sin., one P^3 sin., one P^4 sin., one M^1 sin., one M^2 sin. MNI 1, NISP 18. Shown in figures 3.12I, 3.13.

Description

The material from this locality consists of two mandible fragments (sin. and dext.), isolated anterior lower teeth and some isolated upper molars and fragments of premolars. The wear on the teeth is very similar in the specimens and the material was found from a small pocket, so it most probably represents a single individual. The dimensions of the teeth are somewhat smaller than those in other localities in the Sinap Formation (fig. 3.14; Appendix 3.1).

p_2. p_2 has two roots in one alveolus with some trace of a subdivision (figs. 3.12I, 3.13F,G).

m_1 and m_2. These molars compare well with morphotype 3 (fig. 3.13F,G).

m_3. The hypoconid is separated from the entoconid (fig. 3.13F,G).

M^1. The mesostyles are split and the posterior mesostyle is not branched (morphotype 2). The parastyle is not connected with the paraconid (fig. 3.13J).

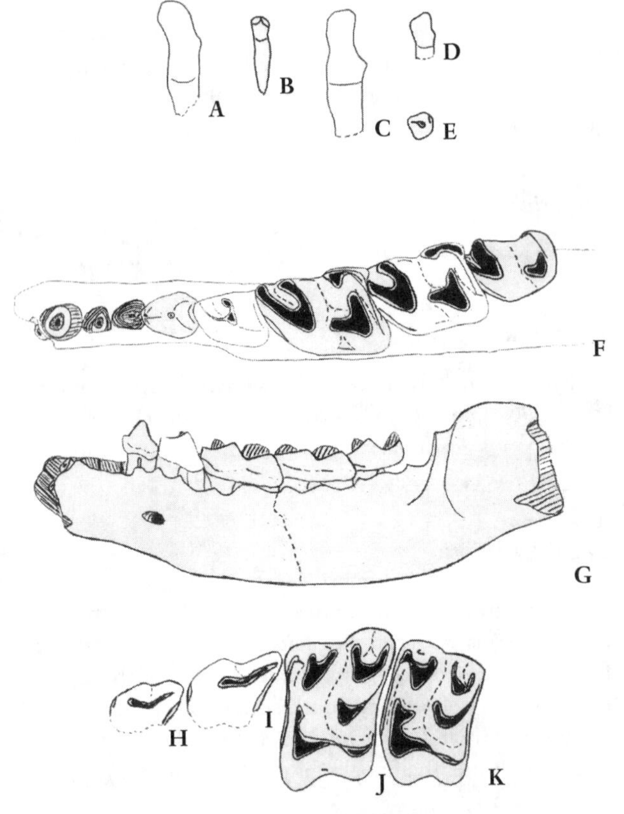

Figure 3.13. *Schizogalerix* cf. *sinapensis* from locality Inönü, Sinap Formation. **(A)** i_1 sin. (invert); **(B)** i_3 sin. (invert); **(C)** $c_{inf.}$ sin. (invert); **(D)** p_1 (?) sin. (invert) buccal view; **(E)** p_1 (?) sin. (invert) occlusal view; **(F)** mandible fragment with p_3–m_3 sin. occlusal view; **(G)** mandible fragment with p_3–m_3 sin. buccal view (magnification 10×); **(H)** P^3 fragment sin.; **(I)** P^4 fragment sin.; **(J)** M^1 sin.; **(K)** M^2 sin. Magnification 15× unless otherwise stated; reproduced at 50%.

Comparison

Specimens of the genus *Schizogalerix* have been found in 26 localities: in Turkey (nine), Greece (six), Austria (three), Germany (one), France (one), Morocco (two), Algeria (two), Moldavia (one), and China (one) (Engesser 1980, Lungu 1981, Bi et al. 1999, NOW database 1999). Ten species of *Schizogalerix* have been described based on these specimens. A comparison of the *Schizogalerix* material from the Sinap Formation with the following species has been made with the help of the literature.

Schizogalerix pasalarensis, Paşalar, Turkey (Engesser 1980) differs from the *Schizogalerix* material from the Sinap Formation in the following ways:

p_4 is bigger and the paraconid is not connected with the protoconid;
in m_1 and m_2 the trigonid is wider, the structure of the talonid is different (in *S. pasalarensis* the hypoconid is connected with the entoconid and the posterior cingulum is separate), and the anterior cingulum is shorter;
m_3 is bigger and the trigonid is wider;
P^3 is bigger;
P^4 is only posterolabially widened; and
in M^1 and M^2 the paracone and the metacone are wider, mesostyles are connected, and the labial cingulum and the parastyle are more pronounced, the protoconule is less pronounced, and M^1 is only posterolabially widened.

The *Schizogalerix* material from the Sinap Formation resembles *S. anatolica*, Yeni-Eskihisar, Turkey (Engesser 1980) in following features:

p_2 is double rooted;
morphotype 1 displayed in m_1 and m_2;
in m_3 the hypoconid connects directly with the entoconid;
morphotype 1 displayed in M^1; and
mesostyles are split in M^2.

Only the upper dentition has been described for *Schizogalerix sinapensis*, locality 8A, Sinap Formation, Turkey (Sen 1990) (see also the discussion of locality 8A in this chapter) and Kastellios Hill, Crete, Greece (van der Made 1996). The upper premolars, M^2 specimens, and the presence of morphotype 3 in the M^1 specimens of the *Schizogalerix* material in other localities of the Sinap Formation correspond with *S. sinapensis*.

The unnamed species *Schizogalerix* nov. sp. Amasya, Turkey (Engesser 1980) is based on a single M^1. It is distinct from the M^1 specimens from the Sinap Formation in that M^1 is more diagonally widened and the branch at the posterior mesostyle is much more pronounced.

Schizogalerix macedonica, Maramena, Greece (Doukas et al. 1995) differs from the *Schizogalerix* material in the Sinap Formation as follows:

there is an extra cuspule next to the hypocone on the M^1 and M^2;
there is an extra cusp next to the entoconid on the m_1 and m_2; and
the teeth are larger.

Schizogalerix zapfei, Kohfidisch, Austria (Bachmayer and Wilson 1970) and *S. atticus*, Pikermi (Chomateri), Greece (Rümke 1976) are considered to be identical by Engesser (1980). The comparison is made based on the original descriptions and the illustrations in Engesser (1980).

The material from Kohfidisch is distinguished from the *Schizogalerix* material in the Sinap Formation by the following characteristics:

the alveolus of i_3 is larger;
p_2 is single-rooted; and
morphotype 3 in m_1 and m_2 resembles *S. zapfei*.

The material from Pikermi is distinguished from the *Schizogalerix* material of the Sinap Formation by the following characteristics:

the paraconid is separated from the protoconid in p_4;
the parastyle of P^3 and P^4 is connected with the protocone by the anterior cingulum;
the protoconule in M^2 is larger;

the anterior arm of the protocone in M^3 is continuous into the anterolabial corner of the tooth; and
morphotype 3 in m_1 and m_2 resembles the material from Pikermi, but only two specimens in the Sinap Formation have a distinct cusp on the posterior arm of the hypoconid.

Schizogalerix voesendorfensis, Vösendorf, Austria (Rabeder 1973) differs from the *Schizogalerix* material in the Sinap Formation as follows:

p_3 is longer and higher with a clear transverse ridge on the posterior cingulum;
in p_4 the cusps of the trigonid are not connected;
the trigonid is broader in lower molars;
m_2 is shorter;
P^3 and P^4 are narrower;
the protocone in P^3 is placed more to the mesial, and the posterior cingulum does not connect with the hypocone;
P^4 and M^1 are widened only posterolabially;
in M^1 mesostyles are less split and form no cusps, the labial cingulum is more pronounced, and the metacone and paracone are broader; and
M^2 is narrower, the metacone and paracone are broader, the protoconule is less pronounced, and the trigon basin is not straight.

Schizogalerix moedlingensis, Eichkogel, Austria (Rabeder 1973) is distinguished from the *Schizogalerix* material in the Sinap Formation in features such as:

p_3 is longer, with a clear transverse ridge on the posterior cingulum;
p_4 has a clear transverse ridge on the posterior cingulum;
in lower molars the trigonid is broader, the anterior cingulum more pronounced, and the connection between the hypoconid and the entoconid more pronounced;
P^4 and M^1 are widened only posterolabially;
P^4 is shorter;
M^1 has a more pronounced labial cingulum, the paracone is broader, the parastyle is more pronounced, the protoconule is Y-shaped, and the trigon basin is more vertical; and
in M^3 the cusps are more pronounced and more distinct and the protoconule is bigger.

The m_1 and M^1 described for *Schizogalerix* nov. sp., Amama II, Algeria (Engesser 1980) are distinct from the *Schizogalerix* material in the Sinap Formation in features such as the teeth are larger and M^1 is anterolingual/posterolabially more widened.

Schizogalerix sarmatica, Buzhor 1, Moldavia (Lungu 1981, NOW database 1999) has the following characters that differ from the *Schizogalerix* material in the Sinap Formation:

p_3 has two distinct cusps;
p_4 is smaller;
P^3 is not posterolabially widened; and
in P^3 and P^4 the labial cingulum is wider.

Schizogalerix duolebulejinensis, Duolebulejin, China (Bi et al. 1999) differs from the *Schizogalerix* material in the Sinap Formation as follows:

the molars are longer;
there is no posterior arm of the hypocone in M^1; and
the metastylid is well developed in m_1 and m_2.

Discussion

Species recognition in the genus *Schizogalerix* has previously been applied to one restricted locality and horizon at a time. The differences in size and morphology between the specimens have been distinct enough so that the identification of the species boundaries and systematics have been straightforward. But as the fossil record improves as a result of a stratigraphically dense and relatively continuous record, discrimination between species becomes more difficult, even as our understanding of their evolution improves (Rose and Bown 1986). This is also the case for the *Schizogalerix* material in the Sinap Formation.

Size Differences

The dimensions of the genus *Schizogalerix* in the Sinap Formation do not show any great variation from sample to sample (Appendix 3.1). The general trend seems to be that the teeth become somewhat narrower the younger the locality of the specimen. This pattern is significant in the cases of p_4, m_1, and M^2 according to a Kruskal-Wallis analysis of variance (fig. 3.14).

The changes in size have a mosaic pattern. The lower fourth premolars from localities 65, 64, 4, and 94 are broader than those from the other localities ($p = 0.002$; $n = 22$). A similar trend is seen in M^2 ($p = 0.009$; $n = 16$). The lower first molars from localities 65 and 64 tend to be broader than those found in other localities ($p = 0.007$; $n = 44$). For single specimens, however, these groupings into older/broader and younger/narrower teeth correlations are only approximate. The dimensions overlap between localities and the difference is not so great (on the order of 0.1–0.2 mm between averages). In other words, no species discrimination can be based on size, because no distinct limit of size can be determined for any single tooth type.

Nonmetric Morphologic Features

The morphologic changes of *Schizogalerix* in the Sinap Formation take place at different rates and at different times in the stratigraphic sequence. The morphology of the anterior teeth, premolars, and M^2 does not show any distinct change between the localities (often there are too few samples of the anterior teeth for comparison). The

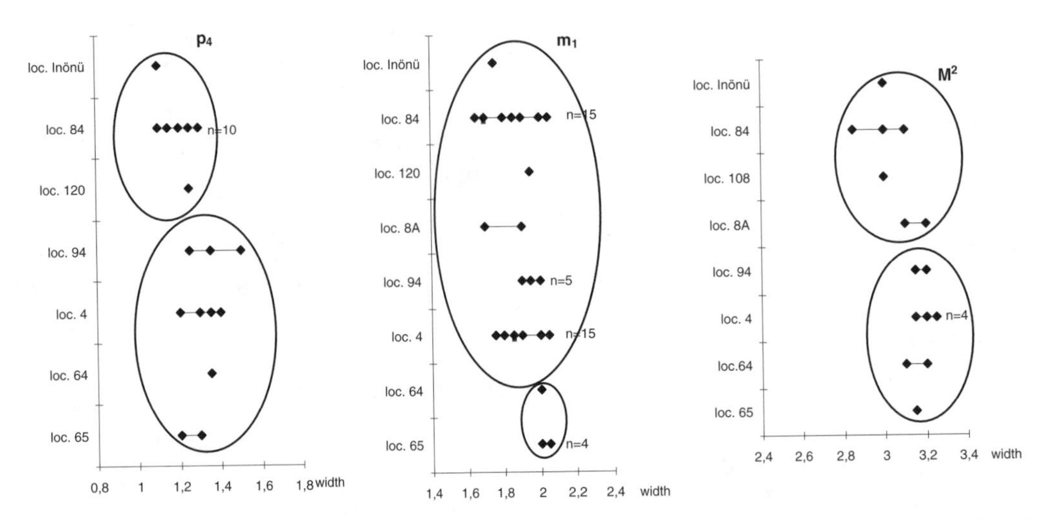

Figure 3.14. The measurements of width (in mm) in p_4, m_1 and M^2 of the genus *Schizogalerix*. Grouping is based on analysis of variance. The number of specimens is given when not clear from the picture.

morphologic differences are best seen in lower molars and in M^1 and M^3. The morphology of m_3 and M^3 is highly variable in every species of *Schizogalerix*. For m_1, m_2, and M^1, three morphotypes can be described (see previous section on the specimens of the genus *Schizogalerix*).

When comparing the material from the Sinap Formation with other *Schizogalerix* species, morphotype 1 in m_1, m_2, and M^1 resembles that found in the species *S. anatolica*. Morphotype 2 in m_1, m_2, and M^1 is morphologically an intermediate form of morphotypes 1 and 3. It does not resemble any known species. Morphotype 3 in M^1 is identical with that in the species *S. sinapensis*. In m_1 and m_2, morphotype 3 resembles that of *S. zapfei* and *S. atticus*, but because there are many other distinct differences in the morphology of these species, I do not think the specimens in the Sinap Formation belong to them (except maybe for locality Inönü; see below). *S. zapfei* and *S. atticus* are also found in younger localities.

The *Schizogalerix* material is presented quantitatively in figure 3.15. The sparse material limits the choice of localities, so that the comparison has been made among four localities only: 65, 4, 94, and 84. These localities cover MN 7/8 (locality 65) and MN 9 (locality 4, 94, and 84).

Figure 3.15 shows that one or more morphologic features are found in one locality and that the relative frequency of the morphotypes changes. The characters that represent locality 65 decrease or vanish in the later localities, and the characters that dominate in locality 84 are unknown in locality 65. Localities 4 and 94 represent two or three morphotypes in more or less equal amounts. The changes in the proportion of the morphologic features seem to accumulate in between localities 65 and 4 (m_1, m_2, and M^1) and between localities 94 and 84 (m_1, m_2, m_3, and M^1). The first marks the MN 7/8–9 boundary as well.

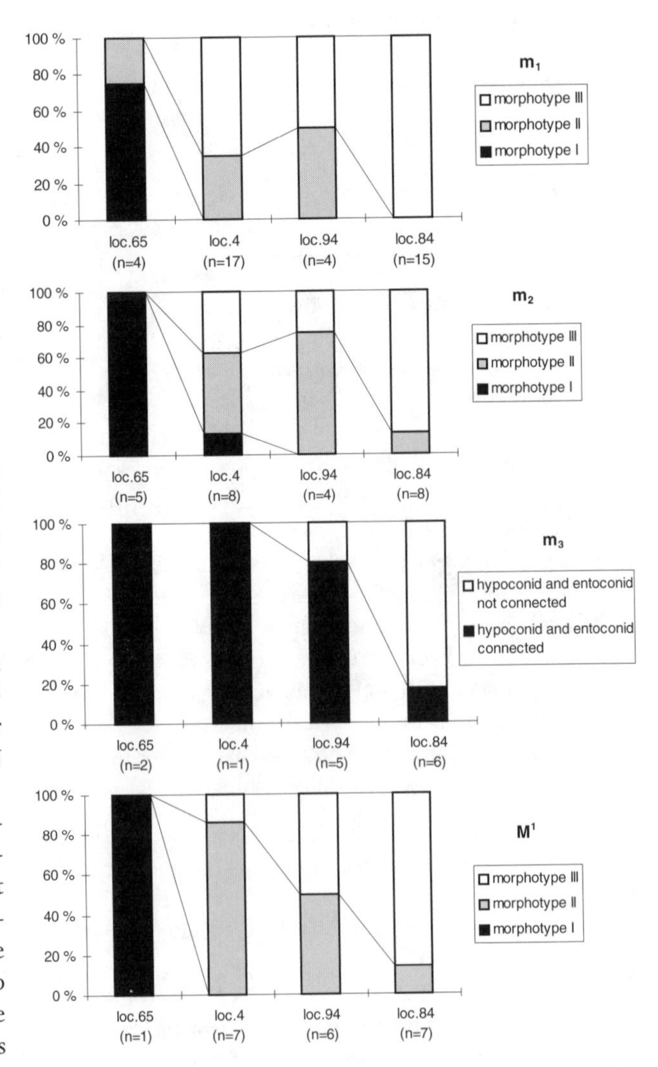

Figure 3.15. The percentage abundances of the morphological features in m_1, m_2, m_3, and M^1 of the genus *Schizogalerix* in localities 65, 4, 94, and 84 in the Sinap Formation.

Species Determination

The *Schizogalerix* material in the Sinap Formation is a good example of gradual change. There is an intermediate form known as morphotype 2 and a continuous variation in the samples without any unambiguous boundaries in morphology or size. The gradual change is also seen in the change of the relative abundance of different morphotypes and in the degree of the difference in specimens within and between the localities.

Longer lineages with sufficient fossil records usually require subdivision into successional species (Rose and Bown 1986). The *Schizogalerix* material in the Sinap Formation could be grouped into three species following the three morphotypes seen in m_1, m_2, and M^1. This classification, however, would place contemporaneous members of the same locality in different taxa, thereby ignoring the biological relationship between them (Rose and Brown 1986). Murphy et al. (1981) assume that all intergrading morphs occurring at a single horizon belong to the same species and, therefore, should have the same name. This is a neomammalogist's approach, in which a species is considered to be a population of individuals, the majority with a basic morphology and the rest with an obvious deviation from it. The genus *Schizogalerix* in the Sinap Formation in each locality can also be considered as one population with one species. This distinction of species based on paleopopulational considerations needs a statistical confirmation (Woodburne 1996).

For example, Agadjanian and Koenigswald (1977) studied some European arvicolid paleopopulations from different time periods and gave species or subspecies names for them based on percentage distributions of the different morphotypes. This approach classifies all varieties of a species together at each stratigraphic level and selects the level in which one variety becomes dominant as the lower boundary of the taxon (Murphy et al. 1981).

In the Sinap Formation, three stages of the *Schizogalerix* material can be distinguished when regarding percentage abundances of morphotypes in m_1, m_2, and M^1 (fig. 3.15): (1) Morphotype 1 is dominant, (2) morphotypes 2 and 3 are almost equally abundant, and (3) morphotype 3 is dominant.

Schizogalerix anatolica and *S. sinapensis* have previously been described only on the basis of morphology. When Engesser (1980) described *S. anatolica,* he included in the species two m_1 specimens with the characteristics of morphotype 2 or 3 (the posterior arm of the hypoconid does not connect with the entoconid/posterior cingulum) and one m_2 with the characteristics of *S. pasalarensis.* Thus there is no reason why localities that have the majority of morphotype 1 (stage 1) should not also be associated with *S. anatolica.*

Schizogalerix sinapensis is known only by its upper dentition. In figure 3.15 it is seen that characteristics such as morphotype 3 in m_1, m_2, and M^1, and the lack of connection between the hypoconid and the entoconid in m_3 dominate in locality 84. Thus these characters should be linked to *S. sinapensis.* Similarly, localities in which the majority of specimens display morphotype 3 (stage 3) should represent *S. sinapensis.*

For stage 2 there is no single species in the majority to identify the stage. To avoid merely giving a genus status to *Schizogalerix* specimens in a locality where morphotypes 2 and 3 are almost equally abundant, it seems appropriate to assign a formal species name (compare Agadjanian and Koenigswald 1977). I suggest a new species name *S. intermedia* for the specimens found in stage 2 localities. This name describes the intermediate nature of the samples and the gradual change from *S. anatolica* to *S. sinapensis.*

New Species Description

Schizogalerix intermedia nov. sp.

Derivatio nominis:	intermedia ("that is between") describes the intermediate status to the species *S. anatolica* and *S. sinapensis.*
Holotype:	left M^1 (AS95-1090); figure 3.9D.
Paratypes:	all the *Schizogalerix* teeth from locality 94 in the Sinap Formation.
Type locality:	locality 94 in the Sinap Formation, province of Ankara, Turkey.
Stratum typicum:	MN 9, near the middle/late Miocene boundary.
Measurements:	length 2.65 mm, width 3.25 mm.

Diagnosis

Morphotypes 2 and 3 of m_1, m_2, and M^1 occur together in almost equal amounts. In m_1 and m_2 the hypoconid is wide and the posterior arm of the hypoconid points obliquely toward the entoconid/posterior cingulum (morphotype 2), or the hypoconid is narrow and the posterior arm is vertically directed toward the middle of the entoconid (morphotype 3). In M^1 the mesostyles are split and in morphotype 3 the posterior mesostyle is branched.

Description of the Holotype

The chosen M^1 is anterolingual/posterolabially widened. It has four main cusps in each corner, of which the hypocone is the lowest. Mesostyles are clearly separated (morphotype 2). The metacone is narrower than the paracone and there is a small labial cingulum next to the paracone. The protoconule is small and connected through a ridge with the protocone and hypocone. The posterior arm of the metaconule stretches to the posterolabial corner of the tooth. The parastyle is fairly big and connected with the paracone. It continues as a cingulum next to the protocone.

Species Distribution

The genus *Schizogalerix* in the Sinap Formation occurs in the following way:

Locality Inönü	*S.* cf. *sinapensis*
Locality 42	*S.* indet.
Locality 84	*S. sinapensis*
Locality 120	*S. sinapensis*
Locality 8A	*S. sinapensis*
Locality 108	*S.* indet.
Locality 94	*S. intermedia*
Locality 4	*S. intermedia*
Locality 64	*S. anatolica*
Locality 65	*S. anatolica*

The species boundaries follow the MN units to some extent. *S. anatolica* occurs in MN 7/8, *S. intermedia* in MN 9, and *S. sinapensis* and *S.* cf. *sinapensis* in MN 9–11.

The material in locality Inönü is somewhat special. It shows resemblance to *Schizogalerix zapfei* (a single alveolus of p_2 and morphotype 3 in m_1 and m_2) and to *S. sinapensis* (morphotype 3 in m_1 and m_2 and the hypoconid and entoconid not connected in m_3). The dimensions of the teeth are somewhat smaller those than from the other localities in the Sinap Formation and clearly smaller than in *S. zapfei*. Although *S. zapfei* has been found from the same age as locality Inönü (suggested as being MN 11), I think that the material more closely resembles *S. sinapensis* than *S. zapfei;* I suggest a species name *Schizogalerix* cf. *sinapensis*, primarily based on size differences.

The disadvantage in species definitions that require the presence of a certain percentage of a given morphology is that isolated teeth cannot be assigned with confidence to one or another species. It is also less precisely applicable from place to place because of inhomogeneties in sampling (Woodburne 1996).

I have also carried out species recognition studies in localities 64, 8A, and 120 based on the quantitative approach described here. The resultant species names might, however, be somewhat misleading because of the sparse material from these localities. The nature of the quantitative approach is such that species recognition becomes increasingly reliable with increasing numbers of specimens from the sites being analyzed. This is particularly true in the case of *Schizogalerix intermedia*. A single specimen with characteristics of morphotype 3 in one locality would be assigned to *S. sinapensis* and a single specimen of morphotype 2 in another locality to *S. intermedia*. In both cases the species determination might change if further specimens are found and the quantitative information could be used with more precision. Thus it should be remembered in future studies of *Schizogalerix* in Turkey that this genus shows a gradual change and the species discrimination should be based not only on the morphologic features but on the statistical aspects of the samples as well.

Evolution of the Genus *Schizogalerix*

After the early middle Miocene, the genus *Schizogalerix* evoled independently from the genus *Galerix*. Four evolutionary lineages are known of *Schizogalerix:* the Turkish lineage begins in the Astracian, the European in the Vallesian, the African lineage is represented in the middle Turolian, and the Chinese in the early middle Miocene (Engesser 1980; Bi et al. 1999).

In Turkey

The Turkish forms of *Schizogalerix* embrace a wide range of stratigraphy: from MN 6 to MN 13. These forms/species undergo a distinct change in morphology during that time.

The oldest known representative of *Schizogalerix* is from Paşalar (MN 6) (Engesser 1980; Engesser and Ziegler 1996). Engesser (1980) described the specimens to *S. pasalarensis* as the most primitive form of the genus *Schizogalerix* known so far (see the morphologic characteristics in table 3.1). The *Schizogalerix* form from Çandir (MN 6) is morphologically between *S. pasalarensis* and *S. anatolica* (Engesser 1980). *S. anatolica* from Yeni-Eskihisar (MN 7/8) Engesser (1980) described as a morphologically intermediate form between *S. pasalarensis* from Paşalar and *S. zapfei* from Kohfidisch (Austria) and Pikermi (Greece). The material from Sofça (MN 7/8) consists of two species: *S.* aff. *pasalarensis* and *S.* aff. *anatolica*. *S.* aff. *anatolica* in this locality is even more advanced than *S. anatolica* from Yeni-Eskihisar (Engesser 1980).

The material from the Sinap Formation has now filled the gap in the Turkish MN stratigraphy from MN 9 to MN 12. The change in the morphology continues through *S. intermedia* to *S. sinapensis*, a more advanced form of *S. anatolica* (Sen 1990). The material from locality Inönü (MN 11 or MN 12) has a form close to *S. sinapensis* and *S. zapfei*. The most evolved form known of the genus *Schizogalerix* in Turkey is a single M^1 from Amasya (*S.* nov. sp.) from MN 13 (Engesser 1980).

The gradual change in *Schizogalerix* is thus not restricted to the forms in the Sinap Formation, but it is seen in other Turkish localities as well. The Turkish forms make a morphologic lineage that relates well to the stratigraphy (species with the most primitive features are found in the oldest localities and the most evolved forms in the youngest; see table 3.1 and fig. 3.16). This morphologic transition series also represents one phylogenetic lineage (Engesser 1980). The new finds from the Sinap Formation confirm the hypothesis of one evolutionary lineage of *Schizogalerix* in Turkey.

Elsewhere

Outside of Turkey, the genus *Schizogalerix* is best represented in Austria and Greece. The lineage is made of species *S. voesendorfensis*, *S. zapfei*, *S. moedlingensis*, and *S. mace-*

Table 3.1. Comparison of the Morphologic Characteristics in Different Species of *Schizogalerix* in Turkey

Tooth	*S. pasalarensis*	*S. anatolica* (morphotype 1)	*S. intermedia* (morphotype 2)	*S. sinapensis* (morphotype 3)	*S.* nov. sp. (Amasya)
m_1, m_2	Wide trigonid	Narrow trigonid	Very narrow trigonid	Very narrow trigonid	—
	Hypoconid is continuous with the entoconid; the posterior cingulum has no connection with them	The posterior cingulum is continuous with the entoconid; the posterior arm of the hypoconid has a weak connection with them	Hypoconid does not join to the posterior cingulum/entoconid, but is still directed obliquely	The posterior arm of the hypoconid more to the centre; very pronounced entoconid	—
M^1	Only posterolabial widened	Diagonally widened	Diagonally widened	Diagonally much widened	Diagonally much widened
	Mesostyles are connected	Mesostyles weakly joined	Mesostyles split	Mesostyles split; posterior mesostyle branched	Branch at the posterior mesostyle more pronounced
	Narrow	Narrow	Broad	Broad	Broad
M^1, M^2	Pronounced parastyle	Less pronounced parastyle	Less pronounced parastyle	Less pronounced parastyle	Less pronounced parastyle
	Clear labial cingulum	Less pronounced labial cingulum	Less pronounced or no labial cingulum	Less pronounced or no labial cingulum	No labial cingulum
	Metacone and paracone very broad	Metacone and paracone broad	Metacone and paracone narrower	Metacone and paracone narrow	Metacone and paracone narrow
M^2	Mesostyles connected	Mesostyles split	Mesostyles split	Mesostyles split	—

donica (fig. 3.16). This lineage is derived from *S. pasalarensis,* but *S. zapfei* may also have descended from *S. anatolica* (Ziegler 1999). The relationship in the morphology between the Turkish and the Austrian-Greek forms is very close, however. This resemblance is seen by Engesser (1980) as a parallel evolution.

In Africa, the morphology of the one M^1 from Amama II (Algeria) is distinct enough to make its own species (*S.* nov. sp.) and evolutionary lineage (Engesser 1980). It is most closely related to the Turkish forms and especially to the M^1 from Amasya.

Schizogalerix doulebulejinensis is the first record of the genus *Schizogalerix* in China. The fauna assemblage in Duolebulejin suggests an age of the early middle Miocene (MN 6), which makes *S. doulebulejinensis* a very early representative of the genus *Schizogalerix.* The relationship of the Chinese lineage to the other *Schizogalerix* lineages is not clear yet because of the sparse material found in China (Bi et al. 1999).

The genus *Schizogalerix* is thus first known to have occurred in Turkey and China during the Astaracian and thereafter seems to have expanded westward from Turkey into Central Europe at the base of the late Miocene. *Schizogalerix* disappears in the late Turolian together with many other mammals in a major turnover event. New insectivores replace *Schizogalerix* in a cooler and more seasonal environment (Engesser 1980; Bernor et al. 1996).

Rate of Evolution

The genus *Schizogalerix* is morphologically and stratigraphically rich and sufficiently well documented to reveal the tempo and mode of its evolution. The total change in the morphology of *Schizogalerix* (from *S. pasalarensis* to *S. macedonica*) happened in ~10 Ma (MN 6–13; Bernor et al. 1996). In the Sinap Formation the gradual change from *S. anatolica* to *S. sinapensis* happened in <1 Ma. MN 10 is not represented in the *Schizogalerix* material from the Sinap Formation, mainly because of its short duration (Bernor et al. 1996).

The frequency and degree of expression have occurred in different teeth at different times, however. In the Sinap Formation the anterior teeth, premolars, and M^2 show stasis through the sequence, but differences are seen when compared with other species in Turkey and Austria. For example, M^2 has already lost the connection between the mesostyles in *S. anatolica,* but is relatively unchanged after that species (table 3.1). m_1 seems to change most rapidly (e.g., in the Sinap Formation morphotype 2 is quite abundant by MN 9, fig. 3.15). p_4 undergoes a distinct molarization (the trigonid becomes more distinct and narrower).

Biostratigraphy

The genus *Schizogalerix* could be used for biostratigraphic correlations: it is fairly abundant over a considerable geographic range and evolves significantly across a wide stratigraphic sequence (MN 6–13). The comparison between the different *Schizogalerix* forms is difficult, however, because there are no common species in Europe and Asia, and the determination of species has been made on different bases. Analysis can thus be based only on the degree of evolution (table 3.1 and fig. 3.16). The geographic distribution and the size of the sample are restrictive elements in the correlation:

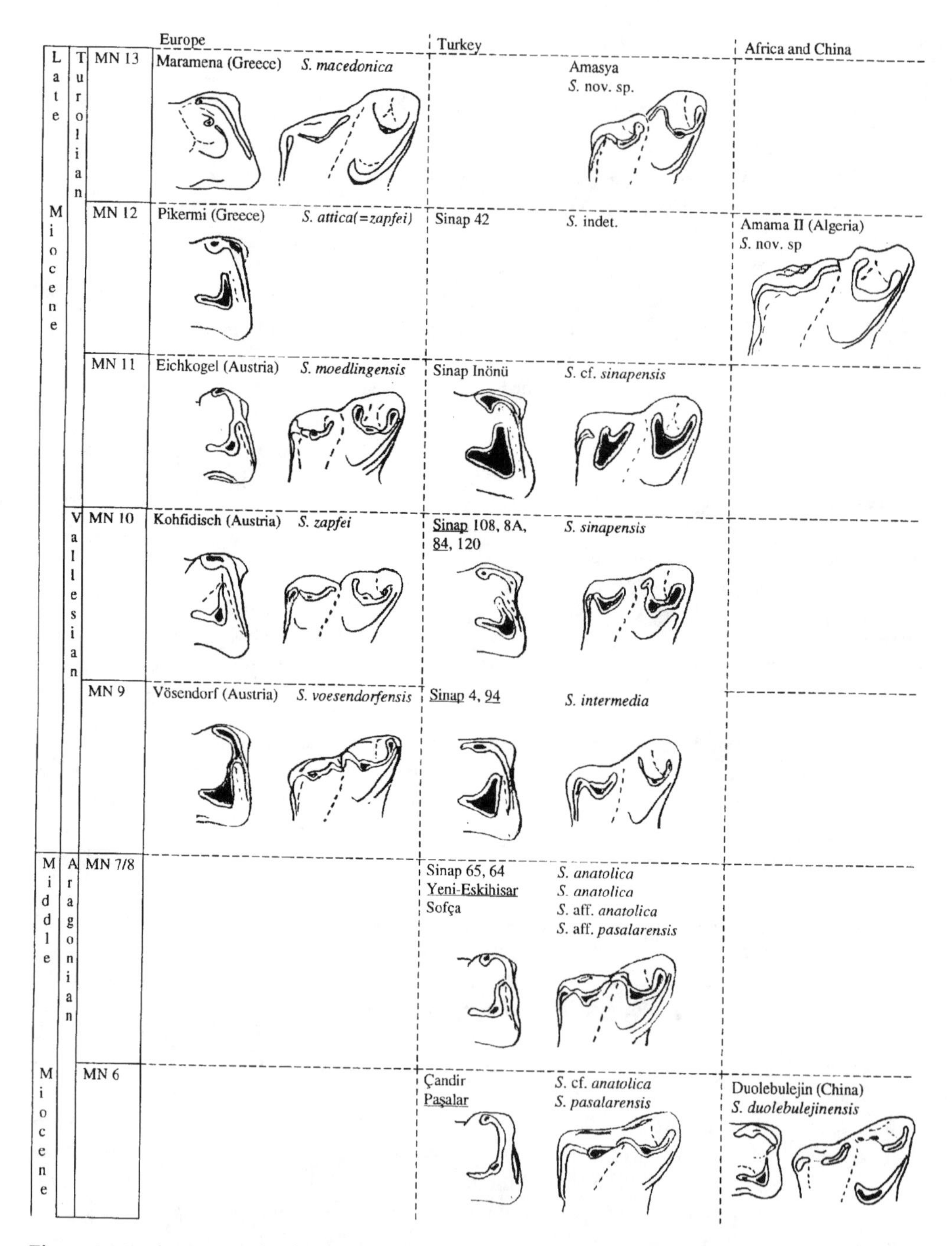

Figure 3.16. The stratigraphic distribution of the genus *Schizogalerix* in Europe, Turkey, Africa, and China. Figures of m_1 and M^1 are shown. MN 10 is not shown in Turkey. If not clear otherwise, the figures are from the underlined localities. Figures from the Sinap Formation are based on the author's drawings; other specimens redrawn after Engesser (1980), Doukas et al. (1995), and Bi et al. (1999).

the evolutionary grade is different in the same stratigraphic unit, for example, in Turkey and Europe, and the determination of the evolutionary grade in a sample is better the more complete the sample.

One *Schizogalerix* species occurs almost solely in one MN unit. However, the species boundaries can be regarded as approximate only, because different morphologic forms are included in one species.

Ecology

Insectivores can provide only scant information about the whole ecological picture in the Sinap Formation. Insectivores have a restricted living area, so they tell more about the immediate area of the locality. Extant Echinosoricinae live in woods and near water and are sometimes even semiaquatic (Engesser 1980). This corresponds well with the geologic interpretations of the Sinap Formation: the area around the Sinap Formation is of fluvial and lacustrine conditions (Lunkka et al., this volume). In addition, the fauna assemblage in the Sinap Formation speaks of a rich vegetation and humid environment in an open woodland (Sen 1990).

Acknowledgments

This chapter is based on my master's thesis, undertaken at Åbo Akademi University, 1998. I gratefully acknowledge the help of my supervisors Dr. Sevket Sen (Muséum National d'Histoire Naturelle, Paris), Professor Mikael Fortelius (University of Helsinki), and Professor Carl Ehlers (Åbo Akademi University). Many thanks to Dr. Burkart Engesser and Dr. Jelle W. F. Reumer for reviewing this chapter. I also thank Dr. Lars W. van den Hoek Ostende for his comments.

Appendix 3.1. The Genus *Schizogalerix* in the Sinap Formation

All dimensions are given in mm; *n* is the number of specimens for the locality.

Table 3A.1. Mandible Height

Locality	Range	Average	*n*
65	2.75–3.3	3.02	3
64			
4			
94	2.8–2.85	2.83	2
108			
8A			
120	2.75–2.85	2.8	2
84	2.55–3.55	2.86	6
Inönü	2.6	2.6	2
Total	2.55–3.55	2.82	15

Table 3A.2. p_2

	Length			Width		
Locality	Range	Average	*n*	Range	Average	*n*
65	1.1–1.25	1.19	4	0.65–0.75	0.7	4
64						
4	1.05–1.2	1.13	2	0.55–0.8	0.68	2
94	1.0–1.2	1.1	3	0.55–0.6	0.58	3
108						
8A						
120	1.05	1.05	1	0.55	0.55	1
84	1.05	1.05	1	0.7	0.7	1
Inönü	0.95	0.95	1	0.65	0.65	1
Total	0.95–1.25	1.08	12	0.55–0.8	0.65	12

Table 3A.3. p_3

	Length			Width		
Locality	Range	Average	*n*	Range	Average	*n*
65	1.3–1.45	1.39	4	0.8–1.0	0.91	4
64	1.35–1.45	1.4	2	0.9–1.0	0.95	2
4						
94	1.4–1.45	1.43	2	0.75–0.9	0.83	2
108						
8A						
120						
84	1.3	1.3	2	0.8–0.9	0.85	2
Inönü	1.15	1.15	1	0.85	0.85	1
Total	1.15–1.45	1.33	11	0.75–1.0	0.88	11

Table 3A.4. p_4

	Length			Width		
Locality	Range	Average	*n*	Range	Average	*n*
65	1.55–1.75	1.63	3	1.2–1.3	1.25	2
64	1.65	1.65	1	1.35	1.35	1
4	1.5–1.65	1.58	4	1.2–1.4	1.31	4
94	1.6–1.9	1.74	5	1.25–1.5	1.35	4
108						
8A						
120	1.6	1.6	1	1.25	1.25	1
84	1.45–1.9	1.61	10	1.1–1.25	1.17	10
Inönü	1.45	1.45	1	1.1	1.1	1
Total	1.45–1.9	1.61	25	1.1–1.5	1.24	23

Table 3A.5. m_1

	Length			Width		
Locality	Range	Average	*n*	Range	Average	*n*
65	2.5–2.95	2.77	3	2.0–2.05	2.04	2
64				2.0	2.0	1
4	2.3–2.85	2.67	9	1.75–2.05	1.87	16
94	2.75–2.8	2.79	5	1.9–2.0	1.93	5
108						
8A	2.55–2.6	2.58	2	1.7–1.9	1.8	2
120	2.55	2.55	1	1.95	1.95	1
84	2.55–2.85	2.74	12	1.65–2.05	1.85	17
Inönü	2.4	2.4	1	1.75	1.75	1
Total	2.3–2.95	2.64	33	1.65–2.05	1.88	45

Table 3A.6. m_2

	Length			Width		
Locality	Range	Average	*n*	Range	Average	*n*
65	2.25–2.3	2.28	3	1.6–2.0	1.78	6
64	2.35–2.45	2.42	3	1.85	1.85	3
4	2.25–2.5	2.33	6	1.5–1.9	1.71	7
94	2.35–2.55	2.42	4	1.7–1.9	1.8	4
108						
8A						
120						
84	2.3–2.45	2.36	8	1.6–2.0	1.74	8
Inönü	2.25	2.25	1	1.6	1.6	1
Total	2.25–2.55	2.34	25	1.5–2.0	1.75	29

Table 3A.7. m_3

	Length			Width		
Locality	Range	Average	*n*	Range	Average	*n*
65	2.0	2.0	2	1.1–1.25	1.18	3
64						
4	1.75–1.9	1.86	4	1.0–1.2	1.08	4
94	1.8–2.25	2.01	5	1.05–1.15	1.12	5
108						
8A						
120						
84	1.85–1.95	1.93	5	0.95–1.05	1	4
Inönü	1.9	1.9	1	1.05	1.05	1
Total	1.75–2.25	1.94	17	0.95–1.25	1.09	17

Table 3A.8. P^2

	Length			Width		
Locality	Range	Average	*n*	Range	Average	*n*
65						
64						
4	0.8–0.85	0.83	2	1.15–1.2	1.18	2
94						
108						
8A	0.82	0.82	1	1.16	1.16	1
120						
84	0.7	0.7	2	0.9–1.0	0.95	2
Inönü						
Total	0.7–0.85	0.78	5	0.9–1.2	1.1	5

Table 3A.9. P^3

	Length			Width		
Locality	Range	Average	*n*	Range	Average	*n*
65	1.7–1.75	1.73	2	1.85–2.05	1.95	2
64	1.7–1.75	1.73	2	1.8–2.0	1.9	2
4	1.7–1.9	1.8	4	1.8–2.0	1.89	4
94	1.9	1.9	1	1.95	1.95	1
108	1.8	1.8	1	1.85	1.85	1
8A	1.96–2.1	2.03	2	2.0–2.1	2.05	2
120						
84	1.75–2.0	1.84	4	1.85–2.0	1.94	4
Inönü						
Total	1.7–2.1	1.83	16	1.8–2.1	1.93	16

Table 3A.10. P[4]

	Length			Width		
Locality	Range	Average	*n*	Range	Average	*n*
65						
64	2.2	2.2	1	2.75	2.75	1
4						
94	2.4	2.4	1	2.7	2.7	1
108						
8A	2.3	2.3	1	2.42	2.42	1
120						
84	2.15–2.45	2.25	5	2.5–2.65	2.61	5
Inönü						
Total	2.15–2.45	2.29	8	2.42–2.75	2.62	8

Table 3A.11. M[1]

	Length			Width		
Locality	Range	Average	*n*	Range	Average	*n*
65	2.55	2.55	1			
64						
4	2.4–2.7	2.55	5	3.05–3.5	3.3	6
94	2.5–2.65	2.56	2	3.25	3.25	2
108	2.0	2.0	1	3.0	3.0	1
8A	2.1	2.1	1	3.1	3.1	1
120	2.4	2.4	1	3.1	3.1	1
84	2.35–2.5	2.44	4	3.05–3.25	3.14	5
Inönü	2.4	2.4	1	3.2	3.2	1
Total	2.0–2.7	2.38	16	3.0–3.5	3.16	17

Table 3A.12. M[2]

	Length			Width		
Locality	Range	Average	*n*	Range	Average	*n*
65	2.1	2.1	1	3.15	3.15	1
64	2.0–2.1	2.05	2	3.1–3.2	3.15	2
4	2.0–2.15	2.05	4	3.15–3.25	3.19	4
94	2.05	2.05	2	3.15–3.2	3.18	2
108	2.0	2.0	1	3.0	3.0	1
8A	2.0–2.1	2.06	3	3.08–3.2	3.13	3
120						
84	2.0	2.0	2	3.0–3.1	3.05	2
Inönü	1.9	1.9	1	3.0	3.0	1
Total	1.9–2.15	2.03	16	3.0–3.25	3.11	16

Table 3A.13. M^3

Locality	Length			Width		
	Range	Average	*n*	Range	Average	*n*
65						
64	1.35	1.35	1	2.0	2.0	1
4	1.25–1.5	1.31	5	2.05–2.15	2.1	5
94	1.25–1.4	1.35	3	2.0	2.0	3
108						
8A	1.05	1.05	1	1.8	1.8	1
120						
84	1.3	1.3	1	1.8	1.8	1
Inönü						
Total	1.05–1.5	1.27	11	1.8–2.15	1.94	11

Literature Cited

Agadjanian, A. K., and W. von Koenigswald, 1977, Merkmalsverschiebung an den oberen Molaren von *Dicrostonyx* (Rodentia, Mammalia) im Jungquartär: Neues Jahrbuch für Geologie und Paläontologie Abhandlungen, v. 153, no. 1, pp. 33–49.

Bachmayer, F., and R. W. Wilson, 1970, Small mammals (Insectivora, Chiroptera, Lagomorpha, Rodentia) from the Kohfidisch Fissures of Burgenland, Austria: Annalen des Naturhistorischen Museums in Wien, v. 74, pp. 533–587.

Bernor, R. L., V. Fahlbusch, P. Andrews, H. de Bruijn, M. Fortelius, F. Rögl, F. F. Steininger, and L. Werdelin, 1996, The evolution of western Eurasian Neogene mammal faunas: A chronologic, systematic, biogeographic, and paleoenvironmental synthesis, *in* R. L. Bernor, V. Fahlbusch, and H-W. Mittmann, eds., The evolution of western Eurasian Neogene mammal faunas: New York, Columbia University Press, pp. 449–469.

Bi S., W. Wu, J. Ye, and J. Meng, 1999, Erinaceidae from the Middle Miocene of North Junggar basin, Xinjiang Uygur autonomous region, China, *in* Wang Y., and Deng T., eds., Proceedings of the 7th annual meeting of the Chinese Society of Vertebrate Paleontology: Beijing: China Ocean Press, pp. 157–165.

Doukas, C. S., L. W. van den Hoek Ostende, C. D. Theocharaopoulos, and J. W. F. Reumer, 1995, The vertebrate locality Maramena (Macedonia, Greece, at the Turolian-Ruscinian Boundary (Neogene), 5. Insectivora (Erinaceidae, Talpidae, Soricidae, Mammalia): Münchner Geowissenschaftlichen Abhandlungen (A), v. 28, pp. 43–64.

Engesser, B., 1980, Insectivora und Chiroptera (Mammalia) aus dem Neogen der Türkei: Sweizerische Paläontologische Abhandlungen, v. 102, pp. 45–149.

Engesser, B., and R. Ziegler, 1996, Didelphids, Insectivores, and Chiropterans from the Later Miocene of France, central Europe, Greece, and Turkey, *in* R. L. Bernor, V. Fahlbusch, and H.-W. Mittmann, eds., The evolution of western Eurasian Neogene mammal faunas: New York, Columbia University Press, pp. 157–167.

Kappelman, J., S. Sen, M. Fortelius, A. Duncan, B. Alpagut, J. Crabaugh, A. Gentry, J.-P. Lunkka, F. McDowell, N. Solounias, S. Viranta, and L. Werdelin, 1996, Choronology and biostratigraphy of the Miocene Sinap Formation of central Turkey, *in* R. L. Bernor, V. Fahlbusch, and H.-W. Mittmann, eds., The evolution of western Eurasian Neogene mammal faunas: New York, Columbia University Press, pp. 78–95.

Lungu, A. N., 1981, Hipparionine fauna of the Middle Sarmatian in Moldavia (Insectivora, Lagomorpha and Rodentia): Kishinev, Georgia, Shtiintsa, 131 pp. (in Russian).

Made, J. van der, 1996, Pre-Pleistocene land mammals from Crete, *in* D. S. Reese, ed., Pleistocene and Holocene fauna of Crete and its first settlers, Monographs in World Archaeology, no. 28: Madison, Wisconsin, Prehistory Press, pp. 69–79.

Murphy, M. A., J. C. Matti, and O. H. Walliser, 1981, Biostratigraphy and evolution of the *Ozarkodina remscheidensis-Eognathodus sulcatus* lineage (Lower Devonian) in Germany and central Nevada: Journal of Paleontology, v. 55, no. 4, pp. 747–772.

NOW database, 1999, Neogene of the Old World, fossil mammal database: www.helsinki.fi/science/now.

Rabeder, G., 1973, *Galerix* und *Lanthanotherium* (Erinaceidae, Insectivora) aus dem Pannon des Wiener Beckens: Neues Jahrbuch für Geologie und Paläontologie Monatshefte, v. 7, pp. 429–446.

Rose, K. D., and T. M. Bown, 1986, Gradual evolution and species discrimination in the fossil record: University of Wyoming Contributions to Geology Special Paper 3, pp. 119–130.

Rümke, C. G., 1976, Insectivora from Pikermi and Biodrak (Greece): Proceedings of the Koninklijke Nederlandse Akademie van Wetenschappen, Amsterdam, series B, v. 79, no. 4, pp. 256–270.

Sen, S., 1991, Stratigraphie, faunes de mammifères et magnétostratigraphie du Néogène de Sinap Tepe, Province d'Ankara, Turquie: Bulletin du Museum National d'Histoire Naturelle, sér. 4e, v. 12, pp. 243–277.

Woodburne, M. O., 1996, Precision and resolution in mammalian chronostratigraphy: Principles, practices, examples: Journal of Vertebrate Paleontology, v. 16, no. 3, pp. 531–555.

Ziegler R., 1999, Order Insectivora, *in* G. E. Rößner, and K. Heissig, eds., The Miocene land mammals of Europe: München, Germany, Verlag Dr. Friedrich Pfeil, pp. 53–75.

4

Hominoidea (Primates)

J. Kappelman, B. G. Richmond, E. R. Seiffert, A. M. Maga, and T. M. Ryan

Hominoid primates were first recovered from the region north of Ankara, Turkey, in the 1950s and 1960s by Ozansoy (1955, 1957, 1965) and were attributed to *Ankarapithecus meteai.* These specimens have figured prominently in ongoing discussions of hominoid evolution—especially those surrounding the possible relationship of *Sivapithecus,* from the Miocene of the Indo-Pakistani Siwaliks, and *Pongo,* the living orangutan. In 1995 additional hominoid cranial and the first postcranial fossils from this species were recovered from the 9.59-Ma locality 12 in the Sinap Formation. The new fossil remains preserve some portions of anatomy not known from the previously discovered specimens and display a mixture of features, including a relatively narrow interorbital region, moderately developed but separate supraorbital tori with an extensive frontal sinus, extensive maxillary sinuses, orbital apertures that are nearly as tall as they are broad, a robust mandibular corpus, upper incisor heteromorphy, and inferred klinorhynchy of the cranium that, when taken together, are features not seen in any other extant or extinct hominoid. Postcranial remains include two phalanges, and a partial radius and femoral shaft that may be attributable to the female (AS95-500) and male (MTA 2125) cranial specimens from locality 12. Analysis of these postcranial remains suggests that *A. meteai* engaged in a range of pronograde quadrupedal behaviors, probably on both arboreal and terrestrial substrates. The postcrania lack forelimb suspensory specializations witnessed in some of the extant apes and cursorial specializations found in terrestrial cercopithecines. Therefore, its morphology challenges the homology of many of the postcranial features shared by modern great apes and humans.

Early Discoveries of Hominoids from the Sinap Formation

Fossil apes are rare but important elements of middle and late Miocene Old World faunas. An interesting element of the Sinap fauna is the fossil ape, *Ankarapithecus meteai,* which is named for the nearby city and capital of Turkey, Ankara, and the acronym "MTA" (Maden Tetkik ve Arama), Turkey's Mineral Research and Exploration Institute. Prior to 1995, there were only two well-known specimens of *A. meteai:* the first, a partial mandible, found in 1955 (Ozansoy 1955, 1957) and described in 1965 (Ozansoy 1965); and the second, a lower face, found in 1967 and fully described in 1980 (Andrews and Tekkaya 1980), with a revised description of both specimens published in 1998 by Begun and Güleç (see table 4.1). These fossils have figured prominently in discussions surrounding the possible relationship between *Sivapithecus* from the Siwaliks of Indo-Pakistan and *Pongo,* the living orangutan (Andrews and Cronin 1982; Pilbeam 1982; Ward and Pilbeam 1983; Brown 1997; Pilbeam 1997; Schwartz 1997; Ward 1997; Begun and Güleç 1998), as well as more general issues in hominoid evolution.

A third specimen from Sinap is enigmatic. Ozansoy (1965) reported an isolated p4 (see Sen, Introduction, this volume) recovered from the lower member of the Sinap Formation, apparently from the region of Sinap Tepe, which he attributed to *Dryopithecus* sp. This specimen was found in association with a fauna that includes *Listriodon* (Ozansoy 1965, pp. 16, 75) but not hipparionines. The only documented occurrence of *Listriodon* in the Sinap Tepe area is at locality 24 in İnönü (see van der Made, chapter 13, this volume), and this locality is thought to date to perhaps 15 Ma (Kappelman et al., chapter 2, this volume). Several other localities in the area surrounding Sinap Tepe include faunas without hipparion but these are also without *Listriodon.* The faunal association suggests that the locality with this primate is probably >11 Ma in age. Unfortunately, this primate specimen was never figured or fully described and its current whereabouts are unknown.

The precise geographic location and stratigraphic position of the two *Ankarapithecus* discoveries were not, unfortunately, given in the early published descriptions; rather,

Table 4.1. Fossil Ape Specimens from the Sinap Formation

ID Number	Description	Discovery Date	Repository[1]	Reference
No number	Partial mandible (type)	1955	MTA	Ozansoy (1955, 1957, 1965); Begun and Güleç (1998)
MTA 2125	Palate	1967	MTA	Andrews and Tekkaya (1980); McHenry et al. (1980); Andrews and Cronin (1981); Begun and Güleç (1998)
AS95.321	Right partial femur	1995	AMM	Discussed briefly in Andrews and Bernor (2000); this chapter
AS95.500	Skull (partial cranium and mandible)	1995	AMM	Alpagut et al. (1996); Begun and Güleç (1998); Andrews and Alpagut (2001)
AS95.501	Distal portion proximal phalanx	1995	AMM	This chapter
AS95.502	Distal portion middle phalanx	1995	AMM	This chapter
AS95.503	Right partial radius	1995	AMM	Discussed briefly in Andrews and Bernor (2000); this chapter
AS95.504	Left upper central incisor (field catalog 12.147, 21.06.95)	1995	AMM	This chapter

[1]Abbreviations: MTA: Maden Tetkik ve Arama, Ankara, Turkey; AS95: Ankara Sinap 1995; AMM: Anadolu Medenyetleri Müzesi, Ankara, Turkey.

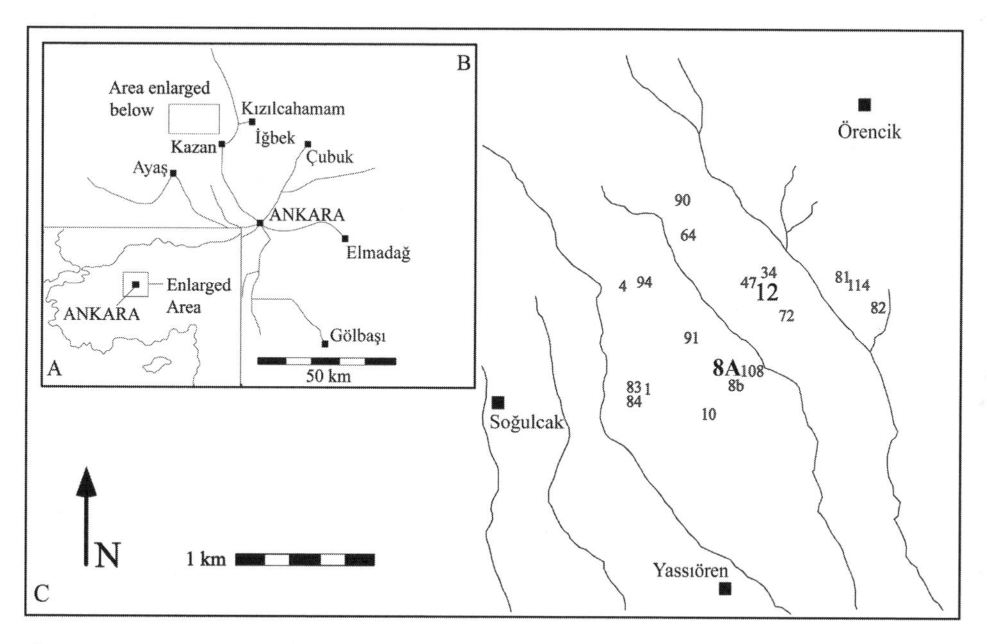

Figure 4.1. Maps of the fossil ape localities in the Sinap region. (**A**) An inset of Turkey. (**B**) The region around Kazan. Lines are major roads. (**C**) Fossil localities 8A and 12 (larger font size) along with several other fossil localities (numbers, smaller font size) in the immediate vicinity of the villages of Yassıören, Örencik, and Soğulcak. Lines are ephemeral stream courses.

they are simply described as being from the "Middle Sinap series" (Andrews and Tekkaya 1980, p. 85). Dr. Sevket Sen, who as a young student of Ozansoy was present at the 1967 discovery, reports that the mandible was found by Ozansoy and the MTA team at our locality 8A (their Loc. I), located on the southeastern flank of Sinap Tepe (fig. 4.1). The mandible preserves the symphysis with left i2–p3 and right i2–c still implanted, and left p4–m3 associated but isolated because the mandibular corpus was destroyed during excavation (Sen, Introduction, this volume). The second specimen, a lower face (MTA 2125), was discovered at our locality 12 (their Loc. II), located about 1 km to the east of locality 8A on the ridge top of Delikayincak Tepe by Dr. Sen (see figs. 4.1, 4.2). The MTA team dynamited and trenched along the southwestern ridge of Delikayincak Tepe and discovered fossils in siltstones below a cobble-and-boulder conglomerate that caps the ridge. Their excavations cut back into the hillside to the maximum extent practical and produced a trench >50-m long and nearly 3-m wide at the foot of a ≤5-m-tall wall consisting of the loosely consolidated cobble and boulder conglomerate (see Sen, Introduction, this volume).

Figure 4.2. A view of Delikayincak Tepe toward the southeast. This ridge exposes a distinctive series of paleosols on its northwestern face. The original excavation of fossil locality 12 (Ozansoy's Loc. II) was at the southwestern edge of the ridge (arrow).

Age of the Sinap Hominoids

As discussed in Kappelman et al. (chapter 2, this volume) and Lunkka et al. (chapter 1, this volume), detailed studies of the chronostratigraphy and sedimentology were undertaken by the Sinap project to provide age estimates for the fauna and reconstruct the depositional settings of the fossil localities. Mapping across the ~1-km-wide valley that separates localities 8A and 12 shows that 8A is at a somewhat lower stratigraphic level than locality 12, and comparisons of the paleomagnetic reversal stratigraphy suggest age estimates of 9.886 Ma for locality 8A and 9.590 Ma for locality 12, for an approximate temporal separation of about 300 Ka (Kappelman et al., chapter 2, this volume). These age estimates are very similar to those for some of the major Vallesian European hominoid localities (Andrews et al. 1996), such as that for *Dryopithecus* from Can Llobateres (9.6–9.7 Ma; Agusti et al. 1996), older than others, such as for the *Oreopithecus* localities of southern Tuscany (7.55 Ma; Rook et al. 2000), and near the younger end of the nearly 5-Ma temporal range that is documented for the Siwalik hominoids (Barry 1986; Kappelman et al. 1991).

Note that in spite of concentrated surface prospecting and detailed excavations at numerous other localities in depositional settings that are essentially identical to those of localities 8A and 12, no new hominoid localities have been discovered in the Sinap Formation beyond the two sites originally discovered by Ozansoy (with the location of the enigmatic third site unknown). We estimate this frequency at <2% of the total number of fossil localities. Sampling bias is always difficult to assess for any rare member of a fauna, but these negative results make it likely that the small number of hominoid fossil localities in the Sinap Formation, along with the limited number of hominoid fossils that have been recovered, probably reflect their relative scarcity in the fauna. Their scarcity suggests that these locations were at the very margins of their habitat range, with their occurrences in the Sinap Formation controlled by local environmental conditions. It is also possible that the very restricted temporal range of the Sinap hominoids is not an artifact of sampling, but this is a point that is even more difficult to assess.

New Discoveries from Locality 12

When the Sinap project began its work in 1989, the trench of the old Delikayincak Tepe excavation was still obvious, even though some 20 years had passed, and it was not uncommon to find fossil fragments on the surface of the old excavation spoil heaps. Preliminary trenching in 1991 at the base of the capping conglomerate revealed that good concentrations of fossils were preserved in the underlying siltstone, but the presence of the thick, loosely cemented, and sometimes overhanging cobble-and-boulder conglomerate made further excavations at this known fossil ape locality too dangerous to undertake. At the end of the 1993 field season a bulldozer was used to remove the capping conglomerate, thus making it possible to safely excavate the site. A platform ~20 m wide by ~50 m long that followed the southeast dipping trend of the sediments was cut across the top of Delikayincak Tepe (fig. 4.3). About 0.5–1.0 m of siltstone was left in place above what was believed to be the bone-bearing level to protect this surface prior to its excavation. Preliminary trenching at the end of the 1993 field season at the northwestern end of the platform revealed some nicely preserved fossils. No fossils were recovered during excavations under the direction of Dr. Peter Andrews early in the 1994 field season, but renewed work during the 1995 field season revealed beautifully preserved and articulated hipparionine postcranials immediately adjacent to the 1994 trenches. Other trenches were opened along the southern portion of the platform and these

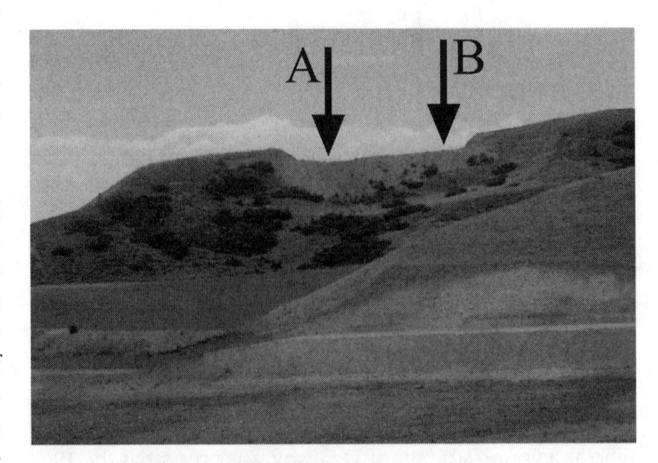

Figure 4.3. A view of Delikayincak Tepe toward the southwest showing a side profile of the ridge and the excavation platform that was bulldozed along the ridge top in 1993. The angle of the excavation platform follows the approximate angle of bedding dip and is at the contact with an overlying cobble to boulder conglomerate that is still present at the northern and southern points of the platform. Arrow A shows the location where the fossil primates were discovered in 1995; arrow B shows the location where associated hipparionine fossils were discovered in 1995.

resulted in the recovery of a large number of fossils, including the fossil hominoid specimens discussed here. After the 1995 field season, Dr. Berna Alpagut elected to carry out fieldwork in the Sinap Formation on her own, thus ending the Sinap project. We understand that Dr. Alpagut has continued to sporadically excavate locality 12 but we are unaware of any reports of new hominoid discoveries. Other excavations carried out by the Sinap project at locality 8A did not produce any new hominoids or fragments of the previously discovered mandible, even though much additional fossil material was recovered from this site.

As noted earlier, the cranial and mandibular specimens of *Ankarapithecus meteai* have been described and discussed in several publications (Ozansoy 1955, 1957, 1965; Andrews and Tekkaya 1980; McHenry et al. 1980; Andrews and Cronin 1982; Alpagut et al. 1996; Begun and Güleç 1998; Andrews and Alpagut 2001). Here we limit our discussion to new observations and interpretations of this material. The postcranial remains have been briefly discussed by Andrews and Bernor (1999) and so here we give this material a much fuller treatment.

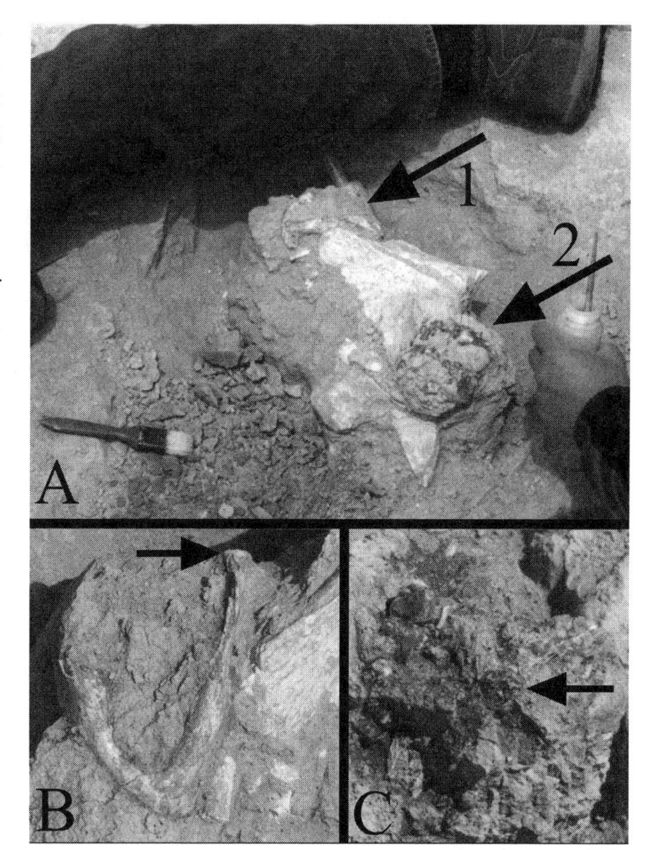

Figure 4.4. (A) The mandible (arrow 1) and partial cranium (arrow 2) of AS95-500 are shown in situ shortly after their discovery. Both specimens are approximately upside down, and a large fragment of proboscidean bone is wedged between them. (B) The inferior margins of the mandibular corpus are preserved whereas the inferior portion of both sides of the rami are broken, and only the right ascending ramus remains. The arrow points to the right mandibular condyle. Note that the small crack seen across the anterior portion of the mandible (see figs. 4.14, 4.17) continues across the mandible and into the sediment and is postdepositional. (C) This view of the partial cranium shows the occlusal surfaces of the molars and premolars. The arrow points to the right M3. The silt pillar that filled the right orbit is visible above this arrow to the right. (See also fig. 4.5 for a posterior stereo view of the partial cranium giving a clearer view of this feature.) Note that the offset between the molars and premolars follows a stepped crack in the sediment and is postdepositional.

Discovery and Observations on the Taphonomy of AS95-500

The partial skull of *Ankarapithecus meteai,* AS95-500, was discovered at locality 12 (40°15′02″N, 32°38′58″E) by Zeynep Bostan on June 20, 1995. Both the mandible and partial cranium were found at the same stratigraphic level in an upside-down position (fig. 4.4) and were prepared and pedestaled for plaster jacketing and removal as a block by Dr. Alpagut and Ms. Bostan. The cranium (entered into the field catalog by Dr. Alpagut as "maxilla" and assigned field number 1995.12.141) and the mandible (field number 1995.12.142) were found in association with numerous other mammal fossil remains. Even though the mandible and partial cranium were not articulated, they are the same size and occlude perfectly. The dentitions display an identical state of eruption and wear, including somewhat more advanced wear on the left upper and lower teeth. These observations demonstrate that the cranium and mandible are almost certainly from the same individual.

A posterior view of the upside-down partial cranium shown prior to final plaster jacketing is shown in figure 4.5. The superior portion of the left frontal preserving a portion of the endocranial contour was inadvertently broken loose during excavation but was subsequently found at the site. The stereo view of the partial cranium in situ in figure 4.5 clearly shows the sediment infilling of the right orbit as a small silt pillar. This mold preserves the smooth contours of this surface along its margins, and a small fragment of bone that forms the inferior wall of the orbit preserves a portion of the optic canal. Most of the left orbital silt infilling was broken loose during excavation, but the small portions that remained also preserve the smooth contours of this surface. Subsequent sieving of what was identified as the spoil dirt from that day's excavation resulted in the recovery of many tiny, eggshell-thin fragments that together formed the walls of the orbit. Unfortunately, the recovery of these fragments was not complete. In addition, the portion of the frontal bone that includes the right supraorbital torus was also recovered from the spoil heap during sieving, but a small portion of the frontal process of the right zygomatic was not found. Although it cannot be known how much of the cranium was originally present, it is clear that a perhaps significant portion of the neurocranium was inadvertently chiseled away in the excavators' attempt to pedestal for removal what they believed to be a maxilla. Furthermore, many of the breaks along the inferior corpus and ascending ramus of the mandible are fresh and are also likely to have occurred during the discovery

Figure 4.5. A posterior stereo view of the AS95-500 partial cranium in situ. The cranium is upside down. The occlusal surfaces of the teeth are in the horizontal plane and the arrow points to the right M3. Note that the silt infilling of the right orbit is preserved in nearly its entirety while that of the right orbit was chiseled away. The right frontal was also chiseled away during preparation in the field but was subsequently discovered during sieving of the spoil dirt. The upper portion of the left frontal is missing in this view, but it was found in the excavation pit. Scale bar in cm.

and initial excavation of this specimen, but subsequent recovery efforts did not succeed in identifying any recognizable fragments of this specimen. Note that these breaks and damage occurred during excavation: future students of these specimens should not mistakenly attribute their origin to taphonomic processes.

It is clear that the mandible and partial cranium of AS95-500 experienced several stages of pre- and post-burial damage beyond that suffered during recovery. For example, a stereo view of the face in figure 4.6 clearly illustrates that the arched contour of the right orbital aperture is undistorted, despite the absence of a small portion of the frontal process of the right zygomatic just inferior to the zygomaticofrontal suture. In contrast, inspection of the left orbit reveals minor breakage and distortion of the frontal that combines to give the left orbital aperture a flattened and more sharply angular appearance along its superior and lateral margin. The external surface of the bone of the left torus differs from that of the right in being very rough and uneven; it appears to preserve evidence of a healed fracture to this part of the upper face (fig. 4.7). As noted earlier, there is a somewhat more advanced degree of wear on the left side of the dentition, and it is possible that this healed fracture to the face produced a moderate degree of malocclusion, which in turn contributed to uneven wear on the dentition. This possibility of a healed fracture to the left orbit also has implications for the interpretation of the configuration of the frontal sinus (see below).

Other damage to AS95-500 clearly occurred after death but before burial. For example, note that the left upper central incisor is missing from the partial cranium and a left central upper incisor (field catalog number 12.147) was recovered at approximately the same depth as the cranium from an adjacent excavation square. This left upper central incisor is nearly identical in size to the implanted right central incisor, and both teeth show nearly identical degrees of wear. The alveolar bone that once surrounded the external surface of the left central incisor is missing from the cranium. Because of these similarities, we are confident that this isolated left upper incisor belongs to the partial cranium. These observations reveal that the cranium experienced some damage, perhaps due to rolling or transport, before it became encased in the surrounding sediments, and that the left central incisor was broken loose from the cranium and this in turn damaged the subnasal region. It is,

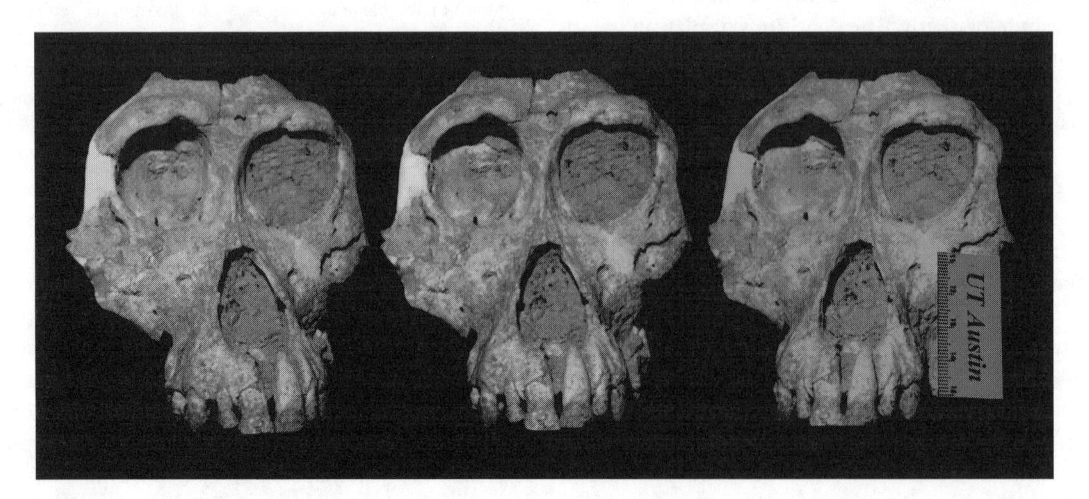

Figure 4.6. Stereo views of the AS95-500 partial cranium in anterior aspect. Note the extreme incisor heteromorphy, modest canine fossa, keeled nasals, recessed lacrimals, distinct supraorbital tori that are not joined across glabella, damage to the left torus, and the right temporal line. The small hole to the left of glabella enters into the frontal sinus (see fig. 4.9). The somewhat vertical and left lateral offset of the molars is clearly visible in the center image.

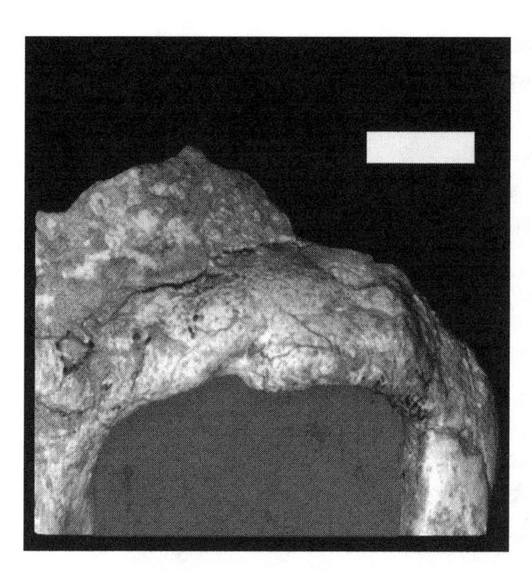

Figure 4.7. An anterior close-up of the left torus, showing the rough, uneven surface that probably points to a healed fracture. The broken surface also emphasizes that this thin table of bone forms the outer surface of the extensive frontal sinus. Scale bar = 1 cm.

however, unlikely that this sort of damage to the cranium was very severe, because the thin bones of the orbits and at least a portion of the neurocranium were present prior to discovery. The close proximity of the mandible and cranium of a single individual also argues for minimal transport or disturbance before burial. Furthermore, other fossil specimens from locality 12 are sometimes preserved as associated and articulated specimens, offering additional evidence for minimal transport. The observation that all of the hollow spaces and foramina of AS95-500 were filled with sediment suggests that burial was relatively quick and complete. Note also that the bone and enamel surfaces of AS95-500, like those of many of the other fossil specimens from this site, are deeply etched by numerous small root traces, but the bone surface does not generally show advanced stages of weathering. When taken together, this evidence suggests that burial in the encasing pebbly siltstone, although rapid, occurred at a relatively shallow depth, well within the root zone of the overlying and subsequently poorly developed paleosol.

Remaining damage to AS95-500 clearly occurred after burial but before recovery. For example, a small crack that runs through the anterior portion of the mandible between the right p4 and m1 (slightly posterior to the right mental foramen) and the left p3 and p4 (directly through the left mental foramen) is also seen in the block of sediment in which the mandible was found (see fig. 4.4). Similarly, it has been previously noted that the posterior portion of the palate that includes the maxillary molars is displaced left laterally and slightly superiorly (Alpagut et al. 1996); this is apparent from a photograph of the palate in Andrews and Alpagut (2001 fig. 9.1B). This step is also apparent in the block of sediment in which the face was discovered and follows the impression of another bone that was preserved on top of the palate. The damage also seems to have occurred after burial but before recovery.

Features of the Cranium

The partial skull of AS95-500 represents a probable female that is conspecific with the lower face of MTA 2125, a probable male, which was discovered at locality 12 about 30 years earlier (Ozansoy 1965). The two specimens preserve different aspects of the facial and palatal anatomy but are very similar in those features that are preserved in both specimens. Many of these features have been described in Ozansoy (1965), Andrews and Tekkaya (1980), McHenry et al. (1980), Andrews and Cronin (1982), Alpagut et al. (1996), and Begun and Güleç (1998), but several of them deserve additional comment and discussion.

Supraorbital Region

The region of the upper face of fossil hominoids has received a considerable amount of attention even though it is only rarely completely preserved in the fossil record. In AS95-500, the supraorbital region over each orbital aperture is inflated and forms moderately developed supraorbital tori. These are clearly visible in stereo view in figure 4.6. (As noted above, the left supraorbital region is damaged, so our comments refer primarily to the undamaged right side.) The tori begin laterally just superior to the tubercle (only preserved on the left side) near the zygomaticofrontal suture. The tori rise superiorly in a gentle arch toward midorbit and descend sharply from the supraciliary notch toward the midline in one continuous structure. The left and right tori are not joined by an inflated torus across glabella as is the case in *Pan* and *Gorilla*. Rather, glabella is a somewhat lower-lying region that separates the tori into left and right structures (fig. 4.8). Slight crushing at glabella resulted in the loss of some of the table bone at midline and across the left torus, exposing the left frontal sinus (see below). Although this damage at midline makes the region look somewhat depressed, in reality it appears that this area was flat rather than concave in life. This flat contour is maintained upward across the frontal squama. The region immediately superior to the tori is slightly depressed and forms a poorly developed sulcus or furrow. This slight supraorbital sulcus gives the tori a somewhat projecting configuration.

The anatomy of the supraorbital region in AS95-500 differs from that of GSP 15000, the best-preserved specimen of *Sivapithecus sivalensis*. As discussed in detail by Ward and Brown (1986), the supraorbital region of GSP 15000 is best described as a thickened costa supraorbitalis (see Clarke 1977), or supraorbital rim, which rises from a weakly developed glabella and arches over the orbit to terminate at the zygomaticofrontal suture. This flattened bar of bone is fused to the frontal squama and there is no evidence of a supratoral sulcus. These features are shared between GSP

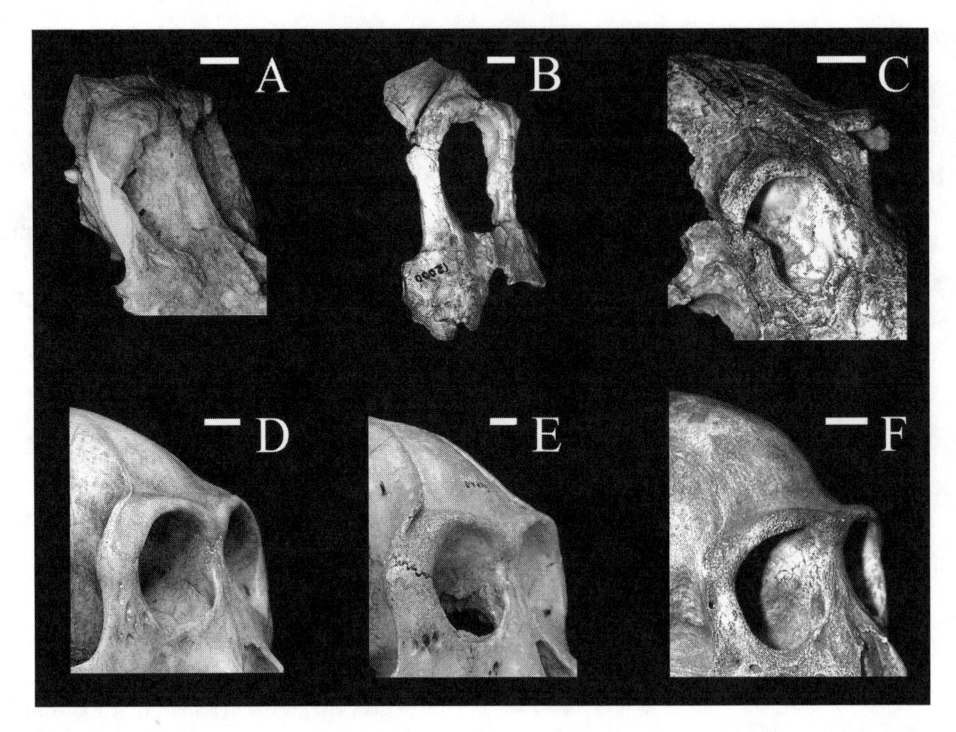

Figure 4.8. Right lateral-posterior oblique views of several extinct and extant species to demonstrate the range of variation in the morphology of the supraorbital region. **(A)** AS95-500, *Ankarapithecus meteai*, has rather bulbous supraorbital tori that are not connected across glabella. The temporal line lies posterior to the torus until it diverges along its posterior path. A shallow furrow is posterior to the tori. **(B)** GSP 15000, *Sivapithecus indicus*, has a much less protrusive supraorbital region that is best described as a "supraorbital rib" (Ward and Brown 1986). The temporal line is fused to the supraorbital rib. This view of the left side was mirrored as a right for easier comparison. **(C)** KNM ER 16950, *Turkanapithecus kalakolensis*, has a supraorbital region that is not inflated with a temporal line that is fused to the posterior portion of the rim. **(D)** A female *Pongo pygmaeus* (Wenner-Gren cast no. 4-CF14/11) has a supraorbital rib with a temporal line that is fused to its posterior margin. These ribs drop across glabella but do not form a continuous torus. **(E)** The male *Pongo pygmaeus* (UTA no.L/111) is similar to the female in **(D)** but shows less distinct ribs and a stronger temporal line. **(F)** A bonobo, *Pan paniscus* (UCB cast), shows a thin torus that connects the two tori across glabella with a supratoral sulcus. Scale bars = 1 cm.

15000 and *Pongo* and differ from both the supraorbital anatomy of *Pan* and *Gorilla* as well as that of AS95-500 as described here.

The configuration and orientation of the temporal line in AS95-500 also contributes to the supraorbital structure. The temporal line is preserved on the right side of AS95-500. It is well developed and is confluent with the lateral posteriosuperior rim of the orbit beginning near the point of the zygomaticofrontal suture. The temporal line diverges from the torus superiorly just before the point of midorbit and continues to rise on its posterior course along the small part of the frontal squama that is preserved. The temporal line of AS95-500 lies at about the same level as the orbital torus through the region where the two are fused; in contrast, the well-defined temporal line of GSP 15000 lies superior to the orbital rim and continues a sharp posterior rise along the small portion of the vertically inclined frontal squama that is preserved. The configuration in GSP 15000 is very similar to that seen in *Pongo* and differs fundamentally from that of *Pan* and *Gorilla*. The temporal line in both *Pan* and *Gorilla* is also fused to the lateral rim of the tori beginning near the zygomaticofrontal suture but does not, however, rise above the orbital rim as in *Pongo;* rather, the temporal line lies posterior to the rim and follows the dip of the supratoral sulcus and rise of the frontal squama as it continues posteriorly. Although the extent of the fusion of the temporal line and the orbit in AS95-500 somewhat resembles that seen in *Pongo*, there are also similarities to the configuration seen in *Dryopithecus* (CL1-18000 and RUD-44) and *Ouranopithecus* (XIR-1) (Cameron 1997). The configuration of the supraorbital region of AS95-500 is unique and probably reflects the presence of a frontal sinus and a different facial profile than other extant and extinct apes (see below).

Frontal Sinus and Interorbital Region

The presence, or absence, and development of a frontal sinus are clearly integral to the overall structure of the supraorbital region. Its presence in *Pan* and *Gorilla* and absence in *Pongo* has suggested to many workers that this

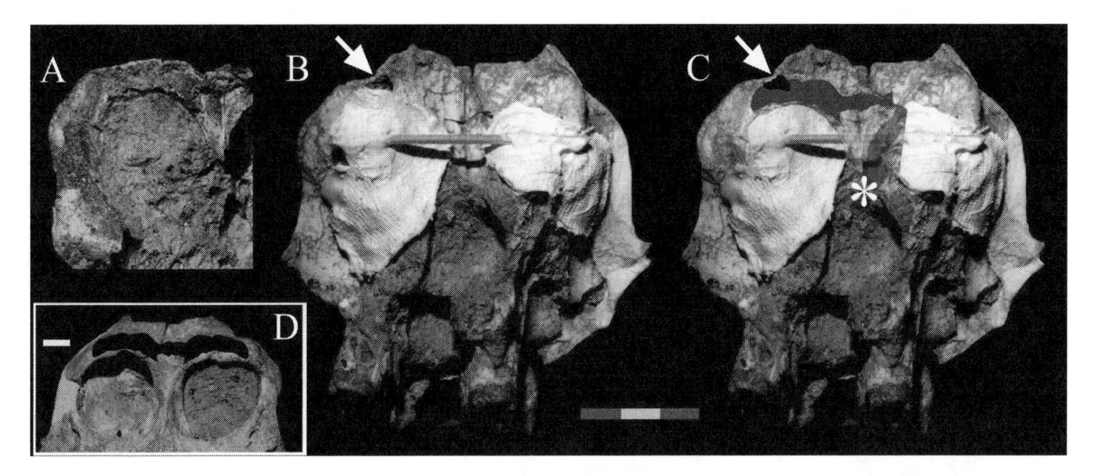

Figure 4.9. The frontal sinus on the left side of AS95-500. (**A**) A close-up of the broken posterior orbit without the covering frontal bone (see fig. 4.5). (**B**) The reconstructed specimen that includes the frontal. The arrow marks the small hole that opens into the sinus. (**C**) The composite image includes an image of the orbit and a tracing of the frontal sinus from (**A**) overlaid on (**B**). The frontal sinus is shown to extend across nearly the full extent of the supraorbital region. The approximate position of the cribriform plate is indicated by the star (*) and is at the base of the frontal crest in AS95-500 (see fig. 4.13). (**D**) The inset photo of an anterior view of the cranium includes mirrored overlays of the frontal sinus and shows what the extent of the sinus would be if it is symmetrical across the midline. An X-ray should reveal the extent of the sinus's inferior extent. Scale bar in cm.

feature might be of some utility in sorting through the relationships of Miocene hominoids. One of the interesting features of AS95-500 is the presence of a frontal sinus. As noted in Alpagut et al. (1996, p. 350), "computed tomography of AS95-500 reveals an invasive frontal sinus that extends 24 mm laterally from the midline." These computed tomography (CT) data are currently in the hands of Dr. Alpagut and are not yet published or available. Photographs of the partial cranium in situ as well as after preparation have been used in an attempt to illustrate the extent of the sinus. As noted above, a portion of the left frontal bone was inadvertently broken loose from the partial cranium during excavation (see fig. 4.5) and this in turn revealed the extent of the left frontal sinus. A close-up of a posterior view of the left orbital region is illustrated in figure 4.9A, along with a posterior view of the reconstructed face in figure 4.9B and a view of the left frontal overlaid on top of the face in figure 4.9C. The left frontal preserves a small portion of both the external surface of the squama (see fig. 4.6) and the endocranial surface. This latter surface is also present on the right side and, together with the left side, preserves a long frontal crest, whose inferior margin ends near what must have been the position of the ethmoid notch near the cribriform plate. The inferior position of the frontal crest lies about one-third of the way above the base of the orbits. The image of the specimen in situ (fig. 4.9A) without the left frontal in place reveals that the sinus extends from the midline through the full extent of the torus. This region has been added as a gray overlay in figure 4.9C. There is an obvious hole (arrow, fig. 4.9B,C) at the point of breakage along the lateral superior margin of the left frontal near midorbit. This hole represents the lateral margin of the left frontal sinus and shows that the sinus fills the inflated space of the torus and serves to separate the external surface of the frontal squama from its preserved endocranial surface, thus placing the frontal lobe in a posterior position. At midline this region is also inflated and the sinus extends inferiorly between glabella and the frontal crest, but its full extent can only be revealed by radiography. This portion of the frontal sinus is also visible in the frontal view of the partial cranium, as noted by the small indentation and hole just to left of midline at glabella (see fig. 4.6). If the sinus is symmetrical on the left and right sides, it would resemble the reconstruction in frontal view given in the inset in figure 4.9D.

The development and expression of the anthropoid frontal sinus has a long history of study. Ward and Brown (1986) note that confusion has arisen from the difficulty in identifying the homologies of the pneumatic spaces present in the hominoid face. Although a pneumatic space is sometimes seen in *Pongo,* this space is derived from an aggressive invasion of the maxillary sinus (Cave and Haines 1940) and, when present, it extends only into the base of the frontal squama (see Ward and Brown 1986, fig. 5a,b). In contrast, *Pan* and *Gorilla* generally have large frontal sinuses that often invade much of the frontal squama (including a lateral expansion through the tori), and these sinuses are always derived from the ethmoid air cells and not the maxillary sinus. The presence of a pneumatic space in the frontal squama has been noted in several late Miocene hominoids, including GSP 15000, RUD-44 (Ward and Brown 1986), and CL1 18000-1 (Moyà-Solà and Köhler 1995), but in each of these instances the sinus is restricted to the inferior portion of the frontal squama, and its lateral extent is very limited.

The correlation between the configuration and extent of pneumatic spaces and the width of the interorbital region appears to be fairly robust. This region in *Pongo* is narrow in comparison to its biorbital breadth, whereas those of *Pan* and *Gorilla* are generally wider. The wider region in the latter two taxa accommodates the ethmoid air cell complex; the narrow region of *Pongo* completely lacks an ethmoid labyrinth. Coolidge (1933) notes that the pygmy chimpanzee, *Pan paniscus,* differs from the common chimpanzee in having a somewhat narrower interorbital region; he provides an X-ray of a single individual with an especially narrow region and minimal development of the ethmoid air cells and frontal sinus. Ward and Brown (1986) and Cramer (1977), however, both note that bonobos have ethmoid air cells and a frontal sinus. The conclusion that must be drawn from these observations is that a relatively narrow interorbital region can have a fully developed ethmoid-frontal sinus complex. The presence of a frontal sinus may be plesiomorphic within Catarrhini; it is found in the stem catarrhines *Aegyptopithecus* (Rossie et al. 2002) and *Anapithecus* (Kordos and Begun 2000), as well as the early Miocene catarrhines *Proconsul* (KNM-RU 7290; Le Gros Clark and Leakey 1951), *Afropithecus* (KNM-WT 16999; Leakey and Leakey 1986a; Leakey et al. 1988a), and *Turkanapithecus* (KNM-WT-16950; Leakey and Leakey 1986b, Leakey et al. 1988b) but is present only as a structure restricted to the midline in the latter two genera. Its absence in *Pongo* is assumed to represent a secondary loss for this genus. A frontal sinus is not found in *Victoriapithecus* (Benefit and McCrossin 1993, 1997).

The interorbital region of AS95-500 displays sharply keeled nasals and anteriorly exposed lacrimals (fig. 4.10), characteristics shared with *Ouranopithecus* (XIR-1) and *Gorilla.* Keeled nasals and anteriorly exposed lacrimals are also sometimes seen in the common chimpanzee, whereas the early Miocene *Afropithecus* has anteriorly exposed lacrimals. All of the taxa noted above have relatively wide interorbital regions (fig. 4.11). The value for AS95-500 falls within the area of overlap between the lower range of variation seen in *Pan* and the upper range of *Pongo* and is greater than that for GSP 15000. It is not clear if the region that would contain the ethmoid air cells is preserved in AS95-500, but CT might provide some information as to whether it is present. Given the large size of the frontal sinus that is present in AS95-500, it seems likely that it developed from an ethmoid air cell complex in spite of the intermediate breadth of its interorbital region. Another possibility is that the left sinus was hollowed out of the frontal bone as a consequence of the probable injury to the left side of the face discussed earlier. This possibility, however, seems remote because the contours of both the left and right frontal squama and endocranial surfaces are identical.

Orbital Aperture

Orbital shape has received a considerable amount of attention in Miocene hominid phylogenetics since the recovery of a facial skeleton of *Sivapithecus* (Pilbeam 1982) that exhibits an orbital aperture morphology most similar to *Pongo* among extant hominids (e.g., Shea 1985; Brown and Ward 1988; Ward 1997). More recently, orbital shape (or some measure of orbit proportions) has been employed as a character in phylogenetic analyses of living and extinct hominids (e.g., Moyà-Solà and Köhler 1995; Begun et al. 1997; Cameron 1997) and various authors have discussed this feature's possible phylogenetic significance (e.g., Benefit and McCrossin 1997; Schwartz 1997). Only the lower portion of the right orbit is preserved in MTA 2125 and so AS95-500 provides the first evidence for this region in *Ankarapithecus.* As noted previously, because the left orbit of AS95-500 is damaged, our comments are focused on the right orbit. The orbits of AS95-500 are about as broad (31.8 mm) as they are tall (30.8 mm), and this breadth: height index is greater than the relatively lower values seen in *Pongo* (Benefit and McCrossin 1993; Moyà-Solà and Köhler 1995) and *Sivapithecus* (Pilbeam 1982; Brown and Ward 1988).

A simple proportion of orbital breadth to height does not, however, provide much information about the complex shape of the aperture. Seiffert and Kappelman (2001) provided a multivariate morphometric analysis of orbital aperture shape in living and extinct catarrhines, including *A. meteai.* They used three different morphometric techniques and resampling analyses of distance statistics derived from these morphometric variables to gauge the consistency with which single fossil specimens could be assigned alternative orbital shape character states. On average, *Ankarapithecus* clustered with the early Oligocene stem catarrhine *Aegyptopithecus,* the early Miocene catarrhine *Afropithecus,* and *Pongo* and *Sivapithecus,* whereas the early Miocene catarrhine *Turkanapithecus,* hylobatids, and *Ouranopithecus, Paranthropus, Pan,* and *Gorilla* formed another cluster. Support for these alternative groupings was surprisingly weak, however—the two primary clusters described here were recovered in only 36–57% of the cluster analyses, depending on morphometric technique. These results suggest that there is a great deal of intraspecific variation in orbital aperture shape among extant hominoids, a conclusion that is further supported by Seiffert and Kappelman's (2001) exact randomization analysis, which revealed that 19 of the 21 distance statistics between the extinct catarrhines included in the study (many of which are often assigned different orbital shape character states) can be observed within a single extant hominid species. Such broad intraspecific shape variation—and the difficulties inherent in categorizing such complex, continuous shape variations into distinct character states—leaves the phylogenetic utility of orbital aperture shape open to debate. These data further suggest that the orbital aperture shape of single individuals, as representative of fossil taxa, should be used with appropriate caution in phylogenetic studies. For example, given only a single complete orbit of *A. meteai,* it is impossible to determine whether the range of orbital shape variation in this species (which, by extrapolation from extant taxa, we may presume was fairly large) extended further into the range of *Pan* and *Gorilla* or into that of *Pongo.* Individuals can be found within each of these extant gen-

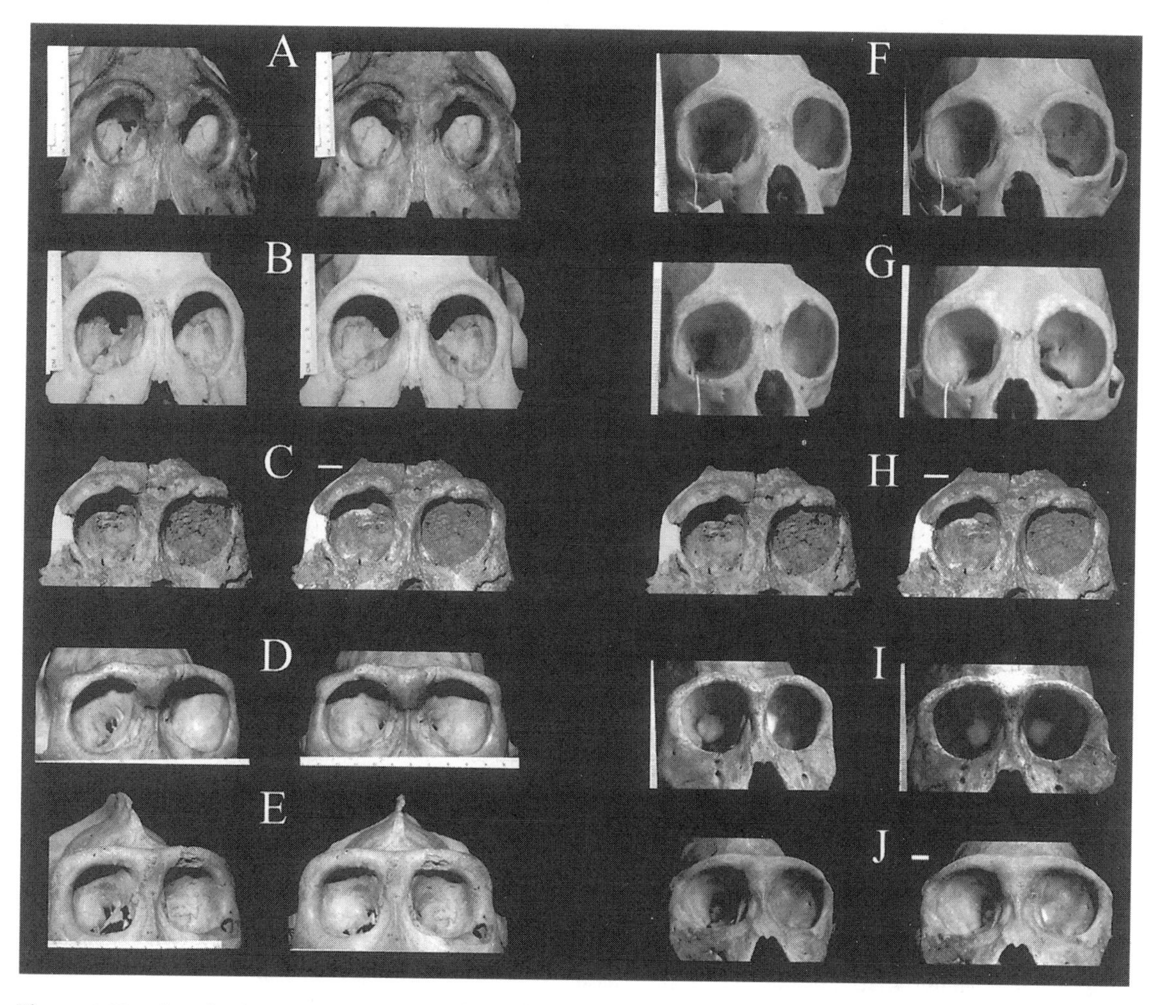

Figure 4.10. The orbital region shows a wide range of variability in topographic relationships among the extant apes. (**A, B**) These stereo images show both wide and narrow interorbital regions in *Pongo pygmaeus* (**A**, male BMNH 1892.11 5 3; **B**, female BM 1976.1434) that are generally smooth, and independent supraorbital ribs that are fused to the frontal bone. The lacrimal fossae are barely visible. (**C**) This morphology in AS95-500 (see also H) shows the narrow region but a keeled set of nasal bones that reveal the lacrimal fossae. The supraorbital tori are positioned forward from the frontal bone and are more inflated than those of *Pongo*. (**D, E**) This same region in *Gorilla gorilla* (**D**, female CMNH 1798; **E**, male CMNH 1795) shows generally broader interorbital septa that has keeled nasals in these two specimens. The keeling of the nasal bones serves to narrow this region and reveal the lacrimal fossae. The supraorbital torus in both specimens is positioned well forward from the frontal squama. (**F, G**) This region in the siamang, *Hylobates syndactylus* (**F**, male BMNH 20.1.26.1; **G**, female BMNH 20.1.26.2) shows a supraorbital rib and a wide region that is flat, not keeled. (**H**) Another view of AS95-500 (see **C**). (**I, J**) The chimpanzee, *Pan troglodytes schweinfurthii* (**I**, female BMNH 20.4.13.2; **J**, male BMNH 27.1.4.1), is similar to the gorilla (see **D, E**) in the configuration of the supraorbital torus and its position relative to the frontal squama, but the torus drops inferiorly in this female, whereas the bar in the male is generally horizontal. Both specimens have keeled nasals, but these are more rounded than the sharp keels in the gorilla. The keels are more obvious in the female and again show the lacrimal fossae. Rulers and scale bars in cm.

era that exhibit orbital shapes similar to those of AS95-500. Orbital shape and its variable expression within hominid species is sure to be under the influence of a multitude of factors (e.g., allometry, sexual dimorphism, facial kyphosis, supraorbital and sinus development, head posture, masticatory biomechanics), few of which are well understood at present. A more complete understanding of the multifactorial nature of orbital development will be required before meaningful functional or phylogenetic inferences can be garnered from the orbital morphology of poorly known fossil hominids such as *A. meteai.*

Facial Architecture

Extant pongines and hominines exhibit strikingly different patterns of upper facial architecture. As discussed previously, pongines lack an ethmofrontal sinus, have poorly developed supraorbital structures, and generally have relatively narrow interorbital regions and dorsoventrally ovoid orbital apertures. In contrast, nonhuman hominines have large, mediolaterally invasive ethmofrontal sinuses that pneumatize large supraorbital tori defined caudally by distinct post-toral sulci; they generally have a relatively wider

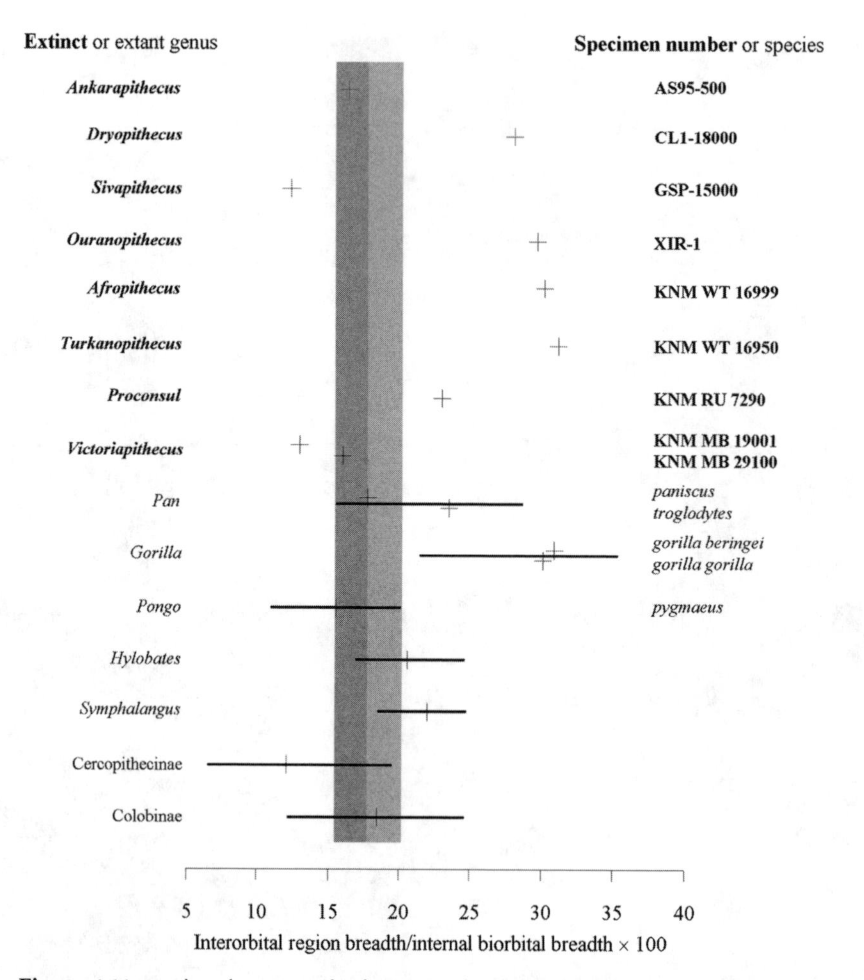

Figure 4.11. A plot of an interorbital region index (interorbital region breadth/internal biorbital breadth × 100) shows that AS95-500 falls within the region of overlap between the lower extent of *Pan* (dark vertical bar) and the upper region of *Pongo* (light vertical bar). *Sivapithecus* (GSP 15000) falls well below this region, whereas the other Miocene apes fall above. Data are from Benefit and McCrosin (1993), Moyà-Solà and Köhler (1995), and Pilbeam (1982).

interorbital region (see fig. 4.11) and orbital apertures that are wider than tall.

Shea (1985, 1988) invoked hypotheses of structural invariance in mammal craniofacial features that had previously been proposed by Enlow and colleagues (e.g., Enlow and Azuma 1975; Enlow 1982) to suggest that the two suites of upper facial features observable in extant nonhuman hominids may be intimately tied to fundamental differences in facial position relative to the braincase. This hypothesis stems from the observation that the facial skeleton of *Pongo* is airorhynch (sensu Hofer 1952), displaying a splanchnocranium that is dorsally rotated with respect to the basicranium; hominines exhibit the opposite, klinorhynch pattern, in which the facial skeleton is more ventrally deflected. The one-to-one correspondence between different patterns of facial kyphosis and different suites of upper facial features among extant nonhuman hominids and *Sivapithecus* seemed to lend considerable support to Shea's (1985, p. 336) argument that "a number of internal and external features of the orangutan skull appear to be related to the position of the face on the braincase. These features should thus be treated as an interrelated complex rather than as a series of independent characters."

The possibility that simple morphogenetic "rules" may (or may not) be driving the various morphological differences observable in hominid craniofacial form has since been of considerable interest for paleoanthropologists, and the airorhynchy/klinorhynchy issue has subsequently been discussed by a number of other authors (Brown and Ward 1988; Begun 1992; Dean and Delson 1992; Ross and Ravosa 1993; Ross and Henneberg 1995; Pickford et al. 1997; Schwartz 1997). However, more recent research on anthropoid craniofacial form has suggested that many of the allegedly invariant architectural relationships identified by Enlow and colleagues and discussed by Shea (1985, 1988) actually do not meet the statistical requirements of structural invariance (at least in extant cercopithecids; Ravosa and Shea 1994). These data have left it less clear how the cranial base may affect various aspects of hominid facial form (e.g., Lieberman et al. 2000). Furthermore, all known hominids have highly derived facial architecture and a number of unique (and still unexplained) craniofacial peculiarities (e.g., Ross and Henneberg 1995; Lieberman et al. 2000), and so it cannot be assumed that craniofacial trends observable in nonhominid anthropoids necessarily hold for living and extinct hominid species. The limited taxonomic and morphological diversity observable in extant hominid species could be hindering our understanding of whether simple morphogenetic influences in the basicranium may be governing details of hominid facial morphology, and whether different craniofacial features of hominids are structurally and/or developmentally independent (an issue that is of great importance for hominid phylogenetics).

Fortunately, this problem can potentially be informed by the study of extinct hominids that preserve unique combinations of craniofacial features not observable in extant taxa. *Ankarapithecus meteai* is one of the first extinct hominids now known to exhibit just such a "unique" pattern of upper and lower facial features (Alpagut et al. 1996), and it is thus of great interest to determine whether the face of AS95-500 was structurally airorhynch or klinorhynch. Pickford et al. (1997) have already hypothesized that *Ankarapithecus, Dryopithecus,* and *Ouranopithecus* exhibited "reduced klinorhynchy"—a stage they hypothesized to be intermediate between what they envisioned as the plesiomorphic condition within Hominoidea ("slight klinorhynchy") and the most derived condition within the clade ("strong airorhynchy," a state that they attributed to *Pongo, Sivapithecus,* and *Lufengpithecus*). These inferences are evidently based on details of facial morphology that are assumed to be correlated with airorhynchy or klinorhynchy, for these patterns of facial kyphosis cannot be gauged without positional information from the basicranium.

Unfortunately, the nature of the preserved facial skeleton of AS95-500, with localized distortion and few pertinent landmarks preserved, does not allow one to estimate the numerous craniofacial angles that have come to be of interest in hominid craniofacial biology (see Lieberman et al. 2000, for a review). We investigated the possibility that the anteroposterior plane from the inferior margin of the orbital aperture to the glenoid fossa (IO-GF) could serve as a reliable estimate for the external plane of the basioccipital-basisphenoid (this plane is of interest because it can also be reconstructed—although less reliably—for *Sivapithecus* [GSP-15000]). In the case of AS95-500, the glenoid fossa is not preserved, so the apex of the undistorted mandibular condyle was taken as an estimate for glenoid fossa position in the anteroposterior plane. Seventy-one hominid crania (table 4.2; 37 *Pan troglodytes,* 23 *Gorilla gorilla,* and 11 *Pongo pygmaeus*) from the Cleveland Museum of Natural History (CMNH) were photographed in lateral view, with two planes identified: (1) the plane of the external basioccipital and basisphenoid rostral to porion, and (2) the plane of the posterior aspect of the hard palate (together forming our angle of facial kyphosis [AFK]). The planes were identified by using metal plates that were affixed to these surfaces with soft clay. The angle formed by these plates, as well as the angle formed by IO-GF and the plane of the hard palate

Table 4.2. Estimated Angle of Facial Kyphosis (AFK)

Species and Measures	*n* total (male, female, unknown)	AFK (°)	
		Estimate	Actual
Pan troglodytes	37 (15, 21, 1)		
Mean		12.81	13.78
Standard deviation		7.19	6.87
Minimum		−2.78	−0.09
Maximum		24.75	30.46
Gorilla gorilla	23 (9, 13, 1)		
Mean		11.65	10.20
Standard deviation		5.80	6.92
Minimum		−4.78	−1.94
Maximum		28.19	25.42
Pongo pygmaeus	11 (4, 7, 0)		
Mean		1.04	−4.38
Standard deviation		5.99	8.66
Minimum		−9.51	−14.59
Maximum		11.24	11.67

(our "estimate of facial kyphosis," [AFKest]) were digitized using SigmaScan software.

A bivariate comparison of AFK and AFKest reveals that the two variables are significantly positively correlated ($r = 0.833$, $p = 0.01$, fig. 4.12). An ordinary least squares (OLS) regression of AFK on AFKest provides a prediction equation of:

$$\text{AFK} = -1.053 + 1.023 \cdot \text{AFKest},$$

whereas a model II reduced major axis (RMA) regression provides a prediction equation of:

$$\text{AFK} = -3.224 + 1.228 \cdot \text{AFKest}.$$

These results suggest that our estimate of facial kyphosis can be useful in identifying differences in facial form.

Damage to the palate of AS95-500 precludes a precise estimate of the posterior plane of the hard palate, but the oral aspect of the maxilla deepens posteriorly (palatal depth at P4 is 7.3 mm and 12.6 mm at M2). Taking this into account, our estimates for AS95-500's angle of facial kyphosis range from ~14° to 19°. An angular estimate of 14° predicts an AFK of 13.3° (OLS) and 14.0° (RMA), whereas the maximum angular estimate of 19° predicts an AFK of 18.4° (OLS) and 20.1° (RMA). Even with large confidence intervals (95% CI of ± 10.35° in the OLS regression), these data strongly suggest that AS95-500 had a relatively klinorhynch lower facial skeleton, and was thus more similar to extant hominines than to pongines (or hylobatids [Shea 1988]) in its pattern of facial kyphosis (fig. 4.13).

The observation that *Ankarapithecus meteai* is likely to have had a klinorhynch lower face as in hominines, yet exhibits a fairly narrow interorbital region that falls within

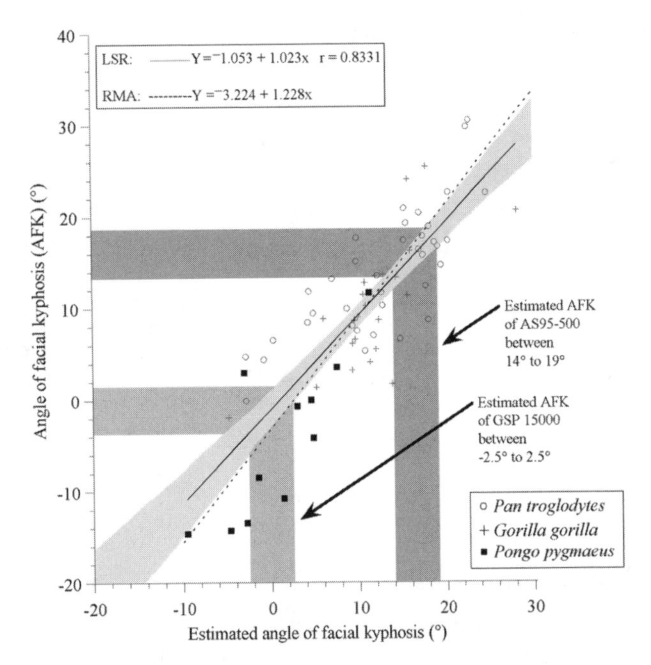

Figure 4.12. Plot of regression equations shows that the estimated angle of facial kyphosis (AFK) can be used as a reliable guide to estimate the true AFK. The estimated AFK of 14°–19° (dark shaded area) for AS95-500 falls within the region of overlap between *Gorilla gorilla* and *Pan troglodytes* and outside that of *Pongo pygmaeus*, while the estimated angle of −2.5°–2.5° (medium shaded area) for GSP 15000 (*Sivapithecus*) falls squarely within the range of values for *Pongo pygmaeus*. LSR is least squares regression for the AFK (y in the figure) as a function of the estimated AFK (x). The light shaded area represents the 95% confidence interval for the LSR. RMA is reduced major axis for the AFK as a function of the estimated AFK. Statistics for the sample given in table 4.2.

the range of values for *Pongo* and *Pan*, robust but distinct tori, and a poorly developed post-toral sulcus may be of great importance for our understanding of the evolutionary morphology of the hominid face. It is tempting to suggest that the lack of a more distinct supraorbital torus in *Ankarapithecus* may be an artifact of allometry, but as noted by Begun and Kordos (1997), other Miocene taxa with well-developed glabellar regions (*Ouranopithecus* and *Dryopithecus*) have smaller supraorbital structures than *Gorilla* and *Pan*, even though these taxa overlap in size. The apparent African apelike klinorhynchy in the face of *Ankarapithecus* conflicts with the suggestion that increased klinorhynchy underlies the increased development of supraorbital structures in extant hominines (Begun and Kordos 1997). The facial morphology of *Ankarapithecus* suggests that some other morphogenetic mechanism or aspect of facial architecture may be responsible for the extreme supraorbital development of nonhuman hominines. That the supraorbital torus is not necessarily the result of size or palatal kyphosis argues for the two features being somewhat independent and thus probably of use as single characters (and not a single complex) in phylogenetic analysis.

Importantly, however, recent research indicates that the orbital region may be more directly integrated with the cranial base than with the palate (Ross and Ravosa 1993; Ravosa and Shea 1994; Lieberman et al. 2000). If this observation holds for living and extinct hominids, then the possibility exists that the palate of *Ankarapithecus meteai* is ventrally deflected independent of the orbits and that the orbits may in turn be more airorhynch and thus less disjunct from the neurocranium than in *Gorilla* or *Pan* (a condition that would be expected to result in less well-

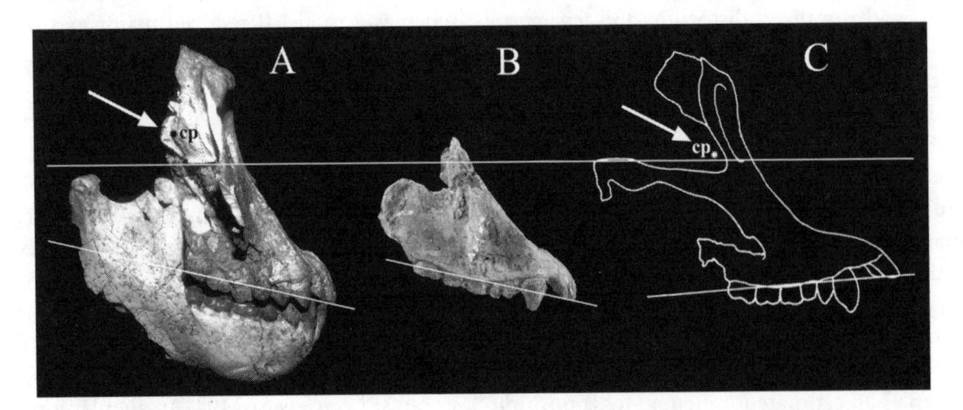

Figure 4.13. A comparison of AS95-500, MTA 2125, and GSP 15000. **(A)** AS95-500 in lateral view with specimen oriented in approximate Frankfurt Horizontal (FH, horizontal line). Determining the exact position of the alveolar plane in AS95-500 is complicated by the superior and left lateral offset of the molars that serves to prevent occlusion with the mandibular molars. **(B)** An image of the undistorted left side of the male lower face of *A. meteai*, MTA 2125, mirrored and reduced to the same size of AS95-500 to approximate this plane. **(C)** GSP 15000 in the same lateral view as AS95-500, also oriented in approximate Frankfurt Horizontal (FH, horizontal line). Note that GSP 15000 shows upward dorsal flexion of the face whereas AS95-500 shows downward flexion. The position of the cribriform plate (arrow, cp) is estimated for both specimens. The position of the cribriform plate of GSP 15000 lies close to the FH in a position similar to that of *Pongo*; the higher position of the cribriform plate in AS95-500 is in a position more like that of *Gorilla* and *Pan*. Right image of GSP 15000 after Brown and Ward (1988, fig. 18-5) is mirrored for this comparison. Image of MTA 2125 is of a cast.

developed supraorbital structures). This possibility cannot be investigated in detail without additional morphological information from the anterior cranial fossa of *A. meteai,* but our observations do indicate that, relative to the external plane (or estimated external plane) of the basicranium, the dorsoventral plane of *A. meteai's* orbital aperture appears to be slightly more airorhynch than the apertures of most *Pan* and *Gorilla* individuals in our sample and generally appears to be intermediate between the condition observable in hominines and that observable in *Pongo.*

Mandible

Because the type mandible of *Ankarapithecus meteai* preserves only a small portion of the corpus and symphysis, the discovery of AS95-500 provides the first complete view of this part of the skull. Various views of the mandible are shown in figure 4.14. It is similar in overall size to some of the smallest Siwalik mandibles (e.g., GSP 4622) but generally differs from them in having absolutely larger teeth, a less divergent tooth row, and a corpus that decreases in depth posteriorly. The incisors are vertically implanted and crowded, thus contributing to a narrow intercanine distance, which contrasts with the mandible's broad bicondylar breadth. The subincisor and subcanine planum is somewhat hollowed and moderately angled along the inferior–superior and mediolateral dimensions to form a moderately developed mental crest (fig. 4.14A–C).

The posterior portion of the corpus broadens buccolingually, and the overall breadth is augmented in the region of the molars by a strong external oblique line that extends anteriorly and inferiorly from the leading edge of the ramus to form a large prominence along the lateral corpus. The initial inferior and subsequent superior sweep of the oblique line from the ramus anteriorly to where it merges with the modest canine eminence imparts a distinct hollowing to the corpus above this line, whereas the corpus

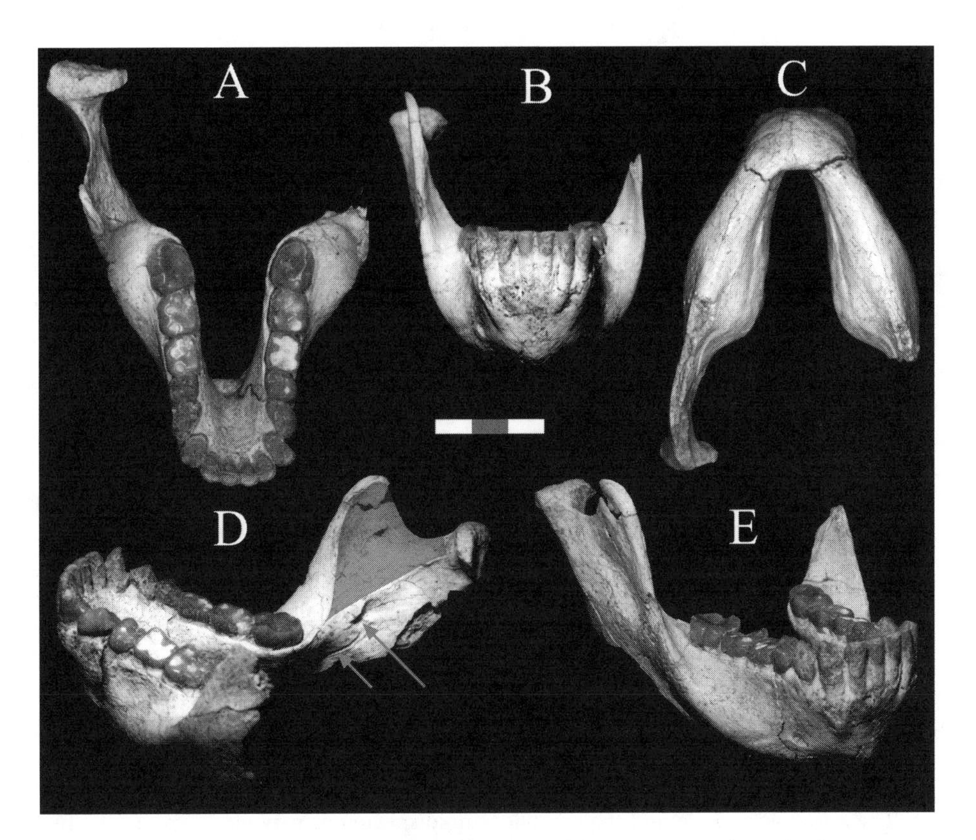

Figure 4.14. Various views of the mandible of AS95-500. (**A**) Superior view shows the distinct superior and inferior tori and the slightly divergent tooth row. The buccinator groove is large and preserved on both right and left sides, and the oblique line runs from the anterior edge of the ramus to drop below the tooth row at about the position of m1/m2. The triangular fossa is also visible on the lingual side of the ramus, as are the enclosing endocoronoid and endocondyloid crests. Note the large, asymmetrical shape of the condyle. (**B**) Anterior view shows the narrow intercanine breadth and broad bicondylar breadth. Note the sweep of the oblique line and the slightly superior orientation of the condyle toward midline. (**C**) Inferior view shows the strong buttressing where the corpus and ramus are joined. Note also the strong groove for the mylohyoid nerve. (**D**) Left lateral oblique view shows the triangular fossa highlighted in gray. The long arrow points to the mandibular foramen and the short arrow points to the posterior position of the grove for the mylohyoid nerve. The hollowing of the corpus lateral to the tooth row is clearly visible. (**E**) Right lateral oblique view shows the full sweep of the oblique line and the modest canine eminence and mental crest. Scale bar in cm.

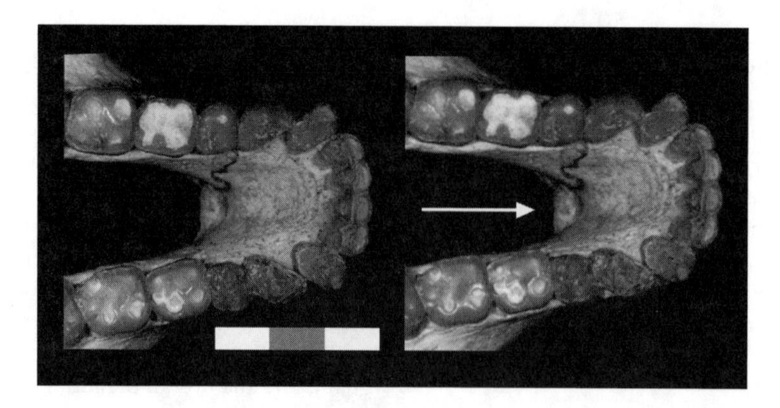

Figure 4.15. Stereo view of the mandibular symphysis of AS95-500 shows the well-defined superior and inferior transverse tori along with the genial fossa. The symphysis is narrow. The arrow points to inferior transverse torus. Scale bar in cm.

below is convex (fig. 4.14D,E). The superior line that defines the attachment area of the buccinator muscle is much less distinct. The overall configuration of this region, and especially the hollowing of the superior lateral surface of the corpus, is similar to that seen in some of the larger *Sivapithecus* specimens (e.g., GSP 15000, GSP 9564) but it is even more extreme in AS95-500. Perhaps the most distinct aspect of this portion of the mandible is the large, mediolaterally broad trough that is lateral and posterior to the second and third molars (fig. 4.14A,B). Again, this configuration is similar to that witnessed in the largest *Sivapithecus* specimens and it is likely that it had the effect of increasing the lever arms of the medial pterygoid and deep temporalis muscles (Ward and Brown 1986).

The mandibular ramus of AS95-500 is partially preserved on the right side only (see fig. 4.13). It is anteroposteriorly broad, but its complete configuration is unknown because most of its inferior portion near the gonial angle is missing. The angle that the leading edge of the ramus makes with the base of the mandible often approximates the gonial angle, and in AS95-500 this is between 109° and 113°. This angle makes AS95-500 more similar to the rearward sloping configuration of the mandibular ramus commonly found in chimps than the nearly right angle configuration of orangutans and some gorillas, but this is a highly variable feature that deserves further detailed analysis, especially with regard to the effect of decreasing mandibular height. The coronoid process and the condyle are preserved in their entirety. The sweep of the sigmoid notch is rather symmetrical with its deepest extent near its midpoint. The condyle is asymmetrical, with its greatest anteroposterior dimension located along its lateral edge, and it is slightly superiorly inclined medially (fig. 4.14B). This is another feature that appears to be quite variable and deserving of more detailed study, but many orangutans show an opposite inferior inclination of the condyle medially, whereas chimpanzee condyles are often more horizontally oriented. The medial surface of the ramus above the mandibular foramen is well preserved (fig. 4.14D, long arrow). The endocoronoid crest drops inferiorly and then posteriorly to wrap around the retromolar fossa as a sharp crest. It is joined by a moderately distinct endocondyloid crest to form the triangular fossa (fig. 4.14D, shaded area). The torus triangularis is distinct and its apex occupies a low position at about the levels of the mandibular foramen and the occlusal surface of the third molar. The relative position of this apex appears to be even lower than that seen in *Sivapithecus* and *Pongo* (Ward and Brown 1986). A distinct groove for the mylohyoid nerve lies inferior to the apex (fig. 4.14D, short arrow), and the mylohyoid line lies anterior to this.

The mandibular symphysis is preserved in its entirety in AS95-500. It has distinct superior and inferior transverse tori (figs. 4.14A, 4.15, 4.16) separated by a well-defined genial fossa. The base of the inferior transverse torus projects posteriorly to about the position of p4/m1, a position more anterior than what appears to be the case for the partially preserved symphysis of MTA 2125 (Begun and Güleç 1998). The superior transverse torus projects posteriorly to about the position of p3/p4, but this portion of the lingual surface of the MTA 2125 symphysis is not preserved. Brown (1997) notes that large and presumably male *Sivapithecus* specimens often have inferior tori that extend further posteriorly than smaller and presumably female specimens, and this is also the case for the extant apes. Although MTA 2125 is only partially preserved, the differences noted between large and small *Sivapithecus* as well as extant hominoid specimens may also prove to be the case for *Ankarapithecus.* Cross-sectional geometries for a variety of extant and extinct species are shown in figure 4.16. AS95-500 is most similar to GSP 4622 and *Pongo* in possessing a robust superior transverse torus and a posteriorly projecting inferior transverse torus, although the latter condition is also seen in some female chimps. It has been suggested that the shapes of the symphysis and corpus represent adaptations for countering the various stresses that are imparted to the mandible during mastication, but there is also a clear relationship to body size (see Brown 1997). The presence of both superior and inferior transverse tori in an individual as small as AS95-500, along with a robust mandibular corpus,

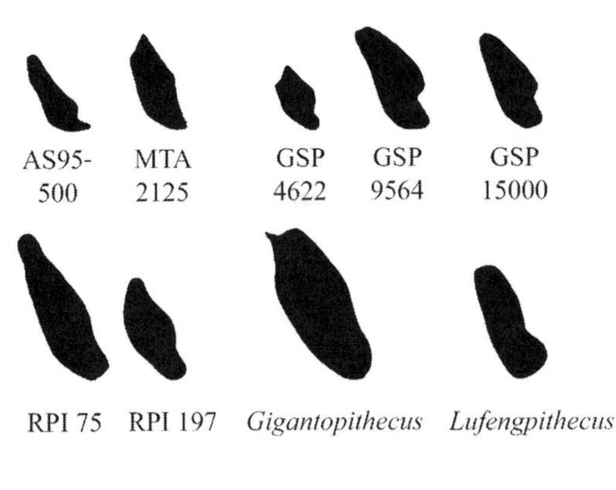

Figure 4.16. Cross-sectional geometries of the mandibular symphysis from a variety of extinct and extant species. The more posteriorly positioned inferior transverse torus of AS95-500 is most similar to that of GSP 4622, a presumably female *Sivapithecus*, and *Pongo*. The inferior transverse torus of MTA 2125 is missing from the original fossil and so it is not known if this specimen had an inferior transverse torus. MTA 2125 in first row after Begun and Güleç (1998), GSP specimens in top row after Ward and Brown (1986), and second and third rows after Brown (1997).

suggests that mastication and/or anterior tooth biting were powerful (see Begun and Güleç 1998; Andrews and Alpagut 2001).

One major difference between *Pongo* versus *Pan* and *Gorilla* is the loss of the anterior digastric muscles. Ward and Brown (1986) and Brown and Ward (1988) note that *Sivapithecus* shows clear evidence of distinct digastric impressions and in this feature differs from *Pongo.* Although this portion of the partial mandible of MTA 2125 is poorly preserved, Begun and Güleç (1998) note the presence of crescent-shaped anterior digastric impressions. This portion of the mandible of AS95-500 is somewhat weathered, but it too shows similar impressions (fig. 4.17).

Estimates of Body Mass, Megadontia, and Sexual Dimorphism

Recent investigations of the orbital aperture have shown that this feature can be used to provide fairly reliable estimates of body mass (Aiello and Wood 1994; Kappelman 1996; but see Delson et al. 2000). The area of the right orbital aperture of AS95-500 was measured following the methods given in Kappelman (1996). The orbital aperture of AS95-500 was calculated to be 727.65 mm^2, and measurements from other Miocene apes were also taken by the same method (table 4.3). These values were next included

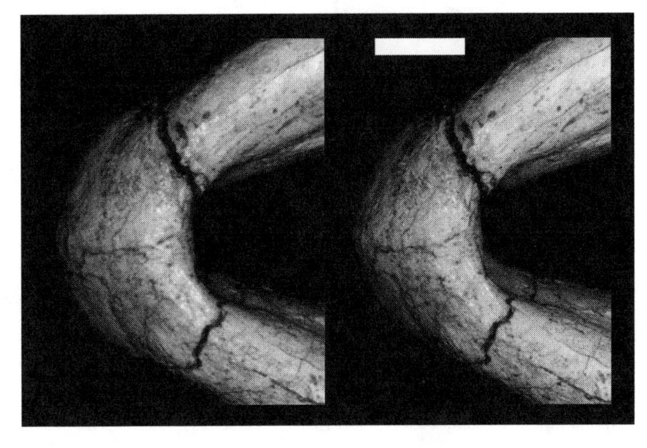

Figure 4.17. Inferior view of the mandible of AS95-500 in stereo shows slight impressions of the anterior digastric muscles that follow the crescent-shaped contour of the inferior lingual surface of the symphysis. Scale bar = 1 cm.

in the two least squares regression equations from Kappelman (1996, tables 1,2):

1. ALLP (all primates, n = 34 males and females from 18 species and subspecies):

$$\log_{10} (\text{body mass in g}) = 2.284 \times \log_{10} (\text{orbital area in mm}^2) - 2.239,\ r = 0.987,\ \text{SEE} = 0.085;$$

2. HOMS (hominoids only, n = 18 males and females from 10 species and subspecies):

$$\log_{10} (\text{body mass in g}) = 2.258 \times \log_{10} (\text{orbital area in mm}^2) - 2.176,\ r = 0.987,\ \text{SEE} = 0.076.$$

Both equations provide corrected body mass estimates of about 20 kg for AS95-500 (see table 4.3). Other calculations by Andrews and Alpagut (2001) use the equations for orbital height and orbital breadth from Aiello and Wood (1994) and provide estimates of 23 and 29 kg, respectively, which are about 15–45% higher than those presented here. It should, however, be noted that the equations shown here that use orbital aperture area are deemed to be somewhat more reliable than those of Aiello and Wood (1994), because these produce lower mean percentage prediction errors and larger percentages of individuals whose estimated mass is within ±20% of their actual body mass (see Kappelman 1996, tables 2, 3). The body mass estimate of about 20 kg for AS95-500 places it well below the female means of 33.2 kg for *Pan paniscus*, 33.7 kg for *P. troglodytes schweinfurthii*, and 35.6 kg and 35.8 kg for *Pongo pygmaeus abelli* and *P. p. pygmaeus*, respectively (Smith and Jungers 1997).

There has been much debate about the degree of megadontia in late Miocene apes (e.g., see Kelley 2001) but the general absence of associated postcranial elements with dental remains has usually precluded any accurate estimates of body mass. New estimates of body mass as derived from the area of the orbital aperture presented above facilitate a reexamination of this issue. The literature

Table 4.3. Least Squares Regression (LSR) Predicted Body Mass Estimates for Fossil Hominoids

Species and Specimen Number	Sex	Geologic Age (Ma)	Orbital Area (mm^2)	Predicted HOMS Body Mass: Corrected Body Mass (g)	Predicted HOMS Body Mass: 95% Confidence Interval (g) Lower	Predicted HOMS Body Mass: 95% Confidence Interval (g) Upper	Predicted ALLP Body Mass: Corrected Body Mass (g)	Predicted ALLP Body Mass: 95% Confidence Interval (g) Lower	Predicted ALLP Body Mass: 95% Confidence Interval (g) Upper
Ankarapithecus meteai AS95-500	F	9.6	727.65	19,754	13,753	28,375	20,096	13,991	28,866
Ouranopithecus macedoniensis XIR-1	M	~9	1080.45	48,222	33,541	69,330	49,570	34,478	71,268
Sivapithecus sivalensis GSP 15000	M	9.3	1235.70	65,295	45,251	94,219	67,358	46,680	97,196
Turkanapithecus kalakolensis KNM WT16950	?M	16 to 18	460.87	7045	4823	10,292	7081	4847	10,345
Afropithecus turkanensis KNM WT 16999	M	16 to 18	927.63	34,177	23,822	49,033	34,991	24,389	50,201
Proconsul heseloni KNM RU 7290	F	18	589.14	12,264	8491	17,714	12,407	8590	17,920

Notes: Least squares regression equations from Kappelman (1996: table 2). HOMS: $y = 2.258x - 2.176$, where $y = \log_{10}$ body mass (g), and $x = \log_{10}$ orbital area (mm^2); $r = 0.987$; SEE = 0.076. ALLP: $y = 2.284x - 2.239$, where $y = \log_{10}$ body mass (g), and $x = \log_{10}$ orbital area (mm^2); $r = 0.987$; SEE = 0.085. Corrected predicted body mass is the detransformed predicted body mass times the correction factor of 1.025 for HOMS and 1.014 for ALLP as calculated from (SE + RE)/2 from Kappelman (1996: table 2).

on body mass is fraught with difficulties (see Smith 1996), not the least of which is that only a few of the museum specimens from which osteologic and dental measurements are taken have associated body mass values. With these cautions in mind, it is still useful to plot the general relationship between body mass and the area of the molars. Figure 4.18 plots the mean body masses for the living male and female great apes as drawn from the literature against M1–3 area (table 4.4). Separate least square regressions are shown for all apes, males only, and females only, but these equations should be viewed with caution, because the latter two groups include the values for only three means. In each species pair, the male demonstrates a larger molar area than its conspecific female, with the greatest differences witnessed in *Pongo* and *Gorilla,* where body mass dimorphism can exceed a 2:1 male:female ratio (see Kelly and Xu 1991; Kelley 1993; Kappelman 1996). There is a general relationship between mean body mass and mean molar area ($n = 6$, $r = 0.78$) that is most apparent within the male means ($n = 3$, $r = 0.95$). It is, however, interesting to note that the general relationship of increasing molar area with body mass does not hold for the female means as it does for male means; rather, the mean for the female orangutan displays a much larger molar area for mean body mass than would be predicted, suggesting a degree of megadontia (also seen, to a lesser extent, in the male orangutan). The fossil apes plotted in figure 4.18 include presumed males (GSP 15000, XIR-1, and KNM WT 16999) and females (AS95-500, KNM RU 7290, and KNM WT 16950). Only the two largest male specimens fall firmly within the range of extant ape mean body mass prediction equations and of these, *Ouranopithecus* (XIR-1) appears to be somewhat more megadont than *Sivapithecus* (see Kelley 2001). The two other specimens that plot close to the limits of the

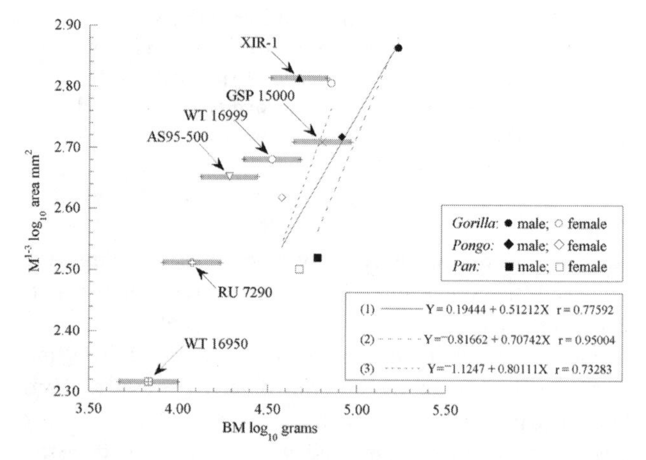

Figure 4.18. Plot of mean body mass ($\log_{10}$ in g) against the sum of upper molar area (M1–3 $\log_{10}$ in mm^2) for the extant apes and several fossil specimens (see table 4.4). Each male from each of the three species is larger than its conspecific female in both body mass and molar area, and the males show a trend of increasing molar area with increasing body mass (LSR: $r = 0.95$). This latter relationship is also true for *Pan* and *Gorilla,* but the female orangutan has larger molars than would be predicted by a regression of body mass against molar area for females only. Several fossil specimens with orbital apertures that are sufficiently well preserved to permit calculations of body mass are also plotted (see table 4.3). GSP 15000, *Sivapithecus,* plots close to the male orangutan, whereas both the presumably male XIR-1, *Ouranopithecus,* and female AS95-500, *Ankarapithecus,* have very large molar areas relative to their estimated body masses. Symbol: +, 95% confidence limits of body mass estimates for fossils from table 4.3. XIR-1, *Ouranopithecus;* GSP 15000, *Sivapithecus;* WT 16999, Kenya National Museum *Afropithecus;* AS95-500, *Ankarapithecus;* RU 7290, Kenya National Museum *Proconsul;* WT 16950, Kenya National Museum *Turkanapithecus.* LSR equations give the $\log_{10}$ (M1–3 area in mm^2) as a function of $\log_{10}$ (body mass in g) for extant: (1) all males and females; (2) males only; and (3) females only.

Table 4.4. Mean Maxillary Canine and Molar Measurements for Extant and Extinct Apes

Species (Fossil Specimen Number)	Sex	Tooth	*n*	Area (mm^2) Mean	Standard Deviation	Maximum	Minimum	Body Mass (g)
Gorilla gorilla gorilla	M	C	20	339.20	38.37	389.76	255.36	169,500
		M1	20	231.86	24.60	290.58	198.65	
		M2	20	266.77	27.79	312.93	217.36	
		M3	20	236.71	27.56	285.60	188.50	
		Sum M1–M3	20	735.33	70.71	858.28	614.54	
	F	C	20	162.34	16.83	182.85	128.75	71,500
		M1	20	208.22	18.91	243.32	178.22	
		M2	20	232.05	25.13	283.92	194.54	
		M3	20	203.14	27.60	247.86	163.80	
		Sum M1–M3	20	643.41	66.41	772.10	542.16	
Pan troglodytes troglodytes	M	C	13	157.25	27.90	205.92	118.08	60,000
		M1	11	110.05	9.94	124.02	95.68	
		M2	12	119.86	12.45	142.08	105.60	
		M3	10	104.62	11.43	120.51	86.10	
		Sum M1–M3	9	332.72	24.74	366.98	300.28	
	F	C	11	99.83	6.03	110.40	92.02	47,400
		M1	11	109.53	9.15	124.02	93.06	
		M2	11	113.20	10.15	130.54	98.94	
		M3	9	98.08	10.80	123.60	87.00	
		Sum M1–M3	9	318.72	25.80	358.52	292.70	
Pongo pygmaeus pygmaeus	M	C	11	237.29	39.09	312.17	186.88	81,700
		M1	11	168.52	20.30	206.55	143.45	
		M2	11	178.69	21.79	212.52	147.63	
		M3	10	174.25	21.22	200.43	143.64	
		Sum M1–M3	10	525.98	53.68	608.70	451.49	
	F	C	8	128.86	17.93	151.39	109.25	37,800
		M1	8	137.26	13.39	156.86	115.47	
		M2	8	143.51	18.15	163.20	123.76	
		M3	8	131.01	15.80	152.32	103.96	
		Sum M1–M3	8	411.77	37.48	456.87	372.27	
Ouranopithecus macedoniensis (XIR-1)	M	C		184.31	*			48,222
		M1		187.99	*			
		M2		226.45	*			
		M3		236.91	l			
		Sum M1–M3		651.35	*			
Ouranopithecus macedoniensis (NKT-89)	F	C		129.80	*			—
		M1		149.16	r			
		M2		173.24	r			
		M3		191.06	r			
		Sum M1–M3		513.46	*			
Sivapithecus sivalensis (GSP 15000)	M	C		188.30	*			65,295
		M1		162.50	l			
		M2		181.76	*			
		M3		171.28	*			
		Sum M1–M3		515.54	*			

(continued)

Table 4.4. Mean Maxillary Canine and Molar Measurements for Extant and Extinct Apes (*continued*)

Species (Fossil Specimen Number)	Sex	Tooth	*n*	Area (mm^2) Mean	Standard Deviation	Maximum	Minimum	Body Mass (g)
Ankarapithecus meteai (AS95-500)	F	C		100.31	*			19,754
		M1		133.33	*			
		M2		166.53	*			
		M3		149.92	*			
		Sum M1–M3		449.77	*			
Ankarapithecus meteai (MTA 2125)	M	C		196.68	l			—
		M1		161.66	l			
		M2		193.05	l			
		M3		213.15	l			
		Sum M1–M3		567.86	l			
Turkanapithecus kalakolensis (KNM WT 16950)	M	C		79.00	l			7045
		M1		52.50	l			
		M2		79.12	r			
		M3		75.65	l			
		Sum M1–M3		207.27				
Afropithecus turkanensis (KNM WT 16999)	M	C		313.20	l			34,177
		M1		149.50	l			
		M2		168.00	l			
		M3		162.50	l			
		Sum M1–M3		480.00	l			
Proconsul heseloni (KNM RU 7290)	F	C		82.98	*			12,264
		M1		75.26	*			
		M2		95.50	*			
		M3		81.38	*			
		Sum M1–M3		252.14	*			

Notes: For fossils, l = left; r = right; * = mean of left and right. *n* = Number of specimens. Data for extant specimens from Pilbeam (1969) and Mahler (1973).

extant data set, *Ankarapithecus* and *Afropithecus,* also appear to be megadont. Both KNM RU 7290 and KNM WT 16950 fall too far outside the limits of the regression equation to assess their relative degree of megadontia.

The simple comparison of the partial crania of AS95-500 and MTA 2125 shown in figure 4.19 suggests that *Ankarapithecus* was marked by a high degree of body mass sexual dimorphism that was probably on par with that witnessed in *Gorilla* and *Pongo.* It is important to remember that both specimens were recovered from the same fossil locality at the same stratigraphic horizon, so it is likely that these specimens belonged to the same general population.

Relative differences in canine size between males and females have frequently been used to investigate sexual dimorphism in primates (Plavcan and van Schaik 1992, 1994; Plavcan 1998; Kelley 2001). Deciding which dental metric to use is complicated because many canines are heavily worn, thus making it difficult to obtain sufficient samples. This difficulty has lead some workers to use canine area (de Bonis and Koufos 1993), whereas others argue for the use of canine height (Kelley 2001), even though the latter approach can limit sample size. The same sample shown in figure 4.18 is included in figure 4.20 and plots mean body mass against the mean maxillary C area. As expected, the larger male of each species demonstrates a larger canine area than does its conspecific female, with the chimpanzee showing the smallest difference between the sexes. There is a stronger correlation between increasing mean body mass and mean canine area across all sexes from each taxon ($n = 6$, $r = 0.93$) than seen with mean molar area in figure 4.18. Although the female gorilla mean now falls within the trend of increasing mean tooth size with increasing mean body mass, it is interesting to note that the female orangutan mean once again does not do so. Of the fossil specimens, GSP 15000 plots nearest the regression line for males while XIR-1, AS95-500, and especially KNM WT 16999 have canine areas that are actually larger than would be predicted for their estimated body masses. As noted in table 4.3, the CIs for the estimated body masses as predicted by orbital area for all of the fossils are quite large, and if the upper 95% CI value of 69,330 g for XIR-1 is used (see "+" and horizontal lines in fig. 4.20), its canine area plots on the regression line for male primates, thus confirming Kelley's (2001) demonstration that, as based on

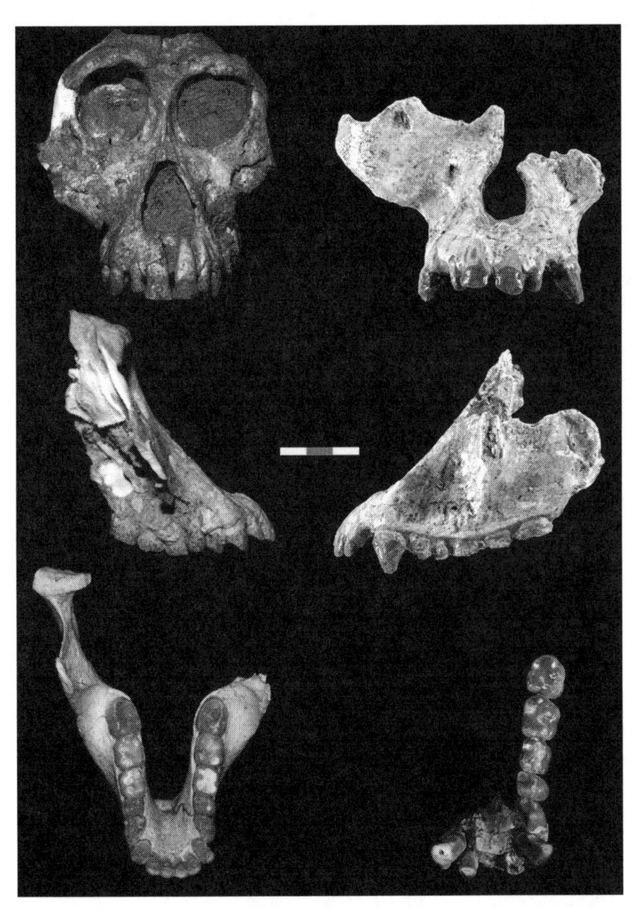

Figure 4.19. Anterior (top row) and lateral (middle row) views of the partial cranium of AS95-500 (female, left column) and lower face of MTA 2125 (male, right column, cast) at the same scale illustrates the degree of sexual dimorphism in *Ankarapithecus meteai*. Bottom row compares occlusal views of the mandible of AS95-500 (female, left) and the type mandible (male, right). Scale bar in cm.

canine height, *Ouranopithecus* does not appear to display any canine reduction for its estimated body mass.

Without a good estimate of body mass for MTA 2125, it is difficult to quantify precisely the degree of sexual dimorphism in *Ankarapithecus* that is suggested by figure 4.19. Unfortunately, dental metrics do not offer highly reliable estimates of body mass (see Smith 1996) because different species can show varying relationships between body mass and tooth size that may ultimately be related to species-specific dietary adaptations. Still, an examination of relative molar size and canine size can be informative because males and females of the same species should demonstrate a similar relationship.

Figure 4.21 plots the summed maxillary molar area against maxillary canine area and shows that there is a strong relationship between these variables within this sample of extant male and female apes (n = 76; see table 4.4). That is, for a particular canine size, the correlation within a particular sex for summed molar area is greater than r = 0.91, and the two groupings by sex are distinct and show no overlap.

The fossil taxa illustrated in figure 4.21 allow some interesting observations that may inform the discussion of sexual dimorphism. In figure 4.21, the presumably male *Ouranopithecus* specimen (XIR-1) displays a molar area near the border between the ranges of male and female gorillas but a canine area that is much closer to that of female gorillas. A presumably female *Ouranopithecus* specimen (NKT-89) plots just outside the lower end of the range for female gorillas in both molar and canine areas. In contrast, the presumably male *Ankarapithecus* (MTA 2125) has a molar area that is intermediate between the mean values for male orangutans and female gorillas but a canine area that is near the lower limit of male orangutans. The female *Ankarapithecus* (AS95-500) plots near the mean value for female orangutans in molar area but has a canine area that is below that of female orangutans and closer to the female chimpanzee mean. Although these two extinct species are limited to two specimens only, these data suggest a high degree of both canine and molar area sexual dimorphism in *Ankarapithecus* that is close to that seen in the living *Pongo* and *Gorilla* and perhaps a somewhat lower degree of canine and molar sexual dimorphism in *Ouranopithecus*. The remaining fossil specimens plot within or at the margins of the range of values seen for *Pongo* (KNM WT16999 and GSP 15000) and near or below the range of values for female chimps (KNM RU 7290 and KNM WT 16950).

Description and Analysis of Sinap Postcranial Fossils

One of the most fundamental problems in hominoid evolution concerns the evolutionary history of the postcranial similarities shared among extant apes. The long-held view (e.g., Huxley 1863; Keith 1903; Gregory 1930) that the hominoid adaptations for orthograde climbing and suspension are a product of their shared ancestry was challenged with the discovery that the humeri of *Sivapithecus* were adapted for pronograde quadrupedalism (Pilbeam et al. 1990). This discovery led the authors to question the relationship between *Sivapithecus* and *Pongo* (Pilbeam et al. 1990). The humeral evidence requires that either the craniofacial similarities between these taxa or the postcranial similarities between *Pongo* and the African apes arose in parallel (Pilbeam 1996; Larson 1998). The evidence is, however, also consistent with the interpretation that *Sivapithecus* forms a clade with *Pongo* and the pronograde postcranial adaptations of *Sivapithecus* are secondarily derived (Ward 1997; Richmond and Whalen 2001). Clarification of sivapithecin relationships and postcranial evolutionary morphology requires the recovery and analysis of additional fossil evidence.

As a possible stem great ape (Alpagut et al. 1996), or sister taxon to a *Pongo-Sivapithecus* clade (Begun and Güleç 1998), *Ankarapithecus* is critical to the reconstruction of the postcranial adaptations of the last common ancestor of great apes and humans. Because the morphology of *Ankarapithecus* may more closely approximate that of the common ancestor of the great ape and human clade than other known late Miocene taxa, the postcranial remains of

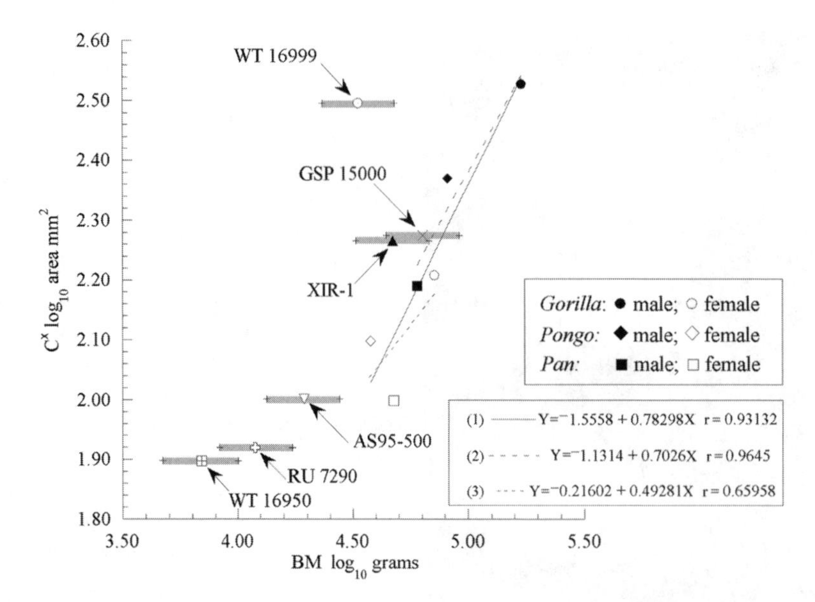

Figure 4.20. Plot of mean body mass ($\log_{10}$ in g) against upper canine area (C $\log_{10}$ in mm^2) for the extant apes and several fossil specimens (see table 4.4). As in the comparison with molar area in figure 4.18, each male from the three species is larger than its conspecific female in both body mass and canine area, but the trend of increasing canine area with increasing body mass shows a much higher correlation (LSR: $r = 0.93$) than does molar area. The fossil specimens from figure 4.18 are also included here. Both GSP 15000, *Sivapithecus*, and XIR-1, *Ouranopithecus*, plot very close to the regression lines, whereas WT 16999, *Afropithecus*, lies far beyond the lines. It is interesting that AS95-500, *Ankarapithecus*, appears to have a larger canine area than would be predicted by any of the LSR equations. Symbols as in figure 4.18. LSR equation for extant: (1) all males and females; (2) males only; and (3) females only.

Ankarapithecus may bear on the problem of homology versus homoplasy in the great ape and human skeleton.

The postcranial remains attributable to *Ankarapithecus* include a nearly complete diaphysis of a right femur, a virtually intact right radius, as well as the distal segments of proximal and middle phalanges (see table 4.1).

Materials and Methods

Comparative skeletal measurements were collected on specimens housed at the Field Museum of Natural History (Chicago) and the Smithsonian Institution's National Museum of Natural History (Washington, D.C.). Comparative taxa were selected to represent a range of locomotor adaptations and include *Pan troglodytes, Pongo pygmaeus, Gorilla gorilla berengei,* and mixed species samples of the genera *Papio, Alouatta,* and *Presbytis.* Each taxon sample consisted of 10–11 adult wild-shot individuals, with near equivalent numbers of males and females in each sample.

Linear measurements were collected using digital calipers. Most measurements of the fossils were collected directly on the original specimens. A few measurements were taken on casts. Measurements collected on both the original fossils and casts differed by <2%.

The following measurements were collected on the radii: head width (ML) and depth (AP), medial and lateral lips of the radial head, neck length, head-neck angle, and width (ML) and depth (AP) at the rugosity of the proximal portion of the interosseous crest. Measurements of the head follow those figured in Rose (1988, fig. 2). Neck length was taken as the average of the lengths from the radial head to the proximal and distal ends of the tuberosity. The angle of the head relative to the neck (degree of anterolateral tilt) was measured with the radius oriented with the tuberosity facing the viewer.

Femur (AS95-321)

The size and morphology of the femur shaft, AS95-321, indicate that it is attributable either to a hominoid or to a nimravid, an extinct group of carnivores. Although the specimen is too large to belong to the same individual from which the skull and/or hominoid radius came, its size is compatible with the larger *Ankarapithecus meteai* face (MTA 2125) previously recovered from the same locality. Our work at locality 12 did not yield nimravid craniodental remains, but Geraads and Güleç (1997) described a previously discovered skull of a nimravid, *Barbourofelis piveteui,* from an unknown locality, which may be the same as our locality 12. The size of the femur is consistent with the possibility of it belonging to *B. piveteui* (S. Viranta, pers. comm.).

Without either of the epiphyses, it is premature to rule out either taxonomic possibility. AS95-321 closely resembles modern and fossil hominoid femora in many respects,

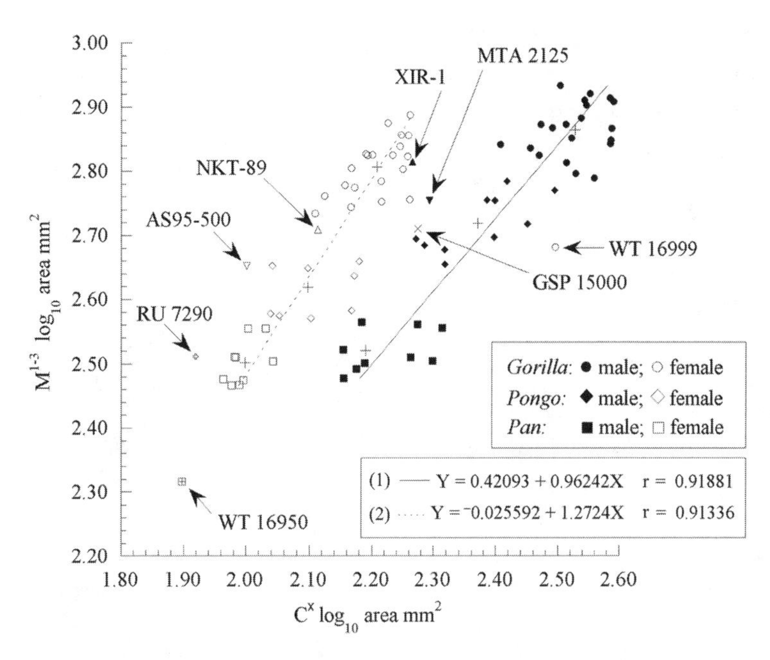

Figure 4.21. Plot of upper canine area (C $\log_{10}$ in g) against the sum of upper molar area (M1–3 $\log_{10}$ in mm^2) for the extant apes and several fossil specimens (see table 4.4). This comparison shows that there is a close relationship between canine area and the summed area of the molars within each sex across the three species, with the correlation in both sexes $>r = 0.91$. Although sample sizes for each species are modest, there is no overlap seen among the sexes from these three species, and there is good separation between the sexes. Three of the presumed male fossils, WT 16999, MTA 2125, and GSP 15000, plot very close to the limits of the extant male distribution, whereas XIR-1, also presumed to be male, plots within the range of extant females. Because its canine area is about what one would expect for its body mass (see fig. 4.20), this plot suggests that it is highly megadont. All of the presumably female fossil specimens plot either within or close to the range of extant females. The fossil female and male belonging to *Ankarapithecus* (AS95-500 and MTA 2125) show a greater separation than the *Ouranopithecus* pair, and this difference suggests that *Ankarapithecus* is more sexually dimorphic than *Ouranopithecus*. Symbols as in figure 4.18, with the exception that + is the mean value for the sex in each extant species. LSR equations for extant: (1) males only; (2) females only.

including the relatively low position and medial orientation of the lesser trochanter, the morphology of the intertrochanteric crest, the size and position of the gluteal tuberosity, and the robusticity and general shape of the shaft. However, some traits are unusual for hominoids, including a relatively narrow anteroposterior diameter at the base of the neck, a slight anterior curve of the proximal anterior surfaces of the neck and greater trochanter, and a somewhat triangular shaft cross-section proximally and distally. In some of these respects, it more closely resembles the femur of the Miocene nimravid *Sansanosmilus palmidens* from Europe (Ginsburg 1961). However, AS95-321 also differs from the nimravid femoral material in several ways. For example, AS95-321 has a more distally located lesser trochanter, a more continuous intertrochanteric crest, and lacks the distinct protuberance at the level of the gluteal tuberosity seen in *S. palmidens*. It also has a more globular greater trochanter, like those of anthropoids and unlike the sharp anterolateral crest in the nimravid. This combination of features does not currently offer a clear taxonomic attribution. Because it might belong to *Ankarapithecus meteai*, AS95-321 is described and illustrated here.

Specimen AS95-321 is as large as the femora of large common chimpanzees and small gorillas (figs. 4.22–4.25). The bulk of the shaft is well preserved, from the base of the neck inferiorly to the distalmost portion of the shaft as it widens to meet the condyles (fig. 4.22). The maximum preserved length is 244 mm along the proximodistal axis. No portions of the proximal or distal articular surfaces remain.

At the proximal end, the head and much of the neck are missing, as are the proximal and posterior portions of the greater trochanter. The base of the neck suggests a fairly high neck-shaft angle. The maximum measurable breadth (AP) of the greater trochanter is 30.4 mm, but it was broader prior to being damaged. The trochanteric fossa extends well below (distal to) the top of the neck, and ends at a ridge running inferolaterally from the base of the neck to the posterior margin of the greater trochanter (fig. 4.23).

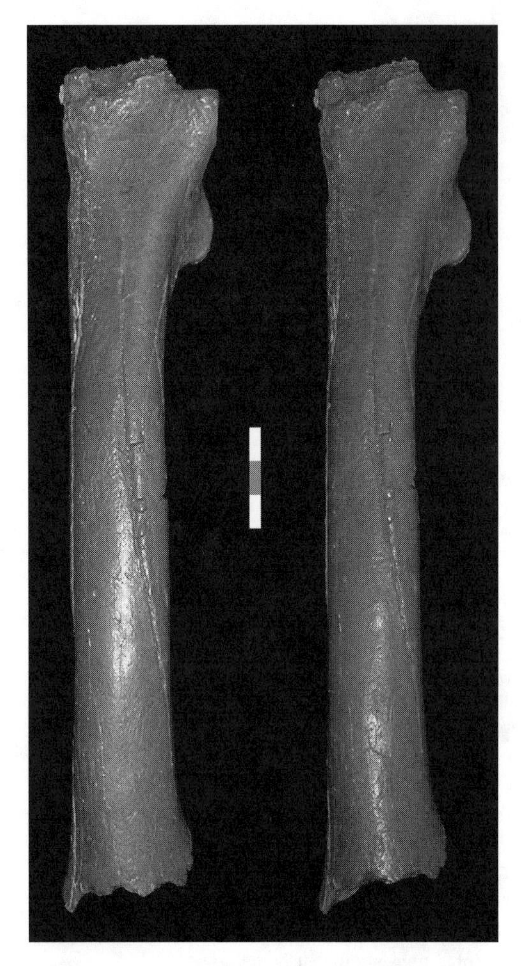

Figure 4.22. Stereo view of *Ankarapithecus* femur, AS95-321 cast. Almost the entire shaft is preserved. Note the robusticity of the shaft. Scale bar in cm.

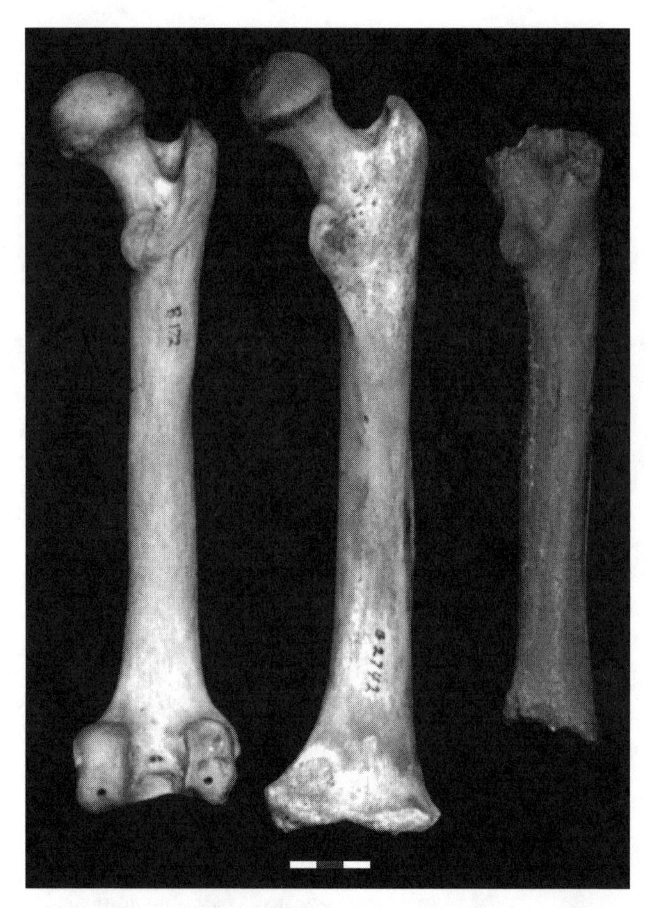

Figure 4.23. Posterior view of *Pongo* (left, adult male, CMNH B172), *Gorilla* (middle, subadult female, CMNH B2742), and AS95-321 (right, cast). AS95-321 is as large and robust as femora of some female gorillas and male orangutans. The trochanteric fossa of AS95-321 extends distally, as in femora of cercopithecoids, and unlike most but not all hominoids (see *Pongo* at left). Like hominoids, the intertrochanteric crest of AS95-321 is discontinuous. The neck appears to form a high angle relative to the shaft. Like African ape and cercopithecoid femora, the lesser trochanter of AS95-321 faces relatively posteriorly. Scale bar in cm.

The intertrochanteric crest is not a continuous bar of bone; instead, the greater trochanter tapers inferomedially into a valley separating the two trochanters. With the exception of the gluteal tubercle, most muscle insertions are not discernible. There is no distinct linea aspera along the posterior surface.

The ovoid, well-defined lesser trochanter is 21.0 mm long and 12.8 mm wide (minimum breadth, slightly tilted relative to ML plane) and projects posteromedially out of the shaft. Just below the lesser trochanter (~4 mm inferior), where the slope of the lesser trochanter disappears into the shaft, the shaft is 33.1 mm wide (ML) and 24.6 mm deep (AP). Using the all-hominid regression equation from McHenry (1992), these subtrochanteric dimensions yield a body mass of 73.6 kg. (using LSR; major axis and RMA regressions yield 75.6 kg and 75.3 kg, respectively).

At this subtrochanteric level, the shaft is roughly triangular in cross section, with fairly flat surfaces facing medially, posteriorly, and anterolaterally. Therefore, the medial side of the shaft projects farthest anteriorly at this level. Near midshaft, the cross section has an anteroposteriorly compressed ovoid shape (30.5 mm ML, 25.2 mm AP, taken on cast; an estimate because the precise location of midshaft cannot be identified). At the distal end, the shaft broadens (ML) and once again assumes a more triangular cross-sectional shape, with the center of the shaft projecting the farthest anteriorly. At approximately 20% of the reconstructed length (26 mm from the distalmost end of the preserved shaft), the shaft is 31.8 mm wide (ML) and 28.1 mm deep (AP). Along its length, the shaft forms an anteriorly convex curve (fig. 4.24).

Like most Miocene hominoid skeletal remains, AS95-321 displays apelike and more primitive features. Like hominoid femora, the base of the neck is consistent with a fairly high neck-shaft angle, but the preserved morphology does not permit a reliable estimate. The intertrochanteric crest is not a continuous ridge of bone that connects the greater and lesser trochanter posteriorly, as seen in cercopithecines. However, the discontinuous intertrochanteric crest is probably primitive, as it is characteristic of both hominoids and most platyrrhines (Rose et al. 1992).

The proximal portion of AS95-321 is unlike extant hominoid femora in some respects. The neck does not expand

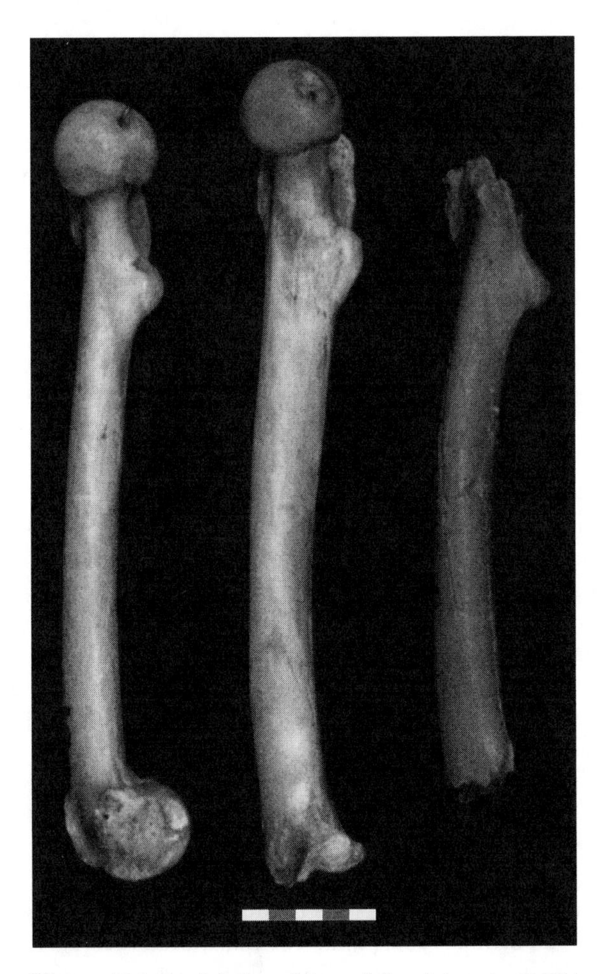

Figure 4.24. Medial view of *Pongo* (left, adult male, CMNH B172), *Gorilla* (middle, subadult female, CMNH B2742), and AS95-321 (right, cast). The shaft of AS95-321 is more curved (anteriorly convex) than most hominoid femora, more closely resembling nonhominoid pronograde quadrupeds. Scale bar in cm.

anteroposteriorly as it meets the greater trochanter and more closely resembles nonhominoid catarrhines in this regard (Rose et al. 1992). The neck appears relatively narrow anteroposteriorly. The trochanteric fossa is extensive and extends distally beyond the top of the neck. In this way, AS95-321 more closely resembles the nonhominoid anthropoid condition rather than the typically restricted and proximally positioned trochanteric fossa. It also resembles other Miocene hominoids, including *Morotopithecus* (MUZM 80), *Proconsul*, and *Equatorius* in this regard (Ward et al. 1993; McLatchy et al. 2000). However, distally-extended fossae are sometimes observed in hominoid femora as well (fig. 4.23).

The lesser trochanter of AS95-321 is positioned low compared with femora of *Pan* and *Papio*, but not unlike the morphology of *Gorilla* femora (fig. 4.23). In the transverse plane, the lesser trochanter is oriented somewhat posteriorly (~55°) relative to the midline of the greater trochanter and femoral neck. In this feature, AS95-321 resembles cercopithecoids and African apes (range, ~45–60°), and is unlike the more medial orientation (~20–45°) in *Pongo*, *Hylobates*, and several Miocene catarrhines, including *Pliopithecus vindobobensis*, *Proconsul nyanzae*, *Equatorius africanus*, *Morotopithecus bishopi*, and the Eppelsheim femur (Ward et al. 1999; MacLatchy et al. 2000).

The shaft at subtrochanteric and near midshaft levels is compressed anteroposteriorly relative to its breadth, similar to great ape (and *Alouatta*) femora (fig. 4.25). The shaft is also quite robust relative to its apparent length, like those of great apes (figs. 4.23, 4.24). The level of shaft robusticity in AS95-321 is consistent with size-related expectations (Jungers et al. 1998). AS95-321 is more curved than femora of suspensory taxa (e.g., *Pongo*, *Hylobates*, *Ateles*), instead resembling femora of anthropoids that frequently use their hindlimbs in some form of pronograde locomotion (Swartz 1990).

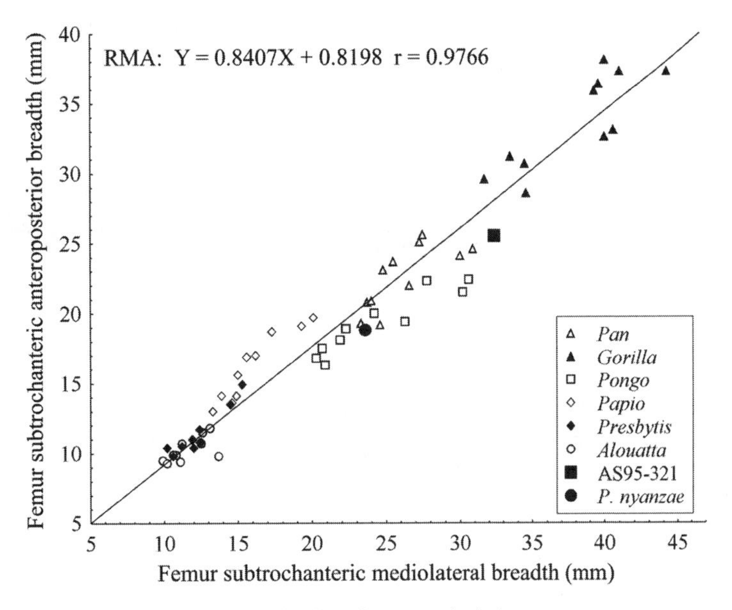

Figure 4.25. Plot of femoral subtrochanteric shaft dimensions, anteroposterior versus mediolateral. Note that the shaft of AS95-321 is anteroposteriorly flattened, as is typical of *Pongo* femora and some femora of *Pan*, *Gorilla*, and *Alouatta*. Note also the large size of the femur.

What can be concluded about the function of AS95-321 is unfortunately limited by the absence of the epiphyses. The only feature suggesting use of the hindlimb in abducted postures is the possibly high neck-shaft angle, but too little of the neck is preserved to obtain a reliable measure. Furthermore, hip mobility is a product of neck angle and neck length, among other (e.g., soft tissue) factors (MacLatchy et al. 2000). Shaft shape has also been related to limb postures. Relatively broad shafts of *Pongo* femora may be interpreted as structural adaptations to resist mediolateral bending incurred by abducted hindlimbs during climbing (MacLatchy et al. 2000). However, the femur of *Proconsul nyanzae* is quite platymeric, but the skeleton is clearly adapted for pronograde quadrupedalism, albeit unlike that of extant cercopithecoids (Ward et al. 1999). The cross-sectional shape of AS95-321 is not as rounded as in cercopithecoids. Instead, it is comparable to femora of *Pan, Pongo, Alouatta,* and *Proconsul nyanzae.* Thus, the shaft structure of AS95-321 is consistent with its use in pronograde quadrupedalism.

The lesser trochanter of AS95-321 is posteriorly oriented like those of African apes and cercopithecoids. Although the functional implications of this feature and others (e.g., extensive trochanteric fossa, greater anterior shaft curvature) are not well understood, their prevalence in pronograde quadrupeds and the frequently terrestrial African apes implies similar functional use. The relatively triangular cross section of the distal shaft is almost certainly related to condyle shape just distal to it (MacLatchy et al. 2000). If so, the anterior projection of the center of the shaft may indicate a relatively narrow patellar groove and deep condyles compared with extant apes. These features are associated with stereotypical parasagittal knee movements (Tardieu 1981).

In sum, the morphology of the femur is somewhat generalized like that of many other Miocene hominoids (Rose 1993b; Ward et al. 1993). The femoral anatomy is consistent with use in pronograde and climbing behaviors, but lacks terrestrial specializations seen in cercopithecoids. The large size of the femur suggests that *Ankarapithecus* spent at least part of its time on the ground.

The size of AS95-321 also suggests that it belonged to a male rather than female, and on the basis of size, it could be associated with the male palate, MTA 2125, which was also found at locality 12.

Radius AS95-503

The radius specimen AS95-503 is comparable in size with the radii of female chimpanzees and large male baboons (fig. 4.26). It is not large enough to belong to the same individual as the femur specimen, AS95-321 (e.g., compared with other primates, the radial head width of AS95-503 is unusually small relative to the femoral subtrochanteric dimensions of AS95-321) and might belong with the female partial skull, AS95-500. Using the all-hominoid regression equations published by McHenry (1992), the transverse breadth of the radial head provides body mass estimates of ~31 kg (between 30.9 and 31.1 kg for least squares, major axis, and reduced major axis equations). It is determined to be a right radius based on the anterior position of the radial tuberosity relative to the interosseous crest and the anterolateral tilt of the radial head.

Almost the entire radius is preserved, with only the distalmost end missing. The maximum length of the preserved portion is 220.8 mm. The radial head is abraded at the lateral (and slightly anterior) and posterior (and slightly medial) margins. A chip of bone is missing on the posterior aspect of the head, leaving an edge ~1 mm deep at the distal end of the damage (fig. 4.26). The anterior surface of the shaft is crushed from the distal end to ~40 mm from the end.

The radial head is moderately oval in outline. The widest dimension (21.2 mm) is along an anterior (and slightly medial) to posterior (and slightly lateral) axis. The

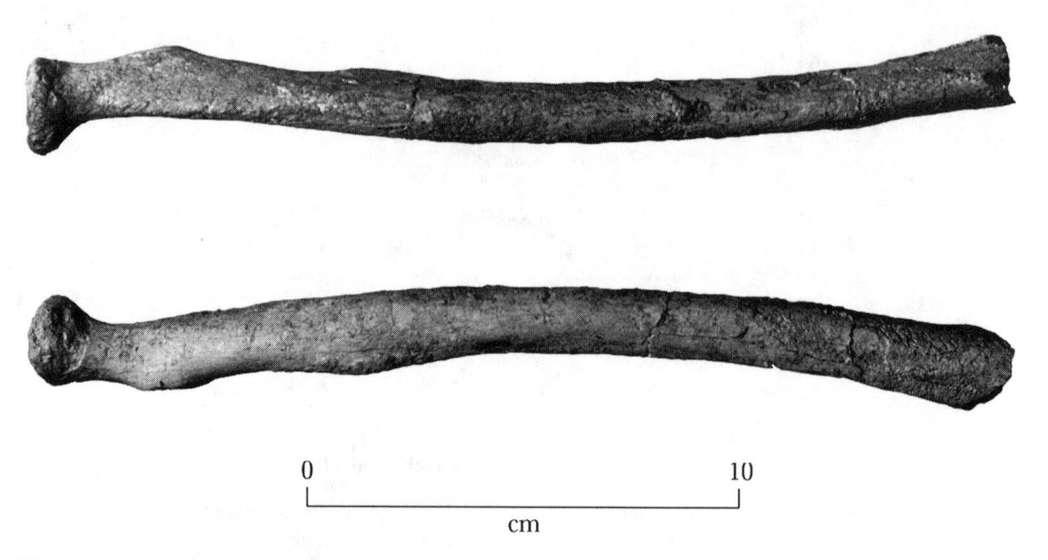

Figure 4.26. Anterolateral (top) and posterior (bottom) views of AS95-503. Note the moderate longitudinal shaft curvature.

abrasion to the lateral side of the head interferes with a minimum head diameter. Without correction, this dimension measures 18.8 mm. It is estimated from the contours on either side of the abrasion that no more than 1 mm is missing. The fovea is quite deep (2.7 mm). The rim of the head is beveled, and the head is anterolaterally angled relative to the neck and shaft.

Neck length is moderate (20.3 mm from proximal head to proximal end of tuberosity, 39.8 mm to distal end of tuberosity). The tuberosity is distinct and oval in outline (19.5 mm, 11.5 mm). A deep groove runs proximodistally within the proximal half of the tuberosity. The neck angles laterally only very slightly relative to the shaft. Running distally from the tuberosity, the oblique line is weakly developed.

The interosseous crest is well separated from the tuberosity, taking origin from the shaft ~14 mm distal to the tuberosity's distal margin. The crest is very sharp, giving the shaft a teardrop cross section with straight, almost concave sides. The proximal part of the crest exhibits a strongly developed prominence. The most projecting portion of this prominence is ~74 mm from the distalmost point on the head. At this point, the shaft is much broader (18.9 mm) than deep (11.4 mm). Distal to the tuberosity, the shaft is moderately curved mediolaterally and is straight anteroposteriorly. Along the distal portion of the shaft, the dorsal surface is smoothly rounded.

AS95-503 exhibits a mixture of apelike and nonapelike features. Perhaps the most notable apelike feature is the distinct beveling of the sides of the radial head (figs. 4.27, 4.28), a feature that corresponds with the presence of a zona conoidea on the distal humerus (Rose 1988). Abrasion prevents an assessment of the extent around the entire circumference of the radial head of the articular surface for the zona conoidea. However, Rose (1988) notes that, in nonhominoids and most fossil hominoid specimens (e.g., *Dendropithecus* and, to a lesser extent, *Proconsul*), the bevel for articulation with the zona conoidea is restricted to the posterolateral side of the head. In AS95-503, the bevel is well expressed on the anterior and medial sides of the head. Thus, *Ankarapithecus* most likely had a distal humerus with the spool-shaped trochlea characteristic of hominoids. AS95-503 also resembles apes in having a deep capitular fovea, unlike the shallow fovea of many pronograde quadrupeds, such as *Papio* (fig. 4.29). The moderate shaft curvature of AS95-503 resembles that seen in great apes and baboons of comparable size (fig. 4.26).

From a proximal view (fig. 4.29), the head is neither as strongly asymmetric as the radial heads of most anthropoid arboreal quadrupeds nor as evenly rounded as the radial heads of extant hominoids (Rose 1988). The abrasion on the lateral margin of the head precludes a direct measurement of minimum head diameter. The radial head contours on either side of the abrasion suggest that the head was mildly asymmetric, comparable with the radii of *Papio*.

In most respects, AS95-503 resembles nonhominoid anthropoids. For example, the "lateral" (anterolateral) lip in AS95-503 is somewhat larger than its medial (postero-

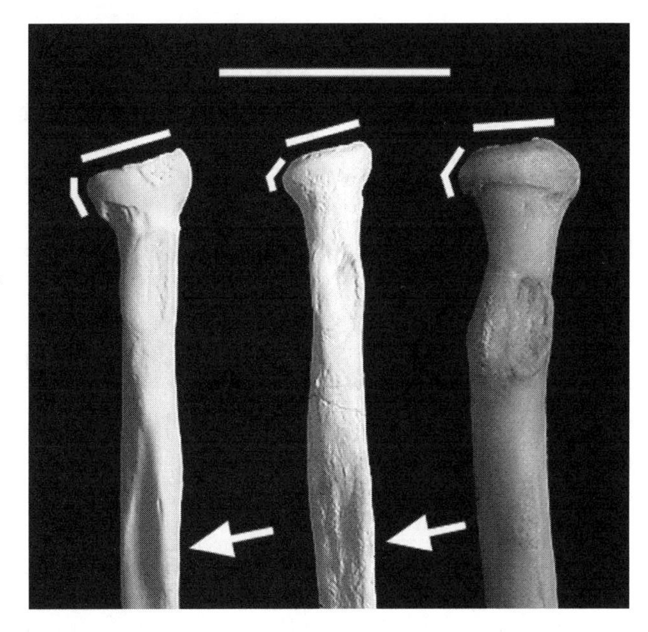

Figure 4.27. Proximal right radii of *Papio* (left), AS95-503 (center, cast), and *Pan* (right) viewed perpendicular to the radial tuberosity. The head of AS95-502 is beveled on its anterior (left and toward viewer) and medial margins, like that of *Pan*, in contrast to the more continuously sloping margin in *Papio*. The beveled surface articulates with the humeral zona conoidea in hominoids. The head of AS 95-503 is tilted anterolaterally (toward the left), resembling nonhominoid anthropoids, including *Papio*. The AS95-503 neck is intermediate in length (relative to head size) between the long necks in hominoids and *Alouatta*, and the short necks in *Papio*. AS95-503 has a sharp interosseous crest (arrows) like that of *Papio* and unlike the more rounded shafts of other anthropoids. The radial tuberosity of AS95-503 is somewhat more anteriorly located compared to the more medial location in hominoids. Scale bar at top = 5 cm.

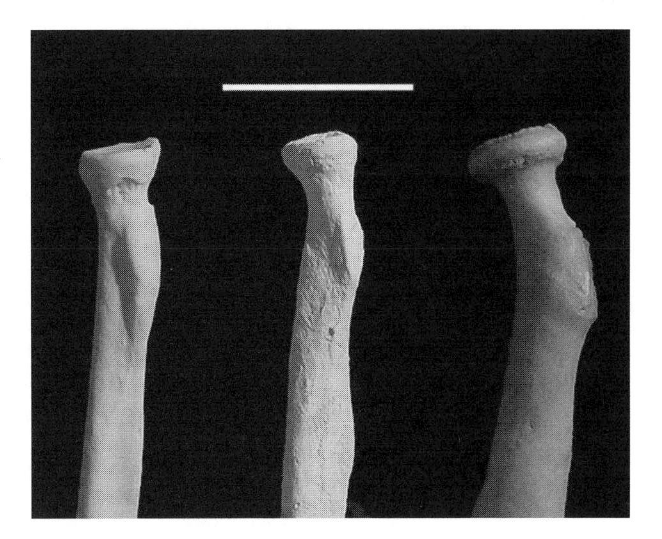

Figure 4.28. Proximal right radii of *Papio* (left), AS95-503 (center, cast), and *Pan* (right) in anterior view. Note the apelike bevel of the head of AS95-503 (see right side of head). The neck of AS95-503 is shorter than those of hominoids, but longer than those of *Papio* (note especially the small distance between the head and proximal end of the tuberosity). Note also the strongly developed prominence at the proximal end of a sharp, *Papio*-like, interosseous crest in AS95-503. Scale bar at top = 5 cm.

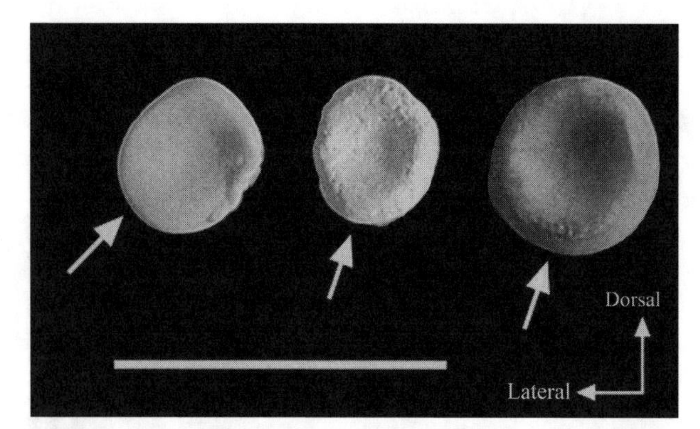

Figure 4.29. Proximal view of right radii of *Papio* (left), AS95-503 (center, cast), and *Pan* (right). Note the apelike deep capitular fovea in AS95-503. Despite slight erosional damage to the lateral margin of the head, it is clear that the head of AS95-503 was not strongly asymmetric, as seen in most pronograde arboreal quadrupeds. However, the lateral lip (arrows) of AS95-503 is more strongly developed than in hominoids (see *Pan*, and fig. 4.30), suggesting that the proximal radioulnar joint is particularly stable in pronation. As is the case in hominoids (Rose, 1993a,b), the lateral lip of AS95-503 is rotated anteriorly, in contrast with its more lateral position in nonhominoid anthropoids (see *Papio*, left). Scale bar at bottom = 5 cm.

medial) complement (figs. 4.29, 4.30), a characteristic of pronograde quadrupeds that habitually employ pronated hand postures. Rose (1988, p. 209) notes that, in nonhominoid anthropoids, the radial head has a "position of particular stability" in pronation because, during pronation, the long edge of the oval radial head makes the greatest contact with the radial notch on the ulna and zona conoidea on the humerus. Specifically, the anteromedial margin of the longer lateral lip rotates medially and achieves a "close-packed" position with the radial notch (Rose 1988, 1993a). The head lip asymmetry in AS95-503 most closely resembles the condition in *Papio*, intermediate between the more extreme asymmetry in radii of *Alouatta* and *Presbytis* and the relative symmetry in *Pan* and *Pongo* radii (fig. 4.29). However, there is a great deal of intraspecific variability in this morphology (fig. 4.30). It should be noted that the lateral lip is located more anteriorly in hominoids (including AS95-503) than other anthropoids, owing to a pronated twisting of the head and neck relative to the remainder of the radius, such as the "reference plane" of the "(generally) flattened anterior surface of the radial shaft" (Rose 1993a, p. 78).

The tilt of the head in most nonhominoid anthropoids may also be involved in this mechanism providing stability in pronation (Le Gros Clark and Thomas 1951; Conroy 1976; Rose 1988, 1993a). Like nonhominoid anthropoids, the head of AS95-503 angles anterolaterally relative to the

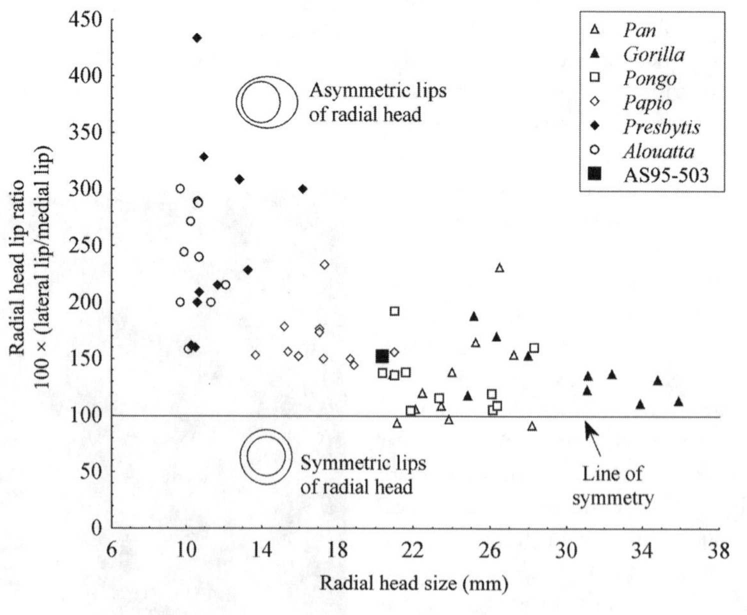

Figure 4.30. Plot of radial head lip asymmetry versus radial head size (mm; square root of the product of anteroposterior and medial head widths) in AS95-503 and a sample of extant anthropoids. In hominoids, including AS95-503, the head is rotated such that the lateral and medial lips are positioned, respectively, more anteriorly and posteriorly. Nonhominoid anthropoids generally have enlarged lateral lips that help provide particular stability at the proximal radioulnar joint in pronated hand postures (Rose 1988). Lateral lips are especially enlarged in pronograde arboreal quadrupeds, such as *Alouatta* and *Presbytis*. AS95-503 most closely resembles the morphology typical of the radii of *Papio* (but also observed in great ape radii) in having only a moderately enlarged lateral lip. The line of symmetry represents equivalent lip widths.

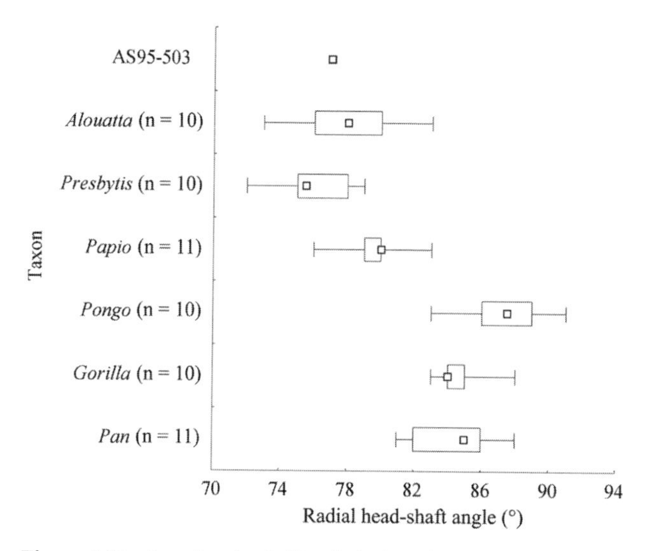

Figure 4.31. Box plot of radial head-shaft angle, reflecting the degree of anterolateral tilt of the head. Like the radii of nonhominoid anthropoids, the head of AS95-503 is strongly tilted anterolaterally, a feature associated with pronograde quadrupedalism. Head angle was measured relative to shaft with tuberosity oriented toward viewer. Symbols: Open squares = median; box = central 50% of the data; whiskers = observed range.

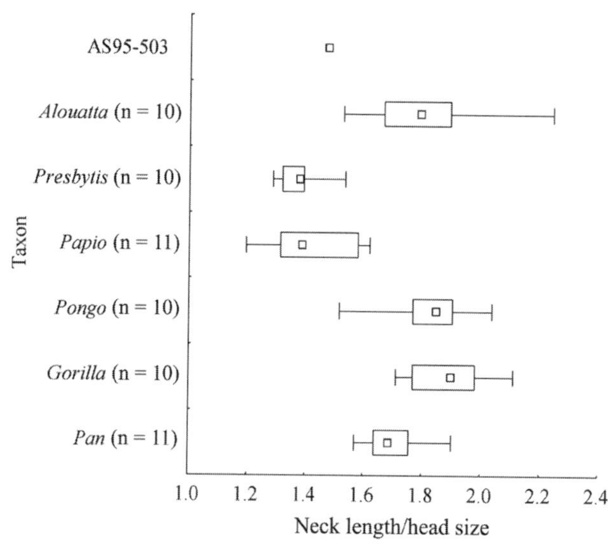

Figure 4.32. Box plot of radial neck length (mm) relative to radial head size (mm; square root of the product of anteroposterior and mediolateral head widths). *Alouatta, Pan,* and especially *Pongo* and *Gorilla* have relatively long radial necks. *Papio* and *Presbytis* have relatively short necks. AS95-503 has an intermediate to short neck length, suggesting the biceps brachii moment arm designed more for speed than power, a feature associated with pronograde quadrupedalism. Symbols: Open squares = median; box = central 50% of the data; whiskers = observed range.

shaft (figs. 4.27, 4.31). The most proximally extended area of the nonhominoid anthropoid radial head contacts the radial notch during pronation (Rose 1988).

Relative to the size of the head, the radial neck is intermediate in length between the long radial necks in great apes and *Alouatta* and the short necks in *Papio* and *Presbytis* (figs. 4.27, 4.32). This suggests that, in *Ankarapithecus,* the leverage for biceps brachii—a feature that increases power in elbow flexion and is thought to be advantageous in climbing—was not as great as it is in extant apes. A short neck, however, offers the potential for greater speed in elbow joint flexion, a trait often characteristic of cursorial mammals (Hildebrand 1995). However, to properly assess muscle leverage, neck length should be compared against radius length. For example, although *Pongo* radii have long radial necks compared with radial head width (fig. 4.32), their radii are so long that their relative neck lengths (and associated biceps brachii power arms) are shorter than those for African apes and humans (Heinrich et al. 1993). Thus, the biomechanical significance of the intermediate neck length relative to head width in AS95-503 must be interpreted with caution.

The sharpness of the interosseous crest of AS95-503 gives the shaft cross section a teardrop shape as extreme as that of *Papio* (fig. 4.27), as opposed to the oval or rounded cross section typical of *Pan, Pongo, Presbytis,* and *Alouatta.* Figure 4.33 shows the mediolateral and anteroposterior diameters of the radial shaft at the rugosity of the proximal end of the interosseous crest. AS95-503 has an anteroposteriorly flattened cross section that most closely resembles that of baboons. Distally, however, the shaft lacks the sharp dorsal crests characteristic of *Papio* and *Erythrocebus* radii.

Taken together, the morphological features of AS95-503 suggest that pronograde quadrupedalism was a major component of the locomotor repertoire of *Ankarapithecus.* In particular, the tilted radial head, with an enlarged lateral lip, indicates that the greatest stability occurred in pronated hand postures, like those employed by pronograde quadrupeds. Although it is a hominoid radius, AS95-503 resembles baboon radii in aspects of shaft shape.

Proximal Phalanx AS95-501

AS95-501 is the distal portion (approximately half) of a proximal phalanx that compares well with female chimpanzees in its size (width). The shaft appears to be fairly well preserved, but the distal articular surface has suffered some erosion. It can be identified as a proximal phalanx based on the relatively palmarly oriented trochlea, semicircular cross section, and the distal portions of the flexor ridges (fig. 4.34). It appears to be a manual phalanx because the trochlea is relatively deep (anteroposteriorly) and the shaft is relatively shallow and broad. In palmar view, the specimen has a trapezoidal shape (like all the taxa examined here except *Papio* and *Erythrocebus*) that suggests it is from a left hand.

One of the most notable features of AS95-501 is its lack of significant longitudinal curvature (fig. 4.35). Species that regularly employ suspensory postures typically have more curved proximal phalanges than those that use above-branch postures and terrestrial supports (Stern et al. 1995).

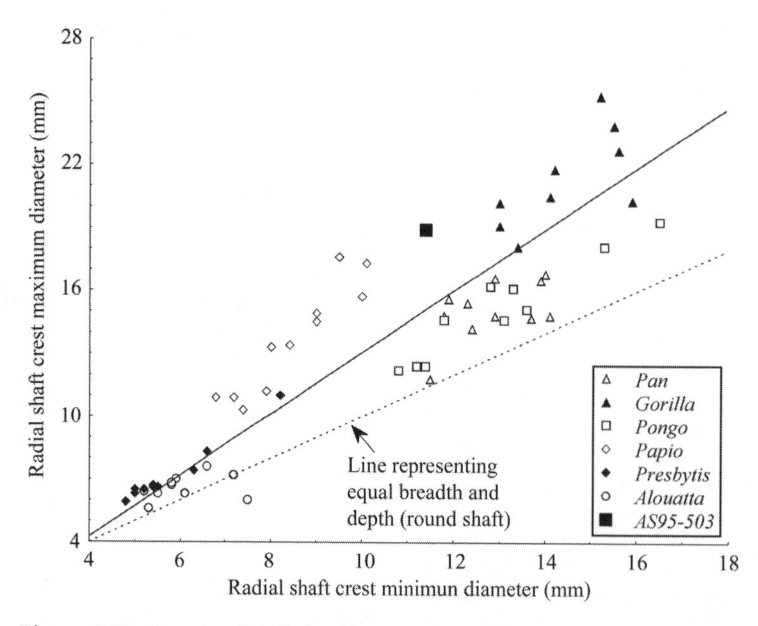

Figure 4.33. Plot of radial shaft crest maximum and minimum diameters. Radii with rounded shafts lie near the line indicating equal breadth and depth; those above the line have more anteroposteriorly compressed shafts. AS95-503 most closely resembles the radii of *Papio* and some gorillas in having a compressed shaft and very well-developed interosseous crest (see also fig. 4.27).

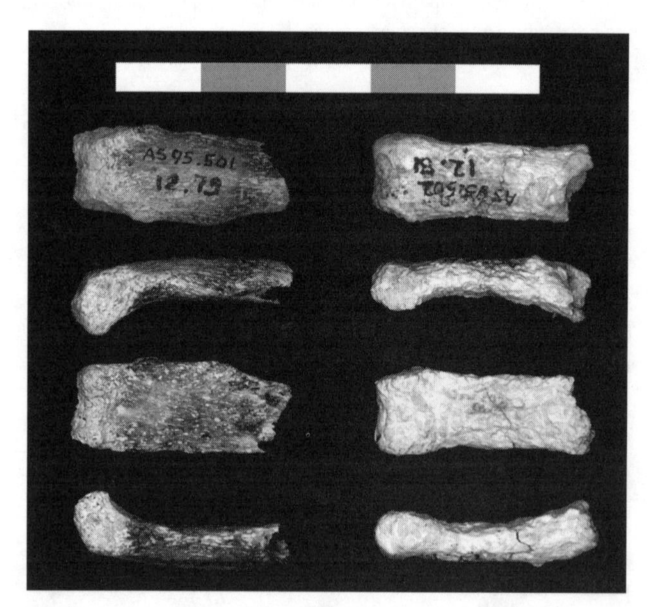

Figure 4.34. Dorsal, ventral, and side views of AS95-501 (left) and AS95-502 (right). Scale bar in cm.

The distal portion of the phalanx is typically more curved than the proximal half. AS95-501 is less curved than the proximal phalanges of *Pongo, Pan,* and *Hylobates* but more curved than the fingers of most *Papio* and *Erythrocebus* individuals. In its curvature, AS95-501 most closely resembles specimens of *Alouatta* and *Presbytis.* The slightly curved shaft of AS95-501 contrasts with the more curved proximal phalanges of *Sivapithecus* (Rose 1986; Richmond and Whalen 2001) and the highly curved proximal phalanges of *Dryopithecus* (Begun 1993; Moyà-Soyà and Köhler 1996).

The trochlea appears dorsovolarly compressed (relative to its width), like those of terrestrial and some arboreal quadrupeds, and less rounded than in hominoid phalanges. However, trochlear depth may have been affected by postdepositional erosion. Despite erosional damage, the trochlea projects farther volarly than observed in proximal phalanges of terrestrial quadrupeds (fig. 4.35).

The shaft, which survived relatively unscathed, exhibits a semicircular cross section at the break (near or just distal to midshaft). It is less rounded than the shafts of the terrestrial cercopithecoids, but lacks the inverted "U" shape of most extant hominoid phalanges. Cross-sectional shape influences the bone's mechanical behavior in resisting bending stresses but is also related to the development of the flexor sheath ridges. Those taxa that have powerful extrinsic flexor musculature, such as the extant hominoids (Tuttle 1969), also have pronounced ridges for the attachment of the flexor sheath, through which the flexor digitorum superficialis and flexor digitorum profundus tendons exert their force on the proximal phalanx. Thus, AS95-501 does not have apelike flexor ridges, and probably lacked the emphasis on extrinsic flexor musculature characteristic of extant hominoids.

In sum, the proximal phalanx AS95-501 provides evidence for pronograde quadrupedal locomotion. It most closely resembles phalanges of arboreal quadrupeds, but its morphology is consistent with a mixed use on both terrestrial and arboreal supports.

Middle Phalanx AS95-502

This specimen is the distal portion (probably more than half) of a middle phalanx (fig. 4.34). Its size is consistent with the

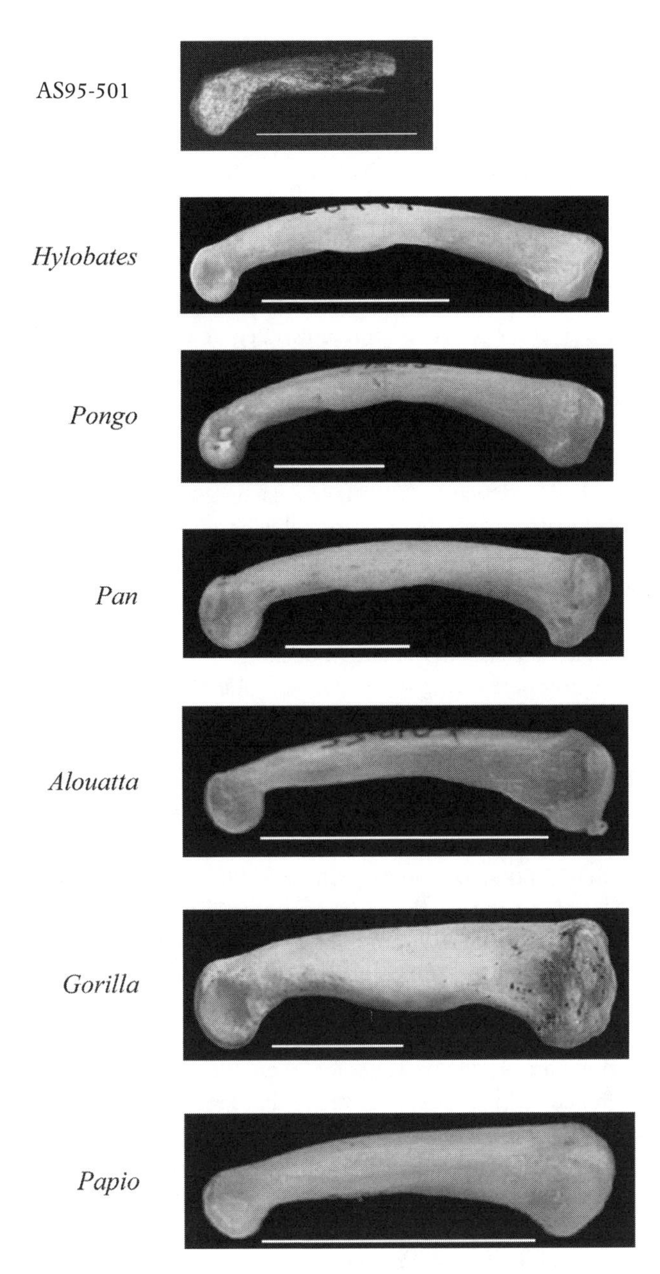

Figure 4.35. Side views of AS95-501 and the third manual proximal phalanges of *Hylobates, Pongo pygmaeus, Pan troglodytes, Alouatta, Gorilla gorilla,* and *Papio.* The distal portion of the shaft of AS 95-501 is more curved than that of *Papio,* but less curved than the suspensory apes. Scale bars = 2 cm.

possibility that it and the proximal phalanx AS95-501 belong to the same individual. AS95-502 has also suffered some erosion, and may have been digested by a carnivore. The trochlea is not palmarly directed but sits squarely on the distal end of the shaft, and the distal shaft is less deep, thus identifying it as a middle phalanx and distinguishing it from proximal phalanges. It is not possible to determine whether it is a manual or pedal phalanx.

AS95-502 is less curved than the phalanges (manual and pedal) of *Pongo, Pan, Presbytis,* and the manual phalanges of *Hylobates* and *Alouatta.* Its curvature resembles that found in pedal phalanges of *Hylobates* and *Alouatta,* as well as hand and foot phalanges of baboons and patas monkeys. However, unlike the terrestrial monkeys, the shaft of AS95-502 does not have a keyhole shape in volar view, with a broad head, narrow neck, and broad proximal shaft. Instead, AS95-502 maintains a fairly even width along its preserved length, like the middle phalanges of *Presbytis* and some *Alouatta* (this is also the case for gibbons and orangutans, but their morphology is distinct from other anthropoids).

The shaft of AS95-502 lacks a depression on the dorsal shaft just proximal to the head, typical of the manual middle phalanges of *Papio* and *Erythrocebus* and the pedal middle phalanges of *Homo* (often shallow). It is likely that this depression is related to hyperextension at the distal interphalangeal (DIP) joint during locomotion, a joint posture used by these taxa to improve frictional contact of the distal phalanx with the ground. The absence of this feature in AS95-502 suggests that the DIP joint was not typically hyperextended as in other taxa adapted for terrestrial digitigrade progression (as in *Papio* and *Erythrocebus*).

In sum, the middle phalanx is not specialized for suspension or digitigrade terrestriality, and most closely resembles those of arboreal quadrupeds.

Discussion

The new material from *Ankarapithecus meteai* serves to greatly expand our knowledge of this late Miocene species and especially of its postcranial skeleton. The cranial material reveals combinations of features that are not seen in any other fossil ape. For example, although AS95-500 does have an absolutely narrow interorbital region, it is not as narrow relative to the biorbital breadth of either *Sivapithecus* or *Pongo;* in fact, it falls easily within the range of values seen in *Pan,* especially in *Pan paniscus.* The configuration of the supraorbital region is not identical to either the supraorbital rib of *Sivapithecus* (GSP 15000) and *Pongo* or the continuous supraorbital torus of *Pan* and *Gorilla.* Rather, in *Ankarapithecus,* this region consists of independent supraorbital tori that are separated across glabella, protrude away from the frontal squama, and are invaded by an extensive frontal sinus that appears to extend across the entire length of the supraorbital region. Such an extensive frontal sinus is unknown from any other late Miocene ape. The shape of the orbital aperture is in some respects similar to that of *Sivapithecus* (GSP 15000) and *Pongo* but does not show the same degree of superior-inferior elongation seen in the latter taxa. The cranium of AS95-500 appears to be klinorhynch and thus differs from the more airorhynch condition seen in *Pongo* and *Sivapithecus.*

In many of its dental (see Andrews and Tekkaya 1980; Begun and Güleç 1998) and mandibular characteristics, *Ankarapithecus* resembles *Sivapithecus* and other late Miocene genera more closely than it resembles *Pongo.* Begun and Güleç (1998) conducted an extensive phylogenetic analysis of *Ankarapithecus* and other hominids and concluded that the most parsimonious cladogram places *Ankarapithecus* as the sister taxon of a *Sivapithecus-Pongo* clade, to the exclusion of all other living and extinct hominid species. Their analysis is largely restricted to the features preserved in the type mandible and lower face specimen MTA 2125, and this latter specimen does not include many of the most important facial features seen in AS95-500. The features they list that serve to link *Sivapithecus* with *Pongo* are compelling (see Begun and Güleç 1998, table 7), but the additional features preserved in AS95-500, such as evidence for klinorhynchy and the existence of an extensive frontal sinus that was probably derived from an ethmoid air cell complex are features that could be interpreted as removing *Ankarapithecus* from an exclusive relationship with the *Sivapithecus-Pongo* clade. *Ankarapithecus* could reasonably be placed as either a sister group to all middle Miocene-Recent Eurasian hominids or perhaps within this group but outside the *Sivapithecus-Pongo* clade. Perhaps the most important lesson to draw from this exercise, however, is that the fossil record is still woefully inadequate to definitively reach robust phylogenetic conclusions. What were once thought to be clearly coupled character complexes have been unraveled by the discovery of the new *Ankarapithecus* material. Because most of the other Eurasian fossils are much less complete than AS95-500, new discoveries are certain to shed more light on these questions.

The discovery of the first postcranial material of *Ankarapithecus* further complicates its phylogenetic reconstruction. Together, the postcranial fossils are most consistent with a generally pronograde mode of locomotion that probably included some amount of terrestriality. In size, *Ankarapithecus* postcrania are comparable with those of anthropoids ranging from large baboons to small gorillas. Large primates typically distribute their weight among multiple branches when moving in trees (Fleagle 1985). Suspensory specializations are lacking in *Ankarapithecus*, especially in the radius (e.g., slanted radial head, strong interosseous crest) and phalanges (fairly straight, weakly developed flexor ridges), and the absence of these features strongly contrasts with the morphology of *Oreopithecus* and the new partial skeleton of *Dryopithecus* from Can Llobateres (Moyà-Solà and Köhler 1996). Combined with large body mass, it is likely that *Ankarapithecus* spent considerable amounts of time on the ground, but it lacked some of the cursorial specializations seen in terrestrial cercopithecines. It may be premature, therefore, to refer to *Ankarapithecus* as a "habitually terrestrial form" (contra Andrews and Bernor 1999, p. 476).

Given the possible phylogenetic position of *Ankarapithecus* as either a stem great ape or perhaps, following Begun and Güleç (1998), a sister taxon to the *Sivapithecus-Pongo* clade, the postcranial evidence for pronograde quadrupedalism challenges the idea that the last common ancestor of great apes and humans was suspensory and the interpretation that quadrupedal adaptations in *Sivapithecus* are secondarily derived (Ward 1997; Richmond and Whalen 2001). It therefore also challenges the homology of many of the postcranial features shared by extant great apes. Unless *Sivapithecus* and *Ankarapithecus* are sister taxa to the exclusion of *Pongo*, the evidence supports the notion that many of the locomotor features shared by living great apes are the product of parallel evolution (Larson 1998).

Conclusions

The late Miocene record of large-bodied fossil hominoids is now among the most complete of that for any primate clade. Teasing apart homoplasy from homology will continue to be one of the most troubling stumbling blocks for our understanding of Miocene hominid evolution, but it is here that a more complete and better-dated fossil record offers great assistance. However, nearly every new discovery taps some previously unknown degree of diversity and reveals ever increasing numbers of unique combinations of character states. Note also that this record is at its richest in Europe and Asia; thus far, the African continent has produced only a handful of specimens from the late Miocene (see Hill and Ward 1988; Haile-Selassie 2001; Senut et al. 2001; Brunet et al. 2002), presumably because there are few geologic outcrops of the appropriate age. Given that the African continent today preserves the greatest specific diversity of living, large hominoids, it is possible that this was also the case in the late Miocene.

A comparison between the new *Ankarapithecus* specimens and other fossil apes from the same general time slice provides an interesting view of hominoid evolution during the late Miocene. The best represented of other species includes *Ouranopithecus macedoniensis* from Greece (de Bonis et al. 1990), *Sivapithecus sivalensis* and *S. parvada* from the Indo-Pakistan subcontinent (Pilbeam 1982; Kelley 1988), *Lufengpithecus lufengensis* from China (Wu 1987), *Oreopithecus bambolii* from Italy (Harrison 1987), *Dryopithecus laietanus* from Can Llobateres (Moyà-Solà and Köhler 1996), and the newly described *Orrorin tugenensis* from Kenya (Senut et al 2001), and *Sahelanthropus tchadensis* from Chad (Brunet et al. 2002). Together, these specimens reveal a high generic and specific diversity that greatly exceeds that of the living hominoids. These recent discoveries force a continuing reassessment of hominoid phylogeny. For example, the description of the relatively complete *Sivapithecus sivalensis* face from Pakistan (Pilbeam 1982) revealed many character states of the upper face that appear to be uniquely shared with the extant hominoid *Pongo*, which in turn led many workers to conclude that *Sivapithecus* is an early member of the *Pongo* clade. However, the subsequent recovery of postcranial elements including the humerus that are attributed to *Sivapithecus* and interpreted to represent a morphology unlike that of *Pongo* have led Pilbeam et al. (1990) and Pilbeam (1996) to question whether the shared facial characters are in fact synapomorphies. However, Moyà-Solà and Köhler (1996) have interpreted features of the postcranial skeleton of *Dryopithecus laietanus* as synapomorphies that are shared with *Pongo*, supporting its inclusion in this clade, whereas Begun and Güleç (1998) view this taxon as a sister group to the extant African apes. Deciding what is and is not homologous or homoplasious will depend on descriptions of all relevant material, including that from older sites. Indeed, the very rich but largely undescribed hominoid collection from the middle Miocene site of Paşalar in Turkey (J. Kelley, pers. comm.) may hold an important key to unraveling phylogenetic issues for the

Eurasian species. It is also likely that the large excavation platform bulldozed into the top of Delikayincak Tepe will continue to produce fossil ape specimens.

Acknowledgments

The Sinap project (1989–1995) enjoyed the efforts of a large number of colleagues and the results presented here reflect the collective efforts of the entire team. The project was supported by the General Directorate of Antiquities, T. C. Ministry of Culture and Tourism, and we thank the ministry for its seven years of support. We thank L. Martin and B. Alpagut, with whom we initiated this project, and I. Temizsoy, director, Museum of Anatolian Civilizations, who held the excavation permits and greatly facilitated our efforts in the field. We thank all members of the field crews for their efforts, and especially Z. Bostan, who found AS95-500, and Sevket Sen for discussions about earlier work at Sinap. W. Sanders and H. Çelebi assisted recovery in the field, and Mr. Çelebi and W. Lindsay carried out the laboratory preparation of the skull. The assistance of A. Kaakinen with molds and casts is greatly appreciated. The Sinap project received support from the National Science Foundation (EAR 9304302); the L.S.B. Leakey Foundation; the University of Texas at Austin; Motorola, Inc.; Autodesk, Inc.; the Academy of Finland; the University of Helsinki; the Sigma Xi Foundation; the Boise Fund; the Doctoral Program in Anthropological Sciences (Stony Brook University); Ankara University; and the Museum of Anatolian Civilizations. Comments by J. Kelley and Steven Ward greatly improved the manuscript.

Literature Cited

Agustí, J., M. Köhler, S. Moyà-Solà, L. Cabrera, M. Garcès, and J. M. Parés, 1996, Can Llobateres: The pattern and timing of the Vallesian hominoid radiation reconsidered: Journal of Human Evolution, v. 31, pp. 143–155.

Aiello, L. C., and B. A. Wood, 1994, Cranial variables as predictors of hominine body mass: American Journal of Physical Anthropology, v. 95, pp. 409–426.

Alpagut, A., P. Andrews, M. Fortelius, J. Kappelman, İ. Temizsoy, H. Çelebi, and W. Lindsay, 1996, A new specimen of *Ankarapithecus meteai* from the Sinap Formation of central Anatolia: Nature, v. 382, pp. 349–351.

Andrews, P., and B. Alpagut, 2001, Functional morphology of *Ankarapithecus meteai, in* L. de Bonis, G. D. Koufos, and P. Andrews, eds., Phylogeny of the Neogene hominoid primates in Eurasia. Hominid evolution and climatic change in Europe, Volume 2: Cambridge, Cambridge University Press, pp. 213–230.

Andrews, P., and R. L. Bernor, 1999, Vicariance biogeography and paleoecology of Eurasian Miocene hominoid primates, *in* J. Augusti, L. Rook, and P. Andrews, eds., The evolution of Neogene terrestrial ecosystems in Europe. Hominid evolution and climatic change in Europe, Volume 1: Cambridge, Cambridge University Press, pp. 454–487.

Andrews, P., and J. Cronin, 1982, The relationship of *Sivapithecus* and *Ramapithecus* and the evolution of the orangutan: Nature, v. 197, pp. 541–546.

Andrews P., and I. Tekkaya, 1980, A revision of the Turkish Miocene hominoid *Sivapithecus meteai:* Paleontology, v. 23, pp. 86–95.

Andrews, P., T. Harrison, E. Delson, R. L. Bernor, and L. Martin, 1996, Distribution and biochronology of European and southwest Asian Miocene catarrhines, *in* R. L. Bernor, V. Fahlbusch, and H.-W. Mittman, eds., The evolution of western Eurasian Neogene mammal faunas: New York, Columbia University Press, pp. 168–207.

Barry, J., 1986, A review of the chronology of Siwalik hominoids, *in* J. C. Else, and P. C. Lee, eds., Primate evolution: Cambridge, Cambridge University Press, pp. 93–106.

Begun, D. R., 1992, Miocene fossil hominids and the chimp-human clade: Science, v. 257, pp. 1929–1933.

Begun, D. R., 1993, New catarrhine phalanges from Rudabanya (Northeastern Hungary) and the problem of parallelism and convergence in the hominoid postcranial morphology: Journal of Human Evolution, v. 24, pp. 373–402.

Begun, D. R., and E. Güleç, 1995, Restoration and reinterpretation of the facial specimen attributed to *Sivapithecus meteai* from Kaylncak, (Yassiören), central Turkey: American Journal of Physical Anthropology Supplement, v. 20, pp. 63–64.

Begun, D. R., and E. Güleç, 1998, Restoration of the type and palate of *Ankarapithecus meteai:* Taxonomic and phylogenetic implications: American Journal of Physical Anthropology, v. 105, pp. 279–314.

Begun, D. R., and L. Kordos, 1997, Phyletic affinities and functional convergence in *Dryopithecus* and other Miocene and living hominids, *in* D. R. Begun, C. V. Ward, and M. D. Rose, eds., Function, phylogeny, and fossils: Miocene hominoid evolution and adaptations: New York, Plenum Press, pp. 291–316.

Begun, D. R., C. V. Ward, and M. D. Rose, 1997, Events in hominoid evolution, *in* D. R. Begun, C. V. Ward, and M. D. Rose, eds., Function, phylogeny, and fossils: Miocene hominoid evolution and adaptations: New York, Plenum Press, pp. 389–415.

Benefit, B. R., and M. L. McCrosin, 1993, Facial anatomy of *Victoriapithecus* and its relevance to the ancestral cranial morphology of old world monkeys and apes: American Journal of Physical Anthropology, v. 92, pp. 329–370.

Benefit, B. R., and M. L. McCrossin, 1997, Earliest known Old World monkey skull: Nature, v. 388, pp. 368–371.

Bonis, L. de, and G. D. Koufos, 1993, The face and mandible of *Ouranopithecus macedoniensis:* Description of new specimens and comparisons: Journal of Human Evolution, v. 24, pp. 469–491.

Bonis, L. de, G. Bouvrain, D. Geraads, and G. Koufos, 1990, New hominoid skull material from the late Miocene of Macedonnia in northern Greece: Nature, v. 345, pp. 712–714.

Brown, B., 1997, Miocene hominoid mandibles: Functional and phylogenetic perspectives, *in* D. R. Begun, C. V. Ward, and M. D. Rose, eds., Function, phylogeny, and fossils: Miocene hominoid evolution and adaptations: New York, Plenum Press, pp. 153–171.

Brown, B., and S. C. Ward, 1988, Basicranial and facial topography in *Pongo* and *Sivapithecus, in* J. H. Schwartz, ed., Orang-utan biology: Oxford, Oxford University Press, pp. 247–260.

Brunet, M., F. Guy, D. Pilbeam, H. Taisso Mackaye, A. Likius, D. Ahounta, A Beauvilain, C. Blondel, H. Bocherens, J.-R. Boisserie, L. De Bonis, Y. Coppens, J. Dejax, C. Denys, P. Duringer, V. Eisenmann, G. Fanone, P. Fronty, D. Geraads, T. Lehmann, F. Lihoreau, A. Louchart, A. Mahamat, G. Merceron, G. Mouchelin, O. Otero, P. Pelaez Campomanes, M. Ponce De Leon, J.-C. Rage, M. Sapanet, M. Schuster, J. Sudre, P. Tassy, X. Valentin, P. Vignaud, L. Viriot, A. Zazzo, and C. Zollikofer, 2002, A new hominid from the Upper Miocene of Chad, Central Africa: Nature, v. 418, pp. 145–151.

Cameron, D. W., 1997, A revised systematic scheme for the Eurasian Miocene fossil Hominidae: Journal of Human Evolution, v. 33, pp. 449–477.

Cave, A.J.E., and R. W. Haines, 1940, The paranasal sinuses of the anthropoid apes: Journal of Anatomy, v. 72, pp. 493–523.

Clarke, R. J., 1977, The cranium of the Swartkrans hominoid, SK 847 and its relevance to human origins [Ph.D. thesis]: University of Witswatersrand, Johannesburg, South Africa.

Conroy, G. C., 1976, Primate postcranial remains from the Oligocene of Egypt: Contributions to Primatology, v. 8, pp. 1–134.

Coolidge, H. J. Jr., 1933, *Pan paniscus.* Pygmy chimpanzee from south of the Congo River: American Journal of Physical Anthropology, v. 18, pp. 1–57.

Cramer, D. L., 1977, Craniofacial morphology of *Pan paniscus:* Contributions to Primatology, v. 10, pp. 1–64.

Dean, D., and E. Delson, 1992, Second gorilla or third chimp?: Nature, v. 359, pp. 676–677.

Delson, E., C. J. Terranova, W. L. Jungers, E. J. Sargis, N. G. Jablonski, and P. C. Dechow, 2000, Body mass in Cercopithecidae (Primates, Mammalia): Estimation and scaling in extinct and extant taxa: Anthropological Papers of the American Museum of Natural History 83, 159 pp.

Enlow, D. H., 1982, Handbook of facial growth, 2nd edition: Philadelphia, Saunders, 486 pp.

Enlow, D. H., and M. Azuma, 1975, Functional growth boundaries in the human and mammalian face, *in* J. Langman, ed., Morphogenesis and malformations of the face and brain: New York, A. R. Liss, pp. 217–230.

Fleagle, J. G., 1985, Size and adaptation in primates, *in* W. L. Jungers, ed., Size and scaling in primate biology: New York, Plenum Press, pp. 1–19.

Geraads, D., and E. Güleç, 1997, Relationships of *Barbourofelis piveteaui* (Ozansoy, 1965), a late Miocene nimravid (Carnivora, Mammalia) from central Turkey: *Journal of Vertebrate Paleontology,* v. 17, pp. 370–375.

Ginsburg, L., 1961, La faune des carnivores Miocenes de Sansan: *Memoires du Museum National D'Histoire Naturelle* Série C. Tome XI, pp. 1–190.

Gregory, W. K., 1930, Basic patents in evolution: Scientific American, v. 143, pp. 112–113.

Gros Clark, W. Le, and L.S.B. Leakey, 1951, The Miocene Hominoidea of east Africa: Fossil Mammals of Africa, v. 1, pp. 1–117.

Gros Clark, W. E. Le, and D. P. Thomas, 1951, Associated jaws and limb bones of *Limnopithecus macinnesi:* Fossil Mammals of Africa, v. 3, pp. 1–27.

Haile-Selassie, Y., 2001, Late Miocene hominids from the Middle Awash, Ethiopia: Nature, v. 412, pp. 178–181.

Harrison, T., 1987, A reassessment of the phylogenetic relationships of *Oreopithecus bambolii* Gervais: Journal of Human Evolution, v. 15, pp. 541–583.

Heinrich, R. E., M. D. Rose, R. E. Leakey, and A. C. Walker, 1993, Hominid radius from the Middle Pliocene of Lake Turkana, Kenya: American Journal of Physical Anthropology, v. 92, pp. 139–148.

Hildebrand, M., 1995, Analysis of vertebrate structure: New York, John Wiley & Sons, 657 pp.

Hill, A., and S. C. Ward, 1988, Origin of the Hominidae: The record of African large hominoid evolution between 14 Myr and 4 Myr: Yearbook of Physical Anthropology, v. 31, pp. 49–83.

Hofer, H.O., 1952, Der Gestaltwandel des Schädels der Säugetiere und Vögel mit besonderer Berücksichtigung der Knickungstypen der Schädelbasis: Verhandlungen Anatomischen Gesellschaft, v. 50, pp. 102–113.

Huxley, T.H., 1863, Evidence as to man's place in nature: London, Williams and Norgate, 159 pp.

Jungers, W. L., D. B. Burr, and M. S. Cole, 1998, Body size and scaling of long bone geometry, bone strength, and positional behavior in cercopithecoid primates, *in* E. Strasser, J. G. Fleagle, A. Rosenberger, and H. M. McHenry, eds., Primate locomotion: Recent advances: New York, Plenum Press, pp. 309–335.

Kappelman, J., 1996, The evolution of body mass and relative brain size in fossil hominids: Journal of Human Evolution, v. 30, pp. 243–276.

Kappelman, J., J. Kelley, D. Pilbeam, K. A. Sheikh, S. Ward, M. Anwar, J. C. Barry, B. Brown, P. Hake, N. M. Johnson, S. M. Raza, and S. M. I. Shah, 1991, The earliest occurrence of *Sivapithecus* from the middle Miocene Chinji Formation of Pakistan: Journal of Human Evolution, v. 21, pp. 61–73.

Kappelman, J., S. Sen, M. Fortelius, A. Duncan, B. Alpagut, J. Crabaugh, A. Gentry, J.-P. Lunkka, F. McDowell, N. Solounias, S. Viranta, and L. Werdelin, 1996, Chronology and biostratigraphy of the Miocene Sinap Formation of central Turkey, *in* R. L. Bernor, V. Fahlbusch, and H.-W. Mittman, eds., The evolution of western Eurasian Neogene mammal faunas: New York, Columbia University Press, pp. 78–95.

Keith, A., 1903, The extent to which the posterior segments of the body have been transmuted and suppressed in the evolution of man and allied primates: Journal of Anatomy and Physiology, v. 37, pp. 18–40.

Kelley, J., 1988, A new large species of *Sivapithecus* from the Siwaliks of Pakistan: Journal of Human Evolution, v. 17, pp. 305–324.

Kelley, J., and Q. Xu, Q., 1991, Extreme sexual dimorphism in a Miocene hominoid: Nature, v. 352, pp. 151–153.

Kelley, J., 1993, Taxonomic implications of sexual dimorphism in *Lufengpithecus, in* W. H. Kimbel and L. B. Martin, eds., Species, species concepts, and primate evolution: New York, Plenum Press, pp. 429–458.

Kelley, J., 2001, Phylogeny and sexually dimorphic characters: Canine reduction in *Ouranopithecus, in* L. de Bonis, G. D. Koufos, and P. Andrews, eds., Phylogeny of the Neogene hominoid primates in Eurasia. Hominid evolution and climatic change in Europe, Volume 2: Cambridge, Cambridge University Press, pp. 269–283.

Kordos, L., and D. R. Begun, 2000, Four catarrhine crania from Rudabanya: American Journal of Physical Anthropology, Supplement 30, pp. 199–200.

Larson, S. G., 1998, Parallel evolution in the hominoid trunk and forelimb: Evolutionary Anthropology, v. 6, pp. 87–99.

Leakey, R.E.F., and M. G. Leakey, 1986a, A new Miocene hominoid from Kenya: Nature, v. 324, pp. 143–146.

Leakey, R.E.F., and M. G. Leakey, 1986b, A second new Miocene hominoid from Kenya: Nature, v. 324, pp. 146–148.

Leakey, R.E.F., M. G. Leakey, and A. C. Walker, 1988a, Morphology of *Turkanapithecus kalakolensis* from Kenya: American Journal of Physical Anthropology, v. 76, pp. 277–288.

Leakey, R.E.F., M. G. Leakey, and A. C. Walker, 1988b, Morphology of *Afropithecus turkanensis* from Kenya: American Journal of Physical Anthropology, v. 76, pp. 289–307.

Leakey, M. G., C. S. Feibel, I. McDougall, and A. Walker, 1995, New four-million-year-old hominid species from Kanapoi and Allia Bay, Kenya: Nature, v. 376, pp. 565–571.

Lieberman, D. E., C. F. Ross, and M. J. Ravosa, 2000, The primate cranial base: Ontogeny, function, and integration: Yearbook of Physical Anthropology, v. 43, pp. 117–169.

MacLatchy, L., D. Gebo, R. Kityo, and D. Pilbeam, 2000, Postcranial functional morphology of *Morotopithecus bishopi,* with implications for the evolution of modern ape locomotion: Journal of Human Evolution, v. 39, pp. 159–183.

Mahler, P. E., 1973, Metric variation in the pongid dentition. Ph.D. dissertation, Ann Arbor, MI, University of Michigan, 467 pp.

McHenry, H. M., 1992, Body size and proportions in early hominids: American Journal of Physical Anthropology, v. 87, pp. 407–431.

McHenry, H. M., P. Andrews, and R. S. Corruccini, 1980, Miocene hominoid palatofacial morphology: Folia Primatologica, v. 33, pp. 241–252.

Moyà-Solà, S., and M. Köhler, 1995, New partial cranium of *Dryopithecus* Lartet, 1863 (Hominoidea, Primates) from the upper Miocene of Can Llobateres, Barcelona, Spain: Journal of Human Evolution, v. 29, pp. 101–139.

Moyà-Solà S., and M. Köhler, 1996, A *Dryopithecus* skeleton and the origins of great-ape locomotion: Nature, v. 379, pp. 156–159.

Ozansoy, F., 1955, Sur les gisements continenteaux et les Mammifères du Néogén et Villafranchien d'Ankara (Turquie): Comptes Rendus de l'Académie des Sciences de Paris, v. 240, pp. 992–994.

Ozansoy, F., 1957, Faunes de Mammifères du Tertiaire de Turquie et leurs révisions stratigraphiques: Bulletin of the Mineral Research Exploration Institute Turkey (Foreign Ed.), v. 49, pp. 29–48.

Ozansoy, F., 1965, Étude des gisements continent aux et de Mammifères du Cénozoïque de Turquie: Mémoires Societe Géologique de France (nouvelle série), v. 44, pp. 1–92.

Pickford, M., S. Moyà-Solà, and M. Köhler, 1997, Phylogenetic implications of the first African middle Miocene frontal bone from Otavi, Namibia: Comptes Rendus de l'Académie des Sciences de Paris II, v. 325, pp. 459–466.

Pilbeam, D. R. 1969. Tertiary Pongidae of East Africa: Evolutionary relationships and taxonomy. Bulletin of the Peabody Museum of Natural History 31:1–185.

Pilbeam, D., 1982, New hominoid skull material from the Miocene of Pakistan: Nature, v. 295, pp. 232–234.

Pilbeam, D. R., 1996, Genetic and morphological records of the Hominoidea and hominid origins: A synthesis. Molecular Phylogenetics and Evolution, v. 5, pp. 155–168.

Pilbeam, D. R., 1997, Research on Miocene hominoids and hominid origins: The last three decades, *in* D. R. Begun, C. V. Ward, and M. D. Rose, eds., Function, phylogeny, and fossils: Miocene hominoid evolution and adaptations: New York, Plenum Press, pp. 153–171.

Pilbeam, D. R., M. D. Rose, J. C. Barry, and S.M.I. Shah, 1990, New *Sivapithecus* humeri from Pakistan and the relationship of *Sivapithecus* and *Pongo:* Nature, v. 348, pp. 237–239.

Plavcan, J. M., 1998, Correlated response, competition, and female canine size in primates: American Journal of Physical Anthropology, v. 107, pp. 401–416.

Plavcan, J. M., and C. P. van Schaik, 1992, Intrasexual competition and canine dimorphism in anthropoid primates: American Journal of Physical Anthropology, v. 87, pp. 461–477.

Plavcan, J. M., and C. P. van Schaik, 1994, Canine dimorphism: Evolutionary Anthropology, v. 2, pp. 208–214.

Ravosa, M. J., and B. T. Shea, 1994, Pattern in craniofacial biology: Evidence from the Old World monkeys (Cercopithecidae): International Journal of Primatology, v. 15, pp. 801–822.

Richmond, B. G., and M. Whalen, 2001, Forelimb function, bone curvature, and phylogeny of *Sivapithecus, in* L. de Bonis, G. D. Koufos, and P. Andrews, eds., Phylogeny of the Neogene hominoid primates in Eurasia. Hominid evolution and climatic change in Europe, Volume 2: Cambridge, Cambridge University Press, pp. 326–348.

Rook, L., P. Renne, M. Benvenuti, and M. Papini, 2000, Geochronology of *Oreopithecus*-bearing succession at Baccinello (Italy) and the extinction pattern of European Miocene hominoids: Journal of Human Evolution, v. 39, pp. 577–582.

Rose, M. D., 1986, Further hominoid postcranial specimens from the Late Miocene Nagri Formation of Pakistan: Journal of Human Evolution, v. 15, pp. 333–367.

Rose, M. D., 1988, Another look at the anthropoid elbow, Journal of Human Evolution, v. 17, pp. 193–224.

Rose, M. D., 1993a, Functional anatomy of the elbow and forearm in primates, *in* D. L. Gebo, ed., Postcranial adaptation in nonhuman primates: DeKalb, Illinois, Northern Illinois University Press, pp. 70–95.

Rose, M. D., 1993b, Locomotor anatomy of Miocene hominoids, *in* D.L. Gebo, ed., Postcranial adaptation in nonhuman primates: DeKalb, Illinois, Northern Illinois University Press, pp. 252–272.

Rose, M. D., M. G. Leakey, R.E.F. Leakey, and A. C. Walker, 1992, Postcranial specimens of *Simiolus enjiessi* and other primitive catarrhines from the early Miocene of Lake Turkana, Kenya: Journal of Human Evolution, v. 22, pp. 171–237.

Ross, C. F., and M. Henneberg, 1995, Basicranial flexion, relative brain size, and facial kyphosis in *Homo sapiens* and some fossil hominids: American Journal of Physical Anthropology, v. 98, pp. 575–593.

Ross, C. F., and M. J. Ravosa, 1993, Basicranial flexion, relative brain size, and facial kyphosis in nonhuman primates: American Journal of Physical Anthropology, v. 91, pp. 305–324.

Rossie, J. B., E. L. Simons, S. C. Gauld, and D. T. Rasmussen, 2002, Paranasal sinus anatomy of *Aegyptopithecus:* Implications for hominoid origins: Proceedings of the National Academy of Sciences, U.S.A., v. 99, pp. 8454–8456.

Schwartz, J. H., 1997, *Lufengpithecus* and hominoid phylogeny: Problems in delineating and evaluating phylogenetically relevant characters, *in* D. R. Begun, C. V. Ward, and M. D. Rose, Function, phylogeny, and fossils: Miocene hominoid evolution and adaptations: New York, Plenum Press, pp. 363–388.

Seiffert, E. R., and J. Kappelman, 2001, Morphometric variation in the hominoid orbital aperture: A case study with implications for the use of variable characters in Miocene catarrhine systematics: Journal of Human Evolution, v. 40, pp. 301–318.

Senut, B., M. Pickford, D. Gommery, P. Mein, K. Cheboi, and Y. Coppens, 2001, First hominid from the Miocene (Lukeino Formation, Kenya): Comptes Rendus de l'Académie des Sciences de Paris, Sciences de la Terre et des planets, v. 332, pp. 137–144.

Shea, B. T., 1985, On aspects of skull form in African apes and orangutans, with implications for hominoid evolution: American Journal of Physical Anthropology, v. 68, pp. 329–342.

Shea, B. T., 1988, Phylogeny and skull form in the hominoid primates, *in* J. H. Schwartz, ed., Orang-Utan biology: London, Oxford University Press, pp. 233–245.

Simons, E. L.,1987, New faces of *Aegyptopithecus* from the Oligocene of Egypt, Journal of Human Evolution, v. 16, pp. 273–289.

Simons, E. L., and D. T. Rasmussen, 1989, Cranial morphology of *Aegyptopithecus* and *Tarsius* and the question of the Tarsier-Anthropoidean clade: American Journal of Physical Anthropology, v. 79, pp. 1–24.

Smith, R. J., 1996, Biology and body size in human evolution: Current Anthropology, v. 37, pp. 451–481.

Smith, R. J., and W. L. Jungers, 1997, Body mass in comparative primatology: Journal of Human Evolution, v. 32, pp. 523–559.

Stern, J. T., W. L. Jungers, and R. L. Susman, 1995, Quantifying phalangeal curvature: An empirical comparison of alternative methods: American Journal of Physical Anthropology, v. 97, pp. 1–10.

Swartz, S. M., 1990, Curvature of the forelimb bones of anthropoid primates: Overall allometric patterns and specializations in suspensory species: American Journal of Physical Anthropology, v. 83. pp. 477–498.

Tardieu, C., 1981, Morpho-functional analysis of the articular surfaces of the knee-joint in primates, *in* A. B. Chairelli, and R. S. Corruccini, eds., Primate evolutionary biology: Berlin, Springer-Verlag, pp. 68–80.

Tuttle, R. H., 1969, Quantitative and functional studies on the hands of the anthropoidea: I. The Hominoidea: Journal of Morphology, v. 128, pp. 309–363.

Ward, S. C., 1997, The taxonomy and phylogenetic relationships of *Sivapithecus* revisited, *in* D. R. Begun, C. V. Ward, and M. D. Rose, Function, phylogeny, and fossils: Miocene hominoid evolution and adaptations: New York, Plenum Press, pp. 269–290.

Ward, S. C., and B. Brown, 1986, The facial skeleton of *Sivapithecus indicus, in* D. Swindler, and J. Erwin, eds., Comparative primate biology, Volume 1: Systematics, evolution, and anatomy: New York, Alan R. Liss, pp. 413–452.

Ward, S. C., and D. R. Pilbeam, 1983, Maxillo-facial morphology of Miocene hominoids from Africa and Indo-Pakistan, *in* R. L. Ciochon, and R. S. Corruccini, eds., New interpretations of ape and human ancestry: New York, Plenum Press, pp. 211–238.

Ward, C. V., A. C. Walker, M. F. Teaford, and I. Odhiambo, 1993, Partial skeleton of *Proconsul nyanzae* from Mfangano Island, Kenya: American Journal of Physical Anthropology, v. 90, pp. 77–111.

Ward, S. C., B. Brown, A. Hill, J. Kelley, and W. Downs, 1999, *Equatorius:* A new hominoid genus from the Middle Miocene of Kenya: Science, v. 285, pp. 1382–1386.

Wu, R., 1987, A revision of the classification of the Lufeng great apes: Acta Anthropologica Sinica, v. 6, pp. 265–271.

Muridae and Gerbillidae (Rodentia)

S. Sen

This chapter describes the remains of two muroid rodent families, the Muridae and Gerbillidae, from several localities of the Sinap Formation, north of Kazan town in central Anatolia. The material was collected during the field seasons 1991–1995 of the International Sinap Neogene project.

In Turkey, late Miocene murids are poorly documented. Their occurrence is mentioned in prelimanary faunal lists of a dozen localities (Sickenberg et al. 1975; Ünay and de Bruijn 1984; Sümengen et al. 1990, de Bruijn et al. 1996). The sole specimen known from the Vallesian is an M1 from Bayraktepe II (northwestern Turkey) that Ünay (1981) attributed to *Progonomys* sp. Recently, de Bruijn et al. (1999) described some murids (*Apodemus* sp. 1 and 2, *Hansdebruijnia neutrum,* and *"Karnimata" provocator*) from the middle Turolian locality of Düzyayla in the Sivas basin. In the neighboring countries, late Miocene murids are recorded from Greece only.

Our knowledge of the family Gerbillidae is little better. Remains of *Myocricetodon* and *Pseudomeriones* have been described from several Turkish localities (Wessels et al. 1987; Sen 1994) The second genus is also known from Greece.

This chapter aims primarily to improve the knowledge of the Muridae and Gerbillidae in Turkey. Moreover, some of the murids described below are the oldest representatives of the family, and thus shed light on the origin of the related European and north African taxa. The localities in the Sinap Formation are well dated, thanks to their rich mammalian assemblages and magnetostratigraphic studies (Sen 1991; Kappelman et al. 1996, this volume; Lunnka et al. 1999); they encompass the middle–late Miocene boundary documented in this area by the *"Hipparion* Datum."

The specimens described in this chapter are kept in the collections of the University of Ankara (DTCF). Casts of most specimens are available in the Laboratoire de Paléontologie du Muséum in Paris.

The cusp terminology follows Michaux (1971) and Van de Weerd (1976) for the Muridae and Tong (1988) for the Gerbillidae. The measurements have been taken with a Mitutoyo measuroscope from the occlusal surface of teeth as the maximum length and width and are given in mm. All teeth are illustrated as if they are from the left side; if inverted, their numbers on the illustrations are underlined.

Family Muridae Gray, 1821

The systematics of late Miocene murids have been intensely debated during the past decade, and there is still no general concensus on the status of some genera (*Progonomys, Occitanomys, Parapodemus*) and the generic attribution of several species. Disagreements also concern the species identification of populations from several European and north African localities. It is of interest here to provide some information about this debate to better establish the identifications of the murids from the Sinap Formation.

Mein et al. (1993, p. 42) consider that *"Progonomys* . . . is clearly a paraphyletic genus, since it houses species that have been brought together on the mere basis of sharing plesiomorphic characters." Based on this assumption, they only retained two species, *P. cathalai* Schaub, 1938, and *P. woelferi* Bachmayer and Wilson, 1970, and excluded from *Progonomys* the species *hispanicus* Michaux, 1971, *debruijni* Jacobs, 1978, and *clauzoni* Aguilar et al., 1986, as well as several Vallesian murid populations from Europe and north Africa that were previously referred to *P. cathalai.* Freudenthal and Martin Suarez (1999) agree with this classification.

However, de Bruijn et al. (1996), Aguilar et al. (1996), Michaux et al. (1997), and van Dam (1997) maintain the species excluded by Mein et al. (1993) in the genus *Progonomys,* arguing for inclusion on the basis of the morphologic similarities of their dentitions with the type species *P. cathalai.*

The systematic approach of Mein et al. (1993) seems rather typologic because it does not accept some morphologic variability of dental features and geographic variability of species. Aguilar and Michaux (1996) and Michaux et al. (1997) showed that the large samples from two Montredon localities display great morphologic variation. These assemblages contain morphotypes that characterize the populations excluded from *Progonomys cathalai* by Mein et al. (1993). Van Dam (1997) includes some populations (e.g., that of Biodrak in Greece) in *P. hispanicus* that Mein et al. (1993) retained as *P. cathalai*. This shows that the mostly plesiomorphic features and a few derived features of Vallesian murids do not allow identification of clear evolutionary lineages and consequently—contrary to what is suggested by Mein et al. (1993)—prevent distinction of genera. The "paraphyletic" status of these species (*hispanicus, debruijni,* and *clauzoni*), together with *cathalai* and *woelferi*, as interpreted by Mein et al. (1993), supposes that they belong to different lineages.

The oldest member of the family Muridae is *Antemus chinjiensis* from the middle Miocene of Pakistan. This genus potentially possesses all characters required to be the origin of all Vallesian and later murids, and there is no other candidate to which the Vallesian murids can be rooted. The idea that *Antemus* is ancestral to all other murids is generally accepted by specialists. In Europe, Turkey, and north Africa, murids appear as immigrants somewhat after the middle–late Miocene transition. Their first occurrence in different areas is apparently diachronic from east to west (Sen 1997). The oldest representatives of this group in Eurasia and north Africa still possess the plesiomorphic characters of *Antemus:* small size, lingual cusps on upper molars lower than the other cusps and situated far back, weak connections between the lingual and central cusps, central and labial cusps paired, strong posterior cingulum (t12), etc. These Vallesian murids are obviously originated from the same ancestral stock. Whether they represent one or more genera is a matter of personal taste. Their oldest representatives have no clear derived features allowing the recognition of distinct lineages and distinct genera, hence their attribution to one genus—*Progonomys.* However, during the late Vallesian, evolutionary trends become increasingly expressed in populations leading to the different tooth morphologies of *Huerzelerimys, Occitanomys,* and *Parapodemus* in Europe and Turkey and to *Karnimata* in southern Asia and the eastern Mediterranean area.

Recently van Dam (1997) described the oldest representatives of *P. hispanicus* from the localities of Masia de la Roma in the Teruel basin in Spain. In many respects, the morphology of these teeth is similar to that of *P. debruijni* from Pakistan: on M1, t9 is lateral to t8, t1 and t4 are lower than the other cusps, t1 is situated well back relative to t2–t3, central and labial cusps are paired, and the posterior cingulum is strong. These characters, considered as plesiomorphic by Mein et al. (1993) and Freudenthal and Martin Suarez (1999), are partly shared by other species referred to *Progonomys* in combination with some new characters: t1 less far back, t9 slightly anterior to t8, lingual cusps more strongly connected to the central cusps in *P. cathalai;* larger size, globular cusps, and reduced posterior cingulum in *P. woelferi;* larger size, wide upper molars, and strong connections between cusps in *P. clauzoni.*

The derived features of these species are not numerous and are too poorly enough defined to distinguish different genera. At our present state of knowledge, the best systematic statement that can be made is to include these species in *Progonomys.* This action is not contradictory to the phylogenetic relationships of these species with younger genera and species as suggested by Freudenthal and Martin Suarez (1999). For example, there is general agreement that the genus *Huerzelerimys* is derived from *P. cathalai,* and the genus *Occitanomys* (at least its western European representatives) from *P. hispanicus.*

Specimens referred to the genus *Progonomys* have been recorded from many localities in north Africa. Heissig (1982), for instance, mentioned *P. cathalai* in association with an early (?) Vallesian fauna from the fissure filling of Farafra in Egypt. Jaeger (1977b) attributed to *P. cathalai* the populations from Bou Hanifia 2 (Algeria) and Oued Zra (Morocco) that Mein et al (1993) consider to belong to a different genus and species. Ameur-Chebbeur (1988) described a new species, *P. chougrani,* from Sidi Salem (Algeria), which is very similar in size and morphology to the material from Oued Zra. *P. mauretanicus* Coiffait-Martin, 1991, from eastern Algeria is identical to *P. chougrani* in all respects, and so it is a synonym of the latter. Unfortunately, these two species are not valid, because they were described in the unpublished theses.

Also important is the status of *Parapodemus* Schaub, 1938. Martin Suarez and Mein (1998) transferred all of the *Parapodemus* species to *Apodemus,* except the type species *P. gaudryi* (Dames 1883) restricted to its type material from Pikermi. Their arguments for including the *Parapodemus* species in *Apodemus* are that (1) this group forms a monophyletic lineage from late Vallesian to Recent, and (2) there is a morphologic continuity between the early–middle Turolian species without t7 on M1 and M2 (*Parapodemus*-morphotype) and the late Turolian and later species with t7 on these molars (*Apodemus*-morphotype). Moreover, these authors asserted that the type specimen of *P. gaudryi* is lost and that the locality from which the type specimen of *P. gaudryi* originated is unknown. Both assertions are untrue. The type specimen of *P. gaudryi* can be retrieved from the collections of the Athens University Geology Department, and the locality of Megaloremma at Pikermi is still accessible for future work (C. Doukas, pers. comm.). Lastly, de Bruijn et al. (1999) redescribed the type material together with a rich collection of *P. gaudryi* from Pikermi-Chomateri in the same formation. Their conclusion, with which I agree, is that "the identity of *Mus gaudryi* Dames, 1833 is clear and that the genus *Parapodemus* is valid." They assigned to this genus the following four species: *P. gaudryi* (Dames, 1833); *P. lugdunensis* Schaub, 1938; *P. barbarae* Van de Weerd, 1976; and *P. meini* Martin Suarez and Freudenthal, 1993.

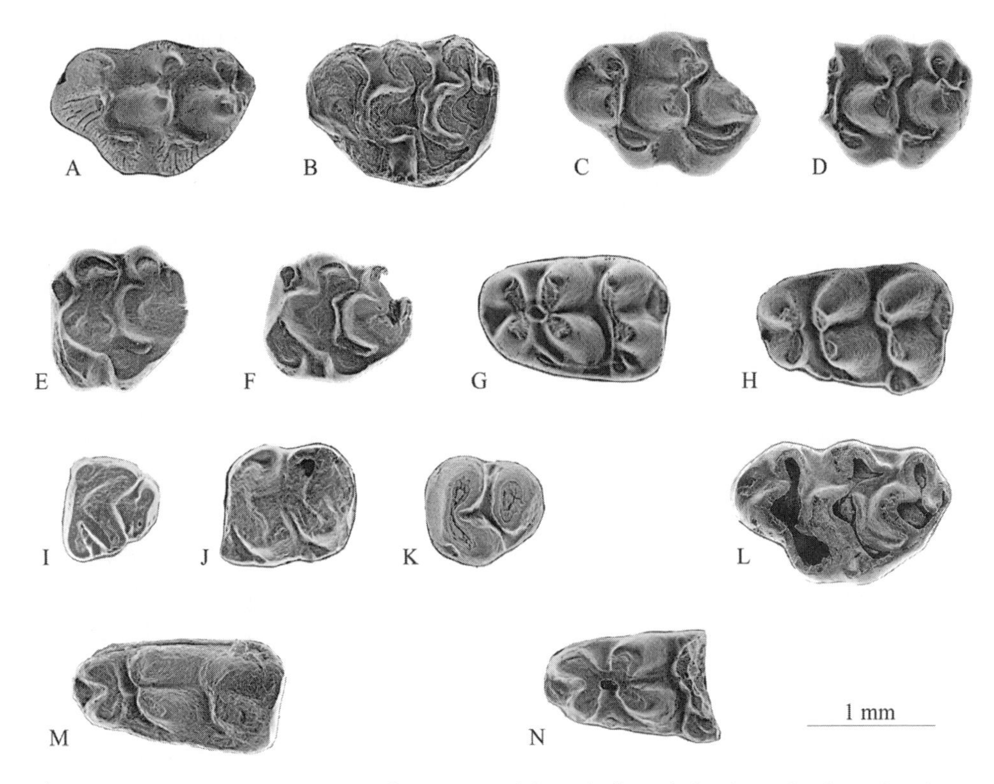

Figure 5.1. *Progonomys minus* sp. nov. from Loc. 8A: **(A)** M1 (Holotype); **(B–D)** M1; **(E, F)** M2, **(G, H)** m1; **(I)** M3; **(J)** m2; **(K)** m3. *Sinapodemus* sp. from Loc. 8A: **(L)** M1; **(M, N)** m1.

Genus *Progonomys* Schaub, 1938
Progonomys minus n. sp.

Holotype. M1 sin (1.58 × 1.08), ST8A-53 (fig. 5.1A).

Hypodigm. Four M1, three M2, one M3, three m1, one m2, and one m3 (ST8A-54–ST8A-66).

Type locality. Loc. 8A, which is situated on the eastern slopes of Sinap Tepe, in the ravine of Igdelik, 1.5 km north-northeast of Yassiören village, Kazan, Ankara.

Etymology. In reference to its small size.

Measurements. See table 5.1 and figure 5.2.

Diagnosis. Small murid. M1 with low lingual cusps that are connected to the central cusps by low ridges, with separated t6 and t9. The m1 has a tiny anterocentral cusp, without connection between the anteroconid and metaconid, terminal heel placed far back.

Differential Diagnosis. *Progonomys minus* differs from *P. cathalai, P. woelferi,* and *P. clauzoni* by its smaller size, lower lingual cusps on M1, and the lack of connection between the anteroconid and metaconid on m1. It differs from *P. hispanicus* and *P. debruijni* by the less far back position of the lingual cusps on upper molars and by the lack of a connection between the first and second chevrons on m1.

Description. Locality 8A yielded 18 murid molars, two-thirds of them being M1 and m1. According to the morphology of M1 and m1, this material clearly represents two distinct taxa. The second and third molars (six specimens in all) are not so well differentiated, and thus cannot be securely attributed to one or the other of these two taxa. These six teeth (three M2 and one each of M3, m2, and m3) are described here as *Progonomys,* but some of them may well belong to *Sinapodemus* sp. Because their identification is not certain, they are not considered in the comparison of the Loc. 8A murids with other species.

The M1 has an amygdaloid outline and the crown is very low. The lingual cusps t1 and t4 are crescent-shaped, notably lower than the others, and connected to the central cusps by low ridges. The central and labial cusps are higher than the lingual ones and strongly connected in pairs. The t6 and t9 are separated by a wide mesosinus. In one specimen, the t6 has a small posterior spur. The ridge connecting the t4 to the t8 is low. The t9 is situated next to the t8 or slightly anteriorly to it. The posterior cingulum (t12) is strong. M1 has three roots.

The M2 has cusps disposed as on M1. The t1 is large, kidney-shaped, and labially connected to the base of the t5. The t3 is small. One specimen with a posterior spur on the t6 may belong to *Sinapodemus* sp. (see later in this chapter). The other specimens lack this spur. The ridge between the t4 and t8 is low. The t12 is strong. There are three roots.

The unique M3 is very small. The t1 is connected to the t5; the t3 is reduced. The t4, t5, and t6 form a symmetric

Table 5.1. Measurements of Upper and Lower Molars of Species Referred to the Genera *Progonomys* and *Sinapodemus* from Various Localities in Europe and Asia

Taxon	Age	Locality	Molar	Length Range	Length Mean	n[1]	Width Range	Width Mean
Progonomys minus sp. nov	MN9	Loc. 8A	M1	1.61–1.65	1.62	3/5	1.00–1.16	1.08
P. cathalai	MN9	Loc. 84	M1	1.63–1.85	1.75	20/29	1.05–1.18	1.11
P. cathalai	MN10	Montredon	M1	1.73–2.06	1.89	33/35	1.12–1.35	1.22
P. hispanicus	MN10	Masia del Barbo	M1	1.64–1.91	1.77	43/45	1.00–1.26	1.13
P. debruijni	~9.5 Ma	YGSP-182A	M1	1.55–1.82	1.66	22	0.88–1.12	1.02
Sinapodemus sp.	MN9	Loc. 8A	M1	—	1.80	1	—	1.11
S. ibrahimi sp. nov	MN9	Loc. 84	M1	1.79–2.03	1.87	7/11	1.10–1.23	1.14
P. minus sp. nov	MN9	Loc. 8A	M2	1.08–1.18	1.13	3	1.02–1.12	1.06
P. cathalai	MN9	Loc. 84	M2	1.08–1.31	1.22	25	1.00–1.17	1.11
P. cathalai	MN10	Montredon	M2	1.04–1.43	1.30	58/63	1.05–1.32	1.21
P. hispanicus	MN10	Masia del Barbo	M2	1.04–1.32	1.19	46	1.00–1.28	1.13
P. debruijni	~9.5 Ma	YGSP-182A	M2	0.98–1.10	1.03	6	0.95–1.10	1.00
S. ibrahimi sp. nov	MN9	Loc. 84	M2	1.28–1.36	1.32	2	1.18–1.20	1.19
P. minus sp. nov	MN9	Loc. 8A	M3	—	0.78	1	—	0.80
P. cathalai	MN9	Loc. 84	M3	0.75–0.89	0.80	7	0.80–0.90	0.86
P. cathalai	MN10	Montredon	M3	0.74–0.98	0.89	13	0.87–1.00	0.94
P. hispanicus	MN10	Masia del Barbo	M3	0.71–0.91	0.80	12/13	0.71–0.99	0.84
S. ibrahimi sp. nov	MN9	Loc. 84	M3	0.86–0.87	0.87	2	0.90–0.93	0.92
P. minus sp. nov	MN9	Loc. 8A	m1	1.61–1.67	1.64	2/3	0.84–0.99	0.94
P. cathalai	MN9	Loc. 84	m1	1.46–1.66	1.57	22/23	0.88–1.06	0.96
P. cathalai	MN10	Montredon	m1	1.45–1.90	1.71	70/69	0.90–1.13	1.04
P. hispanicus	MN10	Masia del Barbo	m1	1.50–1.70	1.61	42/45	0.88–1.06	0.97
P. debruijni	~9.5 Ma	YGSP-182A	m1	1.32–1.55	1.40	12	0.78–0.92	0.86
Sinapodemus sp.	MN9	Loc. 8A	m1	1.62–1.63	1.63	2/3	0.90–0.93	0.92
S. ibrahimi sp. nov	MN9	Loc. 84	m1	1.58–1.71	1.67	4	1.00–1.09	1.04
P. minus sp. nov	MN9	Loc. 8A	m2	—	1.11	1	—	0.93
P. cathalai	MN9	Loc. 84	m2	1.08–1.32	1.19	35	0.97–1.15	1.04
P. cathalai	MN10	Montredon	m2	1.12–1.48	1.32	71/70	1.01–1.21	1.14
P. hispanicus	MN10	Masia del Barbo	m2	1.05–1.30	1.17	51/49	0.95–1.19	1.05
P. debruijni	~9.5 Ma	YGSP-182A	m2	0.98–1.28	1.12	10	0.88–1.00	0.94
S. ibrahimi sp. nov	MN9	Loc. 84	m2	1.22–1.30	1.25	3	1.04–1.10	1.08
P. minus sp. nov	MN9	Loc. 8A	m3	—	1.00	1	—	0.85
P. cathalai	MN9/10	Loc. 84	m3	0.88–0.99	0.96	8	0.80–0.87	0.84
P. cathalai	MN10	Montredon	m3	0.88–1.13	1.05	35	0.85–1.02	0.94
P. hispanicus	MN10	Masia del Barbo	m3	0.82–1.00	0.92	33	0.70–0.90	0.82
P. debruijni	~9.5 Ma	YGSP-182A	m3	—	0.88	1	—	0.80

[1]*n*, Number of measurements.

arc. The t8 and t9 are fused to form a transverse loph. There are three roots.

The m1 has a tiny and low anterocentral cusp. The lophs are slightly oblique. There is no connection between the anteroconid and metaconid. The labial cingular margin is continuous but low; it bears a small c1 and a thickening in the place of the c4. The c1 is connected to the hypoconid on one specimen, and isolated on the other. The terminal heel is small, low, ridge shaped, lingually placed and posteriorly inclined. The roots are broken.

The unique m2 is quite worn. The anterolabial cusp is small. The labial cingulum is very weak with an isolated and small c1. The two lophids are not connected. There are two strong roots.

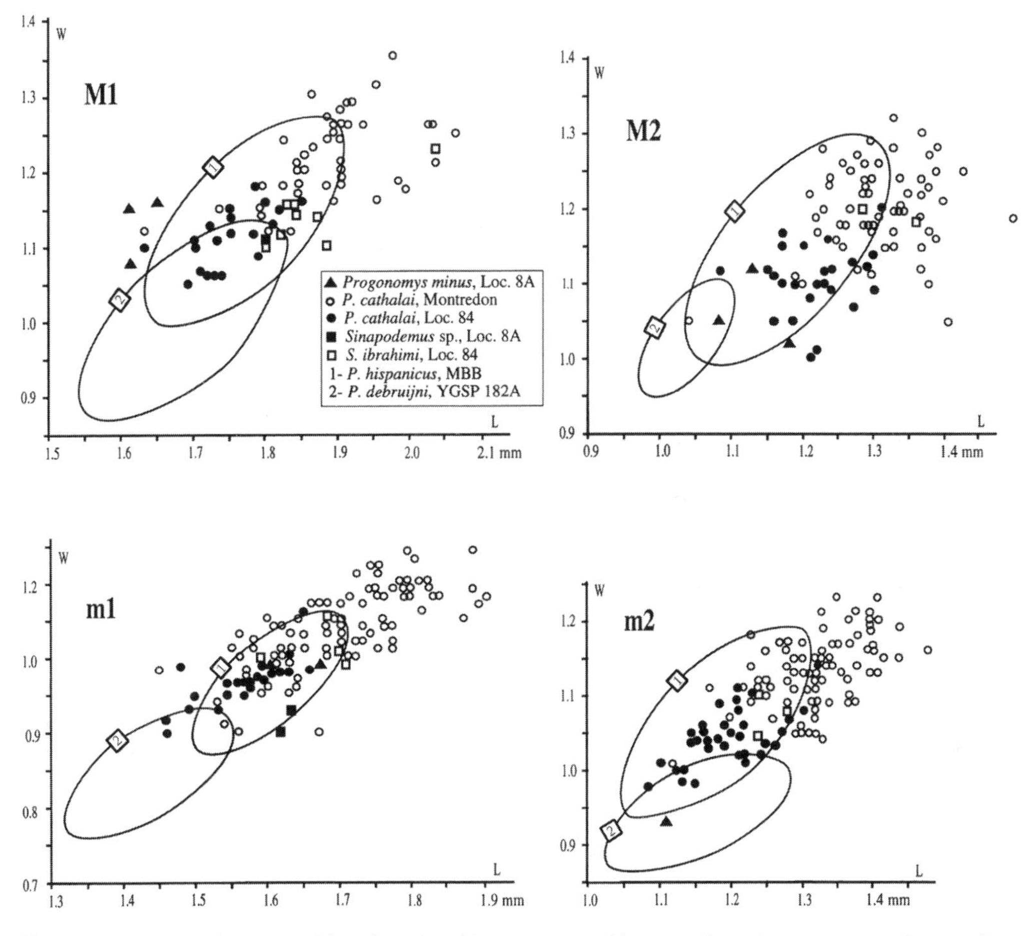

Figure 5.2. Scatter diagrams of length and width in upper and lower molars of *Progonomys* and *Sinapodemus* species. The data for *P. cathalai* from Montredon are after Aguilar (1982); for *P. hispanicus* from Masia del Barbo 2B (MBB), after van Dam (1997); and for *P. debruijni* from the Siwaliks of Pakistan, after Jacobs (1978).

The m3 has a simple pattern. The anterolabial cusp is lost but there is still a remnant of the protosinusid. The posterior lophid is reduced to a single cusp with an oval outline. There are two roots.

Comparison. The material from Loc. 8A contains two species of Muridae as evidenced by the morphology of M1 and m1. The second and third molars (six specimens in all) are not sufficiently differentiated to be securely attributed to one or to the other species. Therefore, the comparison is based on the first molars.

I compared these specimens with Vallesian murids housed in the collections of the Université de Montpellier, notably with those from Montredon (the type locality of the genotype *Progonomys cathalai*), Castelnou 1b, Lo Fournas 6 and 7, and Can Llobateres 1.

Progonomys cathalai Schaub, 1938, from Montredon differs from the Loc. 8A specimens by its larger size (fig. 5.2). The lingual cusps of the M1 of *P. cathalai* are higher, the ridges connecting the lingual and central cusps are higher and stronger, the labial cusps (t3, t6, and t9) are more inclined posteriorly, the anterior valley between the t2 and t3 is deeper, and the t6 and t9 are situated closer or even connected by a ridge on 23% of specimens (Aguilar 1982). Most M1 specimens from Montredon have a t9 situated anteriorly to the t8. On m1 from Montredon, the anteroconid is generally connected to the metaconid, the cingular margin is stronger, the terminal heel is not protruding backwards, and it forms an ellipsoid cusp.

Progonomys castilloae Aguilar and Michaux, 1996, from Lo Fournas 7 (MN10, Pyrénées Orientales, France) is similar in size to *P. cathalai*. According to Aguilar and Michaux (1996, p. 36), this species differs from *P. cathalai* in "having no or a reduced anterocentral cusp on m1, a more reduced cingular margin, and a less developed terminal heel." Considering all these differences as intraspecific variations, Freudenthal and Martin Suarez (1999) proposed to synonymize *P. castilloae* with *P. cathalai*. J. P. Aguilar (pers. comm., 2000) does not agree with this statement and mentions that he is preparing a note to analyze the systematics of Vallesian murids from southern France. Whatever the status of this species, the Lo Fournas 7 material belongs to a species larger in size and having many more derived features than does *P. minus*.

Progonomys minus is close in size to *P. debruijni* from Pakistan, and they share many plesiomorphic features: the molars are low crowned, the lingual cusps on M1 and M2 are lower than the others and have weak connections with

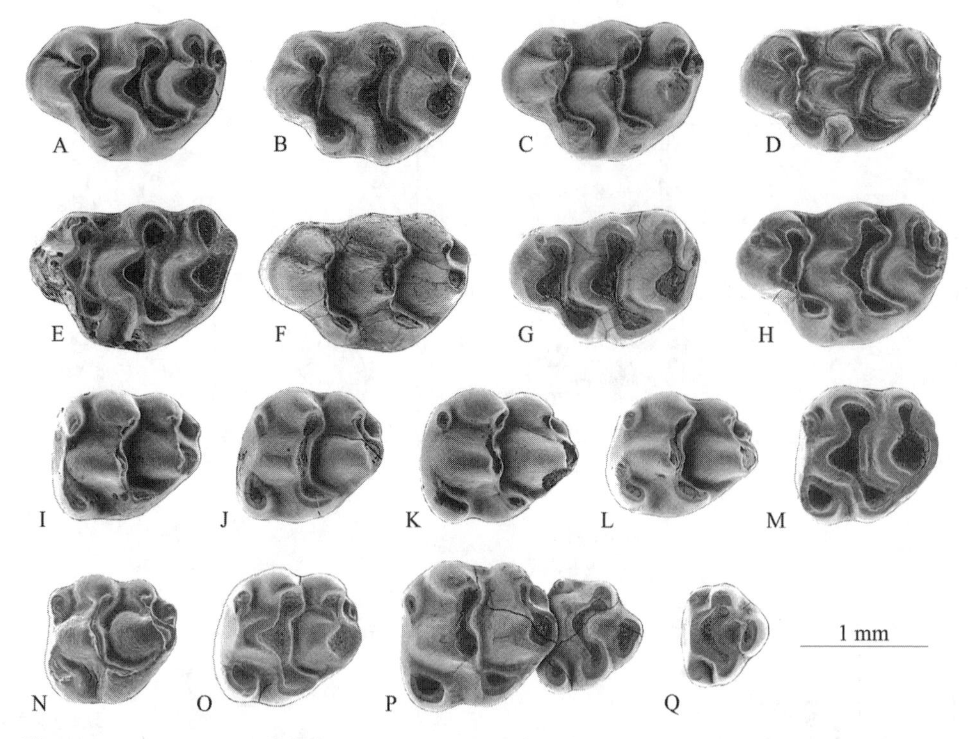

Figure 5.3. *Progonomys cathalai* from Loc. 84, upper molars: (**A–H**) M1; (**I–O**) M2; (**P**) M2–M3; (**Q**) M3.

the central cusps, the t12 is strong, and the labial cingulum is weak on m1 and m2. However, *P. debruijni* differs from *P. minus* in having M1 with an elongate t1 that is situated very far back relative to the t2, a t3 next to the t2, a t9 next to the t8, and m1 without anterocentral cusp and with an X-shaped connection between the anteroconid and the second cusp row. In this respect, *P. debruijni* is more derived than *P. minus* and *P. cathalai*. The tooth pattern of *P. debruijni* resembles that of *P. hispanicus* from some Spanish localities, such as Masia de la Roma 4B and 4C and Masia del Barbo 2B (type locality; Van Dam 1997). *P. hispanicus* is on average smaller than *P. cathalai*, and in particular its upper molars are narrower. I did not see the original materials referred to *P. debruijni* and *P. hispanicus*, but the comparison of measurements, descriptions, and illustrations given by Jacobs (1978) (*P. debruijni*) and by Michaux (1971) and Van Dam (1997) (*P. hispanicus*) show that these species are very similar. In particular, the oldest representatives of *P. hispanicus* from Masia de la Roma seem identical in size and morphology to *P. debruijni* from the locality YGSP I82A from Pakistan. In *P. hispanicus*, the molars, in particular M1 and M2, are larger on average than those of *P. minus*.

The species *Progonomys woelferi* and *P. clauzoni* are larger in size, and their upper molars are wider and have strongly connected globular cusps that form rather transverse chevrons. Their lower molars possess a thicker labial cingulum with several cusps.

Progonomys cathalai Schaub, 1938

Material from Loc. 84. Thirty M1, 26 M2, eight M3, 24 m1, 36 m2, and nine m3 (ST84-109–ST84-235).

Measurements. See table 5.1 and figures 5.3, 5.4.

Description. The M1 has a rather rounded outline, and the mean width:length ratio is 0.63. The t1 is situated clearly back with respect to the t2, whereas the t3 is at the level of the t2 (five specimens) or slightly back from it (20 specimens). The t2 and t3 are well separated by a narrow anterior valley. The indentation between the t1 and t2 (the anterolingual indentation) is smooth. Two specimens out of 24 have a weak posterior spur of the t3. With respect to the t5, the t4 is situated more to the back than is the t6. The t4–t5 connection is weaker than the t5–t6 connection. In eight specimens, the t6 has a weak posterior spur directed toward the t8–t9 junction. The tip of the t9 is situated slightly anteriorly to that of the t8 or is almost at the same level. The ridge between t4 and t8 is absent on seven specimens, weak on 13 others, and prominent on five specimens. The t12 is strong but never cusp-shaped. All central cusps are inclined posteriorly; the lingual and labial cusps are inclined posteriorly and centrally. All M1s have three roots.

The M2 is slightly longer than it is wide. The t1 is large and connected to the anterior face of the t5. The t3 is small and isolated or connected to the base of the t5 by a short ridge. The t4 and t6 are situated back from the t5. The ridge between the t4 and t8 is absent or very weak (two specimens), weak (14) or high (four specimens). The t9 is situated next to the t8 or its tip may be slightly anterior to that of the t8. Only four specimens of 21 have a trace posterior spur of the t6. The posterior cingulum is present in all specimens. The inclination of cusps is as on M1. There are three roots on 15 specimens and four roots on three others.

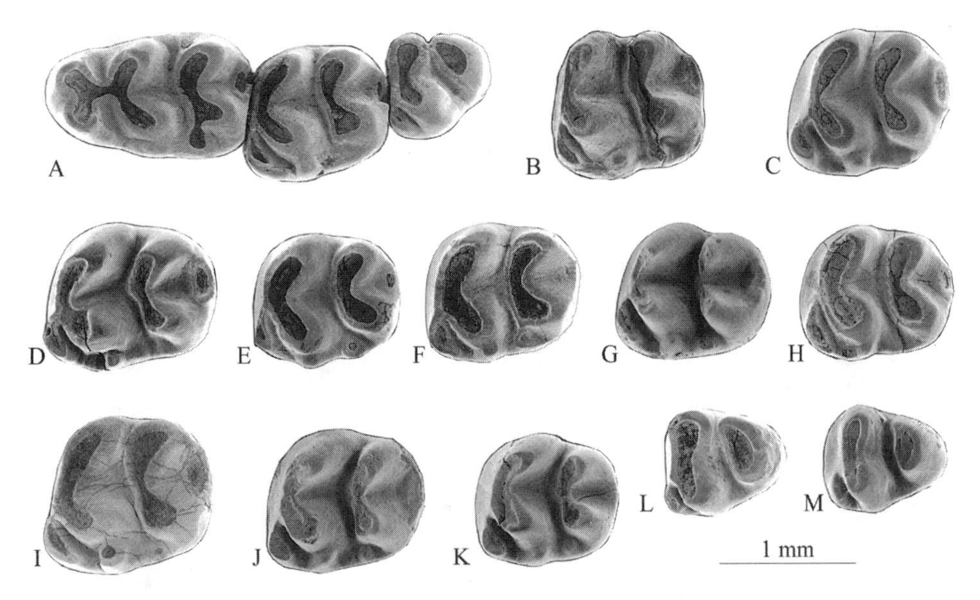

Figure 5.4. *Progonomys cathalai* from Loc. 84, lower molars. **(A)** m1–m3; **(B–K)** m2; **(L, M)** m3.

The M3 has a triangular outline. The t1 is connected to the base of the t5. The t3 is missing on two specimens and small on six others. The posterior cusp is ellipsoidal in shape and isolated or connected to the t4 by a low ridge. There are three roots, but the anterior ones are very close to one another.

The m1 has a small anterocentral cusp on three specimens, a labial cingulum with an isolated c1 (and in some cases with one or two small tubercles), and a low and ellipsoidal small terminal heel, which is situated lingually. The lophs are slightly oblique. There is a low connection between the lingual anteroconid and metaconid. The m1 has two roots, and has a small central radicle in one-third of the specimens.

The m2 has a rather small anterolabial cusp on 22 specimens and large one on 15 others. The anterior face is almost at right angles or slightly oblique to the longitudinal axis on 25 specimens; in 15 specimens the obliquity of the anterior face is much stronger. The degree of this obliquity is correlated to the size of the anterolabial cusp. The labial cingulum is generally weak and it usually bears one anterior cusp, but a second, smaller cusp exists in one-third of the specimens. The terminal heel is as on m1. There are two roots.

The m3 has a small labial anteroconid. The anterior chevron is almost straight, and the posterior chevron is reduced to a single cusp. There are two roots.

Comparison. Direct comparisons show that the specimens from Loc. 84 are generally smaller than *Progonomys cathalai* from Montredon, but there is a large overlap between the populations (table 5.1 and fig. 5.2). The morphology of the dentition is also somewhat different: at Montredon, the t3 is situated generally more to the back; the t6 often has a posterior spur (which is connected to the t9 in 23% of M1 and 8% of M2 specimens) (Aguilar 1982); and M1 and M2 have narrower mesosinuses. The lower molars do not present any consistent differences. The characters noted here are interpreted as indicative of a higher degree of evolution in the Montredon population, which is consistent with the ages as inferred from magnetostratigraphic studies for Loc. 84 (10.1 Ma, after Lunkka et al. 1999; Kappelman et al., this volume) and the late Vallesian age of Montredon (bracketed between 8.7 and 9.7 Ma, after Steininger 1999). However, Michaux et al. (1997) suggested a correlation of Montredon with a time interval between 10 and 11 Ma, using a chronologic scheme in which the middle–late Miocene boundary is put at 12 Ma.

P. woelferi and *P. clauzoni* are larger than *P. cathalai*. The t1 in the M1 is situated close to the t2, the upper molars are wide and in *P. clauzoni,* they have rather straight chevrons. The lower molar has a strong labial cingulum that bears several cusps in both species.

Material from Loc. 49 (Igbek). One M2 dex (1.20 × 1.05), ST49-1 (fig. 5.5I).

Description and Discussion. This tooth is fresh and well preserved. Its dimensions are within the range of *P. cathalai* from Loc. 84. Its morphology does not show any significant difference from the material of the latter locality, except that is has a smaller posterior cingulum and a rather ellipsoidal shape of the t1 that is only present in a few specimens from Loc. 84. The t6 and t9 are well separated. The t9 is situated slightly anterior to the t8. There is a low ridge connecting the t4 to the t8. The roots are broken.

The occurrence of this species in Loc. 49 is interesting for dating this locality. Loc. 49 yielded a rich large-mammal fauna and very few small mammals. The students of different groups hesitated between correlating this assemblage to either the late Vallesian or early Turolian (Kappelman et al., this volume). The time range of *P. cathalai* over all Eurasia does not exceed the late Vallesian (Freudenthal and Martin Suarez 1999); thus, this single tooth favors the first hypothesis (i.e., the late Vallesian age of this locality).

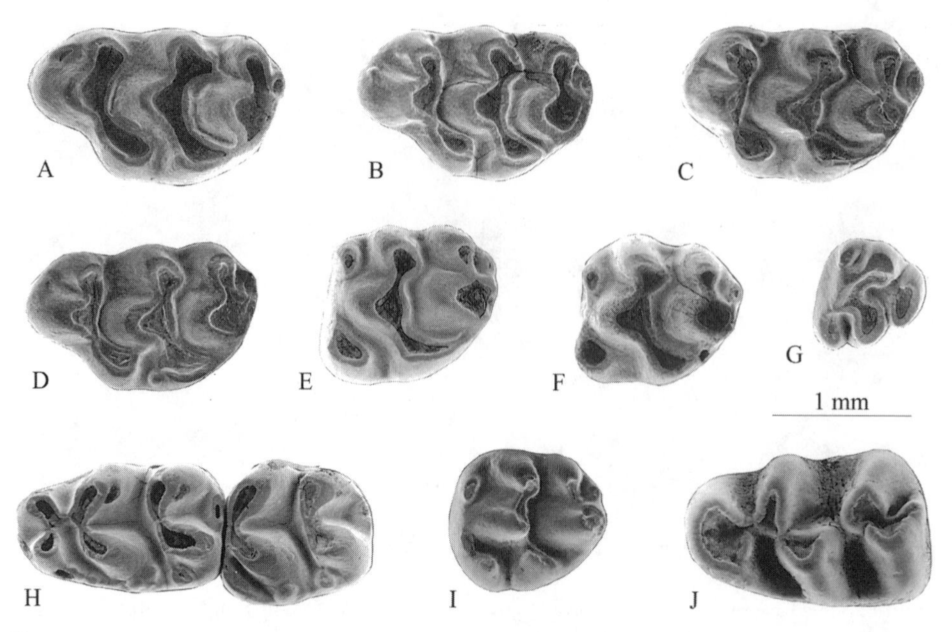

Figure 5.5. *Sinapodemus ibrahimi* gen. and sp. nov. from Loc. 84: **(A)** M1, Holotype; **(B–D)** M1; **(E, F)** M2; **(G)** M3; **(H)** m1–m2. *Progonomys cathalai* from Loc. 49 (Igbek): **(I)** M2; Myocricetodontinae indet. from Loc. 49 (Igbek): **(J)** m1.

Progonomys sp.

Material from Loc. 108. Fragment of M1 (— × 1.14) and m2 (1.14 × 0.99). Specimen numbers ST108-1 and 2 (fig. 5.6A,B).

Description and Comparison. This is the oldest murid record in the Sinap Formation. On the M1, the t6 and t9 are separated, the t4 is situated posterior to the t6, there is a ridge between the t4 and t8, the tip of the t9 is anterior to the t8, and the t12 is strong.

The m2 is small. Its labial cingulum is incomplete and low. The anterolabial cusp is crest shaped. The terminal heel is flattened anteroposteriorly and developed as a posterolophid.

The anterior position of the t9 relative to the t8, as observed on the M1, is considered a derived feature. This assumption is based on observations of primitive murids in western Europe: in the older representatives of *"Progonomys" hispanicus,* the t9 is lateral to the t8, and "the orientation of the t8–t9 pair is more or less at right angles to the longitudinal axis of the molar" (Van Dam 1997, p. 52). In the earlier *Antemus* from Pakistan, this configuration is also present. In *P. cathalai,* the position of the t9 is a little anteriorly to that of the t8; in the *Parapodemus-Apodemus* group, the t9 is clearly shifted toward the t6. The M1 from Loc. 108 has a t9 that is situated anteriorly to the t9, even somewhat more so than in the *Progonomys* species from the younger Sinap localities. This feature does not fit in the evolutionary trend in *Progonomys.*

On the m2, the crest-shaped labial anteroconid and the weak labial cingulum without cusp are primitive characters shared with *Antemus* and *Progonomys debruijni.* The other features of the Loc. 108 specimens are as in *Progonomys.*

Sinapodemus gen. nov.

Type species. *Sinapodemus ibrahimi* sp. nov.

Etymology. From the word "Sinap," which is the name of the hill containing the localities for this genus, and from "*Apodemus,*" in reference to the resemblence of some dental features with those of the *Parapodemus-Apodemus* group.

Diagnosis. The outline of M1 is elongated as in *Parapodemus,* with a well-marked anterolingual indentation. The t4 is elongated and connected to the t8 by a high ridge. There is no connection between the t6 and t9, but a strong spur issues from the t6. The anterocentral cusp of m1 is absent and the connection between the lingual anteroconid and metaconid is strong.

Differential Diagnosis. *Sinapodemus* differs from *Progonomys* in having an elongate M1, the t3 situated next to the t2, the t6 having a posterior spur, and having a high ridge between t4 and t8. The m1 of *Sinapodemus* differs from that of *Progonomys* by its elongate outline and the strong connection between the lingual anteroconid and metaconid. *Sinapodemus* differs from *Parapodemus* by the lack of a connection between t6 and t9 in the upper molars. In *Huerzelerimys,* the upper molars are wider, there is a connection between t6 and t9 in the upper molars, and the ridge between t4 and t8 is absent or very low.

Sinapodemus ibrahimi sp. nov.

Holotype. M1 dex, ST84-236 (fig. 5.5A).

Hypodigme. Ten M1, two M2, two M3, four m1, and three m2 (ST84-237–ST84-257). See figure 5.5A–H.

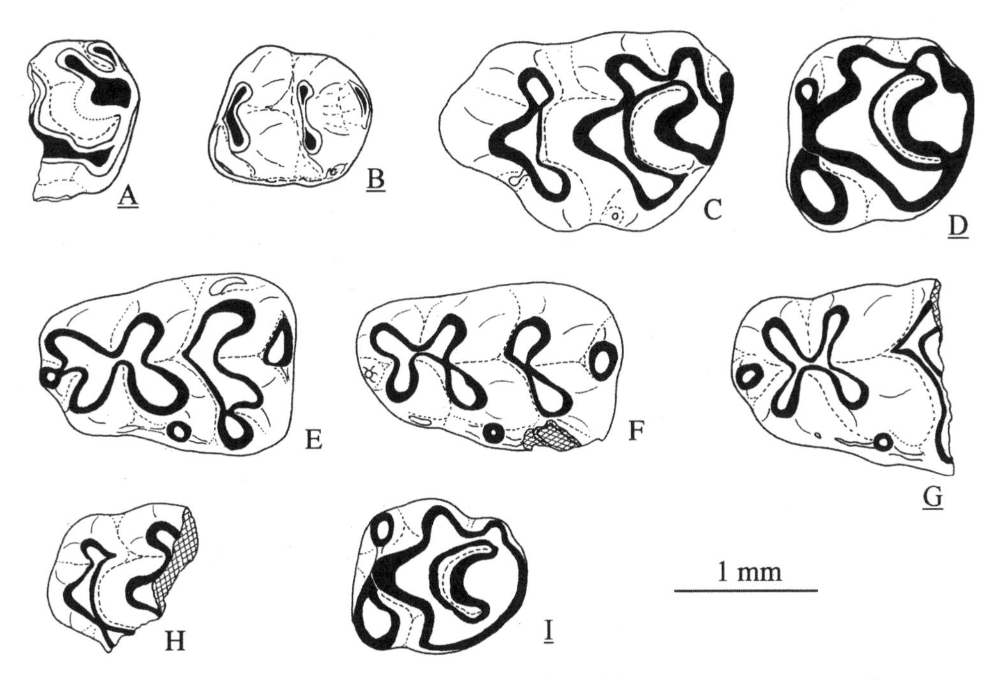

Figure 5.6. *Progonomys* sp. from Loc. 108: (A) Fragment of M1; (B) m2. "*Karnimata*" *provocator* from Loc. 42 (Çobanpinar): (C) M1; (D) M2; (E–F) m1. Muridae g. and sp. indet. from Loc. 42 (Çobanpinar): (G) fragment of m1. *Hansdebruijnia* cf. *neutrum* from Loc. 42 (Çobanpinar): (H) fragment of M1; (I) M2.

Type Locality. Loc. 84, which is situated on the western slopes of Sinap Tepe, 1.5 km north of Yassiören village, Kazan, Ankara.

Etymology. Named in honor of Ibrahim Abaci from the Yassiören village, for his taking interest in our collecting in the Sinap Tepe area.

Measurements. See table 5.1.

Diagnosis. As for the genus.

Description. The M1 outline is amygdaloid. The anterolingual indentation is well marked between the t1 and t2. The t1 is situated well back with respect to t2, whereas the t3 is situated at the same level as the t2 (ten specimens) or slightly in back of it (two specimens). The t3 has a trace of a posterior spur on two-thirds of the M1 specimens. The t4 is placed posteriorly to the t5 and t6, its shape is elongate posteriorly and it is connected to the t8 by a high ridge. The t6 has a well-developed posterior spur directed toward the t8–t9 junction. The t9 is situated next to the t8. The t12 is small and ridge shaped. There are three roots and one central rootlet.

The M2 of *Sinapodemus* is longer than that of *Progonomys.* The t1 is larger than the t3, and the t4 is posteriorly elongate (almost round or ellipsoid in *Progonomys*), situated closer to the t8, and more firmly connected to this cusp. Other characters are as for M1. There are three roots.

The M3 has the t1 connected to the anterior face of the t5. The t3 is quite strong. The t4, t5 and t6 are firmly connected. There are three roots.

The m1 is more elongate than that of *Progonomys* from the same locality, and the posterior angles of the chevrons are narrower. The anterocentral cusp is absent, but in its place there is a low cingulum. The connection between the anteroconid and metaconid is quite central. The labial and lingual anteroconids are situated almost at the same level, forming a transverse crest. The labial cingulum is strong and possesses three accessory cusps. The terminal heel is oval and situated almost centrally on the longitudinal axis. There are two roots.

In the three m2 specimens referred to this species, the labial anteroconid is larger and the chevrons have a narrower angle than in *P. cathalai* from this locality. This molar has two roots.

Comparison. Compared to the oldest species of *Progonomys* (*cathalai, debruijni, hispanicus* and *minus* sp. nov.), *Sinapodemus* from Loc. 84 has some derived features: elongate outline of M1 (width:length ratio is 0.61 versus 0.63 in *P. cathalai* from Loc. 84, 0.64 in *P. cathalai* from Montredon and *P. hispanicus* from Masia del Barbo 2B, and 0.67 in *P. minus* sp. nov. from Loc. 8A), the well-marked anterolingual indentation, the occurrence of a posterior spur on the t6, the elongate shape of the t4 and the high ridge between the t4 and t8 on M1 and M2, elongate shape, narrow chevrons, symmetrically placed anteroconids and the developed labial cingulum on m1. These characters prevent the attribution of this material to the genus *Progonomys.* This murid also differs from *Parapodemus* by the absence of a connection between the t6 and t9, the weak posterior spur of the t3 on M1 and M2, and the lack of the anterocentral cusp on the m1. These differences and the older age of this species in comparison with the first Eurasian murids distinguishes this Loc. 84 murid as a new genus and new species. It should be noted that its differences from *Progonomys* are in fact the occurrence of some new characters pointing to a *Parapodemus* dental pattern. However, I am aware that, together with its derived features, this form shares many primitive characters of *Progonomys,* and consequently raises the question of whether it is wise to define a new genus for a form that is in many respects morphologically intermediate between *Progonomys* and *Parapodemus.* However, such a taxon cannot be securely attributed to the one of these genera.

Sinapodemus sp.

Material from Loc. 8A. One M1 and three m1 (ST8A-67–ST8A-70). See figure 5.1L–N.

Measurements. See table 5.1.

Description and Comparison. The unique M1 is more elongate than the M1 of the other species from Loc. 8A. There is a well-marked anterolingual indentation as in *Sinapodemus ibrahimi.* The t1 is situated far back relative to the t2–t3. The t4 is connected to the t8 by a high ridge. There is a strong posterior spur on the t6. The t12 is developed. There are three roots.

The m1 is relatively longer and narrower than that of *Progonomys minus.* There is a low crest in the position of the anterocentral cusp. The lingual anteroconid is connected to the metaconid. The angle of the second and third chevrons is narrow. There are two roots.

All these features characterize *Sinapodemus ibrahimi* sp. nov. from Loc. 84. However, the present specimens are somewhat smaller and the lingual cusps are much lower in the unique M1 than is true for *S. ibrahimi.*

Genus *Karnimata Jacobs, 1978*

Because Mein et al. (1993) have shown that the type species of this genus, *Karnimata darwini,* is a junior synonym of *Progonomys woelferi,* there is no generic name for other species included in the genus *"Karnimata"* from the late Miocene of Pakistan and Afghanistan (species *huxleyi, minima,* and *intermedia*) and from the Aegean area (*"Occitanomys" provocator*).

"Karnimata" provocator (de Bruijn, 1976)

Material from Loc. 42 (Çobanpinar). One M1 (2.02 × 1.38), one M2 (1.38 × 1.33), three m1 (1.81 × 1.25, 1.84 × 1.16, and a fragment) (Ç42-1–Ç42-5). See figure 5.6C–F.

Description and Comparison. The t1 and t3 of the M1 are situated symmetrically and lack the posterior spur. The t1bis is present but weak. The t6 and t9 are weakly connected. The ridge between t4 and t8 is low and tenuous. The t12 is quite strong. The M1 has three roots, of which the lingual one has two cavities.

The unique M2 is worn. The t1 is large and oval in outline, whereas the t3 is small. The t4 is situated quite close to the t8. The connection between t6 and t9 is weak. The t12 is present, although the tooth is severely worn. The roots are broken.

The anterocentral cusp of m1 is isolated in one specimen but connected to the lingual anteroconid in two others. The anteroconid is lingually connected to the metaconid. The main cusps are slightly alternating. The longitudinal crest is complete in one m1, but is like a spur in two others. The labial cingulum is continuous and bears the c1 and two other tubercles. The m1 has two roots.

The general pattern of these specimens is similar to that of *Occitanomys.* This genus is well known in western Europe, but rather rare in eastern Europe and Turkey. Martin Suarez and Mein (1991), Mein et al. (1993), Martin Suarez and Freudenthal (1993), and Freudenthal and Martin Suarez (1999) discussed the systematics of this genus and the phylogenetic relationships of species previously referred to it or to some other genera. They recognize three subgenera: *O.* (*Occitanomys*) with the species *brailloni, alcalai, clauzoni,* and *faillati, O.* (*Rhodomys*) with *hispanicus, sondaari, adroveri,* and *debruijni,* and *O.* (*Hansdebruijnia*) with the unique species *neutrum.* The last subgenus is recognized by de Bruijn et al. (1999) as an independent genus. The status of ?*Occitanomys provocator* de Bruijn 1976 from Chomateri (Greece) remains open. Michaux et al. (1996) and Van Dam (1997) do not agree with the attribution of *Progonomys hispanicus* to *Occitanomys* and prefer to maintain it as a congeneric species of *P. cathalai.* I agree with these authors that the differences between *P. hispanicus* and the typical species of *Occitanomys* (*O. brailloni* Michaux, 1969) are clear and speak against its inclusion in the genus *Occitanomys,* even though, as many students have demonstrated, *O. sondaari* is probably derived from *P. hispanicus.* Consequently, the latter species may well be the ancestor of all *Occitanomys,* unless this genus is polyphyletic.

The western European Vallesian and Turolian species *Occitanomys sondaari, O. alcalai, O. clauzoni,* and *O. faillati* are smaller than *"Karnimata" provocator.* The Loc. 42 specimens are similar in size to the common western European species *O. adroveri.* However, the M1 and M2 of this species have strong connections between t1 and t5, t3 and t5, and t6 and t9, the M1 has a deep anterolingual incision, and the t1bis is present in most M2.

Martin Suarez and Freudenthal (1993) and Freudenthal and Martin Suarez (1999) included *Occitanomys adroveri* into the subgenus *Rhodomys,* the type species of which is *O.* (*Rhodomys*) *debruijni* from the early Ruscinian locality of Maritsa on the Island of Rhodes, Greece. This species differs from the Loc. 42 specimens by its smaller size (mean values of M1 are 1.83 × 1.27), the t1 being very far back, t1 and t3 with strong posterior spurs (which are often connected to the t5) on M1, and by having a developed longitudinal crest and a weaker labial cingulum on lower molars (de Bruijn et al. 1970).

Two other species from Greece were included in the genus *Occitanomys: O.* (*Hansdebruijnia*) *neutrum* and ?*O.*

provocator, both initially described from Chomateri (de Bruijn 1976). The first species was later described from Maramena (northern Greece; Storch and Dahlman 1995), whereas the second has become known from Samos S3 (Black et al. 1980) and Düzyayla in central Anatolia (de Bruijn et al. 1999). *H. neutrum* differs from the Loc. 42 specimens by its smaller size, more slender upper and lower molars, the presence of a strong posterior spur of t3 on M1, and a more voluminous and anteriorly protruding anterocentral cusp on m1. The scarce material referred to ?*O. provocator* from Chomateri, Samos S3, and Düzyayla is similar in size and morphology to the Loc. 42 specimens. The few differences I noted are that the specimens from these three localities have upper molars with a weaker t6–t9 connection, and the m1 is narrower and has an incomplete or very weak longitudinal crest. These differences are interpreted here as indicative of a more progressive stage of evolution in the Loc. 42 specimens, and consequently of the younger age of the Çobanpinar locality.

Genus *Hansdebruijnia* Storch and Dahlman, 1995
Hansdebruijnia cf. *neutrum* (de Bruijn, 1976)

Material from Loc. 42. One anterior fragment of M1 and one M2 (1.21 × 1.09) (Ç42-7–Ç42-8). See figure 5.6H,I.

Description and Comparison. A fragment of M1 and a worn M2 indicate the presence of a small murid at Çobanpinar. On the M1, the t1 is situated in back of the t3 and pinched laterally. The t3 has a posterior spur. On the M2, the t1 is large and rather rounded in outline. The t3 is small but not especially reduced. The t6–t9 and t4–t8 connections are quite strong. The roots are broken.

These molars are quite similar in size to that of small species of *Parapodemus* and *Occitanomys,* and more particularly to that of *Hansdebruijnia neutrum* from Chomateri (Greece) and Düzyayla (Turkey). The two teeth from Çobanpinar show no differences from this species, but the material is too poor for a certain attribution.

Muridae gen. and sp. indet.

Material from Loc. 42. One fragmentary right m1 (Ç42-6). See figure 5.6G.

Description and Comparison. This tooth belongs to a murid larger than *"Karnimata" provocator.* The anterocentral cusp is large and isolated. The connection between the anteroconid and the second chevron is centrally placed and strong. The longitudinal crest is developed as a spur in front of the entoconid. The labial cingulum has a high cusp next to the protoconid. This tooth is wider posteriorly than the m1 of *"Karnimata" provocator.* In size and some other characters it resembles some large species of *Paraethomys,* but the presence of a strong anterocentral cusp prevents its attribution to this genus. *Paraethomys* is known in the Aegean area from Chalkoutsi (?*Paraethomys* sp.) and from Maritsa (*P. anomalus*). The first locality is referred to the Turolian (de Bruijn 1976) and the second to the early Ruscinian (de Bruijn et al. 1970). The m1 from Çobanpinar is similar in size to *P. anomalus,* but this species lacks the anterocentral cusp. The Chalkoutsi material is poor, smaller in size, and has a residual anterocentral cusp.

Family Gerbillidae Gray, 1825; Subfamily Myocricetodontinae Lavocat, 1961

Genus *Myocricetodon* Lavocat, 1952
Myocricetodon eskihisarensis Wessels, Ünay and Tobien, 1987 (fig. 57)

Material from Sinap Localities. From the oldest to the youngest: Loc. 65: two m2 (ST65-10,11); Loc. 64: six M1 and two molar fragments (ST64-1–ST64-8); Loc. 4: one each M1, M2, m1, and three m2 (ST4-8–ST4-13); Loc. 8A: two M1, four m1, and five m2 (ST8A-71–ST8A-81). See figure 5.7.

Measurements. See table 5.2.

Description. Because the specimens from all the Sinap Tepe localities mentioned do not show consistent differences, they will be described together.

The M1 is elongated. The width:length ratio varies between 0.51 and 0.63. The lingual face is more or less convex, with an indentation between the anterocone and protocone in some specimens. The labial face is concave. The anterocone is generally simple and triangular; one M1 from Loc. 64 and another from Loc. 4 have the anterocone indistinctly divided by a wide and shallow anterior valley. Two specimens out of eight have an anterior accessory cusp. Other accessory cusps, more or less developed, may occur near the edges of the protosinus and anterosinus. The anterocone-protocone connection is always present but low. However, the protolophule and the "new longitudinal crest" (see Tong 1988; Wessels 1996) between the paracone and hypocone are strong. A short metalophule connects the metacone to the posterior arm of the hypocone. All M1 specimens, except one specimen from Loc. 64, have a strong lingual cusp (enterocone), which is connected to the posterior arm of the protocone. The mesoloph is short or absent. The posterosinus is lost. The M1 has three roots and a small central radicle.

The M2 is known only from Loc. 4. Its outline is trapezoidal. The labial anteroloph is strong whereas the lingual one is very weak. The anteroloph-protocone connection is formed of a short, low ridge. The paracone is triangular; its posterior spur peters out in the mesosinus. The enterocone is voluminous; a ridge issued from its top is directed toward the paracone. The metacone and hypocone are fused, and their anterior and posterior arms encounter an enamel island between these cusps. There are three roots.

The m1 is represented by four specimens from Loc. 8A and one half-tooth from Loc. 4. The anteroconid is central, moderately elongated, and anteriorly rounded. It has no lingual anterolophid, but its labial anterolophid is a thick,

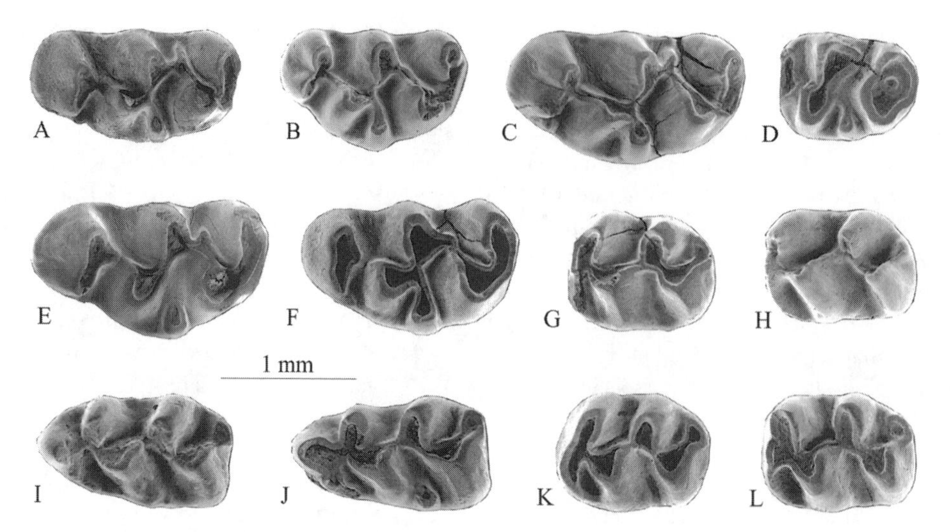

Figure 5.7. *Myocricetodon eskihisarensis* from four localities of the Sinap Formation. **(A–C)** M1; **(D)** M2; **(E, F)** M1; **(G, H)** m2; **(I, J)** m1; **(K, L)** m2. Specimens **A, B,** and **E** are from Loc. 64; **C, D,** and **L** from Loc. 4; **F, G, I, J,** and **K** are from Loc. 8A; **H** from Loc. 65.

long crest, which in one specimen from Loc. 8A bears a large cusp. The longitudinal crest is continuous from the anteroconid to the hypoconulid. The main cusps clearly alternate. There is a large accessory cusp or a thick ridge in the sinusid next to the posterolabial face of the protoconid. This molar has two roots.

The m2 is represented by ten specimens from three localities. The labial anterolophid is strong, whereas the lingual one is lost. The other characters are similar to those of m1, except for the weaker accessory cusp in the sinusid and the lingually closed posterosinusid. The posterior part of this tooth is narrower than the anterior part. There are two roots.

Comparison. Jaeger (1977a,b) distinguished three groups (A, B, and C) among the north African middle- and late Miocene Myocricetodontinae. Wessels (1996) retained this classification with the addition of new discoveries from Pakistan and Turkey. This classification is based on the degree of division of the anteroconid of m1, the development of the longitudinal crest, and the presence or absence of a lingual accessory cusp on M1 and M2. The stage of evolution of these characters defines the systematics of taxa and enlightens their phylogenetic relationships. It is interesting that the species referred to the genus *Myocricetodon* by Jaeger (1977a,b) as well as by Wessels (1996) are included in Group A or Group B.

At present, twelve species are referred to *Myocricetodon;* eight of them come from the middle–late Miocene of north Africa (Jaeger 1977a,b). In Group A (*M. cherifiensis, M. irhoudi, M. ouedi,* and *M. ternanensis*) the cricetid-type of longitudinal crest is retained and the cusps are arranged in transverse rows. In Group B (*M. parvus, M. parvus intermedius, M. seboui, M. ouachi, M. trerki, M. eskihisarensis, M. sivalensis,* and *M. chinjiensis*), the cricetid-type of longitudinal crest is replaced by a new longitudinal crest, accessory cusps are developed in the upper and eventually in the lower molars, and the cusps alternate, especially on the first and second molars.

All *Myocricetodon* remains from the Sinap Formation, as well as those from other Turkish localities, fit in Group B. Therefore, the comparison will be limited to species in this group. Species referred to this group are known from north Africa, Turkey, and Pakistan (Jaeger 1977a,b; Wessels et al. 1987; Lindsay 1988; Wessels 1996).

In table 5.2, the dimensions of first and second upper and lower molars of *Myocricetodon* from Turkish, north African, and Pakistani localities are given. A close examination of the values shows that:

The two north African species *M. trerki* and *M. ouachi* (known by a single M1) are larger than the other species. All other species are quite similar with respect to the size of their first and second molars.

The specimens from the Sinap Formation are in the range of variation of those of the other Turkish localities (Yeni Eskihisar, Kalamis, and Sofça) (Wessels et al. 1987; Rümmel 1998).

The species *M. parvus* (Beni Mellal), *M. parvus intermedius* (Pataniak 6), and *M. seboui* (Oued Zra and Jebel Semmene) have dimensions similar to those of the Sinap Formation material, except for the lengths of M2 and m2, which are shorter. Indeed, these north African species have first molars enlarged, whereas the second and third molars are relatively reduced.

I compared the Sinap material with the specimens from Beni Mellal and Pataniak 6. *Myocricetodon cherifiensis* differs from the Sinap material by the lack of longitudinal crests, the accessory cusp on M1 and M2, and the undivided anterocone of M1. In *M. parvus* from Beni Mellal and Pataniak 6, the M1 anterocone is also simple, the lingual accessory cusp of M1 and M2 is occasionally present (19% of samples;

Table 5.2. Measurements of Upper and Lower Molars of Species Referred to the Genus *Myocricetodon* from Various Localities in Turkey, Pakistan, and North Africa

				Length			Width	
Taxon	Age	Locality	Molar	Range	Mean	*n*[1]	Range	Mean
Myocricetodon eskihisarensis	MN9	Loc. 8A	M1		1.67	1		1.01
M. eskihisarensis	MN9	Loc. 4	M1		1.78	1		1.00
M. eskihisarensis	MN8	Loc. 64	M1	1.50–1.80	1.66	6	0.86–1.05	0.96
M. eskihisarensis	MN8	Yeni Eskihisar	M1	1.41–1.70	1.58	38	0.79–1.02	0.93
M. cf. *eskihisarensis*	MN8/9	Kalamis	M1	1.51–1.71	1.62	4	0.88–0.95	0.92
M. parvus	MN6	Beni Mellal	M1	1.41–1.68	1.53	45/48	0.81–0.99	0.90
M. p. intermedius	MN7/8	Pataniak 6	M1	1.39–1.63	1.51	50	0.77–0.96	0.86
M. seboui	MN10	Oued Zra	M1	1.49–1.63	1.55	6/7	0.78–0.97	0.88
M. cf. *seboui*	MN9	Jebel Semmene	M1	1.38–1.52	1.47	5	0.86–0.99	0.90
M. ouachi	MN11	Khendek el Ouaich	M1		1.84			1.05
M. trerki	MN10	Oued Zra	M1			0/2	1.36–1.42	1.39
M. sivalensis	~16.2 Ma	YGSP 592	M1	1.58–1.72	1.66	4/8	0.93–1.15	0.99
M. chinjiensis	~11.5 Ma	YGSP 76	M1	1.70–1.88	1.79	2/3	0.98–1.08	1.02
M. eskihisarensis	MN9	Loc. 4	M2		1.12	1		0.79
M. eskihisarensis	MN8	Yeni Eskihisar	M2	1.03–1.19	1.11	15	0.82–0.96	0.88
M. cf. *eskihisarensis*	MN8/9	Kalamis	M2	1.10–1.15	1.13	2	0.89–0.92	0.91
M. parvus	MN6	Beni Mellal	M2	0.97–1.11	1.02	17	0.81–0.95	0.88
M. p. intermedius	MN7/8	Pataniak 6	M2	0.86–1.02	0.94	28	0.74–0.92	0.81
M. seboui	MN10	Oued Zra	M2	0.93–1.04	1.00	13	0.80–0.89	0.84
M. cf. *seboui*	MN9	Jebel Semmene	M2	0.90–1.02	0.97	4	0.76–0.88	0.82
M. trerki	MN10	Oued Zra	M2	1.25–1.35	1.32	3	1.17–1.30	1.22
M. sivalensis	~16.2 Ma	YGSP 592	M2	1.18–1.25	1.21	6	0.96–1.12	1.08
M. chinjiensis	~11.5 Ma	YGSP 76	M2	1.05–1.16	1.11	3	0.93–0.93	0.92
M. eskihisarensis	MN9	Loc. 8A	m1	1.46–1.52	1.49	4	0.88–0.89	0.88
M. eskihisarensis	MN8	Yeni Eskihisar	m1	1.21–1.49	1.37	13	0.74–0.86	0.79
M. cf. *eskihisarensis*	MN8/9	Kalamis	m1	1.39–1.53	1.48	6	0.77–0.92	0.84
M. parvus	MN6	Beni Mellal	m1	1.13–1.42	1.29	62/59	0.74–0.89	0.81
M. p. intermedius	MN7/8	Pataniak 6	m1	1.11–1.47	1.26	86	0.69–0.91	0.80
M. seboui	MN10	Oued Zra	m1	1.30–1.39	1.35	6	0.76–0.83	0.80
M. cf. *seboui*	MN9	Jebel Semmene	m1	1.27–1.36	1.31	7	0.69–0.83	0.76
M. trerki	MN10	Oued Zra	m1	1.76–1.86	1.81	3	1.04–1.17	1.14
M. sivalensis	~16.2 Ma	YGSP 592	m1	1.33–1.49	1.42	8	0.84–1.00	0.94
M. chinjiensis	~11.5 Ma	YGSP 76	m1	1.46–1.48	1.47	3	0.82–0.92	0.87
M. eskihisarensis	MN9	Loc. 8A	m2	1.13–1.19	1.16	3	0.89–0.91	0.90
M. eskihisarensis	MN9	Loc. 4	m2	1.15–1.17	1.16	2/3	0.86–0.93	0.89
M. eskihisarensis	MN8	Loc. 65	m2	1.13–1.26	1.20	2	0.88–0.93	0.91
M. eskihisarensis	MN8	Yeni Eskihisar	m2	1.07–1.21	1.12	19	0.78–0.92	0.85
M. cf. *eskihisarensis*	MN8/9	Kalamis	m2	0.96–1.10	1.08	3	0.75–0.83	0.79
M. parvus	MN6	Beni Mellal	m2	0.96–1.13	1.05	27	0.76–0.98	0.88
M. p. intermedius	MN7/8	Pataniak 6	m2	0.86–1.09	0.97	45	0.77–0.96	0.85
M. seboui	MN10	Oued Zra	m2	0.93–1.09	1.01	5	0.91–0.94	0.92
M. cf. *seboui*	MN9	Jebel Semmene	m2	0.92–1.04	0.99	9	0.78–0.89	0.83
M. trerki	MN10	Oued Zra	m2	1.20–1.36	1.28	2	1.20–1.21	1.21
M. sivalensis	~16.2 Ma	YGSP 592	m2	1.20–1.35	1.25	12/13	0.95–1.08	1.03
M. chinjiensis	~11.5 Ma	YGSP 76	m2	1.12–1.29	1.17	8/9	0.92–1.08	0.98

[1]*n*, Number of measurements.

Jaeger 1977a) and rarely connected to the protocone, and the anteroconid of m1 is shorter.

The specimens from Sinap Tepe compare better with *Myocricetodon seboui* from Oued Zra (Morocco; Jaeger 1977b). Except for the M2 and m2, which are proportionally shorter than in the Turkish specimens, the tooth pattern of this species is very similar to that of *M. eskihisarensis.* A detailed comparison of these species was given by Wessels et al. (1987, p. 78). The main differences noted by these authors are the high percentage of M1 from Oued Zra with a divided anterocone and the more frequent occurrence of the new longitudinal crest. However, note that *M. seboui* is known by only six M1. Because of the great similarities in morphology and size, the synonymy of these two species cannot be excluded.

Myocricetodon sivalensis from Pakistan is the most primitive species referred to *Myocricetodon.* Its M1 has a slightly divided anterocone, lacks the enterocone, and often has the cricetid type of longitudinal crest. The alternation of cusps is much less pronounced than in the younger representatives of Group B. The anteroconid is short as in *M. parvus* (Lindsay 1988; Wessels 1996). In fact, it seems better to group this species with Group A.

Myocricetodon chinjiensis is only known from the late middle Miocene of the Potwar plateau in Pakistan. Lindsay (1988) initially referred this species to a new genus, *Paradakkamys,* which Wessels (1996) correctly synonymized with *Myocricetodon.* This species shares the following characters with *M. eskihisarensis:* size, the shape of the anterocone of M1, the alternation of cusps, the forward-directed protolophule, and the weakly developed longitudinal crest (Wessels 1996). However, in *M. chinjiensis,* the cusps of upper molars are strongly inclined, the lingual accessory cusp of M1 is not at all or only weakly connected to the protocone and its size is somewhat larger.

This comparison provides support for the attribution of the Sinap *Myocricetodon* to *M. eskihisarensis,* although the possible synonymy of this species with *M. seboui* and/or *M. chinjiensis* cannot be excluded.

Family Gerbillidae Gray, 1825; Subfamily Gerbillinae Gray, 1825

Genus *Pseudomeriones* Schaub, 1934
Pseudomeriones cf. *rhodius* Sen, 1983

Material from Loc. 42. Two M1 (2.87 × 1.69, a damaged specimen), two M3 (0.82 × 1.10, 0.85 × 1.12), three m2 (1.59 × 1.60, 1.48 × 1.55, 1.52 × 1. 44) (Ç42-9–Ç42-15). See figure 5.8.

Description. The M1 is large and quite elongate. The labial sinuses are deep and slightly curved backward. The lingual sinuses are rather transverse, and the protosinus is very shallow. A small notch indicates that a tenuous posteroloph was present on fresh M1. There are three roots.

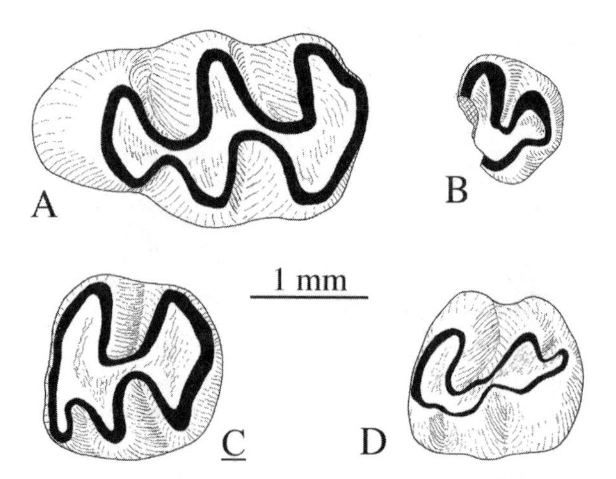

Figure 5.8. *Pseudomeriones* cf. *rhodius* from Loc. 42. **(A)** M1; **(B)** M3; **(C–D)** m2.

The M3 is very reduced. It has a shallow lingual sinus and a deep labial sinus, which may transform to an enamel island when the tooth is worn. Its two roots are fused well below the crown.

The occlusal outline of m2 is almost square. The mesosinusid is proverse in one fresh m2, whereas it is rather transverse in two others, which are worn. The anterosinusid is short, and the labial anterolophid is crest shaped. There are two roots but the anterior one has two cavities in one specimen.

Comparison. These molars have lower crowns and are proportionally wider than in *Pseudomeriones tchaltaensis* from the MN15 locality of Çalta (Sen 1977). The Loc. 42 specimens resemble in size *P. pytagorasi* from Samos S3, *P. rhodius* from Maritsa, and *P. abbreviatus* from Pul-e Charkhi (Black et al. 1980; Sen 1983). However, in *P. pytagorasi* and *P. abbreviatus,* M1 is proportionally wider and its protosinus is deeper. Moreover, the molars in the Samos S3 form are lower crowned, a character that indicates an older stage of evolution in this group. From Maritsa only one M1 of *P. rhodius* was recorded; it is longer and narrower than the Loc. 42 M1. However, in the material from Monasteri (northern Greece, late Turolian), also referred to *P. rhodius* (de Bruijn 1989), the M1 is not so long as the one from Maritsa. Our material is similar in size and morphology to the specimens from Monasteri. Consequently, the material from Loc. 42 is referred to *P. rhodius* with some doubt because of slight differences with the type material.

Gerbillinae g. and sp. indet.

Material from Loc. 49 (Igbek). One m1 dex (1.87 × 1.15) (S49-2). See figure 5.5J.

Description and Discussion. This molar belongs to a young individual. The anteroconid is massive and slightly angular in shape; its labial anterolophid partly closes the protosinusid. The main cusps alternate slightly. The longitudinal ridge is low between the anteroconid and the metaconid and even lower between the protoconid and hypoconid. With respect to these two characters, this specimen resembles *Myocricetodon* species included in the lineage of *M. parvus–M. seboui* (Jaeger 1977b). However, it differs from *Myocricetodon* in having a higher crown and less alternating cusps. The posterior arm of the hypoconid is thickened and forms a small hypoconulid. The posterosinusid is present and is as deep as half of the crown height. All these characters fit with those of the Gerbillinae.

The occlusal pattern of this tooth has all the morphologic structures of the genus *Pseudomeriones*. In the early representatives of this genus, such as *P. pytagorasi* Black et al., 1980, and *P. latidens* Sen, 1998, the cusps are less integrated into lophs than in younger species, the crown is less high, and a well-developed hypoconulid and posterosinusid are present (Sen 1983, fig. 76). However, these species are notably larger, the longitudinal crest is higher and thicker, and the cusps are more strongly connected into pairs.

This tooth from Loc. 49 indicates the presence of a primitive gerbil with an occlusal pattern that is intermediate between *Myocricetodon* and *Pseudomeriones*. This enlightens the phylogenetic relationship of these two genera.

Conclusions

The assemblages of small mammals from the Sinap Formation are dominated by the families Cricetidae (mainly *Byzantinia*) and Erinaceidae (mainly *Schizogalerix*). Other families (except the Muridae, which are relatively well represented in Loc. 84) are generally represented by only a few specimens. The percentage of Muridae shows a gradual increase through time in the Vallesian localities: 7.7% of rodents in Loc. 108, 10.6% in Loc. 8A, and 11.5% in Loc. 84.

In the Sinap Formation, the Muridae are represented by the genera *Progonomys, Sinapodemus* g. nov., *"Karnimata,"* and *Hansdebruijnia*. We have to add the genera described from the late Ruscinian locality of Çalta, which is in the same formation (*Occitanomys, Centralomys, Orientalomys,* and *Apodemus;* see Sen 1977). The material is rich enough in the Vallesian localities (Locs. 8A and 84) to allow the identification to species level. The limited amount of sediment we washed from the Turolian locality of Çobanpinar (Loc. 42) provided scarce remains of *"Karnimata" provocator, Hansdebruijnia neutrum,* and one indeterminate taxa.

The Gerbillidae are of special interest because the genus *Myocricetodon* occurs in four successive localities (Locs. 65, 64, 4, and 8a), and then suddenly disappears. This event is dated as late as MN9. The rich material from the younger Loc. 84 does not contain any trace of Gerbillidae. Later, the Gerbillidae reappear in the Sinap Formation with an indeterminate taxon from Loc. 49 (latest Vallesian) and with *Pseudomeriones* during the Turolian and Ruscinian.

Acknowledgments

The author is grateful to all participants of the International Sinap Neogene project, especially to its co-leaders B. Alpagut, M. Fortelius, and J. Kappelman, for taking an interest in the collecting of small mammals and for their help in several phases of field and laboratory studies. The field work was possible thanks to grants from the National Science Foundation and the Academy of Sciences of Finland, and from Centre National de la Recherche Scientifique (UMR 8569) for travel expenses. J. P. Aguilar allowed me to compare the Sinap murids with those from many localities of southern France. J. P. Aguilar, H. de Bruijn, J. van Dam, P. Mein, D. Russell, and W. Wessels have kindly read the manuscript and suggested many improvements. The scanning microscope photographs were taken by C. Chancogne and retouched by H. Lavina and P. Loubry.

Literature Cited

Aguilar, J. P., 1982, Contributions à l'étude des Micromammifères du gisement Miocène supérieur de Montredon (Hérault). 2—Les rongeurs: Palaeovertebrata, v. 12, pp. 81–117.

Aguilar, J. P., and J. Michaux, 1996, The beginning of the age of Murinae (Mammalia: Rodentia) in southern France: Acta Zoologica Cracoviensia, v. 39, pp. 35–45.

Aguilar, J. P., M. Calvet, and J. Michaux, 1986, Découvertes de faunes de micromammifères dans les Pyrénées Orientales (France) de l'Oligocène supérieur au Miocène supérieur; espèces nouvelles et rèflexion sur l'étalonnage des échelles continentale et marine: Comptes Rendus de l'Académie des Sciences de Paris, sér. II, v. 303, pp. 755–760.

Ameur-Chebbeur, R., 1988, Biochronologie des formations continentales du Néogène et du Quaternaire de l'Algérie: Contribution des Micromammifères. Thèse d'Etat, Université d'Oran, 480 pp.

Black, C. C., L. Krishtalka, and N. Solounias, 1980, Mammalian fossils of Samos and Pikermi. I. The Turolian rodents and insectivores of Samos: Annals of the Carnegie Museum, v. 49, pp. 359–378.

Bruijn, H. de, 1976, Vallesian and Turolian rodents from Biotia, Attica and Rhodes (Greece): Proceedings of the Koninklijke Nederlandse Akademie van Wetenschappen, ser B. v. 79, pp. 361–384.

Bruijn, H. de, 1989, Smaller mammals from the Upper Miocene and Lower Pliocene of the Strimon Basin, Greece. Part I—Rodentia and Lagomorpha: Bolletino della Societa Paleontologica Italiana, v. 28, no. 2/3, pp. 189–195.

Bruijn, H. de, M. Dawson, and P. Mein, 1970, Upper Pliocene Rodentia, Lagomorpha and Insectivora (Mammalia) from the Isle of Rhodes (Greece): Proceedings of the Koninklijke Nederlandse Akademie van Wetenschappen, ser. B, v. 73, pp. 535–584.

Bruijn, H. de, J. van Dam, G. Daxner-Höck, V. Fahlbusch, and G. Storch, 1996, The genera of Murinae, endemic insular forms excepted, of Europe and Anatolia during the late Miocene and early Pliocene, *in* R. L. Bernor, V. Fahlbusch, and H.-W. Mittmann, eds., The evolution of western Eurasian Neogene mammal faunas: New York, Columbia University Press, pp. 253–260.

Bruijn, H. de, G. Saraç, L. W. Hoek Ostende, and S. Roussiakis, 1999, The status of the genus name *Parapodemus* Schaub, 1938: New data bearing on an old controversy, *in* J.W.F. Reumer, and J. de Vos, eds., Elephants have a snorkel! Papers in honor of Paul Y. Sondaar, Rotterdam, DEINSEA 7: Annual of the Natural History Museum of Rotterdam pp. 95–112.

Coiffait-Martin, B., 1991, Contribution des rongeurs du Néogène d'Algérie à la biochronologie mammalienne d'Afrique nord-occidentale. Thèse d'Etat, Université de Nancy, 389 pp.

Dam, J. van, 1997, The small mammals from the Upper Miocene of the Teruel-Alfambra region (Spain): Paleobiology and paleoclimatic reconstructions: Geologica Ultraiectina, v. 156, pp. 1–204.

Dames, W., 1883, Hirsche und Mäuse von Pikermi in Attika: Zeitschrift Deutschen Geologischen Gesellschaft, v. 35, pp. 92–100.

Freudenthal, M., and E. Martin Suarez, 1999, Family Muridae, *in* G. E. Rössner, and K. Heissig, eds., The Miocene land mammals of Europe: Verlag Dr. F. Pfeil, München, pp. 401–409.

Heissig, K., 1982, Kleinsäuger aus einer obermiozänen (Vallesium) Karstfüllung Ägyptens: Mitteilungen der Bayerischen Staatssammlung für Paläontologie und Historiche Geologie, v. 22, pp. 97–101.

Jacobs, L. L., 1978, Fossil rodents (Rhizomyidae and Muridae) from Neogene Siwalik deposits, Pakistan: Bulletin of the Museum of Northern Arizona Press, ser. 52, pp. 1–103.

Jaeger, J. J., 1977a, Rongeurs (Mammalia, Rodentia) du Miocène de Beni Mellal: Palaeovertebrata, v. 7, no. 4, pp. 91–125.

Jaeger, J. J., 1977b, Les rongeurs du Miocène moyen et supérieur du Maghreb: Palaeovertebrata, v. 8, no. 1, pp. 1–166.

Kappelman, J., S. Sen, M. Fortelius, A. Duncan, B. Alpagut, J. Crabaugh, A. Gentry, J.-P. Lunkka, F. McDowell, N. Solounias, S. Viranta, and L. Werdelin, 1996, Chronology and biostratigraphy of the Miocene Sinap Formation of central Turkey, *in* R. L. Bernor, V. Fahlbusch, and H.-W. Mittman, eds., The evolution of western Eurasian Neogene mammal faunas: New York, Columbia University Press, pp. 78–95.

Lindsay, E. H., 1988, Cricetid rodents from Siwalik deposits near Chinji village. Part I: Megacricetodontinae, Myocricetodontinae and Dendromurinae: Palaeovertebrata, v. 18, no. 2, pp. 95–154.

Lunkka, J.-P., M. Fortelius, J. Kappelman, and S. Sen, 1999, Chronology and mammal faunas of the Miocene Sinap Formation, Turkey, *in* J. Agusti, L. Rook, and P. Andrews eds., Hominoid evolution and climatic change in Europe, Volume 1: New York, Cambridge University Press, pp. 238–264.

Martin Suarez, E., and M. Freudenthal, 1993, Muridae (Rodentia) from the Lower Turolian of Crevillente (Alicante, Spain): Scripta Geologica, v. 103, pp. 65–118.

Martin Suarez, E., and P. Mein, 1991, Revision of the genus *Castillomys* (Muridae, Rodentia): Scripta Geologica, v. 96, pp. 47–81.

Martin Suarez, E., and P. Mein, 1998, Revision of the genera *Parapodemus, Apodemus, Rhagamys* and *Rhagapodemus* (Rodentia, Mammalia): Geobios, v. 31, pp. 87–97.

Mein, P., E. Martin Suarez, and J. Agusti, 1993, *Progonomys* Schaub, 1938 and *Huerzelerimys* gen. nov. (Rodentia): Their evolution in western Europe: Scripta Geologica, v. 103, pp. 41–64.

Michaux, J., 1971, Muridae (Rodentia) néogènes d'Europe sud-occidentale. Évolution et rapports avec les formes actuelles: Paléobiologie Continentale, Montpellier, v. 2, pp. 1–67.

Michaux, J., J. P. Aguilar, S. Montuire, A. Wolff, and S. Legendre, 1997, Les Murinae (Rodentia, Mammalia) Néogènes du sud de la France: Évolution et paléoenvironnements: Geobios, v. 20, pp. 379–385.

Rümmel, M., 1998, Die Cricetiden aus dem Mittel—und Obermiozän der Türkei: Documenta Naturae, v. 123, pp. 1–300.

Schaub, S. 1938, Tertiäre und Quartäre Murinae: Abhandlungen des Schweizerischen Paläontologische Gesellschaft, v. 61, pp. 1–38.

Sen, S., 1977, La faune de rongeurs Pliocènes de Çalta (Ankara, Turquie): Bulletin du Muséum National d'Histoire Naturelle, Paris, sér. Sciences de la Terre, v. 61, pp. 89–172.

Sen, S. 1983, Rongeurs et lagomorphes du gisement Pliocène de Pul-e Charkhi, bassin de Kabul, Afghanistan: Bulletin du Muséum National d'Histoire Naturelle, sér. C, v. 5, pp. 33–74.

Sen, S., 1991, Stratigraphie, faunes de mammifères et magnétostratigraphie du Néogène de Sinap Tepe, Province d'Ankara, Turquie: Bulletin du Muséum National d'Histoire Naturelle, Paris, sér. C, v. 12, pp. 243–277.

Sen, S., 1994, Les gisements de Mammifères du Miocène supérieur de Kemikli Tepe, Turquie: 5. Rongeurs, Tubulidentés et Chalicothères: Bulletin du Muséum National d'Histoire Naturelle, sér. 4e, v. C 16, pp. 97–112.

Sen, S., 1997, Magnetostratigraphic calibration of the European Neogene mammal chronology: Palaeogeography, Palaeoclimatolology, Palaeoecology, v. 133, pp. 181–204.

Sickenberg, O., J. D. Becker-Platen, L. Benda, D. Berg, B. Engesser, W. Gaziry, K. Heissig, K. A. Hünermann, P. Y. Sondaar, N. Schmidt-Kittler, K. Staesche, U. Staesche, P. Steffens, and H. Tobien, 1975, Die Gliederung des höheren Jungtertiärs und Altquartärs in der Turkei nach Vertebraten und ihre Bedeutung für die internationale Neogen-Stratigraphie: Geologische Jahrbuch, v. 15, pp. 1–167.

Steininger, F. F., 1999, Chronostratigraphy, geochronology and biochronology of the Miocene "European Land Mammal Mega-Zones" (ELMMZ) and the Miocene "Mammal-Zones" (MN-Zones), *in* G. E. Rössner and K. Heissig, eds., The Miocene land mammals of Europe: München, Germany, Verlag Dr. F. Pfeil, pp. 9–24.

Storch, G., and T. Dahlman, 1995, The vertebrate locality Maramena (Macedonia, Greece) at the Turolian-Ruscinian boundary (Neogene). 10. Murinae (Rodentia, Mammalia): Münchner Geowissenschafliche Abhandlungen, v. A 28, pp. 121–132.

Sümengen, M., E. Ünay, G. Saraç, H. de Bruijn, I. Terlemez, and M. Gürbüz, 1990, New Neogene rodent assemblages from Anatolia (Turkey), *in* E. H. Lindsay, V. Fahlbusch, and P. Mein eds., European Neogene mammal chronology: New York, Plenum Press, pp. 61–72.

Tong, H., 1988, Origine et évolution des Gerbillidae (Mammalia, Rodentia) en Afrique du Nord: Mémoire de la Société Géologique de France, v. 155, pp. 1–120.

Ünay, E., 1981, Middle and Upper Miocene rodents from the Bayraktape section (Çanakkale, Turkey): Proceedings of the Koninklijke Nederlandse Akademie van Wetenschappen, ser. B, v. 84, pp. 217–238.

Ünay, E., and H. de Bruijn, 1984, On some Neogene rodent assemblages from both sides of the Dardanelles, Turkey: Newsletter on Stratigraphy, v. 13, pp. 119–132.

Weerd, A. van de, 1976, Rodent faunas of the Mio-Pliocene continental sediments of the Teruel-Alfambra region, Spain: Utrecht Micropaleontological Bulletins Special Publication 2, pp. 1–217.

Wessels, W., 1996, Myocricetodontinae from the Miocene of Pakistan: Proceedings of the Koninklijke Nederlandse Akademie van Wetenschappen, ser. B, v. 99, pp. 253–312.

Wessels, W., E. Ünay, and H. Tobien, 1987, Correlation of some Miocene faunas from northern Africa, Turkey and Pakistan by means of Myocricetodontidae: Proceedings of the Koninklijke Nederlandse Akademie van Wetenschappen, ser. B, v. 90, pp. 65–82.

6

Spalacidae (Rodentia)

N. Sarica and S. Sen

Among muroid rodents, the family Spalacidae forms a specialized group that is adapted to a fossorial mode of life. The fossil spalacids are known by three genera (*Debruijnia, Heramys,* and *Pliospalax)* and nine species (seven of which have been included in *Pliospalax)*. The taxonomic status of living spalacids is more controversial. Although some authors (Topachevski 1969; de Bruijn 1984) recognize several genera (*Microspalax, Nannospalax, Spalax,* and others), others place all living species in one genus, *Spalax* (Sen 1977; Savic and Nevo 1990; Nevo et al. 1995; Ünay 1996).

The past and present distribution of the family Spalacidae remains restricted in the area from Crimea to Sinai and from the Balkans to Iran. They are believed to be derived from a cricetid stock. However, the avaliable documentation does not allow the designation of a particular group as ancestral to this family.

The fossil documentation, although constantly increasing during the past decade, is still quite scarce. Except in some Pliocene and Pleistocene localities, spalacids are quite rare in Neogene faunas. This is also the case in the Sinap Formation localities. Nevertheless, this region yielded well-preserved specimens in several localities that have a well-controlled stratigraphic order. The spalacid remains from the Sinap Formation localities are sufficiently numerous to document the spalacid diversity during the Astaracian–Vallesian transition in this region. The present chapter describes the spalacid remains collected in the Sinap Formation localities between 1989 and 1995. It includes a new genus and several species. As a result of this study, we redefine the status and the species level content of the genus *Pliospalax* and discuss the phylogenetic relationships between the spalacid genera.

Materials and Methods

The spalacid material studied herein includes isolated cheek teeth and mandibles. They were collected by one of us (S. Sen) from six localities throughout the Sinap Formation at Sinap Tepe and its surroundings (Ankara). From the oldest to the youngest, these are Loc. 65, Loc. 4, Loc. 120, Loc. 41, Loc. 12, and Loc. 84. Most specimens have been collected by using a screen wash technique during the field seasons of 1991–1995, in the context of the Sinap Neogene project. These specimens are currently housed in Muséum National d'Histoire Naturelle, Paris, but they will be transferred, after study, to the University of Ankara (DTCF). The drawings of teeth were prepared using a camera lucida. For ease of comparison, all specimens were illustrated as if they were from the left-hand side; if the original is a right tooth, the number concerned is underlined on the figures.

The nomenclature for molars (fig. 6.1) is adapted from that of Mein and Freudental (1971) for cricetids, and for mandibles from Topachevski (1969) and Korth (1994). If the homology of a structure is not evident, it is discussed in the descriptions. The maximum length and maximum width of specimens were measured by a double axis Mitutoyo measuroscope and the measurements are given in mm. To give a reference frame for the size of specimens, the minimum–maximum lengths and widths observed for the family are divided into three fractions, small (S), medium (M), and large (L). Then for each molar, the length–width diagrams are gridded into the size-range areas of S, M, L, (figs. 6.2–6.7). Additionally, to assess to the length variation ranges of each molar, a length variation table was prepared by subtracting maximum–minimum lengths of the specimens reported from all localities (table 6.1). Upper molars are abreviated as M1, M2, M3 and lower molars as m1, m2, m3.

Systematics

Family Spalacidae Gray, 1821
Genus *Heramys* Hofmeijer and de Bruijn, 1985
Heramys anatolicus n. sp.

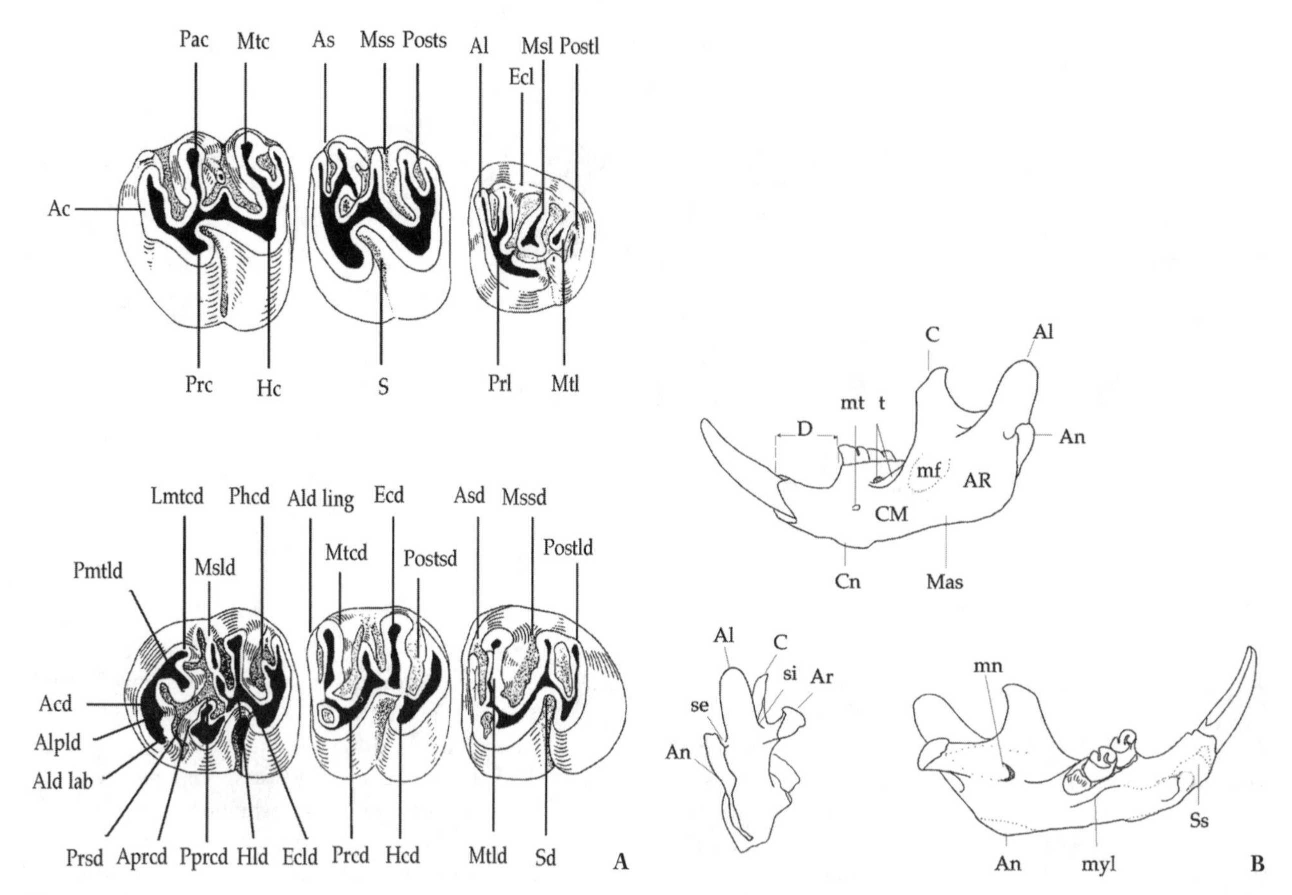

Figure 6.1. Nomenclature of (A) spalacid upper and lower molars and (B) mandible. Abbreviations for upper molars: Ac: anterocone, Al: anteroloph, As: anterosinus, Ecl: ectoloph, Hc: hypocone, Msl: mesoloph, Mss: mesosinus, Mtc: metacone, Mtl: metalophule, Pac: paracone, Postl: posteroloph, Posts: posterosinus. Prc: protocone, Prl: protolophule, S: sinus. Abbreviations for lower molars: Acd: Anteroconid, Ald lab: labial anterolophid, Ald ling: lingual anterolophid, Alpld: anterolophulid, Aprcd: anterior arm of protoconid, Asd: anterosinusid, Ecd: entoconid, Ecld: ectolophid, Hcd: hypoconid, Hld: hypolophulid, Lmtcd: lingual arm of metaconid, Msld: mesolophid, Mssd: mesosinusid, Mtcd: metaconid, Mtld: metalophulid, Phcd: posterior arm of hypoconid, Pmtld: posterior metalophulid, Postld: posterolophid, Postsd: posterosinusid, Pprcd: posterior arm of protoconid, Prcd: protoconid, Prsd: protosinusid, Sd: Sinusid. Abreviations for mandible: Al: alveolar process, An: angular process, AR: ascending ramus, Ar: articular process, C: coronoid process, CM: *corpus mandibulae*, Cn: chin process, D: diastema, Mas: masseteric ridge, mf: masseteric fossa, mn: mandibular foramen, mt: mental foramen, myl: mylohyoid line, se: saddle exterior, si: saddle interior, Ss: symphyseal surface, t: *temporalis* muscle scar.

Type Species. *Heramys eviensis* Hofmeijer and de Bruijn, 1985

Holotype. One m1 sin (2.31 × 1.93) (ST4-1). See figure 6.8A.

Type Locality. Sinap Tepe Loc. 4, Kazan, Ankara.

Etymology. Derived from Anatolia, where a *Heramys* species was first reported.

Paratypes. A right mandible with incisor but without molars, S89/555, one M2 sin (2.07 × 1.90), one M3 sin (1.62 × 1.56), one m1 dex (2.39 × 1.95), one m2 sin (2.34 × 2.43), two m3 dex (2.24 × 2.01, 2.27 × 2.12) (ST4-2–ST4-7). See figures 6.8–6.10.

Diagnosis. Small to medium-sized spalacid with low crowned cheek teeth; ridges relatively thin and high. The m1 is elongated, with a rounded anterior edge, anteroconid entirely incorporated into anterolophid, anteriorly oblique protosinusid, enlarged metaconid, protoconid with a distinct posterior arm, posteriorly directed oblique sinusid. On M2, the anterosinus invades deeply, paracone with ectoloph, the protoloph is long and transverse, and the mesoloph short. M2 and M3 have the metalophule connected to the posterior arm of the hypocone. M3 has a rounded outline and is without sinus.

Differential Diagnosis. *Heramys anatolicus* n. sp. differs from *H. eviensis* in being larger, having a shallower protosinusid, a central anterolophulid, and an open posterosinusid on m1. With reference to the M2, *H. anatolicus* differs from *H. eviensis* in having a protolophule connected to the endoloph, a shorter mesoloph, and a metalophule connected to the posterior arm of the hypoconid. *H. anatolicus* differs from *Debruijnia arpati* in having higher crowned and more lophodont cheek teeth. It differs from all species of the genus *Sinapospalax* in having m1 with a less hypsodont crown, anteriorly oblique and very shallow protosinusid, presence of the central anterolophid, and a very oblique sinusid. *Heramys anatolicus* differs from *Pliospalax* by having

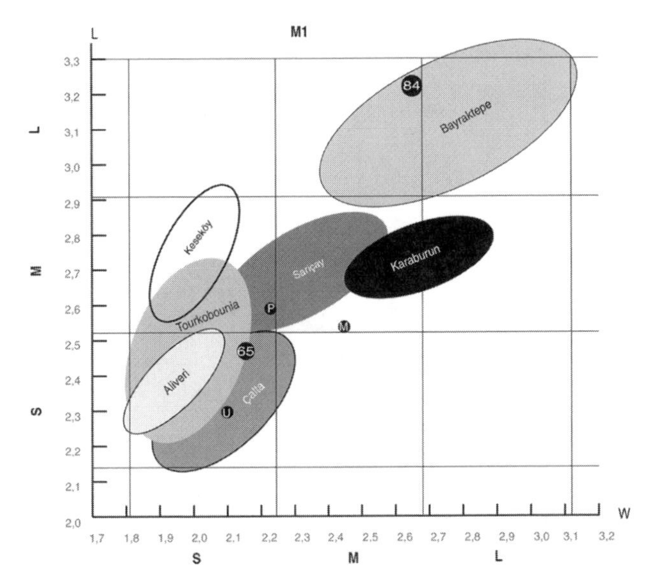

Figure 6.2. Length-width scatter diagram of the M1 of Spalacidae from different localities. Aliveri: *Heramys eviensis;* Bayraktepe: *Sinapospalax canakkalensis;* Çalta: *Pliospalax macoveii;* Karaburun: *Spalax odessanus;* Keseköy: *Debruijnia arpati;* Maritsa (M): *Pliospalax sotirisi;* Pasalar (P): *Sinapospalax marmarensis;* Sariçay: *Sinapospalax primitivus;* Tourkobounia (T): *Pliospalax tourkobouniensis;* U (Ukraina): *Pliospalax compositodontus.* Numbers refer to Sinap Formation localities: 4: *Heramys anatolicus;* 12: *Sinapospalax* n. sp. 41a: *Sinapospalax incliniformis;* 41b: *Sinapospalax* sp. 3, 65: *Sinapospalax canakkalensis;* 65i: Spalacidae indet., 84: *Sinapospalax sinapensis;* 120: *Sinapospalax* sp. Size ranges: S: small, M: medium, L: large.

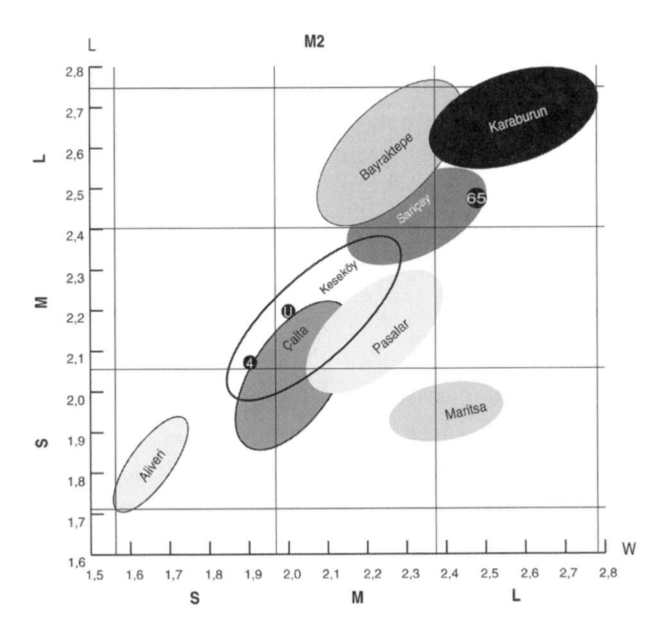

Figure 6.3. Length-width scatter diagram of the M2 of Spalacidae from different localities. See figure 6.2 for legend.

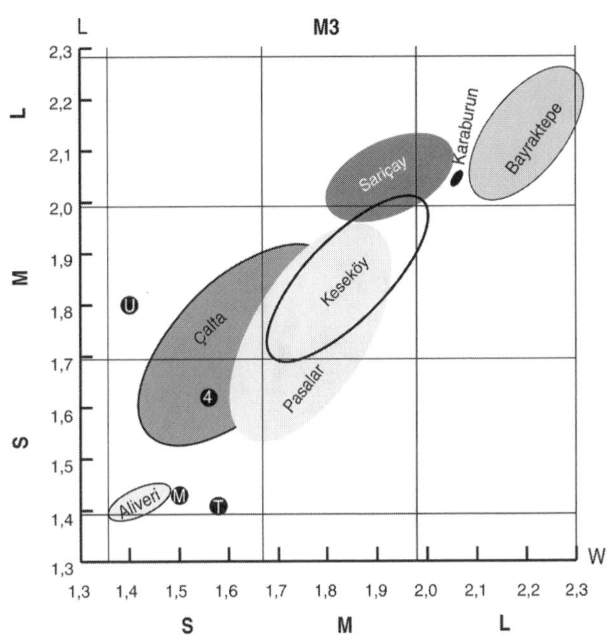

Figure 6.4. Length-width scatter diagram of the M3 of Spalacidae from different localities. See figure 6.2 for legend.

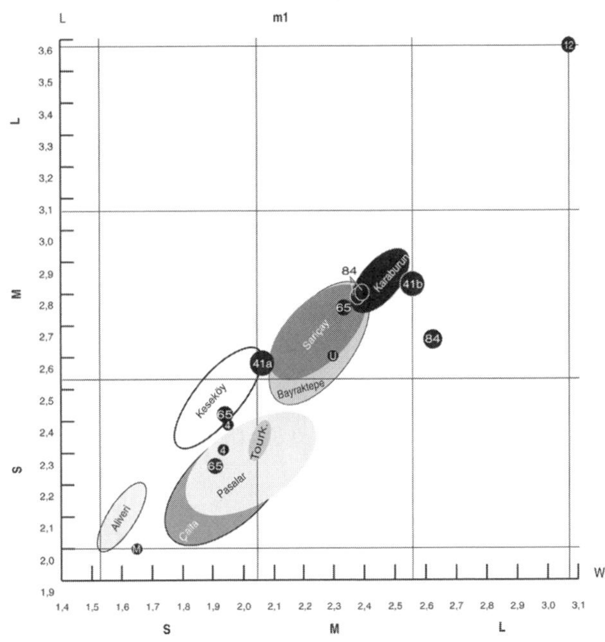

Figure 6.5. Length-width scatter diagram of the m1 of Spalacidae from different localities. See figure 6.2 for legend.

lower crowned molars, anteriorly oblique protosinusid on m1, with higher length:width ratio of m1 and more complex occlusal pattern of molars.

Description. The mandible is broad and deep (6.98 mm in height below m1). The molars are missing. The diastema is shallow. The incisor has three ridges on its anterior face. The mental foramen is situated below the anterior root of m1. On the labial face, the dorsal and ventral ridges of the masseteric scar fuse below m2; there is an inflation anterior to this fusion. A narrow triangular muscle scar (probable insertion of the *temporalis* muscle) is situated below m1–m2. The masseteric fossa is U-shaped and depressed; it extends anteriorly below the roots of m2 and terminates at the point of the fusion of dorsal and ventral ridges of the masseteric scar. The articular, angular, and chin processes

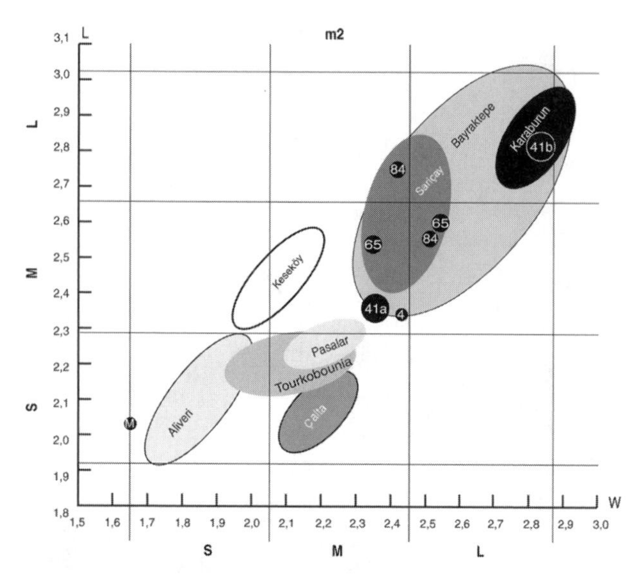

Figure 6.6. Length-width scatter diagram of the m2 of Spalacidae from different localities. See figure 6.2 for legend.

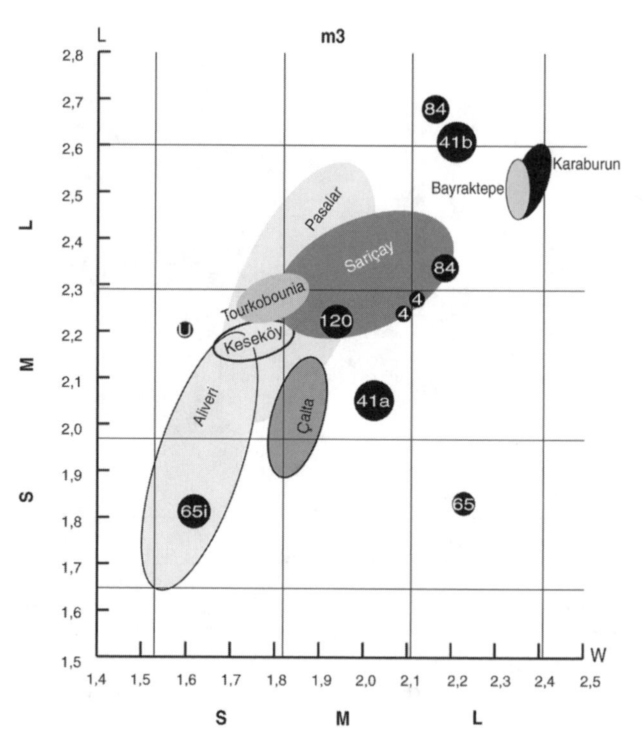

Figure 6.7. Length-width scatter diagram of the m3 of Spalacidae from different localities. See figure 6.2 for legend.

are broken. The coronoid process is well preserved; it is small and posteriorly curved. The ascending ramus originates below m2, rises steeply, obscuring m3; it is thin and high and has a slight groove on the anterior border; on the lingual side, it is separated from the tooth row by a shallow and wide valley. In the posterior part of this valley, between the tooth row and the ascending ramus, there is a slight wedge-shaped scar (probable insertion of the extension of the *pterygoideus lateralis* muscle). The lingual face of the *corpus mandibulae* has a faint mylohyoid line inferior to the molar alveoli. The lingual side of the angular process has a groove parallel to the ventral border. From the posterior view, the broken angular process can be traced as slightly deviating from the vertical plane of the *corpus mandibulae.*

The M2 is longer than it is wide. The protocone is fused with the anteroloph, which is very strong and labially connected to the base of the paracone. The transverse protolophule is connected to the endoloph. On the posterior end of the paracone there is an ectoloph. The oblique mesoloph is relatively short. It is derived from the endoloph, posterior to the protolophule. The metalophule is connected to the posterior arm of the hypocone. The

Table 6.1. Length Variation of Spalacid Molars

Locality	$L_{max}-L_{min}$ (mm)					
	m1	m2	m3	M1	M2	M3
Ubeidiya	0.35 (270)	—	—	0.39 (305)	—	—
Tourkobounia	0.12 (2)	0.07 (3)	0.06 (3)	*0.52* (3)	—	—
Çalta	0.32 (8)	0.20 (3)	0.25 (4)	0.38 (14)	0.34 (6)	0.33 (7)
Maritsa	—	—	—	—	0.03 (3)	—
Sinap Tepe Loc. 84	0.15 (3)	—	—	—	—	—
Sinap Tepe Loc. 65	*0.49* (3)	0.06 (2)	—	—	—	—
Sinap Tepe Loc. 41	0.26 (2)	0.46 (2)	0.46 (2)	—	—	—
Sinap Tepe Loc. 4	0.08 (2)	—	0.03 (2)	—	—	—
Bayraktepe	0.35 (3)	0.50 (11)	0.12 (2)	0.38 (2)	0.33 (5)	0.23 (2)
Sariçay	0.25 (5)	0.45 (4)	0.15 (2)	0.30 (6)	0.15 (5)	0.10 (3)
Pasalar	0.23 (8)	0.09 (4)	0.53 (12)	—	0.26 (6)	0.39 (13)
Aliveri	0.20 (7)	0.64 (7)	0.55 (7)	0.29 (5)	0.23 (3)	0.50 (2)
Keseköy	0.30 (6)	0.28 (8)	0.04 (3)	0.38 (4)	0.36 (9)	0.30 (10)

Notes: Numbers in parentheses indicate sample size. Length variability range of specimens from Ubeidiya (*Spalax ehrenbergi:* Tchernov 1986) is chosen as a reference for the family. Values that exceed the reference value are shown in italics.

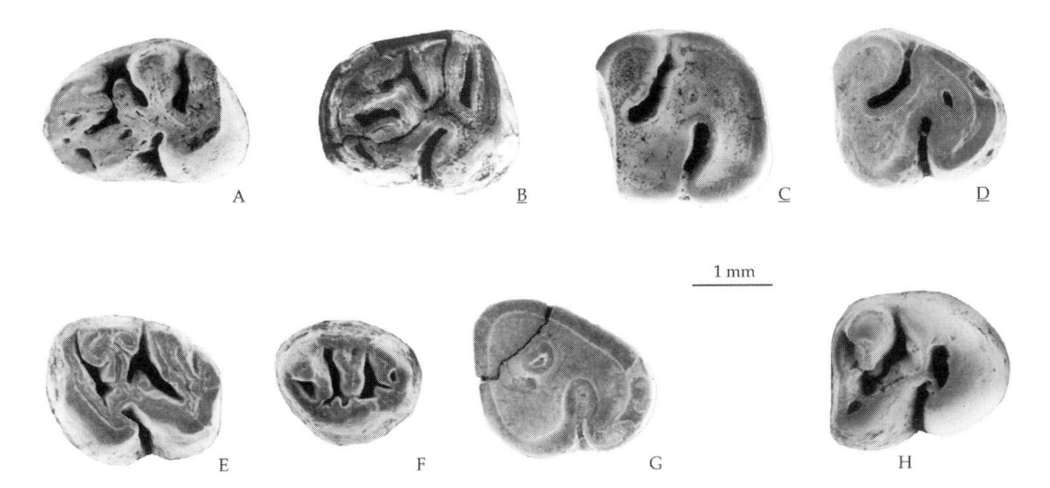

Figure 6.8. *Heramys anatolicus* n. sp. from Sinap Tepe Loc. 4. Occlusal view of molars. (**A, B**) m1; (**C**) m2; (**D**) m3; (**E**) M2; (**F**) M3; (**G**) m3. *Sinapospalax* sp. from Sinap Tepe Loc. 120: (**H**) m3.

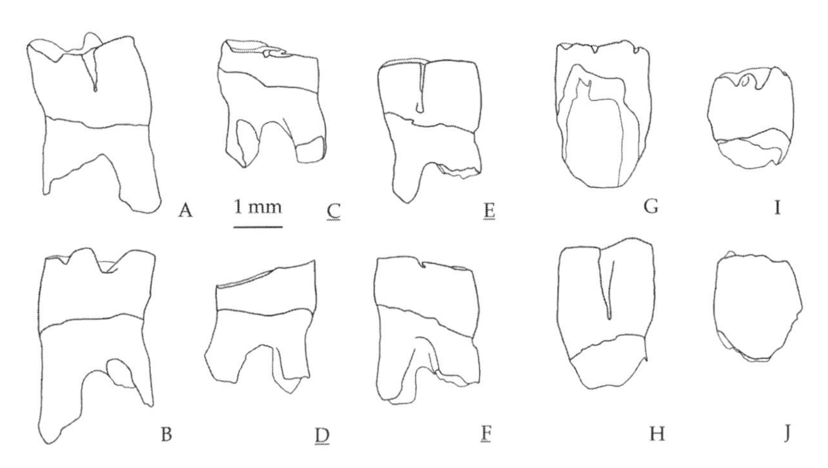

Figure 6.9. *Heramys anatolicus* n. sp. from Sinap Tepe Loc. 4. Labial (top) and lingual (bottom) views of molars. **A, B:** m1; **C, D:** m2; **E, F:** m3; **G, H:** M2; **I, J:** M3.

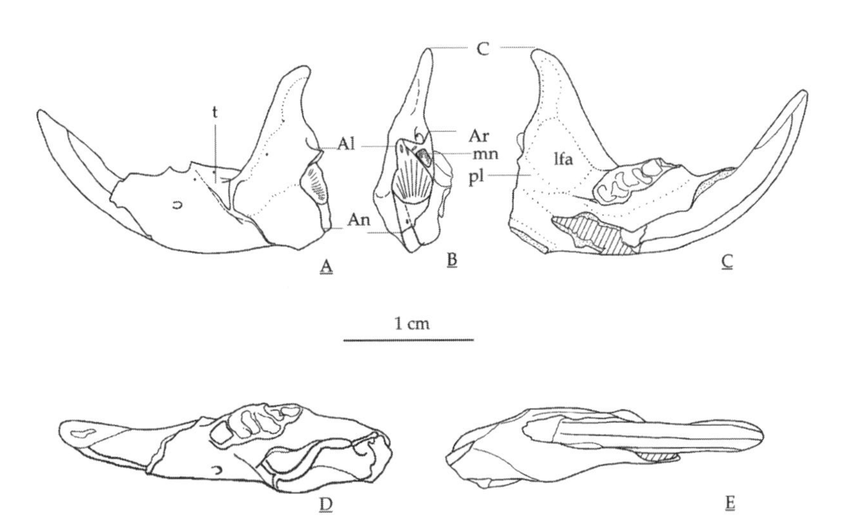

Figure 6.10. Mandible of *Heramys anatolicus* n. sp. from Sinap Tepe Loc. 4. Views of mandible. (**A**) Labial; (**B**) posterior; (**C**) lingual; (**D**) dorsal; (**E**) ventral. Abbreviations: Al: alveolar process, An: angular process, Ar: articular process, mn: mandibular foramen, pl: insertion of the *pyterygoideus lateralis* muscle, t: insertion of the temporalis muscle.

posteroloph is long and extends towards the base of the metacone. The sinus is oblique anteriorly.

The outline of M3 is rounded. There is no sinus. The paracone is the most prominent cusp and is situated anterolabially. The labial anteroloph is connected to the paracone, enclosing a short anterosinus. The protolophule is curved and joins the anteroloph just anterior to the protocone. The mesoloph is long and bifurcated at the labial border; it issues from the neo-endoloph and divides the mesosinus into two valleys. The metalophule is connected to the posterior arm of the hypocone. The posteroloph is strongly curved toward the metacone, thus enclosing the posterosinus. The hypocone is reduced.

The m1 is longer than it is wide, and is narrower and more lophodont anteriorly than posteriorly. The anterior border is rounded. The anteroconid is incorporated into the anterolophid. The labial spur of the anterolophid reaches the labial border, delimiting the protosinusid. The protosinusid is oblique anteriorly and quite shallow. The metaconid is large and placed anterolingually. The posterior metalophulid is longitudinal and reaches the mesolophid. There is a strong spur issuing from the anterolophid and extending back. Because the cricetid pattern is not homologous to this spur, we prefer to term it the central anterolophid. The posterior arm of the protoconid joins the mesolophid, enclosing an enamel island. The mesolophid is strong but does not reach the labial border; it issues from the ectolophid in front of the hypolophulid. The posterolophid is transverse. The sinusid is directed backward, almost parallel to the lingual border and its depth is two-thirds of the crown height on the less-worn m1. There are two roots.

The m2 material includes a very worn specimen. Because of the advanced degree of attrition, it is slightly wider than it is long. The long mesolophid is fused to the entoconid, and they enclose the posterior branch of the mesosinusid as an enamel island. The anterior branch of the mesosinusid is still open. The sinusid is oblique backward. There are two roots.

The m3 is longer than it is wide and has a triangular outline. The anterolophid and metaconid are fused by wear. The mesolophid is apparently absent. The strongly curved posterolophid joins the entoconid and encloses an enamel island. The transverse hypolophulid is connected to the ectolophid, anterior to the hypoconid. The sinusid is transverse and deep. The tooth has two roots.

Comparison. The cheek teeth from Loc. 4 share a number of characters with *Heramys eviensis*, such as elongated form of m1 and M2; small M3 without sinus; and m1 with (1) an anteroconid incorporated into the anterolophid; (2) a strong labial spur of the anterolophid, which delimits the protosinusid; (3) an anteriorly oblique protosinusid; and (4) a posteriorly oblique sinusid. Therefore, the specimens from Loc. 4 are attributed to this genus, although they are larger than those of *H. eviensis*. In figures 6.2–6.7, m2 and m3 from Loc. 4 plot in the M–L size range, whereas M2, M3, and m1 plot in the size ranges S and S–M. This may suggest the presence of two different taxa, but the larger size of these particular m2 and m3 specimens is probably due to their high degree of attrition.

The genus *Heramys* is known by the unique species *H. eviensis* from the early Miocene (MN4) of Aliveri, Greece. It differs from the material of Sinap Tepe by having teeth with smaller size; m1 with more elongated shape, a deeper protosinusid, a slight central anterolophid, and stronger posterior arm of the hypoconid; and M2 with a weaker anteroloph and anterosinus, a protolophule connected to the anteroloph, a stronger mesoloph reaching the labial border, a metalophule connected to the hypocone, and a more transverse posteroloph.

Some heterogeneity in size and morphology of the specimens from Loc. 4, Aliveri, Bayraktepe I, and Loc. 65 raises the question of co-occurence of the genus *Heramys* and *Sinapospalax* in these localities. The Aliveri material (Hofmeijer and de Bruijn 1985) includes one m2, which is larger and morphologically different from the other Aliveri specimens (Hofmeijer and de Bruijn 1985, pl. 1, fig. 2, p. 187). This m2 shares some characteristics of *Sinapospalax* (see later in this chapter), such as larger size and lower length:width ratio. However, it is a worn specimen, and there is no other molar that shares the characteristics of *Sinapospalax*. This is an insufficient reason to assume that *Sinapospalax* is also present at Aliveri. However, the range of the length in m2, m3, and M3 from Aliveri exceeds the normal maximum variation limits (~0.40 mm) as known from other spalacid populations. This observation raises again the question of the homogeneity of the Aliveri material. As noted by Ünay (1981), the material from Bayraktepe I (Çanakkale, Turkey) contains some specimens smaller than those attributed to *"Pliospalax" canakkalensis*. One M2, which is illustrated by Ünay (1981, pl. I, fig. 2), shares some characteristics of *Heramys:* small size; higher length:width ratio; anteriorly directed, very oblique sinus; transverse metalophule; and transverse posteroloph. Because these characters cannot be observed in the new genus *Sinapospalax* in which this species is included, we think that this specimen, and perhaps some others that Ünay (1981) did not illustrate, belong to the genus *Heramys*. The high variation in the length of m2 from Bayraktepe I (fig. 6.6) reinforces this hypothesis.

Sinap Tepe Loc. 65 yielded one m1 that is smaller than the other specimens from the same locality. Unfortunately this m1 is worn and the pattern is completely changed; the only information available about the morphology comes from the outline of the tooth, which is anteriorly rounded and reduced; this morphology is shared by the species of *Heramys*. Although still equivocal, our study supports the presence of the genus *Heramys* in some other localities with a time range of MN4–MN9.

The mandible from Loc. 4 is edentulous. It is referred to *Heramys anatolicus* because the form of the m1 alveolus is compatible with the form of the roots of an isolated m1, and also its dimensions indicate a small to medium-sized spalacid. The mandible of *H. eviensis* is not known. Consequently, we only can compare the mandible from Loc. 4

with that of *Sinapospalax incliniformis* n. gen. n. sp. from Loc. 41. The mandible of *S. incliniformis* differs from that of *H. anatolicus* by having a slender, shallower and narrower *corpus mandibulae,* thinner incisor, an ascending ramus originating below the anterior root of m2 (i.e., slightly more anterior than in *H. anatolicus*), a less lingually shifted alveolus of m3, a depressed area for the probable insertion of extension of *pterygoideus lateralis* muscle, a weaker ventral ridge of the masseteric scar, a more pronounced inflation anterior to the masseteric ridge fusion, and a larger triangular muscle scar between the dorsal and ventral ridges of the masseteric scar. However, they share a number of characters: an abruptly and linearly rising ascending ramus that is separated from the tooth row by a wide and shallow valley, the presence of the triangular muscle scar between the dorsal and ventral ridges of the masseteric scar, U-shaped and deep masseteric fossa, mental foramen situated inferior to the anterior root of m1, and a triple-ridged incisor that rises above the abrasion plane of the molars and posteriorly reaches the mandibular condyle.

Genus *Sinapospalax* n. gen.

Type Species. *Sinapospalax canakkalensis* (Ünay 1981)

Etymology. From Sinap Tepe where this genus is recorded in several levels.

Diagnosis. Medium-sized to large spalacids with medium crowned semihypsodont cheek teeth. Upper molars have one strong lingual and two weaker labial roots; lower molars have one anterior and one posterior root of equal strength. M1, m1, m3 are longer than they are wide; m2, M2, M3 have length:width ratios of ~1. All molars have strong mesoloph(id)s except M1 and m3. The m1 has a subrounded or square anterior border, metaconid is in the same plane with anteroconid or slightly back of it; their conicle shapes are often preserved because of posterior and/or anterior grooves on the anterior ridge; the protoconid has short (or remnant) anterior and distinct posterior arms; spurlike anterolophulid is fused with the anterior arm of the protoconid by wear; the protosinusid is generally deep. On m2 and m3 there are distinct anterosinusid and anterolophid in young individuals; the protosinusid is transformed into an enamel island even in unworn teeth. The M1 has an anterolingual depression distinguishing the protocone from the anterocone. The M2 is almost symetrically divided into two parts by a strong mesoloph; the metalophule is connected to posteroloph; the anterosinus and posterosinus are weak or enclosed. The M3 is subtriangular to subrounded in outline with a complete pattern of 4–5 lophs. The mandible is broad, the ascending ramus originates anteriorly below m2 and rises linearly, there are robust masseteric ridges, strongly depressed masseteric fossa, and a well-defined triangular muscle scar on the lateral face of mandible between the intersection of dorsal and ventral masseteric ridges; there are three ridges on the lower incisor.

Differential Diagnosis. *Sinapospalax* differs from *Debruijnia* Ünay, 1996, in having a lower length:width ratio in all molars and higher crowned and more lophodont cheek teeth with thicker ridges. It differs from *Heramys* Hofmeijer and de Bruijn, 1985, by its larger size; lower length:width ratios of m1, m2, m3; having more transverse sinusid and transversely elongated protoconid on m1; and the absence of protosinusid on m2 and m3. *Sinapospalax* differs from *Pliospalax* Kormos, 1932, by being larger, m1 having a more complex trigonid pattern and higher length:width ratio, and having lower incisors with three ridges. In *Sinapospalax,* m1 is frequently square anteriorly or subrounded and its anteroconid and metaconid still preserve their conicle forms (whereas in *Pliospalax* it is strongly rounded because of the total fusion of the aforementioned cusps). In addition, the presence of a strong mesolophid and the posterior arm of the protoconid, a deeper protosinusid, and a longer hypolophulid on m1 differentiate it from *Pliospalax. Sinapospalax* also has M1 with broken-arc-shaped anterior part caused by an anterolingual depression; on M1, the metalophule is directed toward the posterior arm of the hypocone in the former (whereas in *Pliospalax,* it joins the posteroloph very close to the labial border). Another difference is that in *Sinapospalax,* M1 and M2 have a strong mesoloph, and M2 and M3 have an anterosinus. *Sinapospalax* clearly differs from *Spalax* by having lower crowned cheek teeth with a more complex occlusal pattern.

Species included in *Sinapospalax* n. gen. are:

Pliospalax canakkalensis Ünay, 1981 (type species); Bayraktepe I (Çanakkale, Turkey);
Pliospalax primitivus Ünay, 1978; Sariçay (Milas, Turkey);
Pliospalax marmarensis Ünay, 1990; Paşalar (Bursa, Turkey);
Sinapospalax incliniformis n. gen. n. sp.; Loc. 41 (Kazan, Ankara, Turkey);
Sinapospalax sinapensis n. gen. n. sp.; Loc. 84;
Sinapospalax n. sp., Loc. 12;
Sinapospalax sp. 1, Loc. 120;
Sinapospalax sp. 2, Loc. 41; and
Sinapospalax sp. 3, Loc. 41.

Sinapospalax canakkalensis (Ünay 1981)

Holotype. One M1 dex (Ünay 1981, pl. II, fig. 2). See figures 6.11, 6.12.

Type Locality. Bayraktepe I (Çanakkale, Turkey).

Emended Diagnosis. Medium-sized to large spalacid. On m1, there is a complex trigonid, the anteroconid and metaconid are anteriorly in the same plane, there is a spurlike anterolophulid behind the anteroconid, a long fragmented mesolophid connects to the posterior metalophulid and/or to the posterior arm of the protoconid, the hypolophulid and posterolophid are long, and the posterosinusid is deeply invaded. On m2, in juvenile specimens there are thin, wedgelike lingual anterolophid and slitlike anterosinusid.

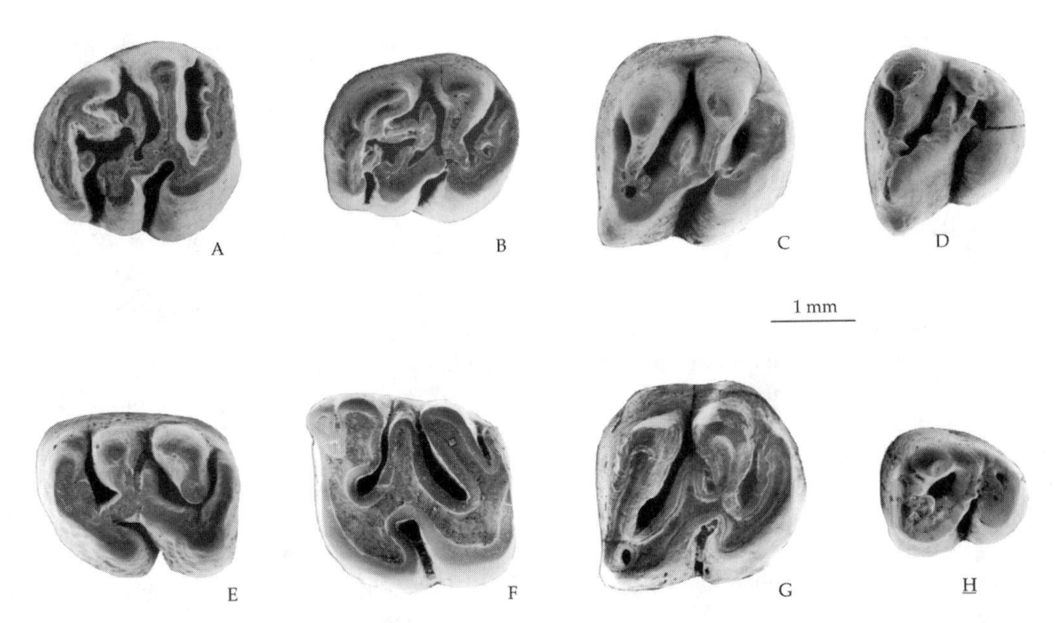

Figure 6.11. *Sinapospalax canakkalensis* from Sinap Tepe Loc. 65. Occlusal view of molars. **A, B:** m1; **C, G:** m2; **D:** m3; **E:** M1; **F:** M2. Spalacidae indet., **H:** m3.

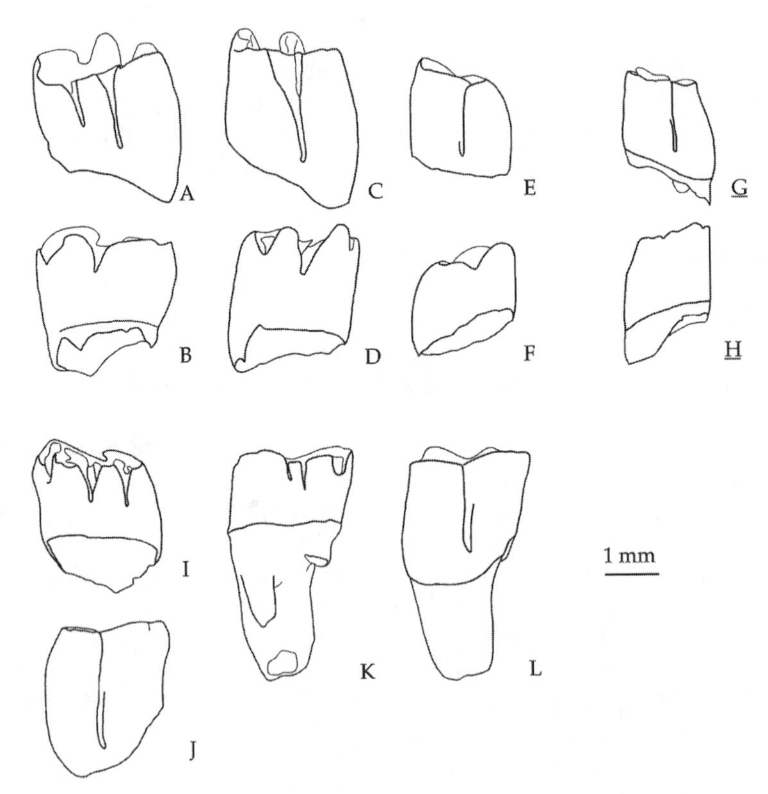

Figure 6.12. *Sinapospalax canakkalensis* from Sinap Tepe Loc. 65. Labial and lingual views of molars. **A, B:** m1; **C, D:** m2; **E, F:** m3; **I, J:** M1; **K, L:** M2; **G, H:** Spalacidae indet., m3.

The M1 has an interrupted metalophule; a labially directed, distinct anterior arm of the protocone enclosing an enamel island behind the anteroloph; and an anteriorly oblique sinus. The M2 has a paracone with a double connection. The M3 is small and has a deep sinus confluent with the posterolabial sinus.

Differential Diagnosis. *Sinapospalax canakkalensis* differs from *S. marmarensis* (Ünay 1990) in having a deeper protosinusid and an anterior lobe connected to the posterior part on m1; narrower mesosinus on M2, and a strong mesolophid on m3. *S. canakkalensis* differs from *S. primitivus* (Ünay 1978) in the presence of a deeper protosinusid on m1, a more anteriorly directed sinus on M1 and M2, more transverse mesoloph on M1, open anterosinus and posterosinus, and a paracone with double connection on M2. *S. canakkalensis* differs from *S. incliniformis* n. gen. n. sp. in having transverse lophids on m1 versus oblique lophids in the latter species. It differs from *S. sinapensis* n. gen. n. sp. in having an oblique sinus on M1 and a long mesolophid on m3.

Material from Loc. 65. One M1 (2.47 × 2.16), one M2 (2.48 × 2.47), three m1 (2.75 × 2.35, 2.26 × 1.91, 2.42 × 1.94), two m2 (2.54 × 2.35, 2.6 × 2.54), one m3 (1.83 × 2.23) (ST65-1–ST65-8). See figures 6.11–6.12.

Description. The M1 is longer than it is wide. Just behind the anterocone, there is a well-developed loph that is difficult to homologize. Because this transverse loph is positioned behind the anterocone and issues from the protocone, it may be considered as a well-developed anterior arm of the protocone; it encloses an enamel island joining the anteroloph. The anterosinus is wide. The paracone is connected to the endoloph by a transverse protolophule. The mesoloph is medium sized and transversal. The metacone is isolated. The posteriorly directed metalophule is separated from the endoloph and posteroloph. However, an anterolingual spur from the hypocone can be considered as a relic of the metalophule-endoloph connection. The strong posteroloph is slightly curved toward the posterior face of the metacone. The depth of the anteriorly directed sinus is five-sixths of the crown height. From the labial side, the anterosinus and mesosinus exceed in depth one-half the crown height, whereas the posterosinus is less deep. Roots are not preserved.

The M2 has a square occlusal outline. The anterosinus is closed by wear and thus the labial anteroloph and the paracone are completely fused. The strong and transverse mesoloph reaches the labial border, dividing the mesosinus into two valleys. The metalophule is directed posteriorly and fused with the posteroloph. The posteroloph joins the posterior wall of the metacone. The sinus is as deep as three-quarters of the crown height (slightly shallower than on M1). There are two small labial and one strong lingual roots.

The m1 is represented with three specimens from Loc. 65, one of which is worn and small. In two unworn specimens, the anteroconid and metaconid are strongly connected, but they still preserve a cusp shape because of the discrete anterior and posterior grooves. The metaconid is situated slightly in back of the anteroconid; it is connected to the mesolophid by a thick crest that we interpret as being the posterior metalophulid. This crest is longitudinal in one specimen and a little oblique in the other. The anterolophulid forms a cuspid between the anteroconid and protoconid. The posterior arm of the protoconid is strong and connected to the mesolophid. The mesolophid is strong but does not reach the lingual border. The hypolophid is transverse. The posterior arm of the hypoconid is weak. The posterosinusid is open. In both specimens, the sinusid is deep—about two-thirds of the crown height—and the protosinusid reaches about one-half of this height. There are two roots.

The m2 is almost square in outline. The lingual branch of the anterolophid is distinct and extends to the metaconid. The metalophulid is connected to the anterolophulid. The protosinusid is preserved as an enamel island. The mesolophid is transverse, medium sized, and strongly connected to the ectolophid. In a slightly worn specimen, a trace of the ectomesolophid is preserved. The hypolophulid is transverse and connected to the ectolophid, anterior to the hypoconid. The hypoconid has no posterior arm. The lingual edge of the posterolophid is curved to join the entoconid. From the labial view, the sinusid is deep, reaching 80% of the crown height.

The m3 is represented by two specimens, one of which is smaller in size. The larger specimen, with respect to its size and morphological similarity to m2, fits the other specimens from this locality. The smaller m3 (fig. 6.11H) is too small to correspond to the other specimens described here. We believe that this smaller specimen belongs to another taxon and it is described below as Spalacidae indet. The larger m3 has a strong lingual anterolophid but lacks the anterolabial enamel island. The mesolophid reaches the lingual border. The posterior portion is reduced. The posterolophid forms a continuous ridge between hypoconid and entoconid. The sinusid is transversal and almost as deep as the crown height. There are two roots.

Comparison. *Sinapospalax canakkalensis* from Loc. 65 is similar in size and cheek teeth morphology to the specimens from Bayraktepe I (Çanakkale, Turkey). Their principal common characters are: m1 with deep protosinusid, fragmented anterolophulid, strong lingual arm of the metaconid, fragmented mesolophid connected to the posterior arm of the protoconid and to the posterior metalophulid, protoconid having distinct posterior and anterior arms, long hypolophulid and posterolophulid; M1 with an enamel island on the anteroloph and an interrupted metalophule. The other species included in this genus differ from *S. canakkalensis* in having complete mesolophids on m1, a stronger lingual anterolophid on m2, and more transverse sinuses on M1 and M2. *S. marmarensis* and *S. primitivus* differ from *S. canakkalensis* in having a more anterior outline in m1. *S. incliniformis* n. sp. differs from *S. canakkalensis* in having inclined lophids, particulary in m1. Additionally, m1 of *S. marmarensis* and *S. incliniformis* n. sp. lacks the anteroconid-protoconid connection. *S. primitivus* m1 has a shallower protosinusid and a shorter mesolophid. *S. sinapensis* n. sp. differs from *S. canakkalensis* in having a more robust anterior edge, a weaker lingual arm of the metaconid on m1, and a more anteriorly directed metalophulid on m2.

This comparision allows the attribution of the material from Loc. 65 to *Sinapospalax canakkalensis*. This assignment is also supported by the similar ages (MN 7/8) of Loc. 65 and Bayraktepe I (Çanakkale, Turkey).

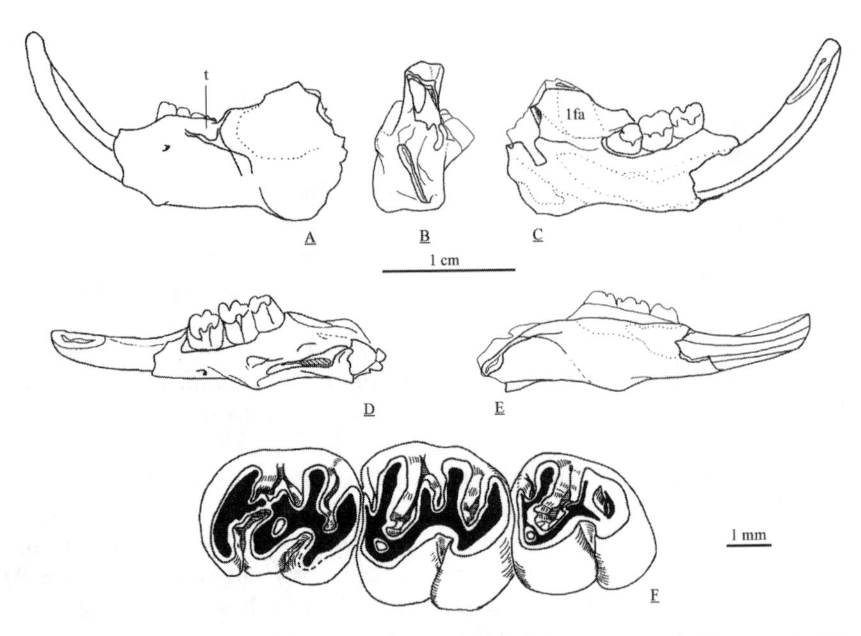

Figure 6.13. *Sinaposspalax incliniformis* n. sp. from Sinap Tepe Loc. 41. Labial (**A**), posterior (**B**), lingual (**C**), dorsal (**D**) and ventral (**E**) views of mandible. (**F**) Occlusal view of molars.

Sinapospalax incliniformis n. gen. n. sp.

Holotype. One left mandible with m1 (2.58 × 2.05), m2 (2.36 × 2.35), and m3 (2.05 × 2.02) (mandible S89/556). See figure 6.13.

Type Locality. Loc. 41.

Etymology. Derived from the shape of lophids, which are oblique, particularly on m1.

Diagnosis. Medium-sized spalacid. Slender mandible has sharp ascending ramus, obscuring labially m3; masseteric ridges are high; masseteric fossa U-shaped; mental foramen situated below the anterior root of m1; there are three ridges on the lower incisor. On the trigonid of m1 lophids, sinusids and crests are all posteriorly inclined, anteroconid and metaconid are strongly fused, metalophulid and posterior arm of the protoconid connected to the mesolophid, and the protosinusid is deep. The mesolophid is medium sized in all lower molars. The m2 is square shaped and has a deep transversal anterosinusid; the metalophulid and hypolophulid are anteriorly directed, and mesolophid is transverse. There is no remarkable size change from m1 to m3.

Differential Diagnosis. *Sinapospalax incliniformis* n. sp. differs from *Heramys eviensis, H. anatolicus,* and *Debruijina arpati* in having a lower length:width ratio in all molars and higher-crowned cheek teeth. The m1 of *S. incliniformis* differs from those of all the other species of the genus by having posteriorly oblique lophids and sinusids. Moreover, *S. incliniformis* differs from *S. marmarensis* in having a mesosinusid-protosinusid connection interrupted by metalophulid-mesolophid connection on m1 (confluent in the latter species) and anteriorly directed metalophulid on m2 and m3. *S. incliniformis* differs from *S. primitivus, S. canakkalensis,* and *S. sinapensis* by the lack of the anterolophulid on m1 and in having a well-developed lingual anterolophid and a wider anterosinusid on m2.

Description. Loc. 41 yielded three mandibles and one lower incisor. It is significant that these mandibles are well differentiated in size and morphology: they cannot be attributed to one species. Can such a large variation in size and morphology be observed in a living or fossil species? The largest populations that we examined (10 mandibles of *Spalax ehrenbergi* from Turkey and seven mandibles of *Pliospalax macoveii* from the Pliocene of Çalta, Turkey) do not show such a variation. Consequently, we decided to describe the specimens separately as *Sinapospalax incliniformis* n. sp. (left mandible S89/556), *S.* sp. 2 (left mandible S89/558) and *S.* sp. 3 (right mandible S89/557 and left lower incisor S89/559).

The mandible S89/556 has a well preserved tooth row. It is broad and deep (6.23 mm in height below m1). Part of the ascending ramus, the apex of the angular processes, and the anterior portion of the chin process (i.e., the digastric eminence) are broken. The diastema is shorter than the tooth row. The chin process is weak but distinct. The mental foramen is placed below the anterior root of m1. The ascending ramus anteriorly originates below the anterior root of m2 and steeply rises, obscuring m3; lingually, it is separated from the tooth row by a wide, shallow valley. The ventral masseteric ridge is elevated in the center of the *corpus mandibulae;* it obliquely runs across the ramus from the angular process to below the interroot area of m1; the dorsal masseteric ridge joins the ventral ridge below m2, where the *corpus mandibulae* inflates; there is a triangular muscle scar below m1–m2 (probable insertion of some of the fibers of the *temporalis* muscle). The masseteric fossa is

U-shaped and delimited anteriorly by the dorsal and ventral ridge fusion. On the labial side, between the anterior origination of ascending ramus and the tooth row, there is a particular small depression below m2 (probable insertion of some of the fibers of the *temporalis* muscle). The lingual face has a well-defined mylohyoid line under the alveoli of molars; the insertion of the *mylohyoideus* muscle is indicated by a wedge-shaped ridge that originates posteriorly below the mandibular foramen and runs through the incisor; just above the posterior part of this ridge, there is another wedge-shaped scar (probable insertion of the extension of *m. pterygoideus lateralis*), which is relatively deep and tapers anteriorly behind the tooth row. The angular process is deep dorsoventrally; it deviates slightly from the vertical plane of the mandible toward the labial side. The lingual side of the angular process has a quite deep fossa (the insertion of the muscles *masseter superficialis* and *pterygoideus medialis*). The tip of the incisor is higher than the abrasion plane of the molars and posteriorly it reaches the area of the mandibular condyle; three ridges adorn the anterior face of the incisor.

The m1 is characterized by its lophids more obliquely oriented than in any other species of *Sinapospalax*. The anteroconid is situated anterior to the metaconid; these two cusps are strongly fused by wear, forming an oblique ridge. The posterior metalophulid is a high, thick crest, slightly oblique; it joins the mesolophid. The anterolophulid is absent; consequently, the protoconid is not connected to the anteroconid and the protosinusid is very deep. The posterior arm of the protoconid is curved backward and connected to the mesolophid, enclosing an enamel island. The mesolophid is medium sized and derived obliquely from the ectolophid in front of the hypolophulid. The sinusid is simple in shape and strongly directed backward; in this it resembles *Heramys*. The hypolophulid is short and directed anteriorly. The swollen form of the posterolophid may indicate the presence of the posterior arm of the hypoconid, which is fused by wear. The sinusid is deep, extending three-quarters of the crown's maximum height.

Mandibular m2 is as long as m1 but much wider. The lingual anterolophid is distinct and joins the anterior wall of the metaconid. The protosinusid is preserved as an enamel island. The metalophulid is anteriorly directed; it is connected to the anterior arm of the protoconid. There is a small longitudinal spur on the metalophulid. The transverse mesolophid is long but does not reach the lingual border. The hypolophulid is anteriorly directed. The posterolophid is long and joins the base of the entoconid. The sinusid is deep and directed posteriorly.

Mandibular m3 is longer than wide and posteriorly narrow. The lingual anterolophid is short and its internal wall has a small spur. The protosinusid is preserved as an enamel island. The mesolophid is short and directed toward the metaconid. The hypolophulid is transverse. The entoconid and the posterolophid are fused so that they enclose a large enamel island. The sinusid is deep and transversally oriented.

Comparison. This species is defined as a new species of the genus *Sinapospalax* because it has a number of characters that cannot be observed in other species of *Sinapospalax:* on m1, lophids and sinusids have an oblique orientation, the hypolophulid anteriorly directed, the sinusid is far back and simple in shape, there is no protoconid-anteroconid connection, and the hypolophulid is short; on m2, the metalophulid is connected to the protoconid instead of the anterolophid. Some characters of this species are shared with *Heramys,* such as the oblique sinusid on m1 and the connection of the metalophulid to the protoconid. However, *Heramys* differs from this species in having a higher length:width ratio for molars, a rounded anterior part, and a complex trigonid pattern on m1, which is also characterized by its transverse main lophids. The m1 of *S. marmarensis* and *S. primitivus* shows some similar characters to *S. incliniformis,* such as the anteroconid situated quite anteriorly to the metaconid (plesiomorphic character), shallow protosinusid (from the labial side), absence of anteroconid-protoconid connection, and elongated outline; however, *S. marmarensis* differs by having a larger anteroconid, robust posterior metalophulid (which does not reach the mesolophid), transverse sinusid, and transverse hypolophulid. *S. primitivus* differs from *S. incliniformis* in its rather transverse lophids and sinusids on m1 and m2.

Loc. 41 also yielded two other poorly preserved mandibles. They are more robust and larger than the mandible described here as *S. incliniformis* and also have visible morphologic differences; they are described later in this chapter as *Sinapospalax* sp. 2 and *Sinapospalax* sp. 3. The morphologic differences between *S. incliniformis* and these specimens are also described later.

Sinapospalax sinapensis n. gen., n. sp.

Holotype. One m1 sin (2.79 × 2.38) (ST84-101). See figure 6.14A.

Type Locality. Loc. 84, Sinap Tepe (Ankara).

Etymology. Derived from the name of Sinap Tepe, the main hill across the Sinap Formation (Ankara).

Paratypes. Two m1 sin (2.65 × 2.62, 2.80 × 2.39), two m2 sin (2.55 × 2.51, 2.75 × 2.42), one m3 sin and one m3 dex (2.33 × 2.18, 2.68 × 2.16), one M1 sin (3.22 × 2.65) (ST84-102–ST84-108). See figures 6.14, 6.15.

Diagnosis. Medium-sized to large spalacid. Molars with sinus(ids) slightly directed forward (backward). The m1 with the square-shaped anterior part has the anteroconid and metaconid in the same plane. The m2 has an open anterosinusid even in adults, a distinct lingual anterolophid, and a short metalophulid anteriorly curved and connected to the anterolophid; the mesosinusid reaches the anterolophid.

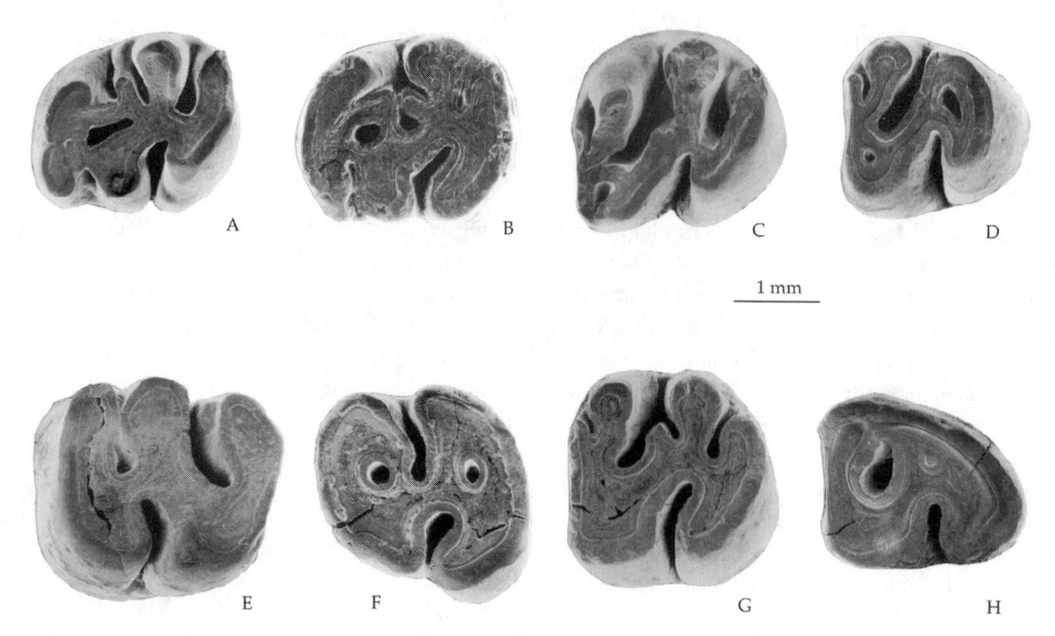

Figure 6.14. *Sinaposalax sinapensis* n. sp. from Sinap Tepe Loc. 84. Occlusal view of molars. (**A, B, F**) m1; (**C, G**) m2; (**D, H**) m3; (**E**) M1.

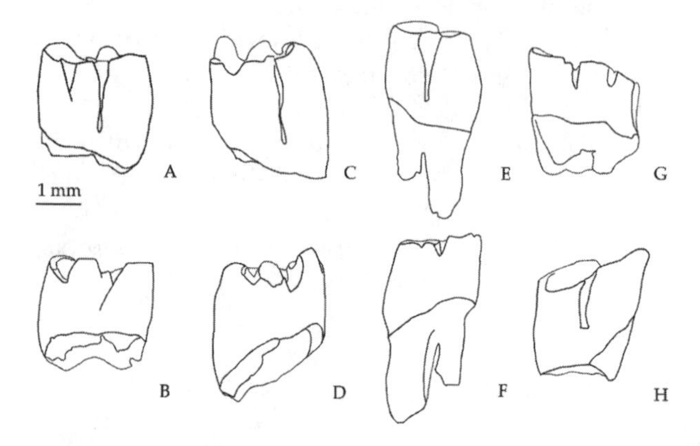

Figure 6.15. *Sinapospalax sinapensis* n. sp. from Sinap Tepe Loc. 84. Labial and lingual views of molars. (**A, B**) m1; (**C, D**) m2; (**E, F**) m3; (**G, H**) M1.

Differential Diagnosis. *Sinapospalax sinapensis* differs from all other species of this genus by its m2 having and L-shaped mesosinusid, which is delimited anteriorly by the anterolophid but not by the metalophulid as in other species; its mesolophid is shorter and anteriorly directed whereas in all the other *Sinapospalax* species it is longer and transverse. This species differs from the *S. marmarensis* and *S. incliniformis* n. sp. in having the anteroconid-metaconid couple in the same anterior plane and a strong anteroconid-protoconid connection on m1. *S. sinapensis* differs from *S. primitivus* by its m1, which is comparatively more square and wider anteriorly and has a deeper protosinusid than in the latter species.

Description. Maxillary M1 is the unique upper molar found at Loc. 84. Compared with the other specimens, its size is slightly larger. It is longer than it is wide. Because the specimen is worn, it is difficult to identify unambiguously which molar it is. The straight anterior face suggests that the tooth is an M2. However, the absence of the anterior contact and the central position of the paracone suggest that it is an M1. The mesoloph is absent or totally fused with the protolophule due to the attrition. This is also the case for posteroloph and metalophule. The anteroloph, protolophule, and metalophule are transverse. There is a small tubercule attached to the anterior wall of the metacone. The transversal sinusid slightly exceeds half of the crown height. The roots are not preserved.

The mandibular m1 is longer than it is wide. The metaconid is fused with the anteroconid; a shallow groove is present on the anterior face between these cusps in a slightly worn specimen. This groove is connected to the protosinusid, so that the anteroconid and metaconid seem to be separated by a very superficial valley. The protosinusid is interrupted by the fusion of the anterior arm of the protoconid and anterolophulid. The posterior metalophulid is either longitudinal or oblique but directed posteriorly; it forms a bridge connecting the metaconid to the mesolophid. The protoconid has a well-developed posterior arm in one specimen; it is absent in the other. The mesolophid is short and oblique. In a slightly worn specimen, there is a mesostylid delimiting the mesolophid in the lingual border. The hypolophulid is transverse. The posterolophid is long and separated from the entoconid. The sinusid is set slightly back. From the labial side, on the less worn specimen, the protosinusid extends one-half of the depth of maximum crown height and the sinusid is three-quarters of the crown height in its depth. There are two equally developed roots.

The mandibular m2 is slightly longer than it is wide. The lingual anterolophid is still present and straight. In occlusal view, the anterosinusid is quite wide and deep, hence it is not closed even in an advanced stage of wear.

The metaconid is a large, prominent cusp. The metalophulid is directed anteriorly to join the anterolophid. The anterolophulid and metalophulid are separated by a slender mesosinusid. Posteriorly, the anterolophulid joins the protoconid and delimits the mesosinusid and the protosinusid. The protosinusid is transformed into a quite large enamel island. The mesolophid is short. The hypolophulid is transverse. The posterolophid is long and ends with a conule at the lingual side in the slightly worn specimen. The sinusid is three-quarters of the crown height in depth and slightly directed posteriorly. The m2 has two equally developed roots.

The mandibular m3 is longer than it is wide. The anterolophid is short and straight. The metalophulid is connected to the anterolophid, as on m2. The protosinusid is preserved as an enamel island. The mesolophid is absent. The oblique hypolophulid is connected to the ectolophid in front of the sinusid. The posterolophid is strongly curved, joining the entoconid and enclosing the posterosinusid as an enamel island. The depth of the sinusid is slightly more than two-thirds of the maximum crown height. There are one compressed anterior and one cylindrical posterior roots.

Comparison. *Sinapospalax sinapensis* is remarkable with its more transverse sinus(id) in all molars and with m1 having a more robust anterior part than in the other species of the genus, except the giant species from Loc. 12 (see later in this chapter). The complex trigonid pattern of m1 of *S. sinapensis* resembles that of *S. canakkalensis,* but the former has a metaconid without any pronounced lingual arm; *S. canakkalensis* has m2 with a weaker anterolophid, its anterosinusid easily disappears by attrition, and its mesosinusid is almost straight compared with the L-shaped mesosinusid of the *S. sinapensis. S. marmarensis* and *S. incliniformis* n. sp. are different from *S. sinapensis* because their m1 lacks the anteroconid-protoconid connection and has a metaconid that is situated posteriorly to the anteroconid. *S. primitivus* from Sariçay differs from *S. sinapensis* in the shape of the m1, which is rounded and narrower anteriorly, and its M1 has a posteriorly directed mesoloph. Thus the material of Loc. 84 is described as a new species of the genus *Sinapospalax.*

Sinapospalax n. sp.

Locality. Loc. 12 (Ankara).

Material. One m1 dex (3.58 × 3.07) (ST12-1). See figure 6.16.

Description and Remarks. The only specimen from Loc. 12 is a right m1. This is the largest spalacid tooth that has been found to date. The strongly fused anteroconid and metaconid are separated from the posterior part by the mesosinusid and protosinusid. The protoconid is directed anterolabially but is not connected to the anteroconid; it has a weak posterior arm but lacks the anterior arm. There is a large cusp between the anteroconid-metaconid complex

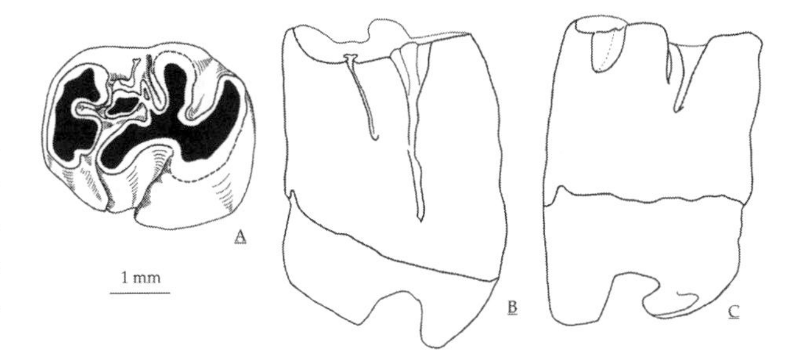

Figure 6.16. *Sinapospalax* n. sp. from Sinap Tepe Loc. 12. Occlusal (**A**), labial (**B**) and lingual (**C**) views of m1.

and mesolophid; it seems isolated on the occlusal surface but unites with the anteroconid and mesolophid at the base. Thus, this cusp is considered as a segmented posterior metalophulid. The mesolophid is long and narrow; it reaches the labial border and there joins the anterolabial base of the entoconid. The hypolophulid is transverse. The posterolophid is separated from the entoconid by a wide posterosinusid. The sinusid depth is about three-quarters of the crown height. The protosinusid does not reach one-half of the maximum crown height in its depth. This tooth has one cylindrical anterior and one compressed posterior root.

Compared with other spalacids, one of the most prominent features of this specimen is its large size, which is striking. Moreover, this m1 is relatively higher crowned and has thicker ridges. However, this tooth shares many characters with other species referred to *Sinapospalax* and differs clearly from all species of *Debruijnia* and *Heramys. S.* n. sp. differs from *S. marmarensis* in having a segmented metalophulid that interrupts the mesosinusid, and deeper proto- and mesosinusid from the labial view. *S.* n. sp. differs from *S. primitivus, S. canakkalensis,* and *S. sinapensis* by the presence of a long mesolophid that reaches the lingual border.

Together with the gigantic size, all of the above-mentioned differences suggest that this specimen should belong to a new species—perhaps even a new genus. However, we would not erect a new taxon based on a unique m1. One M1 and one m3 of gigantic spalacids are recorded at Düzyayla I (MN 11, Anatolia; Ünay 1996) and Çalta (MN 15, Ankara, Anatolia; undescribed material collected by S. Sen in 1995). These remains may indicate that a rare, very large spalacid existed in Turkey during the late Miocene and Pliocene. The m1 from Loc. 12 shares the following characters with *Sinapospalax* to justify its inclusion in this genus: high crown, square-shaped anterior part, anteroconid and metaconid situated in the same plane and preserving their conical shapes, deep protosinusid, and fragmented posterior metalophulid forming a bridge between the metaconid and mesolophid.

Sinapospalax sp. 1

Locality. Loc. 120.

Material. One m3 sin (2.22 × 1.93) (ST120-1). See figure 6.8H.

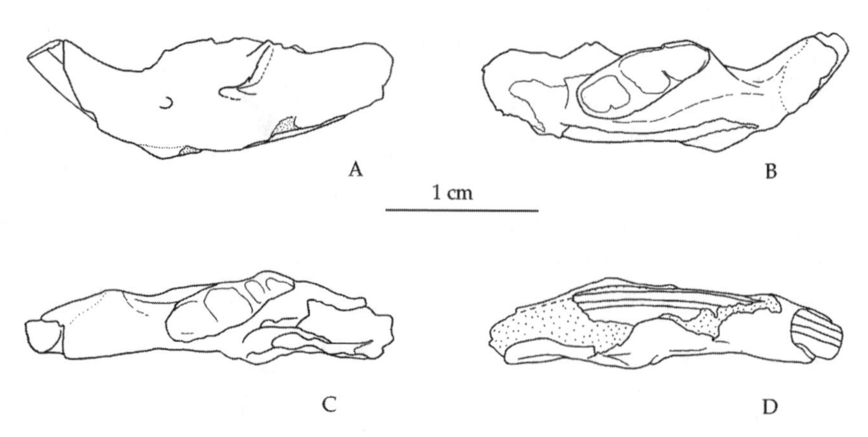

Figure 6.17. *Sinapospalax* sp. 2, from Sinap Tepe Loc. 41. Labial (**A**), lingual (**B**), dorsal (**C**) and ventral (**D**) views of mandible.

Description and Remarks. The only specimen from this locality is a well-preserved left m3. The lingual anterolophid is slightly curved and joins the metaconid. The metalophulid is short, fragmented, and connected to the anterolophulid. The metaconid seems to be isolated due to the fragmented metalophulid. The protosinusid is preserved as an enamel island. The mesolophid is absent. The hypolophulid is transverse and connected to the ectolophid. The posterolophid is also transverse but its lingual end is strongly curved and joins the metaconid, enclosing the posterosinusid as an enamel island. The sinusid is deep and transverse.

Sinapospalax sp. 2

Locality. Loc. 41 (Kazan, Ankara).

Material. One left mandible (S89/558). See fig. 6.17.

Description and Remarks. This mandible is poorly preserved and lacks the molars. The mandibular condyle, the coronoid process, and the angular process are missing. The *corpus mandibulae* is relatively shallow (7.13 mm in height below m1) but broad. The diastema is deep and symetrical; it is shorter than the tooth row. The mental foramen is large and situated below the anterior root of m1; it is at the same level as the anterior end of the masseteric ridge. The ascending ramus originates below the interroot area of m2 and is separated from the *corpus mandibulae* by a narrow valley. The masseteric scar is bordered by elevated dorsal and distinct ventral ridges that join each other below m2. The masseteric fossa is limited anteriorly by the fusion of dorsal and ventral ridges of the masseteric scar. Labially below the tooth row, there is a small slight depression situated between the base of the roots of m2. Lingually below the tooth row, there is a slight mylohyoid line. The chin process (digastric eminence) has a distinct groove at its ventral limit, parallel to the *corpus mandibulae.* The incisor rises above the abrasion table of the molars. There are three ridges on the anterior face of the incisor.

This mandible differs from all the other mandibles from the Sinap Tepe localities by having a shallow *corpus mandibulae* and a narrow valley between the ascending ramus and the *corpus mandibulae* (in these characters, it resembles the genus *Spalax*). But, as in *Sinapospalax incliniformis* n. gen. n. sp. from Loc. 41, the *corpus mandibulae* is broad, the masseteric ridges are elevated, the ascending ramus originates below the anterior root of m2, and there are three ridges on the incisor surface. Thus, it may be included in the genus *Sinapospalax.*

Sinapospalax sp. 3

Locality. Loc. 41.

Material. One right mandible (S89/557). See figure 6.18.

Description and Remarks. This mandible is deep and broad; the *corpus mandibulae* is 8.54 mm in height. The diastema is deep. The mental foramen is relatively small and situated below the posterior half of the diastema and anteriorly at the same level of the fusion of masseteric ridges. The ascending ramus is thick anteriorly and more slender posteriorly; there is a strong groove on the anterior border; it originates anteriorly, below the limit of m1–m2 and rises abruptly, obscuring the posterior part of m3; lingually, it is separated from the tooth row by a wide, deep valley such that the declination of the tooth row from the horizontal plane of the mandible is high; there is a deep fossa situated on the ventral part of the ascending ramus delimited by this valley. The ventral ridge of the masseteric scar is slight but more prominent than the dorsal ridge; the dorsal limit of the masseteric scar fuses with its ventral limit medially on the mandible below m1–m2 border, making a V-shape; thus, the *corpus mandibulae* is inflated anterior to this fusion. The masseteric fossa is V-shaped and depressed. On the lingual side of the mandible, the mylohyoid line is faintly defined; there is an elevated wedge-shaped ridge between the posterior part of the tooth row and the mandibular foramen (probable insertion of the extension of the *musculus pterygoideus lateralis*). The tooth row is elevated and recumbent lingually. The incisor posteriorly extends at least to the area of the mandibular condyle. There are three ridges on the anterior surface of

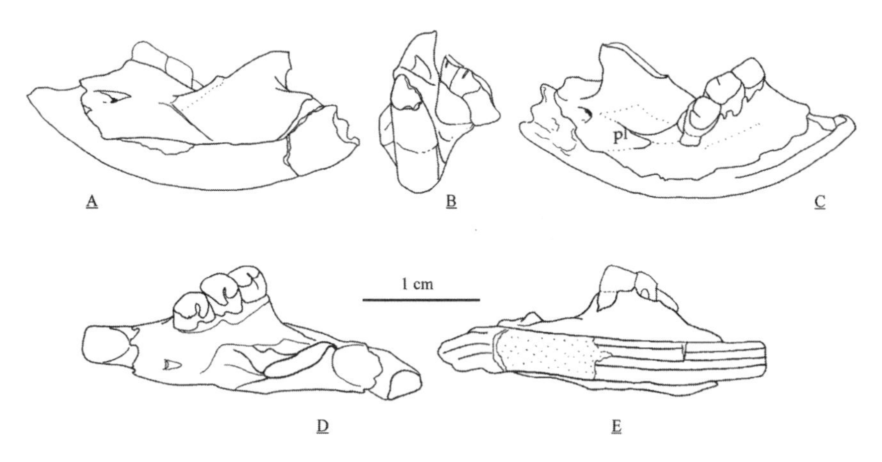

Figure 6.18. *Sinapospalax* sp. 3 from Sinap Tepe Loc. 41. Views of mandible. (**A**) Labial; (**B**) posterior; (**C**) lingual; (**D**) dorsal; (**E**) ventral. Abbreviation: pl: insertion of the *pyterygoideus lateralis* muscle.

the incisor. The molars are worn, but it is still possible to see a trace of the protosinusid, indicating that it was deep.

This mandible is described as *Sinapospalax* sp. 3 because it is large, the *corpus mandibulae* is broad, the first molar has a deep protosinusid, and the ascending ramus rises abruptly. All these characters are shared with the new genus *Sinapospalax* (see also *Pliospalax* later in this chapter).

This mandible is different from the other mandibular material of the Sinap Formation localities by being larger and more robust, and having a high tooth row declination angle, an elevated wedge-shaped ridge between the tooth row and the mandibular foramen (here interpreted as being the insertion area of the extension of *pterygoideus lateralis*), and a mental foramen situated anterior to the anterior root of m1.

Spalacidae gen. and sp. indet.

Locality. Loc. 65.

Material. One m3 dex (1.81 × 1.62) (ST65-9). See figure 6.11H.

Description and Remarks. This specimen is notably smaller than the other m3 specimens from Loc. 65 referred to *Sinapospalax canakkalensis*. The lingual anterolophid is cusplike. The protoconid has a lingual spur. The metaconid is a high cusp connected posteriorly to the mesolophid but anteriorly separated from the anterolophid by a narrow anterosinusid. The hypolophulid is short and directed anterolingually; it is almost fused with the mesolophid. The posterolophid curves through the lingual border and joins the hypolophulid. The posterosinusid is preserved as a small enamel island. The sinusid descends almost to the base of the crown. The roots are broken.

This m3 is smaller than that of other Miocene and Pliocene spalacids, except for some m3 specimens of *Heramys eviensis* from Aliveri. It may be considered as unusual; however, its semihypsodont crown, occlusal pattern, lophid-sinusid ordering, and the ease of the recognition of character homologies with spalacid m3 suggest that this specimen belongs to the Spalacidae. Unfortunately, m3 is not sufficiently characteristic for further determination.

Remarks on the Systematics of the Spalacidae

In previous studies on fossil Spalacidae, their Miocene and Pliocene representatives are grouped in three genera: *Debruijnia, Heramys,* and *Pliospalax.* Here, we add a new genus, *Sinapospalax,* which includes some middle and late Miocene species previously referred to *Pliospalax.* Therefore, some brief characterizations are needed to clarify the main features and the species content of these genera.

Debruijnia *Ünay, 1996*

This genus is known only from its type species *Debruijnia arpati* Ünay, 1996. Its molars are low crowned and still retain many cricetid characters. Thus, the Spalacidae would appear to be derived from a primitive member of the paraphyletic taxon Cricetidae s.l. The main features of its molar pattern are high lophs, narrow valleys, m1 and m2 with strong posterior arms of protoconid and hypoconid, ectomesolophid occasionally present, m1 with an anteroconid situated on the longitudinal axis of the crown, M1 with protosinus and sometimes with an anterocone, which may be bicuspid interiorly, and M3 with lingually closed sinus.

Heramys *Hofmeijer and de Bruijn, 1985*

The type species is *Heramys eviensis* Hofmeijer and de Bruijn, 1985, from Aliveri (MN 4, Greece). Locality 4 of Sinap Tepe yielded a new species, *H. anatolicus.* This genus is characterized by small to medium-sized cheek teeth, elongated m1 with anteriorly oblique protosinusid and posteriorly oblique sinusid, a very large metaconid with respect to the anteroconid, and anteroconid-metaconid connection by an anterolophid.

Sinapospalax *n. gen.*

As shown earlier, we retain *Sinapospalax canakalensis* (Ünay 1981) as its type species. Taking into account their morphological similarities, the following species are grouped in *Sinapospalax: S. primitivus, S. marmarensis, S. icliniformis,* and *S. sinapensis.* The characteristics of this genus and of species referred to are given above.

Pliospalax *Kormos, 1932*

The type species is *Pliospalax macoveii* (Simionescu 1930). Subsequently, several other newly described species were included in this genus. Here, we retain some of them as belonging to this genus (see later in this chapter) and attribute others to the new genus *Sinapospalax.* Kormos (1932) based this genus on the presence of two labial and two lingual reentrant folds on unworn m1 and used these characters to distinguish a Pliocene spalacid from the extant genus *Spalax.* However, the new discoveries show that such reentrant folds (labially protosinusid and sinusid, lingually mesosinusid and posterosinusid) are plesiomorphic characters shared with many muroid rodents and recently described spalacid genera: *Debruijnia, Heramys,* and *Sinapospalax.* Ünay (1990, 1996), added the following traits to the original diagnosis of the genus: M1 protocone is entirely incorporated into the anteroloph and an anterolabial sinusid is developed as a lake even in unworn m2 and m3.

The description of the new genus *Sinapospalax* and the inclusion in this genus of some species previously referred to *Pliospalax* lead us to redefine the main characteristics of the genus *Pliospalax;* hence the emended diagnosis and differential diagnosis given here.

Emended Diagnosis. Small to medium-sized spalacid. High crowned molars have lower length:width ratios, loph(id)s are short, thick, and reduced in number; all molars tend to acquire S-Z shape pattern in early stages of attrition. The m1 generally has a rounded anterior part, a very shallow protosinusid, and an oblique and weak mesolophid. On m2, the protosinusid is preserved as a lake in young individuals, but disappears in adults; a very weak anterosinusid is present in juvenile specimens; the mesolophid is frequently absent or weak. The M1 has the protocone entirely incorporated into the anteroloph; sometimes a weak mesoloph is present. The M2 has the paracone incorporated into the anteroloph; the mesoloph is fused with the paracone, thus enclosing the mesosinus as a transverse enamel island; the metalophule is incorporated into the posteroloph. The M3 pattern is incomplete, anterior and posterior lophs are fused. *Corpus mandibulae* has a parabolically rising ascending ramus that obscures the posterior part of m3 and originates inferior to m1–m2. There is a strong groove on the anterior border of the ascending ramus. A dorsoventrally deep angular process rises steeply to the area of mandibular condyle. The lower incisors have a smooth surface.

Differential Diagnosis. *Spalax* Guldenstaedth, 1770, differs from *Pliospalax* by having higher crowned cheek teeth, simplified molar pattern with Z-S-shape organization of loph(id)s even in young individuals, wide loph(id)s and narrow reentrant folds, the absence of protosinusid and mesolophid on m1 and the absence of anterosinusid on m2, and by having upper molars that tend to show one semicylindrical root and a narrower *corpus mandibulae.* *Sinapospalax* n. gen differs from *Pliospalax* by having larger molars; its m1 has a cusplike anteroconid and metaconid separated by discrete posterior and/or anterior grooves, and by having a protoconid with a posterior arm, strong mesolophid, generally deep protosinusid, long hypolophulid, and a higher length:width ratio. Its M1 has a protocone distinct from the anteroloph, a deep mesosinus, slightly oblique protolophule and metalophule; its M2 has a narrow but deep anterosinus, a strong mesoloph reaching the labial border, a paracone with anterior or posterior or double connection; and its M3 has a subrounded outline and complete pattern. *Sinapospalax* also has a broad *corpus mandibulae* with an abruptly and more steeply rising ascending ramus with more posterior origination, more elevated masseteric ridges, more depressed masseteric fossa, and three ridges on the lower incisors. *Heramys* Hofmeijer and de Bruijn, 1985, differs from *Pliospalax* by lower crowned molars with higher length:width ratio, m1 has a very large metaconid with respect to its anteroconid, an anteriorly oblique protosinusid and posteriorly oblique sinusid, long mesolophid, and protoconid and hypoconid with posterior arms. Its M1 has a distinct protocone and transverse lophs, M2 and M3 has a five lophed complete pattern. *Debruijnia* Ünay, 1996, differs from *Pliospalax* by having lower crowned molars with a higher length:width ratio and by retaining the cricetid pattern.

Species included in the genus *Pliospalax* are:

Pliospalax macoveii (Simionescu 1930), type species;
Pliospalax compositodontus Topachevski, 1969;
Pliospalax sotirisi (de Bruijn, Dawson and Mein 1970);
Pliospalax tourkobounensis de Bruijn and van der Meulen, 1975.

Spalax *Guldenstaedth, 1770*

There is presently no consensus on whether the extant spalacids belong to a single genus or several genera. Although some workers include all living species and their late Pliocene-Pleistocene representatives in one genus, *Spalax,* other specialists (Mehely 1913; Topachevski 1969; Gromov and Baranova 1981; Carleton and Musser 1984; de Bruijn 1984; Gromov and Erbajeva 1995), recognize three genera: *Microspalax, Spalax,* and *Nannospalax.* Because of the karyotype heterogeneity in living species, it has been proposed that the single genus *Spalax* would be a convenient nomen to avoid confusion until a thorough taxonomic revision of the family is done (Savic and Nevo 1990; Nevo et al. 1995). This idea is followed by Ünay (1996, 1999), who uses generic nomen *Spalax* for all of the younger spalacids usually characterized by two lingual and one labial reentrant folds on the m1. In addition to these characters, de Bruijn and Van der Meulen (1975, p. 330) recognize two other dental characteristics: "m1 and m3 relatively short, while the posterior loph of m3 is not isolated from the anterior part." It is not within the scope of this chapter to define the generic status of living spalacids and their Pliocene-Pleistocene relatives. For the purpose of comparison, all the living spalacids are grouped under the nomen *Spalax* s.l. The living species of *Spalax* as well as those from late Pliocene (Karaburun and Odessa) and Pleistocene localities in Greece, Turkey, Israel, and the like share a number of derived molar characters defined by simplified occlusal pattern with S-Z shape, increased semihypsodonty, and lophodonty.

Phylogenetic Status of *Sinapospalax*

This part of our study addresses the phylogenetic relationships of the Spalacidae based on molar morphology and tests the specific hypothesis that *Sinapospalax* n. gen. is a member of this group.

First, we propose a set of discrete characters, each with two or more states. We assume that topologically related structures are homologous. The characters used in this study are listed in figure 6.1A. Characters are expressed as discrete binary (0, 1) or multistate (0, 1, 2, 3, etc.), according to their morphologic variation and are then entered into a data matrix (see table 6.2). In this table, when a character was found to be polymorphic, the predominant state is used. When a character was clearly variable, it is expressed as "0&1" or any other potential set of states. The states that are not applicable to certain taxa are entered as a dash (—). We analyzed the dataset by using PAUP version 3.1 (Phylogenetic Analysis Using Parsimony; Swofford 1993). Character states were optimized with ACCTRAN (Accelerated Transformation) analysis. Characters were polarized using outgroup comparison.

Outgroup Taxa

Two genera—*Prokanisamys* and *Spanocricetodon*—were chosen as outgroup taxa that are thought to be sufficiently related to the ingroup so that character homologies can be established without difficulty.

Prokanisamys *de Bruijn, Hussain and Leinders, 1980*

This form is known from the early–middle Miocene of the Murree Formation (Banda Daud Shah, Pakistan). It was defined as a new genus and species, *Prokanisamys arifi,* based on isolated cheek teeth and included in the Rhizomyidae (de Bruijn et al. 1980, p. 73). In the differential diagnosis, two groups of characters are indicated: "transversely to posteriorly directed hypolophulids and the position of mesolophids which arise from the protoconid and not from the mure" as primitive and "the height of the ridges in *Prokanisamys* relative to the cusps (almost as high as the cusps), the strong connection between the protocone and the anteroloph in M1 and the small anterior cingula in M2 and M3" as derived. These authors state that most of the dental features are also diagnostic for *Kanisamys* Wood, 1937. However, *Kanisamys* is characterized by the lophate pattern of cheek teeth and a relatively large M3, which are accepted as the most diagnostic features for this genus. For de Bruijn et al. (1980), *Prokanisamys* should be the direct ancestor of *Kanisamys.* At the family level, the similarity in dental morphology of Rhizomyidae, Tachyoryctidae, and Spalacidae is regarded as the result of adaptative convergence; therefore, they are considered to be separate families.

Flynn et al. (1985) insisted on the neccessity of identifying apomorphies in the definition of families. Considering all Muroidea, the myomorphy and a well-developed anterior protoconid-metaconid connection on m1 are proposed as apomorphies for Cricetidae. Early rhizomyids (*Kanisamys, Prokanisamys*), which are myomorphous but lack well-developed anterior and posterior protoconid-metaconid connections on m1 (Flynn 1982), are excluded from Cricetidae as a separate family; the same is true for the spalacids, which are primitively hystricomorphous and lack this connection. Flynn et al. (1985, p. 608) also stated that Rhizomyidae and Spalacidae have separate origins: "these traits indicate that spalacids evolved from histricomorphous muroids; that is spalacids evolved from more primitive muroids than did rhizomyids." We retain *Prokanisamys* for the present phylogenetic analysis because it represents, among all muroids, the morphologicaly most similar genus to the oldest Spalacidae that is proposed to have muroid heritage (Flynn et al. 1985).

Spanocricetodon *Li, 1977*

This genus was first described with *Spanocricetodon ningensis* by Li (1977) from the early Miocene of China. De Bruijn et al. (1980) reported two more species, *S. khani* and *S. lii,* from the early Miocene Murree Formation in Pakistan. The diagnostic features of this genus, as reported by de Bruijn et al. (1980, p. 81) are: "*Spanocricetodon* comprises small cricetid species whose lower cheek teeth may or may not have a short mesolophid. Mesolophs of first upper molars usually short and directed towards the metacone rather than towards the labial margin. Metalophule of M1 usually transverse and lingually connected to hypocone. The simple anterocone of M1 usually bears a 'ledge' or cingulum on the anterior face." We retain this genus for the present phylogenetic analysis as a primitive representative of Asian Cricetidae because the spalacids are presumably

Table 6.2. Data Matrix for Family Spalacidae

	Spalacidae, matrix	1 Pr	2 Sp	3 Spx	4 Plx	5 Snx	6 Her	7 Deb
1	m1 anteroconid	0	0	—	1	1	1	0
2	m1 metaconid	0	0	—	1	1	1	1
3	m1 anteroconid-metaconid	0	0	3	3	2	1	1
4	m1 anteroconid-metaconid size	0	0	—	0	0	1	0
5	m1 protoconid-metaconid	0	0	1	1	1	1	1
6	m1 protoconid	0	0	—	1	1	0	0
7	m1 mesolophid	1	0	0	2	1	1	1
8	m1 labial anterolophid	0	0	2	1	1	1	1
9	m1 anterolophulid	0	0	4	3	2	0	1
10	m1 hypolophulid orientation	1	0	1	1	1	1	1
11	m1 posterior arm of protoconid	0	0	0	0	1	1	1
12	m1 posterior arm of hypoconid	0	0	0	0	0	1	1
13	m1 ectomesolophid	0	1	1	1	1	1	0
14	m1 number of lingual reentrant fold	0	0	1	0	0	0	0
15	m1 protosinusid	0	0	0	0	0	1	0
16	m1 depth of protosinusid	0	0	3	2	1	1	1
17	m1 anterior shape of tooth	0	0	3	2&3	2	3	1
18	m2 lingual anterolophid	1	0	2	2	1	1	0
19	m2 posterior arm of hypoconid	0	0	0	0	0	0	1
20	m2 protosinusid occurrence	0	0	2	1	1	0	0
21	m2 posterosinusid occurrence	0	0	1	1	0	0	0
22	m3 protosinusid occurrence	0	0	2	1	1	0	0
23	m3 anterosinusid occurrence	0	0	2	0	0	0	1
24	M1 position of metacone	0	0	0	0	1	0	0
25	M1 metalophule orientation	2	0	2	2	1	1	0
26	M1 mesoloph	0	0	1	1	0	0	0
27	M1 anterior arm of protocone	0	1	0	0	0	1	0
28	M1 ectoloph	0	0	1	1	1	1	0
29	M1 protosinus occurrence	1	0	3	3	2	2	1
30	M2 Paracone	0	0	2	2	1	0	0
31	M2 Anterolophule	0	0	1	1	0	0	0
32	M2 Protolophule	0	0	1	1	0	0	0
33	M2 Mesoloph	0	0	2	2	1	0&1	1
34	M2 Metalophule	0	0	1	1	0	0	0
35	M2 Ectoloph	0	1	1	1	0	0	0
36	M2 Posterosinus	1	0	2	2	1	0	0
37	M3 Hypocone-protocone	1	0	3	3	3	2	1
38	M3 Endoloph	0	0	2	2	1	2	0
39	M3 Lingual branch of anteroloph	1	0	1	1	1	1	1
40	M3 Labial ends of lophs	1	0	3	3	2&3	1	1
41	M3 Anterosinus	0	0	2	2	1	1	0
42	M3 Number of labial sinus	0	0	2	2	1	1	0
43	M3 Shape of tooth	0	0	2	2	1	0	0
44	Endoloph (ectolophid) position	1	0	—	1	1	1	1
45	Loph and cone relation	0	0	2	1	1	1	1
46	Crown height	0	0	2	2	2	1	1
47	Hypsodonty of upper molars	0	0	1	1	1	1	1
48	Sinus(id)s width	0	0	1	1	1	1	1
49	M1 and m1 anteocone(id)	1	0	2	1&2	1	1	1
50	Loph(id)s height	1	0	1	1	1	1	1
51	Number of roots	0	1	1	1	1	1	1
52	M1/M3 length ratio	0	0	1	1	1	1	1

Notes: Taxa choosen as outgroup, Pr: *Prokanisamys*, Sp: *Spanocricetodon*. Ingroup taxa: Spx: *Spalax*, Plx: *Pliospalax*, Snx: *Sinapospalax*, Her: *Heramys*, Deb: *Debruijnia*. &, indicates polymorphic states; —, states not applicable to taxon.

originated from muroid-cricetid stock (Hofmeijer and de Bruijn 1985; Flynn et al. 1985; Ünay 1996, 1999).

Results

Only one shortest tree among 945 trees resulted from the analysis (length 97, consistency index [CI]: 0.91 [0.84, excluding uninformative characters], retention index [RI]: 0.83). In figure 6.19, only the characters that do not vary as a function of optimization are presented. The cladogram generates four monophyletic groups at nodes A, B, C, and D.

Node A: (*Debruijnia* (*Heramys* (*Sinapospalax* (*Pliospalax, Spalax*)))) form a monophyletic group based on 14 synapomorphies, 12 of which do not vary with optimization (see fig. 6.19). The unambiguous synapomorphies are represented by one homoplaseous (11(1)) and 11 autapomorphic characters (2(1), 3(1), 5(1), 8(1), 16(1), 33(1), 45(1), 46(1), 47(1), 48(1), 52(1)).

Node B: (*Heramys* (*Sinapospalax* (*Pliospalax, Spalax*))) are monophyletic based on 11 synapomorphies, nine of which do not vary with optimization. All the unambiguous synapomorphies are unique (characters 1(1), 13(1), 18(2), 25(1), 28(1), 29(2), 38(2), 41(1), 42(1)).

Node C: (*Sinapospalax* (*Pliospalax, Spalax*)) are sister taxa based on 10 synapomorphies, five of which do not vary as a function of optimization. All of the unambiguous synapomorphies are unique (characters 6(1), 20(1), 22(1), 40(3), 46(2)).

Node D: (*Pliospalax, Spalax*) are sister groups based on 21 synapomorphies, 11 of which do not vary with optimization. All of the unambiguous synapomorphies are unique (characters 21(1), 25(2), 26(1), 29(3), 31(1), 32(1), 33(2), 34(1), 35(1), 41(2), 42(2)).

The new genus *Sinapospalax* is a discrete monophyletic taxon based on four unambiguous autapomorphies (characters 17(2), 24(1), 37(1), 38(1)): (1) the blunt anterior shape of m1; (2) isolated metacone on M1; (3) closely spaced or slightly connected hypocone-protocone on M3 (homoplaseous character); and (4) when present, segmented endoloph on M3.

In the cladogram, all other terminal taxa except *Pliospalax* appeared as monophyletic with certain unambiguous unique automorphies: four for *Debruijnia* (characters 9(1), 12(1), 19(1), 23(1)), four for *Heramys* (characters 4(1), 15(1), 12(2), 27(1)), and six for *Spalax* (characters 8(2), 14(1), 20(2), 22(2), 23(2), 45(2)). However, the autapomorphies of *Pliospalax* vary as a function of optimization.

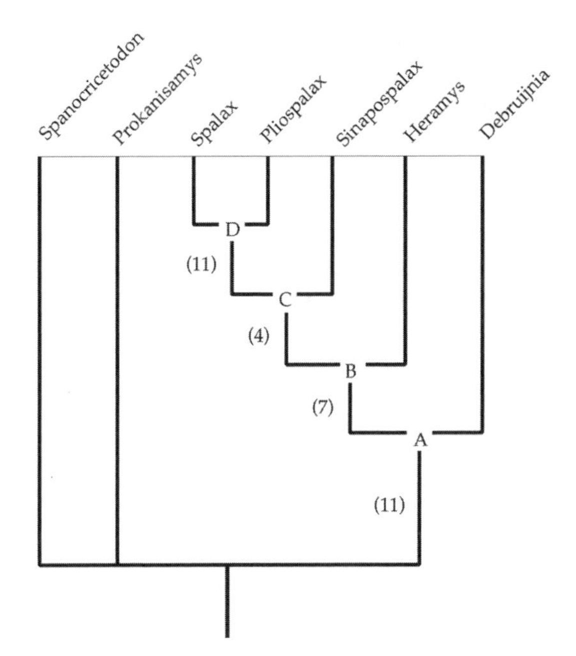

Figure 6.19. Cladogram showing the relationship of the new genus *Sinapospalax* in the family Spalacidae. Characters numbered as in Appendix 6.1. Characters that change unambiguously at nodes are: Node A: 2, 3, 5, 8, 11, 16, 33, 45, 46, 47, 48, 52: (0–1); Node B: 1, 13, 18, 25, 28, 41, 42: (0–1), 29(1–2), 38(0–2); Node C: 6, 20, 22:(0–1), 40(1–3), 46(1–2); Node D: 21(0–1), 25(1–2), 26(0–1), 29(2–3), 31(0–1), 32(0–1), 33(1–2), 34(0–1), 35(0–1), 41(1–2), 42(1–2). Numbers in parentheses indicate Bremer branch support values (Bremer 1994) calculated using PAUP (version 3.1; Swofford 1993).

Discussion

The hypothesis of relationships inferred in figure 6.19 shows that all the monophyletic groupings within spalacids are supported by mostly unique synapomorphies. For many years, only one genus, *Pliospalax,* was recognized to include many of the middle–late Miocene spalacid species. The information presented here leads us to nominate the new genus *Sinapospalax.* In fact, our data do not support the recognition of the genus *Pliospalax* as a monophyletic taxon. Current information suggests that *Pliospalax* is a paraphyletic group.

Conclusion

Here, for the first time, we have described spalacid remains from six successive horizons spanning the late Astaracian (MN 8) to late Vallesian (MN 10) (~9–11.5 Ma) of Anatolia. In previous studies, only discoveries from isolated localities were described.

The continuity of the documentation allowed us to observe the evolutionary trends inside the new genus *Sinapospalax* described in this chapter. Based on new observations and considering all previously described taxa, we undertook a phylogenetic analysis. From a systematic point of view, Sinap Tepe localities yielded five spalacid species belonging to two genera: *Heramys* and *Sinapospalax* n. gen. Other fragmentary remains are also described.

In Sinap Tepe localities, as well as in other Miocene localities in Turkey, spalacids remain a discrete element of rodent faunas, whereas they are more abundant in Pliocene faunas of the region (e.g., at Çalta 15.12% of rodents). This may be due to mode of life, climate, ecology, and/or taphonomy. Spalacids are known as inhabitants of steppic environments and grasslands with soft soil.

The first phylogenetic analysis presented here shows that all the monophyletic groupings within spalacids are supported by mostly unique synapomorphies. For many years, only one genus, *Pliospalax,* was recognized to include many of the middle–late Miocene spalacid species. The systematic studies and phylogenetic analyses lead us to recognize a new genus, *Sinapospalax,* that includes some of the species previously referred to *Pliospalax.* Also, if these analyses demonstrate the monophyly of the genera *Debruijnia, Heramys, Sinapospalax,* and *Spalax,* our data do not support the recognition of the genus *Pliospalax* as a monophyletic taxon. Current information suggests that *Pliospalax* is a paraphyletic group.

Acknowledgments

The material studied in this chapter was collected in the framework of the Sinap Project (1989–1995). We thank its organizers, B. Alpagut, M. Fortelius, and J. Kappelman, for allowing us to participate in this project. The field work was done with the authorization of the General Directorate of Antiquities, Ministry of Culture. Carefully reading the manuscript, Raymond L. Bernor and Lawrence J. Flynn suggested many technical and linguistic improvements. The SEM photography was done by Christiane Chancogne. Henri Lavina and Philippe Loubry helped with the composition of the figures. We are grateful to all of these individuals.

Appendix 6.1

The following is a list of characters and their states presented in the data matrix (table 6.2) and figure 6.19.

1. m1 anteroconid
   0: anterior; 1: shifted labially
2. m1 metaconid
   0: lingual; 1: anterior
3. m1 anteroconid-metaconid
   0: separated; 1: connected by lophid; 2: confluent, with posterior and/or anterior grooves; 3: fused forming a loph
4. m1 anteroconid-metaconid size
   0: almost equal in size; 1: metaconid strikingly bigger
5. m1 protoconid-metaconid
   0: connected by metalophule; 1: not connected
6. m1 protoconid
   0: longitudinally elongated; 1: transversely elongated
7. m1 mesolophid
   0: absent; 1: well developed; 2: weakly developed
8. m1 labial anterolophid
   0: connected to the base of protoconid; 1: not connected; 2: absent
9. m1 anterolophulid
   0: complete; 1: complete or fragmented; 2: fragmented; 3: weak tubercule or absent; 4: absent
10. m1 hypolophulid orientation
    0: posteriorly directed; 1: transverse
11. m1 posterior arm of protoconid
    0: absent; 1: present
12. m1 posterior arm of hypoconid
    0: absent; 1: long; 2: short
13. m1 ectomesolophid
    0: present; 1: absent
14. m1 lingual reentrant folds
    0: two; 1: one
15. m1 protosinusid
    0: almost transverse; 1: anteriorly oblique
16. m1 depth of protosinusid
    0: very deep; 1: deep; 2: shallow; 3: absent
17. m1 anterior shape of tooth
    0: sharp; 1: blunt; 2: square; 3: rounded
18. m2 lingual anterolophid
    0: long; 1: short; 2: absent
19. m2 posterior arm of hypoconid
    0: absent; 1: present
20. m2 protosinusid occurrence
    0: open; 1: enamel island; 2: absent
21. m2 posterosinusid occurrence
    0: open; 1: enamel island
22. m3 protosinusid occurrence
    0: open; 1: enamel island; 2: absent
23. m3 anterosinusid occurrence
    0: enclosed; 1: open; 2: absent
24. M1 position of metacone
    0: not isolated; 1: isolated
25. M1 metalophule orientation
    0: directed to hypocone; 1: directed to posterior arm of hypocone; 2: directed to posteroloph
26. M1 mesoloph
    0: present; 1: absent
27. M1 anterior arm of protocone
    0: present; 1: absent
28. M1 ectoloph
    0: present; 1: absent
29. M1 protosinus occurrence
    0: deep; 1: reduced; 2: depression state; 3: absent
30. M2 paracone
    0: labial; 1: shifted anteriorly; 2: incorporated into anteroloph
31. M2 anterolophule
    0: present; 1: absent
32. M2 protolophule
    0: present; 1: absent
33. M2 mesoloph
    0: stays in the area of mesosinus; 1: reaches the labial border; 2: fused with paracone
34. M2 metalophule
    0: present; 1: absent
35. M2 ectoloph
    0: present; 1: absent

36. M2 posterosinus
    0: well developed; 1: weak; 2: absent
37. M3 hypocone and protocone connection
    0: separated; 1: slightly connected or closely spaced; 2: fused
38. M3 endoloph
    0: continuous; 1: if present, segmented; 2: absent
39. M3 lingual anteroloph
    0: present; 1: absent
40. M3 labial ends of lophs
    0: separated; 1: close to each other but separated; 2: all connected by ectoloph; 3: some connected leaving one sinus open
41. M3 anterosinus
    0: open; 1: enclosed by anteroloph and paracone; 2: absent
42. M3 number of labial sinus
    0: four to three; 1: three to two; 2: two to one
43. M3 shape of tooth
    0: triangular; 1: slightly rounded; 2: rounded
44. endoloph (Ectolophid) position
    0: central; 1: labially shifted
45. loph and cone relation
    0: all cones recognisable, lophs weak; 1: all cones recognisable, lophs strong; 2: cones included in lophs
46. crown height
    0: low; 1: medium; 2: high
47. hypsodonty of upper molars
    0: labially and lingually equal; 1: lingually high;
48. sinus(id)s width
    0: wide; 1: narrow
49. M1 and m1 anteocone(id) shape
    0: distinct cone; 1: retracted loph-like; 2: included completely in the anterior loph
50. loph(id) height
    0: lower than cone(id) height; 1: equal to cone(id) height
51. number of roots
    0: upper molars (M1, M2, M3) = 3, 4, 3 and lower molars (m1, m2, m 3) = 2, 3, 2; 1: upper molars (M1, M2, M3) = 3, 3, 3 and lower molars (m1, m2, m 3) = 2, 2, 2
52. M1/M3 length ratio
    0: high; 1: low

Literature Cited

Bremer, K., 1994, Branch support and tree stability: Cladistics, v. 10, pp. 295–304.

Bruijn, H. de, 1984, Remains of the mole-rat *Microspalax odessanus* Topachevski, from Karaburun (Greece, Macedonia) and the family Spalacidae: Proceedings of the Koninklijke Nederlandse Akademie van Wetenschappen, ser. B, v. 87, pp. 417–425.

Bruijn, H. de, and A. J. Van Der Meulen,1975, The early Pleistocene rodents from Tourkobonia-1 (Athens, Greece): Proceedings of the Koninklijke Nederlandse Akademie van Wetenschappen, ser. B, v. 78, pp. 314–338.

Bruijn, H. de, M. R. Dawson, and P. Mein, 1970, Upper Pliocene Rodentia, Lagomorpha and Insectivora (Mammalia) from the isle of Rhodes (Greece): Proceedings of the Koninklijke Nederlandse Akademie van Wetenschappen, ser. B, v. 73. pp. 535–584.

Bruijn H. de, S. T. Hussain, and J. J. M. Leinders, 1980, Fossil rodents from the Muree formation near Banda, Daud Shah, Kohat, Pakistan: Proceedings of the Koninklijke Nederlandse Akademie van Wetenschappen, ser. B, v. 84, pp. 71–99.

Carleton, M. D., and G. G. Musser, 1984, Muroid rodents, *in* S. Anderson, and J. Knox, eds., Orders and families of recent mammals of the world: New York, John Wiley and Sons, pp. 289–379.

Flynn, L. J., 1982, Systematic revision of Siwalik Rhizomyidae (Rodentia): Géobios, v. 15, pp. 327–389.

Flynn, L. J., L. L. Jacobs, and E. H. Lindsay, 1985, Problems in muroid phylogeny: Relationship to other rodents and origin of major groups, *in* W. P. Luckett, and J. L. Hartenberger, eds., Evolutionary relationships among rodents: A multidisciplinary analysis: New York, Plenum Press, pp. 589–616.

Gromov, I. M., and G. I. Baranova, 1981, Catalogue of mammals of USSR: Leningrad, Russia, Nauka, 456 pp.

Gromov, I. M., and M. A. Erbajeva, 1995, The mammals of Russia and adjacent territories: Lagomorphs and Rodents: St. Petersburg, Russia, Russian Academy of Sciences, Zoological Institute, 521 pp.

Hofmeijer, G. K., and H. de Bruijn, 1985, The mammals from the Lower Miocene of Aliveri (Island of Evia, Greece). Part 4: The Spalacidae and Anomalomyidae: Proceedings of the Koninklijke Nederlandse Akademie van Wetenschappen, ser. B, v. 88, pp. 185–198.

Kormos, T., 1932, Neue pliozäne Nagetiere aus der Moldau: Paläontoligische Zeitschrift, v. 14, pp. 193–200.

Korth, W. W., 1994, The Tertiary record of rodents in North America: New York, Plenum Press, 319 pp.

Li, C. K., 1977, A new Miocene cricetodont rodent of Fangshan, Nanking: Vertebrata Palasiatica, v. 15, pp. 67–75.

Mehely, L., 1913, Species generis *Spalax:* Die Arten der Blindmäuse in systematischer Beziehung: Mathematische und Naturwissenschaftliche Berichte aus Ungarn, v. 28, pp. 1–385.

Mein, P., and M. Freudenthal, 1971, Les Cricetidae (Mammalia, Rodentia) du Néogéne moyen de Vieux-Collonges. Partie 1: Le genre *Cricetodon* Lartet, 1851: Scripta Geologica, v. 5, pp. 1–37.

Nevo, E., M. G. Filippucci, C. Redi, S. Simson, G. Heith, and A. Beiles, 1995, Karyoptype and genetic evolution in speciation of subterranean mole rats of the genus *Spalax,* Turkey: Biological Journal of the Linnean Society, v. 54, pp. 203–229.

Savic, I. R., and E. Nevo, 1990, The Spalacidae: Evolutionary history, speciation and population biology, *in* E. Nevo, and O. A. Reig, eds., Evolution of subterranean mammals at the organizmal and molecular levels: New York, Alan R. Liss, pp. 129–143.

Sen, S., 1977, La faune de rongeurs Pliocènes de Çalta (Ankara, Turquie): Bulletin du Muséum National d'Histoire Naturelle, ser. Sciences de la Terre, v. 61, pp. 89–172.

Swofford, D. L., 1993, PAUP, Phylogenetic Analysis Using Parsimony, Version 3.1: Computer program distributed by the Illinois Natural History Survey, Champaign, Illinois.

Tchernov, E., 1986, The rodents and lagomorphs from Ubeidiya Formation: Systematics, paleoecology, and biogeography, *in* E. Tchernov, ed., Les Mamiféres du Pléistocéne Inférieur de la Vallée du Jourdan à Oubeidiyeh MTJ 5: Paris, Association Paleorient, pp. 235–350.

Topachevsky, V. O., 1969, Fauna of the USSR: Mammals, molerats, Spalacidae: Akademia Nauk USSR, n.s., v. 99, pp. 1–247.

Ünay, E., 1978, *Pliospalax primitivus* n. sp. (Rodentia, Mamalia) and *Anomalomys gaudryi* Gaillard from the *Anchitherium* fauna of Sariçay: Bulletin of the Geological Society of Turkey, v. 21, pp. 121–128.

Ünay, E., 1981, Middle and Upper Miocene rodents from the Bayraktepe section (Çanakkale, Turkey): Proceedings of the Koninklijke Nederlandse Akademie van Wetenschappen, ser. B, v. 84, pp. 217–238.

Ünay, E., 1990, A new species of *Pliospalax* (Rodentia, Mammalia) from the Middle Miocene of Pasalar Turkey: Journal of Human Evolution, v. 19, pp. 445–453.

Ünay, E., 1996, On fossil Spalacidae (Rodentia), *in* R. L. Bernor, V. Falbusch, and H. W. Mittman, eds., The evolution of western Eurasian Neogene mammal faunas: New York, Columbia University Press, pp. 246–252.

Ünay, E., 1999, Family Spalacidae, *in* G. E. Rössner, and K. Heissig eds., The Miocene land mammals of Europe: München, Germany, Verlag Dr. F. Pfeil, pp. 421–425.

7

Lagomorpha

S. Sen

Sinap Tepe is one of the major reliefs some 6 km northwest of the town of Kazan, between the villages of Yassiören, Sogucak, and Örencik. Along the slopes of this hill and in its surrounding area, several mammal localities were discovered in 1950s by Ozansoy (1965 and references therein). This author described Neogene large mammals from several localities, including an anthropoid primate, *Ankarapithecus meteai* Ozansoy 1957. The first small mammals from Sinap Tepe were collected in 1970s and described much later (Sen 1991).

The material described in this chapter was collected between 1989 and 1995 by members of the International Sinap project. The location of mammal localities and their stratigraphic context are given in Kappelman et al. (this volume).

Lagomorph remains are recorded in seven localities, which are, from the oldest to the youngest, Locs. 120, 8A, 12, 84, Inönü, Loc. 45, and Kömürlük Dere. Specimens from the Locs. 12, 45, and Kömürlük Dere are in fact surface findings, whereas those from the other localities were collected by washing-screening of sediments. The unique tooth from Loc. 12 represents a leporid. The material from the other localities belongs to the ochotonids.

The age of these localities, as based on their mammal faunas, spans the early Vallesian to the Ruscinian. It is of interest that the earlier Sinap localities 65, 64, 4, and 94, which yielded rich small mammal associations, do not contain any trace of lagomorph. They are all situated below Loc. 120, which is the oldest lagomorph locality of the Sinap Tepe area. The localities 65 and 64 are dated as latest Astaracian and the two others (Locs. 4 and 94) as earliest Vallesian. The latter two localities also contain the earliest hipparion record in the Sinap Tepe area.

The specimens here described are stored in the collections of the Dil Tarih ve Cografya Fakültesi, Ankara, Turkey.

The terminology used here to describe tooth structures is adapted from Lopez Martinez (1989). Instead of being simply descriptive, this terminology proposes an interpretation of the lagomorph dental pattern (fig. 7.1). Using the wear facets of the occlusal pattern, Lopez Martinez (1989) recognized cusp homologies and the other structures (e.g., flexa, fossets, lophs) of the mammal molar pattern. Another terminology was developed by White (1991) for leporid tooth morphology; it is rather descriptive and only recognizes folds but not cusps and ridges. It cannot be easily applied to ochotonids to name adequately all characteristic structures.

The measurements have been taken from the occlusal surface of the teeth, outside from enamel to enamel; they indicate in mm the maximum length and width of each tooth. The specimens displayed in the figures are all presented as if they were from the left side; if the original is from the right, the number of the illustration is underlined. Capital letters (I, P, M) are used for upper teeth, and small letters (i, p, m) for lower teeth.

Systematics of the Ochotonidae

Before describing the ochotonid remains from the Sinap Tepe area localities, I must clarify the systematics of some genera and species. In Western European literature, all ochotonids later than the Eocene/Oligocene boundary are included in the family Ochotonidae (see Lopez Martinez 1989, and references therein). Russian specialists, however, recognize three families: Palaeolagidae, Prolagidae, and Ochotonidae (Gureev 1964; Erbajeva 1988, 1994). According to Erbajeva (1988), who published the most recent synthesis on lagomorph systematics, the Palaeolagidae includes the Eurasian genera *Amphilagus, Piezodus, Titanomys,* and *Gymnesicolagus;* Prolagidae includes *Prolagus* and *Ptychoprolagus;* all other Eurasian and African genera (see below) were referred to Ochotonidae.

A complete revision of ochotonid systematics is beyond the scope of this chapter. However, some remarks about

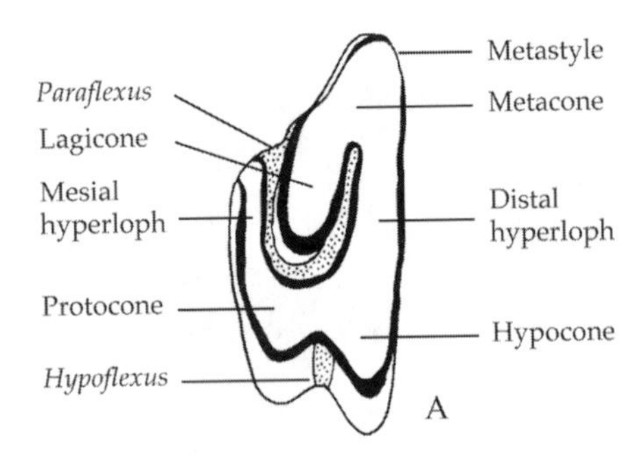

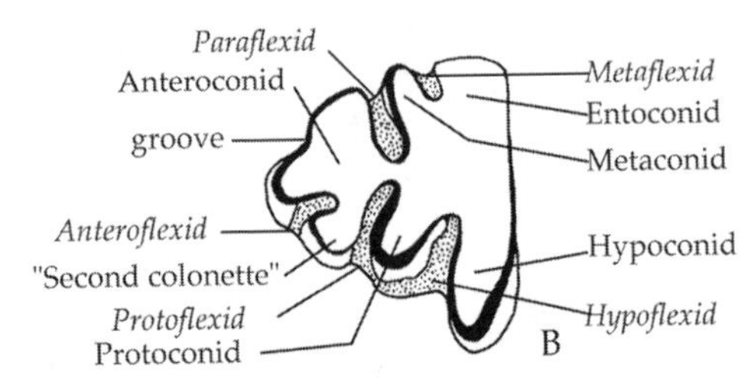

Figure 7.1. Dental terminology of the ochotonid cheek-teeth. **(A)** P3 sin; **(B)** p3 sin in occlusal view. Blackened lines represent enamel; white, dentine; dotted areas are cement infill. Modified after Lopez Martinez (1985).

the systematics of some genera are necessary to compare the Sinap ochotonids to other taxa.

Concerning the attribution of ochotonids to one or another family, note that all lagomorph students recognize that this group includes several phylogenetic lineages that are well differentiated in their dental morphology. Most of these lineages can be followed as evolving independently since at least late Oligocene or even earlier. Therefore, their classification in different families is more appropriate than grouping them together in a single family. However, the family grouping of some genera as proposed by Erbajeva (1988) is not fully satisfactory, because the genera belonging to one evolutionary lineage are included in two different families (e.g., *Titanomys* and *Piezodus* in Palaeologidae, *Prolagus* and *Ptychoprolagus* in Prolagidae). As shown by Lopez Martinez (1974, 1989) and Tobien (1975), these four genera represent a single European evolutionary lineage and are better included in the same family, the Prolagidae. The genera grouped in Palaeolagidae (Gureev 1964; Erbajeva 1988) are in fact the primitive representatives of ochotonids; as understood at present, this family is paraphyletic because it includes taxa ancestral to many lineages. A concerted effort is currently underway by M. A. Erbajeva and me to better resolve the systematics of ochotonid lagomorphs and enlighten their phylogeny.

The genera grouped by Erbajeva in Ochotonidae represent in fact several evolutionary lineages, among which at least three can be easily recognized. First is the *Marcuinomys-Lagopsis-Paludotona* lineage that is restricted to the early–middle Miocene of Western Europe and, for *Paludotona*, to the late Miocene of Italy. This lineage was intensively analyzed by Tobien (1963, 1974), Bucher (1982), and Lopez Martinez (1986, 1989). Some students suggested that this lineage may be derived from the late Oligocene western European *Titanomys*. Erbajeva (1988) classified this lineage in the family Ochotonidae, which I retain here, faute de mieux.

Another lineage is formed by the genera *Albertona* and *Alloptox* (Lopez Martinez 1986). The origin of this group and its phylogenetic relationships with other taxa are not well known, probably because of the lack of fossil record.

Ochotonids included in the genera *Bellatona, Proochotona, Pliolagomys, Ochotonoides, Ochotonoma,* and *Ochotona* have similar dental morphologies and probably form a third monophyletic lineage that I term here the "*Ochotona* group." This group is characterized by some synapomorphies that include (1) the posterolingual process of M2, (2) well-defined anteroconid with round or more or less triangular outline on p3, and (3) the presence of the opposite paraflexid and protoflexid on p3. Moreover, the genera included in this group preserve some plesiomorphic characters, such as the lack of centroflexid and metaflexid on p3 and one anterior flexus on P2.

The first apomorphic character (posterolingual process of M2) appears in the early–middle Miocene genus *Bellatona* Dawson 1961 from central and eastern Asia. This process is missing in all other ochotonids. Three species were referred to this genus: *B. forsythmajori* Dawson 1961 (type species), *B. yanghuensis* Zhou 1988, and *B. kazakhstanica* Erbajeva 1988. The latter species is known from the early Miocene of Kazakhstan; it is notably smaller than the others, and according to Erbajeva (1988, p. 60), its M2 lacks the posterolingual process. *Bellatona yanghuensis* is known by only one individual (skull and lower jaw) from the middle Miocene locality of Yanghu in the Shanxi Province in China, and its M2 has no posterolingual process. The type species *B. forsythmajori* is represented with well-described, rich material. In this species, the posterolingual process may be absent, tenuous, or quite strong. Dawson (1961, p. 11) notes that "the lack of uniformity of the process in *Bellatona* may indicate that the character was recently initiated in this line of ochotonid." The latest occurrence of *B. forsythmajori* is at the Moergen V horizon of the Tung Gur Formation in Inner Mongolia, correlated by Qiu (1996) to the latest middle Miocene. In this locality, M2 has a quite well-developed posterolingual process. In later representatives of this group, this process is strong and separated lingually from the second loph by a deep groove all along the shaft of M2.

Another evolutionary trend of the *Ochotona* group is the progressive complication of the p3 anteroconid pattern. The p3 of *Bellatona kazakhstanica* lacks an anteroconid or has a trace bulge in its place. In the single p3 known of *B. yanghuensis*, there is no anteroconid. In *B. forsythmajori*, it may be absent, or there may be a small cuspid delimited posteriorly by a deep protoflexid and a shallow paraflexid (Dawson 1961; Qiu 1996) (see also fig. 7.8A). In later representatives of the *Ochotona* group, the anteroconid of p3 is well developed or even complicated by some enamel folds and plications.

Late Miocene to Recent species of this group are referred to the genera *Proochotona, Pliolagomys, Ochotonoma, Ochotonoides,* and *Ochotona.* In a recent paper (Sen 1998), I examined the dental features characterizing each of these genera and stated their morphological differences. Most of them have dental morphologies—in particular, the pattern of P2 and p3—sufficiently differentiated to characterize accurately each genus with some apomorphies.

All living species are included in *Ochotona.* Its oldest occurrence is with *O. lagreli* Schlosser 1924 in the latest Miocene of Inner Mongolia (Qiu 1987). It is significant that among the *Ochotona* group, the genus *Ochotona* includes the most "primitive" species, with a dental pattern much less differentiated than in other genera. For example, in species such as *O. pusilla, O. transcaucasica,* and *O. rufescens,* p3 has a primitive pattern, and in that it resembles *Bellatona forsythmajori:* the anteroconid is small and rounded in outline, the protoflexid and mainly paraflexid are shallow and transversal, the bridge is wide, the metaconid-entoconid complex is large, the hypoflexid is short, and on P2 the occlusal outline is round with a short paraflexus. In some other species of *Ochotona*—in particular *O. daurica* (type species), *O. tibetana,* and *O. antiqua*—the pattern of p3 is more specialized in having a large and triangular anteroconid, a narrow bridge, deep and posteriorly directed paraflexid and protoflexid, and shorter metaconid-entoconid complex. For these species, P2 is wide and has a deep paraflexus.

The other genera of the *Ochotona* group are all extinct. However, they have a more derived dental pattern than found in most species of *Ochotona* (see Erbajeva 1988; Sen 1998). Among these genera, *Proochotona* has a p3 pattern similar to that of *Ochotona,* and for that reason, its synonymy with *Ochotona* was suggested by several authors (Argyropulo and Pidoplichka 1939; Qiu 1987; Sen 1991 1998). Because many samples from the Sinap localities resemble in some characters the species included in *Proochotona,* I provide here a detailed analysis of this genus.

The genus *Proochotona* was erected by Khomenko (1914) with *P. eximia* based on a lower jaw fragment with p3–m1 from the late Miocene locality of Taraklia in Moldavia. This species is characterized by its larger size compared with *Ochotona* species. Khomenko considered this new taxon as intermediate between *Lagopsis* (early–middle Miocene of western Europe) and the living *Ochotona.* As we now know, with more fossil evidence at hand, *Lagopsis* belongs to a different evolutionary lineage not closely related to the *Ochotona* group. Khomenko (1914) also compared this new taxon with the extant species *O. rutila* Severtzov, 1873 and *O. nepalensis* Hodgson, 1841 (which is synonymized with *O. roylei* Ogilby, 1839; see Erbajeva 1988, p. 138) from which it differs, according to Khomenko (1914, p. 17) "by the presence of a groove on the lingual face of the anteroconid." As we will see later in the chapter, this feature is in fact quite common in several species of *Ochotona,* even in some individuals of *O. daurica* (Pallas, 1776) which is the type species of the genus. In *P. eximia* from Taraklia, the p3 anteroconid is triangular shaped and has lateral depressions; the protoflexid and paraflexid are equally deep and posteriorly oblique; the bridge is central and narrow; the hypoflexid does not reach the half-width; and the lingual length of the metaconid-entoconid complex is more than two-thirds of the total length. The measurements given by Khomenko (1914, p. 15) are p3: 2.2 × 2.0 mm; p4: 2.2 × 2.4 mm, and m1: 2.4 × 2.5 mm.

Argyropulo and Pidoplichka (1939) recognized *Ochotona (Proochotona) eximia* from the Odessa fissure fillings (Ukraine) and the Karboliya beds (Moldavia), both correlated to the "middle Pliocene." Unfortunately, their descriptions and illustrations are insufficient to be certain of the species assignment.

In his monograph on lagomorphs, Gureev (1964) retained *Proochotona* as a valid genus, and added three other species from the Pliocene of the Ukraine and central Asia: *P. gigas* (Argyropoulo and Pidoplishka 1939), *P. kurdjukovi* sp. nov., and *P. kirgizica* sp. nov. The first species was later referred to as the type species of a new genus *Pliolagomys* Erbajeva, 1988. The latter two species are based on few mandibles, without p3, from the Pliocene localities in Kyrgyzstan (Dmitrieva and Nesmeyanov 1982). The lack of p3 does not allow a secure systematic assignment, and thus *P. kurdjukovi* and *P. kirgizica* should be considered as nomen vanum. Gureev (1964, p. 223) proposed a new generic diagnosis for *Proochotona* based on the dimensions of the tooth-row (p3–m3 length, 10–13 mm) and of the lower jaw (thickness in symphysis > 4 mm, and the depth of the mandible below m2, 8 mm). These measurements are taken on specimens attributed to *P. gigas, P. kurdjukovi,* and *P. kirgizica,* and thus they cannot be reliably used to characterize the genus *Proochotona.* Nevertheless, it is true that the type specimen of *P. eximia* from Taraklia belongs to—according to the dimensions given by Khomenko (1914)—a large individual, as large as the largest fossil and living species of *Ochotona.*

Lungu (1981) described *Proochotona kalfense* from the Vallesian locality of Kalfa, Moldavia. He compared it with the other species that Gureev (1964) referred to *Proochotona,* but he did not discuss its generic status. We see later in this chapter that this species belongs in fact to a new genus.

Mats et al. (1982) referred to *Proochotona* sp. some isolated teeth from the middle or late Pliocene of Olkhon island in Lake Baikal. The dimensions given (length of M2, 1.9–2.0 mm, $n = 4$; length of p3, 2.45 and 2.50 mm, $n = 2$) indicate a moderately large ochotonid. The pattern of the unique p3 they illustrated is similar to that of *Ochotona lagreli* from Inner Mongolia (Qiu 1987).

From three late Miocene (?late Turolian) localities in Kazakhstan, well-preserved specimens were briefly described and referred to *Proochotona* cf. *eximia.* These localities are Kanal Irtysh and Gucinii Perelet near Pavlodar (Agadjanian and Erbajeva 1983) and Ecekartgan near Tekes town ~250 km east of Alma Ata (Tleuberdina 1982). Moreover, in faunal lists from several Russian and central Asian localities, some ochotonids were referred to *Proochotona.* More detailed studies of these materials are needed for a better systematic assignment.

Finally, Erbajeva (1988) revised the family Ochotonidae and retained the genus *Proochotona* with two species: *P. eximia* and *P. kalfense.* The emended generic diagnosis given by this author is: "length of the p3 is 2.2 mm and its width 2.0 mm. The anterior segment of p3 is very large, with trilobed configuration, and disposed symmetrically in respect to the posterior segment. These two parts are connected by a narrow enamel bridge, around which there are equally deep labial and lingual folds filled with cement."

The characters retained as diagnosic by Erbajeva (1988, p. 71) are all shared by several fossil and extant species of *Ochotona,* and among them particularly by *O. antiqua* from the early Pliocene of Odessa (Argyropulo and Pidoplichka 1939) and the living *O. daurica* (type species), *O. alpina,* and *O. tibetana.* These observations and the great similarities of the dentition in species referred to *Proochotona* with that of *Ochotona* agree well with the suggestion of the previous authors, that is, that *Proochotona* is a junior synonym of *Ochotona.*

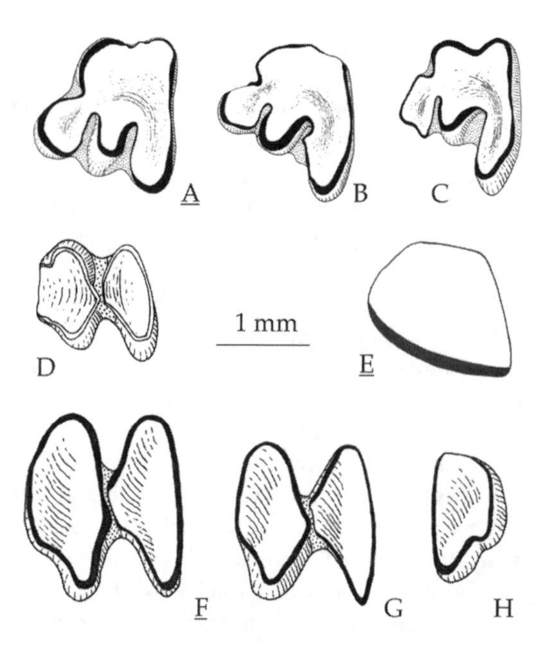

Figure 7.2. *Bellatonoides eroli* gen. and sp. nov. from Loc. 8A. Lower teeth in occlusal view. **(A)** p3 (holotype); **(B, C)** p3; **(D)** d4; **(E)** lower incisor; **(F, G)** lower check teeth; **(H)** m3. Blackened lines represent enamel; white, dentine; dotted areas are cement infill.

Systematics

Order Lagomorpha Brandt, 1855
Family Ochotonidae Thomas, 1897
Subfamily Ochotoninae Thomas, 1897
Genus *Bellatonoides* g. nov.

Type Species. *Bellatonoides eroli* sp. nov.

Etymology. Because of its similarities with *Bellatona forsythmajori* Dawson, 1961.

Diagnosis. Small ochotonids with p3 having anteroconid generally rounded and situated labially, lacking emanel fold, posteriorly delimited by a deep protoflexid, a shallow paraflexid, and wide central bridge; P2 has short paraflexus and rounded metastyle; M2 has a strong posterolingual process.

Included Species. *Bellatonoides eroli* sp. nov. and *B. kalfense* (Lungu 1981).

Bellatonoides eroli g. and sp. nov.

Synonymy. cf. *Ochotona* sp., Sen 1991 (p. 255, fig. 3).

Holotype. One p3 dex (ST8A-21) from Sinap Tepe Loc. 8A. See figure 7.2A.

Other Material from Loc. 8A. Seven I1, one D3, one D4, one P2, five P3, 12 P4 and M1, four M2, one i1, one d4, six p3, eight p4 or m1 or m2, four m3 (ST8A-1–ST8A-52). All are isolated teeth. See figures 7.2–7.4.

Type Locality. Loc. 8A, Igdeli Dere, southeastern slopes of Sinap Tepe, Yassiören village, Kazan, Ankara.

Age. Early Vallesian.

Etymology. In honor of Prof. Dr. Oguz Erol, who discovered the first Sinap Tepe mammals on September 7, 1951.

Measurements. See table 7.1.

Diagnosis. Small ochotonid. The p3 has rounded anteroconid or labially angular, situated labially; paraflexid short and wide and not always filled with cement. The P2 has a short paraflexus and round metastyle. The M2 has a strong posterolingual process separated from the second loph by a wide lingual angle.

Differential Diagnosis. This species differs from *Bellatona forsythmajori* in having a large anteroconid on p3 and stronger posterointernal process on M2. It differs from *B. kalfense* by its smaller size, a shorter and wider paraflexid, and more labially displaced anteroconid on p3.

Description of Loc. 8A Material. The permanent teeth are rootless, whereas the milk teeth have short roots.

The lower incisor has its anterior face slightly curved; its section has an almost triangular outline; the enamel is thin and only covers the anterior face. The upper incisor is well curved; it is divided into two parts by an anterior longitudinal furrow; the thin enamel only covers the anterior face; the medial border is straight whereas the lateral border is rounded.

The D3 occlusal pattern has a V-shaped paraflexus and a deep hypoflexus. The D4 has two lophs separated by a deep

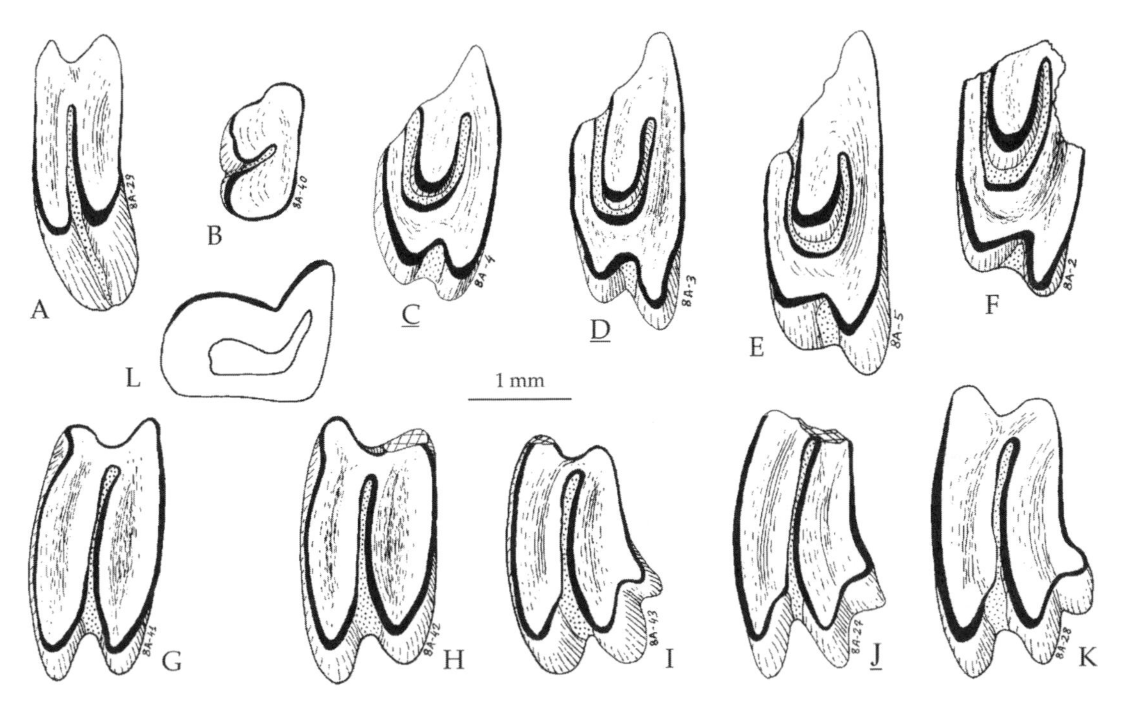

Figure 7.3. *Bellatonoides eroli* gen. and sp. nov. from Loc. 8A. Upper teeth in occlusal view. **(A)** D4; **(B)** P2; **(C–F)** P3; **(G–H)** P4 or M1; **(I–K)** M2; **(L)** upper incisor. Blackened lines represent enamel; white, dentine; dotted areas are cement infill.

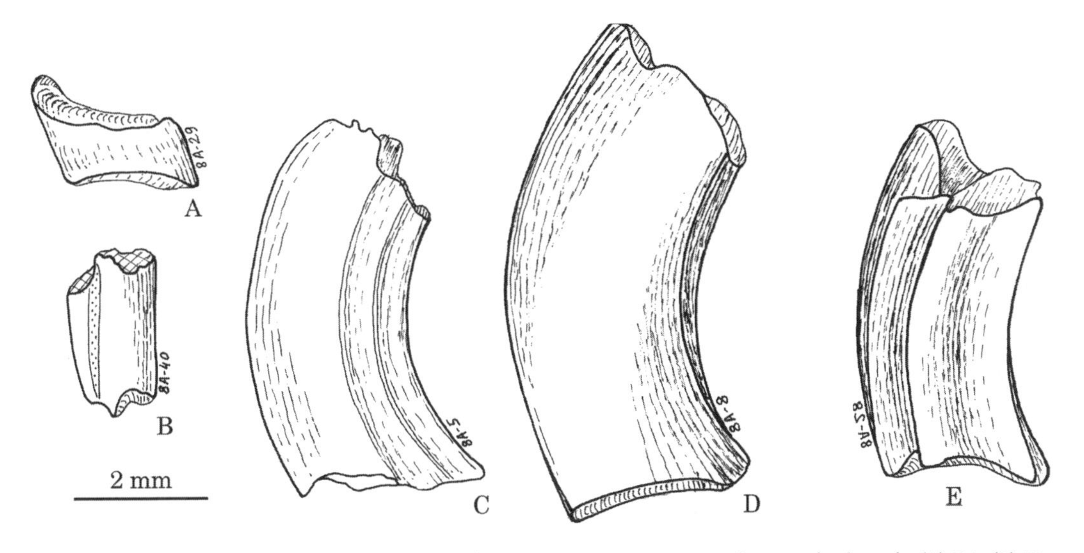

Figure 7.4. *Bellatonoides eroli* gen. and sp. nov. from Loc. 8A. Anterior view of upper cheek teeth. **(A)** D4; **(B)** P2; **(C)** P3; **(D)** P4 or M1; **(E)** M2.

hypoflexus. The D4 crown is low, as on D3. The P2 has a simple occlusal pattern; the paraflexus is narrow and labially directed and the metastyle is rounded.

The P3 shaft is strongly curved and has a anterolabial torsion. The occlusal outline is trapezoidal in shape. The paraflexus is U-shaped. The protoloph is not differentiated from the protocone by a constriction, as it is in some ochotonids. The hypoflexus is shallow and wide on three specimens, but narrow in two others. The metastyle is sharp. The thickness of the enamel is not much differentiated, but it is absent on the labial face.

The P4–M1 shaft has a similar curve and torsion to that of the P3; the occlusal pattern is formed of two lophs separated by a deep hypoflexus. The M2 is characterized by the presence of a posterolingual process on the posterior loph. The lingual angle between this process and the posterior loph is wide on two M2 samples and at right angle on two others, giving to the process a posteriorly directed orientation. Other M2 characters are as on the previous teeth.

The d4 crown is formed of two rounded lophs; the trigonid has a anterolingual notch. This tooth has four roots that spread outward.

Table 7.1. Measurements of the Upper and Lower Teeth of *Bellatonoides eroli* g. and sp. nov. from Tepe Loc. 8A

		Length (mm)		Anterior Width (mm)		Posterior Width (mm)	
Tooth	*n*	Mean Value	Range	Mean Value	Range	Mean Value	Range
I upper	6/5	1.51	1.38–1.80	1.97	1.79–2.17		
D3	0/1	—	—	1.59	—		
D4	1	1.02	—	1.99	—	2.02	—
P2	1	0.72	—	1.40	—		
P3	5/3	1.25	1.17–1.33	2.76	2.39–3.06		
P4–M1	11	1.46	1.26–1.71	2.56	1.96–3.01	2.53	2.10–3.08
M2	4	1.51	1.31–1.60	2.36	2.11–2.60	2.16	1.95–2.36
I lower	1	1.55	—	1.62	—		
d4	1	1.34	—	1.07	—	1.08	—
p3	5	1.61	1.55–1.66	1.65	1.56–1.83		
p4–m2	4/6/5	1.67	1.55–1.76	1.72	1.48–1.96	1.73	1.59–1.85
m3	4	0.74	0.69–0.81	1.19	1.15–1.25		

Notes: n is the number of measures of length and anterior and posterior widths, respectively.
—, No measure.

The p3 shaft is gently curved with a lingual concavity. The hypoflexid and the protoflexid are deep. The metaflexid is absent, but three specimens of six have a marked groove lacking cement all along the shaft. The paraflexid is very shallow and filled with a little cement in two specimens, but in four others the paraflexid has no cement. The anteroconid is small and labially displaced; its outline is round in five specimens but rather triangular in two others.

The p4–m2 shaft is gently curved backward. The lengths and widths of both the trigonid and talonid are roughly equal. The enamel is a little thicker on the posterior borders of the lophs than on their anterior borders. The labial and lingual triangles are almost symmetrical. In all specimens, the trigonid possesses a marked anterolabial groove, which corresponds to the paraflexid.

The m3 is reduced in size. The shaft is more strongly curved backward than are the other teeth described here. Its occlusal outline is rather oval. The paraflexid groove is present, but less marked than in p4–m2.

Material from Loc 120. A left maxillary with P3–M2, one isolated P2, one P3, two M2 (one damaged), one lower incisor, two p3 and three fragmentary lower cheek teeth, cataloged as ST120-2 to 11. See figure 7.5.

Age. Early Vallesian.

Measurements. See table 7.2.

Description of Loc. 120 Material. This locality was not well explored. However, the material contains a few remains of an ochotonid very similar to that of the Loc. 8A. The teeth are similar or a little larger than those of Loc. 8A. In morphology, the P2 is a little wider and its paraflexus deeper than seen in the Loc 8A specimens. The two p3 from Loc. 120 have rounded anteroconid and shallow paraflexid, as in most of the p3 from Loc. 8A.

Comparison. The specimens from Locs. 8A and 120 resemble the youngest species of *Bellatona* (*B. forsythmajori* from Tairum Nor and Tung Gur in Inner Mongolia; Dawson 1961; Qiu 1996) in having a rounded P2 with one shallow flexus, the posterolingual process on M2, and a small anteroconid and shallow paraflexid on p3. However, in Sinap specimens, these features are much more developed and indicate a more advanced degree of evolution. As an example, in *B. forsythmajori,* the p3 anteroconid is just initiated by the occurrence of shallow protoflexid and paraflexid, whereas it is a well-separated cusp in Sinap specimens. Also note that *B. forsythmajori* has a little larger cheek teeth than *Bellatonoides* from Sinap localities.

Among the *Ochotona* group, *Ochotonoides* (late Pliocene to middle Pleistocene) differs from the Sinap specimens by its several derived characters: large size, p3 with widened and folded anteroconid, deep protoflexid and paraflexid and enamel plications, P2 with a very deep and very oblique paraflexus, and the like. Thus, we can definitely exclude all comparison of the Sinap ochotonids with *Ochotonoides.*

Pliolagomys Erbajeva 1983 is known in late Pliocene localities from an area extending from Moldavia to Lake Baikal (Agadjanian and Erbajeva 1983; Erbajeva 1988, 1994). The species referred to this genus have p3 with a narrow enamel bridge connecting obliquely the anteroconid to the protoconid; in other genera of the *Ochotona* group, as well as in the Sinap ochotonids, the bridge is central and straight and joins the anteroconid to the protoconid-metaconid complex. This unusual anteroconid-protoconid connection in

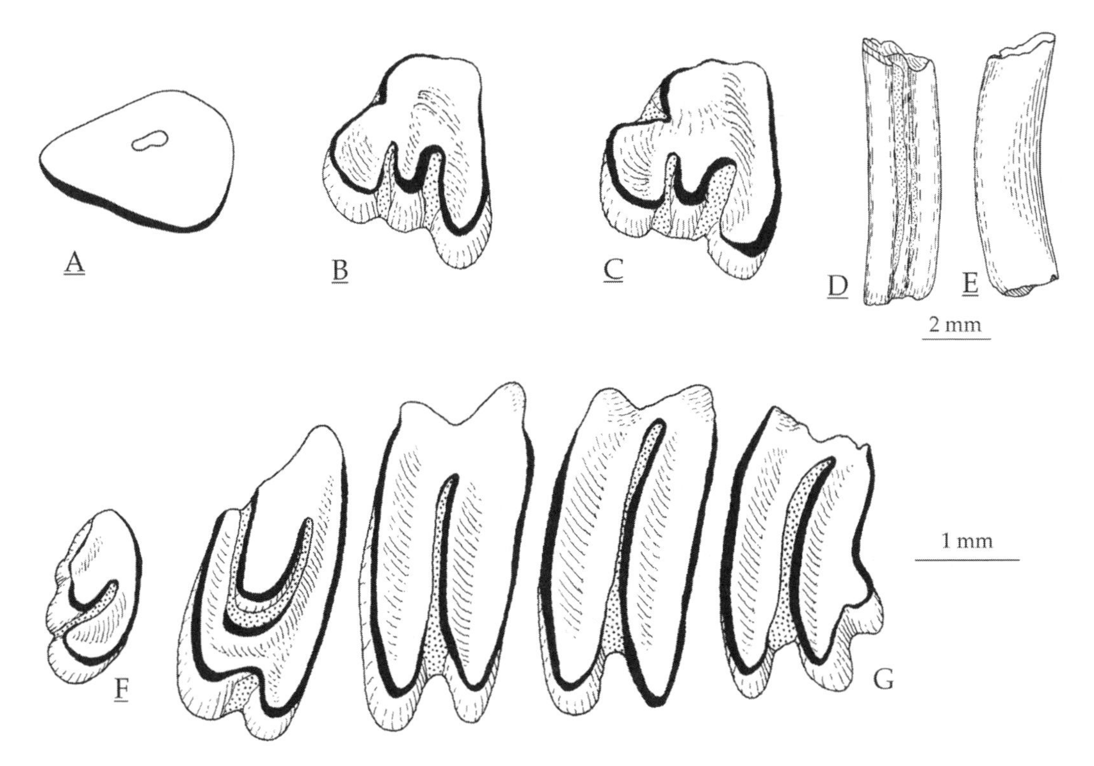

Figure 7.5. *Bellatonoides eroli* gen. and sp. nov. from Loc. 120. **(A)** lower incisor; **(B, C)** p3; **(D, E)** labial and posterior views of p3; **(F)** P2; **(G)** P3–M2. The 2 mm scale concerns only the lateral views. Blackened lines represent enamel; white, dentine; dotted areas are cement infill.

Table 7.2. Measurements of the Upper and Lower Teeth of *Bellatonoides eroli* g. and sp. nov. from Loc. 120

		Length (mm)		Anterior Width (mm)		Posterior Width (mm)	
Tooth	*n*	Mean Value	Range	Mean Value	Range	Mean Value	Range
P2	1	0.74	—	1.52	—	—	—
P3	2	1.22	1.14–1.30	2.68	2.55–2.81	—	—
P4	1	1.61	—	2.82	—	3.02	—
M1	1	1.55	—	2.86	—	3.04	—
M2	1	1.60	—	2.54	—	2.20	—
I lower	1	1.40	—	1.83	—	—	—
p3	2	1.65	1.57–1.72	1.79	1.71–1.87	2.20	2.14–2.33
p4–m2	0/1/2	—	—	1.85	—	1.94	1.84–2.04

Notes: n is the number of measures of length and anterior and posterior widths, respectively. See also table 7.1. —, No measure.

Pliolagomys is a derived character and it creates a dyssymmetric shape in the occlusal pattern of p3. Also, because of this anteroconid-protoconid connection, the paraflexid becomes markedly deeper than the protoflexid.

Ochotonoma Sen, 1998, is known from Pliocene localities of Turkey, Greece, Romania, and Hungary (Sen 1998; Ünay and de Bruijn 1998). It is characterized by the triangular-shaped anteroconid on p3 and the presence of labial and/or lingual additional flexids of the anteroconid, which are filled with cement. These derived features, later complicated in *Ochotonoides,* do not suggest any comparison with *Bellatonoides.*

As discussed earlier in the chapter, the oldest representatives of the genus *Ochotona* are from the late Miocene of Moldavia (*O. eximia*), Inner Mongolia, and Tibet (*O. lagreli* Schlosser, 1924 and *O. minor* Bohlin, 1942). Subsequently, several other species occur in Asia, Europe, and North America (Erbajeva 1988). The type specimen of *O. eximia* from Taraklia (Moldavia; a fragmentary lower jaw with p3–m1) is larger than the Sinap specimens and its p3 has a

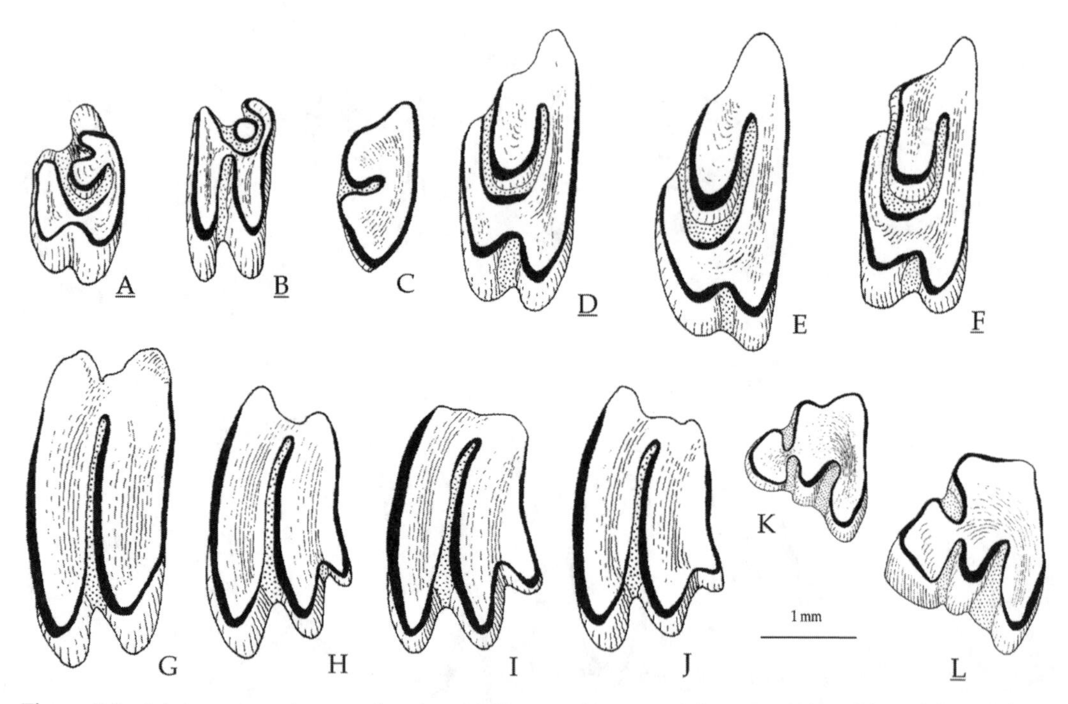

Figure 7.6. *Ochotona ozansoyi* sp. nov. from Loc. 84. Upper and lower teeth in occlusal view. (**A**) D3; (**B**) D4; (**C**) P2; (**D–F**) P3; (**G**) P4 or M1; (**H–J**) M2; (**K–L**) p3. Blackened lines represent enamel; white, dentine; dotted areas are cement infill.

triangular anteroconid and deep paraflexid. Specimens described as *Proochotona* cf. *eximia* by Agadjanian and Erbajeva (1983) from some late Miocene localities in Kazakhstan present the same differences. *O. lagreli* Schlosser, 1924 and *O. minor* Bohlin, 1942 co-occur in Ertemte (Inner Mongolia), which is tentatively correlated with the late Turolian (MN 13) of the European Neogene Mammal Chronology (Qiu 1987). *Ochotona guizhongensis* Ji et al. 1980 has been reported from the late Miocene Guizhoing deposits in Tibet and is only known by an upper jaw; Qiu (1987) believes that this taxon may be the junior synonym of *O. lagreli*. In this species as well as in several other representatives of the genus, the paraflexid and the protoflexid of p3 are equally deep, opposite one to the other, and directed backward. In *O. lagreli*, the anteroconid generally has a triangular outline, sometimes with tenuous lateral depressions. On the P2, the paraflexus is deep and labially oriented, the metastyle is sharp, and the lingual border has tenuous enamel folds in some specimens. All these characters prevent any reliable comparison of the Sinap specimens with *Ochotona*. It is obvious that the dental pattern of fossil and living species of *Ochotona* is more evolved than that of *Bellatonoides*.

"*Proochotona*" *kalfense* Lungu, 1981 from the Vallesian of Moldavia is better referred to *Bellatonoides:* its p3 has a rounded and labially displaced anteroconid and a shallow paraflexid. A. N. Lungu kindly lent me five isolated p3 from Kalfa (see discussion later in this chapter). Their comparison with the Sinap specimens shows that the Moldavian and Turkish species are very similar in the characters mentioned here. The Moldavian species is better assigned to the new genus *Bellatonoides* than to any other genus of the *Ochotona* group. However, its size is notably larger than *B. eroli*, and the p3 has a deeper paraflexid, probably indicating a more advanced stage of evolution. Kalfa is also known to be younger than Sinap Loc. 8A.

Genus *Ochotona* Link, 1795
Ochotona ozansoyi sp. nov.

Holotype. One p3 dex (ST84-1). See figure 7.6L.

Hypodigm. Six upper incisors, one D3, one D4, one P2, three P3, five P4 and M1, four M2, one lower incisor, two other p3, two p4 or m1 or m2, one m3 (ST84-2–ST84-28). All are isolated teeth. See figure 7.6.

Type Locality. Loc. 84 along the southwestern slopes of Sinap Tepe, to the north of the Yassiören village, Kazan, Ankara.

Age. Base of the late Vallesian.

Etymology. In honor of Prof. Dr. Fikret Ozansoy, who first explored Sinap mammal localities in 1951.

Measurements. See table 7.3.

Diagnosis. Small ochotonid. The p3 has a triangular shaped anteroconid with shallow lateral grooves and a narrow enamel bridge; the paraflexid is as deep as the protoflexid, both are transverse; the metaconid-entoconid complex is short; P2 has a short paraflexus; M2 has a strong posterior process whose lingual angle is ~90° or less.

Table 7.3. Measurements of the Upper and Lower Teeth of *Ochotoma ozansoyi* sp. nov. from Loc. 84

		Length (mm)		Anterior Width (mm)		Posterior Width (mm)	
Tooth	*n*	Mean Value	Range	Mean Value	Range	Mean Value	Range
I upper	5/6	1.49	1.34–1.63	2.00	1.65–2.31		
D3	1	1.00	—	1.28	—		
D4	1	0.94	—	1.50	—	1.54	—
P2	1	0.90	—	1.78	—		
P3	3	1.31	1.21–1.39	2.76	2.63–2.96		
P4–M1	5/3/5	1.57	1.42–1.73	2.83	2.66–3.03	2.74	2.54–2.99
M2	4	1.69	1.62–1.78	2.48	2.41–2.52	2.20	2.14–2.33
I lower	1	1.22	—	1.68	—		
p3	2	1.49	1.35–1.63	1.59	1.37–1.80		
p4–m2	2	1.79	1.78–1.79	1.94	1.88–1.99	1.96	1.91–2.00
m3	1	0.74	—	1.18	—		

Notes: n is the number of measures of length and anterior and posterior widths, respectively. See also table 7.1. —, No measure.

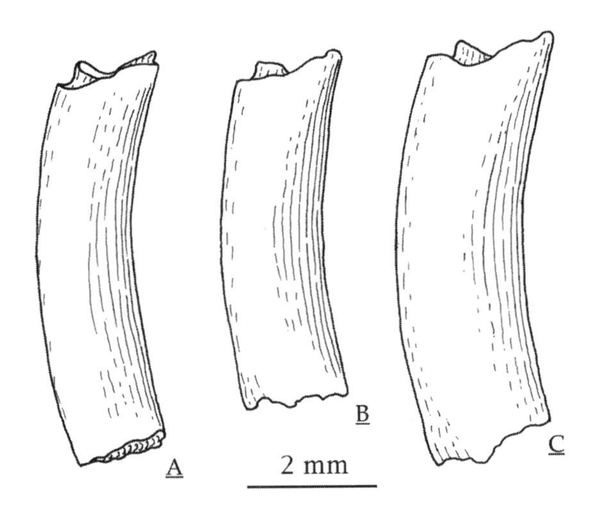

Figure 7.7. Comparison of p3 posterior views. **(A)** *Bellatonoides eroli* from Loc. 8A. **(B)** *Ochotona ozansoyi* from Loc. 84. **(C)** *Bellatonoides kalfense* from Kalfa.

Differential Diagnosis. *Ochotona ozansoyi* differs from *O. eximia* by its smaller size, transversely oriented paraflexid (posteriorly oblique in *O. eximia*), and deeper hypoflexid. It differs from *Bellatonoides eroli* in being larger and in having a deep paraflexid, triangular anteroconid, and narrow enamel bridge on p3 (fig. 7.8). It differs from *B. kalfense* by its smaller size, triangular anteroconid, transversal protoflexid and paraflexid, and deeper hypoflexid.

Description. Upper and lower incisors are as in *Bellatonoides eroli* from Locs. 8A and 120. The P2 is represented by one specimen, which has a trapezoidal occlusal outline, sharp metastyle, and short paraflexus. The P3 has an occlusal pattern similar to that for *B. eroli*. This is also the case for P4 and M1. However, these three teeth, as well as M2, are slightly larger in Loc. 84. The M2 is known from Loc. 84 with four specimens; on two of them, the posterolingual process has a lingual angle approaching 90°, but in two others, this angle is clearly sharper. One specimen has weakly developed cement in this angle.

The p3 is represented by three specimens; one of them belongs to a young individual (small size, thin enamel, and quite rounded anteroconid). The other two specimens (one fragmentary) belong to adult individuals and have larger dimensions, a triangular anteroconid, and thick enamel along the external angles. The paraflexid and protoflexid are equally deep and transversely oriented. The enamel bridge is narrow. The hypoflexid reaches almost the middle width of the tooth.

The other lower cheek teeth have the common features of the *Ochotona* group and are a little larger than those from Locs. 8A and 120.

Comparison. This species has cheek teeth slightly larger than those of *Bellatonoides eroli* from Locs. 8A and 120. It differs from *B. eroli* in that its P2 is wider and has a deeper paraflexus, its p3 is triangular with a centrally situated anteroconid, the paraflexid and protoflexid are equally deep, the bridge is narrower, and the hypoflexid deeper. Loc. 84 samples have p3 with a shorter metaconid-entoconid complex (less than two-thirds of the total tooth length). On the M2, the angle between the second loph and the posterior process is generally narrower than for specimens from Locs. 8A and 120.

Bellatonoides kalfense from Moldavia differs from this species by its notably larger size and in having p3 with a rounded and labially displaced anteroconid, shallower paraflexid, and wider enamel bridge. The similarities of this

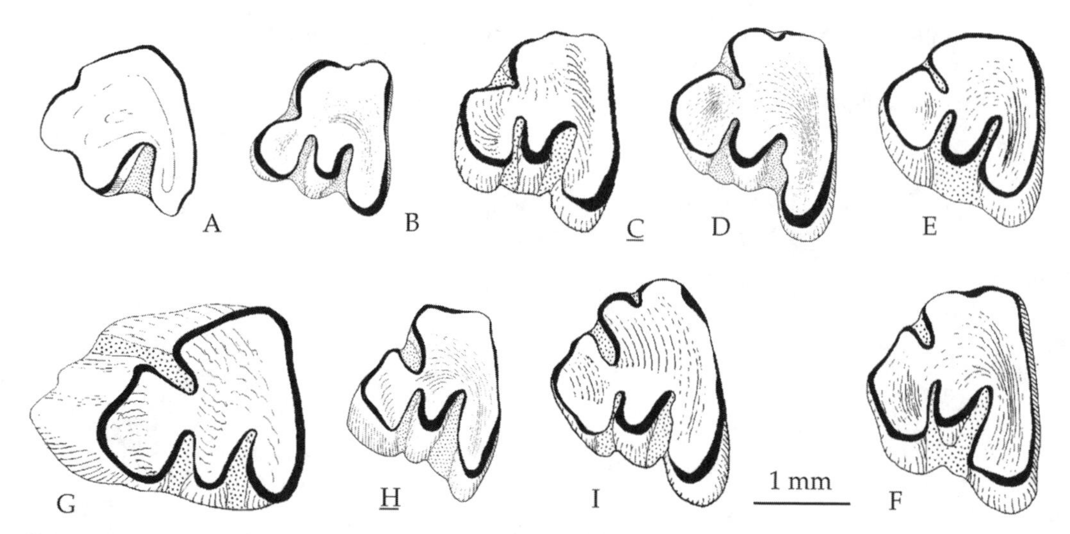

Figure 7.8. Comparison of p3 occlusal pattern in some species of the *Ochotona* group. (**A**) *Bellatona forsythmajori* from Moergen V in Inner Mongolia (after Qiu 1996). (**B**) *Bellatonoides eroli* g. and sp. nov. from Loc. 8A. (**C**) *B. eroli* g. and sp. nov. from Loc. 120. (**D–F**) *B. kalfense* from Kalfa in Moldavia. (**G**) *Ochotona eximia* from Taraklia in Moldavia (redrawn after Khomenko 1914). (**H**) *O. ozansoyi* sp. nov. from Loc. 84. (**I**) *O.* cf. *eximia* from Loc. 45. Blackened lines represent enamel; white, dentine; dotted areas are cement infill.

species to the samples from Sinap Locs. 8A and 120 are discussed above.

The Loc. 84 ochotonid compares most closely with *Ochotona eximia* from Taraklia, (Moldavia, MN 12), which is known only from a fragmentary lower jaw with p3–m1. The p3 of this species (fig. 7.8G) resembles those from Loc. 84 in having anteroconid with triangular occlusal outline and a shallow groove on its lingual face, equally deep protoflexid and paraflexid, a narrow bridge, and short metaconid-entoconid complex. The main differences are the larger size of the Taraklia species (p3, 2.2 × 2.0 mm [after Khomenko 1914], versus 1.63 × 1.80 mm for the largest p3 from Loc. 84), the larger anteroconid of p3, and the paraflexid directed backward on this tooth. The poor material described by Khomenko (1914) does not allow further comparison with the Loc. 84 specimens.

Ochotona sp. cf. *O. ozansoyi sp. nov.*

Locality. An unnumbered locality in the Inönü area, 2 km southeast of the Sarilar village, Kazan, Ankara.

Material. One P2 (0.84 × 1.60), one P3 (1.26 × 2.81), one P4 (1.48 × 2.60 × 3.00 for P4 and lower molariform teeth; second and third measures are the widths of anterior and posterior lophs, respectively) and left lower jaw with p3–m3 (p3, 1.59 × 1.52; p4, 1.73 × 1.78 × 1.98; m1, 1.82 × 1.93 × 1.91; m2, 1.79 × 1.97 × 1.82; m3, 0.71 × 1.14) (IN 15–18). See figure 7.9.

Age. Early? Turolian.

Description and Comparison. The P2 is larger and for the most part wider than in the specimens from Locs. 8A, 120, and 84; its paraflexus is deeper and its metastyle is sharper. All of these characters indicate a species more progressive than both *Bellatonoides eroli* and *Ochotona ozansoyi*. The P3 and P4 are similar in size and morphology to the equivalent teeth of *O. ozansoyi* from Loc. 84.

The lower jaw apparently belongs to a young individual, because the enamel is not well differentiated on the occlusal surface borders; this is particularly true for p3. The thickness of the enamel and the general pattern of the occlusal surface resemble those features of the juvenile p3 (S84-17) from Loc. 84. Therefore, the measurements given for this mandible and its teeth should be considered with some caution when estimating the size of this species.

The length of alveoli p3–m3 is 8.7 mm, and the maximum depth of the mandible below p4 is 5.9 mm. The depth of the mandible does not change under the tooth row; in other words, there is no lowering of the mandible depth below m3, as is often the case in species of *Ochotona*. There are two main foramina on the labial face, one anterior to p3 and the other inferior to m2, near the ventral border. Moreover, there is a longitudinal groove along the ventral border that I did not observe in living species of *Ochotona*. The incisor terminates inferior to p4.

The p3 is triangular in outline; its anteroconid forms a rounded triangle, slightly inclined labially; the paraflexid is less deep than is the protoflexid; the hypoflexid reaches the middle width of the occlusal surface. The dimensions of this tooth are a little larger than the juvenile p3 (S84-17) from Loc. 84, but the morphology of the occlusal pattern is the same. Based on the morphological similarities of teeth from this locality with those from Loc. 84, the material from Inönü is referred to *Ochotona* cf. *ozansoyi*.

Ochotoha sp. cf. *O. eximia* (Khomenko 1914)

Locality. Loc. 45, in the Inönü area, 2 km southeast of Sarilar village, Kazan, Ankara.

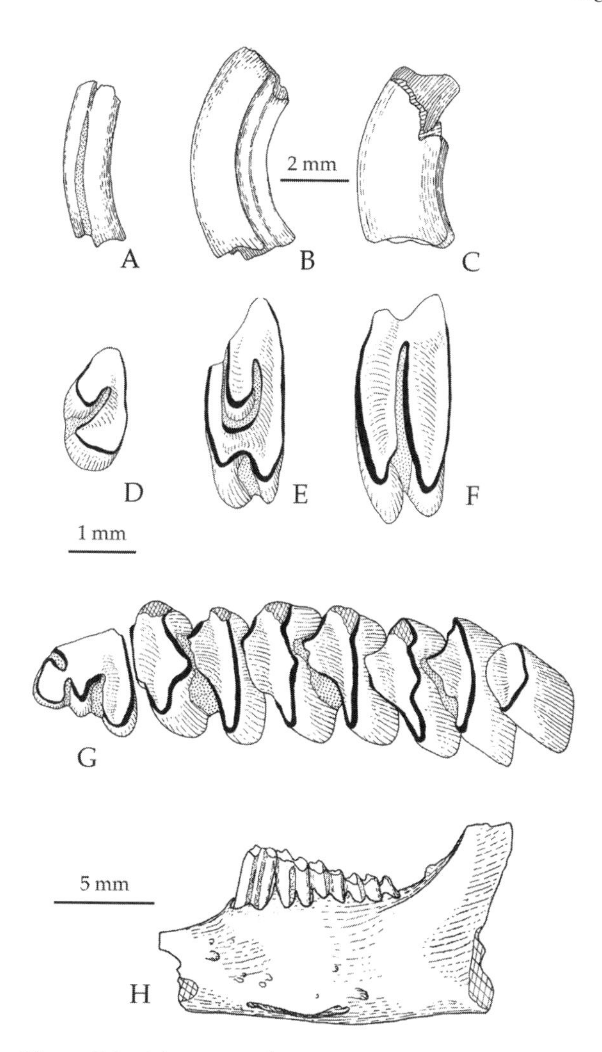

Figure 7.9. *Ochotona* sp. cf. *O. ozansoyi* from Inönü. (**A–C**) Anterior view of P2, P3, and P4. (**D–F**) Occlusal view of the same teeth. (**G**) Occlusal view of p3–m3. (**H**) Lateral view of the mandible. The 2-mm scale is for the anterior view of upper teeth; the 5-mm scale is for the lateral view of the mandible. Blackened lines represent enamel; white, dentine; dotted areas are cement infill.

Material. Fragment of lower jaw with incisor (2.20 × 1.68) and p3–p4, AS92-403 (p3, 1.82 × 2.03; p4, 1.97 × 1.91 × 2.21). See figure 7.10C–E.

Age. Middle? Turolian.

Description and Comparison. Only a portion of the mandible is preserved. The mental foramen is situated far anterior to p3. The incisor ends inferior to p4 and forms a bulge on the lingual face. The incisor cross section describes an irregular triangle. Enamel covers its anterior face and partly covers its labial face.

The shaft of p3 is lingually curved. Its occlusal outline is almost triangular. The anteroconid is relatively short but wide; its labial and lingual faces have shallow grooves. The depth of the protoflexid and paraflexid are similar, and both are directed posteriorly. The enamel bridge is central and quite wide. The metaflexid is present and filled with cement. The hypoflexid does not reach one-half of the tooth width. The p4 has a lozenge-shaped trigonid and a triangular talonid. The trigonid has an anterolabial flexid filled with cement. The talonid is wider than the trigonid.

This specimen is clearly larger than *Ochotona ozansoyi* from Loc. 84. In both localities, the occlusal pattern of p3 is quite similar. However, the Loc. 45 specimen has a wider anteroconid compared with the tooth's length, the paraflexid and protoflexid are directed backward, and the protoconid is much more strongly built. The Loc. 45 specimen is similar in size to *Bellatonoides kalfense,* but in the latter species, the anteroconid of p3 has a rounded occlusal outline and is labially inclined, and the paraflexid of *B. kalfense* is shallower than is the protoflexid.

Both in size and occlusal pattern, the p3 from Loc. 45 is quite similar to that of *Ochotona eximia* from Taraklia. The main difference is that in the latter species, the anteroconid is more elongated. These are derived features in the *Ochotona* group, probably indicating a younger age for Taraklia. The rarity of specimens both in Taraklia and Loc. 45 does not warrant a more detailed comparison; therefore, the material from Loc. 45 is referred to *O.* cf. *eximia* (Khomenko 1914).

Genus *Ochotonoma* Sen, 1998
Ochotonoma sp. (fig. 7.10A, B)

Locality. Kömürlük Dere, east of Kazan, Ankara.

Material. One p3 dex (1.83 × 1.94) (AS90-317). See figure 7.10A,B.

Age. Late Ruscinian or early Villanyian.

Description and Comparison. This unique specimen is a surface finding in the gully of Kömürlük Dere. The shaft is lingually curved. The occlusal outline is quite round. The anteroconid is large and particularly wide; its labial border (second colonette) protrudes. The anteroconid has a deep labial anteroflexid filled with cement and a lingual shallow groove. The protoflexid and paraflexid have similar depths, so that the enamel bridge between the anteroconid and protoconid-metaconid complex is centrally placed. There is a shallow depression in place of the metaflexid, which becomes progressively obliterated progressing down the shaft. The hypoflexid does not reach the middle width of the crown.

All of these characters fit with those of the genus *Ochotonoma.* This specimen cannot be attributed to *Ochotona* and *Pliolagomys* because of its large and wide anteroconid, deep anteroflexid, widened "second colonette," and the central position of the enamel bridge between the anteroconid and protoconid-metaconid complex.

This specimen is notably larger than *Ochotonoma anatolica* from Çalta (early MN 15; Sen 1998) and Ciuperceni-2 in Romania (MN 15; Terzea 1997), and *O. csarnotanus* Kretzoi, 1962 from Csarnota-2 (late MN 15) in Hungary. The Kömürlük Dere p3 compares better with *Ochotonoma* sp.

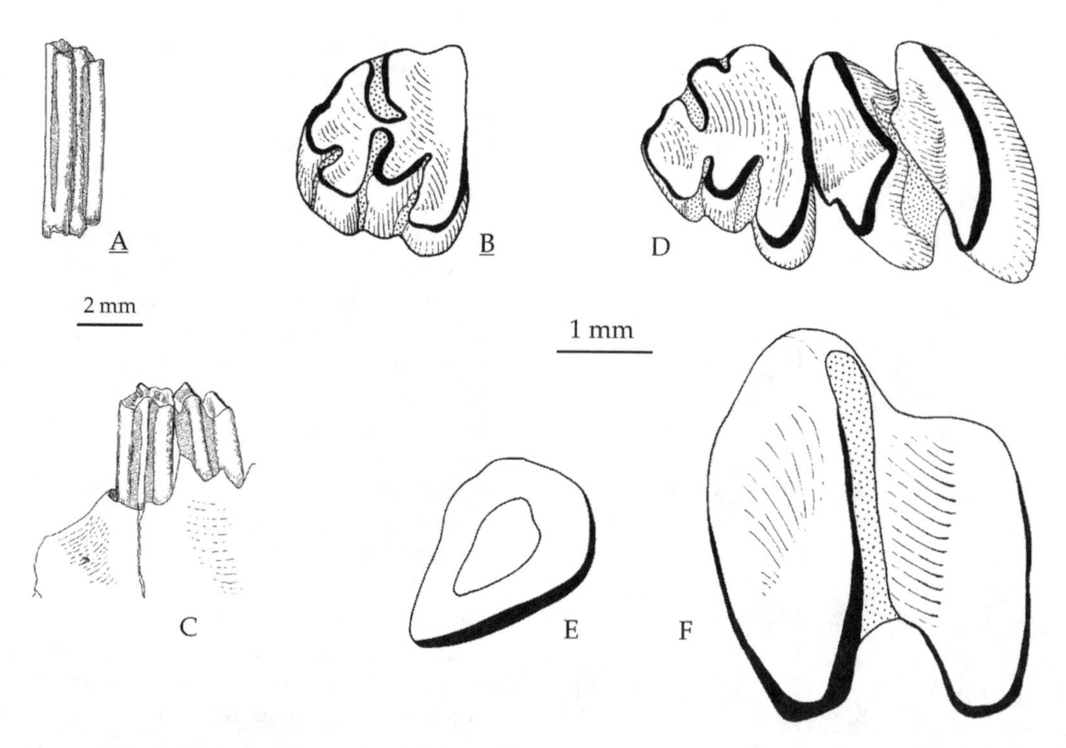

Figure 7.10. *Ochotonoma* sp. from Kömürlük Dere: (A) Labial view of p3. (B) Occlusal view of p3. *Ochotona* cf. *O. eximia* from Loc. 45. (C) Lateral view of the mandible with p3–p4. (D) Occlusal view of the same teeth. (E) Section of the lower incisor. Leporidae indet. from Loc. 12. (F) Occlusal view of a lower cheek tooth. Blackened lines represent enamel; white, dentine; dotted areas are cement infill.

from Apolakkia (Rhodes, Greece) because of its similar occlusal morphology and dimensions. I cannot discern any notable difference between the unique p3 from Apolakkia and the Kömürlük Dere specimen, except the greater width of the latter. Apolakkia has been correlated to late MN 15 (van de Weerd et al. 1982; van der Meulen and van Kolfschoten 1986). The rare material from these two localities has some derived characters (larger size, broadened anteroconid, strong second colonette, deep anteroflexid, and short metaconid-entoconid complex) compared with *O. anatolica* from Çalta. The Apolakkia and Kömürlük Dere specimens probably represent a new species of this genus that cannot be adequately characterized because the material is too limited.

Recently, Ünay and de Bruijn (1998) described a new species from Ortalica (MN 15), north central Anatolia: *Ochotonoides ortalicensis*. The species has tooth dimensions similar to, or a little larger than, *Ochotonoma anatolica* from Çalta. The characteristic teeth P2 and p3 are also similar, albeit with less variable and deeper flexids on the p3 anteroconid of *O. ortalicensis*. In any case, the Ortalica species cannot be referred to *Ochotonoides* because of its smaller size, simple p3 pattern, and shorter metaconid-entoconid complex (see Sen 1998). The specimens from Ortalica are intermediate in size and morphology of P2 and p3 between those from Çalta on the one hand and those from Apolakkia and Kömürlük Dere on the other.

Family Leporidae Gray, 1821
Leporidae, indet.

Material. One lower molar sin (3.48 × 4.08 × 3.12) (S91-353). See figure 7.10F.

Locality. Surface finding at Loc. 12.

Description and Comparison. This leporid remain is unique to the Sinap Tepe area. This lower tooth is large, its shaft is almost straight. It has a maximum crown height of 11.5 mm. The occlusal pattern of leporid lower molars is not diagnostic for genus and species referral. This specimen is clearly larger than *Alilepus annectens* (Schlosser 1924) and *Hypolagus schreuderi* Teilhard, 1940 but smaller than *Trischizolagus dimitrescuae* Radulesco and Samson, 1967. *Oryctolagus lacosti* Pomel, 1853 and the living hares (*Lepus*) have lower molars quite similar in size to the Sinap specimen.

Eurasian leporids emigrated from North America. Their oldest record in Eurasia is apparently at Can Ponsic (MN 9), San Miguel de Taudell, Soblay, and Salmendingen (MN 10), but their remains are always rare and determined as Leporidae indet. (Lopez Martinez 1989). Their presence in Asia during the Vallesian has yet to be documented. During the Turolian, the genus *Alilepus* is a little more common all over Eurasia.

Conclusions

The mammal localities in the Sinap Tepe area yielded the lagomorphs determined as *Bellatonoides eroli* n. g. n. sp. (Locs. 8A and 120), *Ochotonoma ozansoyi* n. sp. (Loc. 84), *O.*

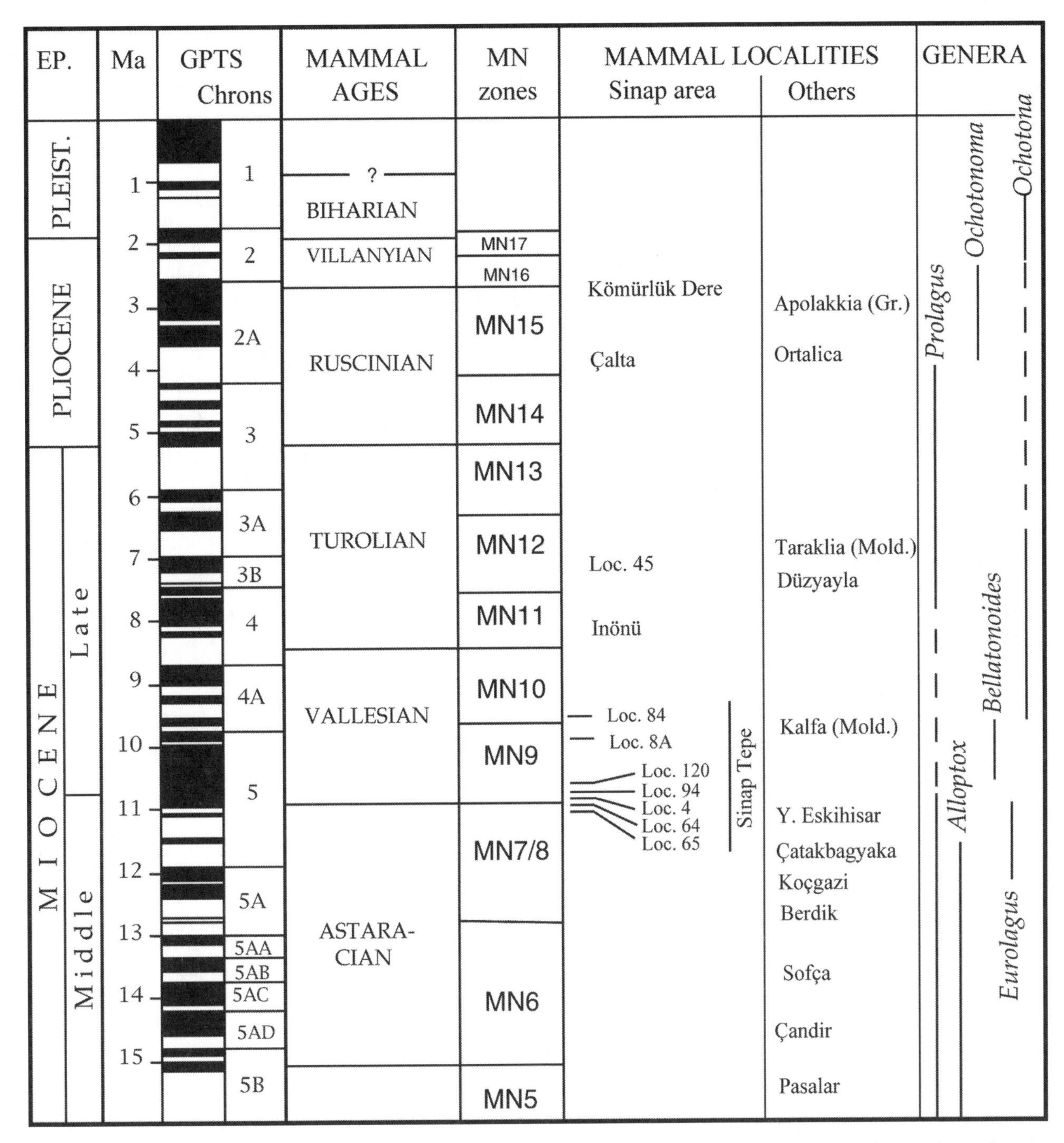

Figure 7.11. Stratigraphic chart of the key localities in the Sinap Formation compared with those of other areas from Turkey and southeastern Europe.

cf. *ozansoyi* (Inönü), *O.* cf. *eximia* (Loc. 45), *Ochotonoma anatolica* (Çalta, Sen 1998), *Ochotonoma* sp. (Kömürlük Dere) and Leporidae indet. (Loc. 12). Except for the leporid, all taxa belong to the *Ochotona* group. The remains of genera such as *Alloptox, Prolagus,* and *Amphilagus,* recorded in other Turkish middle–late Miocene or Pliocene localities, are unknown in the Sinap Tepe area.

Lagomorph remains are quite rare in the Sinap localities, except at Çalta. Late Miocene small mammal associations from this formation are dominated by cricetids and erinaceids, whereas at Çalta (late Ruscinian), the dominant group among the small mammals is the Gerbillidae, followed by the Ochotonidae.

The oldest small mammal localities in the Sinap Formation (Locs. 65, 64, 4, and 94) are dated as latest Astaracian and earliest Vallesian (Kappelman et al., this volume); they did not yield any lagomorph remains, despite the abundance of material we collected by washing-screening techniques. This absence is interesting, because the genera *Alloptox* and *Prolagus* are quite common in several middle Miocene localities of Anatolia, such as Paşalar, Çandir, Sofça, and Berdik, as is the genus *Amphilagus* (or *Eurolagus*) at Koçgazi, Yeni Eskihisar, and Çatakbagyaka (Sickenberg et al. 1975; Sen 1990). Late Miocene small mammals are still poorly represented in Turkey, and before the present study, not a single lagomorph had been described from this period. We know more about them in the Pliocene, because the genera *Prolagus* and *Ochotonoma* are recorded at Ayseki, Igdeli, Ortalica, Kavurca, and Develi (Sickenberg et al. 1975; Ünay and de Bruijn 1998; Sen 1998). The record as we now know it shows that *Alloptox* disappears in Anatolia before the end of the middle Miocene and that there is a gap (late Miocene) in the stratigraphic distribution of *Prolagus*. The presence of a break in the stratigraphic distribution of lagomorphs in Anatolia between the latest Astaracian–early Vallesian was not mentioned in previous studies. However, is it reasonable to posit a "lagomorph vacuum" in Anatolia during this interval? The record is still too poor to propose such an hypothesis.

The early and middle Miocene is a time of Eurasian ochotonid diversification. In central and western Europe, 11 genera of ochotonids with about 20 species were recorded (Lopez Martinez 1989). In central and eastern Asia during the same time interval, there seems to be less generic and specific diversity—only five genera with a dozen species are recognized (Erbajeva 1988). Toward the end of the middle Miocene and later, the diversity of ochotonids decreases in both areas. Late Miocene central and western European faunas have only three ochotonid genera: *Prolagus*, *Eurolagus*, and the Italian endemic *Paludotona*. In central and eastern Asia, the record of late Miocene ochotonids is still poor; the only ochotonids we know are those attributed to the genera *Proochotona* (= *Ochotona*, see above), *Pliolagomys*, and *Ochotona*, all members of the *Ochotona* group, and even they are still poorly documented. There is no record of lagomorphs from the Indian subcontinent prior to the latest Pliocene.

These observations show that:

The late Miocene is a time of decline of ochotonids in Eurasia, probably due to some environmental changes, and also perhaps to competition with leporids, which immigrated from North America to Eurasia in early Vallesian;

The members of the *Ochotona* group appear in central and eastern Asia during the middle Miocene, and in western Asia and eastern Europe during the Vallesian; and

The late Miocene is a time of diversification of modern ochotonids.

The "Lagomorph vacuum" recorded in the Sinap Formation is not well documented elsewhere in Turkey. The late Astaracian localities (MN 7–8), such as Berdik, Çatakbagyaka, Koçgazi, Sariçay, Sofça, Yeni Eskihisar, and Yukari Kizilca, have all yielded remains of *Prolagus*, *Alloptox*, or *Amphilagus* (fig. 7.11). On the other hand, the Vallesian is poorly documented in Turkey, except in the Sinap area. A few localities already known (Bayraktepe II, Dereikebir, Mahmutköy, Tuglu) did not yield any lagomorph.

The new genus *Bellatonoides* appears as an immigrant form derived from the Asiatic *Bellatona*. Its occurrence is later than the *Hipparion* datum in the Sinap Formation, but almost contemporaneous with the murid genus *Progonomys*, which is another Asiatic immigrant in Anatolia and Europe (Sen 1997).

The presence of *Ochotona* group representatives in Neogene faunal associations was often interpreted as indicative of open habitats in plateau grasslands and/or landscapes with relief and rocky cover. This assumption is mainly based on the ecology of living pikas. The faunal composition of the late Miocene Sinap faunas and the environmental data inferred from them do not support such an assumption. Instead, these faunas indicate open woodland environments, and they are generally fossilized in floodplain deposits. Therefore, we can argue that the present-day mode of life of pikas is a recent adaptation in such habitats because of the strong competition with arvicolid rodents and leporids, which became prolific during the Pliocene and upward. The competition with arvicolids and leporids might also have been the main cause of the decline in ochotonid diversity and spatial distribution, and consequently of their adaptation to steppic environments of high plateaus.

Acknowledgments

The material for this study was collected during the International Sinap Project under the leadership of Mikael Fortelius, John Kappelman, and Berna Alpagut, with the authorization of the General Directorate of Antiquities, Ministry of Culture of Turkey. All field participants of this project helped with collecting small mammal remains and with the washing-screening operations. Margarita Erbajeva (Ulan Ude) allowed me to compare this material with the rich collections of ochotonids in her care. A. N. Lungu (Kishinev) lent me five ochotonid p3 from Kalfa, which were useful for comparison with the Sinap material. The suggestions of the referees Raymond L. Bernor, Marguerite Hugueney, and Qiu Zhuding improved this chapter. Philippe Loubry helped with the redrawing of the illustrations. I am grateful to all of these individuals.

Literature Cited

Agadjanian, A. K., and M. A. Erbajeva, 1983, Late Cenozoic rodents and lagomorph of the USSR: Moscow, Akademia Nauk, 189 p.

Argyropulo, A. I., and I. G. Pidoplichka, 1939, Representatives of Ochotonidae (Duplicidentata, Mammalia) in the Pliocene of the USSR: Comptes Rendus (Doklady) de l'Académie des Sciences de l'URSS, v. 24, no. 7, pp. 723–728.

Bucher, H., 1982, Etude des genres *Marcuinomys* Lavocat et *Lagopsis* Schlosser (Lagomorpha, Mammalia) du Miocène inférieur et moyen de France: Implications biostratigraphiques et phylogénétiques: Bulletin du Muséum National d'Histoire Naturelle, ser 4e, v. 4, no. C1/2, pp. 43–74.

Dawson, M. R., 1961, On two ochotonids (Mammalia, Lagomorpha) from the later Tertiary of Inner Mongolia: American Museum Novitates, v. 2061, pp. 1–15.

Dmitrieva, E. L., and S. A. Nesmeyanov, 1982, Mammals and stratigraphy of continental Tertiary deposits from the southeast central Asia: Trudy Paleontological Institute 193: Moscow, Nauka, 138 pp.

Erbajeva, M. A., 1988, *Pishchukhi kainozija* (taxonomia, systematica, philogenia): Moscow, Akademia Nauk, 222 pp.

Erbajeva, M. A., 1994, Phylogeny and evolution of Ochotonidae with emphasis on Asian ochotonids, *in* Y. Tomida, C. K. Li, and T. Setoguchi eds., Rodent and lagomorph families of Asian origins and diversification: National Science Museum Monographs,Tokyo 8, pp. 1–13.

Gureev, A. A., 1964, The rabbits (Lagomorpha): Akademia Nauk, Fauna USSR, Mammals: Moscow, Akademia Nauk, 276 pp.

Ji, H. X., C. Q. Hsu, and W. P. Huang, 1980, The Hipparion fauna from Guizong basin, Xizang: Ser. Scientific Expedition to Qinghai-Sizang Plateau, Xizang Paleontology, v. 1, pp. 18–32.

Khomenko, I. P., 1914, La faune méotique du village Taraklia du district de Bendery: Fissipedia, Rodentia, Rhinocerotinae, Equinae, Suidae, Proboscidae: Trudy Bessarabskoe Obshchestvo Estestvoispytatelei, Kishinev, v. 5, pp. 1–55.

Lopez Martinez, N., 1974, Evolution de la lignée *Piezodus-Prolagus* (Lagomorpha, Ochotonidae) dans le Cénozoïque d'Europe sud-occidentale, Ph.D. thesis, Université Sciences et Techiques du Languedoc, Montpellier, 165 pp.

Lopez Martinez, N., 1986, The mammals from the Lower Miocene of Aliveri (Island of Evia, Greece). VI: The ochotonid lagomorph *Albertona balkanica* n. gen. n. sp. and its relationships: Proceedings of the Koninklijke Nederlandse Akademie van Wetenschappen, ser. B, v. 89, pp. 177–194.

Lopez Martinez, N., 1989, Revision sistematica y biostratigrafica de los Lagomorpha (Mammalia) del Terciario y Quaternario de Espana: Memorias del Museo Paleontologico de la Universidad de Zaragoza, v. 3, no. 3, pp. 1–296.

Lungu, A. N., 1981, Hipparionovaia fauna crednego Sarmata Moldavii: Kishinev, Moldova, Ministerstvo Bishevo i Brednevo Spetsialnovo Obrasovania Moldavskoi SSR, 118 pp.

Mats, V. D., 1982, The Pliocene and Pleistocene of middle Baikal: Novossibirsk, Siberia, Science Press, 192 pp.

Meulen, A. J. van der, and T. van Kolfschoten, 1986, Review of the late Turolian to early Biharian mammal faunas from Greece and Turkey: Memoria della Società Geologica Italiana, v. 31 pp. 201–211.

Ozansoy, F., 1965, Etude des gisements continentaux et des mammifères du Cénozoïque de Turquie: Mémoires de la Société Géologique de France, v. 102, pp. 1–92.

Qiu, Z., 1987, The Neogene mammalian faunas of Ertemte and Harr Obo in Inner Mongolia (Nei Mongol), China. 6: Hares and pikas—Lagomorpha: Leporidae and Ochotonidae: Senkenbergiana Lethaea, v. 67, pp. 375–399.

Qiu, Z., 1996, Middle Miocene micromammalian fauna from Tunggur, Nei Mongol: Science Press, Beijing, 216 pp.

Schlosser, M., 1924, Tertiary vertebrates from Mongolia: Palaeontologia Sinica, v. 1, no. 1, pp. 1–119.

Sen, S., 1990, Middle Miocene lagomorphs from Pasalar, Turkey: Journal of Human Evolution, v. 19, no. 4/5, pp. 455–461.

Sen, S., 1991, Stratigraphie, faunes de mammifères et magnétostratigraphie du Néogène de Sinap Tepe, Province d'Ankara, Turquie: Bulletin du Muséum National d'Histoire Naturelle, ser. 4e, v. 12, no. C3/4, pp. 243–277.

Sen, S., 1997, Magnetostratigraphic calibration of the European Neogene mammal chronology: Palaeogeography, Palaeoclimatology, Palaeoecology, v. 133, pp. 181–204.

Sen, S., 1998, Pliocene vertebrate locality of Çalta, Ankara, Turkey. 4: Rodentia and Lagomorpha: Geodiversitas, v. 20, no. 3, pp. 359–378.

Sickenberg, O., J. D. Becker-Platen, L. Benda, D. Berg, B. Engesser, W. Gaziry, K. Heissig, K. A. Hünermann, P. Y. Sondaar, N. Schmidt-Kittler, K. Staesche, U. Staesche, P. Steffens, and H. Tobien, 1975, Die Gliederung des höheren Jungtertiärs und Altquartärs in der Türkei nach Vertebraten und ihre Bedeutung für die internationale Neogen-Stratigraphie: Geologische Jahrbuch, v. B15, pp. 1–167.

Terzea, E., 1997, Biochronologie du Pliocène du bord méridional du bassin dacique (Roumanie), *in* J. P. Aguilar, S. Legendre, and J. Michaux eds., Actes du Congrès BiochroM'97, Mémoires et Travaux de l'EPHE, Institut de Montpellier: Montpellier, EPHE, v. 21, pp. 649–660.

Tleuberdina, P. A., 1982, Late Neogene faunas from southeast Kazakhstan: Akademia Nauk Kazakhskoy SSR, Alma Ata, 118 pp.

Tobien, H., 1963, Zur Gebiss-Entwicklung Tertiärer Lagomorphen (Mamm.) Europas: Notizblatt des Hessischen Landesamtes für Bodenforschung zu Wiesbaden, v. 91, pp. 253–344.

Tobien, H., 1974, Zur Gebisstruktur, systematik und evolution der genera *Amphilagus* und *Titanomys* (Lagomorpha, Mammalia) aus einigen Vorkommen im jungeren Tertiar Mittel- und Westeuropas: Mainzer Geowissenschaften Mitteilung, v. 3, pp. 95–214.

Tobien, H., 1975, Zur Gebisstruktur, systematik und evolution der genera *Piezodus, Prolagus* und *Ptychoprolagus* (Lagomorpha, Mammalia) aus einigen Vorkommen im jungeren Tertiar Mittel- und Westeuropas: Notizblatt des Hessischen Landesamtes für Bodenforschung zu Wiesbaden, v. 103, pp. 103–186.

Ünay, E., and H. de Bruijn, 1998, Plio-Pleistocene rodents and lagomorphs from Anatolia, *in* The dawn of the Quaternary: Mededelingen Nederlands Instituut voor Toegepaste Geowetenschappen, v. 60, pp. 431–465.

Weerd, A. van de, J. W. F. Reumer, and J. de Vos, 1982, Pliocene mammals from Apolakkia Formation (Rhodes, Greece): Proceedings of the Koninklijke Nederlandse Akademie van Wetenschappen, v. B85, pp. 89–112.

White, J. A., 1991, North American Leporinae (Mammalia: Lagomorpha) from Late Miocene (Clarendonian) to latest Pliocene (Blancan): Journal of Vertebrate Paleontology, v. 11, no. 1, pp. 67–89.

Zhou, X., 1988, Miocene ochotonid (Mammalia, Lagomorpha) from Xinzhou, Shanxi: Vertebrata Pal Asiatica, v. 26, pp. 139–148.

8

Carnivora

S. Viranta and L. Werdelin

Despite intensive work over the past 150 years, our knowledge of the taxonomy, phylogeny, and paleoecology of Neogene carnivores is still in a state of flux. In a recent synthesis (Werdelin 1996a) it was, for example, impossible even to begin to unravel the taxonomy and relationships of Eurasian Miocene Mustelidae. Other families of Carnivora are somewhat better understood, but no consensus has been attained regarding any of them. In large part, this may be due to the fragmentary nature of the material, with Carnivora generally being known from very few specimens at any given site, presumably due to the small population sizes of Carnivora brought on by the trophic position of the majority of its constituent taxa. Another problem has been the constant theme of iterative evolution characteristic of Carnivora (Martin 1989; Werdelin 1996b), an evolutionary pattern that suggests that a limited number of morphological pathways are available to the group. This has led to a general disregard of Carnivora in many broader studies of paleoecology and faunal change. Herein we shall attempt to place the record of Carnivora from the Sinap Formation in the context of carnivore evolution in western Eurasia in general and that of southeast Europe and western Asia in particular, as the latter regions have comparatively good records of Carnivora, especially in the late upper Miocene (Turolian).

Compared with the many studies of Neogene Carnivora from neighboring regions such as Greece and Iran (e.g., Solounias 1981 and references therein), the carnivoran faunas of Turkey have been relatively little studied. Nevertheless, a number of important investigations have been published by Senyürek (1954, 1957, 1958, 1960), Ozansoy (1965), Gürbüz (1974, 1981), Schmidt-Kittler (1976), de Bonis (1994), and Viranta and Andrews (1995). These have shown that Turkish faunas in general fit well in the evolutionary patterns seen in other parts of western Eurasia. They also, however, include material of a taxon, *Dinocrocuta senyureki,* that is otherwise known only from north Africa (Sahabi; Howell, 1987). Given this evidence, it is possible that Turkey represents the western- or northwesternmost limit to the distribution of certain carnivoran taxa. Study of the Carnivora of the Sinap Formation will therefore add significantly to our understanding of the evolution of carnivorans in western Eurasia. In this chapter, we list the material of Carnivora from the various Sinap Formation localities, describe relevant specimens, and discuss the significance of the material in the context of carnivoran evolution in the later Neogene of western Eurasia in general. The material will be presented by locality and in stratigraphic order within MN zones.

Methods

All specimens identifiable as Carnivora are included in the listings. Only those identified to at least the family level are described. We use dental terms as defined in Werdelin as Solounias (1991). Measurements were taken with calipers and are given to the nearest 0.1 mm. Standard measurements are maximum length (l) and width (w) for teeth (generally given as length × width in the specimen lists) and maximum length (l) and transverse width for proximal (p) and distal (d) ends of long bones (including phalanges). Other measurements are defined as they are used. Measurements for specimens that are broken or otherwise altered by taphonomic processes are given as approximations (~).

MN 6 Localities

Loc. 24

Hemicyon sansaniensis Lartet, 1851

Material. S.89.522, m2 dex (24.4 × 16.9); no number, cranial fragments with I3 (6.7 × 9.4), P2 (~11), P3 (13.8 × 6.6), M1 (l = 23.8), M2 (l = 20.2), p3 (11.2 × 6.7), p4 (17.1 × 8.9), m1 (33.0 × 14.9), m2 (23.5 × 15.8), m3 (14.1 × 11.2).

Description. S.89.522 is a worn m2 with additional heavy postmortem damage, including rootmarks. Most of the morphology is still perceivable, however, and is similar to that of *H. sansaniensis* (see Ginsburg 1961). It has the shape of an elongated rectangle in buccal view. The trigonid is longer than the talonid, but both are well developed. In size, the tooth lies within the range of known m2 specimens of this species.

The unnumbered material consists of an articulated mandibular condyle and mandibular fossa with a part of the masseteric fossa and a fragment of the basicranium with the left zygomaticum arch preserved. The buccal half of the M2 is preserved in place. Most of the mandibular and maxillary bone is lost. Only fragments of the incisors are preserved, and all of the canines are lost.

Of the cheek teeth, the left p2 and m1–m3 and the right p3–p4 and m1–m3 are preserved. The molars are very worn. On the first and second molars the cusps are completely worn down. The first molar has heavy wear on the paraconid and protoconid and lighter wear on the metaconid and hypoconid. The entoconid is very flat and unworn.

Most of the upper teeth are lost. Only the buccal sides of the molars are preserved, and the right upper first molar is missing. The upper carnassials are represented by paraconid fragments, and only P1 and P2 are preserved. The buccal cusps of the upper molars are very worn. They are about equal in size, and have a distinct cingulum around them. The cingulum of the M2 is enhanced at the anterobuccal corner (the facet for the M1), and the cingulum of the M1 is enhanced at both the antero- and posterobuccal corners. The premolars have only one cusp, which is slightly worn. There is also an I3 and an unidentified incisor fragment. They are both worn.

cf. *Hemicyon sansaniensis* Lartet, 1851

Material. AS.92.897, I2 dex (6.3 × 10.8); AS.92.735, canine fragment; AS.92.713, navicular sin.

Description. AS.92.897 is an I2 similar in shape to those known for hemicyonids and probably presents *H. sansaniensis*. Similarly, a fragment of a canine (AS.92.735) is tentatively attributed to this species, based on overall size and what is left of the morphology.

A large and complete navicular (S.92.713) resembles that of a recent brown bear (*Ursus arctos*), although it differs from it in certain respects. The facet for the ectocuneiform is very large, whereas in the brown bear it is smaller (in relation to the other facets). The facet for the mesocuneiform is relatively small. The facet for the entocuneiform in S.92.713 is larger again and concave, whereas in the brown bear it is straight or slightly convex. The articulation for the astragalus is almost round in S.92.713; in the brown bear it is more oval. The specimen S.92.713 is about twice the size of an average navicular of a female brown bear. It has a transverse width of 46.1 mm, and an anteroposterior width of 36.5 mm.

cf. *Pseudaelurus* sp.

Material. S.89.491, ulna dex proximal fragment.

Description. This specimen is a proximal fragment of an ulna, which resembles those of felids in having a relatively open semilunar notch. It is about the size of that of a modern lynx (*Lynx lynx*). The greatest height of the semilunar notch is 16.7 mm and the height of the olecranon process above it is 47.1 mm.

MN 7/8 (9?) Localities

Loc. 64

Mustelidae indet.

Material. AS.92.29, p4 sin (10.2 × 4.5).

Description. The specimen is a premolar of a relatively large mustelid. It consists of four cusps placed along the midline.

*Sansanosmilus jourdani (*Filhol 1881)

Material. AS.92.65 (fig. 8.1), dp4 sin (16.3 × 5.8); AS.92.828, premolar fragment.

Description. The specimen AS.92.65 has three cusps (paraconid, protoconid, and metaconid) placed along the midline of the tooth. It is worn on the buccal side of the paraconid and protoconid. It is similar to the lower carnassials of nimravids (*Sansanosmilus* spp.) in appearance, although it differs from them in having a distinct metaconid. It differs from the known deciduous carnassials of Eocene and Oligocene nimravids in lacking a talonid cusp (Harold Bryant, pers. comm.). No dp4 has been described for *Sansanosmilus* spp. We attribute AS.92.65 to this genus based on its similarities to the lower permanent carnassial of the genus and to the other known nimravid dp4 specimens. *Sansanosmilus jourdani* is the only nimravid species known to occur in Europe in MN 7/8.

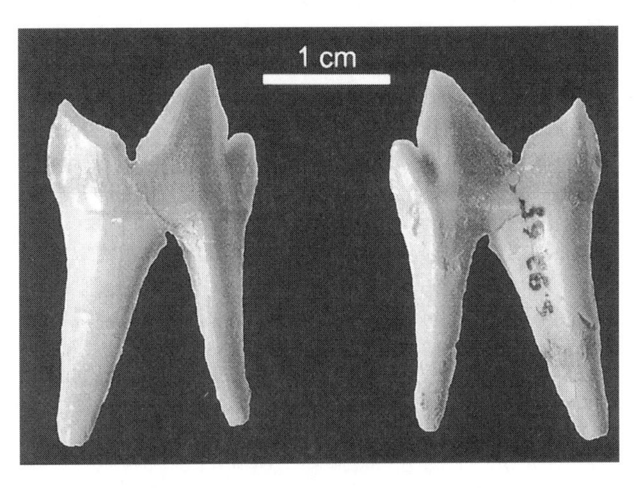

Figure 8.1. AS.92.65, dp4 sin of *Sansanosmilus jourdani* from Loc. 64 in (left) lingual view; (right) buccal view.

The specimen AS.92.828 preserves a partial main cusp and probably the posterior accessory cusp. The cusps are rounded, like those of *Sansanosmilus* spp.

Hyaenidae indet.

Material. AS.92.826, I3 dex (3.0 × 5.1).

Description. This specimen is an unworn hyenid I3. It is larger than that of *P. crassum* from Sofça and Esme Akçaköy (Schmidt-Kittler 1976) and comes from a hyenid of about the size of *Thalassictis montadai* or slightly smaller.

MN 9 Localities

Loc. 4

cf. *Thalassictis montadai* (Villalta Comella and Crusafont-Pairo, 1943)

Material. AS.93.364, medial phalanx (l = 17.7, p = 8.7, d = 8.2); AS.93.82, medial phalanx (l = 17.3, p = 8.7, d = 8.2); AS.93.81, proximal phalanx (l = 22.0, p = 8.2, d = 6.9,); AS.93.97, MC V proximal (proximal articular surface 10.5 × 13.6); AS.93.445, i3 (3.5 × 5.1); AS.93.84, canine fragment; AS.93.199, upper canine fragment.

Description. The phalanges (AS.93.364, AS.93.82, AS.93.81) and MC V (AS.93.97) have morphologies similar to those of later Miocene hyenids, and although no *T. montadai* phalanges are available for comparison, they are of similar size to those expected for that species. This species is known elsewhere at Sinap from this stratigraphic level (Loc. 94).

The i3 AS.93.445 is about the size of the i3 specimens of *T. montadai* and probably belongs to that taxon. The same is true of the two partial canines.

cf. *Hyaenotherium wongii* (Zdansky 1924)

Material. AS.94.792, m2 dex (6.4 × 5.5).

Description. The m2 AS.94.792 is a single-rooted molar with a low crown and three small cusps. It is round and differs in that respect from the m2 of *Protictiherium crassum* and *Thalassictis montadai.* It is similar to the m2s of *Hyaenotherium wongii* but is not sufficiently diagnostic to record the presence of the species in MN 9 of the Sinap area with confidence. The specimen is a little larger than those recorded for *H. wongii* of Samos by Kurtén (1985) and Werdelin (1988).

Felidae indet.

Material. AS.93.96, proximal phalanx (l = 21.7, p = 8.0, d = 6.8).

Description. This phalanx is attributed to a felid on the basis of the distally situated insertion for the flexor digitorum superficialis muscle.

Mustelidae, medium

Material. AS.94.788, calcaneum dex (l = 18.2, maximum w = 10.4).

Description. Specimen AS.94.788 is a calcaneum with characters indicating mustelid affinities, such as a short neck and rounded facets for the astragalus.

Carnivora indet.

Material. AS.93.94, i2 sin (3.6 × 4.8); AS.93.84, canine fragment; AS.93.199, canine fragment; AS.93.327, dental fragment; AS.93.83, dental fragment; AS.92.374, ulna fragment; AS.93.87, tibia fragment; AS.93.423, navicular fragment; AS.93.427, navicular; AS.93.86, navicular; AS.94.801, navicular fragment; AS.93.85, cuboid; AS.93.444, mesocuneiform; S.91.663, MP distal fragment (d = 4.0); AS.93.364, medial phalanx (l = 17.7); AS.94.786, medial phalanx (l = 16.9); AS.94.790, distal phalanx (l = 15.8).

Description. These specimens represent several size classes, but we could not place them in subordinal taxa.

Loc. 94

Thalassictis montadai (Villalta Comella and Crusafont-Pairo, 1943)

Material. AS.92.463 (fig. 8.2), mandible sin with c (12.9 × 10.7), p2 (15.1 × 8.6), p3 (17.4 × 10.0), m1 (23.4 × 10.5); AS.92.464 (fig. 8.2), mandible sin with i2 (3.5 × 4.9), i3 (5.2 × 5.8), c (12.8 × 12.1), p2 (15.7 × 8.1), p3 (17.4 × 10.3), p4 (21.4 × 11.1), m1 (22.8 × 10.1), m2 (5.6 × 5.2); AS.92.526, I2 (3.6 × 6.1); AS.95.312, M1 sin (7.8 × 14.9); AS.92.552, humerus distal fragment; AS.92.465, radius dex (l = 180, p = 20.1, d = 31.5); AS.92.466, tibia dex proximal fragment (p = 36.2).

Description. Both mandibles (AS.92.463, AS.92.464) have heavily worn teeth. Mandible AS.92.463 has lost m2 and the anterior part of p4. There are no alveoli for either of these teeth and they probably were lost early in life and the alveoli closed completely. The upper incisor, AS.92.526, is also heavily worn and may have belonged to the same individual as one of the mandibles.

The incisors of AS.92.464 are worn down quite close to the root margins, and no cuspules or other morphological features remain. The p2 of both mandibles have a main cusp and a posterior accessory cusp. Only a rudiment of an anterior accessory cusp is preserved. The accessory cusps are more distinct in the p3 and in the p4 preserved in AS.92.464. The talonids of the m1 are short and have two low cusps; the entoconid and hypoconid. The metaconids are low but distinct. They are unworn, although the protoconids are worn down to the level of the metaconid. This is especially clear in AS.92.434.

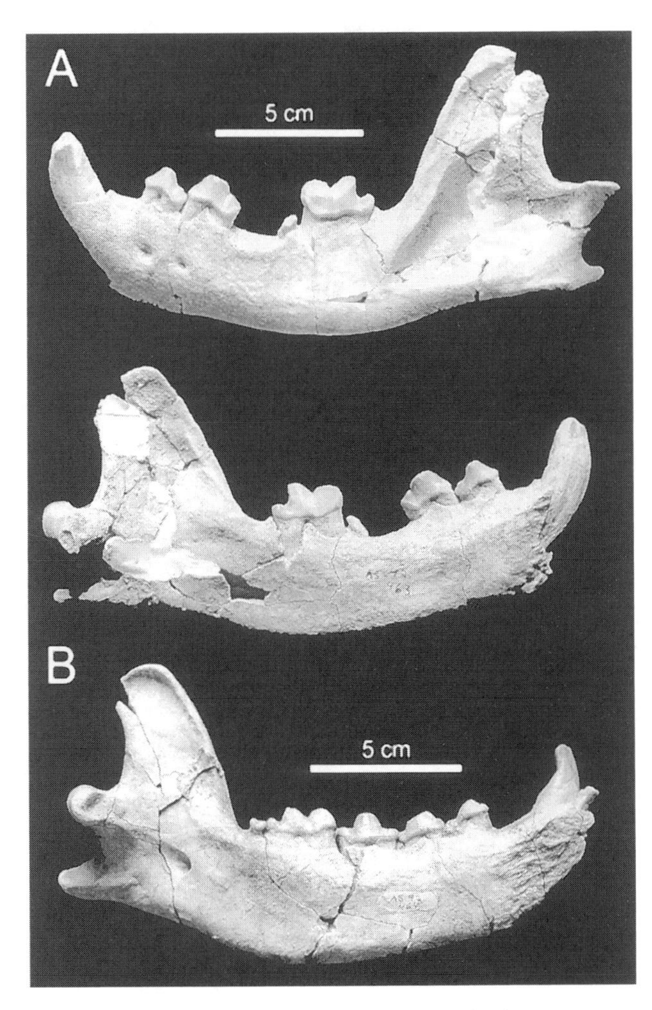

Figure 8.2. Mandibles sin of *Thalassictis montadai* from Loc 94 in lingual view. (**A**) AS.92.463; (**B**) AS.92.464.

The postcranial bones were attributed to *Thalassictis montadai* on the basis of comparisons with *Hyaenotherium wongii* limb bones. They are stouter than those of *H. wongii,* scaling well to the greater body mass estimated for *T. montadai* (Werdelin, unpublished data). These limbs are shorter than those of *H. wongii,* indicating an evolution toward greater cursorial abilities in the latter species.

cf. *Thalassictis montadai* (Villalta Comella and Crusafont-Pairo, 1943)

Material. AS.92.526, incisor (3.6 × 6.1); AS.92.487, distal phalanx (l = 17.4); AS.93.185, MC II dex fragment (p = 14.4).

Description. These specimens have typical hyenid characters and are of the size estimated for these elements in *T. montadai.*

Protictitherium crassum (Depéret, 1892)

Material. AS.92.488, m1 sin (12.6 × 5.7); AS.93.65, m1 sin (12.3 × 5.8).

Description. These two lower carnassials are almost unworn and display the morphology and size of those of *Protictitherium crassum.* They have long, well-developed talonids with three cusps and strong metaconids. The protoconid is the tallest trigonid cusp and is set posterobuccal to the paraconid.

cf. *Protictitherium crassum* (Depéret, 1892)

Material. AS.92.799, p2 dex (9.5 × 4.3); AS.93.122, femur shaft (immature); AS.92.486, ulna dex proximal fragment; AS.93.117, radius shaft (immature); AS.93.600, premolar fragment, AS.92.570, trapezoid; AS.93.186, calcaneum fragment; AS.93.220, calcaneum fragment.

Description. Specimen AS.92.799 is a lower premolar with a main cusp and a posterior accessory cusp. The posterior cingulum is thickened.

The postcranial specimens (AS.93.122, AS.92.486, AS.93.117) have hyenid affinities and are of the size expected of *Protictitherium crassum.* The epiphyses of the long bones have not been fused, indicating that the individual to which they belonged was immature. It follows that these specimens could potentially pertain to a larger species, although not much larger.

Felidae indet., small

Material. AS.92.550, MC II sin proximal fragment (p = 5.0).

Description. This proximal end of an MC II (AS.92.550) has felid characters. It is fairly small in size, about one-half the size of that described for *Pseudaelurus quadridentatus* by Schmidt-Kittler (1976). It may thus belong to one of the smaller species of *Pseudaelurus.*

Carnivora indet.

Material. AS.92.566, tooth fragment; AS.93.167, trapezoid; AS.92.489, medial phalanx (l = 9.2, p = 4.7, d = 4.2); AS.92.525, proximal phalanx proximal fragment (p = 5.1).

Description. These specimens cannot be identified beyond the ordinal level and they represent more than one size category.

Loc. 107

cf. *Protictitherium* sp.

Material. AS.93.98, P4 sin fragment.

Description. The P4 fragment AS.93.98 preserves only the metastyle blade and a partial paracone. The metastyle appears relatively short, as is typical of *Protictitherium* spp.

Loc. 108

cf. *Dinocrocuta senyureki* (Ozansoy, 1965)

Material. AS.93.66, P4 fragment; AS.93.67, AS.93.68, AS.93.69, premolar fragments; AS.93.197, C fragment.

Description. This material consists only of fragments of teeth that match those of percrocutids in morphology. It includes premolar accessory cusps with inflated pyramid shapes. They appear relatively tall. The original sizes of the teeth to which these fragments belonged can be extrapolated to about the size of the teeth of *Dinocrocuta senyureki.*

cf. *D. minor* (Ozansoy, 1965)

Material. AS.93.67, p2 fragment; AS.93.69, p2 fragment.

Description. These premolars are similar in morphology to but smaller than the other percrocutid dental material known from Loc. 108. *Dinocrocuta minor* is smaller in size than *D. senyureki* and has been recorded from Yassiören (Ozansoy 1965).

Carnivora indet., small

Material. AS.94.149, MC IV sin proximal fragment (p = 2.8).

Description. This small and gracile metacarpal fragment is probably of a small mustelid, but attribution to a small viverrid, such as *Semigenetta* spp., cannot be ruled out.

Loc. 72

cf. *Protictitherium crassum* (Depéret, 1892)

Material. AS.93.99, mandible sin fragment with p4 (10.1 × 4.8), m1 fragment.

Description. The m1 preserves only the trigonid, which is tall and has a paraconid that is much lower than the protoconid. The morphology of the three-cusped p4 is similar to that of *Protictitherium crassum.*

Hyaenidae indet.

Material. AS.92.206, mandible sin fragment c (5.3 × 5.6), p1 (3.1 × 1.9), p2 fragment.

Description. The dentition preserved is not sufficient to allow for identification to species. The teeth measure a little above the size range known for *Protictitherium crassum.*

Loc. 91

Ictitherium cf. *intuberculatum* (Ozansoy, 1965)

Material. AS.93.63 maxilla sin fragment with C fragment, P1 (6.1 × 4.1), P2 (12.6 × 6.5).

Description. The maxillary bone is broken posterior to the P2 and continues toward the nasals but does not quite reach them. Anteriorly, it continues to the premaxilla, which is lost. The upper canine is badly damaged. The P1 is small and has a single cusp. The P2 has two cusps. The main cusp of the P2 has a concave anterior margin with no accessory cusp. The posterior margin has a distinct but low accessory cusp. These premolars are equal in size to those of *Ictitherium intuberculatum.* This species has been described from the middle Sinap by Ozansoy (1965), and AS.93.63 is likely to represent that taxon.

Loc. 114

Hyaenidae indet.

Material. AS.94.189, maxillary dex fragment with I1 (l = 3.3), I2 (l = 4.6), I3 (l = 6.3), C (10.0 × 6.4), P2 (14.0 × 6.2), P3 (17.5 × 9.1).

Description. The teeth are unworn. The tip of the canine is broken, whereas the other teeth are complete. Bone is preserved only around the dentition. The material available is not diagnostic within the Hyaenidae.

Loc. 8

Ictitherium intuberculatum Ozansoy, 1965

Material. AS.92.223 (fig. 8.3), maxilla dex fragment with P3 (17.1 × 10.1), P4 (24.2 × 15.3), M1 (10.7 × 16.6), M2 (~11 × ~6); AS.92.276, I2 dex (4.5 × 5.3).

Description. These teeth match those of *Ictitherium intuberculatum,* as described by Ozansoy (1965), in both size and shape. The M2 is small but has three cusps, as in the M1. The metastyle wing of M1 is reduced, although not as much as in *Hyaenotherium wongii.* The metastyle blade is short in relation to the length of the paracone in the upper carnassial, and the protocone is level with the parastyle. The paracone is well developed. The P3 consists of the

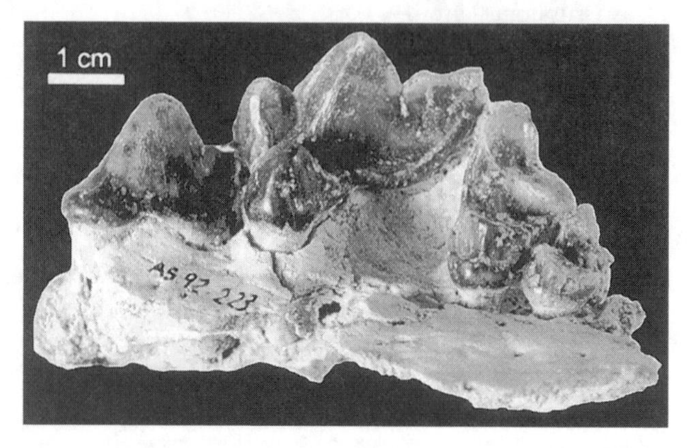

Figure 8.3. AS.92.223, maxilla dex fragment of *Ictitherium intuberculatum* from Loc. 8 in lingual view.

main cusp and a small posterior accessory cusp. The maxillary bone is preserved only around the dentition.

Loc. 12

Indarctos vireti Villalta and Crusafont, 1943

Material. Number 2404 in the MTA Museum, Ankara. maxillary dex fragment with I3 (8.5 × 10.0), C (18.4 × 12.4), P1 (10.0 × 6.0), P2 (9.8 × 6.3), P4 (21.8 × 17.2), M1 (23.2 × 19.4), M2 (26.3 × 19.5).

We were able to study only an incomplete cast of the specimen. The original material was studied by D. Geraads in Ankara, and the description given here is partly based on his notes.

Description. There is a gap between the I3 and canine. The canine is transversely compressed. The first two premolars are very simple, unicuspid teeth. They are situated close to each other and the canine. The P4 has a short metastyle blade, no parastyle blade, and a low but very wide protocone. The molars are transversely compressed and quadrangular in shape. They are so worn that morphological details are not preserved.

The bunodont but transversely slim upper molars of this specimen ally it with *Indarctos* spp. The lengthened muzzle and the uncrowded premolars ally it with *Indarctos vireti*. The size is also within the known range of *I. vireti*. This species has also been tentatively recorded from Yeni Eskihisar (*I.* cf. *vireti* in Wessels et al. 1987).

Dinocrocuta senyureki Ozansoy, 1965

Material. AS.95.413, skull fragment with dex I1–I3, P1 (9.8 × 7.3), P2 (23.8 × 13.9), P3 (28.2 × 18.4) and sin P1 (10.0 × 7.8), P2 (23.4 × 14.5), P3 (27.0 × 17.9), P4 (46.3 × 14.4), M1 (15.7 × 6.3); AS.95.414, C dex (18.9 × 15.6); AS.95.404, P3 sin (28.3 × 20.3); AS.95.602, dex P4 (46.0 × 21.9); AS.95.446, maxillary sin fragment (associated with AS.95.404); AS.95.318, mandible dex with i3 (7.8 × 8.9), c (20.6 × 17.4), p2 (22.9 × 16.3), p3 (25.9 × 17.5), p4 (26.4 × 16.6), m1 (31.8 × 16.6); AS.95.279 (fig. 8.4), mandible sin fragment with m1 (30.8 × 14.9) (associated with AS.95.280); AS.95.280 (fig. 8.4), mandible sin fragment with p3 (26.3 × 6.7), p4 (29.9 × 16.4) (associated with AS.95.279); AS.95.410, mandible dex fragment with i3 (6.5 × 9.8), c sin (20.0 × 17.4), AS.95.347 (fig. 8.5), d4 dex (24.0 × 9.7); AS.95.601 (fig. 8.5), d4 sin (22.3 × 9.2); AS.95.326, atlas (for measurements, see the description); AS.95.434, scapula dex proximal fragment (p = 53.4).

Description. This material represents at least two adult and two juvenile animals. Both the adults have moderate dental wear, and the milk teeth of the juvenile specimens are also somewhat worn.

The partial skull AS.95.413 was found in very close proximity to AS.95.326, an atlas, and they probably belong to the same animal. The skull preserves the maxillary and

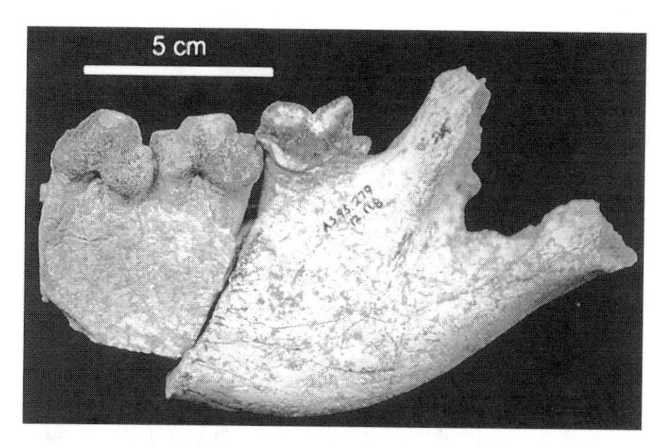

Figure 8.4. AS.92.279 (posterior piece) and AS.95.280 (anterior piece), mandible sin pieces of *Dinocrocuta senyureki* from Loc. 12 in buccal view.

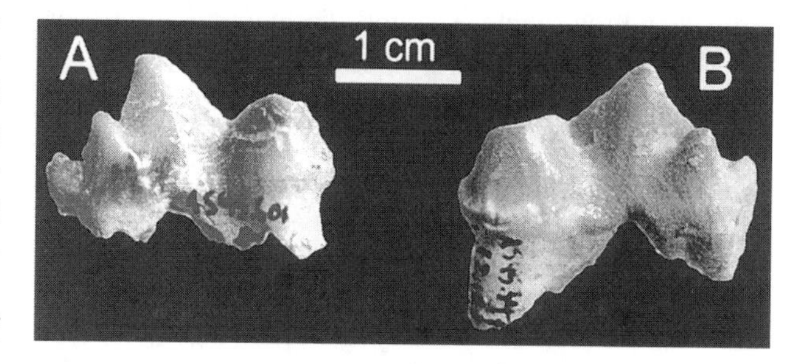

Figure 8.5. Deciduous carnassials of *Dinocrocuta senyureki* from Loc 12. **(A)** AS. 95.601, sin in lingual view; **(B)** AS.95.347, dex in lingual view.

premaxillary bones of both sides, as well as fragments of the nasals. Parts of the palatine are also preserved. The back of the skull has been eroded away and part of the dentition, the right P4 and M1 are lost, as well as most of the incisors and both canines. The right P4 AS.95.602 was found separate from but near AS.95.413. This tooth is likely to have belonged to the skull.

The snout is very short and the cheek teeth densely packed. Large infraorbital foramina are situated 31.5 mm above the lingual margin of the anterior root of P3. The zygomatic arch emerges above and ~10 mm posterior to the foramen. The anterior palatine foramina are large. The rest of the foramina and the sutures on the palatal side are indistinct, due to the fragmented state of the bone. (In addition, there is one cavity on the inferior side of the maxillary bone that is caused by a dental pick used in the excavations.)

The two preserved incisors are so worn that the original morphology is obliterated. The premolars have not worn symmetrically, as those on the right side are more heavily worn than those on the left. The P1 is a unicuspid, round tooth with an excavation of enamel on the posterolingual side. The P2 is also unicuspid. It has, however, enhancements of the cingulum at the anterior and posterior ends. The anterior of these is set lingual to the main axis of the tooth. The same is true of P3, which has a small cusp on this enhancement. This tooth also has a small posterior accessory cusp. The P2 and P3 are not on the same sagittal line, but at an angle of 170° to each other. The parastyle blade, protocone, and metastyle blade of P4 form a long, sagittally oriented cutting blade. The protocone is situated directly lingual to the parastyle. It is reduced, but has heavy wear on both the left and the right sides. The M1 is a diminutive, slim tooth.

The mandibles (AS.95.318 and AS.95.279/AS.95.280) may belong to a single individual. The ramus AS.95.318 is almost complete, lacking the symphysis (along with i1 and i2), whereas at the caudal end, it continues to the anterior part of a broken ascending ramus. The bone is very poorly preserved and broken, and the original traits are mostly lost. Two mental foramina are preserved. They are situated below the anterior and posterior margins of the p2. The symphysis appears to have been strong, as the mandibular bone is thicker in the anterior part and seems to narrow a little behind the canine and short diastema posterior to it. The masseteric fossa was probably deep, as indicated by the preserved anterior part. It reaches anteriorly to just behind the lower carnassial.

The i3 is very low, due to the heavy wear. The canine is similarly worn. There is no alveolus for a p1. The p2 and p3 have small anterior accessory cusps and slightly larger posterior ones. The p4 is more elongated, with both the accessory anterior and posterior cusps present. These cusps are distinct but low. The m1 has no metaconid and a tiny talonid with two very low cusps.

The two deciduous lower carnassials are morphologically exactly like those described for percrocutids by Schmidt-Kittler (1976) and Chen and Schmidt-Kittler (1983). They both have a high protoconid, lower paraconid, and trenchant metaconid, the latter set distinctly posterior to the protoconid. In addition, two caudally inclined talonid cusps are present. Both teeth have vertical wear facets on the buccal side of the para- and protoconids.

The atlas (AS.95.326) is large and robust. The maximum anteroposterior length is 53.8 mm. The maximum transverse width is unknown, because the wings are broken. The total width of the facets for the cranial condyles is 51.1 mm, and the articulation for the axis measures 61.1 mm in transverse width. The length of the body is 24.6 mm on the ventral side and 16.0 mm on the dorsal side. The cranial articular fovea is very small and smooth. The vertebral foramen broadens considerably toward the dorsal margin. In appearance, AS.95.326 is very similar to the atlas of the living spotted hyena, *Crocuta crocuta*.

Ozansoy (1965) described a new species of *"Hyaena," H. senyureki,* based on a right mandible and a partial left maxilla found at Yassiören, middle Sinap. Our specimens from Sinap project Loc. 12 are very similar in appearance and dimensions to the same elements in the species described by Ozansoy (1965).

cf. *Dinocrocuta senyureki* (Ozansoy, 1965)

Material. AS.95.434, scapula dex proximal fragment.

Description. A very small proximal part of a robust scapula is preserved. The neck seems relatively long and narrow. The scapular spine is distinct and the proximal projection is relatively large. The articulation with the humerus is completely preserved and measures 45.6 × 34 mm.

Miomachairodus pseudailuroides Schmidt-Kittler, 1976

Material. AS.95.657, skull sin fragment with P3 (17.1 × 10.3), P4 (32.2 × 11.7), mandible sin with c (fragment l = 17.4), p3 (16.6 × 7.4), p4 (22.3 × 9.4), m1 (26.1 × 11.9); AS.95.411, radius sin (l = 235.0, p=34.7, d=50.2); AS.95.599, radius dex proximal fragment (p = 32.5); AS.95 412, ulna sin (l = 300.0); AS.95.443, pelvis; AS.95.416 fibula sin (l = ~220); AS.92.415, tibia sin (l = ~210 mm); AS.95.576, MC V sin (l = 70.3, p = 20.3, d=16.4); AS.95.432, MT distal fragment (d = 18.2); AS.95.419, a + b MT distal frags. (a: d = 20.8, b: d = 20.3).

Description. At the time AS.95.657 was studied, it was still partly in matrix and very fragmentary, so that no observations on the skull itself were possible. Description of the skull will thus have to wait until the cleaned specimen is studied. The mandible was cleaned and available for measurement and study. The total length of the ramus is 180.1 mm. Most of the ascending ramus is lost, with only the lowermost part preserved. The canine is broken at the tip and has also lost a section from the buccal side. The p3 has a rounded main cusp and very low accessory cusps. The p4 has a similarly low anterior accessory cusp but more distinct and pointed posterior accessory cusp. The m1 has a large and prominent protoconid and a lower paraconid. The talonid is represented by only a small bump on the posterior margin of the protoconid. It is very worn on the buccal side. Both the cusps have vertical wear facets that reach to the alveolus. The morphology and size of this specimen indicate affinities with *Miomachairodus pseudailuroides,* described by Schmidt-Kittler (1976) from Yeni Eskihisar and Esme Akçaköy, Turkey.

Both the dentition and the postcranial bones are of the size of a modern lion, although the distal limb bones from Loc. 12 are shorter than those of a modern lion. Given that the majority of machairodontines had relatively shorter limbs than the modern felids, this is to expected (Turner and Antón 1997). The articular surfaces are of similar size in the lion and our Loc. 12 specimens. Based on this and the general machairodontine aspect of the bones, they are attributed to the same species as AS.94.657.

The two radii (AS.95.411 and AS.95.599) are similar in size and shape. The ulnar sides are very flat, whereas the lateral sides are rounded. The lesser tubercles are broad and distinct, as in extant large felids such as the lion. The lateral tubercles are less prominent. The proximal articulations are oval with lateral depressions. The distal end, as exemplified by AS.95.411, is large with a clearly concave medial side.

The ulna (AS.95.412) was found very close to the left radius (AS.95.411) and probably belongs to the same individual. The olecranon process is slightly broken proximally, but appears to have been relatively tall. The semilunar notch is shallow (craniocaudally) and open. The coronoid process is also quite small. The radial notch is well developed. The shaft is straight. At the distal end the styloid process is distinct, while the articulation for the radius is only slightly free of the shaft.

Specimen AS.95.443 is an almost complete pelvis preserving the ilium on both sides, most of the ischium on the right side, and parts of the acetabulum. The articulation for the femur is round and measures 34.7 × 34.0 mm. The total length of the pelvis is a little more than 200 mm.

The fibula (AS.95.416) and tibia (AS.95.415) were found together. Both specimens lack the proximal end and in the tibia the distal end is also broken. The fibula is quite robust, with a very large tubercle for the peroneus tertius muscle. The caudal tubercle is the smaller. The shaft has a very sharp mediocaudal edge. The tibia is robust with a heavy cranial ridge.

Specimen AS.95.576 is a fifth metatarsal that is smaller than, but very similar to, AS.91.700 (see Loc. 49). The distal fragments of metatarsals (AS.95.433, AS.95.419) also show similar affinities.

cf. *Paramachairodus* sp.

Material. AS.95.572, C sin (23.9 × 13.1).

Description. This upper canine (AS.95.572) is smaller than that expected for AS.95.657 and may be about the size of that of *Paramachairodus* spp.

Metahyaena confector sp. nov.
Genus *Metahyaena* gen. nov.

Etymology. From Latin "meta," conical column at ends of the Roman Circus (for the shape of the p4) and the Latin "hyaena." Feminine.

Type and Only Species. *Metahyaena confector* sp. nov.

Diagnosis. See description of the type specimen of the type species.

Metahyaena confector sp. nov.

Etymology. From Latin "confector," maker or destroyer. In reference to the amount of puncture marks on the bones of other taxa found in the type locality.

Type Specimen. AS.95.417 (fig. 8.6), mandible dex with c (9.8 × 7.2), p2 (11.6 × 6.0), p3 (13.7 × 7.2), p4 (15.4 × 8.1), m1 (17.9 × 8.2).

Type Locality. Loc. 12, Turkey.

Description. The type and only specimen of this new taxon is a right lower ramus with canine and p2–m1 as well as alveoli for i1–i3, p1, and m2. The ramus is complete apart from the ascending branch, which has been broken off about level with the alveolar margins. The ramus as a whole is slender, with a very long symphysis that is nearly exactly aligned with the long axis of the ramus and only presents a very small chin between p2 and p3. The anterior part of the ramus surrounding the incisors and canine is slightly damaged. There is a single large mental foramen located beneath the midline of p2. Posterior to the posterior-most point of the symphysis the ramus curves gently and is only slightly deeper beneath m1 than beneath p3. The masseteric fossa is deep but short and does not quite extend to the posterior end of the alveolus for m2.

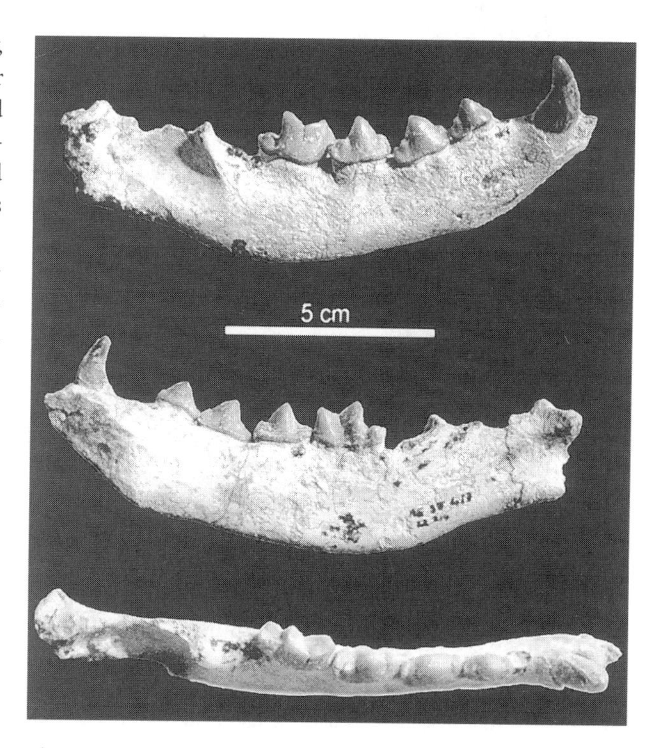

Figure 8.6. AS.95.417, holotype, *Metahyaena confector* gen. et sp. nov., from Loc. 12, mandible dex in (top) lingual view; (middle) buccal view; (bottom) occlusal view.

The incisors are lost but must have been quite small to fit into the space provided for them. The canine, on the other hand, is relatively robust and short and is placed at a slight angle to the anteroposterior axis of the ramus, as is normal in hyenids. The postcanine diastema is very short, 2–3 mm only. The p1 was a small single-rooted tooth set slightly anterolingual to p2. The second and third premolars of this specimen show a very distinctive set of morphological characters. The anterior accessory cusp of p2 is minute, whereas the main cusp is large and stout with strongly concave anterior and posterior margins. The apex of the main cusp is set well anterior to the anteroposterior midline of the tooth. The posterior accessory cusp and posterior shelf are low and relatively indistinct and the posterior end of the tooth tapers gradually posterad. The p3 is very similar in general morphology to p2. The anterior accessory cusp is very small, whereas the main cusp is large and stout with strongly convex anterior and posterior margins. The apex of the main cusp is set anterior to the anteroposterior midline of the tooth, although not as far anteriorly as in p2. The posterior accessory cusp and posterior shelf are low and indistinct, but the posterior end of the tooth tapers less posterad than does p2. The anterior accessory cusp on p4 is small and low. The main cusp is tall and short, whereas the posterior accessory cusp and shelf are low but long, contributing an unusually large proportion to the total length of the tooth. The m1 has a tall trigonid with a protoconid that is slightly taller than the paraconid. The talonid is short, with a distinct entoconid, hypoconid, and hypoconulid. To judge from the alveolus, the m2 must have been relatively small.

This specimen is of the size of the common Miocene hyenids *Hyaenotherium wongii* and *Ictitherium viverrinum,* yet it differs from them in several important respects that instead ally it with more derived hyenids, such as *Palinhyaena reperta.* A distinct such feature is the convex anterior faces of the main cusps of the premolars, in which *Metahyaena confector* is particularly reminiscent of *Belbus beaumonti.* In other respects, however, the new taxon is more primitive than either *P. reperta* or *B. beaumonti,* especially in the narrowness of the premolars.

Protictitherium indet.

Material. AS.95.269, p4 dex (~11.0 × 5.0).

Description. This specimen is complete, but has become broken into two fragments that have been glued together. In this process, the original dimensions of the specimen have been slightly altered. The morphology of the specimen is similar to that of *Protictitherium* spp., but it is larger than P4 of *P. crassum.*

Hyaenidae indet.

Material. AS.95.408, C sin (10.3 × 7.3); AS.95.403, C dex (9.9 × 7.1).

Description. These two upper canines are quite similar in appearance and size and could represent the same species. They appear too large to belong to *Metahyaena confector.*

Pseudaelurus quadridentatus (Blainville, 1841)

Material. AS.95.406, mandible with c dex (9.1 × 6.9), p3 (9.0 × 4.8), p4 (13.3 × 6.5); AS.95.407, m1 sin (16.5 × 7.5); AS.95.409, C sin (13.1 × 8.2).

Description. AS.95.406 preserves the anterior part of the mandible. The symphysis has been lost, along with the incisors and their alveoli. The canine is upright in the typical felid manner and the premolars have three round cusps. The anterior cusp of the p3 is tiny. The sizes of these teeth are within the range known for *P. quadridentatus* premolars. Similarly, the m1 (AS.95.407) and the upper canine (AS.95.409) match those of *P. quadridentatus* from Sansan and La Grive (Heizmann 1973). No *P. quadridentatus* material is previously known from localities younger than MN 8. These specimens were found within a small area and are all only slightly worn and could therefore represent a single individual.

cf. *Pseudaelurus quadridentatus* (Blainville, 1841)

Material. AS.95.481, scapula sin; AS.95.608, humerus sin shaft; AS.95.447, ulna sin shaft; AS.95.609, radius sin shaft.

Description. Scapula AS.95.481 is almost complete, lacking only the distal margin. The total height of the specimen is ~137 mm and the anteroposterior length 77 mm. The humeral articulation measures 24.3 × 20.4 mm, excepting the coracoid process. All three limb bones have broken proximal and distal ends. Specimen AS.95.608 consists of a shaft with both the epiphyses missing. Ulna AS.95.447 preserves the shaft and a part of the olecranon process. The semilunar notch appears wide, although it is not possible to observe it adequately, due to the broken proximal part. The coronoid process is well developed. The radial shaft, AS.95.609 is lacking only the epiphyses. The approximate length is 160 mm and this specimen scales well with AS.95.447.

Carnivora indet.

Material. AS.95.405, c sin (8.3 × 6.0).

Description. This specimen has suffered heavy postmortem damage. All the enamel has worn off, and the dentine has additional damage. The original shape has, however, preserved quite well. It is probably the lower canine of a relatively large mustelid.

Loc. 51

Ursidae indet.

Material. S.90.153, radius dex proximal fragment (p = 45.0 × 33.0).

Description. Specimen S.90.153 preserves the proximal end and a part of the shaft of a right radius. The proximal epiphysis is broken on the medial side. It shows ursid affinities in its distally placed lateral tubercle and conspicuous lesser tubercle.

Loc. 84

Carnivora indet., small

Material. AS.92.776, P4 sin (5.9 × 1.9).

Description. AS.92.776 is a small upper carnassial with a carnassial notch, a small but distinct parastyle, and a protocone, which is partially broken parallel to the lingual margin of the tooth. On the buccal side, there is a narrow cingulum.

Loc. 7

Ictitherium intuberculatum Ozansoy, 1965

Material. S.91.653, p2 sin (11.7 × 6.0); S.91.654, p4 sin (16.8 × 8.5); S.91.655, m1 sin fragment; S.90.221, mandible sin fragment with p2 (11.6 × 6.0), p3 (15.4 × 7.1), p4 (16.8 × 8.2), m1 (19.5 × 9.0).

Description. This material shows clear affinities with *Ictitherium* spp. and is very similar in its dimensions to the material described as *I. intuberculatum* by Ozansoy (1965). The p4 has both the anterior and posterior accessory cusps. In the p3, the anterior cusp is smaller than that of, e.g., *I. viverrinum*. In addition, the premolars are more robust than those of *I. viverrinum*.

Pseudaelurus turnauensis Hoernes, 1882

Material. S.89.1, mandible sin with p3 (6.7 × 3.8), p4 (9.2 × 4.4), m1 (11.9 × 4.8).

Description. These felid teeth are within the size range of *Pseudaelurus turnauensis* teeth (Heizmann 1973). The dentition is unworn and thus preserves the morphology well. The m1 has postmortem damage posteriorly.

Loc. 37

Dinocrocuta senyureki Ozansoy, 1965

Material. S.89.138, p4 sin (31.2 × 18.6).

Description. This is an unworn tooth of very large size. The p4 reported for this species from the Sinap area by Ozansoy (1965) is slightly more slender (32.0 × 16.8). We consider this to represent intraspecific variation.

Loc. 1

Carnivora indet., large

Material. AS.92.74, premolar frag.

Description. This specimen consists of an accessory cusp and partial main cusp. The accessory cusp is relatively low and there is no cingulum. We could not match AS.92.74 to any carnivore premolar known to us.

Carnivora indet., medium

Material. AS.89.22, mandible fragment.

Description. AS.89.22 belonged to a much smaller animal than AS.92.74. This edentulous mandible could not be identified beyond the ordinal level.

Loc. 10

Machairodontinae indet., large

Material. S.89.39, radius sin proximal fragment.

Description. S.89.39 preserves only the anteromedial side of the proximal end of the radius. The radial tuberosity is preserved and is similar to that of felids.

MN 11 Localities

Loc. 49

Percrocutidae indet.

Material. AS.95.86, dP3 sin (l = ~25).

Description. The dP3 AS.95.86 is broken at the paraconid. It has heavy vertical wear on the cusps. The preserved part of the tooth is similar to a tooth identified as a percrocutid dP3 by Schmidt-Kittler (1976).

Machairodontinae indet.

Material. S.91.700, MC V sin (l = 94, p = 29.4, d = 21.9); S.91.752, radius dex proximal fragment (p = 36.2); AS.92.173, astragalus dex.

Description. These specimens can be referred to a very large machairodont felid, such as *Machairodus irtyschensis* Orlov (1936), which is only known from MN 13, or *M. giganteus,* which is known from MN 11 to MN 13 from China to Spain (Beaumont 1975).

The proximal left radius (S.91.752) preserves only a small portion of the caudal part of the radial head. The radial head appears to have been rounded and mediolaterally slightly oval. The groove for the extensor carpi ulnaris muscle is situated above the radial tuberosity on the caudal side. The posterior side of the head has a radial tuberosity (bicipital tuberosity) which is broad and not high or extended. Under the radial tuberosity there is a large and deep fossa. This fossa is situated lateral to a short ridge, which seems to be an extension of the oblique line of the radius.

The posterior side of the radius has very strong muscle scars for the flexor digitorum profundus muscle. The insertion for the supinator brevis muscle is also well developed. The protrusion within the insertion of the supinator muscle on the medial side of the radius is very large, forming a pedicle. The shaft is rounded in cross section. The proximal part of the shaft appears to be straight.

The fifth metacarpal (S.91.700) is complete. In dorsal view, the main articular facet for the unciform is triangular and anteriorly pointed. The facet continues posteriorly and distally on the anterior side of the bone. On the medial side, it has a rather well-developed protrusion for the articulation with the fourth metacarpal. On the lateral side, there are two epicondyles. The proximomedial one for the cuneiform metacarpal ligament is flat and well defined and in caudal view continues posteriorly well below the facet of the unciform. The lateral epicondyle for the extensor carpi ulnaris muscle is large, circular in lateral view, and subdivided by a central ridge. The proximomedial and lateral facets are separated by a groove that is oriented proximolaterally. The anteromedial side of the shaft forms a well-defined ridge, as does the posteromedial side of the shaft. The medial surface of the bone between the two ridges is rather flat.

The anterior side of the entire shaft is flat and the posterior side convex. In lateral and medial views, the shaft appears arched. The posteromedial side of the anterior region preserves the muscle scar for the lateral fibers of the third palmar interosseus. The scar for the flexor digiti muscle is not observable on the lateral side of the posterior aspect. In anterior view, the distal articular condyle appears low and wide. Two enlargements (epicondyles) are observed above the condyle. The lateral epicondyle is the larger of the two. The posterior side of the condyle has a medial ridge that is strongly elevated but does not continue on the palmar side of the phalanx dorsal to the articular facet.

Specimen AS.92.173 is similar to felid astragali, such as that of the lion, in being short and broad. The facet for the tibia is not very deep. The facets for the calcaneum are large, the lateral being larger than the medial. The lateral facet continues all the way to the head. In lateral view, the bone narrows toward the anterior end. There is no foramen on the posterior side of the bone.

Felidae indet., small

Material. AS.92.170, proximal phalanx proximal fragment (p = 7.3); AS.93.700, proximal phalanx (l = 16.2, p = 6.2, d = 5.0).

Description. These relatively small phalanges have characters that indicate felid affinities, such as the relatively distal insertion for the flexor digitorum muscle and the curved overall morphology of the shaft.

Mustelidae

Material. AS.94.380 astragalus dex (15.7 × 21.6).

Description. This is a typical mustelid astragalus with a broad head set at an angle to the body.

MN 11 Localities (Kavakdere)

Above Loc. 34

Pseudaelurus cf. *lorteti* Gaillard, 1899

Material. AS.95.299, m1 dex (l = ~14.7).

Description. This specimen is broken on the posterior side of the protocone. It has the morphology of *Pseudaelurus* spp., and its size approximates that of the m1 of *Pseudaelurus lorteti*.

Loc. 33

Carnivora indet., small

Material. S.91.117, proximal phalanx (l = 13.5, p = 4.7, d = 3.9).

Description. We could not identify this specimen beyond the ordinal level, although it is certainly a carnivore phalanx.

MN 12 Localities

Loc. 42

Ictitherium cf. *intuberculatum* Ozansoy, 1965

Material. "Sc.1" nasals, premaxilla, maxilla dex with P3 (17.5 × 9.5), P4 (26.9 × 14.7), M1 (15.7 × 9.4), M2 (8.4 × 4.3); S.91.854 (fig. 8.7), nasals, premaxilla, maxilla dex with I1 (fragment), I2 (4.2 × 5.5), I3 (6.8 × 4.2), C (12.1 × 8.0), P1 (5.1 × 4.8), P2 fragment, P3 fragment.

Description. Specimen "Sc.1." is from the earlier excavations by Senyurek (Ozansoy 1965) at his locality "Çoban Pinar," a locality that corresponds to Sinap project Loc. 42.

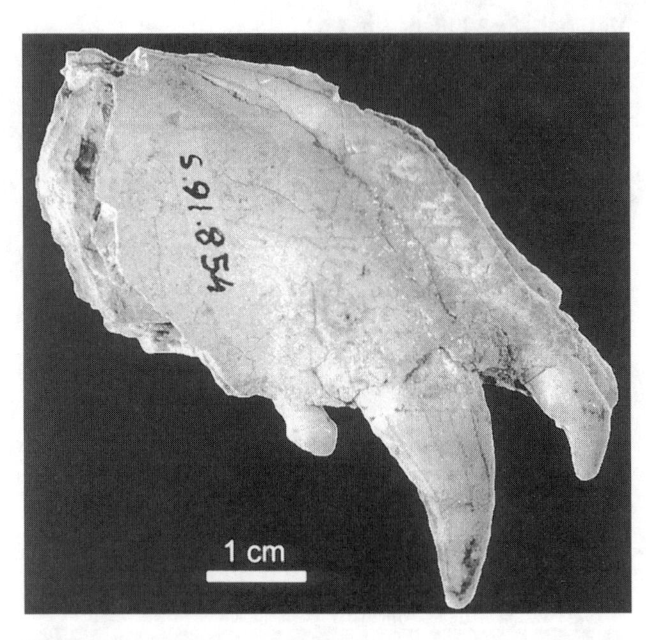

Figure 8.7. AS.91.854, maxilla dex fragment of *Ictitherium intuberculatum* from Loc. 42 in buccal view.

The dentition of this specimen is similar to that of *Ictitherium intuberculatum*. The P3 has well-developed main and posterior accessory cusps, whereas the anterior accessory cusp is reduced to a small enhancement of the cingulum. The P4 is surrounded by a cingulum and has a relatively large protocone. The metastyle blade is relatively short, being about the length of the paracone, as is typical of *Ictitherium* spp. The M1 has a strong, lingually enlarged cingulum and two distinct buccal cusps (para- and metacone).

The teeth preserved in S.91.854 are not very diagnostic. They are, however, of the size of those of *Ictitherium intuberculatum*, and the specimen is tentatively attributed to this species.

cf. *Belbus beaumonti* (Qiu 1987)

Material. AS.93.785, M1 dex (15.5 × 8.8).

Description. AS.93.785 preserves both the paracone and metacone. The latter is lingually shifted and situated about midway to the lingual end of the tooth. The tooth resembles that of *Belbus beaumonti* (cf. Werdelin and Solounias 1991, fig. 51).

Hyaenotherium wongii (Zdansky, 1924)

Material. AS.94.1283, mandible with sin i1 (l = 2.4), i2 (l = 3.5), i3 (l = 5.8), c (8.2 × 11.5), p2 (12.5 × 6.2), p3 (15.6 × 7.5), p4 (18.0 × 8.6), m1 (20.3 × 9.4), m2 (6.2 × 5.8), dex i1 (l = 2.5), i2 (l = 3.5), i3 (l = 5.8), c (8.0 × 11.7), p1 (3.1 × 3.0), p2 (12.5 × 6.2), p3 (15.3 × 7.2), p4 (18.0 × 8.7), m1 (20.4 × 9.4).

Description. The mandible AS.94.1283 is completely preserved on both sides, apart from the ascending ramus, which is lost from both rami. There is only one mental foramen on either side. It is situated below the p2.

The dentition is nearly complete, lacking only the right m2, the left p1, and part of the left canine. The incisors are densely packed in the relatively narrow anterior part of the ramus. The p1 is rudimentary. The p2 and p3 lack anterior accessory cusps and have concave anterior margins of the main cusp. They have distinct posterior accessory cusps and posterior shelves. The p4 has three cusps and a well-developed posterior shelf. It also has a concave anterior margin of the main cusp, and the anterior accessory cusp is free of the main cusp. The m1 has a high trigonid, in which the paraconid and protoconid are of equal height. The metaconid forms a distinct cusp. The talonid is low but long, with three cusps in place (hypoconid, entoconid, and hypoconulid). The m2 is small and round. All the teeth are only slightly worn.

Specimen AS.94.1283 is assigned to *Hyaenotherium wongii* on the basis of its dental characters, such as the shape of premolar cusps and the size of the talonid of the lower carnassial (see Werdelin 1988; Werdelin and Solounias 1991).

cf. *Hyaenotherium wongii* (Zdansky 1924)

Material. AS.93.1111, P4 dex fragment.

Description. AS.93.1111 preserves only the paracone and metastyle blade. These cusps have only slight wear. There is a thick cingulum on the lingual side. The metastyle blade is longer than the paracone, a character that distinguishes *Hyaenotherium* sp. from similarly sized taxa, such as *Ictitherium* spp.

Felidae indet., large

Material. S.91.433, MC IV dex proximal fragment (p = 15.3); AS.93.1110 MP distal fragment (d = 22.6).

Description. These specimens represent a very large felid, probably of similar size to that from Loc. 49, as described above.

Felidae indet., small

Material. S.91.434 calcaneum sin (16.6 × 10.3).

Description. This felid calcaneum is smaller than *Pseudaelurus quadridentatus* as given in Schmidt-Kittler (1976).

Localities of Uncertain Stratigraphic Position

South of Loc. 45

Carnivora indet., medium

Material. AS.92.417, mandible dex with m2 fragment.

Description. AS.92.417 consists of the posterior part of a ramus with the coronoid process broken off. The roots of the m1 and posterior part of the m2 are preserved. The specimen clearly belongs to a carnivore, but further affinities are difficult to elucidate, due to the few preserved diagnostic features.

Loc. 45

Carnivora indet., medium

Material. AS.92.669, calcaneum sin distal fragment.

Description. Too little of this specimen is preserved for identification beyond the ordinal level.

Loc. 78

Carnivora indet., large

Material. S.91.612, proximal phalanx, proximal end (p = 22.0).

Description. This locality is of Turolian age and the large carnivores known to be present are felids, ursids, and percrocutids. This specimen is very large, but cannot be assigned with confidence to either of these families.

Loc. 111

Mustelidae indet.

Material. AS.94.79, P4 dex (7.3 × 3.6); AS.94.80 (fig. 8.8), M1 sin (4.8 × 8.6).

Description. These very small mustelid teeth apparently belong to a single species, probably even to one individual.

Felidae indet., medium

Material. AS.94.89, C dex (9.7 × 5.9).

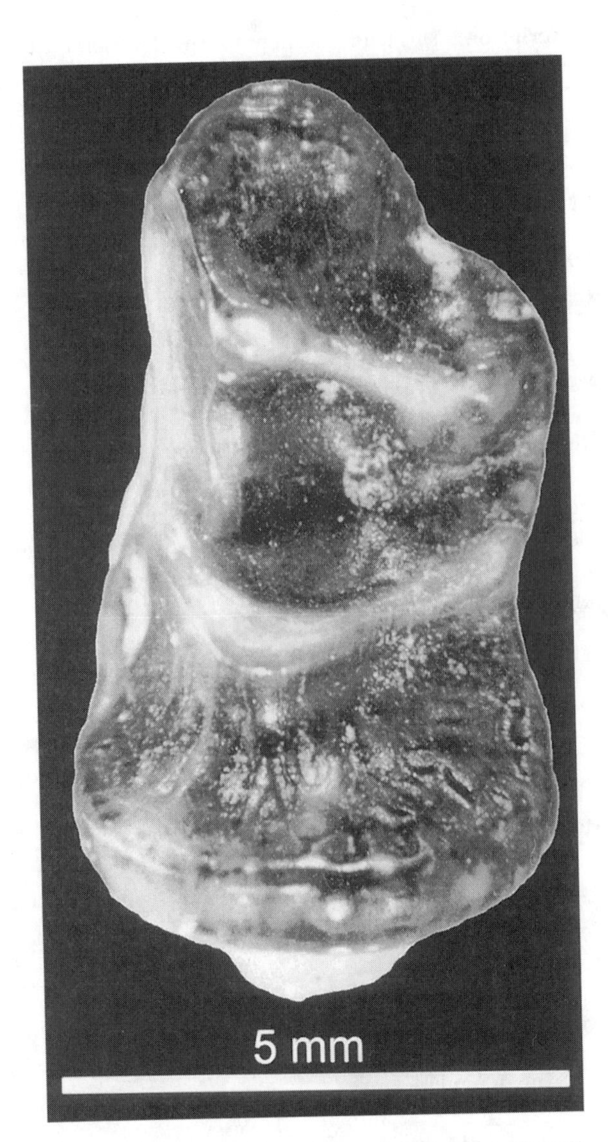

Figure 8.8. AS.94.80, M1 sin. of small mustelid from Loc. 111 in occlusal view.

Description. The size of this felid canine indicates that it could be attributable to *Pseudaelurus lorteti*.

Loc. 118

Carnivora indet., large

Material. AS.94.642, femur distal fragment (condyle); AS.94.1159, ulna fragment (olecranon process); AS.94.1173, medial phalanx (l = 22.4); AS.94.1175, proximal phalanx (l = 20.1).

Description. These specimens could not be identified more precisely than as Carnivora indet.

Anatolian Carnivores of the Middle and Late Miocene Compared with Adjacent Areas

MN 6

The carnivore fauna of the Sinap Formation is not adequately represented until early MN 9 (~10.5 Ma). Only *Hemicyon sansaniensis* is recorded from Loc. 24, which is placed in MN 6 (~15.2 Ma). Earlier workers have also recorded *H. sansaniensis* (Gürbüz 1981) and a small hyenid (*"Ictitherium prius" nomen nudum*) (Ozansoy 1965) from this same locality (Loc. 24 or In-Önü). The next stratigraphic level in the Sinap Tepe area is correlative with MN 8, and we have here recorded a nimravid, *Sansanosmilus jourdani*, from this time.

Fortunately, the middle Miocene is better represented at other fossil localities in Anatolia. Paşalar, ~500 km west of Sinap Tepe, comprises a rich fossil fauna, including carnivores and is placed in MN 6. Çandir near Ankara is also placed in MN 6. In the absence of an exhaustive systematic study, the exact number of carnivore taxa at Paşalar remains unknown. The carnivore material collected from Paşalar is extensive, and a detailed study of the collection would contribute useful information toward a better understanding of carnivore communities of the Neogene.

Paşalar shares many faunal similarities with contemporaneous European localities, such as Sansan (Ginsburg 1961). Large amphicyonid and ursid species are present at both localities. The diversity of mustelids also appears similar at these two localities. Sansan lacks early, civetlike hyaenids like *Protictitherium* spp., whereas these are present at Paşalar. These forms were, however, present in western Europe in MN 4–11 (Werdelin and Solounias 1996) and their absence from the Sansan fauna may be due to local conditions. The percrocutids, however, which are definitely present at Paşalar (*Percrocuta miocenica*), may have been totally absent from western Europe until MN 9. *Percrocuta miocenica* and other percrocutids are present in China from 12.5 Ma (Tunggurian age), which is roughly equivalent to MN 6. They also probably appeared in the Siwaliks around that time (Barry and Flynn 1990).

The Nimravidae are absent from Paşalar, whereas two species (*Sansanosmilus palmidens* and *S. jourdani*) are fairly common in western Europe in the Astaracian. It is of interest that there is a younger record of a nimravid in Anatolia, as *S. jourdani* is found in Sinap Loc. 64, which is placed in MN 8. This is only the second record of a Miocene nimravid in Anatolia (Geraads and Gülec 1997).The family is represented in the Siwaliks between 15.1 Ma and 7.4 Ma (Barry and Flynn 1990).

MN 7/8

In addition to Loc. 64, which records *Sansanosmilus jourdani* and Mustelidae indet., there are two other fossil localities of MN 7/8 age in Anatolia that have produced carnivores: Sofça and Yeni Eskihisar (Schmidt-Kittler 1976). *Agriotheriinae* indet. and *Proticitiherium crassum* are known from Sofça. As the agriotheriine is only identified by a tibia, this referral may be considered doubtful, especially as the group is not generally identified in western Eurasia until MN 11. We have no record of agriotheriines in the Sinap Tepe area. *Proticitiherium crassum* is common in MN 7/8–11 throughout western Eurasia.

Yeni Eskihisar has produced two species of hyaenids, *Thalassictis montadai* and *Protictitherium cingulatum,* a sabertoothed felid, *Miomachairodus pseudailuroides,* and a mustelid, Peruniinae indet. *Thalassictis montadai* is known from Spain (Crusafont-Pairó and Petter 1969) in the interval MN 7/8–10. It is also known from Sinap at localities placed in MN 9.

The Yeni Eskihisar record is the only one known for *Protictitherium cingulatum. Miomachairodus pseudailuroides* is the earliest and most primitive sabertooth felid (Schmidt-Kittler 1976) and is not known outside of Anatolia.

No percrocutids are recorded from Anatolia in MN 7/8. Given the scarcity of the available material from this time, their presence cannot be ruled out, however. The same is true of *Pseudaelurus* spp. and hemicyonine ursids. All these groups were present both before and after MN 7/8, and most likely were there during this period as well. The Amphicyonidae present is a more interesting case. This group is well represented in Anatolia in MN 6, which seems to mark the last appearance of the group in the area, whereas it continues to be common east and west of Anatolia at least until MN 10.

The carnivore record of MN 7/8 of Anatolia is too poorly known for any firm biogeographic or paleoecologic conclusions. However, that two out of the four described taxa have not been identified outside of Anatolia suggests a strong element of endemism for the period.

MN 9

Faunas placed in MN 9 are well represented in the Sinap Tepe area, as well as in some other parts of Anatolia (Esme Akçaköy; Schmidt-Kittler 1976). The period is also well known in the rest of western Eurasia and the Siwaliks.

Loc. 94 is near the level of the first appearance datum of *Hipparion* and probably represents a very early phase of MN 9. Loc. 4 is geographically some 100 m away from Loc. 94 and at the same stratigraphic level. Together, these two localities record eight carnivore taxa, of which two are identified to species (*Thalassictis montadai* and *Proticititherium crassum*) and one as cf. species (cf. *Hyaenotherium wongii*).

Percrocutids have not been found in Locs. 94 or 4, but they are known by two species from Loc. 108, a little higher in the sequence. One is *Dinocrocuta senyureki,* and the other is an indeterminate smaller species (cf. *D. minor*). *Dinocrocuta senyureki* is not recorded from Europe or elsewhere in Asia, but it is known from north Africa (Sahabi, Libya) from younger sediments (MN 13; Howell and Petter 1985). It seems thus to be endemic to Anatolia in MN 9. The same is true of *Ictitherium intuberculatum,* which is also present in this level (Locs. 7 and 8), and is not known outside Anatolia.

Loc. 12, which is placed in the upper part of MN 9, differs from other Sinap localities in its more extensive carnivore record. Six species are recognized. *Dinocrocuta senyureki* and *Protictitherium crassum* continue to be present. In addition, *Miomachairodus pseudailuroides,* known from older sites in Anatolia, and *Pseudaelurus quadridentatus,* a species that already had occurred in Anatolia in MN 6, are found in Loc. 12. *Pseudaelurus quadridentatus* disappears from western Eurasia in MN 9, and this might be its youngest record. In addition to these species, a nimravid represented by cranial material may have been present in loc 12. This material was found by the Senyürek expedition, and according to his notes, this material comes from Kayinçak. Kayinçak, however, refers to two hills that contain several fossil accumulations apparently of different ages (Mikael Fortelius, pers. comm.). The nimravid material was described as *Barbourofelis piveteaui* by Geraads and Gülec (1997) and if it comes from Loc. 12, it is the youngest record of a nimravid in the Old World.

Indarctos vireti, a species otherwise recorded only from Spain, was identified from Loc. 12. Because the status of this species is questionable (Werdelin 1996b), the meaning of our record remains open. *Indarctos* spp. is widely known from western Eurasia. A new species of a hyenid is described from Loc. 12. This species shows affinities to other Western Eurasian and Chinese hyenids.

MN 10

One relatively poor site in the Sinap Tepe area, Loc. 37, is placed in MN 10. The only carnivore present is *Dinocrocuta senyureki.*

MN 11

MN 11 is represented by Loc. 49 (Igbek), which has produced many remains of large mammals. Carnivores are less well presented and only four taxa, of which none is identified to species, are known. A large machairodontine and a

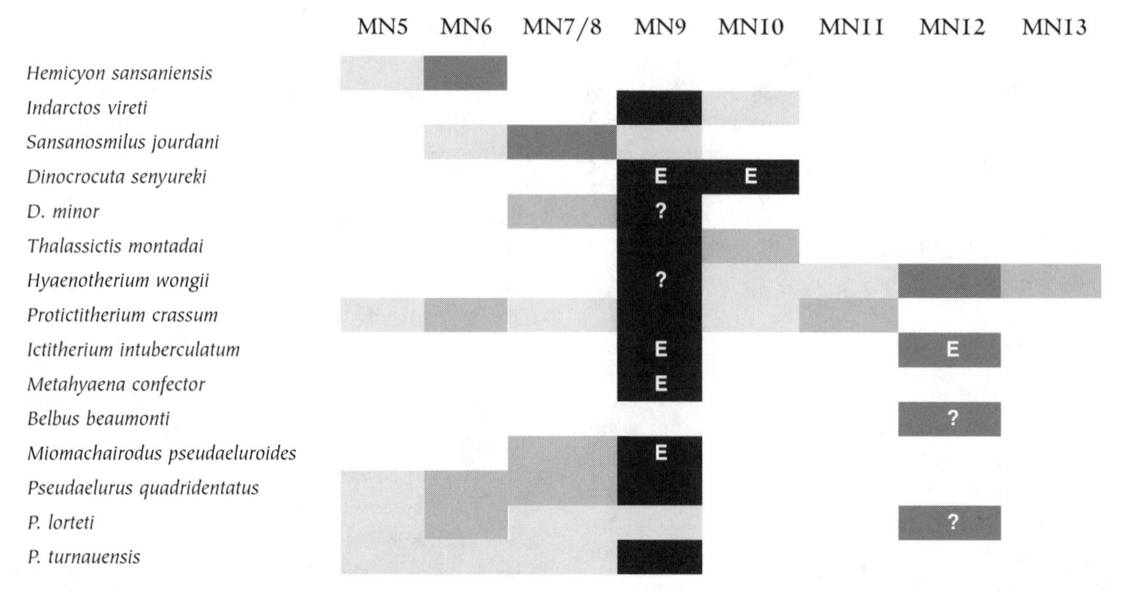

Figure 8.9. Diagram summarizing the carnivore record of the Sinap Formation. Sinap records are given as black bars. Doubtful records are marked with question marks. Endemic taxa marked with E. Grey bars represent inferred ranges of specific taxa.

percrocutid are typical elements in the faunas of western Eurasia during this time. The same is true of the small felid and mustelid that are also recorded from Loc. 49.

MN 12

Loc. 42 (Çoban Pinar in Ozansoy 1965) has yielded a variety of hyaenid species. *Ictitherium intuberculatum* is identified from Loc. 42, which means that the stratigraphic range of this apparently endemic Anatolian hyenid reaches MN 12. *Belbus beaumonti* has been described from Samos, Greece, in MN 12 and *Hyaenotherium wongii* is well known in Eurasia (Spain to China) from MN 10–12.

Conclusions

Figure 8.9 summarizes the Sinap carnivore species and their occurrences. The Sinap sequence shows endemism, as well as incorporating European, Asian, and African elements in its carnivore faunas. Due to a lack of current revisions of ursids, percrocutids, felids, and mustelids, biogeographic reconstructions are not necessarily accurate for these groups.

The presence of percrocutids in MN 6 onward allies Anatolia with the east rather than the west. Percrocutids appeared from Saudi Arabia to Mongolia during the Astaracian (MN 6–7/8), but not earlier than MN 9 in Europe.

Endemism of the Sinap percrocutids and the suggestion that *Dinocrocuta senyureki* occurs later in Africa may be an artefact of the poorly known taxonomy of the group. Relationships of Sinap percrocutids with other western Asian forms should be assessed critically in the light of the new fossil material. Such a study is, however, beyond the scope of this chapter, and we have here relied on the work of Howell and Petter (1985).

Anatolia also differs from European faunas in the absence of nimravids in MN 9 (and perhaps MN 6). Nimravids seem to be rare in Asia in the Miocene, with only one additional record from China in MN 8.

Apart from the endemic species, the most peculiar trait of the Sinap carnivore fauna is the absence of amphicyonids. This group is fairly common in European faunas until MN 10, and similarly occurs in the Siwaliks, China, and even in Africa during the middle and beginning of the late Miocene.

Acknowledgments

We thank Drs. M. Fortelius, J. Kappelman, and B. Alpagut for the opportunity to study Sinap carnivores. Dr. H. N. Bryant's help with the nimravid milktooth and Dr. D. Geraads' help with the *Indarctos* specimens are gratefully acknowledged. We also thank Drs. J. Barry and R. Bernor for reviews and comments that improved the manuscript. The work of SV was supported by the Finnish Cultural Foundation; that of LW by the Swedish Natural Science Research Council.

Literature Cited

Barry, J. C., and L. J. Flynn, 1990, Key biostratigraphic events in the Siwalik sequence, *in* E. H. Lindsay, V. Fahlbusch, and P. Mein, eds., European Neogene mammal chronology: New York, Plenum Press, pp. 557–571.

Beaumont, G. D., 1975, Recherches sur les Félidés (Mammifères, Carnivores) du Pliocène inférieur des sables à *Dinotherium* des environs d'Eppelsheim (Rheinhessen): Archives des Sciences, v. 28, pp. 369–400.

Bonis, L. de, 1994, Les gisements de mammifères du Miocène supérieur de Kemiklitepe, Turquie: 2. Carnivores: Bulletin du Museum National d'Histoire Naturelle, ser. C, v. 16, no. 1, pp. 19–41.

Chen G., and N. Schmidt-Kittler, 1983, The deciduous dentition of *Percrocuta* Kretzoi and the diphyletic origin of the hyaenas (Carnivora, Mammalia): Paläontologisches Zeitschrift, v. 57, pp. 159–169.

Crusafont Pairó, M., and G. Petter, 1969, Contribution à l'étude des Hyaenidae. La sous-famille des Ictitheriinae: Annales de Paléontologie (Vertébrés), v. 55, pp. 89–127.

Geraads, D., and E. Gülec, 1997, Relationships of *Barbourofelis piveteaui* (Ozansoy, 1965), a late Miocene nimravid (Carnivora, Mammalia) from central Turkey: Journal of Vertebrate Paleontology, v. 17, pp. 370–375.

Ginsburg, L., 1961, La faune des carnivores Miocènes de Sansan (Gers): Mémoires du Muséum National d'Histoire Naturelle, ser. C, v. 9, pp. 1–190.

Gürbüz, M., 1974, *Amphicyon major* Blainville discovered in the Middle Miocene beds of Candir: Bulletin of the Mineral Research and Exploration Institute of Turkey, v. 83, pp. 109–111.

Gürbüz, M., 1981, Inönü (KB Ankara) Orta Miyosenindeki Hemicyon sansaniensis (Ursidae) türünün tanimlanmasi ve stratigrafik yayilimi: Bulletin of the Geological Society of Turkey, v. 24, pp. 85–90.

Heizmann, E. P. J., 1973, Die Carnivoren des Steinheimer Beckens. B. Ursidae, Felidae, Viverridae sowie Ergänzungen und Nachträge zu den Mustelidae: Palaeontographica, Supplementband, v. 8, pp. 1–95.

Howell, F. C., 1987, Preliminary observations on Carnivora from the Sahabi Formation (Libya), *in* N. T. Boaz, A. El-Arnauti, A. W. Gaziry, J. de Heinzelin, and D. D. Boaz, eds., Neogene paleontology and geology of Sahabi: New York, Alan R. Liss, pp. 153–181.

Howell, F. C., and G. Petter, 1985, Comparative observations on some Middle and Upper Miocene hyaenids. Genera: *Percrocuta* Kretzoi, *Allohyaena* Kretzoi, *Adcrocuta* Kretzoi (Mammalia, Carnivora, Hyaenidae): Géobios, v. 18, pp. 419–476.

Kurtén, B., 1985, *Thalassictis wongii* (Mammalia: Hyaenidae) and related forms from China and Europe: Bulletin of the Geological Institutions of the University of Uppsala, v. 11, pp. 79–90.

Martin, L. D., 1989, Fossil history of the terrestrial Carnivora, *in* J. L. Gittleman, ed., Carnivore behavior, ecology, and evolution: London, Chapman and Hall, pp. 536–568.

Ozansoy, F., 1965, Étude des gisements continentaux et des mammifères du Cénozoique de Turquie: Mémoires de la Société Géologique de France, v. 102, pp. 1–92.

Schmidt-Kittler, N., 1976, Raubtiere aus dem Jungtertiär Kleinasiens: Paleontographica, v. A155, pp. 1–131.

Senyürek, M., 1954, A study of a skull of *Promephitis* from the Pontian of Kücükyozgat: Belleten Türk: Tarih Kurumu, v. 18, pp. 279–315.

Senyürek, M., 1957, A new species of *Epimachairodus* from Küçükyozgat: Belleten Türk: Tarih Kurumu, v. 21, pp. 1–60.

Senyürek, M., 1958, Adaptive characters in the dentition of *Crocuta eximia* (Roth and Wagner), together with a survey of the finds of Crocuta in Anatolia: Publications of the Faculty of Language, History and Geography, University of Ankara, v. 122, pp. 1–48.

Senyürek, M., 1960, The Pontian Ictitheres from the Elmadag district: Publications of the Faculty of Language, History and Geography, University of Ankara, v. 5, Supplement, pp. 1–223.

Solounias, N., 1981, The Turolian fauna from the island of Samos, Greece, Contributions to Vertebrate Evolution, v. 6, pp. 1–232.

Turner, A., and M. Antón, 1997, The big cats and their fossil relatives: New York, Columbia University Press, 234 pp.

Viranta, S., and P. Andrews, 1995, Carnivore guild structure in the Pasalar Miocene fauna: Journal of Human Evolution, v. 28, pp. 359–372.

Werdelin, L., 1988, Studies on fossil hyaenas: The genera *Thalassictis* (Gervais ex Nordmann), *Palhyaena* Gervais, *Hyaenictitherium* Kretzoi, *Lycyaena* Hensel and *Palinhyaena* Qiu, Hyang & Guo: Zoological Journal of the Linnean Society, v. 92, pp. 211–265.

Werdelin, L., 1996a, Carnivores, exclusive of Hyaenidae, from the later Miocene of Europe and western Asia, *in* R. L. Bernor, V. Fahlbusch, and H.-W. Mittmann, eds., The evolution of western Eurasian Miocene mammal faunas: New York, Columbia University Press, pp. 271–289.

Werdelin, L., 1996b, Carnivoran ecomorphology: A phylogenetic perspective, *in* J. L. Gittleman ed., Carnivore behavior, ecology, and evolution, Volume 2: Ithaca, New York, Cornell University Press, pp. 582–624.

Werdelin, L., and N. Solounias, 1991, The Hyaenidae: Taxonomy, systematics and evolution: Fossils and Strata, v. 30, pp. 1–104.

Werdelin, L., and N. Solounias, 1996, The evolutionary history of hyenas in Europe and western Asia during the Miocene, *in* R. L. Bernor, V. Fahlbusch, and H.-W. Mittmann, eds., The evolution of western Eurasian Neogene mammal faunas: New York: Columbia University Press, pp. 290–306.

Wessels, W., E. Ünay, and H. Tobien, 1987, Correlations of some Miocene faunas from northern Africa, Turkey and Pakistan by means of Myocricetodontidae: Proceedings of the Koniklijke Akademie van Wetenschappen, ser. B, v. 90, no. 1, pp. 65–82.

9

Orycteropodidae (Tubulidentata)

M. Fortelius, S. Nummela, and S. Sen

When present in a fossil assemblage, aardvarks are usually rare, which means that they are mostly found at localities from which large numbers of fossils have been collected. Their apparently spotty occurrence in the Neogene of Eurasia is thus likely to represent poor sampling of a more continuous actual distribution, perhaps limited to the western and southern parts of the continent closest to their biogeographic center in Africa (their known range is from France to Pakistan). The last full-scale revision of fossil aardvarks was by Patterson (1975), and predates several important discoveries. Overviews relevant to the Eurasian Neogene are provided by Pickford (1975), Patterson (1978), Sen (1994), and de Bonis et al. (1994).

The Sinap Formation was named by Ozansoy (1957, 1965). The stratigraphy has subsequently been developed by Öngür (1976), Sen (1991), and Kappelman et al. (1996). It is now considered to span the time from ~15 to ~2.5 Ma, with a densely sampled interval at ~11–9 Ma. This densely sampled interval corresponds to the lower and middle Sinap Members of Ozansoy, their boundary roughly corresponding to the middle/late Miocene boundary, marked by the local appearance of hipparionine horses. For an up-to-date summary of the geology and chronology of the Sinap Formation, we refer the reader to Lunkka et al. and Kappelman et al. (this volume).

Aardvark material from the Sinap Formation was first described by Ozansoy (1965), who referred specimens from the middle Sinap Member to a new species, *Orycteropus pottieri*. Patterson (1975, 1978) argued that fossil aardvark material cannot be correctly assigned at the genus level without good postcranial material. He did, however, include *O. pottieri* in *Orycteropus* sensu stricto. *Orycteropus pottieri* was considered valid by Sen (1991, 1994) and de Bonis et al. (1994). The latter referred material from Pentalophos I (the early late Miocene of Greece) to *O. pottieri*, apparently the first extension of the taxon beyond the type locality. Here we report additional specimens from Sinap localities spanning the range 10.1–9.6 Ma, including the first cranial aardvark material from the Sinap Formation.

Lower Sinap Loc. 64 (~10.8 Ma) has yielded a few teeth distinct from those of *Orycteropus pottieri* and similar to the Anatolian early middle Miocene species *O. seni*. Finally, material too fragmentary for specific identification has been recovered from two "Sinap" localities, one of them Loc. 49, paleomagnetically calibrated at ~9.1 Ma (Kappelman et al., this volume).

Material and Abbreviations

The material studied here is housed in the Museum of Anatolian Civilizations, Ankara. See the Introduction to this volume for a history of the Sinap project.

Measurements: APD = maximum distal anteroposterior diameter, APP = maximum proximal anteroposterior diameter, APS = minimum anteroposterior diameter of shaft, L = maximum length, WD = maximum distal width, WP = maximum proximal width, WS = minimum width of shaft.

Ear Anatomy: ow = oval window, pr = promontorium, rw = round window.

Dental Position: C = upper canine, c = lower canine, M = upper molar, m = lower molar, P = upper premolar, p = lower premolar. Numbers indicate position in series (e.g., m3 = lower third molar).

Other Abbreviations: AS = Ankara Sinap, Ma = megayears.

Catalog of Fossil Material

Orycteropus pottieri *Ozansoy, 1965*

Taxonomy

Order Tubulidentata Huxley, 1872
Family Orycteropodidae Bonaparte, 1852

Figure 9.1. Partial skull of *Orycteropus pottieri* AS.91.424, Loc. 12. (**A–C**) Dorsal, ventral, and lateral views of specimen before the preparation that unfortunately resulted in loss of the right occipital condyle and other parts. (**D–F**) Dorsal, posterior, and lateral views of the specimen after preparation. The unfused occipital suture (**D**) and the possible puncture marks on the right parietal (**D, F**) can be clearly seen. Scale bar = 50 mm.

Genus *Orycteropus* Geoffroy, 1795
Orycteropus pottieri Ozansoy, 1965

Synonymy

1991 *Orycteropus pottieri* Sen
1993 *Orycteropus pottieri* Tekkaya
1994 *Orycteropus pottieri* de Bonis et al.
1996 *Orycteropus* sp. Kappelman et al. (in part)

Sinap Material

Loc. 108. Metatarsal II sin AS.94.241, distal part of metatarsal III or IV AS.93.282.

Loc. 72. Partial mandible dex with p3–m1 AS.92.202.

Loc. 12. Partial skull AS.91.424 (with left incus), maxillary dex fragment with C–P1 AS.91.423, M2 dex AS.95.700, M3 sin AS.92.581, m2 sin AS.92.580, partial scapula sin AS.92.579, metacarpal IV dex AS.91.366, first phalanx III sin AS.95.63, second phalanx AS.91.415, third phalanges AS.91.356, AS.95.176, AS.95.251, AS.95.578

Age. The magnetostratigraphic age estimates for Locs. 108, 72, and 12 span ~10.1–9.6 Ma (Kappelman et al., this volume).

Description and Remarks. The partial skull AS.91.424 (fig. 9.1) is the first described from the Sinap Formation. It is broken at the postorbital constriction and thus represents almost exactly the same portion of the skull as specimen PNT-130 from Pentalophos I described by de Bonis et al. (1994), to which it is also generally quite similar. Judging from the unfused lambdoidal suture (fig. 9.1A,D), the Sinap specimen represents a younger individual, which may explain why the skull roof is smooth, with only very faintly indicated longitudinal crests, as well as its somewhat smaller size (width at ear openings 52.2 mm, height of skull in same plane 37.0 mm, maximum width of braincase 44.3 mm). The occipital crest (along the unfused suture) is strong, as in the Pentalophos specimen. As it is undamaged in the Sinap specimen, the strong anterior bend in the midline of the skull is clearly seen. Between-season preparation in Ankara resulted in damage to the skull, including loss of the right occipital condyle (fig. 9.1B,E). This was fortunately preserved on the Pentalophos specimen, and we refer the reader to de Bonis et al. (1994) for a detailed description of it and other parts of the skull, which we do not wish to duplicate here.

In both ears, the middle ear cavity walls are well preserved (fig. 9.2A), although the tympanic ring in both is absent. In the left cavity, one middle ear ossicle, the left incus (anvil) was found, complete except that the small processus lenticularis, which forms the joint with the stapes, is absent (fig. 9.2B,C). The bone is similar to that of extant *Orycteropus. Crus longum* and *crus breve* are both long and divergent (Doran 1878, *O. afer aethiopicus*), and perpendicular to each other (Fleischer 1973, *O. afer afer*). According to the terminology of Fleischer (1978), the incus

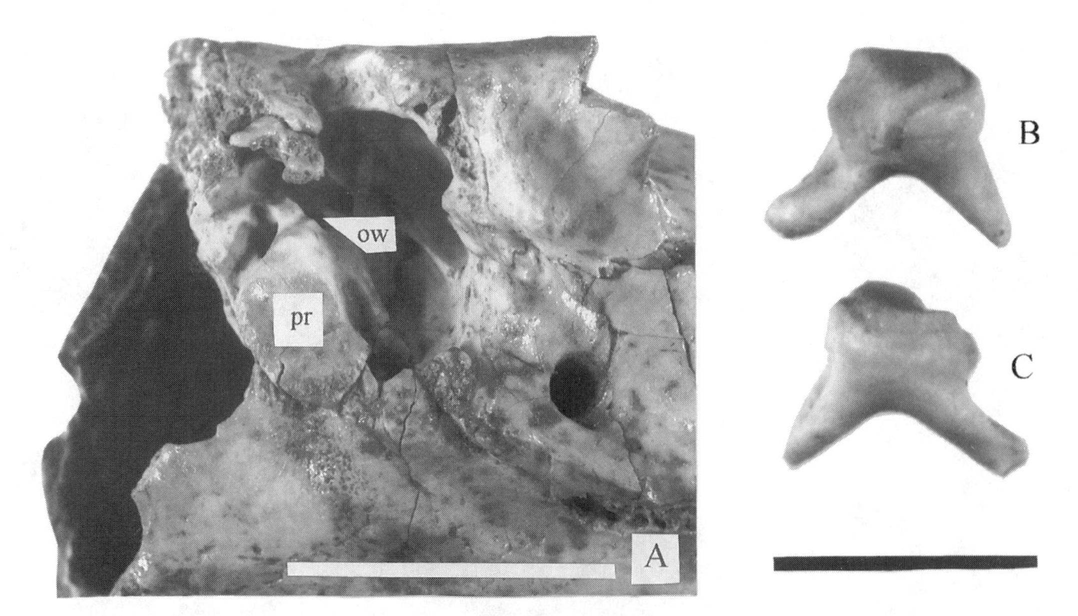

Figure 9.2. (A) Right ventrolateral view of AS.91.424, showing right tympanic cavity and promontorium (pr) with the oval (ow) and round (rw) windows. Scale bar = 20 mm. (B) Frontolateral view of left incus. (C) Distomedial view of left incus. Scale bar = 4 mm.

represents the ancestral-transitional type of the mammal middle ears.

The skull has two punctures on the right side, near the occipital crest (fig. 9.1D,F). It is not possible to say without further study whether they might have been made by a carnivore, but partly digested aardvark molars (described below) suggest that an aardvark predator was present at the locality.

The maxillary fragment AS.91.423 (fig. 9.3A) may well be part of the same individual as AS.91.424, but there is no actual contact between the two pieces. This specimen shows an enlarged canine in front of a much smaller premolar. This feature is also seen in the material from Pentalophos I (de Bonis et al. 1994, fig. 2), but the size difference is more marked in the Sinap specimen, perhaps because it is from a younger individual. The canine has well-developed occlusal facets both mesially and distally, forming a sharp, narrow wedge at the tip. The premolar has a single, well-developed distal facet, forming an acute angle with the mesial wall of the tooth.

There are two partial toothrows, the right C–P1 described above (AS.91.423) and a right mandibular fragment with p3–m1 (AS.92.202; fig. 9.3B,C). There is also an isolated right m2 (AS.92.580), which is probably part of the previous specimen, and a set of partly digested left M2–M3 (AS.95.700 and AS95.581), probably from one individual.

The upper molars both have an emaciated appearance. They are smaller and relatively longer and narrower than the corresponding teeth from Pentalophos I (de Bonis et al. 1994; PNT-127), and are more strongly waisted, all presumably as a result of carnivore digestion, indicated by a porous surface texture and "molten" overall shape.

The lower premolars are small relative to the molars, with a simple, ellipsoidal outline with only the faintest trace of waisting (fig. 9.2B,C). Well-developed occlusal facets mesially and distally meet at an obtuse angle, but with a sharp edge near the middle of the tooth. The lower molars are strongly "8" shaped, with almost equally strong buccal and lingual grooves. The occlusal facets of the molars are more horizontal and less distinct than on the premolars. All lower occlusal facets dip buccally as well as mesially or distally. See Appendix table 9.1 for dental dimensions.

The left scapular fragment AS.92.579 is made up of the glenoid surface, the neck, and the base of the blade (fig. 9.4A,B). The base of the spine is present but broken, and most of the coracoid process has also been lost. As in living *Orycteropus,* the glenoid surface curves strongly downward at the anterior end, whereas the coracoid process is directed almost straight forward. This is in contrast to *O. gaudryi,* in which the glenoid surface is more horizontal and the coracoid process more laterally directed. The basal part of the inferior border also seems to have a gently curved outline as in *O. afer,* rather than the tight curvature resulting from an expanded infraspinatus portion of the shoulder blade, as in *O. gaudryi* (Colbert 1933, figs. 10, 24). As far as can be determined from this fragmentary piece, *O. pottieri* thus had a shoulder joint similar to that of living aardvarks and different from at least one of the other Eurasian fossil species.

The metapodials (fig. 9.4C–G, Appendix table 9.2) are similar to those of *Orycteropus afer* in shape, but are about one-third the size and have slightly more slender proportions. The difference in robustness is about as much as would be expected from basic limb bone allometry and functional similarity and is unlikely to imply any functional difference. This is in contrast to *O. gaudryi,* in which the metapodials and phalanges are clearly more slender than in *O. afer* (Colbert 1933, figs. 13, 14, 17, 18). Of the metapodials,

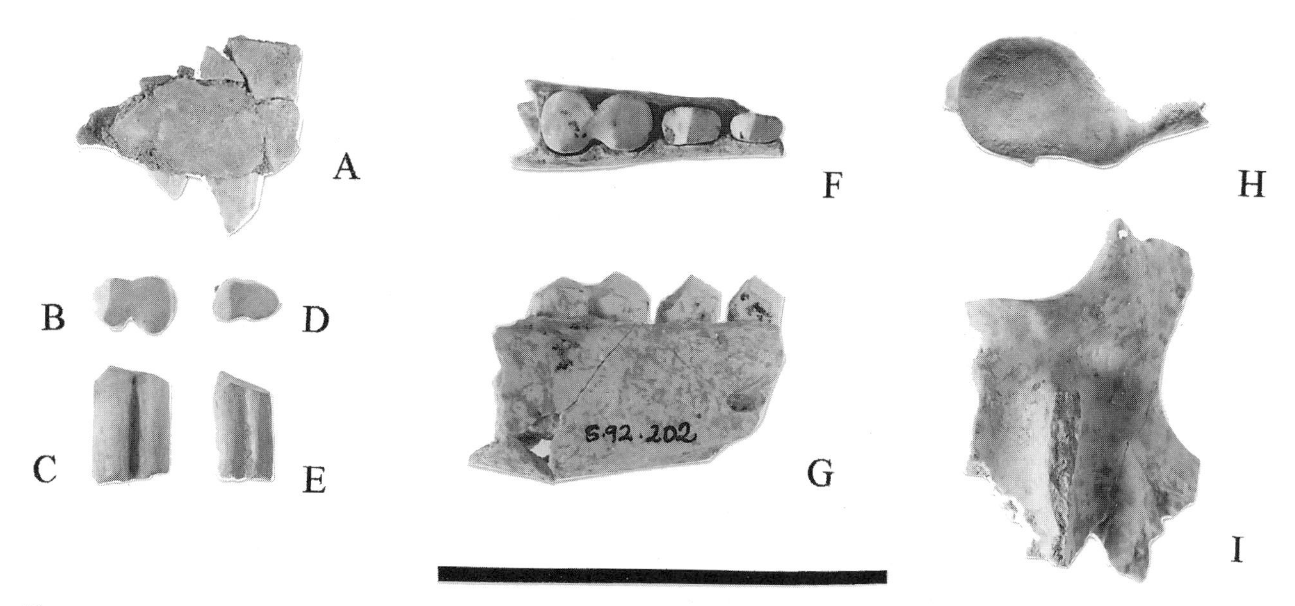

Figure 9.3. (**A**) Right maxillary fragment of *Orycteropus pottieri* AS.91.423 with canine and first premolar, Loc. 12. (**B, C**) Occlusal and buccal views of left M2 AS.92.23 of *Orycteropus* cf. *O. seni*, Loc. 64. (**D, E**) Occlusal and buccal views of right p4 AS.92.811 of *Orycteropus* cf. *O. seni*, Loc. 64. (**F, G**) Occlusal and buccal views of right mandibular fragment of *O. pottieri* AS.92.202 with p3–m1, Loc. 72. (**H, I**) Left partial scapula AS.92.579, distal and dorsal views, Loc. 12. Scale bar = 50 mm.

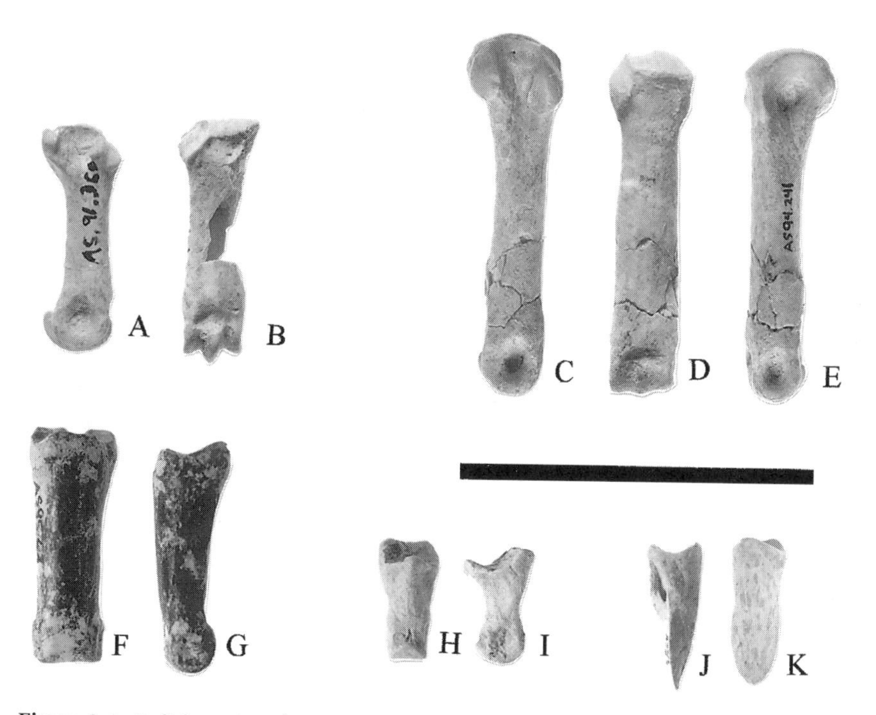

Figure 9.4. Podial remains of *Orycteropus pottieri* from the Sinap Formation. (**A, B**) Right metacarpal IV AS.91.366, lateral and plantar views. (**C–E**) Left metatarsal II AS. 94.241, medial, plantar, and lateral views. (**F, G**) Left first pedal phalanx III AS.95.63, plantar and lateral views. (**H, I**) Second phalanx AS.91.415, plantar and side views. (**J, K**) Third phalanx AS.95.251, side and plantar views. (**C–E**) are from Loc. 108; all others from Loc. 12. Scale bar = 50 mm.

only metacarpal IV is known from both Sinap and Pentalophos I. In both cases, the bone is slightly damaged, but the Sinap specimen (AS.91.360) has fewer breaks than the Pentalophos specimen (PNT-63; de Bonis et al. 1994, fig. 5a). The Sinap bone (fig. 9.4C,D) has a straight shaft, suggesting that the slightly curved shaft of the Pentalophos specimen is an artifact due to damage. Otherwise the bones are closely similar in shape and size.

Like the metapodials, the phalanges are also generally similar but smaller and less robust than those of recent *Orycteropus* (fig. 9.4H–K), but not as slender as in *O. gaudryi* (Colbert 1933, figs. 13–14, 17–18). Judging from their shape and the metacarpal/metatarsal size ratio, the first and second phalanges (fig. 9.4H–K) are probably from the third ray of the pes. The largest of the third phalanges (AS.95.251; fig. 9.4L,M) is also likely to be pedal, whereas the smaller ones are more difficult to place.

Orycteropus *cf.* O. seni *Tekkaya, 1993*

Synonymy

1996 *Orycteropus* sp. Kappelman et al. (in part)

Sinap Material

Loc. 64. M2 dex AS.92.23, M3 sin AS.92.810, p4 dex AS.92.811.

Age. The magnetostratigraphic age estimate for Loc. 64 is ~10.8 Ma.

Description and Remarks. All the teeth are relatively small (Appendix table 9.1). The M2 has a strong buccal and weak lingual groove, separating a larger mesial lobe from a smaller distal one (fig. 9.3D,E). There are two occlusal facets, which meet with a sharp edge near the distal end of the tooth. Neither facet has a significant dip component toward the lingual or buccal side. Apart from the smaller size and presence of a distal facet, the tooth is extremely similar to a left upper molar G1206 from Paşalar (Fortelius 1990, fig. 1), an early middle Miocene locality in western Anatolia. It is distinctly smaller and has a much weaker lingual groove than *Orycteropus pottieri* from Pentalophos I (de Bonis et al. 1994; PNT-127). The M3 is a short tooth with a distinct buccal and a faint lingual groove and a single, nearly horizontal occlusal wear facet. The p4 is distinctly elongated, with a strong lingual groove and a faint buccal one (fig. 9.3F,G). There are two occlusal facets that meet, forming a wedge near the distal end of the tooth; both facets have a significant dip toward the buccal side. The mesial facet is larger than the distal one. The tooth is thus very different from the lower premolars of *O. pottieri* (e.g., AS.92.202, PNT-129), and resembles the description of the smaller African species *O. chemeldoi* (Pickford 1975) in incipient bilobation and the disposition of the wear facets. No other form reviewed by Pickford (1975) or de Bonis et al. (1994) shows this combination of characters. Unfortunately, no upper teeth of *O. chemeldoi* are available for comparison. There are also no elements in common between the material from Loc. 64 and the material of *O. seni* described by Tekkaya (1993) from the central Anatolian middle Miocene locality Çandir, but the lower molars of the type mandible are distinctly elongated, like the p4 from Loc. 64. The type specimen of *O. seni* has lower molars that appear somewhat larger than would be expected from the upper molars and the p4 from Loc. 64 (Appendix table 9.1; Tekkaya 1993, fig. 1, table 1), but the difference is difficult to gauge and interpret. Given the similarity in age and fauna of Çandir and Paşalar, and the similarity between the upper molars from Paşalar and Loc. 64, we tentatively assign the material from Loc. 64 to *Orycteropus* cf. *O. seni.*

Orycteropus sp. indet.

Sinap Material

Loc. 46. Metapodial fragment AS.92.413.

Loc. 49. Partial third phalanx AS.95.1011.

Age. The magnetostratigraphic age estimate for Loc. 49 is 9.1 Ma (Kappelman et al., this volume).

Description and Remarks. The specimens are too fragmentary for identification beyond the genus level (if that). Faunally, Loc. 49 is very similar to Pentalophos I (Fortelius et al., this volume [rhinoceroses]), a locality with *Orycteropus pottieri* as discussed above.

Discussion

Despite the scarcity and mostly fragmentary nature of the aardvark remains from the Sinap Formation, it is clear that two temporally successive taxa can be distinguished: *Orycteropus* cf. *O. seni* from the pre-*Hipparion* Loc. 64 (~10.8 Ma) and *O. pottieri* from the middle Sinap Locs. 108, 72, and 12 (~10.1–9.6 Ma). *Orycteropus pottieri* seems to replace *O.* cf. *O. seni* without overlap or evolutionary change.

As far as the available fossil material shows, *Orycteropus pottieri* appears to be quite similar to recent aardvarks, except for being about one-third smaller and retaining a relatively unreduced canine. The limb bones are proportionately more slender, probably due to a simple scaling effect. The difference between *O. gaudryi* and *O. afer* seems to be considerably greater, the fossil form being distinctly more slender in build and having a differently constructed shoulder joint. Another possibly important character uniting *O. pottieri* with living *Orycteropus* and separating it from all Eurasian fossil forms is the symmetrical "8" shape of the upper molars (de Bonis et al. 1994, fig. 2). The norm is for the upper molars of the extinct Eurasian species to be asymmetrical, with a much weaker lingual than buccal groove (e.g., fig. 9.3D; Sen 1994, p. 104; de Bonis et al. 1994, fig. 11). *Orycteropus pottieri* is only known from Anatolia and Greece, from localities of closely similar age, and could conceivably represent a brief and limited dispersal of an African lineage not closely related to the other Eurasian forms. *Orycteropus* cf. *O. seni,* in contrast, has asymmetrical upper molars of the usual Eurasian type, and could be the basal taxon of the main clade of Eurasian aardvarks, including at least *O. gaudryi* and *O. depereti.* The African middle Miocene *O. chemeldoi* might be the sister taxon of such a clade.

Until more material becomes available, one can have only moderate confidence in such scenarios, however. Still, a relatively robust conclusion seems to be that the Anatolian succession of Neogene aardvarks does not represent a single evolving lineage. Their first dispersal into Eurasia seems to have taken place near the early/middle Miocene boundary, with what may be a separate, more limited dispersal near the middle/late Miocene boundary.

Hearing in a Fossil Aardvark

The incus from specimen AS.91.424 was used to estimate the theoretical high-frequency limit of hearing of *Orycteropus pottieri*. It has a mass of 10.770 mg and a density of 2.81 mg/cm^3 (for the method of density measuring, see Nummela et al. 1999). The approximate mass of the incus in a live animal could be calculated by assuming a density of 2.2 mg/cm^3 for terrestrial mammals (Giraud-Sauveur 1969). Further, the mass of the malleus could be estimated from the incus/malleus allometric relationship for mammals (Nummela 1995). Finally, with these mass values, the theoretical high-frequency cutoff for this species was predicted to be ~35 kHz (see Hemilä et al. 1995).

This, however, tells us nothing about the absolute hearing sensitivity of this species, especially when there are hardly any data available of the hearing abilities of the extant *Orycteropus*. It apparently has a good sense of hearing and smell, but poor vision. In general, the threshold of best sensitivity is quite similar (0 ± 10 dB sound-pressure level) in many different kinds of terrestrial mammals (Fay 1988), and when scaled with middle ear size, the behavioral audiograms become more or less equal (Nummela 1997). Furthermore, there are observations of aardvarks (as well as bears and ratels) using a loud snort and its associated echo to identify cavities before entering (Kingdon 1977). The high-frequency hearing limit of *O. pottieri* is somewhat lower than predicted on the basis of its interaural distance (Masterton et al. 1969; Heffner and Heffner 1990). However, there is evidence for sound propagation in underground tunnels that shows that the least attenuation of sound occurs at low frequencies (Heth et al. 1986). Thus, if low-frequency hearing is adaptive in aardvarks, the shifting of the whole hearing range toward lower frequencies may well have lowered the high-frequency hearing as well (see Fay 1988; Hemilä et al. 1995).

Paleoecology of Eurasian Fossil Aardvarks

It is probable that aardvarks of the genus *Orycteropus* have been specialized feeders on ground-living colonial ants and termites during their entire known history, from the early Miocene to the Recent (Patterson 1975). The loss of dental enamel would be easy to explain as a means of increasing tolerance to formic acid (which attacks enamel more than it does dentine), but according to Patterson (1975), aardvarks lost their dental enamel prior to the evolution of specialized myrmecophagy. Perhaps even a moderate ingestion of acid-producing insects would have been sufficient for an evolutionary loss of enamel. The reason why teeth persist at all in aardvarks is not well understood, as they tend to be lost entirely in other anteating groups (Patterson 1975). The most likely reason seems to be that aardvarks ingest a significant amount of plant foods (Kingdon 1974; Patterson 1975, 1978), an explanation at least consistent with the presence of distinct, slothlike occlusal wear facets on the teeth. The burrowing, solitary lifestyle of living aardvarks seems likely for the genus *Orycteropus*, but is difficult to verify, because the digging adaptations are already present as part of the feeding mode. According to Kingdon (1974), *Orycteropus afer* requires a sufficient supply of ground-living insects and a local source of drinking water. Thus aardvarks are not particularly specific environmental indicators, distinctly less so than many other taxa found at the same localities.

Acknowledgments

We thank Prof. Dr. Berna Alpagut, Prof. Louis de Bonis, Prof. Ann Forstén, Dr. George Koufos, and Dr. Ilhan Temizsoy for permission to study material under their care and for discussions, and fellow members of the Sinap project for all the fun. We are grateful to Mr. Markku Lehtonen for producing the photographs and to Dr. Gerçek Saraç and Ms. Anneli Paldanius for help with obtaining some of the literature. Special thanks to Celâl Metin of Kazan, King among Taxi Drivers.

Appendix

Table 9.1. Dental Measurements

Tooth	Specimen Number	Locality	Length (mm)	Mesial Width (mm)	Distal Width (mm)
			Orycteropus pottieri		
C sup	AS.91.423	12	6.2	3.5	
P1	AS.91.423	12	5.1	2.5	
p3	AS.92.202	12	5.0	2.8	
p4	AS.92.202	12	6.2	3.8	
m1	AS.92.202	12	12.4	5.8	6.5
m2	AS.92.580	12	12.9	6.3	7.0
			Orycteropus cf. *O. seni*		
M2	AS.92.23	64	9.1	5.6	6.5
M3	AS.92.810	64	6.5	5.1	4.0
p4?	AS.92.811	64	7.5	4.9	4.1

Note: Specimens AS.95.700 and AS.95.581 are not measurable owing to digestive removal of dentine.

Table 9.2. Postcranial Measurements

Bone	Specimen Number	Locality	Maximum Extent of Articular Surface (mm)					WD (mm)	APD (mm)
			L	WP	APP	WS	APS		
Scapula	AS.92.579	12						14.1	21.5
Mc IV	AS.91.366	12	33.2	11.8	11.6	—	—	8.3	10.6
Mt II	AS.94.241	108	49.5	12.1	13.4	8.2	7.5	10.2	8.7
Mt III?	AS.93.282	108	—	—	—	8.6	6.6	9.8	8.2
Phalanx 1	AS.95.63	12	32.2	12.2	11.1	8.5	6.5	10.2	7.2
Phalanx 2	AS.91.415	12	16.1	8.6	9.9	6.0	4.6	6.4	6.8
Phalanx 3	AS.95.251	12	20.7	8.1	8.3	—	—	—	—
Phalanx 3	AS.95.176	12	16.9	6.5	7.6	—	—	—	—
Phalanx 3	AS.91.356	12	14.7	5.6	6.5	—	—	—	—

Literature Cited

Bonis, L. de, G. Bouvrain, D. Geraads, G. Koufos, and S. Sen, 1994, The first aardvark from the late Miocene of Macedonia, Greece: Neues Jahrbuch für Geologie und Paläontologie, Abhandlungen, v. 194, no. 2/3, pp. 343–360.

Colbert, E., 1933, A study of *Orycteropus gaudryi* from the island of Samos: Bulletin of the American Museum of Natural History, v. 78, pp. 305–351.

Doran, A. H. G., 1878, Morphology of the mammalian Ossicula auditus: Transactions of the Linnean Society, London, v. 2., ser. 1, Zoology, pp. 371–497.

Fay, R. R., 1988, Hearing in vertebrates: A psychophysics databook: Winnetka, Illinois, Hill-Fay Associates, 621 pp.

Fleischer, G., 1973, Studien am Skelett des Gehörorgans der Säugetiere, einschliesslich des Menschen: Säugetierkundliche Mitteilungen, v. 21, pp. 131–239.

Fleischer, G., 1978, Evolutionary principles of the mammalian middle ear, Advances in Anatomy, Embryology and Cell Biology, v. 55, no. 5, pp. 1–70.

Fortelius, M., 1990, Less common ungulate species from Pasalar, Middle Miocene of Anatolia (Turkey): Journal of Human Evolution, v. 19, pp. 479–488.

Giraud-Sauveur, D., 1969, Recherches biophysiques sur les osselets des Cétacés: Mammalia, v. 33, pp. 285–340.

Heffner, R. S., and H. E. Heffner, 1990, Hearing in domestic pigs (*Sus scrofa*) and goats (*Capra hircus*): Hearing Research, v. 48, pp. 231–240.

Hemilä, S., S. Nummela, and T. Reuter, 1995, What middle ear parameters tell about impedance matching and high frequency hearing: Hearing Research, v. 85, pp. 31–44.

Heth, G., E. Frankenberg, and E. Nevo, 1986, Adaptive optimal sound for vocal communication in tunnels of a subterranean mammal (*Spalax ehrenbergi*): Experientia, v. 42, pp. 1287–1289.

Kappelman, J., S. Sen, M. Fortelius, A. Duncan, B. Alpagut, J. Crabaugh, A. Gentry, J.-P. Lunkka, F. McDowell, N. Solounias, S. Viranta, and L. Werdelin, 1996, Chronology and biostratigraphy of the Miocene Sinap Formation of central Turkey, *in* R. L Bernor, V. Fahlbusch, and H.-W. Mittmann, eds., The evolution of western Eurasian Neogene mammal faunas: New York, Columbia University Press, pp. 78–95.

Kingdon, J., 1974, East African mammals, Volume I: Chicago, University of Chicago Press, 446 pp.

Kingdon, J., 1977, East African mammals, Volume IIIA: Chicago, University of Chicago Press, 476 pp.

Masterton, B., H. Heffner, and R. Ravizza, 1969, The evolution of human hearing: The Journal of the Acoustical Society of America, v. 5, no. 4, pp. 966–985.

Nummela, S., 1995, Scaling of the mammalian middle ear: Hearing Research, v. 85, pp. 18–30.

Nummela, S., 1997, Scaling and modeling the mammalian middle ear: Comments on Theoretical Biology, v. 4, pp. 387–412.

Nummela, S., T. Wägar, S. Hemilä, and T. Reuter, 1999, Scaling of the cetacean middle ear: Hearing Research, v. 133, pp. 71–81.

Öngür, T., 1976, Kizilcahamam, Camlidere, Celtikci ve Kazan dolayinin jeoloji durumu ve jeotermal enerji olanaklari: Maden Tetkik ve Arama Enstitüsü, Ankara, unpublished report.

Ozansoy, F., 1957, Faunes de mammifères du Tertiaire du Turquie et leurs révisions stratigraphiques: Bulletin of the Mineral Resource Exploration Institute of Turkey, Foreign Edition, v. 49, pp. 29–48.

Ozansoy, F., 1965, Études des gisements continentaux et de mammifères du Cénozoïque du Turquie: Mémoires de la Societé Géologique de France, Nouvelle Série, v. 44, pp. 1–92.

Patterson, B., 1975, The fossil aardvarks (Mammalia: Tubulidentata): Bulletin of the Museum of Comparative Zoology, v. 147, pp. 185–237.

Patterson, B., 1978, Pholidota ad Tubulidentata, *in* V. J. Maglio, and H.B.S. Cooke, Evolution of African mammals: Cambridge, Massachusetts, Harvard University Press, pp. 268–278.

Pickford, M., 1975, New fossil Orycteropodidae (Mammalia, Tubulidentata) from East Africa. *Orycteropus minutus* sp. nov. and *Orycteropus chemeldoi* sp. nov.: Netherlands Journal of Zoology, v. 25, pp. 57–88.

Sen, S., 1991, Stratigraphie, faunes de mammifères et magnétostratigraphie du Néogène de Sinap Tepe, Province d'Ankara, Turquie: Bulletin de la Museum National d'Histoire Naturelle, Paris, ser. 4e, v. 12, section C, no. 3/4, pp. 243–277.

Sen, S., 1994, Les gisements de mammifères du Miocène supèrieur de Kemiklitepe, Turquie: 5, Rongeurs, Tubulidentés et Chalicothères: Bulletin de la Museum National d'Histoire Naturelle, ser. 4e, v. 16, section C, no. 1, pp. 97–111.

Tekkaya, I., 1993, Türkiye fosil Orycteropodidae'leri. T.C. Kültür Bakanligi Anitlar ve Müzeler Genel Müdürlügü. VIII. Arkeometri Sonuclari Toplantisi 1992: Ankara, Turkey, T.C. Kültür Bakanligi, pp. 275–289.

10

Proboscidea

W. J. Sanders

Proboscidean fossils were first recovered from the Sinap Formation during paleontologic fieldwork in the early 1950s around the area of Sinap Tepe, north of the village of Yassiören and northwest of Ankara (Ozansoy 1955, 1957, 1965). These fossils were provenanced to the middle member of the formation and allocated among a number of taxa, including several species of *Synconolophus* (Ozansoy 1957, 1965). In addition, a few juvenile specimens were made the type series of a novel species, *Trilophodon (Choerolophodon) anatolicus* (Ozansoy 1965), which was subsequently subsumed into *Choerolophodon pentelici* (Gaziry 1976). More recently, Tassy et al. (1989) and Tassy (1994) formally recognized morphologic distinctions between more advanced *Choerolophodon pentelici* and these specimens by including them in a new subspecies, "*C. pentelici lydiensis.*" Although there are no published descriptions of the *Synconolophus* fossils, it is conceivable that Ozansoy's sample from Sinap Tepe represents a single choerolophodont species, as this genus has since been synonymized under *Choerolophodon* (Gaziry 1976; Tassy 1985).

Other Sinap Formation proboscideans collected during this interval, from sites to the west of Yassiören, were organized into a more bewildering array of taxa. Ozansoy (1955, p. 992) originally recorded *Mastodon angustidens* and *Mastodon* sp. from Inönü, where he reported the occurrence of a "veritable cemetary of mastodonts," but later divided this material among *Synconolophus* sp., *Serridentinus* sp., and *Trilophodon* sp. (Ozansoy 1965). It is possible that some of the Inönü fossils are gomphotheres, as *Mastodon, Trilophodon,* and *Serridentinus* have all been alternatively used to refer to *Gomphotherium* (Tobien 1973a). With regard to gomphotheres, however, *Mastodon* is an inappropriate genus nomen because it is a junior synonym of the American mastodon genus *Mammut,* and *Gomphotherium* has priority over *Trilophodon* (Tobien 1973a; Gaziry 1976). *Serridentinus* is equally invalid (see Tobien 1972). Proboscidean fossils from Kavak Dere were also initially placed in *Mastodon* sp. (Ozansoy 1955, 1957), but in an ensuing publication (Ozansoy 1965, table 5) they were referred to *Synconolophus* sp. 1. In addition, *Trilophodon (Choerolophodon) pentelici, Tetralophodon grandincisivus, Tetralophodon longirostris,* and *Deinotherium* sp. were listed from Çoban Pinar (Ozansoy 1965, table 5). Although Ozansoy (1955, 1957, 1965) believed the localities at Inönü, Çoban Pinar, and Kavak Dere to be stratigraphically lower than the Sinap Formation, for the most part they are correlatable with the middle member of the formation (see Kappelman et al. 1996). The exception is locality Inönü I, which is equivalent to lower member Sinap Locs. 24 and 24A (Kappelman et al. 1996). Because these specimens remain undescribed as well, it is impossible to assess the validity of Ozansoy's taxonomic assignments and interpretation of such proboscidean diversity at these sites. Nonetheless, it seems reasonable to assume that in addition to choerolophodonts, the inaugural proboscidean collection from the Sinap Formation also includes deinotheres and gomphotheres.

More recent paleontologic prospecting and excavation in the Sinap Formation of the Ankara region, between 1989 and 1995, produced a large sample ($n = 205$) of proboscidean remains, from localities at Sinap Tepe, Kavak Dere, Delikayinçak Tepe, and Igbek (see Kappelman et al. 1996). These fossils are chronostratigraphically well provenanced (Kappelman et al. 1996, chapter 2, this volume). The present study provides detailed descriptions of teeth and dentaries from this sample and makes comparative morphometric analyses of the specimens. These form the basis for taxonomic evaluation of the sample. The results of the study document the presence of proboscideans in both the lower and middle members of the Sinap Formation, confirm the occurrence of deinotheres and gomphotheres in the Sinap Formation, and contribute new evidence useful for better understanding the evolution of circum-Mediterranean choerolophodonts during the late Miocene. In addition to their taxonomic utility, the new proboscidean fossils from the Sinap Formation are also informative for paleoecological reconstruction of depositional environments, as well as for biochronologic correlation, of Sinap localities.

Abbreviations and Definitions

Choerolophodont = member of the genus *Choerolophodon* or genus *Afrochoerodon*
dP or dp = deciduous premolar; dP2 refers to an upper second premolar, and dp2 refers to a lower second premolar
H = maximum crown height, measured from the cervix of the crown to the apex of the highest conelet, on a line perpendicular to the base of the crown
Gomphothere = member of the genus *Gomphotherium;* used here only in this restricted sense
L = crown length, measured parallel to the long axis of the tooth
M or m = molar; M1 refers to an upper first molar, and m1 refers to a lower first molar
MC = metacarpal; MC IV refers to the fourth metacarpal
Mesoconelet(s) = the cusp(s) closest to the midline in each half-loph(id) of a molar (Tassy 1996a)
MT = metatarsal; MT III refers to the third metacarpal
P or p = premolar; P3 refers to an upper third premolar, and p3 refers to a lower third premolar
Pretrite = refers to the more worn half of each loph(id), which is buccal in lower molars and lingual in upper molars (Vacek 1877)
Posttrite = refers to the less worn half of each loph(id), which is lingual in lower molars and buccal in upper molars (Vacek 1877)
Principal, or main, cusp = the cusp farthest from the midline in each half-loph(id) of a molar; usually the largest cusp (Tassy 1996a)
W = greatest crown width, measured across the broadest loph(id) perpendicular to the long axis of the tooth, including cementum
+ = indicates a missing portion of a tooth, and that the original dimension was greater
x = indicates presence of an anterior or posterior crescentoid or cingulum not constituting a full loph(id)

Systematics

Deinotherium giganteum *Kaup, 1829*

Taxonomy

Order Proboscidea Illiger, 1811
Suborder Deinotherioidea Osborn, 1921
Family Deinotheriidae Bonaparte, 1845
Genus *Deinotherium* Kaup, 1829
Deinotherium giganteum Kaup, 1829

Referred Specimens. Sinap Formation, middle member, Loc. 49: AS 94.471, dP2–4 dex (fig. 10.1).

Diagnosis. Large species of deinothere. Permanent tooth formula 0-0-2-3/1-0-2-3 (deciduous 0-0-3/1-0-3), with vertical cheek tooth replacement. The dP4/dp4 and M1/m1 trilophodont; P4/p4 and M2–3/m2–3 bilophodont. Molars and dP4/dp4 tapiroid, vertical shearing teeth; premolars and dP2–3/dp2–3 for crushing. Cranium short, low, dorsally flat, with very large, elevated occipital condyles. Nasal opening retracted and substantial. Rostrum long and rostral fossa broad. Mandibular symphyses very long and curved downward, perpendicular to long axis of corpus. Distinguished from *Prodeinotherium* by greater size of dentition and reduced development of posterior cingula in M2–3; from *Deinotherium bozasi* by greater width of rostral trough and size of nasal aperture, less posterior retraction of nares, lower, wider cranium, lesser flexion of rostrum, longer and less abruptly flexed symphysis, more anteriorly situated mental foramina, and lack of anterior median projection on nasals; from *D. indicum* by more elliptical cross-section of dentary at m3, more gracile construction of dentary, and lack of p4–m3 intravalley tubercles (Weinsheimer 1883; Gräf 1957; Sahni and Tripathi 1957; Bergounioux and Crouzel 1962; Harris 1975, 1976; Sarwar 1977; Tobien 1988).

Description. Deinotheres are represented in the new Sinap Formation collection by the right dP2–4 of a single individual, AS 94.471 (fig. 10.1). The dP2 is complete and virtually unworn and retains the remnants of two buccolingually transverse roots. The specimen is relatively elongate; crown length is 49.3 mm, anterior width is 37.9 mm, and posterior width is 45.2 mm. Crown height is greatest at the protocone, 30.0 mm. Small tubercles are aligned along the anterior face of the tooth to form a low, conspicuous cingulum, which projects farthest mesially at the buccal side of the crown. The cingulum is connected to the ectoloph by a weak crest and it continues along the lingual side of the crown as a low, indistinct ridge. Posteriorly, the crown is bordered by a low, almost imperceptible cingulum that is joined to the apex of the hypocone by a steeply rising crest. The protocone and hypocone are large and narrowly separated by a V-shaped transverse valley that opens lingually (fig. 10.1A). On the buccal side of the tooth, the ectoloph runs anteroposteriorly without transverse interruption. Apically, the ectoloph is superficially subdivided into numerous fine mammillons, and displays two prominences equivalent to the paracone and metacone (fig. 10.1C). Distally, the ectoloph terminates as a flange perpendicular to and confluent with the posterior cingulum. The paracone connects transversely with the protocone via a low, anteriorly convex ridge to form the protoloph. A long, deep fovea that is open posteriorly is enclosed anteriorly by the protoloph, buccally by the ectoloph, and lingually by the protocone and hypocone.

The dP3 preserves parts of two large roots and required reconstruction of a portion of the crown at its anterolingual

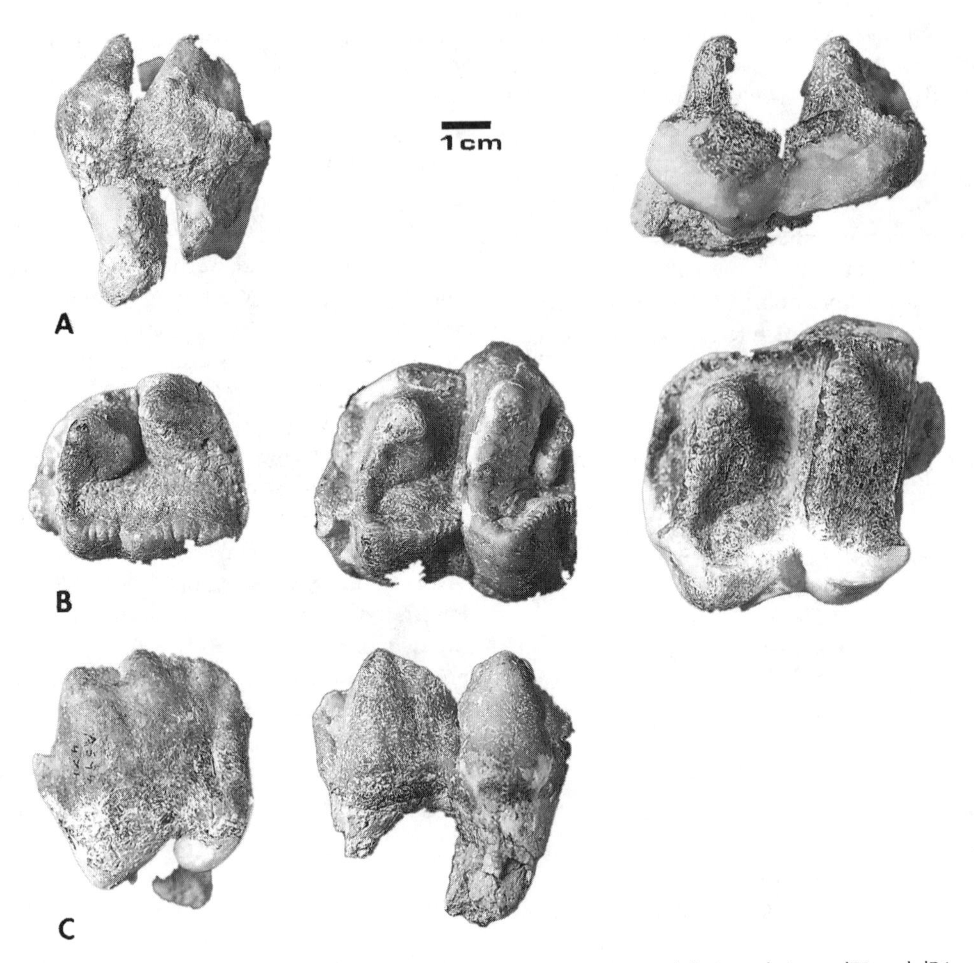

Figure 10.1. Specimen AS 94.471, right dP2–4, *Deinotherium giganteum*. **(A)** Lingual views, dP2 and dP4. Anterior is to the right. **(B)** Occlusal views, dP2–4 (from left to right). Anterior is to the left. **(C)** Buccal views, dP2–3 (from left to right). Anterior is to the left.

edge (fig. 10.1B). This tooth is bilophodont and is considerably larger (L = 57.6 mm) than the dP2. The protocone and paracone, and hypocone and metacone, are connected by sharp, anteriorly convex crests to form the protoloph and metaloph, respectively (fig. 10.1B). These lophs exhibit small, flat vertical wear facets oriented mesially. Constriction of the crown at the interloph (w = 43.9 mm), relative to widths of 49.0+ mm across the protoloph and 57.5 mm across the metaloph, gives the dP3 a figure-eight occlusal shape. Greatest height is 35.5 mm at the hypocone. Anteriorly, the specimen possesses a low cingulum that is most prominent on the buccal side. The cingulum is connected to the paracone by a short ridge and continues on the lingual side of the crown as a shelf that terminates at the base of the hypocone. The anterior cingulum has a small horizontal wear facet on its buccal aspect. The transverse valley between the protoloph and metaloph is open lingually and closed off on the buccal side by a postparacrista running from the apex of the paracone to the base of the metacone. In turn, a postmetacrista runs distally from the apex of the metacone to the posterior cingulum. The posterior cingulum is low, closely appressed to the metaloph, and curves back on itself to occupy the distal fovea. The lophs, cingula, and ridges are all superficially subdivided into numerous fine mammillons. Although interproximal facets are not distinct, the dP3 conforms well with the distal surface of the dP2 and mesial end of the dP4.

The dP4 is incomplete, having pieces of the crown broken away anterobuccally from the bases of the protocone and hypocone and from the buccal side of the metaloph. It is also sharply fractured posterior to the metaloph and missing its tritoloph and posterior cingulum (fig. 10.1B). There is little occlusal wear in evidence, and traces of fine mammillons can be seen along the lophs and cingular margins. The specimen is 63.5+ mm in length, 58.3 mm in width across the protoloph, and ~58.5 mm across the metaloph. Crown height (41.0 mm) is greatest at the hypocone. The anterior cingulum appears as a low shelf that is connected to the paracone by a short ridge, and continues along the lingual aspect of the crown as a narrow fringe (fig. 10.1A); it evidently was most prominent on the buccal side of the mesial face of the crown. The protocone and paracone, and hypocone and metacone, are connected by apically thin crests that are nearly straight transversely, to form the protolophs and metalophs, respectively. The cusps and lophs are anteroposteriorly expanded at their bases and would encompass more extensive occlusal surfaces with wear. A postparacrista runs posteriorly from the apex of the paracone toward the base of the metacone. The transverse valley is narrow, V-shaped, and partially blocked by the postparacrista at its buccal side; lingually, it is more open and U-shaped.

Remarks. In deinotheres, dP2–4 resemble P3–4 and M1, respectively (Harris 1975). However, dP2 is distinguished from P3 by the mesial projection of its anterior cingulum, dP3 differs from P4 in having good separation between its buccal cusps and a broader metaloph than protoloph, and in dP4 the metaloph is wider than the protoloph, inverse to the condition in M1 (Bergounioux and Crouzel 1962; Tobien 1988). These features are present in AS 94.471 (fig. 10.1) and identify the specimen as a deciduous dental series. Deinothere cheek tooth series were bifunctional, with dP2–3/dp2–3 used for crushing and dP4/dp4 for shearing in the early stages of life (Harris 1975, 1976, 1978). The lack of pronounced occlusal wear and incomplete closure of its roots indicate that AS 94.471 was a very young juvenile or an infant.

Because Turkey is at a paleobiogeographic crossroads between Europe, Asia, and Africa (Kappelman et al. 1996), and no criteria have been advanced to morphologically differentiate deciduous teeth among deinothere taxa, specific allocation of AS 94.471 is not necessarily straightforward. Deinotheres are represented by two genera, *Prodeinotherium* and *Deinotherium,* which were successively distributed across Africa, south Asia (Indo-Pakistan), and Europe during the late Tertiary (Sahni and Tripathi 1957; Harris 1973, 1976, 1977, 1978, 1983, 1987; Sarwar 1977; Raza and Meyer 1984). In Europe, *Prodeinotherium bavaricum* first appears in early Miocene mammal faunal zone MN 4 (Antunes 1989; Ginsburg 1989; Tassy 1989), and is replaced by *D. giganteum* in the middle Miocene (Bergounioux and Crouzel 1962). *Deinotherium giganteum* is primarily a late Miocene species, most common from Vallesian and Turolian localities (Tobien 1988); its last reported occurrence is from the middle Pliocene of Romania (Sarwar 1977). Similarly, in south Asia, *Prodeinotherium* is documented from the early Miocene Bugti beds (Cooper 1922; Raza and Meyer 1984) and persists into the middle Miocene Chinji Formation, where it is replaced by *D. indicum* (Sarwar 1977). *Deinotherium indicum* is most prevalent in the late Miocene Dhok Pathan Formation (Sarwar 1977) and disappears from the fossil record ~7 Ma (Barry and Flynn 1989). The earliest documentation of deinotheres is from late Oligocene localities at Chilga, Ethiopia, suggesting an African origin for the group (Sanders and Kappelman 2001). *Prodeinotherium* is reasonably well represented in the early Miocene of Africa, and was succeeded by *D. bozasi* at the beginning of the late Miocene (see Harris 1978; Hill et al. 1985; Nakaya 1993; Leakey et al. 1996). The last record of deinotheres is that of *D. bozasi* from the Kanjera Formation, Kenya, at ~1.0 Ma (Behrensmeyer et al. 1995).

Dentally, *Deinotherium* is distinguished from *Prodeinotherium* in each of these areas by greater crown dimensions and reduced postmetaloph ornamentation on M2–3 (Bachmann 1875; Weinsheimer 1883; Roger 1886; Gräf 1957; Sahni and Tripathi 1957; Bergounioux and Crouzel 1962; Harris 1975, 1978; Sarwar 1977). Specimen AS 94.471 is assigned to *D. giganteum* on the basis of size. The great dimensions of the deciduous premolars of AS 94.471 exclude them from *Prodeinotherium* and place them at the upper end of the range for European *Deinotherium* (fig. 10.2). Of the

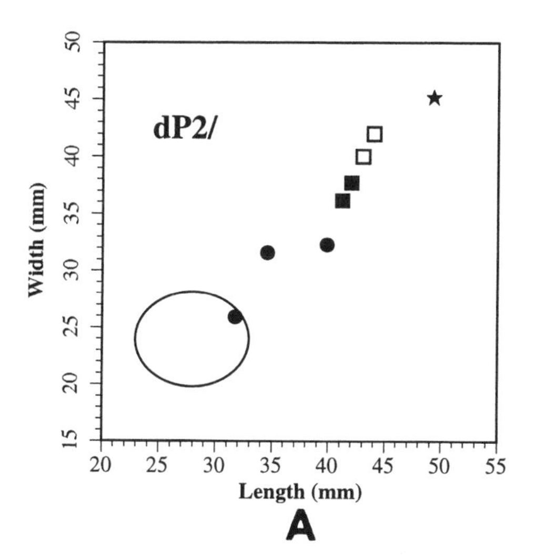

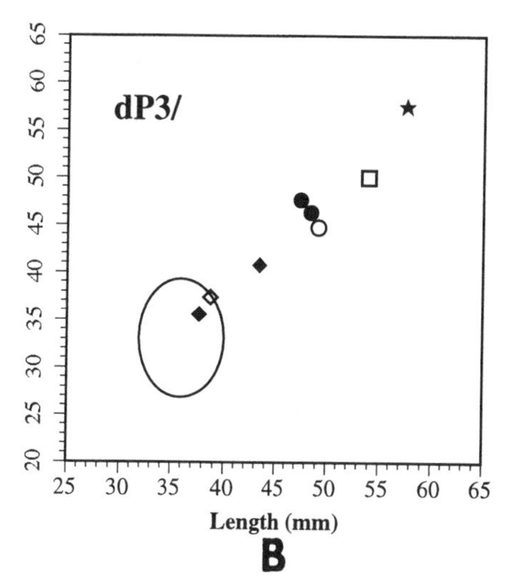

Figure 10.2. Bivariate plots of crown length versus width in deciduous deinothere teeth. Comparative dimensions are from Weinsheimer (1883), Gräf (1957), Gaziry (1976), and Tobien (1988). Mammal faunal units and zonations follow Steininger et al. (1989, 1996), Andrews et al. (1996), Woodburne et al. (1996), and Fejfar et al. (1997). Data points represent individuals of *Deinotherium giganteum*. Large ovals represent the morphospace of *Prodeinotherium bavaricum*, based on dimensions given in Bergounioux and Crouzel (1962). Symbols: ★, AS 94.471, Sinap Formation, Turkey (Vallesian, late MN 10); □, Kayadibi, Turkey (Turolian, earliest MN 11); ●, Eppelsheim, Germany (Vallesian, MN 9); ■, Montredon, France (Vallesian, MN 10); ◇, Esselborn, Germany (Vallesian, MN 9); ◆, Wissberg, Austria (?Vallesian, ?MN 9); ○, Wolfersheim (?Turolian), Germany. (**A**) DP2 dimensions. (**B**) DP3 dimensions.

small sample available for comparison, dP2–3 of AS 94.471 most closely matches in size the deciduous premolars of *D. giganteum* from the Turkish site of Kayadibi (fig. 10.2). Furthermore, the preserved dimensions of dP4 of AS 94.471 (W = 58.5 mm; H = 41.0 mm) are nearly identical to those of Kayadibi (W = 58.0 mm, 59.0 mm; H = 38.0 mm, 43.0 mm) (Gaziry 1976). Specimen AS 94.471, from Loc. 49 and dated paleomagnetically to ~9 Ma (late MN 10) (Kappelman et al. 1996, chapter 2, this volume), is also similar to Kayadibi in age. Kayadibi is correlated with lowermost MN 11 (Steininger et al. 1989, 1996). The dP4 specimens from AS 94.471 and from Kayadibi are larger than those of *D. giganteum* from older localities, such as Eppelsheim, Wissberg, and Montredon (for dates, see Michaux 1988; Steininger et al. 1989, 1996; Andrews et al. 1996; Woodburne et al. 1996; Fejfar et al. 1997; for dimensions, see Weinsheimer 1883; Gräf 1957; Tobien 1988), suggesting an intraspecific trend of increasing tooth size over time with potential utility for biochronologic correlation. The dP4 of AS 94.471 is much greater in size than those of *Prodeinotherium* (Bergounioux and Crouzel [1962] give an upper limit of W = 45.0 mm for *P. bavaricum*). In contrast, the few dP4 specimens known for *D. indicum* and *D. bozasi* (Harris 1976, 1983; Sarwar 1977) are slightly larger than that of specimen AS 94.471. To date, of these *Deinotherium* species, only *D. giganteum* has been identified in fossil localities elsewhere in Turkey and the circum-Mediterranean (Senyürek 1952; Ozansoy 1957, 1965; Symeonidis 1970; Gaziry 1976; Tsoukala and Melentis 1994).

Gomphotherium angustidens *(Cuvier, 1817)*

Taxonomy

Suborder Elephantiformes Tassy, 1988
Family Gomphotheriidae Hay, 1922
Genus *Gomphotherium* Burmeister, 1837
Gomphotherium angustidens (Cuvier, 1817)

Referred Specimens. Sinap Formation, lower member, Loc. 24: AS 92.603, M3 sin (fig. 10.3A,B); AS 92.647, astragalus dex.

Loc. 24A: AS 92.699, dP3 dex (fig. 10.3C,D).

Diagnosis. Advanced gomphothere species. Longirostral mandible; lower incisors have pyriform cross section. Inter-

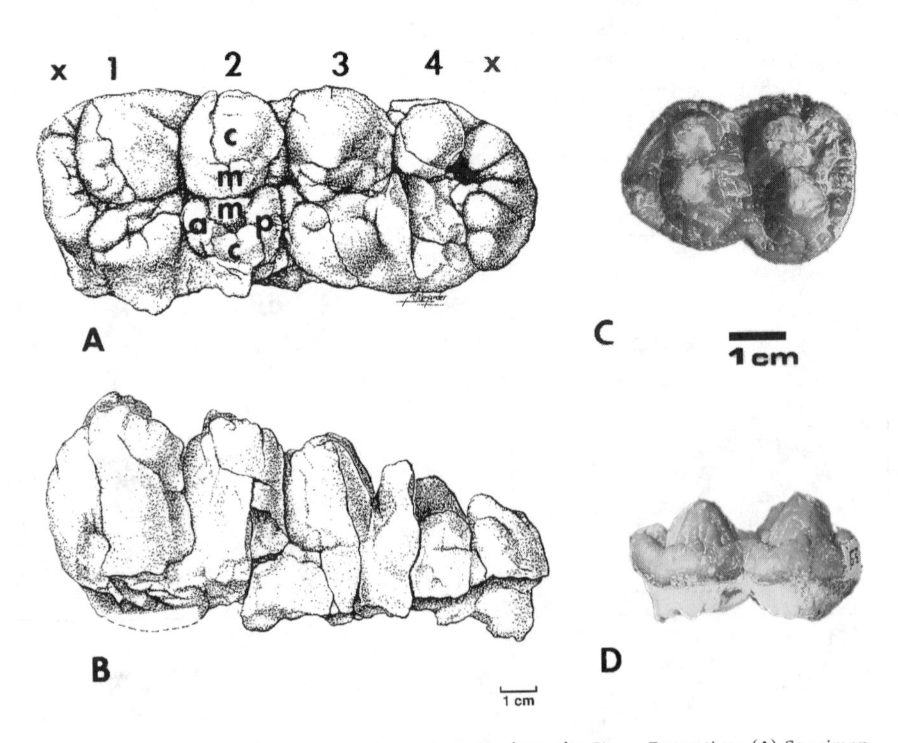

Figure 10.3. *Gomphotherium angustidens* specimens from the Sinap Formation. **(A)** Specimen AS 92.603, left M3, diagram of occlusal view. Anterior is to the left. Abbreviations: m, mesoconelet; c, principal cusp, pre- or posttrite half-loph; a, anterior accessory pretrite conule; p, posterior accessory pretrite conule; z, posterior accessory posttrite conule; 1–4, loph counted from the anterior end of the tooth; x, anterior or posterior cingulum. **(B)** Diagram of lingual view. **(C)** Specimen AS 92.699, right dP3, occlusal view. Anterior is to the left. **(D)** Buccal view.

mediate molars are trilophodont; M3/m3 have four massive loph(id)s and tendency for strong "heel" or incipient fifth loph(id). Transverse valleys are narrow. Pretrite accessory conules are large and high, forming trefoil wear figures at least to loph(id) 3. Half-loph(id)s are separated by median longitudinal sulcus and set in transversely straight line. Principal cusps are larger and slightly higher than mesoconelets. Paracone and hypocone of dP3 are obliquely connected via a low ridge. Distinguished from primitive ("*annectens*-grade") gomphotheres by a stronger expression of accessory conules and greater development of fourth loph(id)s in M3/m3; from *Gomphotherium steinheimensis* by greater height of pretrite conules, generally smaller size, greater relative crown height, and greater asymmetry of trefoil wear figures; from *G. browni* by longer mandibular rostrum, more ovoid cross-section of lower tusks, greater crown height, and thinner molar enamel (Tassy 1985).

Description. Two teeth are referred here to *Gomphotherium angustidens.* The best preserved of these is AS 92.699, an isolated right dP3 (fig. 10.3C,D). Specimen AS 92.699 has two lophs, both worn; a relatively large protruding anterior cingulum with a large parastyle; and a smaller posterior cingulum closely appressed to the last loph. In addition, a low, narrow cingular shelf borders the crown along its lingual and buccal sides. The tooth is small, measuring 43.2 mm in length and having a greatest width across the last loph of 31.5 mm. The first loph is composed of equal-sized protocone and paracone; the hypocone and metacone of the last loph are also similarly sized, and are more sharply demarcated by a narrow median sulcus. A broad V-shaped transverse valley separates the lophs. The crown is strongly notched into an ectoflexus and entoflexus at the buccal and lingual exits of the transverse valley, respectively, giving it a figure-eight shape. The enamel surface is marked by fine vertical striations.

An adult left M3, AS 92.603 (fig. 10.3A,B), is further evidence of a gomphothere in the lower member of the Sinap Formation. Although there is some damage to the pretrite cusps, the molar is unworn and retains its overall occlusal outline and original dimensions. It is relatively narrow, with a length of 163.4 mm and greatest width, at the second loph, of 70.4 mm. Crown height is 60.4 mm at the second loph, yielding a hypsodonty index of 85.8. Loph formula is x4x. The anterior cingulum is compressed against the first loph and highest at the lingual side. It continues as a low shelf on each side of the tooth until the first interloph. The posterior cingulum is formed of five large conelets and forms a substantial "heel." In lateral view, the lophs are bulbous and pyramidal in shape and are interspersed by deep V-shaped transverse valleys (fig. 10.3B). A deep, constricted median sulcus separates the lophs into pre- and posttrite halves. The posttrite half-lophs are each superficially subdivided into a buccally-situated main cusp and one to three mesoconelets. There are very low accessory conules posterior to posttrite half-lophs 1, 2, and 3, and posttrite half-loph 4 may also have had a posterior accessory conule. The pretrite half-lophs are more complex. Each has a lingually-situated main cusp accompanied by anterior and posterior accessory conules; these embrace more centrally located mesoconelets. The effect of this arrangement is to give the unworn pretrite half-lophs an inwardly facing, C-shaped configuration (fig. 10.3A). Fine horizontal wrinkling embellishes the enamel.

Remarks. Proboscidean specimens from the lower member of the Sinap Formation show a suite of features characteristic of *Gomphotherium.* In the M3 (AS 92.603), these include principal cusps and mesoconelets arrayed in nearly straight rows across the crown; principal cusps larger and slightly higher than the mesoconelets; separation of half-lophs by a narrow but distinct median longitudinal sulcus; presence of small, low posterior posttrite accessory conules; and arrangement of anterior and posterior accessory conules around the principal pretrite cusps such that with wear, they would form trefoil figures (see Osborn 1936; Tobien 1973b). Specimen AS 92.699 is typical of the gomphothere dP3: it is bilophodont with strong anterior and posterior cingula; its buccal cusps are respectively furnished with diminutive postparacrista and premetacrista; the hypocone is mesially offset from the metacone, and the paracone and hypocone contact via low, inflated ridges (see Tassy 1984, 1985, p. 354–355). The latter is a synapomorphy of *G. angustidens* + tetralophodont gomphotheres + elephantids and is not found in other elephantoids (Tassy 1985).

Subdivision of the anterior and posterior accessory conules into multiple conelets, as well as full expression of the last half-loph pair, and strong development of the posterior "heel" into an incipient fifth loph well separated from the fourth reveal AS 92.603 to be from an advanced form of *Gomphotherium.* Although a heterogeneous assemblage of gomphotheres from the early late Miocene has traditionally been accommodated in a single species, *Gomphotherium angustidens* (Osborn 1936; Tobien 1973b), more recently the alpha taxonomy and systematics of gomphotheres have been reexamined and extensively revised (Tassy 1985, 1989, 1996b). The most primitive gomphotheres are now gradistically united into a *Gomphotherium* "*annectens* group" by Tassy (1996b), diagnosed only by plesiomorphic traits, such as possession of simple bunolophodont molars with poorly expressed accessory conules and small fourth lophids in m3. Included in this group are the early Miocene species *G. annectens* from the Hiramaki Formation of Japan, *G. cooperi* from the Bugti beds, Pakistan, *G.* sp. from Mfwangano and Mwiti, Kenya, and *G. sylvaticum* from Europe (Cooper 1922; Coppens et al. 1978; Raza and Meyer 1984; Tassy 1986, 1989, 1996c; Antunes 1989; Ginsburg 1989). Clearly, the morphology of AS 92.603 is closer to that of molars from more derived species of *Gomphotherium,* such as *G. browni* of south Asia, *G. steinheimensis,* and *G. angustidens.*

These derived gomphothere species are primarily middle Miocene in age range. Tassy (1983 1985, 1989) appears to have subsumed the derived gomphotheres of the Siwalik Series (see Chakravarti 1957; Sarwar 1977) into a single species, *Gomphotherium browni,* which is probably limited to the Chinji Formation (MN 6–9) (Pilbeam et al. 1996). *Gomphotherium steinheimensis* is known from the localities of Steinheim (Klähn 1931), Massenhausen, Hisrchhorn, Markt Indersdorf, and Groβlappen (Göhlich 1998), faunally correlated with mammal faunal zone MN 7 (Tassy 1985; Göhlich 1998). In Europe, *G. angustidens* is apparently restricted to faunas dated to between the late Orleanian and early Vallesian (MN 5–9) (Gaziry 1976; Tassy 1985, 1989; Alpagut et al. 1989). It is generally thought of as a typical Astaracian taxon (Mazo 1996). It is also documented at the early Miocene (~18–17 Ma) site of Wadi Moghara, Egypt, and possibly from the slightly younger site of Ad Dabtiyah, Saudi Arabia (Sanders and Miller, 2002; see also Fourtau 1920; Gentry 1987). Differences among these species in incisor and mandibular morphology are more easily discerned than the subtle distinctions in the occlusal construction of their cheek teeth (Tassy 1985). Although this makes specific identification of isolated molars difficult, the closest resemblances of the Sinap gomphothere teeth are with early–middle Astaracian specimens of *G. angustidens* (see Tassy 1985; Mazo 1996). Compared with the large sample of gomphothere M3 specimens from En Péjouan, France (MN 7) (Tassy 1985), AS 92.603 is comparable in occlusal morphology but proportionally narrower (fig. 10.4). The bimodal distribution of M3 dimensions of individuals from En Péjouan (fig. 10.4) suggests that *G. angustidens* is sexually dimorphic and that AS 92.603 is from a male. Although AS 92.603 is similar in size to M3 specimens from Paşalar, Turkey (lowermost MN 6; Steininger et al. 1989, 1996) (fig. 10.4), it is morphologically more advanced in having multiple anterior and posterior accessory pretrite conules and a larger, more complex molar "heel." These observations are consistent with an age of ~15 Ma (Kappelman et al., 1996, chapter 2, this volume) and assignment of faunas from Sinap Locs. 24 and 24A to MN 6 (Gürbüz 1981).

Choerolophodon anatolicus *(Ozansoy, 1965)*

Taxonomy

Subfamily Choerolophodontinae Gaziry, 1976
Genus *Choerolophodon* Schlesinger, 1917
Choerolophodon anatolicus (Ozansoy, 1965)

Referred Specimens. Sinap Formation, middle member, Loc. 12: AS 93.844, palatal fragment with dP2–3 sin and dex (fig. 10.5); AS 95.551, dentary fragment sin with m3 (see fig. 10.8A,B); AS 95.569, partial dentary dex with dp4 (fig. 10.6).

Loc. 45: AS 92.410, proximal phalanx sin?, hallux.

Loc. 49: AS 91.735, calcaneum dex; AS 92.136, astragalus; AS 92.144, astragalus; AS 94.494, navicular sin; AS

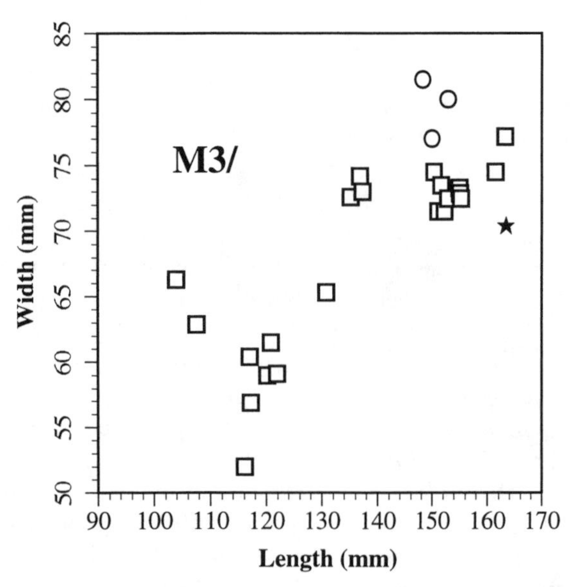

Figure 10.4. Bivariate plot of crown length versus width of M3 in selected samples of *Gomphotherium angustidens* and Sinap specimen AS 92.603 (*G. angustidens*). Comparative dimensions are from Gaziry (1976) and Tassy (1985). Mammal faunal units and zonations follow Tassy (1985), Steininger et al. (1989, 1996), and Kappelman et al. (1996). Symbols: ★, AS 92.603, Sinap Formation, Turkey (Astaracian, lower MN6); □, En Péjouan, France (Astaracian, MN7); ○, Paşalar, Turkey (Astaracian, lowermost MN6).

Figure 10.5. Specimen AS 93.844, left and right dP2–3, *Choerolophodon anatolicus*. Occlusal view. Anterior is to the left.

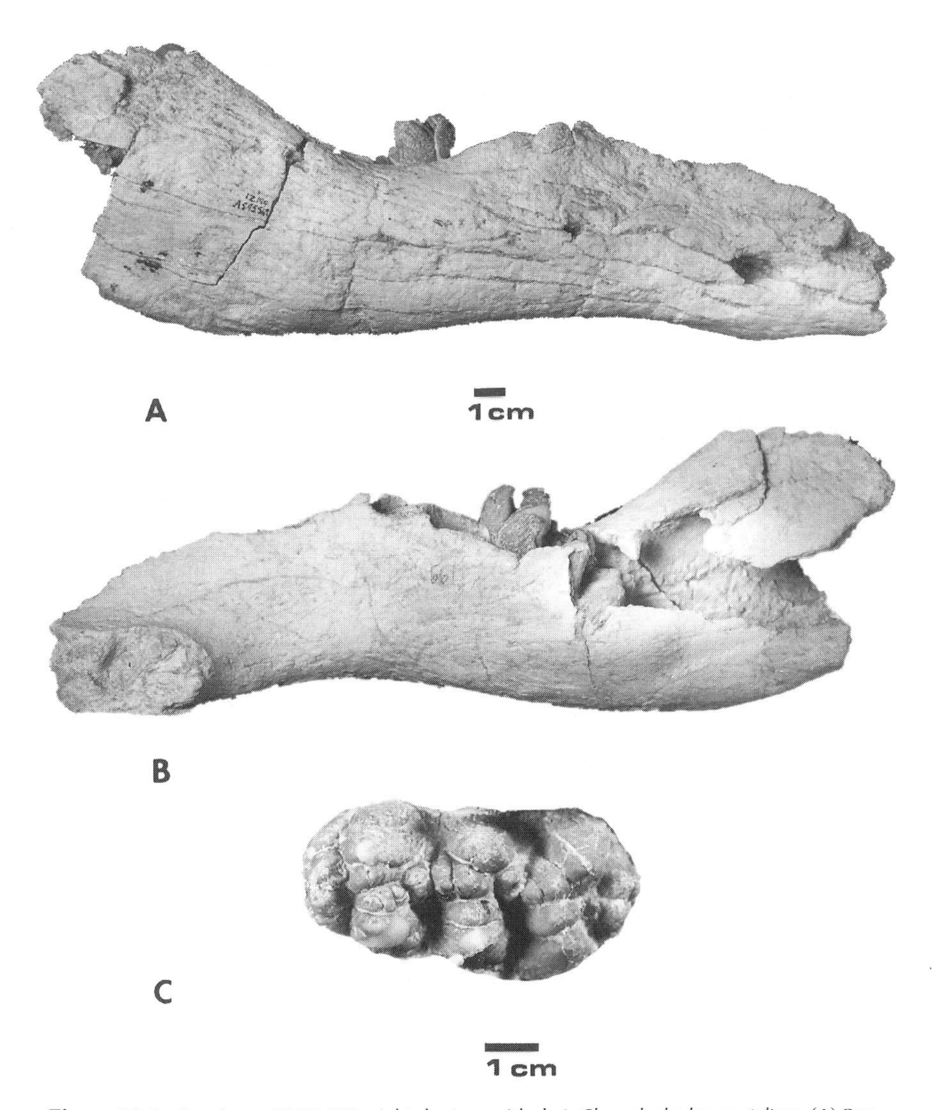

Figure 10.6. Specimen AS 95.569, right dentary with dp4, *Choerolophodon anatolicus*. (**A**) Buccal view. Anterior is to the right. (**B**) Lingual view. Anterior is to the left. (**C**) Occlusal view, dp4. Anterior is to the left.

94.581, associated postcrania including MC II–V sin, proximal phalanx dex (ray II?, manus), middle phalanx sin (ray II?, manus), right proximal and middle phalanges (ray III, manus), right proximal and middle phalanges (ray IV, manus), sesamoid; AS 94.805, calcaneal sin fragment; AS 94.1350, MC III dex; AS 94.1352, unciform sin; AS 94.1359, MC IV dex; AS 94.1361, astragalus sin; AS 94.1375, lunate sin; AS 94.1376, magnum sin; AS 94.1379, manual cuneiform sin; AS 94.1392, trapezoid dex; AS 94.1394, pisiform? dex?; AS 94.1395, magnum dex; AS 94.1420, astragalus dex.

Loc. 51: AS 90.141, patella dex?; AS 90.143, manual cuneiform sin; AS 90.145, lunate sin; AS 90.146, unciform dex; AS 90. 147, calcaneum sin; AS 90.148, MC IV dex; AS 90.149, MC II dex; AS 90.150, MC III dex; AS 90.151, middle phalanx.

Loc. 83: AS 92.605, dentary sin with m2–3, associated m2–3 dex (figs. 10.7, 10.8C).

Diagnosis. Large choerolophodont species. Long, tuskless mandibular symphysis; there is a large, deep symphyseal "gutter;" accessory conules and mesoconelets are high; molar half-loph(id)s posterior to loph(id) 1 are arranged into anteriorly pointing chevrons; ptychodonty and choerodonty are poorly expressed or absent. Distinguished from *Choerolophodon pentelici* by greater degree of downward symphyseal deflection in mandibles, presence of a large retromolar gap between m3 and the mandibular ramus, posterior angulation of the ramus on the corpus, lack of a second entoflexus and third loph in dP3, and generally smaller size of deciduous premolars; from *C. corrugatus* by lesser degree of symphyseal angulation, lesser height and greater posterior angulation of ramus, presence of retromolar gap, smaller size of deciduous premolars, and tendency for lesser size and complexity of molar crowns; from *Afrochoerodon ngorora* by lesser symphyseal angulation (Tobien 1980; Tassy 1985, 1994; Tassy et al. 1989).

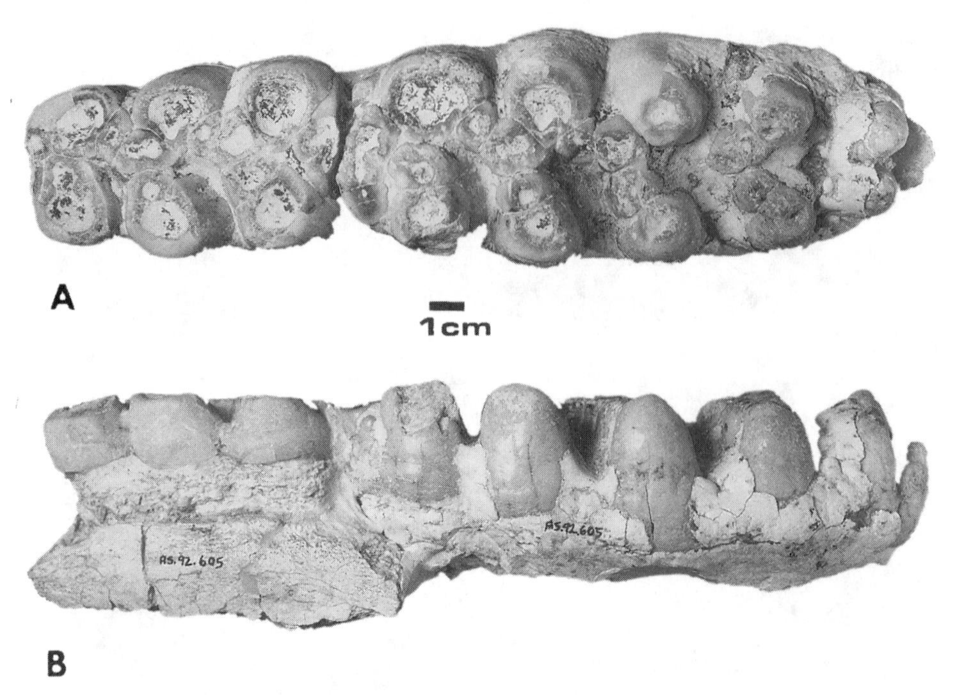

Figure 10.7. Specimen AS 92.605, right m2–3, *Choerolophodon anatolicus*. **(A)** Occlusal view. Anterior is to the left. **(B)** Lingual view. Anterior is to the left.

Description. The craniodental sample of *Choerolophodon anatolicus* from the Sinap Formation is small but preserves sufficient morphological detail for precise taxonomic identification. Among these remains is a partial palate, AS 93.844, with left and right dP2–3 (fig. 10.5). The palatal fragment measures 94.9 mm in length and has a breadth of 59.3 mm between the dP3 specimens. Breakage has occurred anteriorly and posteriorly, and bone is missing near the midline. In cross section, the palate is shallow and concave. Enamel has spalled off of the posterior edge of the right dP2, and the right dP3 is missing its posterobuccal quadrant.

The left dP2 has a length of 24.5 mm, is widest (19.8 mm) across its posterior half, and has a height of +12.2 mm at the paracone. The right dP2 is 23.4 mm long and 21.1 mm wide. Each tooth has a relatively massive paracone with a sharp postparacrista that runs posterobuccally toward a low, diminutive metacone. The protocone is also small and low and is worn flat. It is set widely apart from the metacone, to the lingual side, giving the dP2 a subtriangular occlusal shape. A narrow anterior cingulum continues along the lingual side of the tooth and ends at the protocone.

The left dP3 measures 37.6 mm in length, 32.6 mm in width across its posterior loph and 25.8 mm across the anterior loph and has a height of +20.6 mm. This tooth exhibits a low anterior cingulum, a protoloph and metaloph separated by a narrow transverse valley, and a prominent posterior cingulum that is nearly as high as the posterior loph. Buccally, a strong ectoflexus further demarcates the two lophs. Consequently, occlusal shape is that of an unsymmetrical figure eight. A large accessory tubercle occludes the lingual opening of the transverse valley. The protoloph is moderately worn and although transversely continuous, its wear figure reveals that it is formed from a single pretrite cusp (protocone) and posttrite half-loph subdivided into a mesoconelet and outer cusp (paracone). The metaloph is also worn and is transversely disjunct, as the posttrite mesoconelet and outer cusp (metacone) are posteriorly offset from the pretrite half-loph, which is composed of a mesoconelet and outer cusp (hypocone). Although the posterior cingulum may be thought of as an incipient third loph, it is not isolated from the metaloph by a transverse valley or ectoflexus. Both dP2 and dP3 have enamel ornamented throughout by fine vertical striations.

Juvenile *Choerolophodon anatolicus* is also represented in the sample by a right dentary with an emerging dp4, AS 95.569 (fig. 10.6). Alveoli for dp3 are preserved as well. The specimen is 345 mm long and is 57.5 mm high and 48.3 mm wide at dp4. The symphyseal articulation measures 81.6 mm in length and 28.3 mm in height. When articulated with its antimere, the effect of inflection of the symphysis would have been to form a median "gutter" at the front of the lower jaw. The ramus is angled posteriorly at ~45° relative to the long axis of the corpus, and the dorsal ridge of the symphysis is reflected moderately downward (~25° relative to the alveolar planum) (fig. 10.6A,B). Ventrally, a shallow concavity is formed where the corpus meets the symphyseal portion of the dentary, and there is very slight downward angulation of the ventral surface of the symphysis relative to the long axis of the ventral surface of the corpus. The superior margin of the symphyseal segment narrows to a ridge.

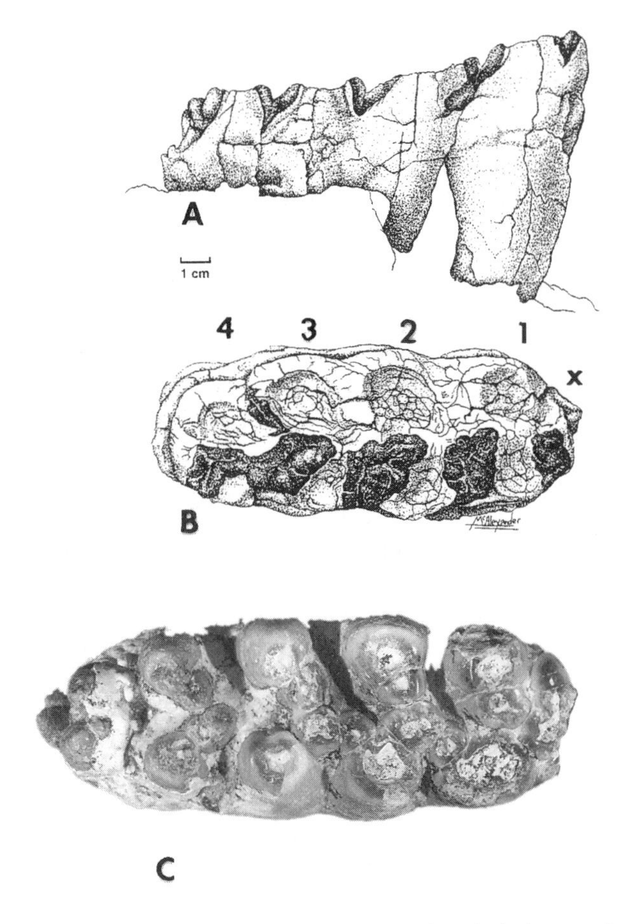

Figure 10.8. Specimen AS 95.551, left m3, *Choerolophodon anatolicus.* (**A**) Diagram of lingual view. Anterior is to the right. (**B**) Diagram of occlusal view. Abbreviations: 1–4, lophid counted from the anterior end of the tooth; x, anterior or posterior cingulum. (**C**) Specimen AS 92.605, right m3, to the same scale for size comparison.

The crown of the dp4 is complete and is comprised of prominent anterior and posterior cingula and three lophids (fig. 10.6C). Dimensions of the tooth are L = 68.3 mm, W = 35.6 mm, and H = 23.2 mm. The anterior cingulum is formed of two cusps and a series of smaller conelets and is fused with the anterior pretrite (accessory) conule of the first lophid. Each lophid is divided by a median sulcus into pretrite and posttrite half-lophids. In turn, the half-lophids are each subdivided into a mesoconelet and outer cone. On the pretrite side, the half-lophs are accompanied by anterior and posterior accessory conules, which are closely appressed to the mesoconelets. Also, a small posterior accessory conule is fused to the posttrite mesoconelet of the first lophid. In the second and especially the third lophid, the mesoconelets are set anterior to the outer cones. As a result, the occlusal outline of the third lophid is that of an anteriorly-pointing "V." The distal cingulum is comprised of three small conelets. As in AS 93.844, the enamel is marked by numerous fine vertical striae.

The most informative dental specimen of an adult is AS 92.605, a left dentary with m2–3 and associated right m2–3 (fig. 10.7). The corpus is damaged anteriorly and is missing its symphysis, and the ramus is incomplete. Greatest thickness (102 mm) and height (157 mm) of the corpus are at the mesial end of m3. Length of the dentary is +707 mm. A large mandibular foramen opens anteriorly 82 mm below m2. There is an extensive retromolar gap between the distal end of m3 and the anterior margin of the ramus. The ramus is angled posteriorly ~45° relative to the long axis of the corpus. Anterior to m2, the superior margin of the dentary is thin and curves slightly outward.

The molars are well preserved and nearly complete. The left m2 has three lophids and a diminutive posterior cingulum, and measures 100.4 mm in length and 61.4 in width (at lophid 3). Corresponding dimensions for the right antimere are +97.2 mm and 60.0 mm. These molars are heavily worn. Enamel is thick (5.2 mm), and traces of cementum are present in the interlophids. The pretrite and posttrite sides of the second and third lophids are each composed of a large main outer cusp and smaller mesoconelet. The mesoconelets are set anterior to the main cusps, particularly on the pretrite side, where they abut closely against anterior pretrite accessory conules. This arrangement forms the lophids into anteriorly-pointing chevrons (fig. 10.7A). Pretrite half-lophids 1 and 2 retain large posterior accessory conules. The posterior cingulum is made up of two moderate sized conelets that are tightly fit between the third lophid and m3.

Each m3 has a well-developed anterior cingulum, four lophids, and a prominent "heel" comprised of an incipient fifth lophid and a low posterior cingulum. Lophid formula can be expressed as x5x. The wear gradient of this tooth is steep, so that the first lophid is heavily worn and truncated but the fifth lophid is unworn. The crown is narrow (W = 73.0 mm, right side; 72.9 mm, left side) relative to length (L = 186.9 mm right side; 191.9 left side). Cementum coats the walls of the interlophids and nearly covers the fifth lophid (fig. 10.7B). The composition of the main lophids is the same as in m2. Lophids 3 and 4 are especially strongly angled into anteriorly pointing chevrons (fig. 10.7A). The anterior cingulum has at least four conelets and the posterior cingulum is made up of two conelets. The enamel exhibits hints of fine horizontal wrinkling and is very thick (~6.0 mm).

The other adult choerolophodont molar in the sample is AS 95.551, a heavily worn left m3 with much of its enamel spalled off, in a dentary fragment (fig. 10.8A,B). Damage to the dentary reveals that the first lophid is supported by a deep, posteriorly curved root, and that the remaining lophids are underlain by an immense compound root. With a length of 141.0 mm and greatest width (at the second lophid) of 57.8 mm, it is considerably smaller than the m3 specimens in AS 92.605 (fig. 10.8B,C). Despite its poor state of preservation, AS 95.551 retains the outline of a large anterior cingulum, four lophids, and a small posterior cingulum. This constitutes a lophid formula of x4x. It is impossible to ascertain the number of conelets in each lophid, but the pretrite and posttrite half-lophids (in lophids 2–4) are clearly angled on one another to form anteriorly pointing "V"s (fig. 10.8B).

A substantial number of choerolophodont postcrania were also recovered from the middle member of the Sinap Formation. Locality 49 in particular yielded a wealth of choerolophodont postcranial fossils, in sharp contrast to the lone deinothere specimen found there. Although their description and comparative analysis are beyond the scope of this chapter, the manus and pes elements have been thoroughly studied and are allocated to *Choerolophodon* on the basis of autapomorphies (e.g., presence of a concave facet on the lateral face of the trapezoid that articulates with a tuberosity on the anterodorsal aspect of the magnum [see Tassy 1986, 1994]) and their conformation with one another. They are robust in construction and very large, similar in size to corresponding elements in modern elephants. The metacarpals are relatively elongate, with lengths about three times greater than mid-shaft breadth, and the calcaneal tuber is also very posteriorly extended. A single astragalus from Loc. 24 is assigned to *Gomphotherium* on general morphological resemblance (see Tobien 1962), but is also not discussed further. Other postcranial elements from the Sinap Formation require additional study and therefore have not been formally allocated to taxa; nonetheless, it is likely that the majority of the samples, especially from Locs. 49 and 51, will prove to belong to *Choerolophodon*.

Remarks. Choerolophodont molars have been characterized as having numerous irregularly arranged conlets (choerodonty) and heavily corrugated enamel (ptychodonty) (Osborn 1936; Tobien 1973a,b). In addition, chevroning of half-loph(id)s appears to be an autapomorphic feature of choerolophodont cheek teeth (Tobien 1973a,b; Tassy et al. 1989; Tassy 1994). As the former two features are variable intra- and interspecifically and are present only in the most derived members of the Choerolophodontinae, the latter trait is the most diagnostic of subfamily members. The development of half-lophids into anteriorly pointed chevrons unambiguously establishes the choerolophodont identity of a number of molar specimens in the new collections from the middle member of the Sinap Formation, even though ptychodonty and choerodonty are poorly expressed or absent. Although choerolophodonts were previously recovered from the middle member of the Sinap Formation (Ozansoy 1955, 1957, 1965; Sen 1990), the new material includes the first described adult choerolophodont molars from the formation and serves to clarify the taxonomic affinity of the sample and to refine the biostratigraphy of Sinap localities.

The first choerolophodont material collected from the Sinap Formation (near Yassiören) was assigned to *Trilophodon (Choerolophodon) anatolicus* by Ozansoy (1965), but was later synonymized under *C. pentelici* (Gaziry 1976). This synonymy was subsequently accepted by Tassy (1985, 1994), Tassy et al. (1989), and Sen (1990). Tassy et al. (1989) also recognized subspecific distinctions within *C. pentelici* and placed the Yassiören material into "*C. p. lydiensis*," along with fossils from Esme Akçaköy, Turkey, to differentiate these specimens from more derived craniodental fossils allocated to "*C. p. pentelici*." Tassy (1994) indicated that choerolophodont fossils from the lower level (lower Turolian) of Kemiklitepe, Turkey, are also distinct from those in "*C. p. pentelici*."

Choerolophodon pentelici is common in late Miocene faunas of the eastern Mediterranean (Tassy 1994), and ranged widely across Eurasia, from Yugoslavia to Iran (Bakalov and Nikolov 1963; Tobien 1973a; Tassy 1985, 1986; Bernor 1986). The species has an especially rich record in Turkey, beyond its occurrence in the Sinap Formation (Senyürek 1952; Viret and Yalçinlar 1952; Yalçinlar 1952; Viret 1953; Ozansoy 1957, 1965; Sickenberg and Tobien 1971; Gaziry 1976; Tassy 1985, 1994; Tassy et al. 1989). In Tassy's (1985, 1994) view, *C. pentelici* is a Vallesian-Turolian species (but see below), documented from faunas correlated with mammal zones MN 9–13 (see also Steininger et al. 1989, 1996; Sen 1991). Older records of choerlophodonts include the middle Miocene species *Afrochoerodon chioticus* from Chios, Greece (Tobien 1980; Tassy 1989; correlated with MN 5, Steininger et al. 1996), early–middle Miocene *A. kisumuensis* from Maboko and Cheparawa, Kenya, and Wadi Moghara, Egypt (Tassy 1986, 1989; Pickford 2001; Sanders and Miller 2002), and the early Miocene species *A. palaeindicus* from the Bugti beds and possibly the Kamlial Formation, Siwalik Series, Pakistan (Tassy 1983, 1985, 1989). *Choerolophodon pentelici* is broadly contemporaneous with *C. corrugatus* from south Asia, which is documented from the Chinji through Dhok Pathan Formations (Sarwar 1977; Tassy 1983, 1985), and *C. ngorora* from the Tugen Hills and Nakali, Kenya (Hill et al. 1985; Tassy 1985, 1986).

The new Sinap choerolophodont fossils were initially placed in *Choerolophodon* sp. (Kappelman et al. 1996), but my findings presented here show there is little reason for such caution. The morphological affinity of the Sinap choerolophodonts is unequivocally with "*C. p. lydiensis.*" As typified by fossil specimens from Esme Akçaköy and lower horizon KTD at Kemiklitepe, "*C. p. lydiensis*" and the Sinap material are morphologically contrasted with "*C. p. pentelici*" in the following features: (1) slight downward angulation of the symphyseal segment of the dentary; (2) low inclination of the ramus; (3) presence of an extensive retromolar gap between the distal end of m3 and the anterior margin of the ramus; and (4) small size of deciduous premolars.

Tassy (1985, 1994) and Tassy et al. (1989) consider the angulation of the symphyseal rostrum on the mandibular corpus to have a strong taxonomic valence. Specimens of "*C. p. lydiensis*" exhibit moderate downward angulation of the symphysis. Specimen AS 95.569 has an angle of 25°, EA-1 from Esme Akçaköy has an angle of 30°, and the symphysis of KTD 66 from Kemiklitepe is reflected downward at 30°. In this measurement, juvenile specimens KB+47 B and C (23° and 28°, respectively) from Kayadibi, Turkey (see Gaziry 1976) also resemble AS 95.569. These are most similar to the condition observed in *Afrochoerodon chioticus* (see Tobien 1980), and contrast with the lack of angulation in mandibles of "*C. p. pentelici*" (e.g., juvenile specimen M 3956 from Maragheh, Iran). Furthermore, the symphyseal reflection in "*C. p. lydiensis*" differs from the strong angulation observed in *C. corrugatus* (e.g., specimen AM 19487 [Tassy 1985]) and *A. ngorora* (e.g., subadult specimen KNM-FT 2788). Although symphyseal angulation may have increased with age, as suggested by Tassy et al. (1989), even juvenile dentaries of *C. corrugatus* show much stronger flexion of the symphysis than those of "*C. p. lydiensis*" and "*C. p. pentelici*" (see Tassy 1985). In this study, the symphyseal flexion was calculated as the angulation of the *dorsal* margin of the symphysis relative to the long axis of the alveolar planum. This is important to note, because in contrast to the condition in *C. corrugatus*, the *ventral* margin of the symphysis in "*C. p. lydiensis*" shows little or virtually no downward reflection and is similar in outline to that in "*C. p. pentelici*." Although Tassy (1985) and Tassy et al. (1989) have argued that there is no symphyseal angulation in the juvenile choerolophodont dentary from Yassiören, the ventral outline of the symphysis is nearly indistinguishable from that of AS 95.569, and the dorsal margin of the symphysis is too broken to permit comparative measurement of angulation (see Tassy 1985). Thus, moderate angulation of the symphysis in the Yassiören dentary and homogeneity of the "*C. p. lydiensis*" sample cannot be ruled out.

The Sinap molars also differ in general from those of "*Choerolophodon p. pentelici*" in having simpler crowns with weaker expression of choerodonty and ptychodonty. In this regard, the Sinap specimens are more reminiscent of molars of *Afrochoerodon ngorora* and the older species *A. chioticus* and *A. kisumuensis*. Mandibular m3 specimens of "*C. p. pentelici*" and *C. corrugatus* tend to be larger than those of *A. ngorora* and the more ancient choerolophodont taxa, but there is a large amount of proportional overlap, possibly due to strong sexual dimorphism (fig. 10.9). The taxonomic limitations of metrical analysis for this tooth are further revealed by the huge size disparity between Sinap m3 specimens AS 95.551 and AS 92.605, which are probably from female and male individuals, respectively. Specimen AS 95.551 is the smallest choerolophodont m3 known; an apparently smaller m3 of *A. kisumuensis* (see Tassy 1985, fig. 261) is more likely an M3. Specimen AS 92.605, however, fits comfortably within the length-width scatter of points for "*C. p. pentelici*" (fig. 10.9). Size appears to be less important than occlusal morphology for taxonomic discrimination of choerolophodont molars.

Comparative morphometric analysis of the deciduous Sinap specimens reinforces the division of *Choerolophodon pentelici* into two distinct groups. The dP3 specimens of the new Sinap palatal fragment are distinguished from those of "*C. p. pentelici*," as from upper horizons KTA and KTB at Kemiklitepe, by the absence of a second entoflexus and lack of a third loph (Tassy 1983, 1994). Conversely, the new deciduous specimens from the Sinap Formation are morphologically identical with upper and lower deciduous premolars previously collected from the Sinap Formation near Yassiören (Ozansoy 1965) and from the locality of Gökdere (Senyürek 1952). Deciduous premolars attibuted to *C. pentelici* from MN 9–lower MN 11 (Vallesian–early Turolian) localities are also generally smaller than those from upper

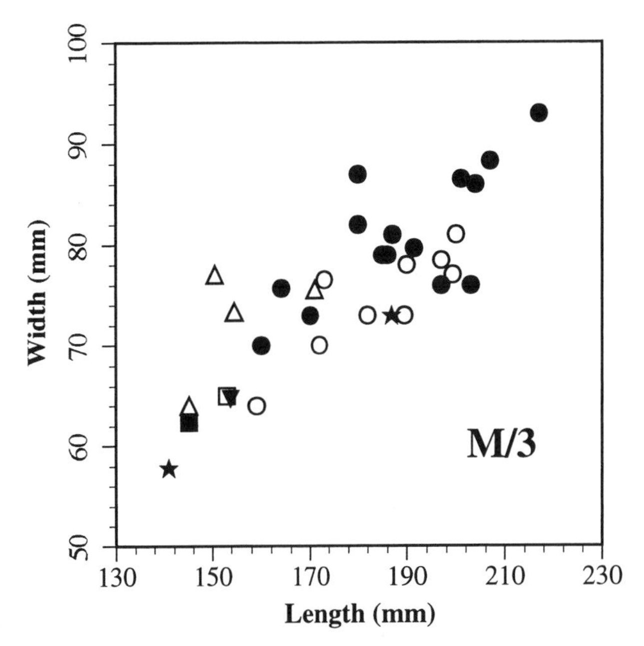

Figure 10.9. Bivariate plot of crown length versus width of m3 in *Choerolophodon* and *Afrochoerodon*. Comparative dimensions are from Gaziry (1976), Tassy (1983, 1985), and Sanders and Miller (2002). Mammal faunal units and zonations follow Steininger et al. (1989, 1996). Chronostratigraphic information is from Bakalov and Nikolov (1963), Tassy (1983, 1985, 1986), Raza and Meyer (1984), Hill et al. (1985), Barry and Flynn (1989), and Miller and Simons (1996). Symbols: ★, new specimens, Sinap Formation, Turkey (Vallesian, late MN 10); ■, *C. palaeindicus*, Dera Bugti, Pakistan (Bugti beds, ?earliest Miocene); ▼, *A. kisumuensis*, Maboko, Kenya (middle Miocene); △, *A. ngorora*, Tugen Hills, Kenya (Ngorora Formation, late Miocene); ●, *C. corrugatus*, Siwaliks, Pakistan (Dhok Pathan Formation, late Miocene); ○, *C. pentelici* (="*C. p. pentelici*"), Samos, Greece (Turolian, MN 12), Sandanski, Bulgaria (?early Pliocene, ?late Miocene), Djebel Hamrin, Iraq (?terminal Vallesian); □, *C. anatolicus*, Esme Akçaköy, Turkey (early Vallesian, MN 9).

MN 11–MN 13 Turolian localities, as well as those of *C. corrugatus* (fig. 10.10; Gaziry 1976, table 1).

The new choerolophodont specimens from the Sinap Formation confirm the morphological distinctions previously noted for older and younger samples of *Choerolophodon pentelici* (see Tassy et al. 1989). Contra Gaziry (1976), Tassy (1985, 1994), Tassy et al. (1989), and Sen (1990), however, the demarcation of these samples should be at the species level. "*Choerolophodon pentelici lydiensis*" is distinguished from "*C. p. pentelici*" by morphometric aspects of the deciduous premolars and a series of mandibular and adult dental traits. These are contrasted at least to the same degree as criteria used to discriminate other choerolophodont species. For this reason, *C. anatolicus* should be recognized as a species in its own right and as encompassing "*C. p. lydiensis*." The taxon includes the choerolophodont material from Esme Akçaköy, Kayadibi; level KTD at Kemiklitepe; the middle member of the Sinap Formation, Gökdere; and possibly Garkin and Çorak-Yerler. The dating of this sample suggests that *C. anatolicus* was replaced by *C. pentelici* sensu stricto at the end of the early

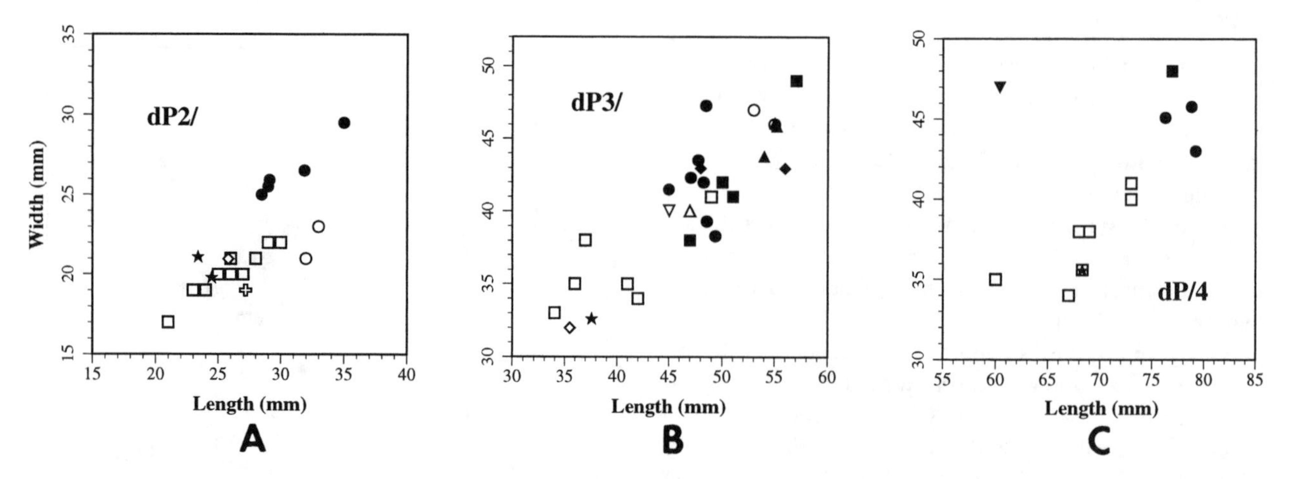

Figure 10.10. Bivariate plots of crown length versus width in deciduous teeth of *Choerolophodon* and *Afrochoerodon*. Comparative dimensions are from Senyürek (1952), Ozansoy (1965), Gaziry (1976), and Tassy (1983, 1994). Mammal faunal units and zonations follow Steininger et al. (1989, 1996) and Sen (1990). Symbols: ★, new specimens, Sinap Formation, Turkey (Vallesian, late MN 10); ◇, *C. anatolicus*, Yassiören, Turkey (=Sinap Formation, early Vallesian, MN 9); □, *C. anatolicus*, Kayadibi, Turkey (Turolian, earliest MN 11); ▽, ?*C. anatolicus*, Garkin, Turkey (Turolian, late MN 11); △, *C. anatolicus*, Gökdere, Turkey (?Turolian); ○, ?*C. anatolicus*, Çorak-Yerler, Turkey (Vallesian, MN 10); ▲, *C. pentelici*, Kemiklitepe KTA + KTB (Turolian, MN12); ■, *C. pentelici*, Maragheh, Iran (Turolian, MN 12); ◆, *C. pentelici*, Pikermi, Greece (Turolian, MN 11/12); ▼, *C. pentelici*, Samos, Greece (Turolian, MN 12); ✚, *A. ngorora*, Ft. Ternan, Kenya (middle Miocene); ●, *C. corrugatus*, Siwaliks, Pakistan (Dhok Pathan Formation, late Miocene). (**A**) DP2 dimensions. (**B**) DP3 dimensions. (**C**) Dp4 dimensions.

Turolian (late MN11), ~7.5 Ma (see Lindsay 1997; Sen 1997).

Discussion

Biochronologic Implications

An important aspect of the proboscidean remains from the Sinap Formation is that they are tied to the absolute time scale. The secure dating of these fossils makes them potentially valuable for helping to establish the timing of biogeographic events and faunal turnover and for biochronologic correlation of more poorly dated faunas elsewhere. Sampling of long sedimentary sections at Kavak Dere, Sinap Tepe, and Delikayinçak Tepe has yielded detailed paleomagnetic reversal stratigraphies for the formation (Sen 1990, 1996; Kappelman et al., 1996, chapter 2, this volume). The combined results of biochronologic correlation of the Sinap mammal faunas (particularly of the small mammals) and of paleomagnetic sampling indicate that the lower member of the formation predates the Vallesian and is probably Astaracian in age, whereas the middle member extends from the lower Vallesian into the Turolian (Sen 1990, 1996; Kappelman et al. 1996, chapter 2, this volume). In addition, abundant volcanic sediments in the area hold promise for precise radiometric dating of the sequence (Kappelman et al. 1996). Although the proboscidean fossils have previously not contributed to the development of the temporal framework of the Sinap Formation, it is a useful exercise to see if they fit with the proposed chronostratigraphy.

Gomphothere specimens from Locs. 24 and 24A are allocated to *Gomphotherium angustidens* and are dated to ~15 Ma, as these localities are stratigraphically close to the boundary between the Sinap Formation and the subjacent Pazar Formation (Kappelman et al. 1996). Potassium/argon dating of a basalt flow underlying the Sinap Formation has produced a maximum age for the fossil deposits of 15.2 Ma (Kappelman et al. 1996). This is supported by assignment of the fauna from these localities to MN 6 (Gürbüz 1981). *Gomphotherium angustidens* as presently constituted is morphologically heterogeneous (Tobien 1973a) and occurs at localities throughout Europe and in Turkey (as well as north Africa and possibly Arabia) that represent a long time span (MN 5–9, late Orleanian–early Vallesian) (Tobien 1973a; Gaziry 1976; Tassy 1985, 1989; Alpagut et al. 1989). The species is certainly in need of revision, and many attributions are problematic (e.g., Mazo's [1996] assignment of gomphotheres from MN 4 localities in Spain to *G. angustidens*). Nonetheless, the new gomphothere M3 from Sinap is more advanced in occlusal morphology than those of Paşalar and is most similar to molars of early–middle Astaracian examples of the species (see Tassy 1985; Mazo 1996). Thus, the new gomphothere specimens from Sinap are biochronologically consistent with current estimations of their temporal context. Additionally, they indicate that correlation of Paşalar to basal MN 6 (Steininger et al. 1989, 1996; Bernor and Tobien 1990) is correct.

The co-occurrence of *Choerolophodon anatolicus* and *Deinotherium giganteum* supports inferences that the middle member of the Sinap Formation is of early Vallesian–early Turolian age. Localities yielding specimens of *C. anatolicus* (Locs. 12, 45, 49, 51, and 83) are dated by paleomagnetic reversal stratigraphy to an interval ~10.5–9.0 Ma (see Kappelman et al. 1996, chapter 2, this volume; Sen 1996), corresponding with mammal faunal zones MN 9–10. These dates are biochronologically consistent with the presence of *C. anatolicus* at other localities in Turkey that are correlated with Vallesian–early Turolian faunas (see Senyürek 1952; Ozansoy 1965; Gaziry 1976; Tassy 1985). The dating of the new choerolophodont material from Sinap reinforces the supposition of a temporal succession from (earlier) *C. anatolicus* to (later) *C. pentelici* in the circum-Mediterranean area at the close of the early Turolian.

Deinotherium giganteum is widely distributed among fossil sites throughout Europe and the circum-Mediterranean, and apparently survived from the end of the middle Miocene until the middle Pliocene (Bergounioux and Crouzel 1962; Sarwar 1977). Although the temporal range of this species implies limited usefulness for biochronologic dating, the morphometric analysis of the present study shows an intraspecific trend for increasing tooth size over time, suggesting correlation of the specimen from Loc. 49 to lowermost MN 11. This assignment accords reasonably well with the calculation of ~9.0 Ma for Loc. 49 (see Kappelman et al. 1996, chapter 2, this volume).

Paleoecologic Implications

The type of bunolophodont occlusal morphology observed in gomphothere M3 AS 92.603 has been functionally associated with grinding-shearing mastication and browsing (Maglio 1973). Conversely, in a number of proboscidean lineages (and particularly among the Elephantidae), reduction of the grinding component of mastication has been adaptively linked with greater emphasis on grazing (Maglio 1972, 1973). Despite the narrowness of the crown in AS 92.603 relative to length, the height:width ratio (or hypsodonty index) is low. In living ungulates, brachyodonty is linked with preferences for feeding on dicotyledonous plants above ground level (Janis 1986). Furthermore, bunolophodonty has been categorized as an adaptation in herbivorous mammals for folivory or folivory/frugivory (Janis 1995). These dietary preferences are most likely to be satisfied in woodland or forested areas, suggesting that such habitats were proximal to the depositional environment of gomphothere Locs. 24 and 24A.

Similarly, the recovery of a deinothere individual from Loc. 49 indicates the presence of wooded or bush conditions. Deinotheres are considered to have been "shearing browsers" adapted for plants located above ground level (Harris 1975, p. 361). Their masticatory function is reconstructed as similar to that in extant tapirs, with the anterior teeth acting to crush food, and occlusion of second and third molars having a strong vertical shearing component but little lateral motion. In this regard, their chewing action differed from the lateral grinding of gomphotheres and the horizontal shearing of elephants (Harris 1975). The faintness of wear scratches on deinothere molars suggests a diet of nongritty, soft vegetation, and use of the unusual downturned lower incisors for stripping bark or other vegetation is indicated by anteromedial wear facets (Harris 1975). However, *Deinotherium giganteum* exhibits a distinctly more digitigrade manus and longer metacarpals than early–middle Miocene *Prodeinotherium* (Tobien 1962; Harris 1978, 1983), possibly reflecting increased cursorial abilities in response to the spread of savannas in Europe during the late Miocene (Harris 1983). Contrasting locomotor and masticatory adaptations in *D. giganteum* suggest the need to range widely to find sufficient browse, perhaps in gallery forest areas (see Harris 1983).

The paleobiology and paleoecology of choerolophodonts from the middle member of the Sinap Formation are more difficult to reconstruct. Their molars are low crowned and bunodont and exhibit a trefoil arrangement of cusps in the pretrite half-loph(id)s. Similar to the condition in gomphotheres, these features have been correlated with a grinding-shearing mastication and browsing (Maglio 1973; Janis 1986, 1995). However, the absence of lower tusks in choerolophodonts could be functionally linked, as in elephants, with greater emphasis on shearing mastication and increased importance of acquiring food from the ground in open country settings (Maglio 1972, 1973). The chevroning of loph(id)s in choerolophodonts would have produced a "double-scissors" shearing action during horizontal fore-aft jaw movements; this chewing mechanism likely would have been less effective, had it been burdened with the weight of lower tusks. Absence of lower tusks and progressive changes in molar occlusal morphology in the proboscidean genus *Anancus* have also been associated with environmental shifts toward more open country habitats (see Mebrate and Kalb 1985; Sanders 1997). It seems logical that the loss of structures like lower tusks, with apparent utility in feeding, should be balanced by some gained advantage in transformed chewing mechanisms for new types of forage.

Interpretation of adaptation to open country settings is supported by cursorial features in *Choerolophodon anatolicus,* including an extended calcaneal heel process and elongate metapodials. Whether this included grazing is uncertain. Although rapid expansion in the Old World of C_4 biomass, including warm-season grasses, postdated the known temporal range of *C. anatolicus* (see Cerling et al. 1993; Ramstein et al. 1997), the possibility of C_3 grasslands (associated with cool growing seasons; Cerling et al. 1993) in the Sinap region during the Vallesian cannot be ruled out. Examination of Miocene rodent faunas from western Europe suggest a climatic shift toward drier environmental conditions during the late Vallesian (Agustí et al. 1997), which, if manifested in central Turkey, would have been favorable to the development of such grasslands. Evidence for the initiation of C_4 grassland expansion in the Old World (Cerling et al. 1993) does coincide temporally (and perhaps ecologically) with the replacement in the circum-Mediterranean area of *C. anatolicus* by *C. pentelici,* with its more elaborate molar occlusal morphology.

Germane to these considerations—and a cautionary note to them—is the probability that proboscidean occlusal structure, masticatory function, and feeding habits are not as tightly correlated as in living ungulates. For example, carbon isotope analysis of brachyodont late Miocene elephants from Kenya shows an unexpected dominance of C_4 plants in their diets, indicating grazing, whereas a large sample of modern African elephants, which are considerably more hypsodont, exhibit isotope signals that suggest a greater preference for browsing (Harris 1995, 1996). It is likely that other mammals of the Sinap Formation are more sensitive paleoecological indicators.

Conclusions

The new proboscidean sample from the Sinap Formation, central Turkey, recovered between 1989 and 1995, is primarily from the middle member of the formation and is dominated by postcranial remains. It is an important collection because it confirms the presence of deinotheres and gomphotheres in the formation and is useful for the classification and systematic study of late Miocene circum-Mediterranean choerolophodonts. In addition, it has utility for paleoecologic reconstruction and biochronologic correlation of Sinap Formation localities.

Morphometric analysis of the dental and mandibular specimens of the new collection provides the following results:

1. *Gomphotherium angustidens* is present in the lower member of the Sinap Formation and is morphologically similar to early–middle Astaracian examples of the taxon;
2. In occlusal morphology and size, the deinothere deciduous premolars from middle member Loc. 49 are most similar to those of *Deinotherium giganteum* from faunas correlated with the lowermost portion of mammal zone MN 11;
3. Choerolophodont specimens from the middle member of the formation have close metric and structural affinity with *"Choerolophodon pentelici lydiensis,"* known from MN 9–lower MN 11 faunas;
4. The Sinap choerolophodont sample reinforces the differences between *"C. p. lydiensis"* and *"C. p. pentelici"* from MN 11–MN 13 faunas, supporting a species level separation between these two groups and resuscitation of the nomen *C. anatolicus;*
5. Previous estimates of an early Astaracian date (~15 Ma) for Locs. 24 and 24A of the lower member, and early and late Vallesian dates for middle member Locs. 12 and 49, respectively, are corroborated; and
6. The occurrence of these taxa is suggestive of woodland or forest habitats in the depositional environment of Locs. 24 and 24A, and expansion of open country conditions at localities higher in the sequence.

Acknowledgments

I am extremely grateful to John Kappelman and Mikael Fortelius for their generous invitation to participate in the 1995 field season and to study the Sinap Formation proboscideans. Appreciation goes to Ilhan Temizsoy, director of the Anatolian Civilizations Museum, and Berna Alpagut, professor of Paleoanthropology, Ankara University, for permission to work with proboscidean specimens from the Sinap Formation and from Paşalar. I am thankful to Jeremy Hooker (The Natural History Museum, London) and Meave Leakey (National Museums, Kenya) for allowing me to study fossil collections in their care. I also thank my Turkish colleagues for their kind hospitality during my stay, and especially Zeynep Bostan for her assistance with the Paşalar collection. Chris Kirk was very helpful in accessing the Sinap collection. Drawings were provided by Jason McAlexander, and Bonnie Miljour helped with layout of the figures. Ray Bernor and Pascal Tassy provided insightful comments on the manuscript. Financial assistance for my stay in Turkey, as well as for subsequent research in Africa and Europe, was provided by John Kappelman, Mikael Fortelius, and a Scott Turner Award in Earth Science from the Department of Geological Sciences, University of Michigan.

Literature Cited

Agustí, J., L. Cabrera, M. Garcés, and J. M. Pares, 1997, The Vallesian mammal succession in the Valles-Penedes basin (northeast Spain): Paleomagnetic calibration and correlation with global events: Palaeogeography, Palaeoclimatology, Palaeoecology, v. 133, pp. 149–180.

Alpagut, B., P. Andrews, and L. Martin, 1989, Miocene paleoecology of Paşalar, Turkey, *in* E. H. Lindsay, V. Fahlbusch, and P. Mein, eds., European Neogene mammal chronology: New York, Plenum Press, pp. 443–459.

Andrews, P., T. Harrison, E. Delson, R. L. Bernor, and L. Martin, 1996, Distribution and biochronology of European and Southwest Asian Miocene catarrhines, *in* R. L. Bernor, V. Fahlbusch, and H.-W. Mittmann, eds., The evolution of western Eurasian Neogene mammal faunas: New York, Columbia University Press, pp. 168–207.

Antunes, M. T., 1989, The proboscideans data, age and paleogeography: Evidence from the Miocene of Lisbon, *in* E. H. Lindsay, V. Fahlbusch, and P. Mein, eds., European Neogene mammal chronology: New York, Plenum Press, pp. 253–262.

Bachmann, I., 1875, Beschreibung eines Unterkiefers von *Dinotherium bavaricum:* Abhandlungen der Schweizerischen Palaeontologischen Gesellshaft, v. 2, pp. 5–19.

Bakalov, P., and I. Nikolov, 1963, Neuer Fund von *Trilophodon (Choerolophodon) pentelicus:* Travaux sur la Géologie de Bulgarie, v. 5, pp. 229–239.

Barry, J. C., and L. J. Flynn, 1989, Key biostratigraphic events in the Siwalik sequence, *in* E. H. Lindsay, V. Fahlbusch, and P. Mein, eds., European Neogene mammal chronology: New York, Plenum Press, pp. 557–571.

Behrensmeyer, A. K., R. Potts, T. Plummer, L. Tauxe, N. Opdyke, and T. Jorstad, 1995, The Pleistocene locality of Kanjera, western Kenya: Stratigraphy, chronology and paleoenvironments: Journal of Human Evolution, v. 29, pp. 247–274.

Bergounioux, F.-M., and F. Crouzel, 1962, Les Déinothéridés d'Europe: Annales de Paléontologie, v. 48, pp. 1–56.

Bernor, R. L., 1986, Mammalian biostratigraphy, geochronology, and zoogeographic relationships of the Late Miocene Maragheh fauna, Iran: Journal of Vertebrate Paleontology, v. 6, pp. 76–95.

Bernor, R. L., and H. Tobien, 1990, The mammalian geochronology and biogeography of Paşalar (middle Miocene, Turkey): Journal of Human Evolution, v. 19, pp. 551–568.

Cerling, T. E., Y. Wang, and J. Quade, 1993, Expansion of C4 ecosystems as an indicator of global ecological change in the late Miocene: Nature, v. 361, pp. 344–345.

Chakravarti, D. K., 1957, A geological, palaeontological, and phylogenetic study of the Elephantoidea of India, Pakistan and Burma: Part 1. Gomphotheriidae: Journal of the Palaeontological Society of India, v. 2, pp. 83–94.

Cooper, C. F., 1922, Miocene Proboscidia [sic] from Baluchistan, Proceedings of the Zoological Society of London, v. 42, pp. 606–626.

Coppens, Y., V. J. Maglio, C. T. Madden, and M. Beden, 1978, Proboscidea, *in* V. J. Maglio and H.B S. Cooke, eds., Evolution of African mammals: Cambridge, Harvard University Press, pp. 336–367.

Fejfar, O., W.-D. Heinrich, M. A. Pevzner, and E. A. Vangengeim, 1997, Late Cenozoic sequences of mammalian sites in Eurasia: An updated correlation: Palaeogeography, Palaeoclimatology, Palaeoecology, v. 133, pp. 259–288.

Fourtau, R., 1920, Contribution a l'étude vertébrés Miocènes de l'Égypte: Cairo, Government Press, 121 pp.

Gaziry, A. W., 1976, Jungtertiäre Mastodonten aus Anatolien (Türkei): Geologisches Jahrbuch, v. 22, pp. 3–143.

Gentry, A. W., 1987, Mastodons from the Miocene of Saudi Arabia: Bulletin of the British Museum (Natural History) (Geology), v. 41, pp. 395–407.

Ginsburg, L., 1989, The faunas and stratigraphical subdivisions of the Orleanian in the Loire Basin (France), *in* E. H. Lindsay, V. Fahlbusch, and P. Mein, eds., European Neogene mammal chronology: New York, Plenum Press, pp. 157–176.

Göhlich, U., 1998, Elephantoidea (Proboscidea, Mammalia) aus dem Mittel- und Obermiozän der Oberen Süßwassermolasse Süddeutschlands: Odontologie und osteologie: Münchner Geowissenschaftliche Abhandlungen, Reihe A, Geologie und Paläontologie, v. 36, pp. 1–246.

Gräf, I. E., 1957, Die Prinzipien der Artbestimmung bei *Dinotherium:* Palaeontographica, v. 108, pp. 131–185.

Gürbüz, M., 1981, Inönü (KB Ankara) Orta Miyosenindeki *Hemicyon sansaniensis* (Ursidae) turunum tanimlanmasi ve stratigrafik yayilimi: Türkiye Jeoloji Kurumu Bülteni, v. C24, pp. 85–90.

Harris, J. M., 1973, *Prodeinotherium* from Gebel Zelten, Libya: Bulletin of the British Museum (Natural History) (Geology), v. 23, pp. 285–350.

Harris, J. M., 1975, Evolution of feeding mechanisms in the family Deinotheriidae (Mammalia: Proboscidea): Zoological Journal of the Linnean Society, v. 56, pp. 331–362.

Harris, J. M., 1976, Cranial and dental remains of *Deinotherium bozasi* (Mammalia: Proboscidea) from East Rudolf, Kenya: Journal of Zoology, v. 178, pp. 57–75.

Harris, J. M., 1977, Deinotheres from southern Africa: South African Journal of Science, v. 73, pp. 281–282.

Harris, J. M., 1978, Deinotherioidea and Barytherioidea, *in* V. J. Maglio and H.B.S. Cooke, eds., Evolution of African mammals: Cambridge, Harvard University Press, pp. 315–332.

Harris, J. M., 1983, Family Deinotheriidae, *in* J. M. Harris, ed., Koobi Fora research project, Volume 2: The fossil ungulates: Proboscidea, Perissodactyla, and Suidae: Oxford, UK, Clarendon Press, pp. 22–39.

Harris, J. M., 1987, Fossil Deinotheriidae from Laetoli, *in* M. D. Leakey, and J. M. Harris, eds., Laetoli, A Pliocene site in northern Tanzania: Oxford, UK, Clarendon Press, pp. 294–297.

Harris, J. M., 1995, Dietary preferences of Lothagam mammals (late Miocene, Northern Kenya): Journal of Vertebrate Paleontology, v. 15, p. 33A.

Harris, J. M., 1996, Isotopic changes in the diet of African proboscideans: Journal of Vertebrate Paleontology, v. 16, p. 40A.

Hill, A., R. Drake, L. Tauxe, M. Monaghan, J. C. Barry, A. K. Behrensmeyer, G. Curtis, B. F. Jacobs, L. Jacobs, N. Johnson, and D. Pilbeam, 1985, Neogene palaeontology and geochronology of the Baringo Basin, Kenya: Journal of Human Evolution, v. 14, pp. 759–773.

Janis, C. M., 1986, An estimation of tooth volume and hypsodonty indices in ungulate mammals, and the correlation of the factors with dietary preference, *in* D. E. Russell, J.-P. Santoro, and D. Sigogneau-Russell, eds., Teeth revisited: Proceedings of the 7th International Symposium on Dental Morphology, Paris: Mémoires, Muséum National d'Histore Naturelle, ser. C, v. 53, pp. 367–387.

Janis, C. M., 1995, Correlations between craniodental morphology and feeding behavior in ungulates: Reciprocal illumination between living and fossil taxa, *in* J. Thompson, ed., Functional morphology in vertebrate paleontology: Cambridge, Cambridge University Press, pp. 76–98.

Kappelman, J., S. Sen, M. Fortelius, A. Duncan, B. Alpagut, J. Crabaugh, A. Gentry, J.-P. Lunkka, F. McDowell, N. Solounias, S. Viranta, and L. Werdelin, 1996, Chronology and biostratigraphy of the Miocene Sinap Formation of central Turkey, *in* R. L. Bernor, V. Fahlbusch, and H.-W. Mittmann, eds., The evolution of western Eurasian Neogene mammal faunas: New York, Columbia University Press, pp. 78–95.

Klähn, H., 1931, Die Mastodonten des Sarmatikum von Steinheim A. Alb.: Palaeontographica, Supplement, v. 7, pp. 1–36.

Leakey, M. G., C. S. Feibel, R. L. Bernor, J. M. Harris, T. E. Cerling, K. M. Stewart, G. W. Storrs, A. Walker, L. Werdelin, and A. J. Winkler, 1996, Lothagam: A record of faunal change in the late Miocene of East Africa: Journal of Vertebrate Paleontology, v. 16, pp. 556–570.

Lindsay, E., 1997, Eurasian mammal biochronology: An overview: Palaeogeography, Palaeoclimatology, Palaeoecology, v. 133, pp. 117–128.

Maglio, V. J., 1972, Evolution of mastication in the Elephantidae: Evolution, v. 26, pp. 638–658.

Maglio, V. J., 1973, Origin and evolution of the Elephantidae: Transactions of the American Philosophical Society, v. 63, pp. 1–149.

Mazo, A. V., 1996, Gomphotheres and mammutids from the Iberian Peninsula, *in* J. Shoshani, and P. Tassy, eds., The Proboscidea. Evolution and palaeoecology of elephants and their relatives: Oxford, UK, Oxford University Press, pp. 136–142.

Mebrate, A., and J. E. Kalb, 1985, Anancinae (Proboscidea: Gomphotheriidae) from the Middle Awash Valley, Afar, Ethiopia: Journal of Vertebrate Paleontology, v. 5, pp. 93–102.

Michaux, J., 1988, Contributions a l'étude du gisement Miocène supérieur de Montredon (Herault). Les grands mammifères. Conclusions generales: Palaeovertebrata, Mémoire Extraordinaire, 1988, pp. 189–192.

Miller, E. R., and E. L. Simons, 1996, Age of the first cercopithecoid, *Prohylobates tandyi,* Wadi Moghara, Egypt: American Journal of Physical Anthropology, Supplement, v. 22, pp. 169–170.

Nakaya, H., 1993, Les faunes de mammifères du Miocène supérieur de Samburu Hills, Kenya, Afrique de l'est et l'environnement des pré-hominidés: L'Anthropologie, v. 97, pp. 9–16.

Osborn, H. F., 1936, Proboscidea. A monograph of the discovery, evolution, migration and extinction of the mastodonts and elephants of the world, Volume I: Moerotherioidea, Deinotherioidea, Mastodontoidea: New York, American Museum Press, 802 pp.

Ozansoy, F., 1955, Sur les gisements continentaux et les mammifères du Néogène et du Villafranchien d'Ankara: Comptes Rendus de l'Académie des Sciences, Paris, v. 240, pp. 992–994.

Ozansoy, F., 1957, Faunes de mammifères du Tertiaire de Turquie et leurs révisions stratigraphiques: Bulletin of the Mineral Research Exploration Institute, Turkey, v. 49, pp. 29–48.

Ozansoy, F., 1965, Étude des gisements continentaux et des mammifères du Cénozoïque de Turquie: Mémoires de la Société Géologique de France, v. 102, pp. 1–92.

Pickford, M., 2001, *Afrochoerodon* nov. gen. *kisumuensis* (MacInnes) (Proboscidea, Mammalia) from Cheparawa, middle Miocene, Kenya: Annales de Paléontologie, v. 87, pp. 99–117.

Pilbeam, D., M. Morgan, J. C. Barry, and L. Flynn, 1996, European MN units and the Siwalik faunal sequence of Pakistan, *in* R. L. Bernor, V. Fahlbusch, and H.-W. Mittmann, eds., The evolution of western Eurasian Neogene mammal faunas: New York, Columbia University Press, pp. 96–105.

Ramstein, G., F. Fluteau, J. Besse, and S. Joussaume, 1997, Effect of orogeny, plate motion and land-sea distribution on Eurasian climate change over the past 30 million years: Nature, v. 386, pp. 788–795.

Raza, S. M., and G. E. Meyer, 1984, Early Miocene geology and paleontology of the Bugti Hills, Pakistan: Memoirs of the Geological Survey of Pakistan, v. 11, pp. 43–63.

Roger, O., 1886, Ueber *Dinotherium bavaricum:* Palaeontographica, v. 32, pp. 215–226.

Sahni, M. R., and C. Tripathi, 1957, A new classification of the Indian deinotheres and description of *D. orlovii* sp. nov.: Memoirs of the Geological Society of India, Paleontologica India, v. 33, pp. 1–33.

Sanders, W. J., 1997, Fossil Proboscidea from the Wembere-Manonga Formation, Manonga Valley, Tanzania, *in* T. Harrison, ed., Neogene paleontology of the Manonga Valley, Tanzania: New York, Plenum Press, pp. 265–310.

Sanders, W. J., and Miller, E. 2002. New proboscideans from the early Miocene of Wadi Moghara, Egypt: Journal of Vertebrate Paleontology, v. 22, pp. 388–404.

Sanders, W. J., and Kappelman, J. 2001. A new late Oligocene proboscidean fauna from Chilga, Ethiopia *in* Scientific programme and abstracts, 8th International Theriological Congress, p. 120.

Sarwar, M., 1977, Taxonomy and distribution of the Siwalik Proboscidea: Bulletin of the Department of Zoology, University of the Punjab, v. 10, pp. 1–71.

Sen, S., 1990, Stratigraphie, faunes de mammifères et magnétostratigraphie du Néogène de Sinap Tepe, Province d'Ankara, Turquie: Bulletin du Museum National d'Histoire Naturelle, ser. 4e, v. 12, pp. 243–277.

Sen, S., 1996, Present state of magnetostratigraphic studies in the continental Neogene of Europe and Anatolia, *in* R. L. Bernor, V. Fahlbusch, and H.-W. Mittmann, eds., The evolution of western Eurasian Neogene mammal faunas: New York, Columbia University Press, pp. 56–63.

Sen, S., 1997, Magnetostratigraphic calibration of the European Neogene mammal chronology: Palaeogeography, Palaeoclimatology, Palaeoecology, v. 133, pp. 181–204.

Senyürek, M. S., 1952, A study of the Pontian fauna of Gökdere (Elmadagi), south-east of Ankara: Belletin Türk Tarih Kurumu, v. 16, pp. 449–492.

Sickenberg, O., and H. Tobien, 1971, New Neogene and lower Quaternary vertebrate faunas in Turkey: Newsletters on Stratigraphy, v. 1, pp. 51–61.

Steininger, F. F., R. L. Bernor, and V. Fahlbusch, 1989, European Neogene marine/continental chronologic correlations, *in* E. H. Lindsay, V. Fahlbusch, and P. Mein, eds., European Neogene mammal chronology: New York, Plenum Press, pp. 15–46.

Steininger, F. F., W. A. Berggren, D. V. Kent, R. L. Bernor, S. Sen, and J. Agusti, 1996, Circum-Mediterranean Neogene (Miocene and Pliocene) marine-continental chronologic correlations of European mammal units, *in* R. L. Bernor, V. Fahlbusch, and H.-W. Mittmann, eds., The evolution of western Eurasian Neogene mammal faunas: New York, Columbia University Press, pp. 7–46.

Symeonidis, N. K., 1970, Ein *Dinotherium*-Fund in Zentralmakedonien (Griechenland): Annales Geologiques des Pays Helliniques, v. 21, pp. 334–341.

Tassy, P., 1983, Les Elephantoidea Miocenes du Plateau du Potwar, Groupe de Siwalik, Pakistan. IIe Partie: Choerolophodontes et Gomphothères: Annales de Paleontologie (Vert.-Invert.), v. 69, pp. 235–297.

Tassy, P., 1984, Le Mastodonte à dents étroits, le grade trilophodonte et la radiation initiale des Amebelodontidae, *in* E. Buffetaut, J.-M. Mazin, and E. Salmon, eds., Actes du Symposium Paléontologique Georges Cuvier: Montbeliard, France, Symposium Paléontologique Georges Cuvier, pp. 459–473.

Tassy, P., 1985, La place des mastodontes Miocènes de l'Ancien Monde dans la phylogénie des Proboscidea (Mammalia): Hypothèses et conjectures [Ph.D. thesis]: Université Pierre et Marie Curie, Paris.

Tassy, P., 1986, Nouveaux Elephantoidea (Mammalia) dans le Miocène du Kenya: Paris, Cahiers de Paléontologie, Éditions du Centre National de la Recherche Scientifique, 135 pp.

Tassy, P., 1989, The "Proboscidean Datum Event": How many proboscideans and how many events?, *in* E. H. Lindsay, V. Fahlbusch, and P. Mein, eds., European Neogene mammal chronology: New York, Plenum Press, pp. 237–252.

Tassy, P., 1994, Les gisements de mammifères du Miocène supérieur de Kemiklitepe, Turquie: 7. Proboscidea (Mammalia): Muséum National d'Histoire Naturelle, Bulletin, ser. 4e, v 16, pp. 143–157.

Tassy, P., 1996a, Dental homologies and nomenclature in the Proboscidea, *in* J. Shoshani and P. Tassy, eds., The Proboscidea. Evolution and palaeoecology of elephants and their relatives: Oxford, UK, Oxford University Press, pp. 21–25.

Tassy, P., 1996b, The earliest gomphotheres, *in* J. Shoshani and P. Tassy, eds., The Proboscidea. Evolution and palaeoecology of elephants and their relatives: Oxford, UK, Oxford University Press, pp. 89–91.

Tassy, P., 1996c, Who is who among the Proboscidea? *in* J. Shoshani and P. Tassy, eds., The Proboscidea. Evolutions and paleoecology of elephants and their relatives: Oxford, UK, Oxford University Press, pp. 39–48.

Tassy, P., S. Sen, J.-J. Jaeger, J.-M. Mazin, and N. Dalfes, 1989, Une sous-espèce nouvelle de *Choerolophodon pentelici* (Proboscidea, Mammalia) à Esme Akcaköy, Miocène supérieur d'Anatolie occidentale: Comptes Rendus de l'Académie des Sciences, Paris, ser. 2, v. 309, pp. 2143–2146.

Tobien, H., 1962, Über carpus und tarsus von *Deinotherium giganteum* Kaup (Mamm., Proboscidea): Palaeontologische Zeitschrift H. Schmidt-Festband, pp. 231–238.

Tobien, H., 1972, Status of the genus *Serridentinus* Osborn 1923 (Proboscidea, Mammalia) and related forms: Mainzer Geowissenschaften Mitteilungsblatt, v. 1, pp. 143–191.

Tobien, H., 1973a, On the evolution of mastodonts (Proboscidea, Mammalia), Part 1: The bunodont trilophodont groups: Notizblatt des Hessichen Landesamtes für Bodenforschung zu Wiesbaden, v. 101, pp. 202–276.

Tobien, H., 1973b, The structure of the mastodont molar (Proboscidea, Mammalia), Part 1: The bunodont pattern: Mainzer Geowissenschaften Mitteilungsblatt, v. 2, pp. 115–147.

Tobien, H., 1980, A note on the skull and mandible of a new choerolophodont mastodont (Proboscidea, Mammalia) from the middle Miocene of Chios (Aegean Sea, Greece), *in* L. L. Jacobs, ed., Aspects of vertebrate history: Flagstaff, Arizona, Museum of Northern Arizona Press, pp. 299–307.

Tobien, H., 1988, Contributions a l'étude du gisement Miocène supérieur de Montredon (Herault). Les Grands Mammifères. Les Proboscidiens Deinotheriidae: Palaeovertebrata, Mémoire Extraordinaire, v. 1988, pp. 135–175.

Tsoukala, E. S., and J. K. Melentis, 1994, *Deinotherium giganteum* Kaup (Proboscidea) from Kassandra Peninsula (Chalkidiki, Macedonia, Greece): Géobios, v. 27, pp. 633–640.

Vacek, M., 1877, Über österreichische Mastodonten und ihre Beziehungen zu den Mastodon-Arten Europas: Abhandlungen der Kaiserlich-Königlichen geologischen Reichenstalt, v. 7, pp. 1–45.

Viret, J., 1953, Observations sur quelques dents de mastodontes de Turquie et de Chine: Université de Lyon, Annales, section C, v. 7, pp. 51–62.

Viret, J., and I. Yalçinlar, 1952, *Synconolophus serridentinoides,* nouvelle espèce de Mastodonte du Miocène supérieur de Turquie: Comptes Rendus de l'Académie des Sciences, Paris, v. 234, pp. 870–872.

Weinsheimer, O., 1883, Über *Dinotherium giganteum* Kaup: Palaeontologische Abhandlungen, v. 1, pp. 207–281.

Woodburne, M. O., R. L. Bernor, and C. C. Swisher III, 1996, An appraisal of the stratigraphic and phylogenetic bases for the *"Hipparion"* datum in the Old World, *in* R. L. Bernor, V. Fahlbusch, and H.-W. Mittmann, eds., The evolution of western Eurasian Neogene mammal faunas: New York, Columbia University Press, pp. 124–136.

Yalçinlar, I., 1952, Les gisements et les *Synconolophus serridentinoides* d'Istanbul: Comptes Rendus de la Société Géologique de France, v. 1852, pp. 227–229.

11

Equidae (Perissodactyla)

R. L. Bernor, R. S. Scott, M. Fortelius, J. Kappelman and S. Sen

The Sinap Formation is located ~40 km northwest of Ankara in central Anatolia. Fossil vertebrates are abundant in the region, with many of the localities found in the area around Sinap Tepe, a small butte or "tepe" located north of the village of Yassiören. The fossils of the Sinap Formation have been investigated by a number of workers since the early 1950s. The Sinap material we describe here is derived from collections made by Ozansoy in the 1950s, by Sen of the Museum National d'Histoire Naturelle (Paris) in 1972, and the most recent Sinap project during the summers of 1989–1995. Previous reports on this material and its stratigraphic context include Ozansoy (1957, 1965), Sen (1986, 1991) and Kappelman et al. (1996). Additional fossil material was recovered by M. Senyürek but remains unpublished. The most recent Sinap project has made a major addition to the fossil material known from this area, but more importantly, it has resulted in the development of a highly resolved magnetochronology of the local biostratigraphic record (Kappelman et al., this volume). Additional background information is given in Temizsoy (Preface, this volume) and Sen (Introduction, this volume). The magnetostratigraphy of the hipparion-bearing horizons is given in table 11.1.

The Sinap Formation is critical to the understanding of Old World hipparion evolution because it preserves continuous sedimentation across the Astaracian-Vallesian (middle-late Miocene) boundary (Lunkka et al., this volume; Kappelman et al., this volume). At Sinap, we can directly address already established hypotheses of hipparion phylogeny, biogeography, and geochronology as they pertain both to the local sequence and broader Old World "Hipparion Datum."

Our analysis marshals all continuous and character state variables that have been found to be pertinent to Old World hipparion evolution. We evaluate the morphologic diversity of the Sinap hipparions as it pertains to their systematics, functional anatomy, and likely ecological preferences. Crucial to our evaluation is a consideration of the evolutionary relationships of the Turkish early Vallesian hipparions to central European *Hippotherium primigenium* and the North American Barstovian-Clarendonian *Cormohipparion* species series, including *C. goorisi, C. quinni* (Woodburne 1996: replaces *C. sphenodus* of previous literature), and *C. "occidentale"* (which itself includes multiple species). Our investigation endeavors to shed further light on the early evolutionary relationships of Old World hipparionine horses.

Materials and Methods

We use both continuous and discrete variables here to analyze the hipparion assemblage under consideration. The continuous variables used follow the 1981 American Museum of Natural History workshop on hipparion research, published and illustrated initially by Eisenmann et al. (1988) and again later by Bernor et al. (1997), adding measurements for some less common postcranial elements and the maxillary and mandibular cheek teeth. These measurements have been used by a number of investigators including Bernor (1985) on the Maragheh hipparionines, Qiu et al. (1988) and Bernor et al. (1990) on separate suites of Chinese hipparionines, on central European populations of *Hippotherium primigenium* from Austria (Bernor et al. 1993a), Hungary (Bernor et al. 1993b), Germany (Höwenegg, Bernor et al. 1997; Dorn Dürkheim, Bernor and Franzen 1997), African hipparions (Bernor and Harris 2003; Bernor and Armour-Chelu 1996, 1999), and in an overall review of Old World hipparion evolution (Bernor et al. 1989).

We further implement the use of 49 morphological character states of the skull, mandible, and dentition, which have been progressively refined through many of the studies cited here and by Bernor and Lipscomb (1991, 1995), as well as those by colleagues who have studied North American hipparion evolution, including MacFadden

Table 11.1. Sinap Mammal Biostratigraphy and Magnetochronology

Locality	Polarity Interval	Correlative Chron from GPTS		Interpolated Age (Ma)		Block Unit	Begin (Ma)	End (Ma)
42	—	—		—		13		
26, 27, 28, 30, 33, 70, KD	R4		C4r.1r	8.121	8.12	12	8.25	8
63, 69	N7	C4r.1n		8.230	8.23			
34	R3		C4r.2r	8.440	8.44	11	8.5	8.25
68	N3	C4An		8.866	8.87	9	9	8.75
49	N5		C4r.1r	9.130	9.13	8	9.25	9
10	N7	C4Ar.1n		9.279	9.28	7	9.5	9.25
1	N7	C4Ar.1n		9.288	9.29			
7	N7	C4Ar.1n		9.295	9.30			
75	N7	C4Ar.1n		9.301	9.30			
84	R7		C4Ar.2r	9.367	9.37			
83	R7		C4Ar.2r	9.452	9.45			
11	R7		C4Ar.2r	9.483	9.48			
12, OZ02	N6	C4Ar.2n–C5n.1n		9.590	9.59	6	9.75	9.5
OZ01, S01	N6	C4Ar.2n–C5n.1n		9.683	9.68			
8A	R6		C5n.1r	9.886	9.89	5	10	9.75
8B	R6		C5n.1r	9.918	9.92			
114	N5	C5n.2n–1n and –2n		9.967	9.97			
108/8	N5	C5n.2n–1n and –2n		9.970	9.97			
91	N5	C5n.2n–1n and –2n		9.977	9.98			
72	N5	C5n.2n–1n and –2n		10.080	10.08	4	10.25	10
108	N5	C5n.2n–1n and –2n		10.135	10.14			
106	N5	C5n.2n–1n and –2n		10.206	10.21			
113	N5	C5n.2n–1n and –2n		10.306	10.31	3	10.5	10.25
89	N5	C5n.2n–1n and –2n		10.406	10.41			
87	R5		C5n.2n–3r	10.452	10.45			
93	N4	C5n.2n–3n		10.488	10.49			
121	N4	C5n.2n–3n		10.526	10.53	2	10.75	10.5
94	N4	C5n.2n–3n		10.551	10.55			
122	N4	C5n.2n–3n		10.577	10.58			
107	N4	Cn5.2n–3n		10.653	10.65			
4	N4	Cn5.2n–3n		10.692	10.69			
88	N3	C5n.2n–4n		10.730	10.73			
64	N3	C5n.2n–4n		10.765	10.77	1	11	10.75
104	N3	C5n.2n–4n		10.868	10.87			
65	N3	C5n.2n–4n		10.899	10.90			
24, 24A	—	—		16–15				

and Skinner (1977), Woodburne and Bernor (1980), Woodburne et al. (1981), MacFadden (1984), Webb and Hulbert (1986), Hulbert (1988), Hulbert and MacFadden (1991), and Woodburne (1996). Utilization of postcranial character states awaits further comparative study across outgroup and sister taxa.

Our metric analysis includes a comparison of several taxa relevant to the Sinap assemblage. Initially, we employ univariate and bivariate plots to gain an understanding of metric comparisons between the most pertinent continuous variables. In bivariate plots, we use two central European populations as standards with which we can compare the Sinap hipparion series and other relevant western Eurasian hipparion populations: Höwenegg (10.3 m.y. [Swisher 1996; Woodburne et al. 1996; Bernor et al. 1997]; Hegau, southern Germany) and Eppelsheim (~10.5 m.y. [Bernor et al. 1996]; Rheinhessen, western Germany). Both these populations are believed to be "biologically uniform," including only a single primitive species, *Hippotherium primigenium*. The Höwenegg population is particularly useful for postcranial comparisons, because it preserves 14 partial to complete skeletons. Eppelsheim is superior for maxillary and mandibular cheek tooth comparisons, because the teeth are most often found without the associated jaws (allowing height measurements) and are more numerous than the Höwenegg population, giving more robust statistical results. We utilize yet a third population of hipparions for skull variable comparisons, the Xmas Quarry sample (North America) of *Cormohipparion "occidentale."* This sample is superior to all western Eurasian hipparion assemblages because of the excellent preservation of several skulls. Moreover, this assemblage, together with the North American localities of the Hans Johnson Quarry, MacAdam's Quarry, Pawnee Creek, and Trinity River Pit, represent successively further removed species of *Cormohipparion,* which represent an important record of early hipparionine horse evolution between 15.5 and 10.0 Ma (Bernor et al. 1989, 1997; Bernor and Lipscomb 1991, 1995; Woodburne 1996). Together, these populations allow us to evaluate the size and proportions of the Sinap hipparions with a population perspective and evolutionary background. All univariate and bivariate plots were computed using Systat 9.0 for Windows (Systat, Inc.: Evanston, Illinois).

Eisenmann (1995, for a review) has pioneered the use of Simpson's log-ratio diagrams for studying cranial and postcranial proportions. Eisenmann (1995 and elsewhere) and we (Bernor et al. 1989, 1999; Bernor and Armour-Chelu 1999; Bernor and Harris, 2003) have found that metacarpal III and metatarsal III are important for understanding hipparion evolution and locomotor adaptation. However, we also believe that metapodial morphology may well be subject to a great deal of homoplasy, and that it is better to incorporate the use of ratio diagrams into a broader analytical research design that considers other anatomical regions and other analytical methodologies. We used Excel 97 to compute the metapodial log-ratio diagrams reported here.

Furthermore, we have elected to employ principal components analysis (PCA) for evaluation of metacarpal III and metatarsal III. Principal components analyses of the covariance matrix for complete third metatarsals and metacarpals were computed using SAS. PCA can be used to identify the major sources of variability in a sample, and plots of principal components can be used to identify potential discrete subsets of a sample. The measurements used in these analyses were M2 (length), M3 (midshaft width), M4 (midshaft craniocaudal depth), M5 (proximal articular facet width), M6 (proximal facet craniocaudal depth), and M10 (width across the distal supraarticular tubercles). Measurements were selected to maximize the sample of complete specimens from Sinap available for analysis and allowed the inclusion of some additional specimens not included in the ratio diagrams. The raw measurements for each element were all divided by the geometric mean of the measurements for that element (GEOMEAN) and these GEOMEAN corrected measurements were used in the PCA (Jungers et al. 1995). PCA included 113 third metatarsals and 76 third metacarpals from Sinap, Esme Akçakoy, Höwenegg, other central European sites, and Xmas Quarry (North America).

Bootstrapping models for principal components 1 and 2 of the PCAs were used to divide the Sinap sample into statistically significant groups ($\alpha = 0.05$, one-tailed test) possibly representing single species. The bootstrapping procedure used here involved randomly sampling with replacement N_1 principal component scores from the Höwenegg sample and N_2 principal component scores from various subsamples from Sinap 10,000 times and calculating the ratio of variances between the Sinap subsample and the Höwenegg sample for each iteration. The number of principal component scores for the Höwenegg sample was N_1, and N_2 was the number of principal component scores for the Sinap subsample. The distribution of variance ratios generated using this procedure can be used to test the null hypotheses that the Sinap subsample under investigation was drawn from a population with the same or less variability for a principal component as the population represented by the Höwenegg sample ($\alpha = 0.05$, one-tailed test). The Höwenegg sample included some left-right pairs of metapodial III samples; therefore, the bootstrapping procedure used here was completed using samples composed of left and isolated metapodial III samples from Höwenegg and then completed again for a sample of right and isolated metapodial III samples from Höwenegg. This bootstrapping approach yields more conservative groupings of the Sinap specimens than does a simple visual interpretation of a plot of principal components, but these groupings have the advantage of being statistically significant. We also used *F*-tests in parallel with the bootstrapping procedure, and in cases where sample sizes were very small, only *F*-tests were performed. The distribution-free bootstrap approach was preferred in the interpretation of results.

The stratigraphic distribution of fossils in Sinap's late Astaracian–Turolian interval readily affords analysis of the hipparion assemblage by stratigraphic interval. Table 11.1 provides a biostratigraphic and magnetochronologic ordering of the hipparion-bearing localities that we analyze here. Table 11.2 lists the stratigraphic order of Sinap hipparion localities by block number and bulk ordering (Stratigraphic Units 1–12 here and in the series of plots that follow). Table 11.3 lists the Eurasian and North American hipparion samples, and their respective codes used in the various bivariate plots.

Abbreviations and Conventions

AMNH—American Museum of Natural History, New York
HLMD—Hessiches Landesmuseum, Darmstadt
MNHN—Museum National d'Histoire Naturelle, Paris
NHMW—Naturhistorisches Museum, Vienna
SMNK—Staatliches Museum für Naturkunde, Karlsrhue

The taxon *Hipparion* has been applied in a variety of ways by different authors. We utilize the following definitions in this work:

Hipparionine or hipparion: horses with an isolated protocone on maxillary premolar and molar teeth

Table 11.2. Stratigraphic Block Ordering of Sinap Hipparion-Bearing Localities

Locality	Block Unit	Begin (Ma)	End (Ma)
42	13		
26, 27, 28, 30, 33, 70, KD	12	8.25	8.00
63, 69			
34	11	8.5	8.25
68	9	9	8.75
49	8	9.25	9.00
10	7	9.5	9.25
1			
7			
75			
84			
83			
11			
12, OZ02	6	9.75	9.50
OZ01, S01			
8A	5	10.00	9.75
8B			
114			
108/8			
91			
72	4	10.25	10.00
108			
106			
113	3	10.50	10.25
89			
87			
93			
121	2	10.75	10.50
94			
122			
107			
4			

and, as far as known, tridactyl feet, including species of the following genera: *Cormohipparion, Neohipparion, Nannippus, Pseudhipparion, Hippotherium, Cremohipparion, Hipparion, "Sivalhippus," Eurygnathohippus* (= senior synonym of *"Stylohipparion"*), *Proboscidipparion, "Plesiohipparion."* Characterizations of these taxa can be found in MacFadden (1984), Bernor and Hussain (1985), Webb and Hulbert (1986), Hulbert (1987), Qiu et al. (1988), Bernor et al. (1988, 1989, 1990, 1996, 1997), Woodburne (1989) and Hulbert and MacFadden (1991), Bernor and Armour-Chelu (1996, 1999).

Hipparion s.s.: The name is restricted to a specific lineage of horses with the facial fossa positioned high on the face (MacFadden 1980, 1984; Woodburne and Bernor 1980; Woodburne et al. 1981; MacFadden and Woodburne 1982; Bernor and Hussain 1985; Bernor 1985; Bernor et al. 1987; Bernor et al. 1989; Woodburne 1989). The posterior pocket becomes reduced and eventually lost, and confluent with the adjacent facial surface (includes Group 3 of Woodburne and Bernor 1980). Bernor's definition departs from some investigators in not recognizing North American species of *Hipparion* s.s. Bernor (1985) and Bernor (in Bernor et al. 1989) have argued that any morphologic similarity between North American "*Hipparion* s.s." and *Hipparion* s.s. is due to homoplasy.

"Hipparion": several distinct and separate lineages of Old World hipparionine horses once considered to be referable to the genus *Hipparion* (Woodburne and Bernor 1980; Bernor et al. 1980; MacFadden and Woodburne 1982; Bernor and Hussain 1985; Bernor 1985; Bernor et al. 1988, 1989).

Measurements are in mm. All postcranial measurements are as defined by Eisenmann et al. (1988) and are rounded to 0.1 mm; all dental measurements are as defined by Bernor et al. (1997). The osteological nomenclature, the enumeration, and/or lettering of the figures have been adapted from Nickel et al. (1986). Getty (1982) was also consulted for morphological identification and comparison. Hipparion monographs by Gromova (1952) and Gabunia (1959) were cited after the French translations. As mentioned previously, we have used SAS, Systat 9.0, and Excel 97 in our statistical analyses.

Geology

The material collected by Ozansoy and Sen is housed in the Museum National d'Histoire Naturelle, Paris, and includes fossils from Sinap Localities OZ01, OZ02, and 108/8. Fossils collected by M. Senyürek are housed in the Faculty of Languages, History and Geography at Ankara University and are from Locality S01. The specimens collected by the Sinap project are stored in the Museum of Anatolian Civilisations, Ankara, and are from the remaining localities. The stratigraphic block ordering of Sinap localities and corresponding stratigraphic units used in our analyses below are included in table 11.2. The sedimentologic context of the fossil localities is given in Lunkka et al. (chapter 1, this volume), and the chronostratigraphic framework is given in Kappelman et al. (1996; chapter 2, this volume).

Description and Analysis of the Material

The Eurasian and North American material with which we compare the Sinap hipparions is listed in table 11.3 by the following categories: locality name, taxa, Bernor's database country (Loc.) and site (Site), site symbol used on bivariate plots (an alphanumeric code, e.g., M = Maragheh, I = Inzersdorf), and PLACE (a numerical code for the various sites

Table 11.3. Localities Used in the Analysis of Sinap Hipparions and Their Corresponding Symbols

Locality	Taxa	DBASE LOC & SITE	General Symbol	Place
Maragheh	Hget, Hpro, Hcam	5, 1	M	7
Inzersdorf	Hpri	6, 1	I	8
Gaiselberg	Hpri	6, 2	G	9
Höwenegg	Hpri	7, 1	H	10
Dorn Duerkheim	Hpri	7, 2	D	11
Eppelsheim	Hpri	7, 4	E	12
Cucuron	Hpro1&2	9, 1	C	13
Hostalets de Pierola	Hcat	10, 1	O	14
Esme Acakoy	Hpri	24, 1	A	15
Sinap	Hank&other	24, 2	S	1
Xmas Q.	Coo1&2	1, 2	X	2
Hans Johnson Q.	Coo1&2	1, 3	J	3
MacAdam's Q.	Coo1&2	1, 4	Q	4
Trinity River Pit	Cog	1, 7	T	5
Pawnee Creek	Cos	1, 8	P	6

used in the database; e.g., 7 = Maragheh, 8 = Inzersdorf). Our analysis considers those structures most commonly used for species distinction and comparison by equid workers: the skull, maxillary and mandibular cheek teeth, calcaneum, astragalus, metacarpal III, and metatarsal III.

Skull

Table 11.4 lists the character state distributions of the 16 available skulls in the Paris and Ankara collections. For reference, we include character states for the primitive central European species *Hippotherium primigenium* (Hoewenegg; Bernor et al. 1997) and North American species *Cormohipparion "occidentale"* (F:AM 71800; MacFadden 1984, fig. 131, p. 164). We draw upon these comparisons further in our Systematics section.

The stratigraphically lowest occurring skull in the assemblage is AS93/826 from Loc. 108 (Block Unit 4 of table 11.2; 10.135 Ma). This skull belongs to a partial skeleton that is critical for understanding the evolutionary relationships of the Sinap hipparion sequence. It has a long POB with the anterior edge of the lacrimal placed more than one-half the distance from the anterior orbital rim to the posterior rim of the fossa (C1 = C). The nasomaxillary fossa is absent (C2 = C) and the POF is subtriangular shaped and apparently anteroventrally oriented (C4 = ?D); both are characters in common with *Hippotherium primigenium* and *Cormohipparion "occidentale."* The POF is deeply pocketed posteriorly (C5 = A), medial depth is great (C6 = A) and the medial wall lacks pits (C7 = A), again characters all in common with *H. primigenium* and *C. "occidentale."* The POF peripheral outline is derived in its moderately delineated outline, which dissipates anterodistally and anteroventrally (C8 = B), but an anterior rim is apparent (C9 = ?A). The infraorbital foramen (IOF) is closely associated with the anteroventral border of POF (C10 = B), and the buccinator fossa is distinct (C11 = B), again as in *H. primigenium* and *C. "occidentale."* The caninus fossa is absent (C13 = A), as is the malar fossa (C14 = A), all features characteristic for *H. primigenium* and *C. "occidentale."* The maxillary cheek teeth are more primitive than *H. primigenium,* exhibiting greater curvature (C17 = AB). Maximum crown height is estimated to be between 40 and 45 mm (C18 = C) and is considered to be primitive for all Old World hipparions. It is interesting that the Loc. 108 specimen shows a slight advance in reduction of the POF peripheral rim but is primitive in cheek tooth curvature.

The next stratigraphic unit represented by skulls is Block Unit 5 and includes specimens AS92/458 and AS93/837 (both from Loc. 91; 9.977 Ma), SEN 1 (Loc. 108/8; 9.970 Ma), AS94/203 (Loc. 114; 9.967 Ma), and AS89/54 and AS92/609 (Loc. 8B; 9.918 Ma). Specimen AS93/837 is poorly preserved and cannot be evaluated for its skull character states. Skull SEN 1 is the only specimen of these three that retains Character 1, POB long with the anterior edge of the lacrimal placed more than half the distance from the anterior orbital rim to the posterior rim of the fossa (C1 = C), which is primitive for Old World hipparions. It also preserves a large lacrimal foramen on the orbital rim (C3 = A). Old World hipparions only infrequently preserve this structure, and when they do so, it is small. In North American *Cormohipparion,* the lacrimal foramen is prominent; this is particularly so in *C. "occidentale."* The POF is subtriangular shaped and anteroventally oriented (C4 = D), it is deeply pocketed posteriorly (C5 = A), and the peripheral outline is moderately delineated (C8 = B) and has a distinct anterior rim (C9 = A). Nasal notch morphology is not preserved.

Specimen SEN1 is similar to AS93/826, *C. "occidentale"* and *H. primigenium* in retaining persistent and functional dP1s (C16 = A). In all critical variables, SEN 1 compares very closely with Loc. 108 specimen AS93/826.

Specimen AS92/458 (Loc. 91) differs substantially from AS93/826 and SEN 1 in having a POF that is strongly reduced in length and dorsoventral height, and in having an anteroposteriorly oriented POF (C4 = F). Furthermore, POF posterior pocketing is absent, but the specimen retains a distinct posterior rim (C5 = C), its peripheral border outline is weakly defined (C8 = C), and the nasal notch closely approaches P2 (C15 = C-). Specimen AS92/458 also exhibits relatively high crowned teeth with the early wear M1 having a crown height of 54 mm and the erupting M2 having a crown height of 57 mm; AS92/458 is advanced in all the characters cited above and compares closely with members of the Old World *Hipparion* s.s. group.

The Loc. 114 specimen AS94/203 is an old adult female in a very advanced stage of cheek tooth wear. This individual has a snout that appears to be somewhat longer, and a POF with a greater dorsoventral height than in the Loc. 91 specimen AS92/458. These differences may be due to the very different ontogenetic stages at which these individuals died (AS92/458 was a juvenile). The POF of AS94/203 has a weakly developed subtriangular shape and anteroposterior orientation (C4 = D/F), a reduced posterior pocketing, and a weakly defined peripheral rim (C8 = C). This specimen has the nasal notch closely approaching P2 (C15 = C). Loc. 8B specimens AS89/54 and AS92/609 are very similar to AS92/203 in their morphology. Specimen AS92/609 preserves a very weakly developed anterior rim (C9 = A-).

Stratigraphic Unit 6 has six MNHN skulls. These include four derived from Locs. OZ01 and S01 (9.683 Ma): MNHNTRQ1064, MNHNTRQ1067, MNHNTRQ9001, and MNHNTRQ1211 (the "type" specimen of Ozansoy's [1965] taxon *Hipparion ankyranum* [Lectotype here]). There are two skull fragments from OZ02 (9.590 Ma): MNHNTRQ1064 and MNHNTRQ1067. These are palates that preserve very little morphological information.

Loc. S01 specimen MNHNTRQ9001 is similar to AS93/826 and SEN 1 in its POF morphology. Its POF is subtriangular shaped and anteroposteriorly oriented (C4 = E); it is posteriorly deeply pocketed (C5 = A), medially deep (C6 = A), its peripheral outline is moderately well developed, but the anterior rim is absent (C9 = B). Nasal notch is advanced in its incision, closely approaching P2 (C15 = C). This specimen is primitive in its strongly curved cheek teeth (C17 = A) and accompanying low crown height (C18 = B/C; maximum crown height estimated to be <45 mm).

The lectotype of *Hipparion ankyranum,* MNHNTRQ1211 (Ozansoy 1965, pl. V, figs. 2, 4 [sic. 1]), is a palate that was collected from Loc. OZ01. It preserves enough of the POF to show that it was dorsoventrally extensive in its dimensions, closely approaching the facial-maxillary crest; it was deeply pocketed posteriorly (C5 = A), deep medially (C6 = A), the medial wall lacked pits (C7 = A), and its peripheral outline may have been strongly developed (C8 = ?A). Placement of the IOF is as in all hipparions listed in table 11.4. The dP1 is persistent and functional, the cheek teeth would appear to have been strongly curved and maximum crown height to have been relatively low.

The two OZ02 skull fragments MNHNTRQ1065 and 1066 differ from other Unit 6 skulls in their absence of a dP1 (C16 = C). This may be ontogenetically related, or a truly advanced character typical of later *Hipparion* s.s. species and other Old World hipparion lineages.

Two specimens from Unit 8 are derived from Locality OZKD: MNHNAKA 15 and 25. These specimens are too fragmentary to be evaluated for more than a minimum of skull characters.

Cheek Teeth

Table 11.5 lists the maxillary and table 11.6 the mandibular cheek tooth character states for the sample that was available for study. Bernor et al.'s (1997, table 4.1, p. 20) study of the Höwenegg population revealed variable stability for various maxillary and mandibular cheek tooth character states. Bernor and Franzen (1997) found that variability in character states was strongly influenced by ontogentic wear stage and that when evaluated at middle stage-of-wear, cheek tooth character states are reasonably stable. Our evaluation of maxillary and mandibular cheek tooth character state distributions follows that of the skulls by highlighting significant differences with *Hippotherium primigenium* and, when known, *Cormohipparion "occidentale."*

Maxillary Cheek Teeth

The lowest stratigraphic unit represented in our maxillary cheek tooth sample is Block Unit 4, Loc. 108 (table 11.1, 10.135 Ma). This sample differs from central European *Hippotherium primigenium* in having less complex cheek tooth plications (C19 = B); in this character, the Sinap unit hipparions compare closely with *Cormohipparion "occidentale"* and most species of Old World *Hipparion* s.s. (Bernor et al. 1989). Pli caballin morphology (C21) varies from being double (A) to complex (C), as can occur in *C. "occidentale"* and less frequently in *H. primigenium.* It is uncommon in *Hipparion* s.s., which usually exhibits single pli caballins. Hypoglyph is consistently less deeply incised (C22 = C) than in *H. primigenium* and *C. "occidentale"* (with the noted exception of a single M3, the one element that often has encircled hypoglyphs due to this tooth's intrinsic mediolateral compression; Bernor and Franzen 1997). This character tracks the *Hipparion* s.s. lineage quite closely. Although protocone morphology (C23) is notoriously variable throughout ontogeny, the middle stage-of-wear of *H. primigenium* is consistently some version of lingually flattened and labially rounded (state C23E); the Sinap hipparions exhibit this character in a more oval-to-rounded form (states C23C, E, and G), anticipating the condition common to *Hipparion* s.s. Character 24, isolation of protocone, exhibits an unusally high incidence of connection to the protoloph (C24 = A), which may be due to its later wear

Table 11.4. Character State Distribution of Sinap Hipparion Skulls

Specimen Number	SPECSHORT	Sex	Characteristic																		Locality	Unit
			1	2	3	4	5	6	7	8	9	10	11	12	13	14	15	16	17	18		
	Hpri ss.		C	C	B	D	A	A	A	A	A	B	B	A	A	A	B	A	B	C		
	Coo		C	C	A	B	A	A	A	A	A	B	B	A	A	A	C	B	B	C		
AS93/826	*C sin* n.s. T	3	C	C	?	?D	A	A	A	B	?A	B	B	?	A	A	?	?	AB	C	108	4
SEN 1	*C sin* n.s.	3	C	C	A	D	A	A	A	B	A	B	?	A	A	A	?	A	AB	C	108/8	5
AS92/458	*H.* sp. 1	2	?	C	?	F	C	A-		C		B	B	A	A	A	C-	A	AB	C	91	5
AS93/837	"*H.*" sp.	3	?	C	?	?	?	?	?	?	?	?	?	?	?	?	?	?	AB	C	91	5
AS94/203	"*H.*" *uzun* T	1	C	C	A	D/F	B/C	A-	A	C		B	B	A	A	A	C	A	AB	C	114	5
AS89/54	"*H.*" *uzun*	3	C	C	A	D/F	B/C	A	?	?	?	?	?	?	?	A	?	?	AB	C	8B	5
AS92/609	"*H.*" *uzun*	3	C	C	A	D/F	B/C	A	A	C	A-	B	B	A	A	A	?	?	AB	C	8B	5
MNHNTRQ1064	*H.* sp.	3											B	A					B		S01	6
MNHNTRQ1067	*H.* sp.	3																A			S01	6
MNHNTRQ9001	"*H.*" *kec* T	3		C		E	A	A	A	B	B	B	B	A	A	A	C		A	B/C	S01	6
MNHNTRQ9002	"*H.*" *kec*	3																	A	B/C	S01	6
MNHNTRQ1211	"*H.*" *anky*	3					A	A	A	?A		B						A	A	B/C	0Z01	6
MNHNTRQ1065	"*H.*" sp.	3					A?	A				B	B					C	B		0Z02	6
MNHNTRQ1066	"*H.*" sp.	3										B							B		0Z02	6
MNHNAKA15	"*H.*" sp.	3																C	B		KD	12
MNHNAKA25	"*H.*" sp.	3																	B		KD	12

Sources: Adapted from Bernor et al. (1989), Bernor and Lipscomb (1991, 1995); Bernor and Armour-Chelu (1999).

Notes: Character states defined as follows.

1. Relationship of lacrimal to the preorbital fossa (POF): A = lacrimal large, rectangular, invades medial wall and posterior aspect of POF; B = lacrimal reduced in size, slightly invades or touches posterior border of POF; C = preorbital bar (POB) long with the anterior edge of the lacrimal placed more than half the distance from the anterior orbital rim to the posterior rim of the fossa; D = POB reduced slightly in length but with the anterior edge of the lacrimal placed still more than half the distance from the anterior orbital rim to the posterior rim of the fossa; E = POB vestigial, but lacrimal as in D; F = POB absent; G = POB very long with anterior edge of lacrimal placed less than half the distance from the anterior orbital rim to the posterior rim of the fossa; H = POB absent.
2. Nasolacrimal fossa: A = POF large, ovoid, and separated by a distinct medially placed, dorsoventrally oriented ridge, dividing POF into equal anterior (nasomaxillary) and posterior (nasolacrimal) fossae; B = nasomaxillary fossa sharply reduced compared to nasolacrimal fossa; C = nasomaxillary fossa absent (lost), leaving only nasolacrimal portion (when a POF is present).
3. Orbital surface of lacrimal bone: A = with foramen; B = reduced or lacking foramen.
4. Preorbital fossa morphology: A = large, ovoid shape, anteroposteriorly oriented; B = POF truncated anteriorly; C = POF further truncated, dorsoventrally restricted at anterior limit; D = subtriangular shaped and anteroventrally oriented; E = subtriangular shaped and anteroposteriorly oriented; F = egg shaped and anteroposteriorly oriented; G = C shaped and anteroposteriorly oriented; H = vestigial but with a C-shaped or egg-shaped outline; I = vestigial without C-shaped outline, or absent; J = elongate, anteroposteriorly oriented; K = small, rounded structure; L = posterior rim straight, with nonoriented medial depression.
5. Fossa posterior pocketing: A = deeply pocketed, < 15 mm in deepest place; B = pocketing reduced, moderate to slight depth, > 15 mm; C = not pocketed but with a posterior rim; D = absent, no rim but a remnant depression; E = absent.
6. Fossa medial depth: A = deep, < 15 mm. in deepest place; B = moderate depth, 10–15 mm in deepest place; C = shallow depth, > 10 mm in deepest place; D = absent.
7. Preorbital fossa medial wall morphology: A = without internal pits; B = with internal pits.
8. Fossa peripheral border outline: A = strong, strongly delineated around entire periphery; B = moderately delineated around periphery; C = weakly defined around periphery; D = absent with a remnant depression; E = absent, no remnant depression.

9. Anterior rim morphology: A = present; B = absent.
10. Placement of infraorbital foramen: A = placed distinctly ventral to approximately half the distance between the preorbital fossa's anteriormost and posteriormost extent; B = inferior to, or encroaching upon anteroventral border of the preorbital fossa.
11. Confluence of buccinator and canine fossae: A = present; B = absent, buccinator fossa is distinctly delimited.
12. Buccinator fossa: A = not pocketed posteriorly; B = pocketed posteriorly.
13. Caninus (= intermediate) fossa: A = absent; B = present.
14. Malar fossa: A = absent; B = present.
15. Nasal notch position: A = at posterior border of canine or slightly posterior to canine border; B = approximately one-half the distance between canine and P2; C = at or near the anterior border of P2; D = above P2; E = above P3; F = above P4; G = above M1; H = posterior to M1.
16. Presence of dP1 (16U) or dP1 (16U): A = persistent and functional; B = reduced and nonfunctional; C = absent.
17. Curvature of maxillary cheek teeth: A = very curved; B = moderately curved; C = straight.
18. Maximum cheek tooth crown height: A = < 30 mm; B = 30–40 mm; C = 40–60 mm; D = 60–75 mm; E = ≥ 75 maximum crown height.
19. Maxillary cheek tooth fossette ornamentation: A = complex, with several deeply amplified plications; B = moderately complex with fewer, more shortly amplified, thinly banded plications; C = little complexity with few, shortly amplified plications; D = generally no plis; E = very complex.
20. Posterior wall of postfossette: A = may not be distinct; B = always distinct.
21. Pli caballin morphology: A = double; B = single or occasionally poorly defined double; C = complex; D = plis not well formed.
22. Hypoglyph: A = hypocone frequently encircled by hypoglyph; B = deeply incised, infrequently encircled hypocone; C = moderately deeply incised; D = shallowly incised.
23. Protocone shape: A = round q-shaped; B = oval q-shaped; C = oval; D = elongate-oval; E = lingually flattened-labially rounded; F = compressed or ovate; G = rounded; H = triangular; I = triangular-elongate; J = lenticular; K = triangular with rounded corners.
24. Isolation of protocone: A = connected to protoloph; B = isolated from protoloph.
25. Protoconal spur: A = elongate, strongly present; B = reduced, but usually present; C = very rare to absent.
26. Premolar protocone/hypocone alignment: A = anteroposteriorly aligned; B = protocone more lingually placed.
27. Molar protocone/hypocone alignment: A = anteroposteriorly aligned; B = protocone more lingually placed.
28. P2 anterostyle (28U) / paraconid (28L): A = elongate; B = short and rounded.
29. Mandibular incisor morphology: A = not grooved; B = grooved.
30. Mandibular incisor curvature: A = curved; B = straight.
31. I3 lateral aspect: A = elongate, not labiolingually constricted; B = very elongate, labiolingually constricted distally; C = atrophied.
32. Premolar metaconid: A = rounded; B = elongated; C = angular on distal surface; D = irregular; E = square; F = pointed.
33. Molar metaconid: A = rounded; B = elongated; C = angular on distal surface; D = irregular; E = square; F = pointed.
34. Premolar metastylid: A = rounded; B = elongate; C = angular on proximal surface; D = irregular; E = square; F = pointed.
35. Premolar metastylid spur: A = present; B = absent.
36. Molar metastylid: A = rounded; B = elongate; C = angular on proximal surface; D = irregular; E = square; F = pointed.
37. Molar metastylid spur: A = present; B = absent.
38. Premolar ectoflexid: A = does not separate metaconid and metastylid; B = separates metaconid and metastylid.
39. Molar ectoflexid: A = does not separate metaconid and metastylid; B = separates metaconid and metastylid; C = converges with preflexid and postflexid to abutt against metaconid and metastylid.
40. Pli caballinid: A = complex; B = rudimentary or single; C = absent.
41. Protostylid: A = present on occlusal surface, often as an enclosed enamel ring; B = absent on occlusal surface but may be on side of crown buried in cement; C = strong, columnar; D, a loop; E, a small, poorly developed loop; F = a small, pointed projection continuous with the buccal cingulum.
42. Protostylid orientation: A = courses obliquely to anterior surface of tooth; B = less oblique coursing, placed on anterior surface of tooth; C = vertically placed, lies flush with protoconid enamel band; D = vertically placed, lying lateral to protoconid band; E = open loop extending posterolabially.
43. Ectostylids: A = present; B = absent.
44. Premolar linguaflexid: A = shallow; B = deeper, V-shaped; C = shallow U shaped; D = deep, broad U-shaped; E = very broad and deep.
45. Molar linguaflexid: A = shallow; B = V shaped; C = shallow U shaped; D = deep, broad U-shaped; E = very broad and deep.
46. Preflexid morphology: A = simple margins; B = complex margins; C = very complex.
47. Postflexid morphology: A = simple margins; B = complex margins; C = very complex.
48. Postflexid invades metaconid/metastylid junction by anteriormost portion bending sharply lingually: A = no; B = yes.
49. Protoconid enamel band morphology: A = rounded; B = flattened.

Table 11.5. Sinap Hipparion Maxillary Cheek Tooth Character States by Stratigraphic Unit

Specimen Number	Species	Bone	Characteristic 19	20	21	22	23	24	25	26	27	28	29	30	31	Locality	Unit
	Hpri	txPM	A	B	AC	B	E	B	BC	B	B	A	A	A	A		
	Coc	txPM	B	B	AB	B	D	B	C	B	B	A	A	A	A		
SINAP																	
AS93/826	*C sin*	txP2	B	A	A	C	G	A	C	B		B				108	4
AS93/826	*C sin*	txP3	B	B	A	C	EG	A	C	B						108	4
AS93/826	*C sin*	txP4	B	B	A	C	C	B	C	B	B					108	4
AS93/826	*C sin*	txM1	B	B	C	C	C	A	C		B					108	4
AS93/826	*C sin*	txM2	B	B	C	C	E	B	C		B					108	4
AS93/826	*C sin*	txM3	B	B	C	A	E	B	C		B					108	4
SEN1	*C sin*	txP2	A-	B	C	B	C	B	C	B		A				108/8	5
SEN1	*C sin*	txP3	A-	A	C	B	C	B	C	B						108/8	5
SEN1	*C sin*	txP4	A-	A	A	B	CE	B	C	B						108/8	5
SEN1	*C sin*	txM1	A-	B	A	C	CE	B	C		B					108/8	5
SEN1	*C sin*	txM2	A-	A	C	C	CE	B	C		B					108/8	5
SEN1	*C sin*	txM3	A-	B	A	A	CE	B	C		B					108/8	5
AS92/458	"*H.*" sp. 1	txdP2	A	B	C	C	E	A	C	B		A				91	5
AS92/458	"*H.*" sp. 1	txdP3	B	B	A	C	C	A	C	B						91	5
AS92/458	"*H.*" sp. 1	txdP4	B	B	A	B	C	B	C	B						91	5
AS92/458	"*H.*" sp. 1	txM1	B	B	B	A	F	B	C		B					91	5
AS92/458	"*H.*" sp. 1	txM2		B			F	B	C		B					91	5
AS94/837	"*H.*" sp.	txP2	AB	A	C	B	J	A	C	B		AB				91	5
AS94/837	"*H.*" sp.	txP3	AB	A	C	B	J	B	C	B						91	5
AS94/837	"*H.*" sp.	txP4	AB	A	C	B	J	B	C	B						91	5
AS94/837	"*H.*" sp.	txM1	AB	A	C	C	J	B	B		B					91	5
AS94/837	"*H.*" sp.	txM2	AB	B	C	B	J	B	C		B					91	5
AS94/837	"*H.*" sp.	txM3	AB	B	C	A	J	B	C		B					91	5
AS94/203	"*H.*" *uzun T*	txI1											A	A		114	5
AS94/203	"*H.*" *uzun T*	txI2											A	A		114	5
AS94/203	"*H.*" *uzun T*	txI3											A	A	A	114	5
AS94/203	"*H.*" *uzun T*	txC														114	5
AS92/609	"*H.*" *uzun*	txP2	A	B	A	B	E	B	C	B		A				8B	5
AS92/609	"*H.*" *uzun*	txP3	A	B	A	B	E	B	C	B						8B	5
AS92/609	"*H.*" *uzun*	txP4	A	B	A	B	E	B	C	B						8B	5

AS92/609	"H." *uzun*	txM1	A	B	B	B	E	B	C		B					8B	5
AS92/609	"H." *uzun*	txM2	A	B	B	B	E	B	C		B					8B	5
AS92/609	"H." *uzun*	txM3	A	B	A	B	EF	B	C		B					8B	5
AS89/54	"H." *uzun*	txP2	D	A	D	D-	G	A	C	B						8B	5
AS89/54	"H." *uzun*	txP3	D	A	D	D-	E	A	C	B						8B	5
AS89/54	"H." *uzun*	txP4	D	B	D	D-	E	A	C	B						8B	5
AS89/54	"H." *uzun*	txM1	D	A	D	D-	G	A	C		B					8B	5
AS89/54	"H." *uzun*	txM2	B	B	D	D	E	A	C		B					8B	5
AS89/54	"H." *uzun*	txM3	B	B	A	C	E	A	C		B					8B	5
MNHNTRQ1211	"H." *anky*	txdP1	B	B	C	B	CD	B	C	B		B				OZ01	6
MNHNTRQ1211	"H." *anky*	txP2	B	B	C	B	CD	B	C	B						OZ01	6
MNHNTRQ1211	"H." *anky*	txP3	C	B	A	B	CE	B	C	B						OZ01	6
MNHNTRQ1211	"H." *anky*	txP4	B	B	C	B	CE	B	C	A						OZ01	6
MNHNTRQ1211	"H." *anky*	txM1	B	B	B	B	CE	B	C		A					OZ01	6
MNHNTRQ1211	"H." *anky*	txM2	B	B	B	B	CE	B	C		B					OZ01	6
MNHNTRQ1211	"H." *anky*	txM3	B	B	B	A	F	B	C		A					OZ01	6
MNHNTRQ9001	"H." *kec*	txI1											A	A		S01	6
MNHNTRQ9001	"H." *kec*	txI2											A	A		S01	6
MNHNTRQ9001	"H." *kec*	txI3											A	A	A	S01	6
MNHNTRQ9001	"H." *kec*	txC														S01	6
MNHNTRQ9001	"H." *kec*	txP2	B	B	A	B	C	B	B	B						S01	6
MNHNTRQ9001	"H." *kec*	txP3	B	B	A	B	C	B	C	B						S01	6
MNHNTRQ9001	"H." *kec*	txP4	B	B	A	B	C	B	C	B						S01	6
MNHNTRQ9001	"H." *kec*	txM1	B	B	A	B	C	B	C		B					S01	6
MNHNTRQ9001	"H." *kec*	txM2	B	B	A	B	C	B	C		B					S01	6
MNHNTRQ9001	"H." *kec*	txM3	C	B	D	D	F	B	A		A					S01	6
MNHNTRQ1064	"H." sp.	txI1											A	A		S01	6
MNHNTRQ1064	"H." sp.	txI2											A	A		S01	6
MNHNTRQ1064	"H." sp.	txI3											A	A	A	S01	6
MNHNTRQ1064	"H." sp.	txC														S01	6
MNHNTRQ1064	"H." sp.	txP2	B	B	A	C	C	A	B	A						S01	6
MNHNTRQ1064	"H." sp.	txP3	B	B	B	C	C	B	B	A						S01	6
MNHNTRQ1064	"H." sp.	txP4	B	B	B	C	C	B	C	A						S01	6
MNHNTRQ1064	"H." sp.	txM1	B	A	C	A	F	B	C		B					S01	6
MNHNTRQ1082	"H." sp.	txM1	B	A	B	A	F	B	C		B					S01	6
MNHNTRQ1071	"H." sp.	txdP4	AB	B	C	A	G	B	C	A						S01	6

(continued)

Table 11.5. Sinap Hipparion Maxillary Cheek Tooth Character States by Stratigraphic Unit *(continued)*

Specimen Number	Species	Bone	Characteristic 19	20	21	22	23	24	25	26	27	28	29	30	31	Locality	Unit
MNHNTRQ9003	"*H.*" sp.	txP4	A	B	C	B	F	B	C	B						S01	6
MNHNTRQ1067	"*H.*" sp.	txdP1														S01	6
MNHNTRQ1067	"*H.*" sp.	txP2	BC	B	C	B	F	B	B	B						S01	6
MNHNTRQ1067	"*H.*" sp.	txP3	BC	A	A	A	E	B	C	B						S01	6
MNHNTRQ1067	"*H.*" sp.	txP4	BC	A	A	A	F	B	B	B						S01	6
MNHNTRQ1065	"*H.*" sp.	txP2	AB	B	C	B	C	A	C	A						OZ02	6
MNHNTRQ1065	"*H.*" sp.	txP3	AB	B	A	B	F	B	C	B						OZ02	6
MNHNTRQ1065	"*H.*" sp.	txP4	AB	B	C	B	F	B	C	B						OZ02	6
MNHNTRQ1065	"*H.*" sp.	txM1	AB	B	A	B	F	B	C		A					OZ02	6
MNHNTRQ1065	"*H.*" sp.	txM2	AB	B	A	B	C	B	C		A					OZ02	6
MNHNTRQ1066	"*H.*" sp.	txP2	B	A	A	B	E	B	C	B						OZ02	6
MNHNTRQ1066	"*H.*" sp.	txP3	B	B	B	B	E	B	C	B						OZ02	6
MNHNTRQ1066	"*H.*" sp.	txP4	AB	B	C	B	E	B	C	B						OZ02	6
MNHNTRQ1066	"*H.*" sp.	txM1	AB	B	A	C	E	B	C		B					OZ02	6
MNHNTRQ1066	"*H.*" sp.	txM2	BA	B	B		E	B	C		B					OZ02	6
MNHNTRQ1066	"*H.*" sp.	txM3	B	B		A	H	B	C		B					OZ02	6
MNHNAKA15	"*H.*" sp.	txP2	B	B	B	D	C	B	C	B						OZKD	12
MNHNAKA15	"*H.*" sp.	txP3	AB	B	A	C	G	B	C	B						OZKD	12
MNHNAKA15	"*H.*" sp.	txP4	AB	B	A	C	G	B	C	B						OZKD	12
MNHNAKA15	"*H.*" sp.	txM1	B	B	B	D	G	B	C		B					OZKD	12
MNHNAKA15	"*H.*" sp.	txM2	AB	B	B	C	G	B	C		B					OZKD	12
MNHNAKA15	"*H.*" sp.	txM3	B	B	C	B	C	B	C		B					OZKD	12
MNHNAKA1069	"*H.*" sp.	txP3	C	A	D	D	C	B	C	B						OZKD	12
MNHNAKA1070	"*H.*" sp.	txP4	C	A	D	D	C	B	C	B						OZKD	12
MNHNAKA42	"*H.*" sp.	txP3	C	A	D	B	F	B	C	B						OZKD	12
MNHNAKA10	"*H.*" sp.	txM1	C	A	B	A	E	B	C		B					OZKD	12
MNHNAKA10	"*H.*" sp.	txM2	D	A	D			B	C		B					OZKD	12
MNHNAKA25	"*H.*" sp.	txP2	AB	B	A	D	C	B	C	B						OZKD	12
MNHNAKA25	"*H.*" sp.	txP3	B	B	B	D	G	B	C	B						OZKD	12
MNHNAKA25	"*H.*" sp.	txP4	B	B	A	C	G	B	C	B						OZKD	12
MNHNAKA25	"*H.*" sp.	txM1	B	B	A	D	G	B	C		B					OZKD	12
MNHNAKA25	"*H.*" sp.	txM2	B	B	A	D	H	B	C		B					OZKD	12
MNHNAKA25	"*H.*" sp.	txM3	B	B	C	B	J	B	C		A					OZKD	12

MNHNAKA24	"*H.*" sp.	txdP2	C	A	D	A	C	B	A	A	OZKD	12
MNHNAKA23	"*H.*" sp.	txdP3	C	B	B	A	C	B	A	A	OZKD	12
MNHNAKA22	"*H.*" sp.	txdP4	C	A	A	A	C	B	C	A	OZKD	12
MNHNAKA58	"*H.*" sp.	txdP3	AB	B	A	B	E	B	C	B	OZKD	12
MNHNAKA54	"*H.*" sp.	txdP1									OZKD	12
MNHNAKA54	"*H.*" sp.	txdP2	B	B	A	C	C	B	C	B	OZKD	12
MNHNAKA56	"*H.*" sp.	txdP3	B	B	B	B	G	B	C	A	OZKD	12
MNHNAKA57	"*H.*" sp.	txdP4	B	A	C	C	E	B	C	A	OZKD	12
MNHNAKA41	"*H.*" sp.	txdP3	C	A	B	A	G	B	C	A	OZKD	12
MNHNAKA49	"*H.*" sp.	txdP4	C	A	C	A	E	B	C	A	OZKD	12

Sources: Adapted from Bernor et al. (1989), Bernor and Lipscomb (1991, 1995); Bernor and Armour-Chelu (1999).

Note: For definition of characters, see table 11.4.

Table 11.6. Sinap Hipparion Mandibular Cheek Tooth Character States by Stratigraphic Unit

Specimen Number	Species	Bone	Character																			Locality	Unit
			31	32	33	34	35	36	37	38	39	40	41	42	43	44	45	46	47	48	49		
	Hpri s.s.	tmpm	A	A	A	A	A	A	B	A	B	A	A	C	B	A	B	B	B	A	A		
AS92/618	*Csin*	tmP4		D		D	B			A		C	B	C	B	B		A	A	B		94	2
AS92/458	"*H.*" sp. 1	tmdP2		A		A	B			A		C	B		B	B		A	A	A	A	91	5
AS92/458	"*H.*" sp. 1	tmdP3		A		A	B			A		C	D	B	B	B		A	A	A	A	91	5
AS92/458	"*H.*" sp. 1	tmdP4		A		A	B			A		C	D	B	A	B		A	A	A	A	91	5
AS92/458	"*H.*" sp. 1	tmM1			AB			AB	B		B	C	B		B		C	A	A	A	A	91	5
AS92/458	"*H.*" sp. 1	tmM2			AB				B				B		B							91	5
AS94/202	"*H.*" sp.	tmP2		C		BE	A			A		A	B		B	A		B	B	A	A	114	5
AS94/202	"*H.*" sp.	tmP3		B		BE	A			A		A	B		B	A		B	B	A	A	114	5
AS94/202	"*H.*" sp.	tmP4		B		BE	A			A		A	B		B	A		B	B	A	A	114	5
AS94/202	"*H.*" sp.	tmM1			B			AB	B		B	B	B		B		B	B	B	A	A	114	5
AS94/202	"*H.*" sp.	tmM2			B			AB	B		B	B	B		B		B	B	B	A	A	114	5
AS94/202	"*H.*" sp.	tmM3			B			AB	B				B		B		B	A	B	A	A	114	5
MNHNTRQ1097	"*H.*" sp.	tmP2		A		A	B			B		B			B	B		A	A	A	A	S01	6
MNHNTRQ1097	"*H.*" sp.	tmP3		A		E	B			A		B	B		B	B		A	A	A	A	S01	6
MNHNTRQ1097	"*H.*" sp.	tmP4		A		A	B			A		B	B		B	B		A	A	A	A	S01	6
MNHNTRQ1097	"*H.*" sp.	tmM1			B			E	B		B	C	B		B		C	A	A	A	B	S01	6
MNHNTRQ1097	"*H.*" sp.	tmM2			B			A	B		B	C	B		B		C	A	A	A	A	S01	6
MNHNTRQ1090	"*H.*" sp.	tmI1																				S01	6
MNHNTRQ1090	"*H.*" sp.	tmI2																				S01	6
MNHNTRQ1090	"*H.*" sp.	tmI3																				S01	6
MNHNTRQ1090	"*H.*" sp.	tmC																				S01	6
MNHNTRQ1090	"*H.*" sp.	tmP2		A		A	A			A		B			B	A		A	A	A	A	S01	6
MNHNTRQ1090	"*H.*" sp.	tmP3				E	A			A		C	B		B	A		A	B	A	A	S01	6
MNHNTRQ1096	"*H.*" sp.	tmP2		B		A	B			A		C			B	B		A	A	A	A	S01	6
MNHNTRQ1096	"*H.*" sp.	tmP3		A		E	B			A		C	A	BC	B	B		B	A	A	A	S01	6
MNHNTRQ1096	"*H.*" sp.	tmP4		A		E	B			A		B	A	BC	B	B		B	A	A	A	S01	6
MNHNTRQ1096	"*H.*" sp.	tmM1			A			E	B		B	B	A	BC	B		C	A	A	A	A	S01	6
MNHNTRQ1098	"*H.*" sp.	tmP2		A		E	B			B		B			B	B		A	A	A	A	S01	6
MNHNTRQ1084	"*H.*" sp.	tmP2																				S01	6
MNHNTRQRLB04	"*H.*" sp.	tmI1																				S01	6
MNHNTRQRLB04	"*H.*" sp.	tmI2																				S01	6
MNHNTRQRLB04	"*H.*" sp.	tmI3																				S01	6

MNHNTRQRLB04	“*H.*” sp.	tmC																			S01	6
MNHNTRQRLB05	“*H.*” sp.	tmP2	A		A	B			A					B	B			A	A	A	S01	6
MNHNTRQRLB05	“*H.*” sp.	tmP3	A		A	B			A		B			B	B		A	A	A	A	S01	6
MNHNTRQRLB05	“*H.*” sp.	tmP4	A		E	B			A		B			B	B		A	A	A	A	S01	6
MNHNTRQRLB05	“*H.*” sp.	tmM1		A			E	B		B	B			B		D	A	A	A	A	S01	6
MNHNTRQRLB05	“*H.*” sp.	tmM2		B			E	B		B	B			B		D	A	A	A	A	S01	6
MNHNTRQRLB05	“*H.*” sp.	tmM3		A			E	B		B	B			B		C	A	A	A	A	S01	6
MNHNTRQ1093	“*H.*” sp.	tmdP2	A		A	B			B		C			A	B		A	A	A	A	OZ02	6
MNHNTRQ1093	“*H.*” sp.	tmdP3	B		A	B			B		C	D	D	A	B		A	A	A	A	OZ02	6
MNHNTRQ1093	“*H.*” sp.	tmdP4	B		A	B			B		C	D	D	A	C		A	A	A	A	OZ02	6
MNHNTRQ1093	“*H.*” sp.	tmM1		B			A	B		B	C	B		B		B	B	A	A	A	OZ02	6
MNHNTRQ1093	“*H.*” sp.	tmM2		A			AC	B		A	C	B		B		B	A	A	A	A	OZ02	6
MNHNTRQ1094	“*H.*” sp.	tmP2	A		AB	B			B		A			B	A	A		A	A	A	OZ02	6
MNHNTRQ1094	“*H.*” sp.	tmP3	E		A	A			A		A	B		B	A		B	A	A	A	OZ02	6
MNHNTRQ1094	“*H.*” sp.	tmP4	E		A	A			A		A	B		B	A		B	A	A	A	OZ02	6
MNHNTRQ1094	“*H.*” sp.	tmM1		B			AE	B	B		C	B		B		B	A	A	A	A	OZ02	6
MNHNTRQ1086	“*H.*” sp.	tmP2	A		B	B			B		B			B	B		A	A	A	A	OZ02	6
MNHNTRQ1086	“*H.*” sp.	tmP3	B		A	A			A		A	B		B	A		B	A	A	A	OZ02	6
MNHNTRQ1086	“*H.*” sp.	tmP4	B		A	A			A		A	B		B	A		B	A	A	B	OZ02	6
MNHNTRQ1086	“*H.*” sp.	tmM1		B			E	B		B	C	B		B		C	B	A	A	A	OZ02	6
MNHNTRQ1086	“*H.*” sp.	tmM2		A			E	B		B	C	B		B		C	A	A	A	A	OZ02	6
MNHNTRQ1087	“*H.*” sp.	tmdP2	A		A	B			B		C	B		B	B		A	A	A	A	OZ02	6
MNHNTRQ1087	“*H.*” sp.	tmdP3	A		A	B			B		C	B		B	B		A	A	A	A	OZ02	6
MNHNTRQ1087	“*H.*” sp.	tmdP4	B		A	B			B		C	B		B	C		A	A	A	A	OZ02	6
MNHNTRQ1087	“*H.*” sp.	tmM1		B			E	B		B	C	B		B		C	A	A	A	B	OZ02	6
MNHNTRQ1087	“*H.*” sp.	tmM2		D			E	B		A	C	B		B		B	A	A	A	B	OZ02	6
MNHNTRQ1088	“*H.*” sp.	tmdP2	A		A	B			A		C			A	B		A	A	A	A	OZ02	6
MNHNTRQ1088	“*H.*” sp.	tmdP3	A		A	B			A		C	D	D	A	B		A	A	A	A	OZ02	6
MNHNTRQ1088	“*H.*” sp.	tmdP4	E		A	B			B		C	D	D	A	B		A	A	A	A	OZ02	6
MNHNTRQ1088	“*H.*” sp.	tmM1		D			A	B		B	C	B		B		A	B	B	A	A	OZ02	6
MNHNTRQ1088	“*H.*” sp.	tmM2		B			E	B		A	C	B		B		A	A	A	A	B	OZ02	6
MNHNTRQ1089	“*H.*” sp.	tmdP2																			OZ02	6
MNHNTRQ1089	“*H.*” sp.	tmdP3							B		C	D	D	A					A		OZ02	6
MNHNTRQ1089	“*H.*” sp.	tmdP4	A		A	B			B		C	D	D	A	D		A	A	A	A	OZ02	6
MNHNTRQ1089	“*H.*” sp.	tmM1		A			A	B		B	C	B	D	B		D	A	A	A	A	OZ02	6
MNHNTRQ1089	“*H.*” sp.	tmM2		A			E	B		B	C			B		B	A	A	A	B	OZ02	6

(continued)

Table 11.6. Sinap Hipparion Mandibular Cheek Tooth Character States by Stratigraphic Unit *(continued)*

Specimen Number	Species	Bone	31	32	33	34	35	36	37	38	39	40	41	42	43	44	45	46	47	48	49	Locality	Unit
			Character																				
MNHNTRQ1091	"*H.*" sp.	tmP2		A		A	B			A		C			B	B				A	B	OZ02	6
MNHNTRQ1091	"*H.*" sp.	tmdP3		E		A	B			B		C	D	D	A	B		A	A	A	A	OZ02	6
MNHNTRQ1091	"*H.*" sp.	tmdP4		E		A	B			B		C	D	D	A	D		A	A	A	A	OZ02	6
MNHNTRQ1091	"*H.*" sp.	tmM1			B			E	B		B	C	A	B	B		D	A	A	A	A	OZ02	6
MNHNTRQ1091	"*H.*" sp.	tmM2			A			A	B		B	B	B	B	B		D	A	A	A	A	OZ02	6
MNHNTRQ1092	"*H.*" sp.	tmdP2		B		A	B			B		C	D	D	A	B		A	A	A	A	OZ02	6
MNHNTRQ1092	"*H.*" sp.	tmdP3		E		AE	B			B		C	D	D	A	B		A	A	A	A	OZ02	6
MNHNTRQ1092	"*H.*" sp.	tmdP4		E		A	B			B		C	D	D	A	C		A	A	A	A	OZ02	6
MNHNTRQ1092	"*H.*" sp.	tmM1			A			A	B		B	C	A	D	B		C	A	A	A	A	OZ02	6
MNHNTRQ1092	"*H.*" sp.	tmM2			B			E	B		B	C	B	B	B		B	A	A	A	B	OZ02	6
MNHNAKA1	"*H.*" sp.	tmdP2		D		E	A			A		B			A	A		A	B	A	B	OzKD	12
MNHNAKA1	"*H.*" sp.	tmdP3		D		E	B			B		B	F	D	A	B		A	A	A	A	OzKD	12
MNHNAKA1	"*H.*" sp.	tmdP4		D		E	B			B		B	F	D	A	B		A	A	A	A	OzKD	12
MNHNAKA46	"*H.*" sp.	tmdP4		B		A	B			B		C	A	D	A	C		A	A	A	A	OzKD	12
MNHNAKA44	"*H.*" sp.	tmP2		A		A	B			A		C	B		B	C		A	A	A	A	OzKD	12
MNHNAKA44	"*H.*" sp.	tmP3		A		E	B			A		B	A	BC	B	C		A	A	A	A	OzKD	12
MNHNAKA44	"*H.*" sp.	tmP4		A		E	B			A		C	A	BC	B	CD		A	A	A	A	OzKD	12
MNHNAKA44	"*H.*" sp.	tmM1			A			A	B		B	C	A	BC	B		D	A	A	A	A	OzKD	12
MNHNAKA44	"*H.*" sp.	tmM2			A			E	B		B	C	A	BC	B		D	A	A	A	A	OzKD	12
MNHNAKA45	"*H.*" sp.	tmP2		B		A	B			A		C			B	A		A	A	A	A	OzKD	12
MNHNAKA45	"*H.*" sp.	tmP3		A		A	B			A		C	B		B	C		A	A	A	A	OzKD	12
MNHNAKA45	"*H.*" sp.	tmP4		A		A	B			A		C	B		B	D		A	A	A	A	OzKD	12
MNHNAKA45	"*H.*" sp.	tmM1			A			A	B		B	C	B		B		D	A	A	A	A	OzKD	12
MNHNAKA45	"*H.*" sp.	tmM2			A			A	B		B	C	B		B		D	A	A	A	A	OzKD	12
MNHNAKA45	"*H.*" sp.	tmM3			A			A	B		B	C	B		B		D	A	A	A	A	OzKD	12
AS93/284	"*H.*" sp.	tmM2		AB				BE	B		B	C	B	C	B		B	B	A	A	A		

Sources: Adapted from Bernor et al. (1989), Bernor and Lipscomb (1991, 1995); Bernor and Armour-Chelu (1999).

Note: For definition of characters, see table 11.4.

stage (crown height <20 mm), but is also caused by the intrinsically low crown height of the early Sinap hipparion (see the beginning of the section on mandibular cheek tooth character state attributes). Characters 25 (state of protoconal spur), C26, and C27 (premolar and molar protocone-hypocone alignment) are essentially the same as for *H. primigenium* and *C. "occidentale."* The Sinap hipparion differs from both *H. primigenium* and *C. "occidentale"* in having an abbreviated P2 anterostyle (C28 = B).

Stratigraphic Block Unit 5 has three skulls, including two from Loc. 91 (AS92/458, a juvenile with dP2–4, M1, and M2 and AS93/837; 9.997 Ma) and one from Loc. 108/8 (SEN 1; 9.970 Ma). From higher within Unit 5, there are two skulls: one from Loc. 114 (AS94/203; 9.967 Ma; still in a partially prepared block with only the incisors and canine available for morphologic evaluation) and another from Loc. 8B (AS89/54; 9.918 Ma). Because character state distributions for deciduous cheek teeth appear to differ *within* species of hipparion, and too little research on them has been published to date, we confine our observations of this and succeeding stratigraphic units to observations on adult cheek teeth. Furthermore, we specifically highlight only those features that differ significantly from *Hippotherium primigenium, Cormohipparion "occidentale,"* and the Unit 4 Sinap sample.

Most of the Unit 5 sample does not appear to have a strongly reduced enamel fossette ornamentation (C19 = A, B), as in the Unit 4 sample (which may well have been due to advanced wear), but it is still less ornate than seen in central European *Hippotherium primigenium*. The exception to this observation is seen in the upper Unit 5 specimen AS89/54 from Loc. 8B that has no plis on P2–M1 (19D), likely due to advanced stage-of-wear. Hypoglyph morphology (C22 = A, B, C, and D) exhibits a range of variability beyond what we see for *Hippotherium primigenium* (and perhaps *Cormohipparion "occidentale"*); this is due only in part to the stage-of-wear in AS89/54. These taxa tend to retain deeper incision (state B) through ontogeny, and in this regard, the Sinap hipparions exhibit an affinity with species of *Hipparion* s.s. Although protocone shape is notoriously variable through ontogeny (C23), the lenticular shape exhibited in the Loc. 91 specimen AS94/837 (C23 = J) is not one common to *Hippotherium primigenium, Cormohipparion "occidentale"* or *Hipparion* s.s. The rounded morphology exhibited in Loc. 8B specimen AS89/54 may partially be due to advanced wear, but may also show an alliance with *Hipparion* s.s. All remaining characters are consistent with observations made for the Unit 4 Sinap hipparions. This unit shows some indication of advances in cheek tooth occlusal morphology: a shift to simpler plication frequencies than found in primitive hipparions.

Stratigraphic Unit 6 includes several individuals: MNHNTRQ1211 (OZ01; 9.683 Ma), MNHNTRQ9001 (S01; 9.683 Ma), MNHNTRQ1064 (S01), MNHNTRQ1082 (S01), MNHNTRQ1071 (S01) MNHNTRQ9003 (S01), MNHNTRQ1067 (S01), MNHNTRQ1065 (OZ02), and MNHNTRQ1066 (OZ02). Maxillary cheek tooth ornamentation exhibits a more consistent simplification in this unit's sample than in Unit 5; in fact, these trends are heralded in upper Unit 5 specimen AS89/54. Maxillary cheek tooth ornamentation is more consistently moderately complex (C19B), pli caballin morphology is variable (21A, B, and C), as is protocone morphology (C23C, E, F, G, and H). Protoconal spur very rarely occurs in this level (C25A, B). Premolar (C26A) and molar (C27A) protocone alignment exhibits a remarkable incidence of primitiveness (anteroposterior alignment) in this unit's sample.

Stratigraphic Unit 12 includes 14 individuals (some represented by a single element) from OZKD, nine of which are deciduous cheek teeth and are not further analyzed here. Protocone shape exhibits an increased incidence of the rounded state (C23 = G). There is no evidence of a protoconal spur (C25 = C) on any of the Unit 8 specimens except the deciduous cheek teeth; this pattern is typical of more advanced members of the *Hipparion* s.s. clade.

In summary, the maxillary cheek teeth reveal an interesting suite of trends in their characteristics, which reflect a more direct derivation from *Cormohipparion "occidentale"* than *Hippotherium primigenium,* and presage the Eurasian *Hipparion* s.s. group. These include relatively low cheek tooth crown height, especially in the lower stratigraphic units; moderate-to-simpler cheek tooth plication; simpler pli caballins; hypoglyphs less deeply incised in middle-to-later wear stages; protocones not so consistently lingually flattened and labially rounded, but with character states divergent from the *Hippotherium primigenium* pattern; protoconal spur not as developed and lost altogether in higher stratigraphic intervals.

Mandibular Cheek Teeth

Bernor et al.'s (1997) and Bernor and Franzen's (1997) studies of the Höwenegg and Dorn Dürkheim populations of *Hippotherium primigenium* showed that mandibular cheek tooth character states are more variable than the maxillary states. Yet, in middle stage-of-wear, there is reasonable stability in many of these states. We have no other comparable data on mandibular cheek teeth for any Old World hipparion or for *Cormohipparion "occidentale,"* which itself is in need of a rigorous taxonomic revision.

Although there is certain Hipparion material from as low as Loc. 4 (table 11.1; 10.692 Ma), the first good diagnostic specimen comes from Loc. 94 (Unit 2; 10.551 Ma). This specimen, AS92/618, is a well-preserved p4 in a very early stage of wear (tooth had just begun to emerge from its crypt and wear). This specimen has a crown height of only 40 mm (very close to its maximum), which is extraordinarily low for an Old World hipparion (see Bernor and Franzen 1997; Bernor et al. 1997; Bernor and Armour-Chelu 1999). Even the chronologically oldest central European members of *Hippotherium primigenium* from Gaiselberg (Austria) have maximum crown heights of 50 mm (NHMW8816, a maxillary M1 has a crown height of 49.8 mm, which is virtually identical to maximum crown heights for all studied populations of Central European *H. primigenium*). We know that this specimen is in early wear because of the early emergent occlusal wear pattern and, in particular, the irregular shape of the metaconid (C32 = D) and metastylid (C34 = D) typical for early wear stages of *H. primigenium* (Bernor and Franzen 1997). The metastylid spur is absent on this specimen (C35 = B), a condition common for an Old World hipparionine p4. As is typical for *H. primigenium* (but not all of the time; Bernor and Franzen 1997) as well as for other Old World hipparions, the premolar ectoflexid does not separate the metaconid and metastylid (C38 = A). However, unlike *H. primigenium,* pli caballinid is absent (C40 = C) and protostylid is absent (C41 = B); these states occur in members of the *Hipparion* s.s. lineage. Specimen AS92/618 does compare closely with *H. primigenium* in protostylid orientation (C42 = C) and its lack of ectostylids (C43 = B), both being plesiomorphies for Old World hipparions. Premolar linguaflexid differs from *H. primigenium* in having a deeper V shape (C44 = B; although *H. primigenium* does variably express this character). Preflexid and postflexid morphology are also unlike *H. primigenium* in that they are simple (C46 = A; C47 = A); this is a state common to *Hipparion* s.s. Postflexid invasion of the metaconid-metastylid junction is primitive, as in most Old World hipparions. On balance, this specimen does not exhibit characters particularly typical of central or western European members of the *H. primigenium* clade. Especially revealing in this regard is the low maximum crown height that is a more primitive state than known to us in western Eurasian hipparionine horses. When specific character states differ with *H. primigenium,* they concur with Old World *Hipparion* s.s.

The next stratigraphic unit represented by mandibular cheek teeth is Unit 5 and includes a juvenile individual from Loc. 91, AS92/458 (with dp2–4, m1–2; associated with the AS92/458 skull) and a mandible fragment with p2–m3, AS94/202, from Loc. 114 (9.967 Ma). Specimen AS92/458 includes deciduous premolars that cannot be adequately compared with any other hipparion population because of the lack of any meaningful comparative study. The molars, however, show strong similarities to AS92/618 in all characters except that the metaconid (C33 = AB) and metastylid (C36 = AB) are rounded to elongate, as is often found in the Dorn Dürkheim hipparions. As is common in *Cormohipparion "occidentale"* and Old World hipparion molars, ectoflexid separates the metaconid from the metastylid (C39 = B); molar linguaflexid agrees with a majority percentage of *Hippotherium primigenium* in having a shallow U shape (C45 = C). Preflexid and postflexid morphology is as in the Unit 2 specimen, and protoconid band morphology is as in all Eurasian hipparions except members of the *"Sivalhippus"* Complex.

The Loc. 114 specimen AS94/202 differs from *Hippotherium primigenium* and/or those individuals from the lower stratigraphic units in the following characteristics: premolar metaconid is elongate to angular on its distal surface (C32 = B or C), premolar metastylid is elongate-square (C34 = BE), and molar pli caballinid is rudimentary or single (C40 = B). Premolar linguaflexid is shallow (C44 = A) as in *H. primigenium.* Likewise, molar linguaflexid (C45 = B), preflexid (C46 = B), postflexid (C47 = B—except m3, a normal variant; C48 = A), and enamel band (C49 = A) morphologies are as in *H. primigenium* rather than as in older Sinap individuals.

Unit 6 has several lower dentitions. Loc. S01 (9.683 Ma) specimens include MNHNTRQ1097, a mandible with p2–m2; MNHNTRQ1090, a mandible with i1–p3; MNHNTRQ1096, a mandible fragment with p2–m1; MNHNTRQ1098, a p2; MNHNTRQ1084, a p2; MNHNTRQRLB04, a mandibular symphysis fragment with i1–c; and MNHNTRQRLB05, a mandible with p2–m3. The stratigraphically slightly higher locality OZ02 (9.590 Ma) also has a number of individuals: MNHNTRQ1093, a juvenile mandible with dp2–m2; MNHNTRQ1094, a mandible fragment with p2–m1; MNHNTRQ1086, a mandible with p2–m2; MNHNTRQ1087, a juvenile mandible with dp2–m2; MNHNTRQ1088, a juvenile mandible with dp2–m2; MNHNTRQ1089, a juvenile mandible with dp2–m2; MNHNTRQ1091, a mandible with p2–m2; and MNHNTRQ1092, a juvenile mandible with dp2–m2.

The Unit 6 sample exhibits the following characteristics, many of which deviate from the *Hipparion* s.s. characterization in table 11.6: premolar metaconid shape is frequently rounded, but also exhibits elongated and square shapes (C32B and E, respectively); molar metaconid exhibits a mostly rounded shape (C33A), but also elongated (C33B) and irregular (C33D) shapes, the latter being due to early stage of wear; premolar metastylid exhibits predominately rounded (C34A) and rarely elongated (C34B) or square (C34E) shapes; premolar metastylid spur occurs (C35A) in less than half the cases, but this is rather high for an Old World hipparion assemblage; molar metastylid is mostly square shaped (C36E), and also exhibits rounded shape (C36A) and rarely has an angular shape on the proximal surface (C36C); molar metastylid spur never occurs (C37B); premolar ectoflexid variably does not separate metaconid from metastylid (C38A) or, alternatively, does (C38B; a primitive character for Old World hipparions); molar ectoflexid mostly separates metaconid and metastylid (C39B); pli caballinid is mostly absent (C40C), occasionally is complex (C40A) or single (C40B); protostylid is mostly absent on the occlusal surface (C41B), but expressed as an enclosed enamel ring (C41A) or loop on a worn tooth (C41D); protostylid orientation varies mostly from being slightly obliquely coursing (C42B), to vertically placed and lying lateral to the protoconid band (C42D); there are no apparent ectostylids on the permanent dentition (C43B); premolar linguaflexid is highly variable expressing shallow (C44A), deeper V shape (C44B), shallow U shape (C44C), and a deep and broad U shape (C44D); molar linguaflexid (C45A, B, and C) exhibits similar variability as seen in the premolars; preflexid morphology is mostly simple (C46A), rarely exhibiting complex margins (C46B); postflexid morphology has even a lower incidence of complex margins (C47B) than seen in the preflexids; postflexid never invades metaconid-metastylid junction by anteriormost portion bending sharply lingually (C48A); protoconid enamel band is mostly rounded (C49A) but rarely flattened (C49B). The variability exhibited in the lower cheek teeth suggest taxonomic diversity at this level.

There is a >1.5 m.y. gap between the youngest Unit 6 locality (Locs. 12 and OZ02, 9.683 Ma) and the Unit 12 sample from OZKD (8.121 Ma). The characterization of the adult cheek teeth from this level are as follows: premolar metaconid shape varies from rounded (C32A) to elongate (C32B); molar metaconid is always rounded (C33A); premolar metastylid varies from rounded to square (C34A and E, respectively); premolar metastylid spur is rarely present (C35A); molar metastylid varies from being rounded (C36A) to square (C36E; rarely with an elongate component, B); molar metastylid spur is always absent (C37B); premolar ectoflexid variably does not (C38A) or does (C38B) separate metaconid from metastylid; molar ectoflexid always separates metaconid from metastylid (C39B); pli caballinid is uniformly absent (C40C); protostylid is mostly absent on the occlusal surface (C41B), but when present is expressed as an enclosed enamel ring (C41A); protostylid orientation is slightly oblique in its orientation (C42BC); ectostylids are absent in the permanent cheek teeth (C43B); premolar linguaflexids are mostly shallowly U shaped (C44C), but also exhibit shallower (C44A) and deeper U-shaped (C44D) morphologies; molar linguaflexids are predominately deeply U shaped (C45D); preflexids (C46A) and postflexids (C47A) have mostly simple margins; postflexid does not have anteriormost portion bending sharply lingually (C48A); and protoconid enamel band is always rounded in adult cheek teeth (C49A).

Statistical Analysis of Skull Continuous Variables

Table 11.7 summarizes some critical skull measurements of the Sinap skulls. Our metric analysis of the skull includes box-and-whiskers plots on several variables. Figure 11.1A is a plot of M9 (length of maxillary cheek tooth row [P2–M3]), which we consider to be a reasonable proxy for size and body mass. The Sinap sample (1) has no outlier, and its median and interquartile range (hereafter IQR) is virtually identical to the Xmas Quarry sample (2) of *Cormohipparion "occidentale."* The Hans Johnson Quarry sample (3) and MacAdams Quarry (4) sample both have an IQR that overlaps with the Sinap and Xmas Quarry samples, but the former has a lower median while the latter has a higher median than Sinap and a potential outlier between the upper inner and outer fences. The Pawnee Creek sample (6) of *C. quinni* (Woodburne 1996), and the Trinity River Pit (5) sample of *C. goorisi,* the oldest and most primitive hipparionine (Bernor et al. 1989; Bernor and Lipscomb 1991, 1995; Woodburne 1996) both have values well below the interquartile ranges of Sinap and all other taxa presented here; they were certainly the smallest bodied horses of this sample. The Sinap assemblage also overlaps in its interquartile range with the Maragheh (7) sample of *Hipparion* s.s. taxa, *H. gettyi, H. prostylum,* and *H. campbelli.* The central European samples of *Hippotherium primigenium* from Inzersdorf (8; Austria) and Höwenegg (10; Germany) have the highest medians and IQRs. Moreover, the Höwenegg sample has a potential outlier between the lower inner and outer fence; this is most likely an older individual in an advanced stage of wear. The Cucuron (13) sample, which includes the Genotype species *Hipparion prostylum,* has a lower median but overlaps with the Sinap series. The one measured specimen from Hostalets de Pierola (14) is elevated compared with the Sinap IQR, but lower than the central European *Hippotherium primigenium* sample. The size similarity of the Sinap assemblage to North American *C. "occidentale"* sample, as well as members of the Old World *Hipparion* s.s. group, is an interesting observation which is further tracked in our ensuing analyses.

Figure 11.1B is a box-and-whiskers plot of M30 (length of the naso-incisival notch from prosthion to the distal limit of the narial opening; effectively, one measurement of snout length). The Sinap sample (1) has a slightly higher median and higher upper quartile boundary (hereafter QU) than both the Xmas Quarry (2) and Hans Johnson Quarry (3) samples; the MacAdams Quarry (4), Pawnee Creek (6), and Trinity River Pit (5) samples have successively lower medians. The Maragheh (7), Inzersdorf (8), and Höwenegg (10) samples all have higher medians than those cited above, but the IQR overlaps between Sinap, Maragheh, and Höwenegg. The Sinap sample is poised as intermediate in this dimension between the North American *Cormohipparion* sample and Old World *Hippotherium primigenium* and Maragheh *Hipparion* s.s. taxa.

Figure 11.1C, showing M31 (cheek length from the posterior limit of the narial opening to the anteriormost point of the orbit), is also revealing. The one Sinap (1) specimen (AS94/203) in this analysis is virtually identical to the median value for the Xmas (2) and Hans Johnson Quarries (3) samples, and only somewhat less than the median for the MacAdam's Quarry (4) sample and somewhat greater than the Pawnee Creek (6) and Trinity River Pit (5) samples. Interestingly, Maragheh (7) has a virtually identical median to the Pawnee Creek (6) sample, whereas Inzersdorf's (8) median closely corresponds to that of Sinap. The Höwenegg (10) median is the highest, but the IQR overlaps with the MacAdam's Quarry (4) sample.

Figure 11.1D plots M32 (the length of the preorbital bar [POB] from anteriormost limit of the orbit to the posteriormost limit of the preorbital fossa [POF]). Sinap (1) has a very restricted IQR, which overlaps with the Xmas (2) and Hans Johnson (3) Quarries samples of *Cormohipparion "occidentale"* and the Maragheh (7) and Cucuron (13) samples of *Hipparion* s.s. The one observation for *"Hippotherium" catalaunicum* (14) is very slightly elevated compared to the median and IQR of Sinap, whereas the central European *Hippotherium* (8 and 10) and the MacAdam's Quarry (4) samples are similarly elevated above Sinap and the other samples under consideration.

Figure 11.1E shows M33 (the maximum length of the POF). Again, Sinap's (1) IQR overlaps with the three *Cormohipparion "occidentale"* quarries (2, 3, and 4), the *C. quinni* sample (6), Maragheh (7), and the lowermost portion of

Table 11.7. Critical Measurements on Sinap Hipparion Skulls

Specimen Number	Species	Locality	Unit	Age (Ma)	M1	M7	M8	M9	M14	M15	M32	M33	M34	M35	M36
AS93/826	*Csin T*	108	4	10.14		76.6	65.8	141.7			41	61			25.6
AS93/837	"*H.*" sp.	91	5	9.98		77.6	64.5	141.8							
AS92/458	*Hns1*	91	5	9.98		85.6			29.6			43.7	51.5	30.2	29.3
SEN 1	*Csin*	108/8	5	9.97		73.8	63.9	136.3			41.2	71.4		23.2	16.5
AS94/203	"*H*" *uzun T*	114	5	9.97	124.7	71.6	59.6	129.4		60	43.4	56.8	66.2	32.5	32.4
AS89/54	"*H*" *uzun*	8B	5	9.92		68.1	55.6	123.8			39.6			47.9	34.2
AS92/609	"*H*" *uzun*	8B	5	9.92		83.4	66	148.9			38.7	59.7	53.7	36.1	31.5
AS92/606	"*H.*" sp.	8B	5	9.92		82.1	78.4	159.1							
MNHNTRQ1211	Hanky T	OZ01	6	9.68		67.5	65.6	141							23.5
MNHNTRQ1064	"*H.*" sp.	S01	6	9.68	94.1	75.9			42.2	56.9					
MNHNTRQ9001	"*H.*" *kecT*	S01	6	9.68	102.1	77.7	64.9	141	43	60		62.8		42	20.2
MNHNTRQ1067	"*H.*" sp.	S01	6	9.68		77.3									
MNHNAKA1066	"*H.*" sp.	OZ02	6	9.59		79.8	62.9	142.2							26.9
MNHNTRQ1065	"*H.*" sp.	OZ02	6	9.59		77.8									24.4
MNHNAKA15	"*H.*" sp.	OZKD	12	8.12		83.9	65.9	148.3							27.5
MNHNAKA25	"*H.*" sp.	OZKD	12	8.12		84.9	72.6	156							

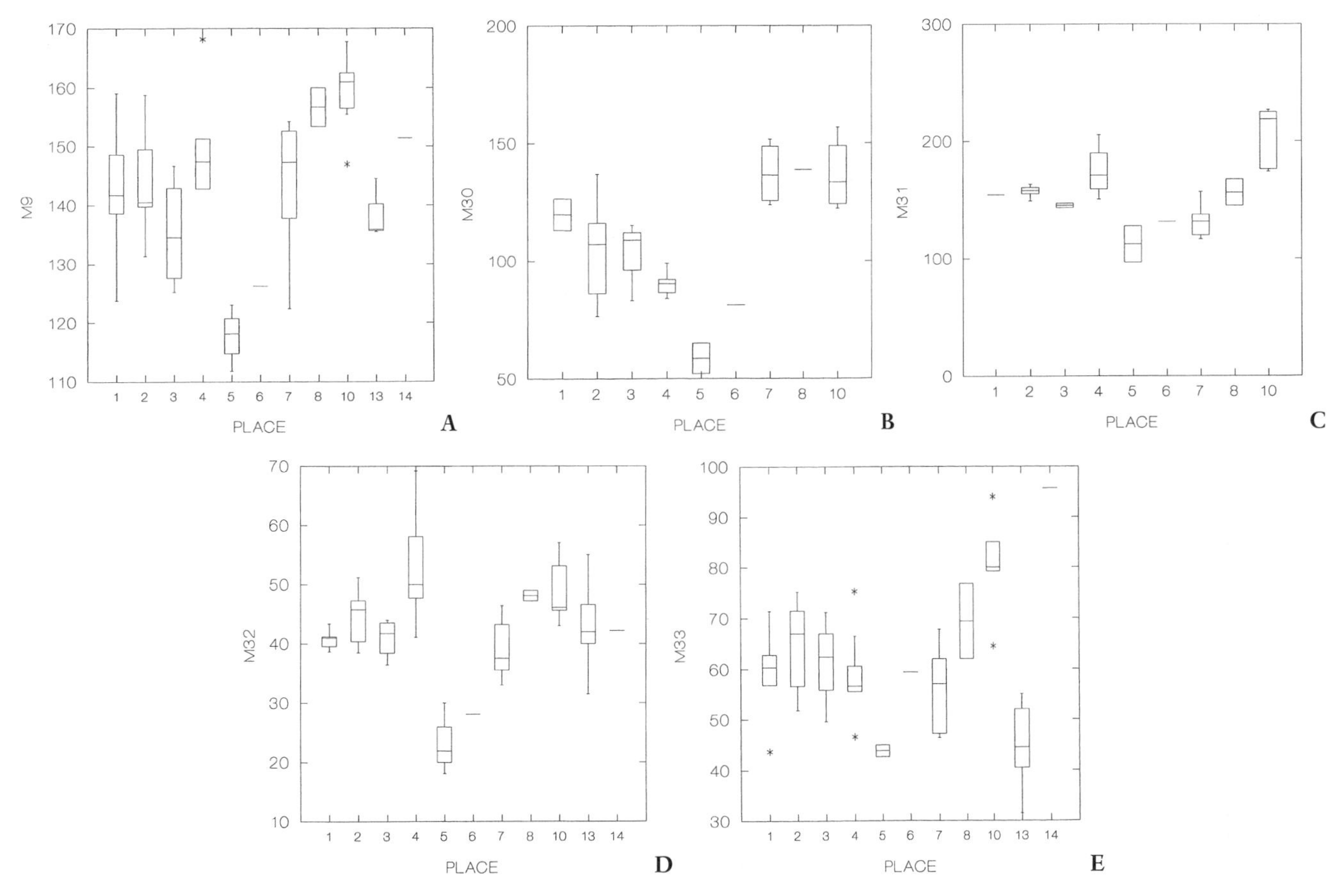

Figure 11.1. Box-and-whiskers plots on skull variables. **(A)** Length of maxillary cheek tooth row (M9). **(B)** Length of the naso-incisival notch from prosthion to the distal limit of the narial opening (M30). **(C)** Cheek length from the posterior limit of the narial opening to the anteriormost point of the orbit (M31). **(D)** Length of the preorbital bar (M32). **(E)** Maximum length of the POF (M33).

the Inzersdorf (8) IQR; there is one Sinap outlier, AS92/458 from Loc. 91. The Höwenegg (10) and Hostalets de Pierola (14) samples have substantially longer POFs (note that the upper and lower fence outliers from Höwenegg may be due to crushing), and the Cucuron population has a lower IQR than Sinap, overlapping with the Maragheh and Trinity River Pit samples.

Figure 11.2A–F includes a number of skull bivariate plots, all with an overlying 95% confidence ellipse calculated from the Xmas Quarry sample. Figure 11.2A compares M1 (muzzle length from prosthion to the midline between the anterior borders of the P2s) versus M9 (maxillary cheek tooth row length); this plot provides an estimate of relative snout length. The Sinap (S) sample includes two individuals, one from Unit 5, and the other from Unit 6 (fig. 11.2B). The Unit 6 individual plots within the ellipse (short snout form: MNHNTRQ1064, Loc. S01), whereas the Unit 5 form with an elongate muzzle plots just outside it (long snout form: AS94/203; Loc. 114). All individuals from the Hans Johnson Quarry (J) lie within the ellipse, whereas the MacAdam's Quarry (Q) sample includes four individuals in the lower part of the ellipse and one individual (with elevated values for M1 and M9) that lies well outside it. The Maragheh sample (M) lies within the ellipse, further arguing for a close size and snout proportion comparison with the *C. "occidentale"* series. The *C. quinni* (P) and *C. goorisi* (T) samples have lower values of M1 and the latter taxon ranges at the lower limit for M9 of the ellipse. The central European *Hippotherium primigenium* populations (H, I) have the longest snouts and have M9 lengths at the upper limit of the ellipse.

Figure 11.2C compares M1 with M15 (muzzle breadth between the posterior borders of I3). Sinap specimens have elevated values of M15 compared with all other hipparions analyzed here. As shown in figure 11.2B, there is a long-snouted form with a broad incisor arcade (AS94/203; Unit 5) that differs significantly from the Paris short-snouted form with a broad incisor arcade (MNHNTRQ9001 and MNHNTRQ1064; Unit 6). Note that all specimens fall at the upper limit of the 95% confidence ellipse for M15, whereas AS94/203 is plotted above the ellipse for M1. The Hans Johnson sample (J) and most of the MacAdams Quarry sample (Q) plot within the ellipse. The Maragheh values are split into a cluster of three specimens at the left side of the ellipse, and two specimens at the right side of the ellipse. Central European *Hippotherium primigenium* values fall within the ellipse's range for M15 but have greatly elevated values for M1. Figure 11.2D again shows that the entire Sinap sample comes from Units 5 and 6.

Figure 11.2E compares M32 (POB length) with M33 (POF length). All Sinap individuals fall within the ellipse.

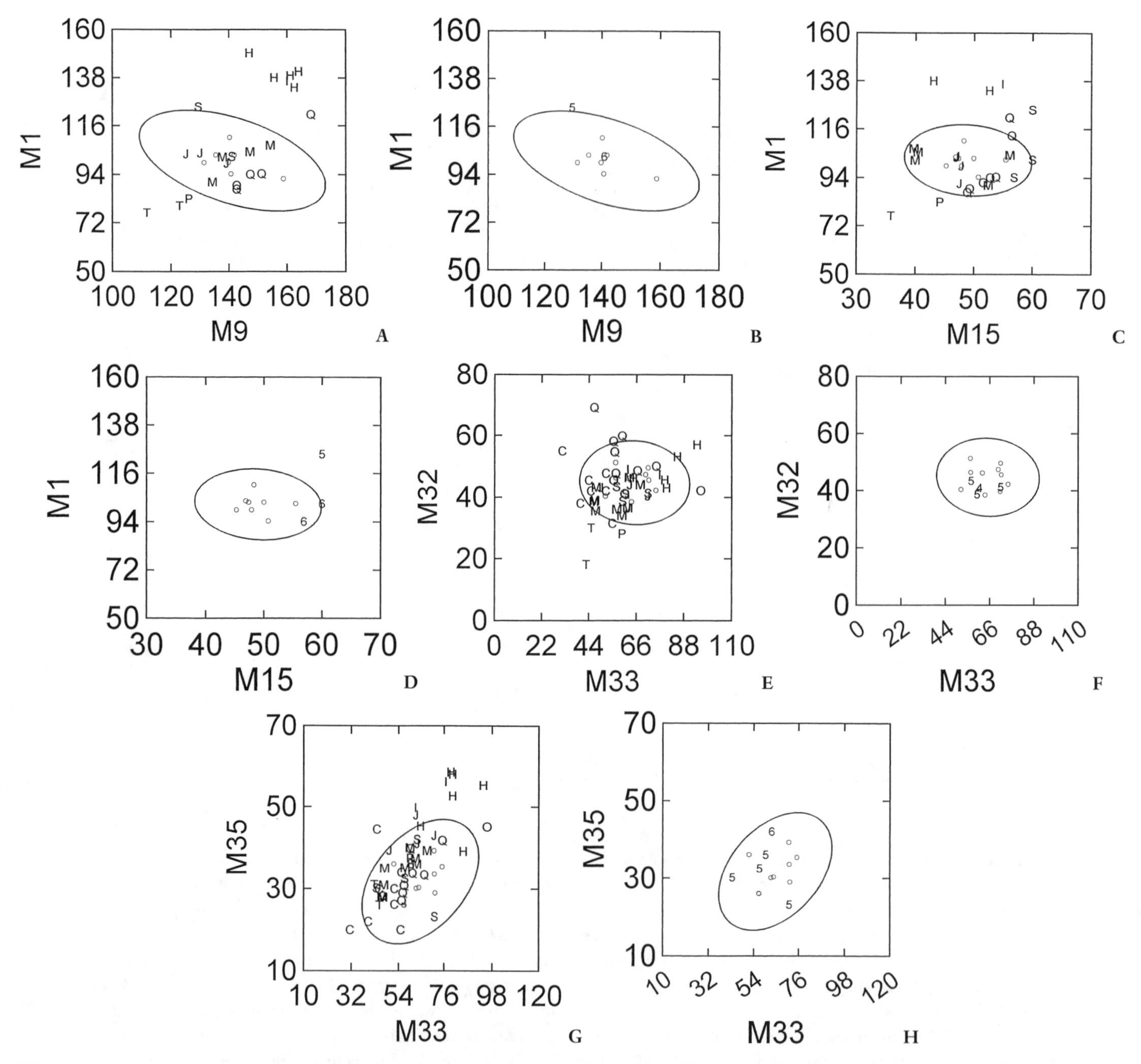

Figure 11.2. Bivariate plots on skull varibles. **(A)** Muzzle length from prosthion to the midline between the anterior borders of the P2s (M1) versus maxillary cheek tooth row length (M9). **(B)** Same as **A**, but with Sinap individuals plotted by stratigraphic unit. **(C)** M1 versus muzzle breadth between the posterior borders of I3 (M15). **(D)** Same as **C**, but with Sinap individuals plotted by stratigraphic unit. **(E)** POB length (M32) versus POF length (M33). **(F)** Same as **E**, but with Sinap individuals plotted by stratigraphic unit. **(G)** M33 versus POF height (M35). **(H)** Same as **G**, but with Sinap individuals plotted by stratigraphic unit.

There is considerable overlap in these dimensions within the entire sample. The only samples found consistently outside the ellipse are those for the smaller horses *Cormohipparion goorisi* (T) and *C. quinni* (P) as well as the one individual of *"Hippotherium" catalaunicum* (O) and one individual of *H. primigenium* from Höwenegg (H), both of which have long POFs. Figure 11.2F shows that the Sinap sample found within the ellipse originates from Stratigraphic Units 4 (AS93/826, Loc. 108) and 5 (SEN 1, AS94/203, and AS92/609), suggesting stability in these variables across this interval of time.

Figure 11.2G,H exhibits the comparison of M33 versus M35 (POF height) for Sinap hipparions from Units 5 and 6 as they compare with the Xmas Quarry sample. Again, the Sinap assemblage compares closely with this sample of *Cormohipparion "occidentale,"* with all five specimens (AS92/458, SEN 1, AS94/203, AS92/609, and MNHNTRQ9001) falling within the ellipse.

Our analysis of the skull leads to some important observations about the evolutionary position of the Sinap sample. First, the Sinap sample compares very closely in size with the North American *Cormohipparion "occidentale"* sample from Xmas Quarry and Hans Johnson Quarry. It likewise compares closely with the Maragheh *Hipparion* s.s. sample. The Sinap skulls exhibit considerable diversity in preorbital fossa morphology and metrics, which suggests that multiple species are present in the assemblage. Other univariate measurements would appear to support an intermediate position of the Sinap sample to all other hipparions under consideration here. The Sinap horses show diversity in M1 within Units 5 and 6 (fig. 11.2B,D), potentially representing two species, but together have the broadest inter-I3 region of the sample under consideration. These specimens' wide gape (high M15 value) support the hypothesis that they were adapted for eating short graze (see further analysis and discussion of later in the chapter; also see Bernor and Armour-Chelu 1999). At the same time, the Sinap hipparions are conservative (like North American *Cormohipparion "occidentale"*) in POB and POF length and height (fig. 11.2G,H).

Cheek Teeth

Maxillary Teeth

Although we analyzed several bivariate dimensions for all the maxillary cheek teeth, we found the results to be largely redundant between tooth classes. We prefer to use P2 to exemplify size trends, because of its certain identification and greater length stability than other maxillary cheek teeth through all wear stages (Bernor and Franzen 1997). All cheek tooth comparisons use the Eppelsheim sample of *Hippotherium primigenium* as the standard for calculating 95% confidence ellipses.

Figure 11.3A shows the bivariate comparison of M3 (occlusal width) versus M1 (occlusal length). The ellipse reflects the great range of length measurements found during ontogeny, rendering these measurements difficult to use for distinguishing taxa. Figure 11.3B reveals that the Sinap sample ranges mostly in the lower half of the ellipse and below the ellipse, meaning only that the Sinap horses have some individuals from Units 5 and 6 that are somewhat smaller than the Eppelsheim and other populations of

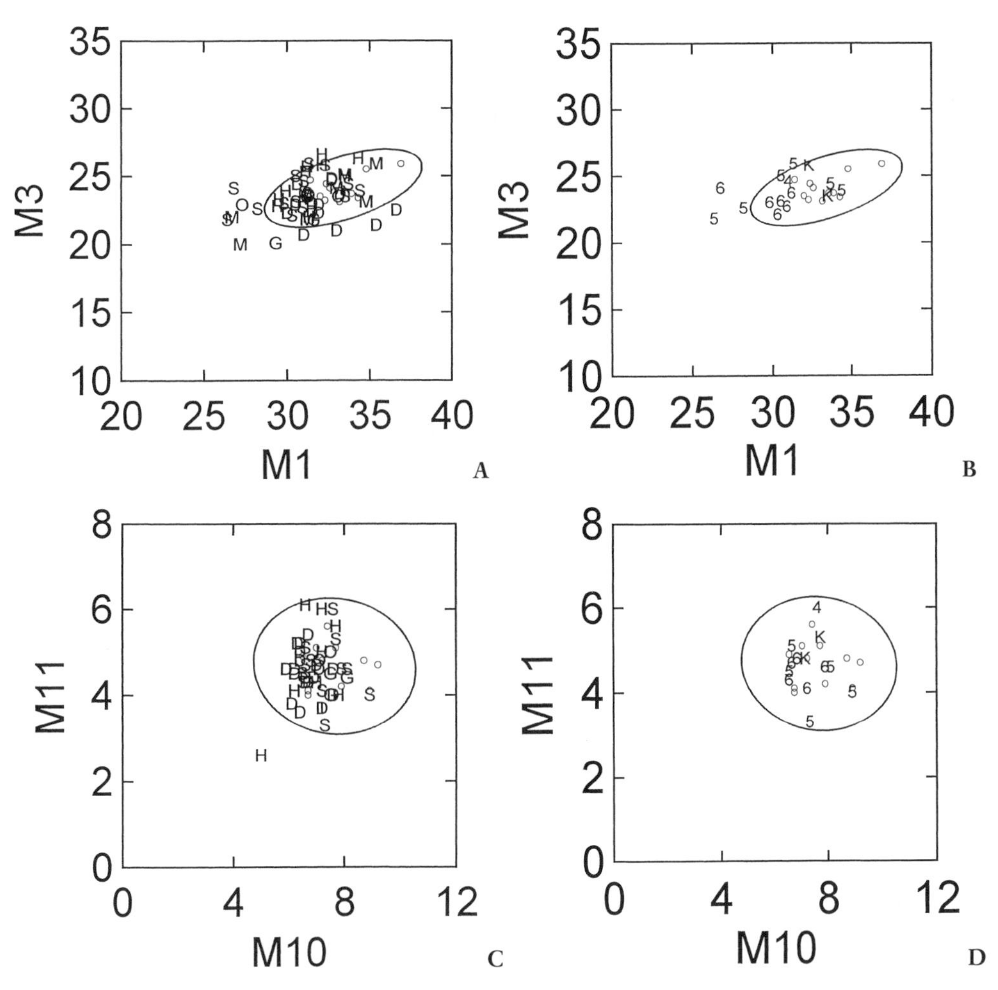

Figure 11.3. Bivariate plots on P2. **(A)** Occlusal width (M3) versus occlusal length (M1). **(B)** Same as **A**, but with Sinap individuals plotted by stratigraphic unit. **(C)** Protocone width (M11) versus protocone length (M10). **(D)** Same as **C**, but with Sinap individuals plotted by stratigraphic unit.

central European *Hippotherium primigenium*. No comparable measurements were available for North American *Cormohipparion*, but we suspect, based on the analysis given above for M9 of the skull, that Sinap would compare most closely with *C. "occidentale."* Figure 11.3C,D compares the European sample P2 protocone width (M11) versus protocone length (M10). The statistical range of this pair of variables is even greater through ontogeny than that of occlusal length and width, and the entire Sinap sample is found to fall within the ellipse (fig. 11.3D).

Mandibular Teeth

We have made a similar analysis of mandibular p2 samples. Figure 11.4A is a plot of M8 (occlusal width across metaconid-protoconid) versus M1 (occlusal length). Sinap, like Dorn Dürkheim, has some of the shorter M1 dimensions in the sample under consideration, being smaller than the central European populations from Höwenegg, Eppelsheim, and Gaiselberg. Figure 11.4B reveals that Sinap Units 6 and K (= Unit 12) exhibit considerable variability in occlusal length versus width. This offers modest support for our previous claims of hipparion species diversity in the Sinap sample, beginning no later than Unit 6.

Figure 11.5A,B showing plots for p4 proves to be similar to those for p2 except that the ellipse reflects greater variability, perhaps due to observer error: p4 is often confused with m1 due to their very similar morphology. The effect of this analysis is to minimize differences between individuals from Units 5, 6, and K.

Postcrania

We conduct bivariate comparisons using the Höwenegg standard 95% confidence ellipses for calcaneum, astragalus, MC III, and MT III. We further plot log-ratio diagrams and principal components analyses for MC III and MT III. We present these results by element.

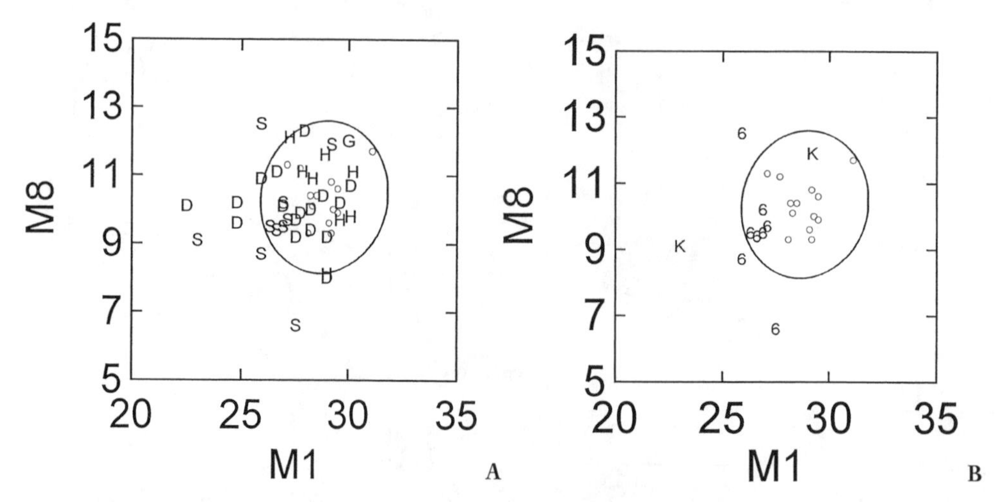

Figure 11.4. Bivariate plots on p2. **(A)** Occlusal width across metaconid-protoconid (M8) versus occlusal length (M1). **(B)** Same as **A**, but with Sinap individuals plotted by stratigraphic unit.

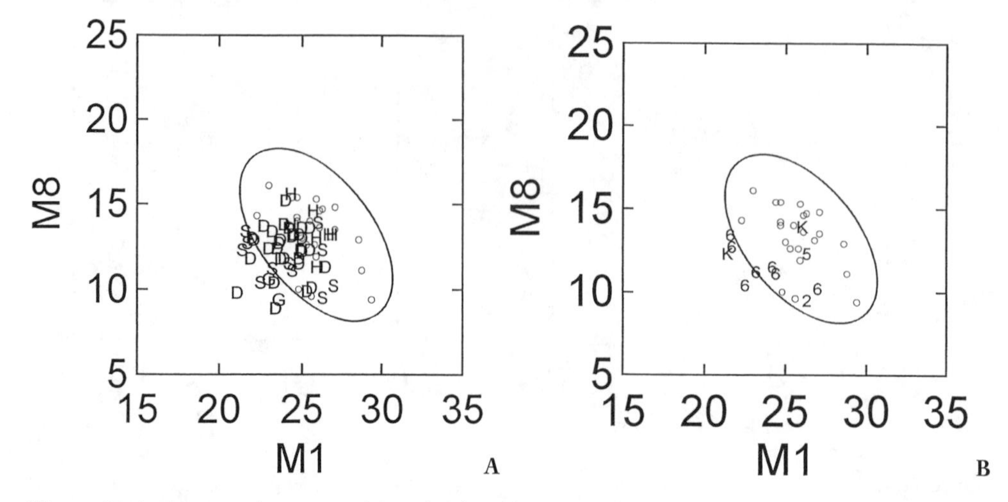

Figure 11.5. Bivariate plots on p4. **(A)** Occlusal width across metaconid-protoconid (M8) versus occlusal length (M1). **(B)** Same as **A,** but with Sinap individuals plotted by stratigraphic unit.

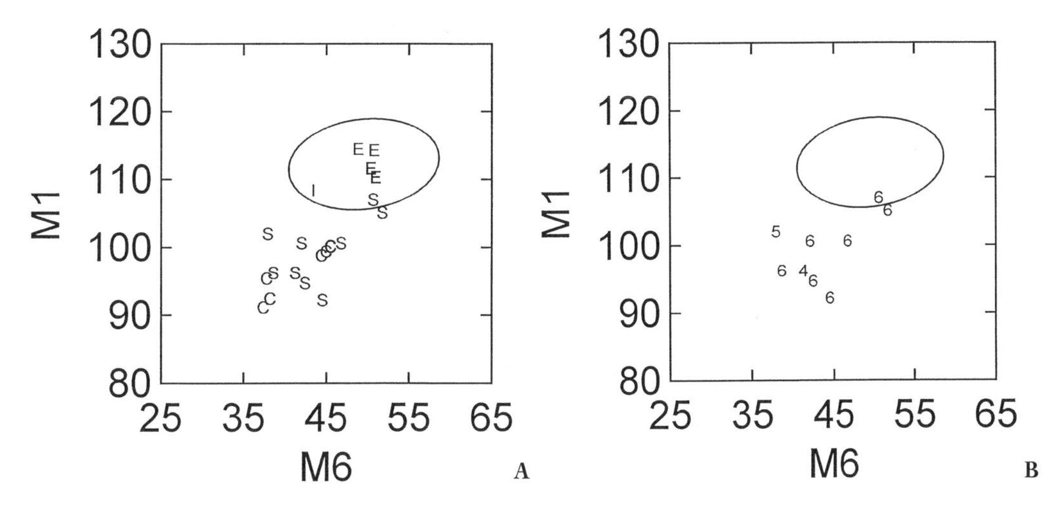

Figure 11.6. Bivariate plots on calcaneum. **(A)** Maximum length (M1) versus distal maximum breadth (M6). **(B)** Same as **A**, but with Sinap individuals plotted by stratigraphic unit.

Calcaneum

Figure 11.6A,B plots M1 (maximum length) versus M6 (distal maximum breadth) of the calcaneum and reveals substantial size diversity in the Sinap and Cucuron samples, with all but two Sinap individuals being well below the Höwenegg ellipse for M1. Figure 11.6B shows that the two individuals found on or adjacent to the lower limit of the Höwenegg ellipse are from Unit 6 and that there is substantial size variability within this unit. Sinap Units 4, 5, and 6 likewise have smaller individuals. Unit 6 has the greatest demonstrable size diversity.

Astragalus

We have calculated four bivariate plots for the astragalus (fig. 11.7A–D). Figure 11.7A,B shows M1 (maximum length) versus M5 (distal articular width). The central European populations of *Hippotherium primigenium* as well as the one individual from Esme Akçakoy (early Vallesian, Turkey) all fall within or close to the limits of the Höwenegg ellipse. The Sinap and Cucuron samples are dispersed from within the ellipse to well below the ellipse, suggesting species diversity at those localities. Figure 11.7B exhibits considerable size variability from Unit K and provides further evidence for size diversity in Sinap Stratigraphic Unit 6. Figure 11.7C,D compares M6 (distal articular depth) versus M5; these dimensions are believed to correspond well with body mass since this is a direct load-bearing joint in a horse. The results of this plot closely reflect results rendered in figure 11.7A,B and reveal that the Sinap and Cucuron hipparions are smaller than their central European counterparts and that there is considerable size diversity in specimens from Units 6 and K.

Metacarpal III

Figure 11.8A–D presents our results on MC III bivariate plots. Figure 11.8A is a plot of MC III maximum length (M1) versus distal articular width (M11). This figure again shows that most of the central European *Hippotherium primigenium* sample and that of Esme Akçakoy fall within or very close to the Höwenegg ellipse, the notable exception being one heavily built specimen from Inzersdorf. Once again, the Sinap sample is very diverse, with some individuals falling within and others decidedly outside the ellipse. Figure 11.8B reveals that Stratigraphic Unit 4 has one specimen of a partial skeleton, AS93/604A, plotting just inside the ellipse, and another outside the ellipse, AS92/289, with an elevated length measurement. Unit 5 has three individuals plotting within the ellipse, and Unit K has one.

By the time of Unit 4's Loc. 72 (10.080 Ma), there is substantial diversity in MC III length (e.g., AS92/289). Morphologic variability increases in Unit 6, plotting one specimen far below the ellipse (MNHNTRQ1129), three at the lower border of the ellipse (AS91/420, MNHNTRQ1125, and AS93/840). Specimen MNHNTRQ1127 is slightly narrower but longer than these three specimens, whereas MNHNTRQ1126 is as narrow but much longer than the cluster of three specimens. There would appear to be multiple taxa in the Sinap Units 4–6 stratigraphic interval, arguably including three or more taxa.

Figure 11.8C,D provides the bivariate dimensions of MC III M6 (proximal articular depth) versus M5 (proximal articular width), a major load-bearing joint of the distal limb. Generally, the same pattern holds true with some interesting distinctions. First, the central European-Esme Akçakoy pattern is the same, except that there are some individuals from Dorn Dürkheim and Inzersdorf with dimensions that lie well outside the Höwenegg ellipse. The lowest dimensions for M5 are found in the Cucuron sample, a decidedly slightly built hipparion. Figure 11.8D reveals that the Sinap sample includes both individuals from Units 4, 5, and K with M6 × M5 dimensions that fall within the Höwenegg ellipse, and several others from those same units outside the ellipse and smaller.

Our log-ratio diagram analysis on MC III is designed to compare the Sinap hipparion metapodial assemblage with

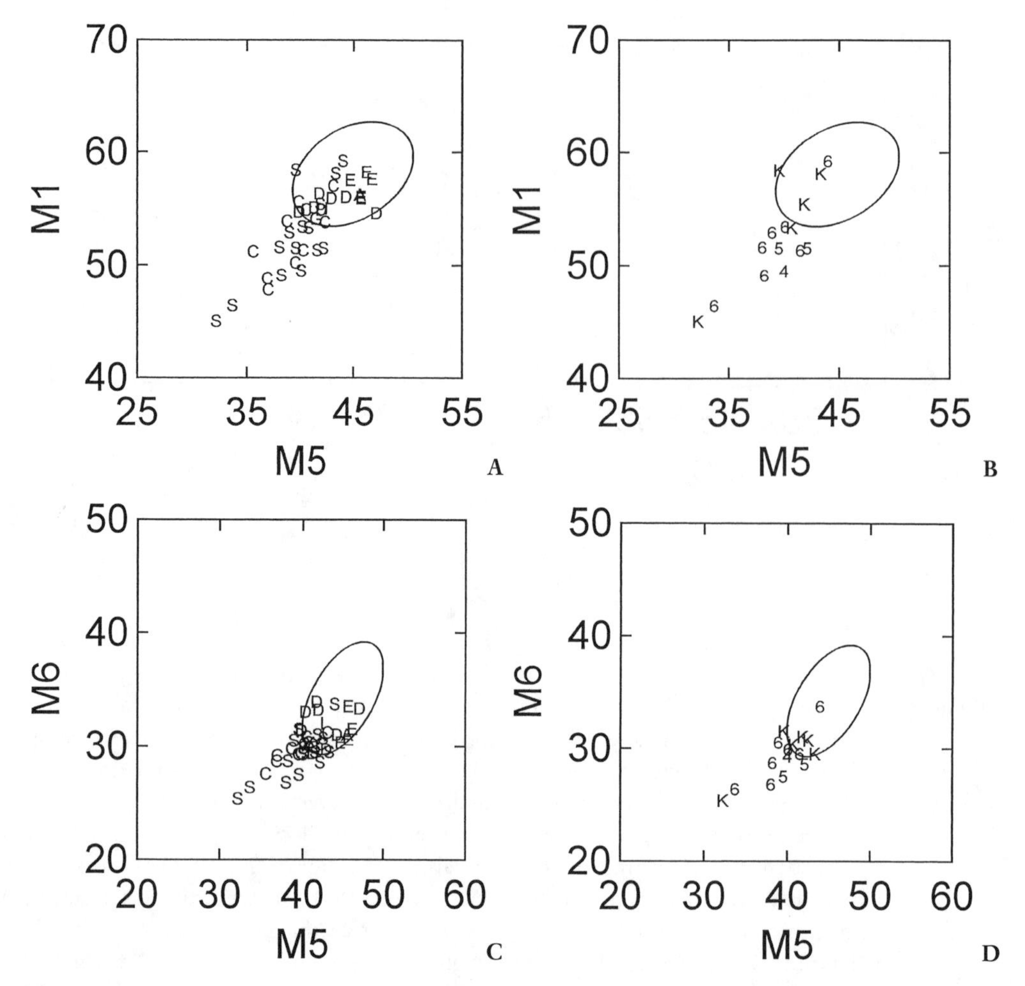

Figure 11.7. Bivariate plots on astragalus. (**A**) Maximum length (M1) versus distal articular width (M5). (**B**) Same as **A**, but with Sinap individuals plotted by stratigraphic unit. (**C**) Distal articular depth (M6) versus M5. (**D**) Same as **C**, but with Sinap individuals plotted by stratigraphic unit.

the splendid samples from Höwenegg, Germany, which we use as our statistical standard, and Esme Açkakoy, Turkey. The log-ratio diagrams are ordered and analyzed by stratigraphic unit to better understand how many taxa are present in each unit, determine how metapodial proportions change between units, and interpret the functional and ecomorphologic significance of the changes.

Figure 11.9A presents a log-ratio diagram for the *Hippotherium primigenium* sample from Höwenegg plotted against Eisenmann's (1995) standard for *"Hipparion"* (= *Cremohipparion* of Bernor and Tobien 1989) *mediterraneum* from Pikermi, Greece. As can be seen from this plot, the Höwenegg hipparion is a more heavily built form, with particularly greater mid-shaft width (M3), proportionally lesser deep mid-shaft depth (M4), similar-to-lesser distal supraarticular width (M10) and articular width (M11). The principal morphological and functional differences depicted here are that *C. mediterraneum* had more slenderly built, elongate metapodials that reflect a more cursorial adaptation than the forest-living *Hippotherium primigenium*.

Figure 11.9B plots Höwenegg MC III against the Höwenegg standard (mean log plot) to provide a visual representation, albeit unconventional, of intrapopulation variability. Figure 11.9A,B supports Bernor et al.'s (1997) conclusion that the Höwenegg sample represents a single species.

Figure 11.9C plots our sample of the Esme Akçakoy MC III samples. This sample has not been described previously, and although the bivariate plots (fig. 11.8A,C) suggest a close similarity to the Höwenegg hipparion, this analysis exhibits a strong difference in some important variables. Most remarkable is the decreased dimension of M3 (mid-shaft width) and strongly contrasting greater dimension of M4 (mid-shaft depth). This contrast is reflected and functionally reinforced by the similarly contrasting M5 (proximal articular width) and M6 (proximal articular depth) dimensions. Interestingly, M10 and M11 are mostly smaller in the Esme Akçakoy specimens. The overall morphological pattern of the Esme Akçakoy sample is different from that of the Höwenegg sample in that the MC III specimens are as elongate but narrower in the various mediolateral dimensions (M3, and variably M10 and M11), but critically greater in mid-shaft depth (M4) and, to a lesser extent, proximal articular depth. We believe that this pattern reflects the functional difference between a primitive forest horse (e.g. Höwenegg; Bernor et al. 1997), in which M3 and

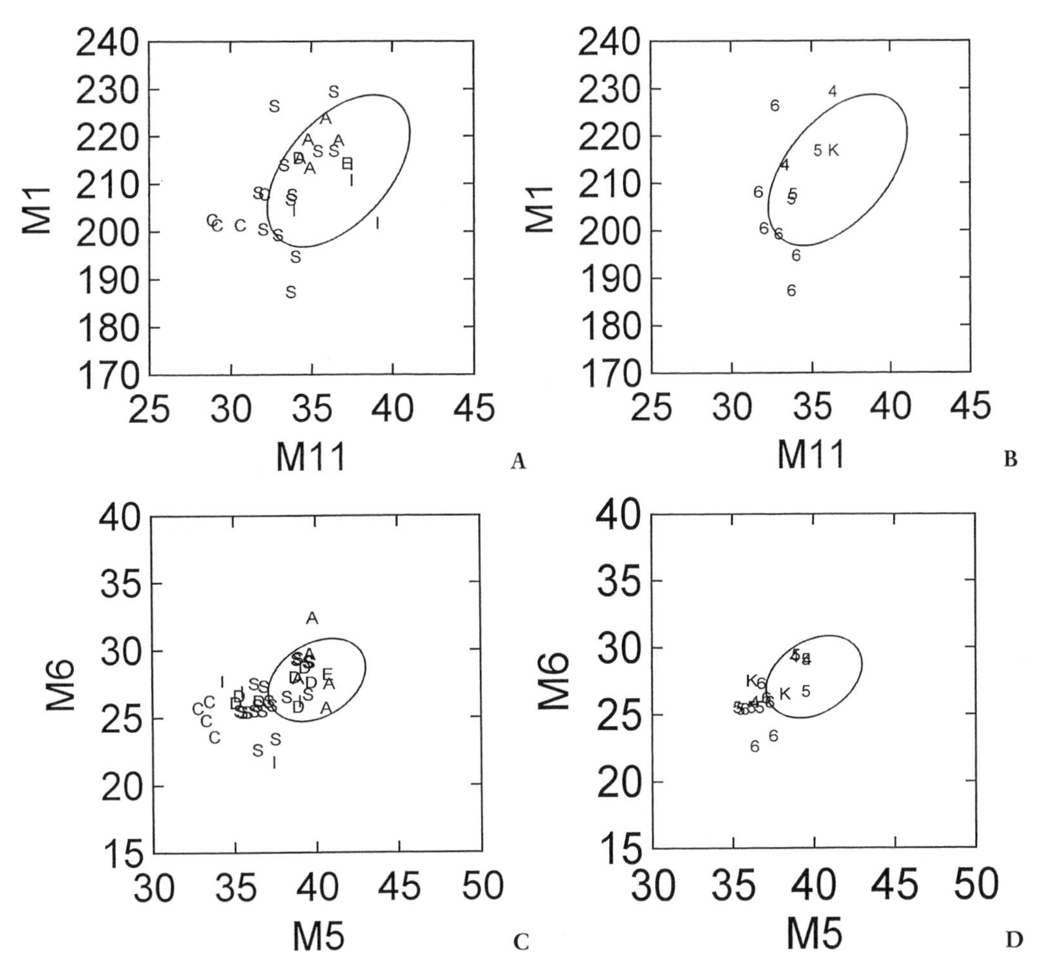

Figure 11.8. Bivariate plots on MC III. **(A)** Maxumum length (M1) versus distal articular width (M11). **(B)** Same as **A**, but with Sinap individuals plotted by stratigraphic unit. **(C)** Proximal articular depth (M6) versus proximal articular width (M5). **(D)** Same as **C**, but with Sinap individuals plotted by stratigraphic unit.

M4 are closely similar to one another, and an open country cursorial lineage, which progressively evolves increased metacarpal length, decreased mid-shaft width and distal articular width, accompanied by sharply increased mid-shaft depth and proximal articular surface depth. This same pattern of relative M3 versus M4 dimensions is seen in figure 11.9A, which contrasts the Höwenegg horse and *Cremohipparion mediterraneum*. Our analysis suggests that the Esme Akçakoy horse is not *Hippotherium primigenium*, but a distinctly different taxon.

Figure 11.9D plots two specimens from stratigraphic Unit 4, AS93/604A from Loc. 108 (10.135 Ma) and AS92/289 from Loc. 72 (10.080 Ma). The younger (Loc. 72) specimen is somewhat longer than the Loc. 108 specimen. It shares with the Loc. 108 specimen and the Esme Akçakoy sample a relatively narrower mid-shaft dimension than is seen in the Höwenegg sample. At the same time, the Loc. 72 specimen has a relatively wider (M5) and deeper (M6) proximal articular surface and larger distal articular dimensions (M10, M11, and M12) than the Loc. 108 specimen. Both specimens compare fairly well with the Esme Akçakoy sample exhibiting the same M3 versus M4 "Esme Akçakoy effect." The Unit 4 sample may be the same species, or closely related species to the Esme Akçakoy sample and different from the Höwenegg horse.

Figure 11.9E compares a single individual, AS93/9, from lower stratigraphic Unit 5 (Loc. 91; ~9.977 Ma). This specimen differs from the Loc. 108 specimen (AS93/604A) and the Esme Akçakoy sample in its greater M3 value as well as the relative size of M3 versus M4, which is reversed here. All of the remaining measurements are as in the Loc. 108 specimen and the Esme Akçakoy sample.

Figure 11.9F includes six specimens from upper Unit 5, Loc. 8B (9.918 Ma): AS92/228, AS92/237, AS92/238, AS92/239, AS92/260, and AS92/275. All of these specimens are similar in length. However, two specimens, AS92/228 and AS92/238, are distinct from the rest of the sample in their very narrow mid-shaft width (M3) and are relatively small compared with the rest of the sample for M5, M6, and M11. Specimen AS92/238 is further distinguished by its very small distal articular width (M11) and contrasting great distal sagittal keel (M12) depth. Specimens AS92/228 and AS92/238 contrast strongly with one another in their distal saggital keel (M12) and distal articular craniocaudal

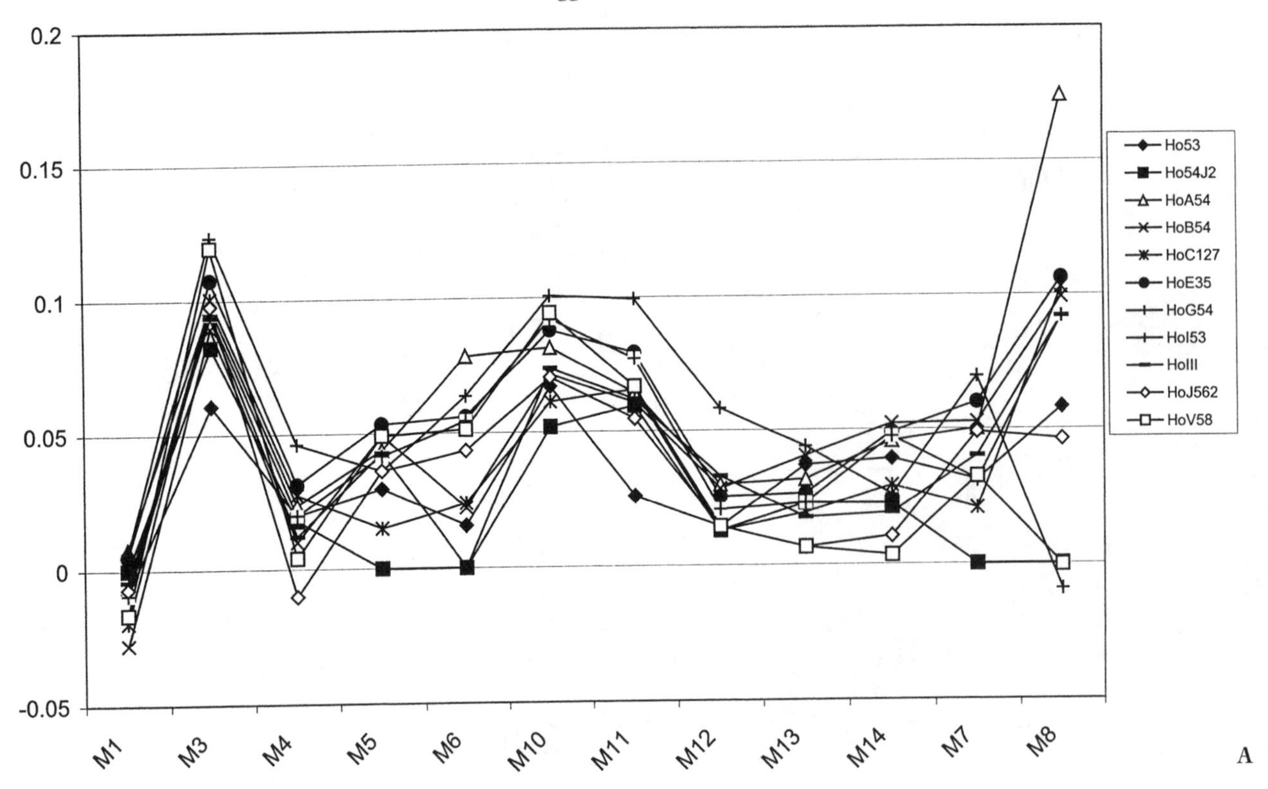

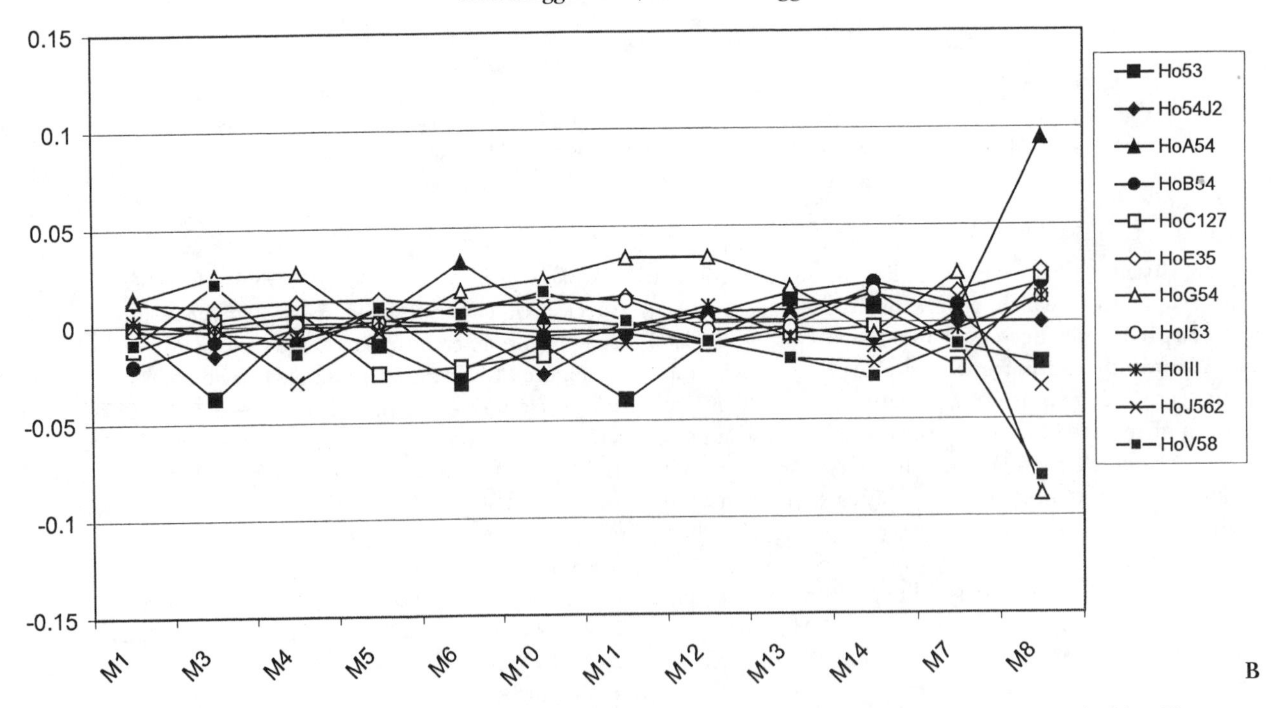

Figure 11.9. Log_{10} ratio plots on MC III. (**A**) Höwenegg MC III sample plotted against *Cremohipparion mediterraneum* standard (see Eisenmann 1995). (**B**) Höwenegg MC III sample plotted against its own mean. (**C**) Esme Akcakoy MC III plotted against the Höwenegg standard. (**D**) Sinap Unit 4 Loc. 108 and 72 hipparion MC III plotted against the Höwenegg standard. (**E**) Sinap Unit 5 Loc. 91 hipparion MC III plotted against the Höwenegg standard. (**F**) Sinap Unit 5 Loc. 8B hipparion plotted against the Höwenegg standard. (**G**) Sinap Unit 6 Locs. 12, OZ02, and OZ01 hipparion plotted against the Höwenegg standard. (**H**) Sinap Unit 12, Locs. 26, 33, and OzKd hipparion plotted against the Höwenegg standard. (**I**) Sinap Unit 13, Loc 42 hipparion plotted against the Höwenegg standard.

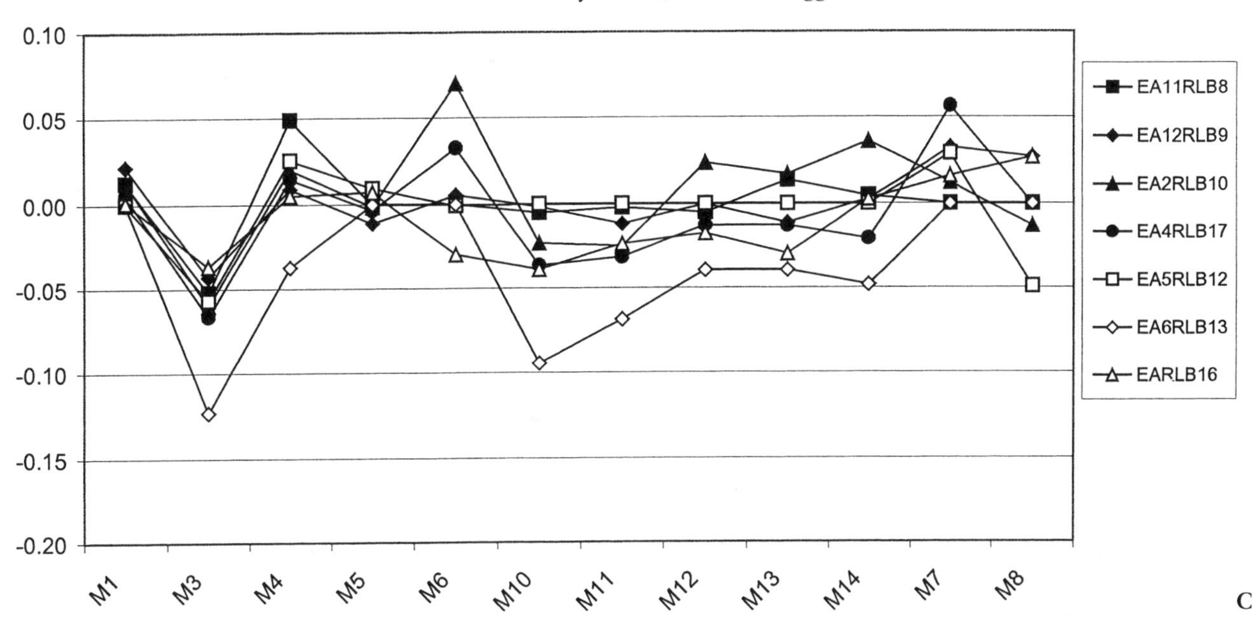

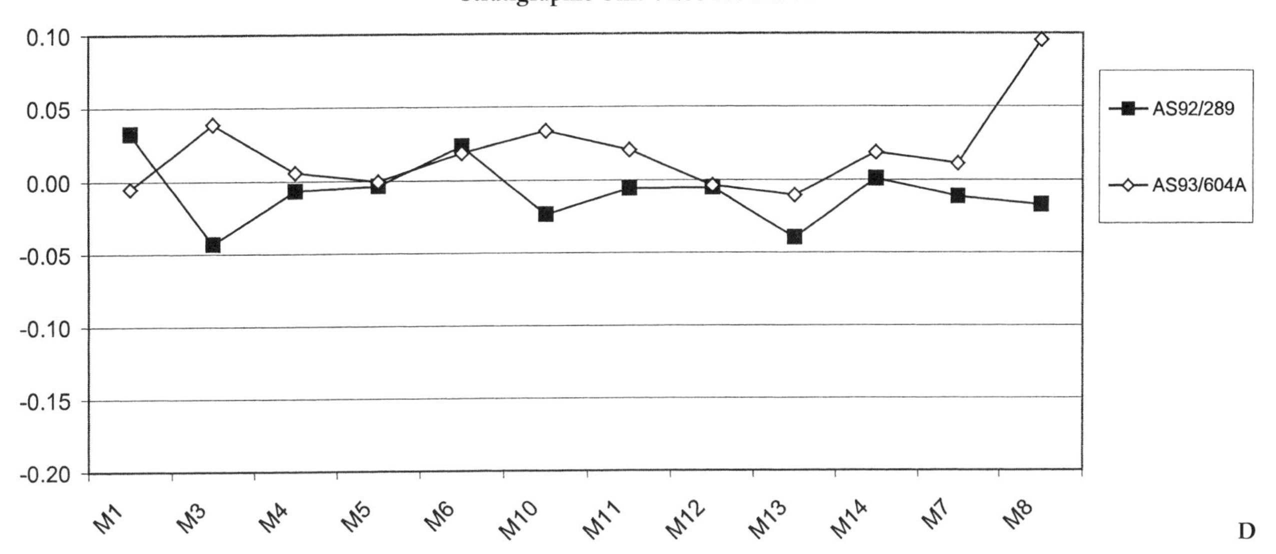

Figure 11.9. *(continued)*

(M13) dimensions. Specimens AS92/237, AS92/239, AS92/260, and AS92/275 would appear to be closely similar to one another in most of their dimensions.

Unit 6 includes seven specimens (fig. 11.9G) from three localities: MNHNTRQ1129 from Loc. OZ01 (9.683 Ma); AS91/420, AS93/840, and AS95/513.15a,b from Loc. 12; and MNHNTRQ1125, MNHNTRQ1126, and MNHNTRQ1127 from Loc. OZ02 (both 12 and OZ02 are 9.590 Ma).

MNHNTRQ1129 (the only specimen from OZ01) exhibits a morphology not seen in the previous samples:

1. It has a very short maximum length (M1; see fig. 11.11A, the shortest dimension on the plot);
2. The M3 and M4 are also small and do not exhibit the strong dimensional contrast seen in the Esme Akçakoy-like forms; and
3. M12 (maximum depth of distal articular keel) and M13 (minimal depth of distal lateral condyle) are only slightly smaller than the Höwenegg standard.

Overall, this morphology suggests a different evolutionary trajectory than seen in the Esme Akçakoy and Sinap hipparions previously described. Functionally, this horse would appear to be goatlike in its shortened length with accompanying slight contrast between mid-shaft width and depth and poorly developed distal sagittal keel, suggesting that it was adapted to locomotion on uneven substrates.

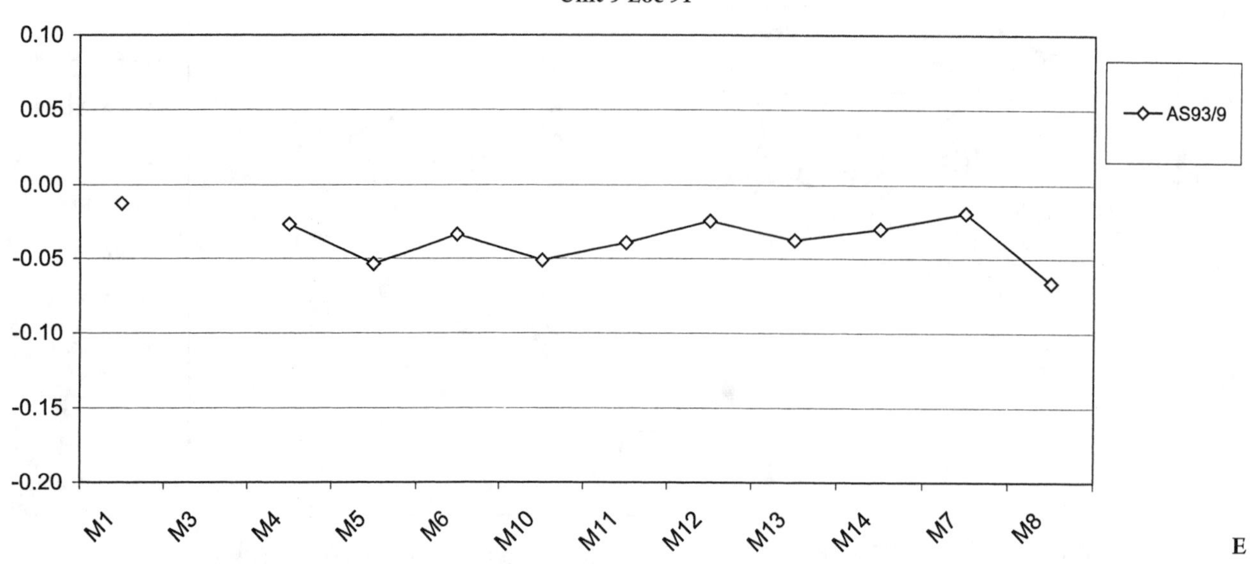

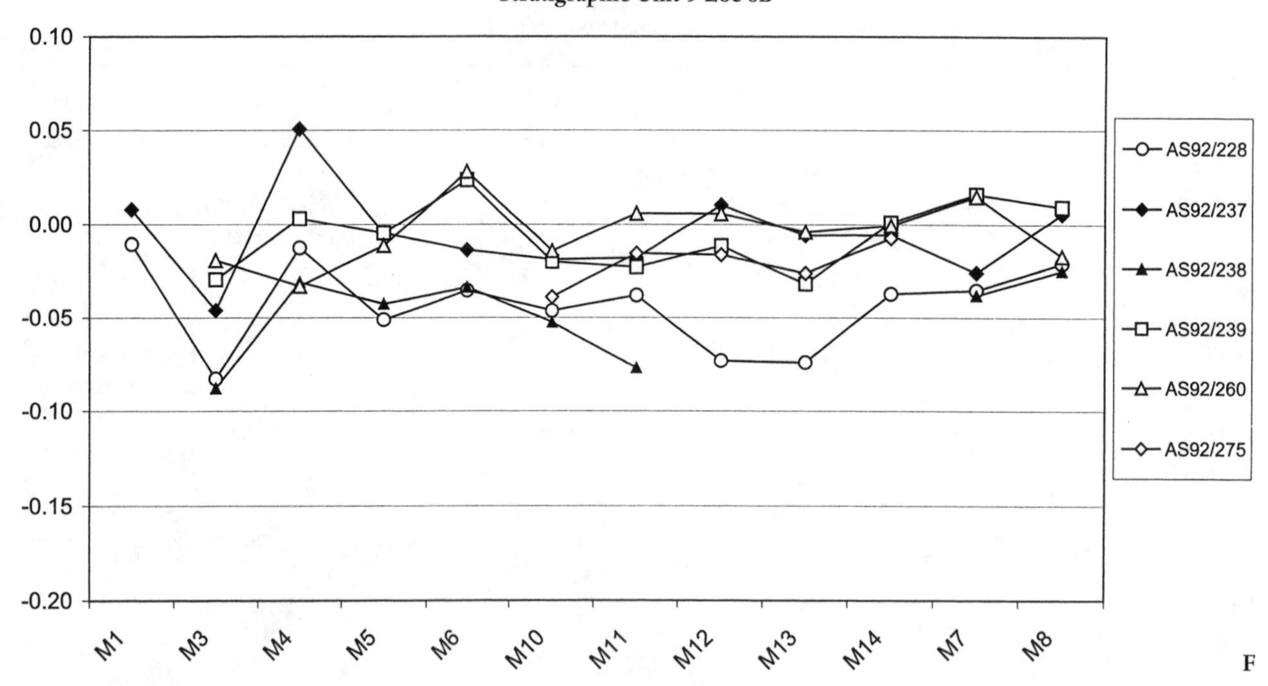

Figure 11.9. *(continued)*

The Locs. 12 and OZ02 specimens show considerable variability among themselves. The two Loc. 12 specimens, AS91/420, and AS93/840 contrast with one another most strikingly in their M3 and M4 dimensions, with AS91/420 being very slender, and in their M6 and M10 dimensions. Specimen OZ02 is at the same stratigraphic horizon as Loc. 12 but is laterally displaced by several meters. Its sample is likewise diverse, with MNHNTRQ1126 being very long (M1), narrow in its mid-shaft dimension (M3), and narrow in its proximal articular surface dimension (M6) compared with the other two specimens. The Unit 6 specimens, differing in age from one another by only about 100,000 years, show remarkable diversity: no less than three morphs would appear in these localities.

Figure 11.9H includes specimens from Stratigraphic Unit 12, Locs. 26, 33, and OZKD. There are seven specimens from this level (= K for Kavak Dere of the bivariate plots): AS89/188, AS89/192, AS89/229, AS91/42, MNHNAKA19, MNHNAKA21, and MNHNAKA36. All individuals except MNHNAKA21 are incomplete. Stratigraphic

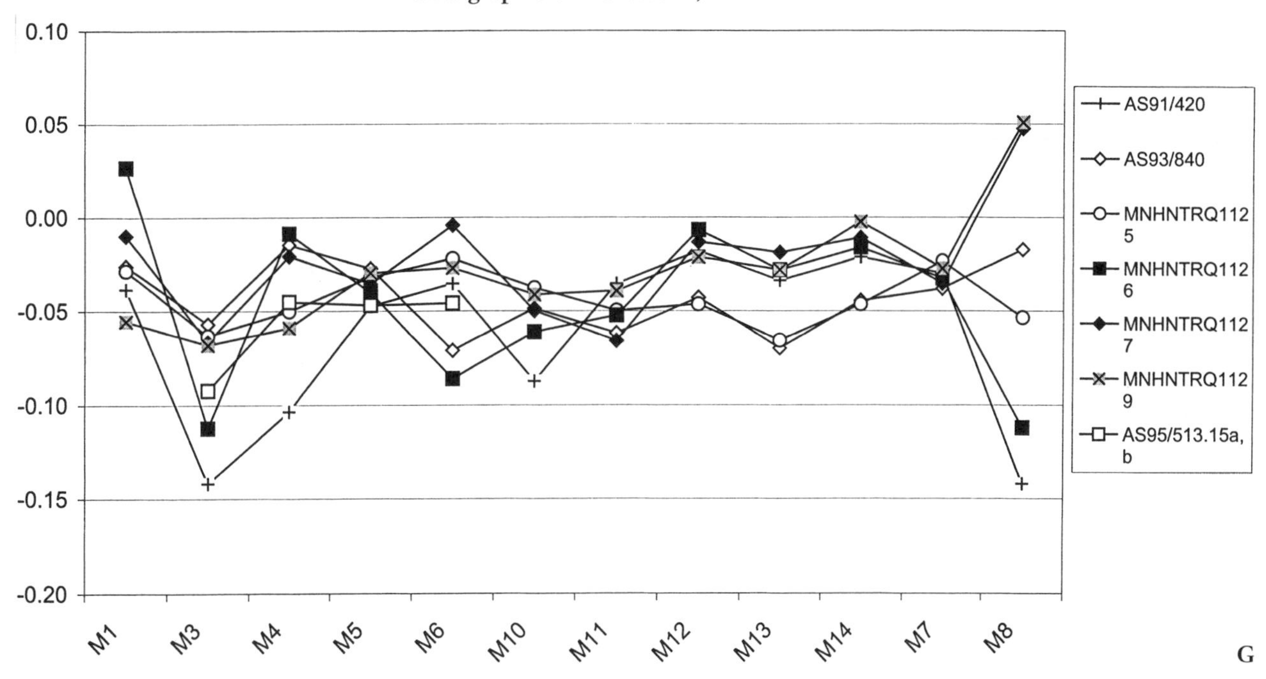

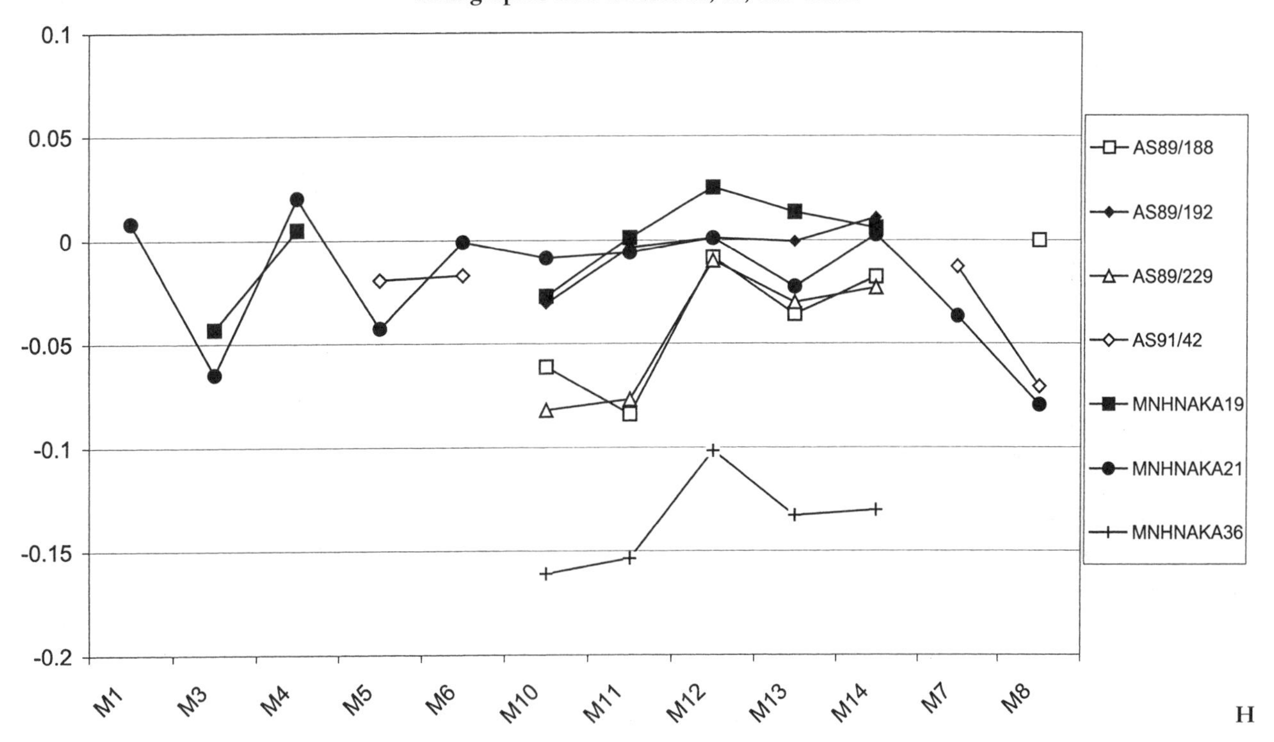

Figure 11.9. *(continued)*

Unit 12 is substantially higher in the section than the previous localities and is much younger in age (~8.121 Ma). Specimen MNHNAKA21 exhibits a similar pattern for its mid-shaft dimensions (M3 versus M4) as found in the Esme Akçakoy lineage. MNHNAKA36 is incomplete but differs sharply from all Sinap hipparion in its strongly reduced M10, M11, M13, and M14 dimensions. Although this skeletal material is incomplete, stratigraphic Unit 12 would appear to include multiple hipparion taxa.

Figure 11.9I includes two partial metapodials from Stratigraphic Unit 13, Loc. 42: AS89/420 and AS93/127. Loc. 42 (Coban Pinar) is of uncertain age, other than it is

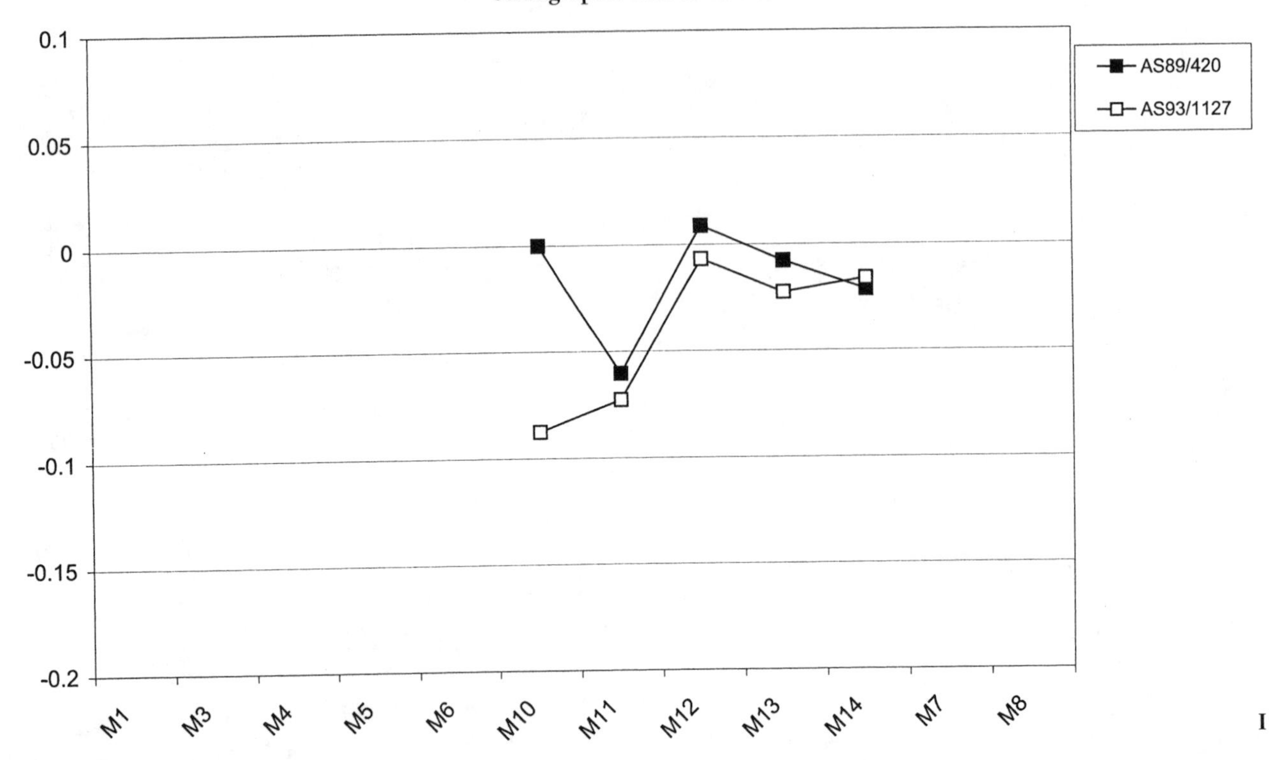

Figure 11.9. *(continued)*

Table 11.8. MC III Morphologies by Stratigraphic Unit[1]

Unit	Pattern 1	Pattern 2 (mc)	Pattern 3	Pattern 4 (mc)	Pattern 6
4	X				
5	X		X		
6	X	X	X	X	X

[1]X, Morphological pattern present.

likely younger than Unit 12 localities, but is still of the latest Miocene age. Not much more can be said except that the distal articular width of these forms is narrow and the distal sagittal keel relatively well developed.

The results for the PCA of MC III (see the appendix and fig. 11.10) show that principal component 1 explains 96% of the variance and this component corresponds closely with M2 divided by the GEOMEAN (e.g., relative length). The second principal component explains only 2% of the variance but may reflect a trend of adaptive significance. Principal component 2 has a positive eigenvector with M3 and M10 and negative eigenvectors with M5 and M6. Thus, positive values for principal component 2 indicate a metacarpal that is relatively broad distally and at midshaft compared to the proximal craniocaudal and mediolateral dimensions. Conversely, a specimen with negative values for principal component 2 would be distally slender. Morphological patterns for MC III revealed by this PCA are listed in table 11.8 by stratigraphic unit.

There are two specimens from Stratigraphic Unit 4: AS93/604A from Loc. 108 and AS92/289 from Loc. 72. Specimen AS93/604A plots with several Xmas Quarry specimens and possibly reflects a primitive MC III morphotype. This morphotype is designated "Pattern 1" (fig. 11.10) and would appear to have a reciprocal "Pattern 1" morphotype for the MT III samples (see below). Stratigraphic Unit 4 includes one specimen from Loc. 72, AS92/289 which the PCA also assigns to Pattern 1, based on its similar principal component scores and in contradiction to the bivariate and log-ratio plots, which suggests that it may be a different taxon. It is apparent from these contrasting analyses that the Loc. 72 specimen is substantially lengthened, but retains a similar basic shape to AS93/604A.

By Unit 5, an additional morphological pattern is evident. Specimen MNHNTRQ1126 from Locality 8B in Unit 5 plots near the range of the Höwenegg sample and away from specimens exhibiting Pattern 1. We assign this

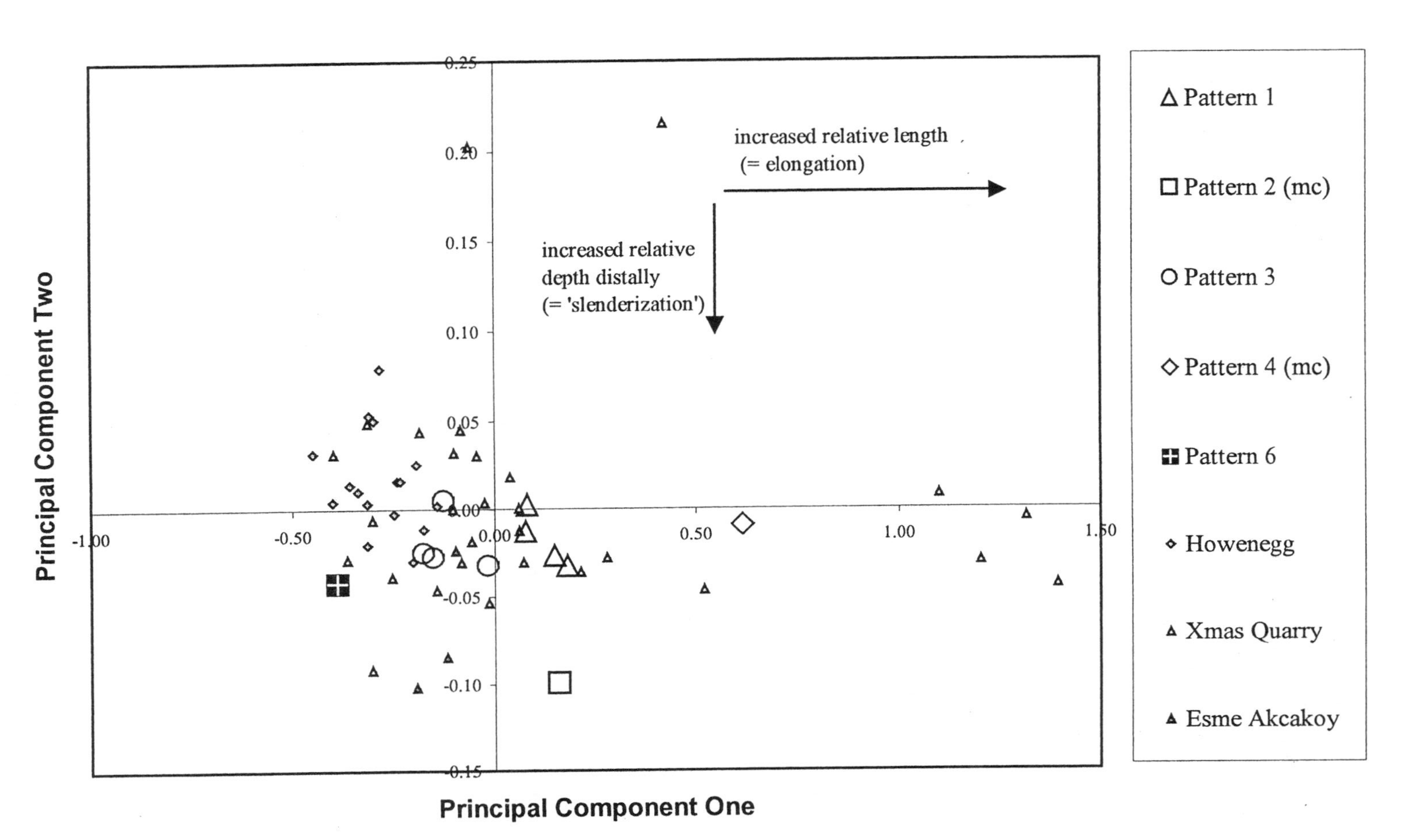

Figure 11.10. Principal components plot of components 1 and 2 for MC III.

specimen to Pattern 3 and suggest that this morphology is parallel to Pattern 3 of the MT III.

Stratigraphic Unit 6 includes three specimens (AS93/840, MNHNTRQ1125, and MNHNTRQ1127) that exhibit the Pattern 3 morphology. Three additional specimens (MNHNTRQ1126, AS91/420, and MNHNTRQ1129) appear to be morphologically unique. Specimens MNHNTRQ1126 and AS91/420 are also smaller in body size (Scott et al., chapter 16, this volume) and either could correspond to the small Pattern 4 MT III specimens defined below. However, it is unlikely that these two specimens both match the Pattern 4 MT III specimens, given their varying scores on both principal components. We refer MNHNTRQ1126 to Pattern 4 (MC III) and note that this particular pattern does not necessarily correspond to MT III Pattern 4 below. Similarly, we refer AS91/420 to Pattern 2 (MC III), which may or may not correspond to MT III Pattern 2a or 2b.

Specimen MNHNTRQ1129 from Loc. OZ01 appears to be morphologically unique in having a markedly short MC III and extreme negative scores for both principal components outside the range of the Höwenegg sample. No parallel morphological pattern is certainly identifiable among the MT III samples from Unit 5; it could be that this specimen represents a species distinct from any represented in the metatarsal sample (see next section). We refer to this morphotype as "Pattern 6" (to prevent confusion with differing metatarsal morphotypes).

The bootstrapping procedure used here indicated that the 11 MC III specimens from Sinap exhibiting either Pattern 1, Pattern 3, or Pattern 6 showed more variation than might be subsumed under a single species ($p < .05$, one-tailed test, see table 11.10). Similarly, the eight specimens from Sinap exhibiting either Pattern 1 or Pattern 3 exhibited more variation than might be subsumed under a single species ($p < .05$, one-tailed test, see table 11.10).

Metatarsal III

The MT III analysis included the same variables analyzed (fig. 11.11A–D) in the MC III analysis. Once again, most of the central European *Hippotherium primigenium* and Esme Akçakoy sample lie within the 95% ellipse, whereas most of the Sinap and Cucuron samples lie outside the ellipse, having lesser M11 dimensions and a greater dispersion of M1 values. Figure 11.11B reveals that Sinap Units 4, 5, and 6 have individuals that plot within the lower part, or just outside the lower part of the Höwenegg ellipse. The Sinap Unit 5 sample exhibits considerable diversity in M1 and M11 dimensions, suggesting that there is a single morphologically variable morph here including: two individuals within the ellipse, AS92/251 (center of ellipse) and AS93/52–53 and AS93/7 (lower left of ellipse; right and left) of a single individual; a narrow-elongate morph, AS92/240 just outside the ellipse. Unit 6 shows a striking increase in morphological diversity with many individuals within the ellipse, below the ellipse, above and to the left of the ellipse and much narrower yet and far and to the left of the ellipse. Distinct clusters include:

1. A short, narrow cluster of four specimens below the ellipse which may be the MT III equivalent of the goatlike specimen identified in the MC III sample (MNHNTRQ1160, 1163, 1164, and 1168);
2. A cluster of five specimens within or just outside the lower left portion of the ellipse that we consider primitive (AS93/1213, AS91/421 left and right, AS91/780, and MNHNTRQ1167);
3. Three specimens, all very narrow and showing some length variability, AS91/373 (shortest), AS93/1193, and MNHNTRQ1169 (longest). Sinap Unit 12 (= K) has an individual with a very slight M11 dimension, whereas Unit 13 (= C) has two individuals strikingly different in their M1 × M11 proportions.

The bivariate comparisons for M6 (proximal articular depth) versus M5 (proximal articular width) have somewhat different results in that the Inzersdorf sample has a number of individuals that fall well below the Höwenegg ellipse (fig. 14C,D). Several of the Sinap specimens lie within the lower half of the ellipse, and several individuals from Units 6, K, and C plot outside the ellipse. The entire Sinap sample from Units 4 and 5 lies within the ellipse except for two Unit 5 specimens that plot on the left-hand border.

Figure 11.12A–C replicates the log-ratio diagram analyses made on MC III for MT III. Figure 11.12A is a plot of the Höwenegg specimens on the *Cremohipparion mediterraneum* standard and reveals a remarkable difference in the ratios: mid-shaft width (M3) versus mid-shaft depth (M4) proportions are reversed, with *Hippotherium primigenium* having a relatively narrower M3 proportion and an equivalent or greater M4 proportion. This suggests that the Esme Akçakoy effect seen in the Turkish hipparion MC III specimens is apparent in those of the Höwenegg horse.

Figure 11.12B, Höwenegg on Höwenegg, exhibits increased individual variability for M3, M4, and M13 in particular suggesting that individual body size/mass differences may have played a role in mid-shaft and distal articular dimensions. Figure 11.12C, the Esme Akçakoy sample plotted on the Höwenegg standard, exhibits less of a proportional contrast between mid-shaft width (M3) and depth (M4) measurements than seen in MC III. Taken together, we interpret these data as revealing that the MT III intra-element dimensions are tracking somewhat differently from those for MC III. This may have a functional significance: MT III would appear to be somewhat less sensitive to differences in mid-shaft proportions than MC III, and perhaps by necessity all hipparions under consideration required MT III with a deep mid-shaft.

Figure 11.12D includes two individuals from Stratigraphic Unit 4, Loc. 108 (10.135 Ma): AS93/332 and AS93/827A. These two specimens track one another closely except for M12, maximum depth of distal sagittal keel, for which AS93/827A exhibits a very high value. With the absence of any other morphologic difference, we suspect that this may reflect a measurement artifact and hesitate to recognize more than one species from this unit on this basis alone. Other than this feature, AS93/827A tracks the

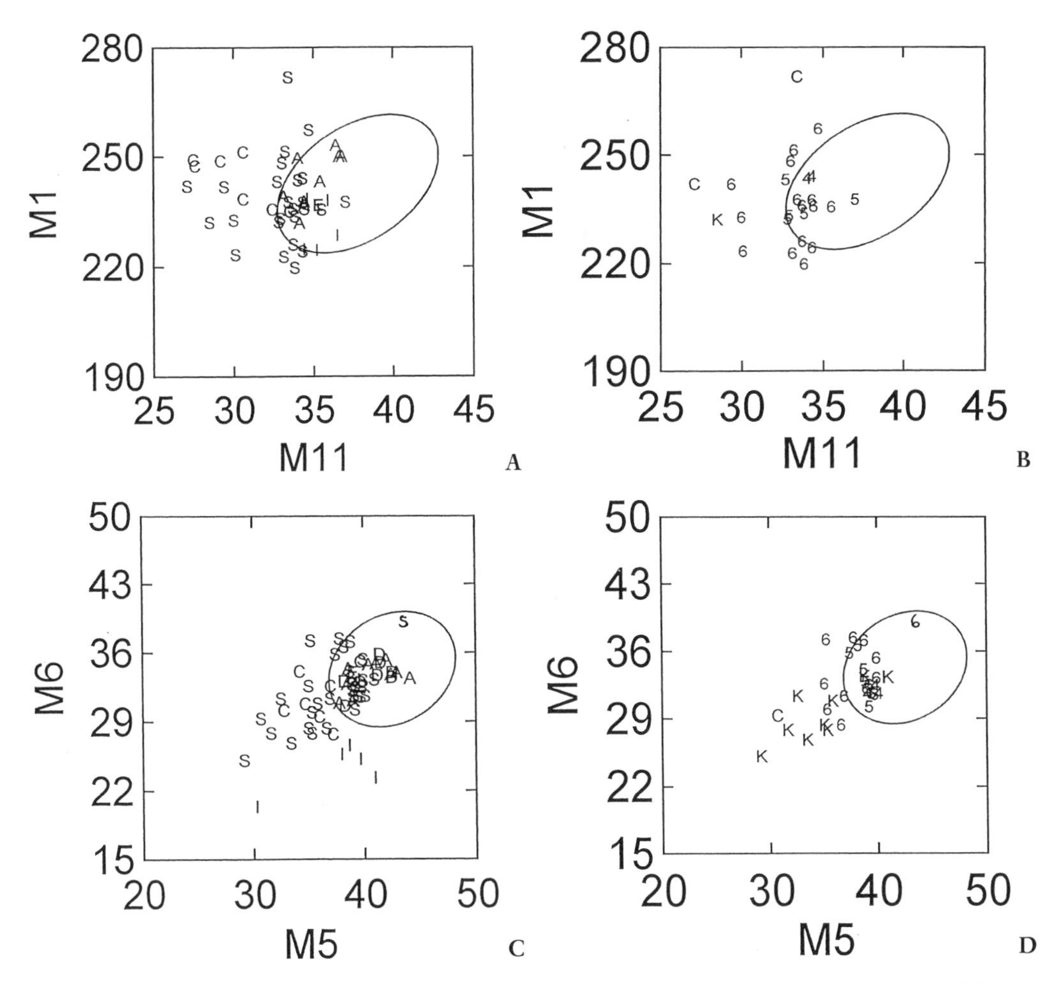

Figure 11.11. Bivariate plots on MT III. **(A)** Maximum length (M1) versus distal articular width (M11). **(B)** Same as **A,** but with Sinap individuals plotted by stratigraphic unit. **(C)** Proximal articular depth (M6) versus proximal articular width (M5). **(D)** Same as **C,** but with Sinap individuals plotted by stratigraphic unit.

Esme Akçakoy sample closely. We conclude that present evidence supports the existence of a single species at Locality 108. This interpretation is supported by figure 11.11B (M1 × M11) and figure 11.11D (M6 × M5), which show that the Unit 4 specimen dimensions are very closely spaced on the outer limits of the Höwenegg ellipse.

There are three specimens from lower Unit 5, Loc. 91 (9.977 Ma). Figure 11.12E depicts two morphologic patterns. AS93/7 exhibits a very strong contrast in M3 versus M4 and M5 versus M6, whereas having the lowest values, albeit marginally, of M10 and M11. Figure 11.11B plots a close clustering of M11 × M1 dimensions at the lower end of the Höwenegg ellipse. As is typical for the Sinap assemblage, these values cluster at the lower left limit of the Höwenegg ellipse revealing smaller and mediolaterally more slender dimensions than *Hippotherium primigenium*.

Figure 11.12F includes two individuals from upper Unit 5 Loc. 8B (9.918 Ma). These specimens are fundamentally similar to each other and to those from lower Unit 5 in their length (M1), midshaft width (M3), and midshaft depth (M4). Specimen AS92/251 is divergent from AS92/240, and the lower Unit 5 sample in its smaller proximal articular depth (M6) and distal articular width (M10 and M11). None of the Unit 5 specimens are particularly long or short (M1).

Figure 11.12G includes four specimens from Locs. OZ01 and S01 (9.683 Ma): MNHNRLB01, MNHNTRQ1161, MNHNTRQ1166, and MNHNTRQ1167. For the most part the ratios are similar, the exception being MNHNTRQ1167's lower values for distal articular dimensions (M12, M13, and M14). The M3 versus M4 size contrast closely corresponds to that of the Esme Akçakoy population. MNHNTRQ1161, 1166, and 1167 exhibit stronger proximal articular dimensions (M5 versus M6) than the entire Unit 4 and Unit 5 sample, except for AS93/7 from lower Unit 5 (9.977 Ma).

Figure 11.12H includes five specimens from Loc. OZ02, slightly higher in Unit 6 (9.590 Ma). This sample contrasts strikingly with the OZ01 and S01 sample in having most of its individuals with shorter M1 dimensions. The one that is not, MNHNTRQ1169, is distinguished from all other Unit 6 individuals by very sharply contrasting midshaft (M3 versus M4) and proximal articular facet (M5 versus M6) dimensions and in having very small distal articular dimensions (M10, M11, M12, and M13). There are clearly two very

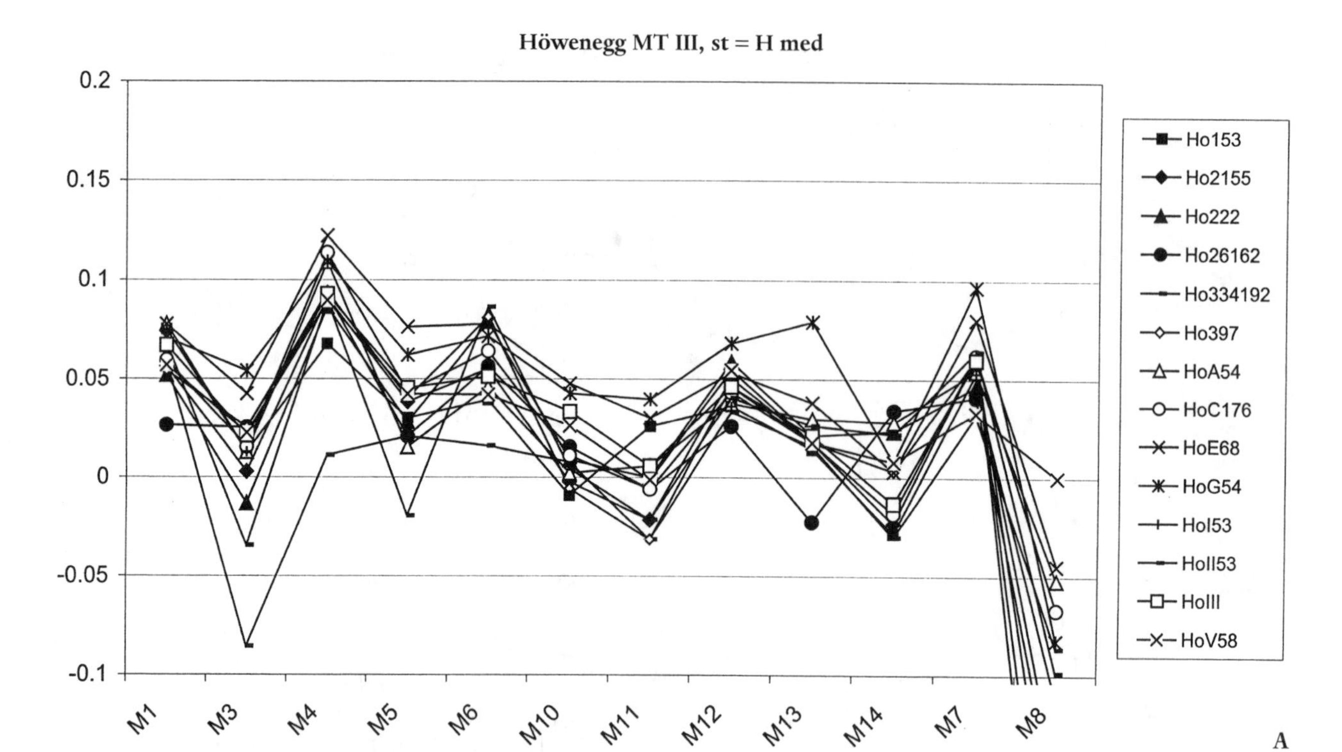

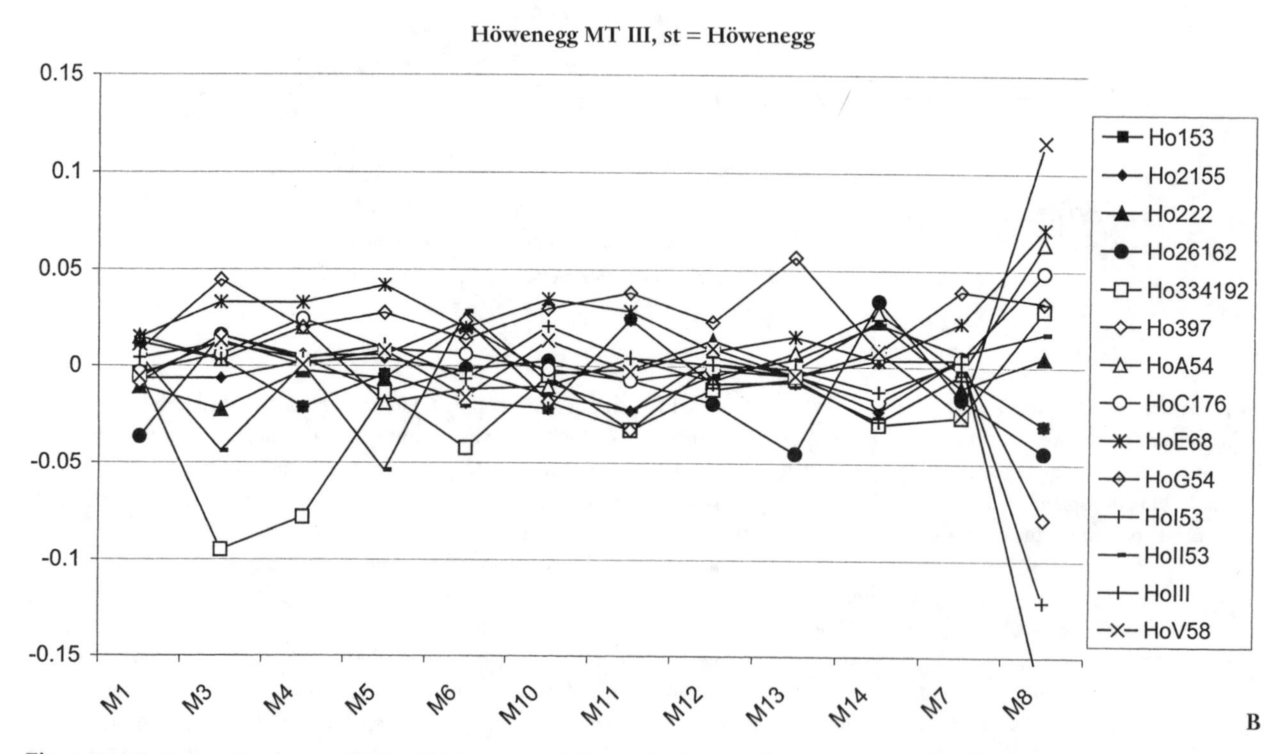

Figure 11.12. Log_{10} ratio plots on MT III. (**A**) Höwenegg MT III sample plotted against *Cremohipparion mediterraneum* standard (see Eisenmann 1995). (**B**) Höwenegg MT III sample plotted against its own mean. (**C**) Esme Akcakoy MT III plotted against the Höwenegg standard. (**D**) Sinap Stratigraphic Unit 4 Loc. 108 hipparion MT III plotted against the Höwenegg standard. (**E**) Sinap Stratigraphic Unit 5 Loc. 91 hipparion plotted against the Höwenegg standard. (**F**) Sinap Stratigraphic Unit 5 Loc. 8B hipparion plotted against the Höwenegg standard. (**G**) Sinap Stratigraphic Unit 6 Locs. Oz01 and S01 hipparion plotted against the Höwenegg standard. (**H**) Sinap Stratigraphic Unit 6 Loc. Oz02 hiparion plotted against the Höwenegg standard. (**I**) Sinap Stratigraphic Unit 6 Loc. 12 hipparion plotted against the Höwenegg standard. (**J**) Sinap Stratigraphic Unit 12 Loc. OzKD hipparion plotted against the Höwenegg standard. (**K**) Sinap Stratigraphic Unit 13 Loc. 42 hipparion plotted against the Höwenegg standard.

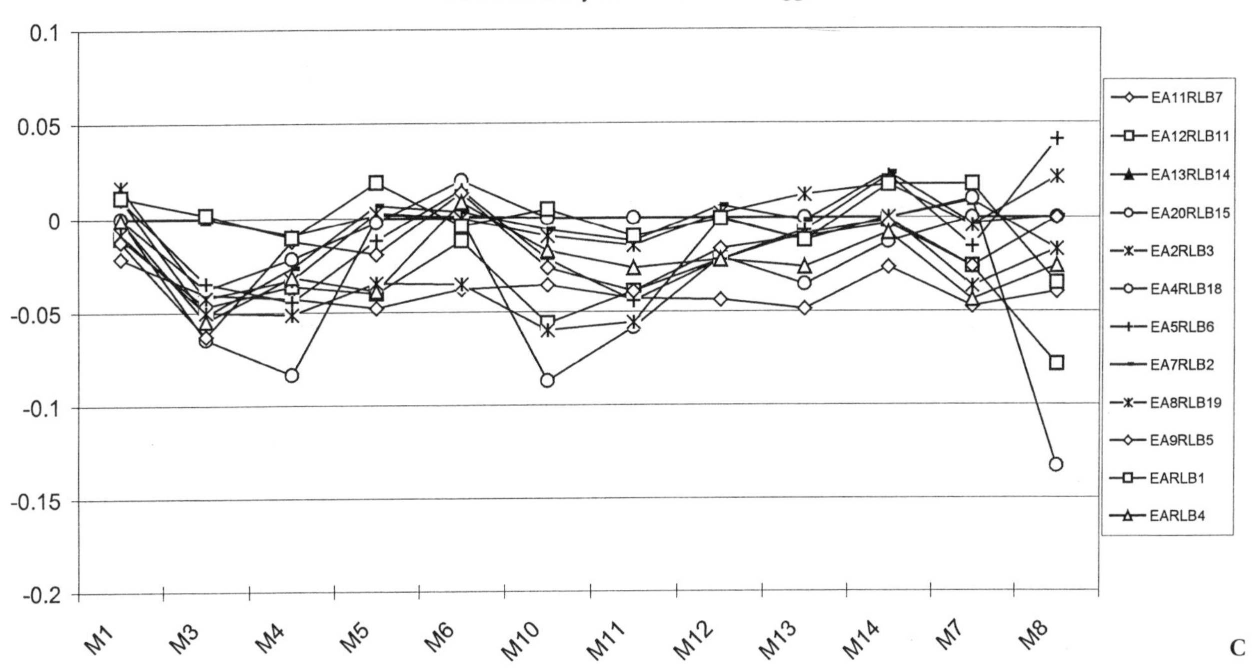

C

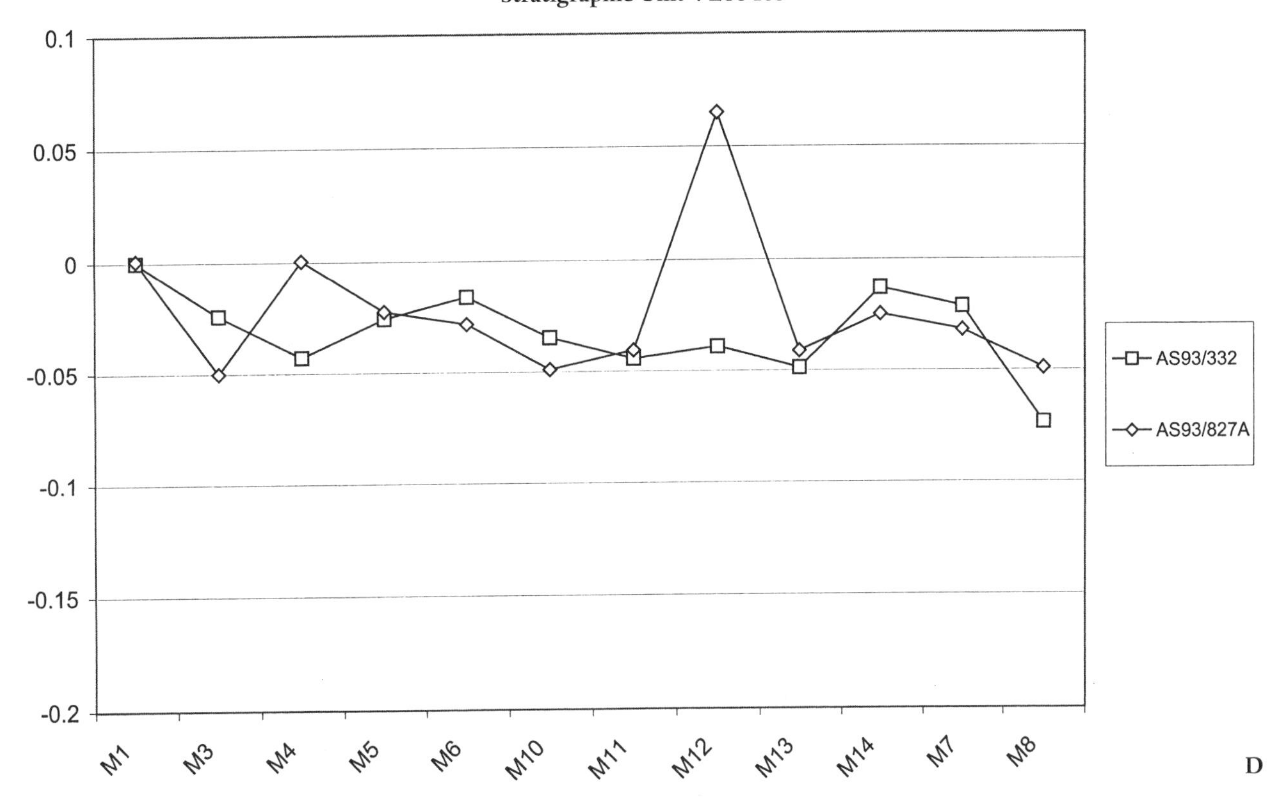

D

Figure 11.12. *(continued)*

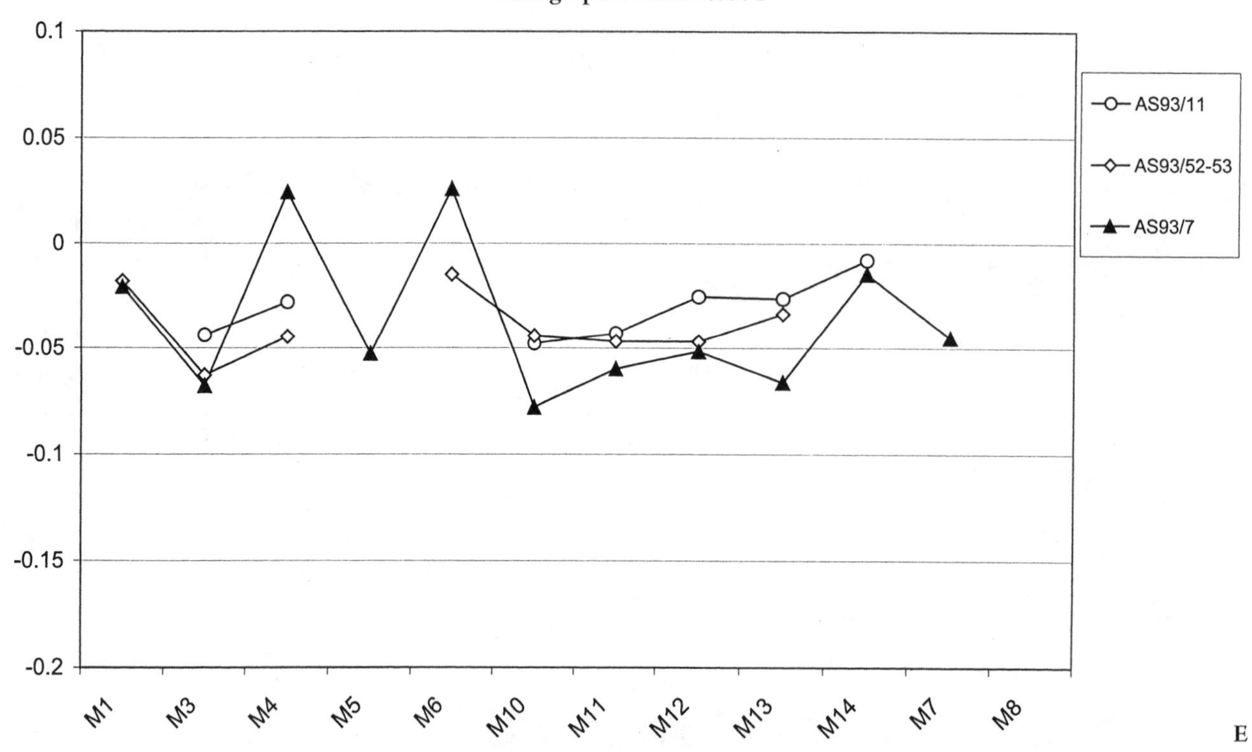

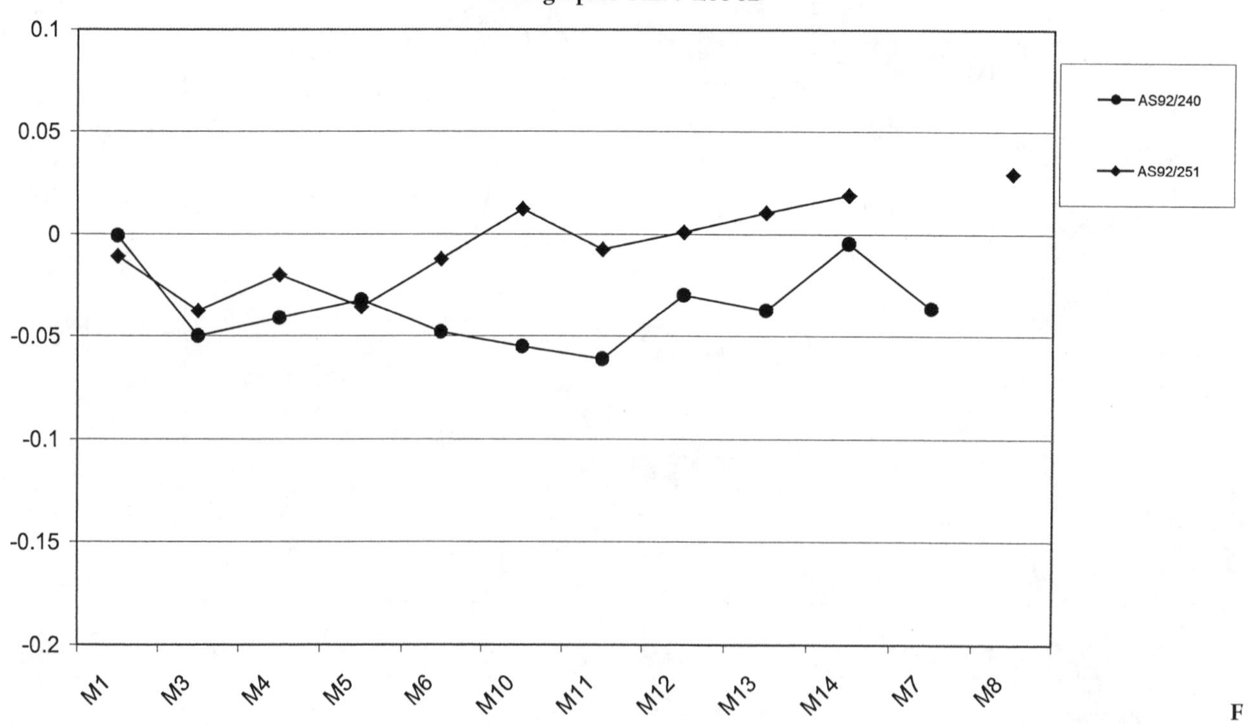

Figure 11.12. *(continued)*

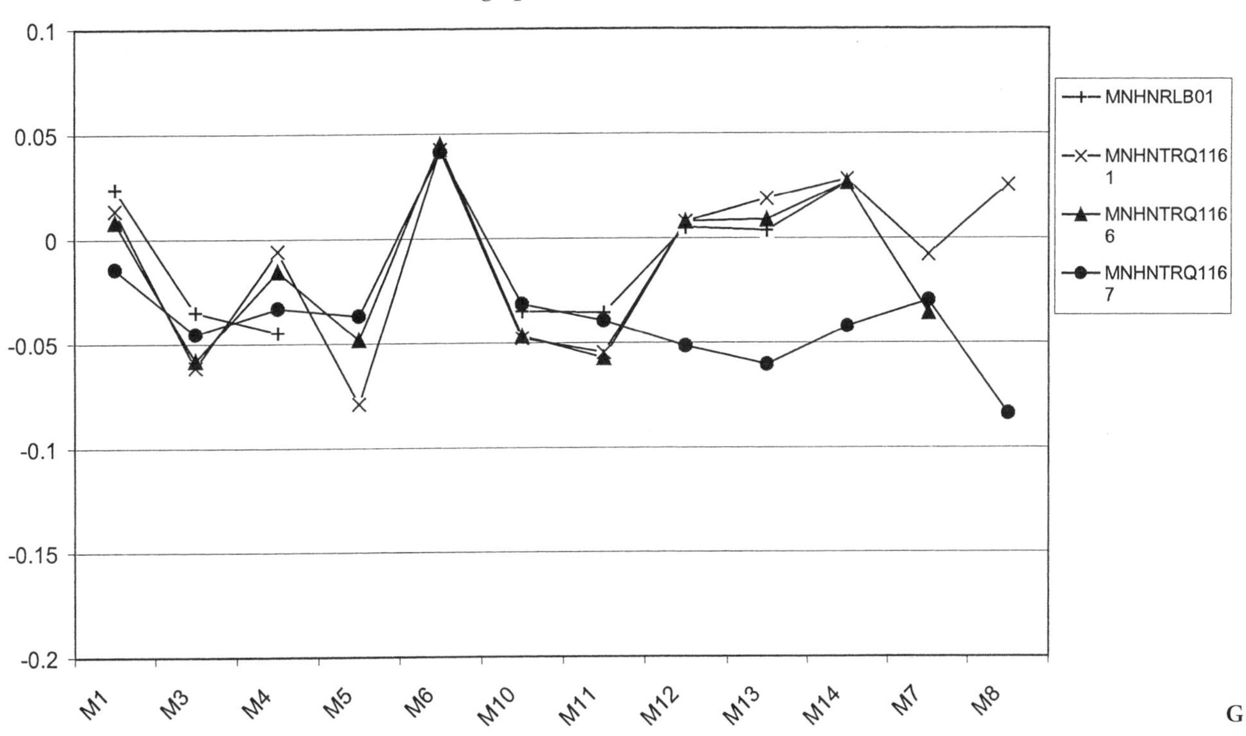

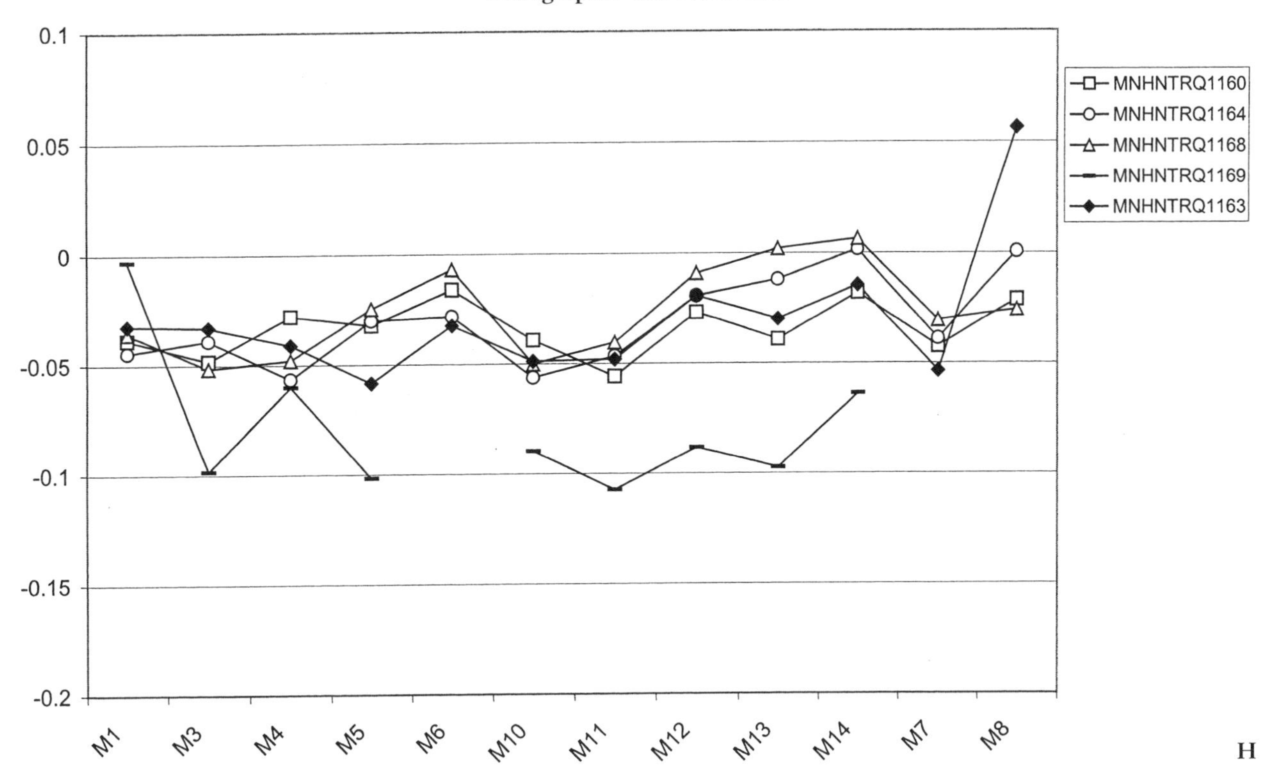

Figure 11.12. *(continued)*

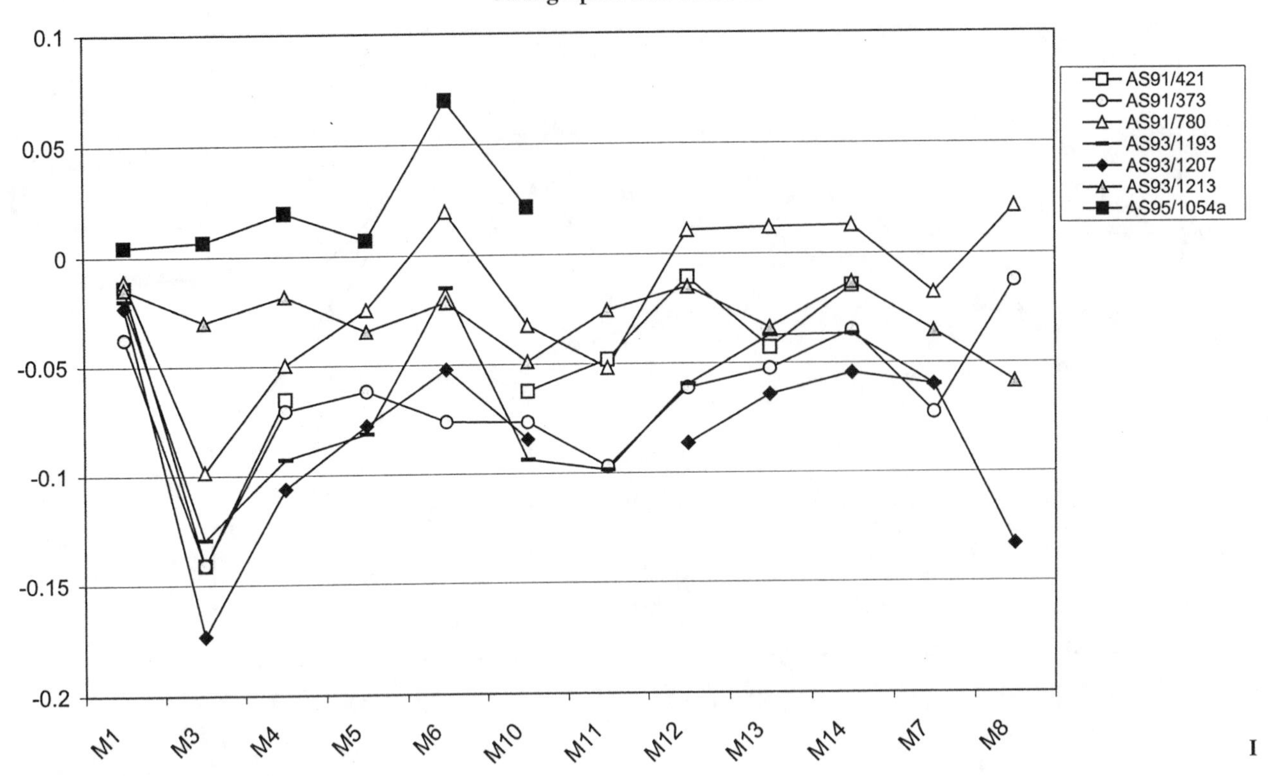

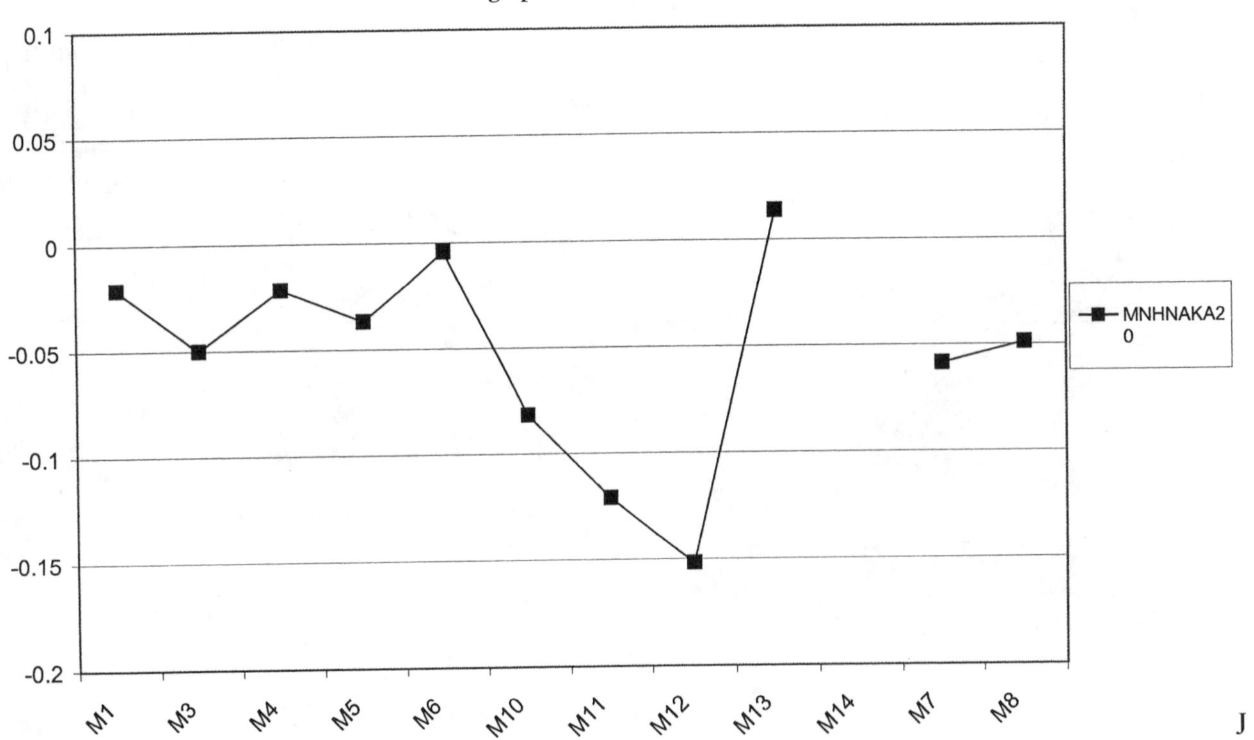

Figure 11.12. *(continued)*

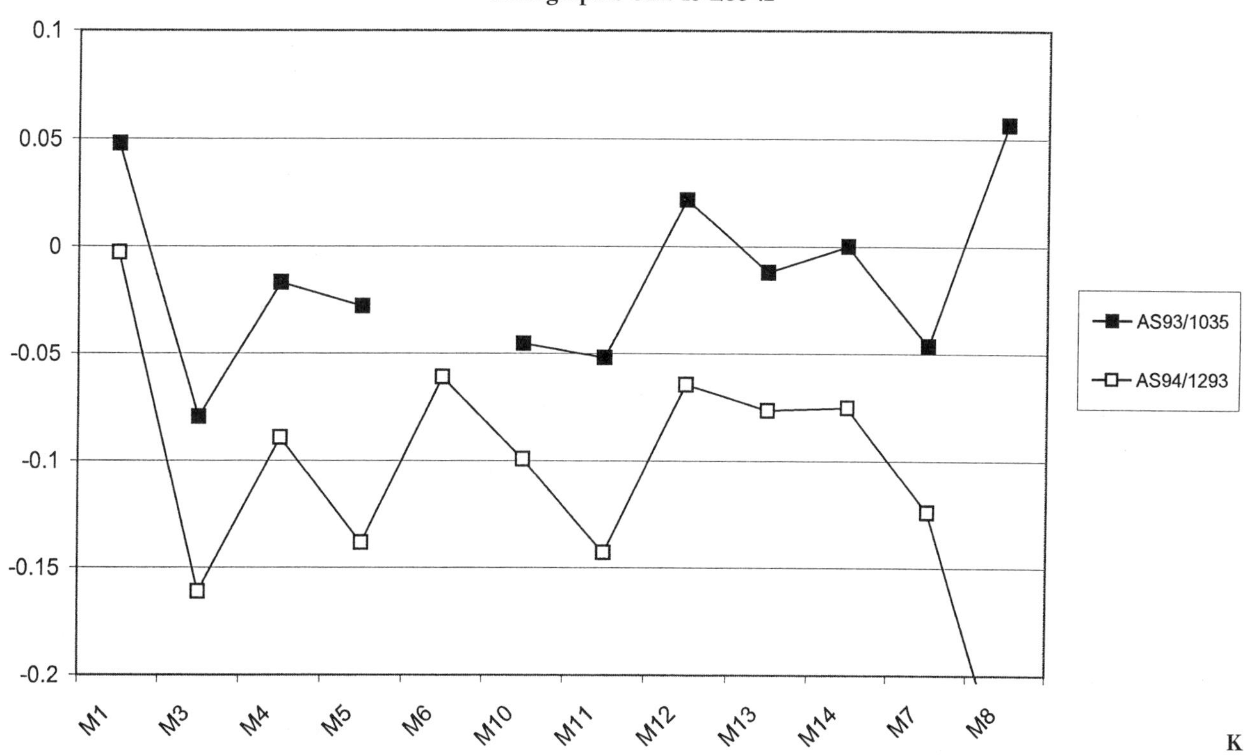

Figure 11.12. *(continued)*

different hipparion morphs in the OZ02 sample, and both of these are different from the OZ01 and S01 samples.

Figure 11.12I includes seven specimens from Loc. 12, at the same stratigraphic horizon in Unit 6 as Loc. OZ02 (9.590 Ma). Specimen AS91/1213 is distinct from the rest of the Loc. 12 sample, and is like the majority of the Loc. OZ02 sample in its slightly contrasting M3 versus M4 dimensions and relatively conservative divergence from the Höwenegg standard in the remaining dimensions. Specimen AS95/1054a is notable for its relatively large size compared with all other specimens from Sinap. The remainder of the sample is distinct in its very small midshaft width (M3) and depth dimensions (M4), very divergent proximal articular surface dimensions (M5 and M6), and contrasting distal articular dimensions (M11, M12, M13, and M14). Clearly, there is considerable variability in this sample built around two themes: the modest divergence in dimensions from the Höwenegg standard seen in AS93/1213 and the remainder of the sample with sharp contrasts in craniocaudal versus mediolateral dimensions.

Referral to the M1 versus M11 plot (fig. 11.11B) reveals tremendous variability in Unit 6 specimens, all within approximately 100,000 years in age of one another. There are no less than four clusters of individuals in these plots: some within the lower left portion of the Höwenegg ellipse, a cluster of four short specimens just below the lower left corner of the ellipse, another cluster of three specimens above the left portion of the ellipse, and three specimens removed far to the left (narrow distal articular width [M11]). Measurements M5 and M6 were available for AS95/1054A and this specimen (fig. 11.11D) plots on the top edge of the ellipse above and to the right of the other specimens from Unit 6. Thus, this specimen suggests a potential fifth type of MT III at Unit 6.

Stratigraphic Unit 12 (= K on the bivariate plots, fig. 11.12J) includes one individual, MNHNAKA20 from Loc. OZKD (8.121 Ma), whereas Unit 13 (Loc. 42; fig. 11.12K) includes two individuals, AS92/1035 and AS94/1293. The Unit 12 specimen has a very narrow midshaft dimension (M3) and proximal articular surface dimension; both strongly contrast with their depth analogues (M4 and M6, respectively). Distal articular width is likewise extremely narrow.

Stratigraphic Unit 13 (C on the bivariate plots) includes two individuals from Loc. 42 (fig. 11.12K): AS93/1035 and AS94/1293. These two specimens have very similar proportions across all measurements but exhibit size variability. The difference in size is strongly reflected in the M1 versus M11 bivariate plot, wherein both have long and slender proportions compared with the rest of the Sinap and the Höwenegg samples. This likely indicates that a single taxon has been sampled at Loc. 42 (Coban Pinar).

The results for the PCA of the MT III are shown in the appendix table 11.13 and the stratigraphic distribution of morphological patterns recognized based on this analysis is summarized in table 11.9. Cumulatively, the first two

Table 11.9. MT III Morphologies by Stratigraphic Unit[1]

Unit	Pattern 1	Pattern 2a	Pattern 2b	Pattern 3	Pattern 4	Pattern 5
4	X					
5	X	X	X	X		
6	X	X	X	X	X	X

[1]X, Morphological pattern present.

principal components explain 98% of the variance and describe two important morphological trends. Principal component 1 corresponds almost exclusively to relative MT III length (M2/GEOMEAN) and explains 96% of the variance in the sample. Principal component 2 explains only 2% of the variance in the sample but illustrates a potentially meaningful and biomechanically significant pattern. This component has positive eigenvectors with the GEOMEAN corrected mediolateral dimensions M3 (midshaft width), M5 (proximal articular surface width), and M10 (distal maximum supraarticular width), and negative eigenvectors for M6 (proximal articular depth). Thus, principal component 2 is mostly a composite measure of the relative mediolateral dimension. Negative values for principal component 2 indicate a relatively deep or slender MT III. Several workers (Eisenmann 1995; Gromova 1952) have noted a functional explanation for differences in relative mediolateral or craniocaudal expansion of the metapodials. According to this explanation, metapodial III forms that are craniocaudally deep are adapted to resist greater loads in the sagittal plane, such as those that might be generated by cursorial locomotion. One prediction of this model is that hipparionines living in open environments and engaging in cursorial locomotion would have craniocaudally deep canon bones, whereas forest dwelling species would have mediolaterally broad canon bones. Thus, we expect that more open country hipparionines will have more negative values for principal component 2, whereas closed habitat dwellers will have more positive values for this component. Similarly, the observation that cursorial forms generally have elongate limbs suggests that hipparionines with high scores for principal component 1 are likely to have low scores for principal component 2.

Previous interpretations based on a variety of anatomical regions and supported by faunal community and paleobotanical information have suggested that *Hippotherium primigenium* was well adapted to subtropical forested habitats (Bernor et al. 1988, 1997). This conclusion is supported by the principal component scores for the Höwenegg sample, which has relatively low values for principal component 1 (decreased relative length, see fig. 11.13A) and relatively high values for principal component 2 (increased mediolateral expansion, see fig. 11.13B) compared with specimens from Sinap. The contrasting trend seen in specimens from Sinap suggests that they may represent hipparionine species adapted to seasonal open country habitats.

The sample of Sinap MT III's is more variable than the Höwenegg sample. Indeed, several morphological patterns are evident for the Sinap sample. Figure 11.13A,B plots the principal component scores for the Sinap specimens by stratigraphic unit relative to the 95% confidence interval for the Höwenegg sample. Figure 11.14 plots principal components 1 and 2 together. The two specimens from Loc. 108 in Stratigraphic Unit 4 appear to be relatively long in contrast to the Höwenegg sample (fig. 11.14, large solid triangles = Pattern 1). However, these specimens have scores for principal component 2 that lie within the 95% confidence interval for the Höwenegg sample. These early Sinap specimens have scores for both principal components similar to the Esme Akçakoy specimens and specimens from the Xmas Quarry sample (fig. 11.14) and may represent a single species. This morphological pattern (Pattern 1) may constitute a primitive MT III morphotype.

The two specimens from Stratigraphic Unit 5, Loc. 91 suggest both diversification and divergence from a possible primitive metapodial morphotype observed for the earlier specimens from Loc. 108 and shared both with *Cormohipparion* specimens from Xmas Quarry and with the Esme Akçakoy sample. Specimen AS93/7 has a negative score for principal component 1, indicating a relative length possibly in keeping with the Höwenegg sample but has a negative score for principal component 2, in sharp contrast with the Höwenegg sample. This particular morphology is referred to as "Pattern 2a" (fig. 11.14); AS93/52 is referred to "Pattern 2b" (fig. 11.14) and contrasts with the Höwenegg sample on both principal components; it may indicate a decidedly cursorial species.

Material from Loc. 8B of Unit 5 includes specimens exhibiting Pattern 1, and a fourth morphological pattern: Pattern 3 (fig. 11.14). Pattern 3 is represented by AS92/251 from Loc. 8B and is most similar to the Höwenegg sample. Pattern 3 specimens have values for principal components 1 and 2 that are within or just outside the 95% confidence interval for the Höwenegg sample. Between Loc. 91 and Loc. 8B, Stratigraphic Unit 5 preserves four different possible MT III morphotypes.

Unit 6 includes specimens exhibiting Patterns 1, 2a, 2b, and 3, as well as two new patterns. The quarry of Loc. 12 has yielded multiple morphologies. Specimen AS95/1054a is a very large individual, which we assign to Pattern 5 (fig. 11.14). This specimen is further notable for its relatively short length and negative score for principal component 2. Locality 12 has two further specimens, AS93/1207a and AS91/373, exhibiting the Pattern 4 (fig. 11.14) morphology combining small body size and highly positive values for principal component 1 (great relative length). Unit 6 Loc.

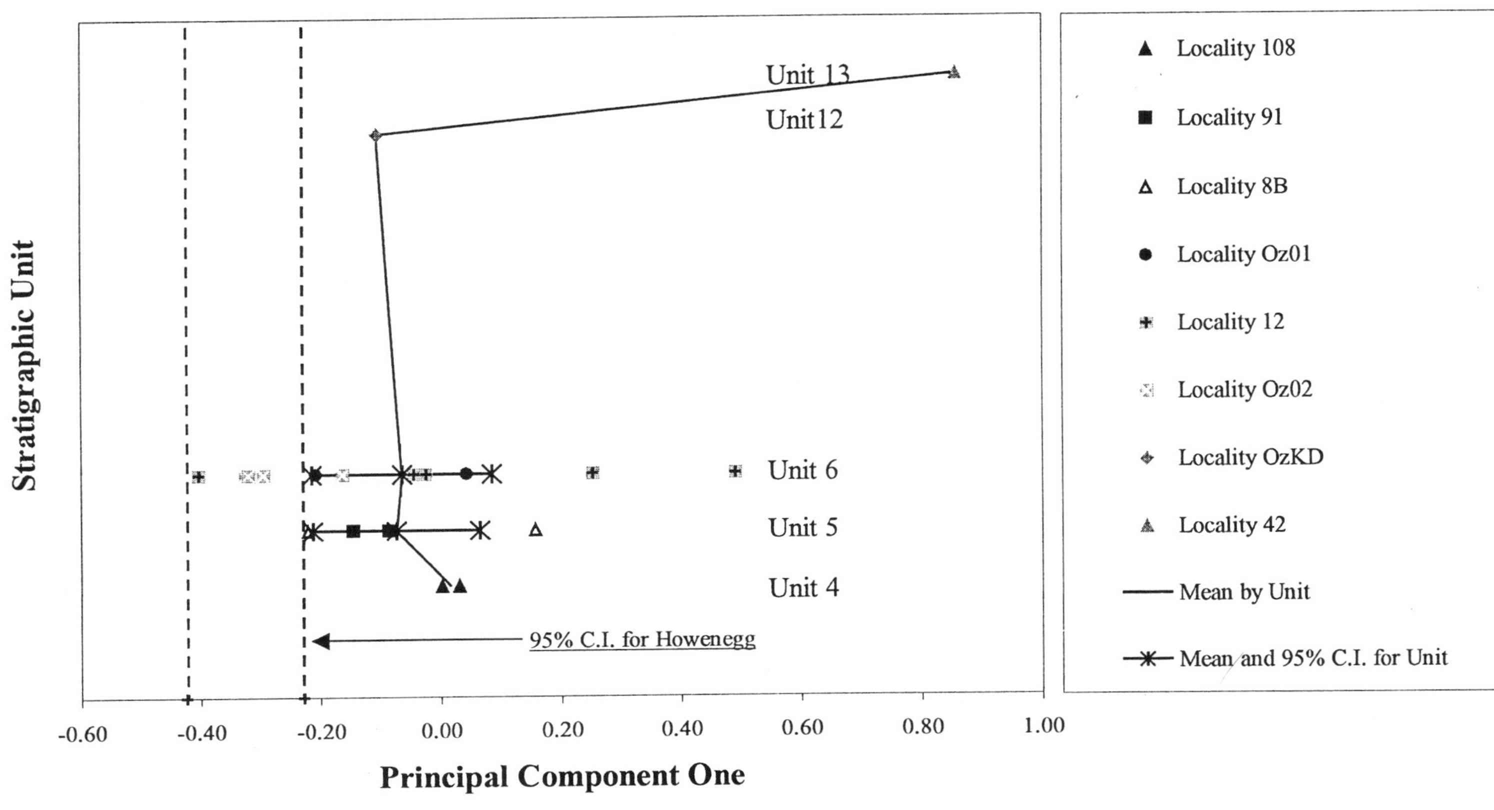

Figure 11.13. Distribution of MT III principal component scores. (A) Principal component 1.

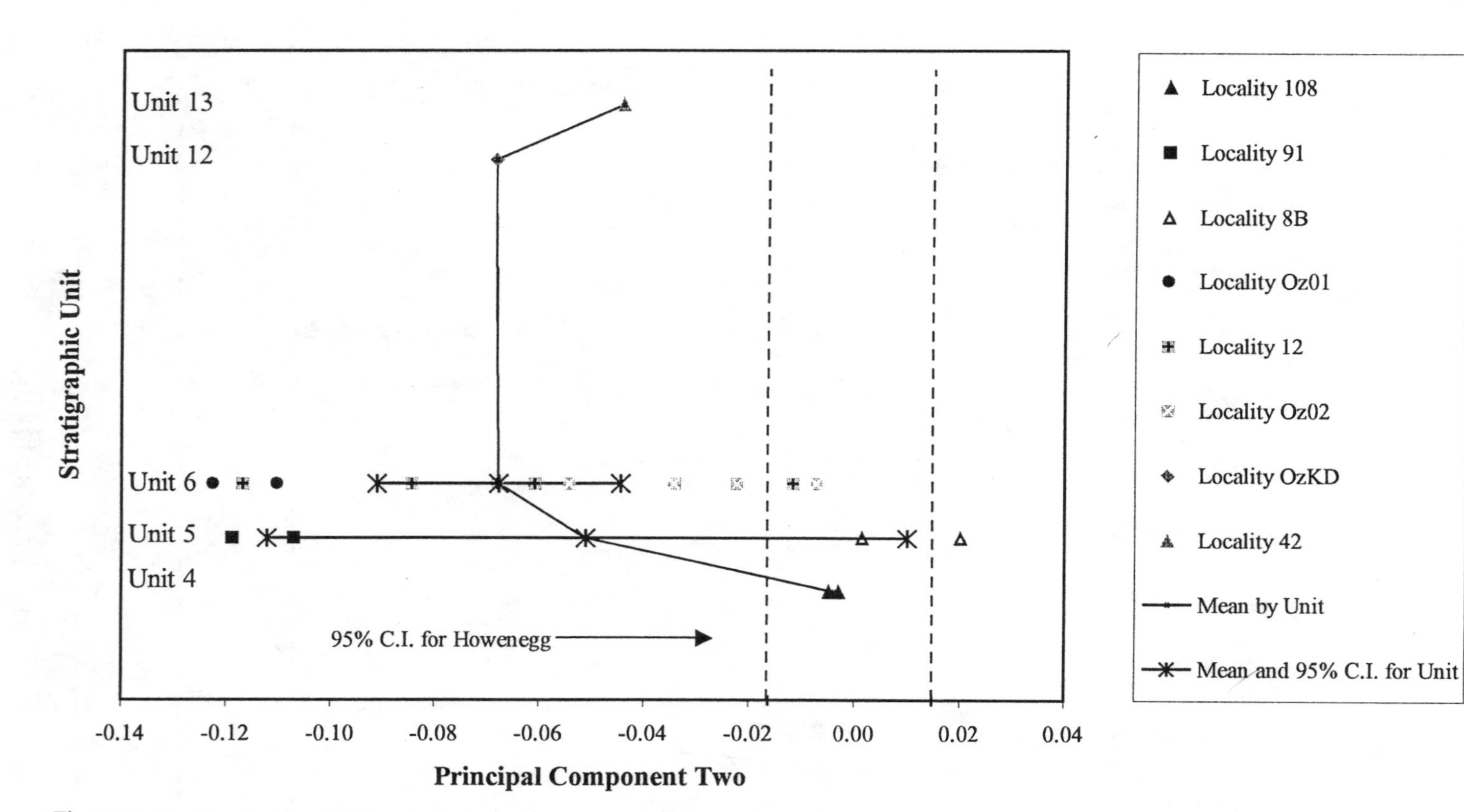

Figure 11.13. Distribution of MT III principal component scores. (**B**) Principal component 2.

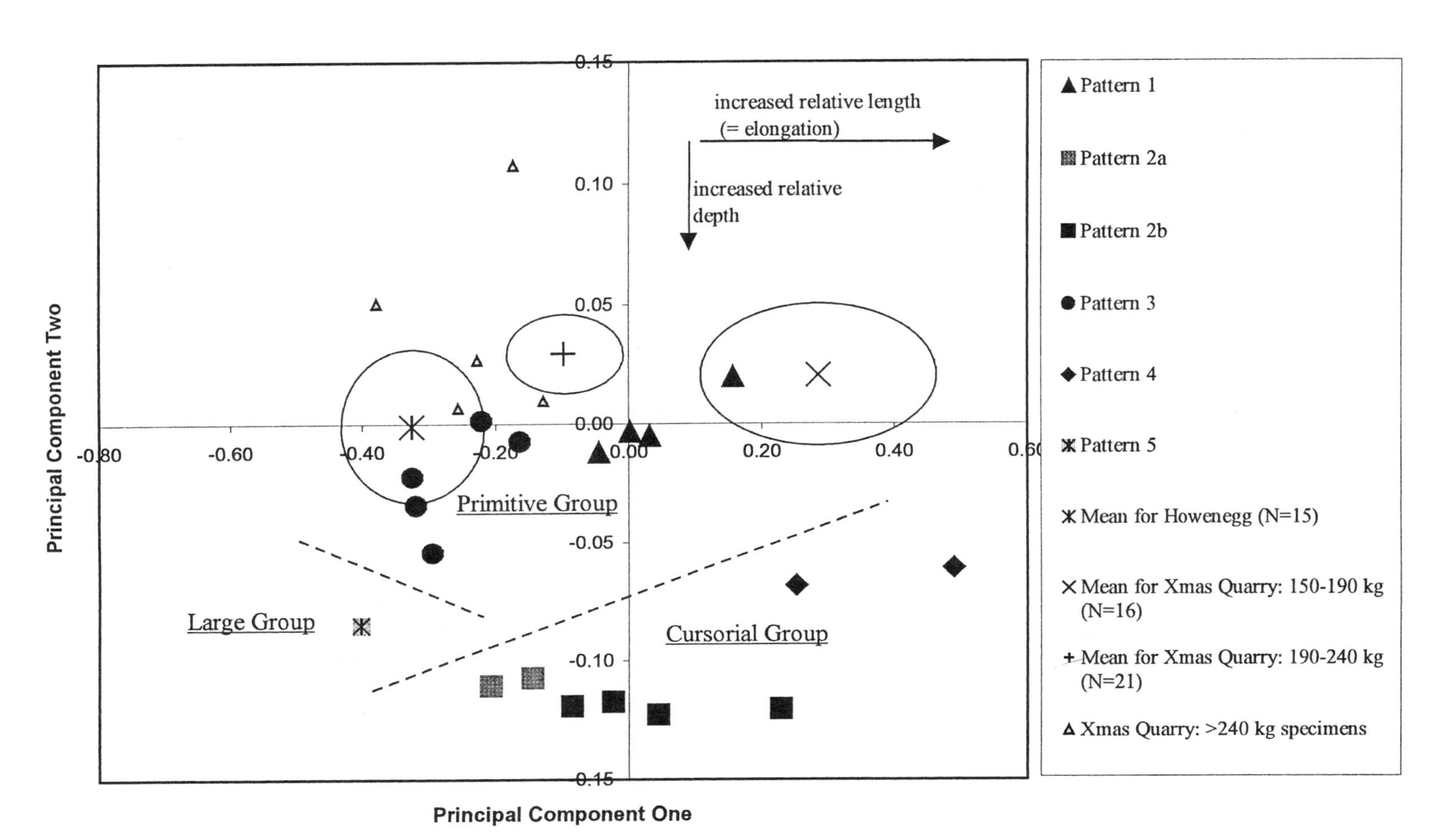

Figure 11.14. Principal components plot of components 1 and 2 for MT III.

Table 11.10. Bootstrap Analysis of Principal Components for MT III and MC III

Element	Group (*i*)	Principal Component	N_i	*p* (*Bootstrap test*)	*p* (*F-test*)	Single Species Hypothesis
MC III	Units 4–6, Sinap Fm.	One	11	**0.0040**	**0.0006**	**Rejected**
MC III	Primitive group	One	9	**0.0134**	**0.0129**	**Rejected**
MC III	Patterns 1 and 3	One	8	**0.0467**	0.0809	**Rejected**
MC III	Patterns 2(mc), 4(mc), and 6	One	3	n/a	<.0001	Rejected
MC III	Patterns 2(mc) and 4(mc)	One	2	n/a	**0.0037**	**Rejected**
MT III	Units 4–6, Sinap Fm.	One	18	0.2547	0.2764	Not rejected
MT III	Moderately large size cohort, Units 4–6	One	15	0.7502	0.7295	Not rejected
MT III	Primitive group	One	9	0.7316	0.6380	Not rejected
MT III	Cursorial group	One	8	0.6767	0.6637	Not rejected
MT III	Quite large and somewhat large size cohorts, Units 4–6	One	3	n/a	**0.0186**	**Rejected**
MT III	Units 4–6, Sinap Fm.	Two	18	**0.0049***$^{0.1454}$	**0.0400***$^{0.2020}$	**Equivocal***
MT III	Moderately large size cohort, Units 4–6	Two	15	**0.0039***$^{0.0730}$	**0.0251***$^{0.1303}$	**Equivocal***

Results are based on comparison of relevant subsample of Sinap specimens [=Group (*i*)] and sample of all left and unpaired right specimens from Höwenegg.

*Results of same comparison using the sample of all right and unpaired left specimens from Höwenegg were not significant. *p*-values for this alternate comparison are shown in the superscript adjacent to the * symbol.

OZ02 includes four specimens all exhibiting Pattern 3 morphology. These Loc. Oz02 specimens, MNHN1160, 1163, 1164, and 1168 can be interpreted as belonging to one of two taxa: (1) the same short, shallow craniocaudal midshaft (M4) represented in Stratigraphic Unit 5 (goatlike form) or (2) a novel form smaller than but similar in its overall morphology to the Höwenegg horse. Pattern 3 may therefore be adapted to either open rocky slopes, or a forested environment. Either interpretation suggests that Loc. OZ02 samples a different paleoenvironment than Loc. 12, with the latter appearing to have harbored more cursorial hipparion species. Locs. OZ02 and 12 are approximately the same age and reveal the distinct possibility of a mosaic of environments at 10.31 Ma.

Table 11.10 provides a boostrapping analysis of the PCA results for both MC III and MT III. The interval of time represented by Units 4–6 is estimated to represent only 1 m.y. but nevertheless shows a great diversification of hipparionines, as revealed from the standpoint of their MT III morphology. Six different morphological patterns (see table 11.9) in the MT III alone could indicate at least six hipparion species. A more conservative approach to determining species number uses the Höwenegg sample and a bootstrap model to divide the Sinap hipparionines from Stratigraphic Units 2–6 into groups with the same or less variation than the Höwenegg sample (see table 11.10). Accordingly, a group including all 18 specimens from Units 4–6 had a significantly greater variance for principal component 2 than the Höwenegg sample ($p < 0.05$, one-tailed test using a sample of 15 left side and unpaired specimens from Höwenegg). Thus, we can reject the hypothesis that all Sinap hipparionines belong to a single population of a single species as represented by the Höwenegg sample. We repeated the test for all Unit-4–6 Sinap hipparionines excluding those exhibiting Patterns 4 and 5 ($n = 15$) and found that this group also had a significantly greater variance for principal component 2 than did the Höwenegg sample ($p < 0.05$, one-tailed test using a sample of 15 left side and unpaired specimens from Höwenegg). Similarly, we found that the Pattern 4 and 5 specimens had significantly more variation on principal component 1 than the Höwenegg sample (*F*-test: $p < 0.05$). The same test was performed for a group composed of specimens exhibiting Patterns 1 and 3 and a group composed of specimens with Pattern 2a or 2b. These results were not significant for either principal component 1 or principal component 2. Thus, we cannot reject the hypotheses that Pattern 1 and 3 MT III specimens belong to a single species and the Pattern 2a and 2b specimens belong to another single species.

The bootstrap analysis of variability for the Sinap third metatarsals suggests at least three species for Units 4–6 that are supported by significant *p* values. According to this conservative grouping of MT III morphs, the Sinap metatarsals from Units 4–6 have been divided into three groups.

Primitive Group. This group includes those specimens with MT III exhibiting Patterns 1 and 3. It appears possible that members of this group were either adapted to rocky slopes or woodland adapted, or a combination of both, and had some similarities with the Höwenegg sample. This group represents a modest derivation from the first appearing North American forms as represented by the Unit 4 specimens and some of the Xmas Quarry sample. Specimens assigned to the Primitive Group by the bootstrap analysis are AS93/1213a, AS93/332, AS93/827a, AS92/240, AS92/251, MNHNTRQ1160, MNHNTRQ1163, MNHNTRQ1164, and MNHNTRQ1168, from Locs. 108, 8B, 12, and OZ02.

Cursorial Group. This group includes specimens exhibiting Patterns 2a, 2b, and 4 and is marked by scores for principal components 1 and 2 suggestive of cursorial behavior. Specimens assigned to the Cursorial Group are MNHNTRQ1161, MNHNTRQ1166, MNHNTRQ1167, AS91/780, AS93/7, and AS93/52, from Locs. 12, 91, and S01/OZ01. Also included in this group but probably representing another taxon (based on body mass) are AS91/373 and AS93/1207A from Loc. 12.

Large Group. The group is represented by AS95/1054a from Loc. 12, with an estimated body mass of 291 kg (Scott 1996).

If we do not recognize Pattern 3 morphs as being the same taxon as the goatlike MC III (MNHN1129 from Unit 5, Loc. OZ01), then this specimen represents yet another taxon identified within the Unit 4–6 stratigraphic interval. We address the species identity of the various metapodials in the section that immediately follows the Systematics section.

Systematics

Taxonomy

Order Perissodactyla Owen 1848
Suborder Hippomorpha Wood 1937
Family Equidae Gray 1821
Subfamily Equinae Steinmann and Doderlein 1890
Genus *Cormohipparion* MacFadden and Skinner, 1977
Cormohipparion sinapensis sp. nov.

Holotype. AS93/826 (fig. 11.15A, B).

Referred Specimens. SEN 1 (fig. 11.15C,D); AS92/618; AS93/9.

Type Locality. Sinap Loc. 108.

Referred Localities. Sinap Locs. 94, 108, 108/8 (between Locs. 108 and 8), and 91.

Age. Late Miocene, early Vallesian (MN 9), 10.551–9.970 Ma,

Geographic Range. Western Anatolia, Turkey.

Diagnosis. A medium-sized hipparionine; POB long with anterior limit of the lacrimal being placed more than one-half the distance from the anterior orbital rim to the posterior rim of the POF; orbital surface of the lacrimal bone with a prominent lacrimal foramen; POF subtriangular shaped and anteroventrally oriented, deeply pocketed posteriorly, deep medially, with moderately delineated peripheral border outline and prominent anterior rim; dP1 persistent and functional; maxillary cheek teeth strongly curved mediolaterally, moderately high crowned having a maximum crown height estimated to be between 40 and 50 mm, fossette ornamentation moderately complex, posterior wall of postfossette variably distinct or confluent with posterior enamel wall, pli caballin varies from complex to double, hypoglyph frequently encircled in earlier wear becoming moderately incised later in wear; protocone shape variably oval to lingually flattened-labially rounded, protocone isolated until later wear frequently becoming connected to the protoloph, protoconal spur known only to be absent, premolar and molar protocone known only to be more lingually placed than hypocone, P2 anterostyle variably elongate to short and rounded; mandibular cheek teeth known to be lacking metastylid spur, premolar ectoflexid not separating metaconid/metastylid; pli caballinid absent; protostylid absent on occlusal surface but vertically placed lying flush with protoconid enamel band; ectostylids absent in adult cheek teeth; premolar linguaflexid with deep V-shape, preflexid and postflexid with simple margins; limbs elongate and slightly built compared to *Hippotherium primigenium.*

Remarks. *Cormohipparion sinapensis* sp. nov. was collected from some of the lowest Sinap stratigraphic units with hipparion, Loc. 94 (ca. 10.551 m.y.), where it is known by a mandibular p4, AS92/618. This individual has a very low crown height (maximum height ~40–45 mm), which is remarkable for an Old World hipparion. Loc. 108, approximately 25 m higher in the section (~10.135 Ma), has produced a partial hipparion skeleton (AS93/826), including a fragmentary skull, which is elected here as the type specimen for *C. sinapensis.* This taxon is further recognized by a skull collected by Senyurek at a locality stratigraphically intermediate between Locs. 108 and 8 (Sen, pers. observ.), which we designate here as Loc.108/8. We further refer a metatarsal, AS93/9 (Loc. 91) to *C. sinapensis.*

Cormohipparion sinapensis is smaller than *Hippotherium primigenium,* as characterized by the Höwenegg, Eppelsheim, and Inzersdorf hipparions, and exhibits some primitive characters compared with that species. *C. sinapensis* mostly shares primitive skull characters with *H. primigenium* and North American *C. "occidentale,"* including Characters C1–C2, C5–C7, and C9–C14 of table 11.4. It shares with *C. "occidentale,"* and *not* with *H. primigenium,* state C3 (=A)—the orbital surface of the lacrimal bone has a large and distinct foramen. It shares with *H. primigenium* and *not* with *C. "occidentale"* states C4 (=D), POF subtriangular shaped and anteroventrally oriented and C16 (= A), dP1 persistent and functional. Woodburne (1996) reports that the dP1 is large and functional in the more primitive *Cormohipparion* species *C. quinni.* In fact, *C. sinapensis* and the Eppelsheim and Höwenegg hipparions all have dP1 morphologies that compare closely with *C. quinni* (Bernor et al. 1997: Höwenegg hipparions; Woodburne 1996: p. 17, fig. 9, type specimen F:AM 71888 from Devil's Gulch Member, Valentine Formation, late Barstovian, Brown County, north central Nebraska, ~13–12.5 Ma). *C. sinapensis* differs from both *C. "occidentale"* and *H. primigenium* in the following characters: the peripheral rim is only moderately delineated

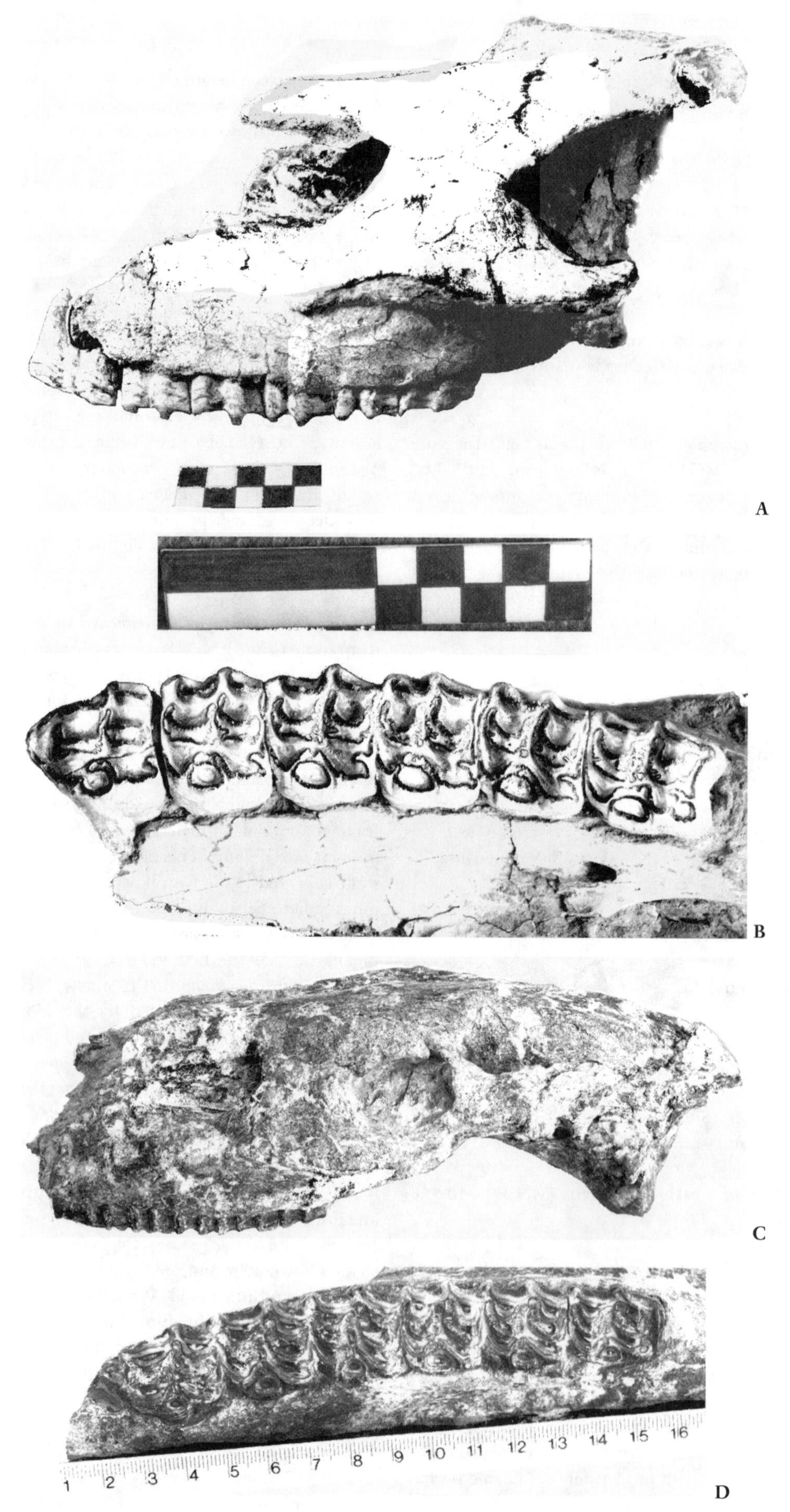

Figure 11.15. *Cormohipparion sinapensis.* (**A**) AS93/826 (Type), skull, lateral view. (**B**) AS93/826 (Type), skull, occlusal view of cheek teeth. (**C**) SEN1, skull, lateral view. (**D**) SEN1, skull, occlusal view of cheek teeth.

(C8 = B), which we believe is an advanced character; the maxillary cheek teeth are lower crowned and more curved mediolaterally (C17 = AB), which is primitive compared to *H. primigenium*.

The maxillary cheek teeth similarly exhibit a mosaic of characters. *Cormohipparion sinapensis* is similar to both *Hippotherium primigenium* and *C. "occidentale"* in characters C25–C27. *Cormohipparion sinapensis* is similar to *H. primigenium* and *not* to *C. "occidentale"* in the variable expression of its pli caballin morphology being either double (C21 = A) or complex (C21 = C). *Cormohipparion sinapensis* is similar to *C. "occidentale"* and *not* to *H. primigenium* in the following characters: maxillary cheek tooth ornamentation is generally less than in *H. primigenium* (C19 = B or sub-A). *Cormohipparion sinapensis* differs from both *H. primigenium* and *C. "occidentale"* in a number of maxillary cheek tooth characters, including: the posterior wall of the postfossette varies considerably in whether it is distinct or merged with the posterior enamel wall of the tooth (primitive) (C20 = A,B); hypoglyph deeply to moderately deeply incised (derived character) (C22 = B,C); protocone with a tendency to become oval (derived character) (C23 = C,E,G); protocone tends to become confluent with the protoloph (primitive character typical of hipparions with lower crowns than *C. "occidentale"* and *H. primigenium*) (C24 = A,B); a tendency to have more abbreviated P2 parastyles than *C. "occidentale"* and *H. primigenium,* typical of the more primitive hipparion *C. quinni* (Woodburne 1996, p. 27, fig. 11, F:AM 71896 from Railway Quarry A, Crookston Bridge Member, Valentine Formation, late Barstovian, Cherry County, north central Nebraska, ~13 Ma; F:AM 71895 from Sawyer Quarry Extension, Devil's Gulch Member of the Valentine Formation, late Barstovian, Brown County, north central Nebraska; ~13 Ma) (C28 = A,B).

Univariate plots of a number of skull measurements (fig. 11.1A–E) support the closest relationship in the various skull parameters calculated between the Loc. 108 skull and the Xmas quarry sample of *Cormohipparion "occidentale."* The bivariate plot on POB length (M32) versus POF length (M33) position, *C. sinapensis* falls within the *C. "occidentale"* ellipse (fig. 11.2E). Maxillary P2 measurements (M3 versus M1 and M11 versus M10; fig. 11.3) place *C. sinapensis* within the Eppelsheim ellipse. Bivariate measurements M1 versus M6 on the calcaneum (fig. 11.6) show that *C. sinapensis* is smaller than *H. primigenium* and among the smallest of the Sinap specimens. Likewise, bivariate plots on the astragalus (M1 versus M5 and M6 versus M5; fig. 11.7A–D) show the Sinap specimens to be smaller than *H. primigenium* and amongst the smaller (but not smallest) of the Sinap horses. Metacarpal III length (M1) versus distal articular width (M11) shows that *C. sinapensis* has an MC III as long, but more slender than the Höwenegg hipparion (fig. 11.8A,B). The MC III log-ratio diagram shows the same long and slender dimensions of *C. sinapensis* but, additionally exhibits the Esme Açkakoy effect of a deeper mid-shaft (M4) than mediolateral width at the same diaphysial level (M3; fig. 11.9D). The PCA analysis establishes the Loc. 108 MC III (fig. 11.10) and MT III (figs. 11.13A,B, 11.14) as being primitive for the sequence with clear morphological linkages to North American *C. "occidentale."*

The various data presented suggest that *Cormohipparion sinapensis* is distinct from central and western European *Hippotherium primigenium* and may well be derived from an earlier member of the *C. "occidentale"* group (sensu Woodburne 1996) than is represented from the Xmas Quarry, Hans Johnson Quarry, and MacAdams Quarry samples, but similar to *C. "occidentale"* type 1 from Xmas Quarry (see Woodburne 1996, p. 47). *Cormohipparion sinapensis'* metapodial proportions are very similar to the Esme Açkakoy horse, but the former is more lightly built.

"Hipparion" ankyranum *Ozansoy, 1965*

Hipparion ankyranum Ozansoy, 1965, pl. V, figs. 2 and 4

Holotype. MNHNTRQ1211 (Ozansoy 1965, pl. V, figs. 2,4; fig. 11.16A–C).

Type Locality. MNHN Loc. OZ01.

Age. Late Miocene, early Vallesian (MN 9), 9.683 Ma.

Diagnosis. A medium-sized hipparionine; POF deeply pocketed posteriorly, medially deep, with strongly delineated ventral peripheral border; dP1 absent; maxillary cheek teeth moderately curved mediolaterally, fossette ornamentation moderately complex, posterior wall of postfossette always distinct, pli caballin varies from complex to single; hypoglyph deeply incised, being encircled only in M3 earlier wear; protocone shape oval-elongate to ovate and remaining separate from protoloph; protoconal spur absent; premolar/molar protocone variably aligned with hypocone, P2 anterostyle short and rounded.

Remarks. Ozansoy (1965; MNHNTRQ1211, pl. V, figs. 2, 4) figured the type specimen of *"Hipparion" ankyranum,* a palate with right and left P2–M3, but mistakenly referred a mandible from another locality (his pl. V, fig. 1) to the taxon and cited the palate (pl. V, fig. 4) to which the left cheek tooth series (pl. V, fig. 2) actually relates, as being derived from Kavek Dere (Unit 12 of this chapter). A review by Sen and Bernor of Ozansoy's Doctor d'Etat (in MNHN, Paris) has revealed that he correctly cited and figured the palate and left cheek tooth series there. There is no doubt that plate V, figures 2, 4 are of the same specimen and represent the type of *"Hipparion" ankyranum* (Bernor and Sen, pers. observ.).

We do, however, restrict the nomen *"Hipparion" ankyranum* to the type specimen, making it a nomen dubium. This is because there is insufficient morphological information contained within the type specimen to assign it to any given species at Sinap or elsewhere; the lack of a complete snout disallows accurate taxonomic discrimination from other Sinap hipparions. Moreover, the type material is insufficient to determine whether *"H." ankyranum* is referable to the genus *Hipparion* or to some other Old World superspecific taxon.

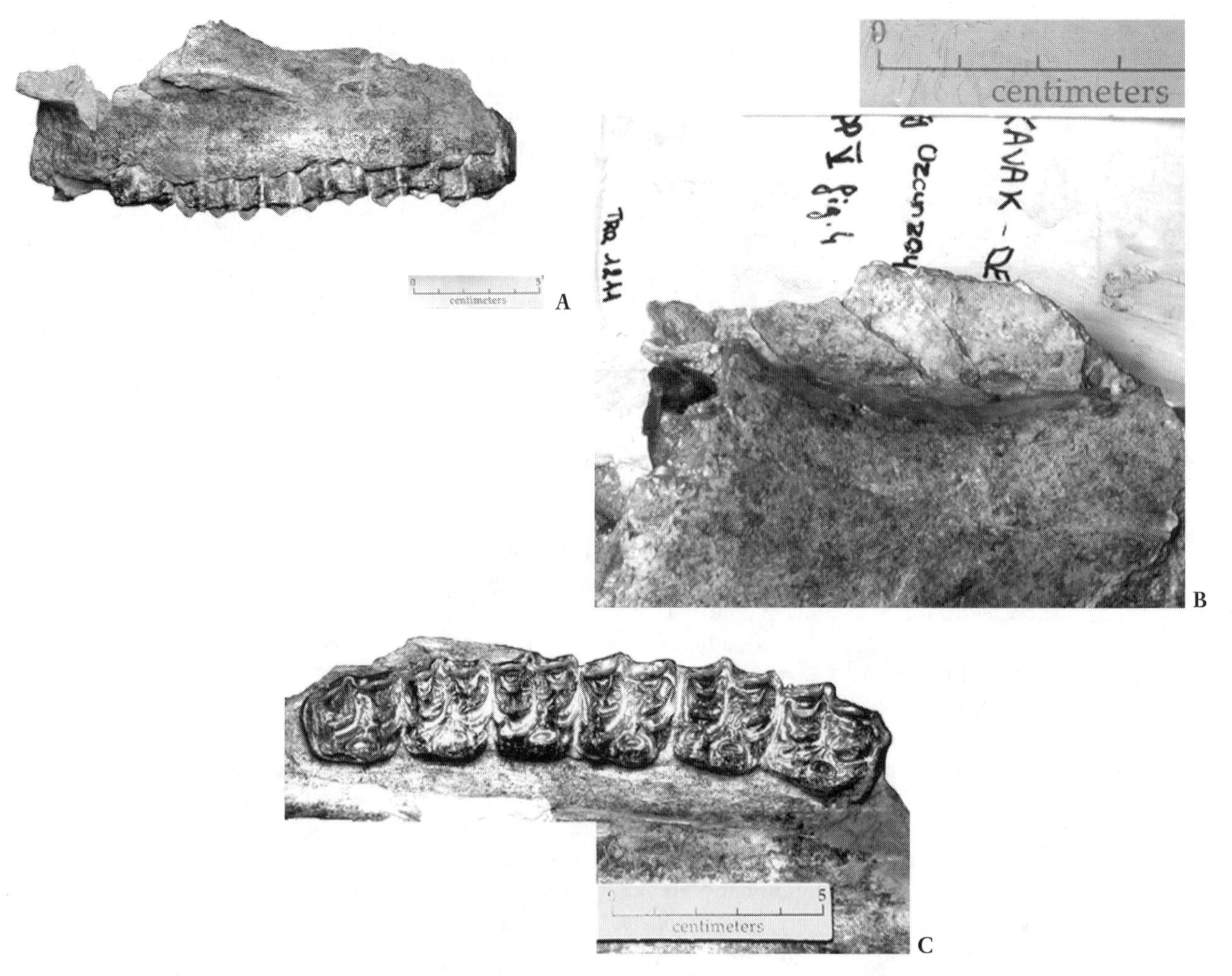

Figure 11.16. *"Hipparion" ankyranum.* (**A**) MNHNTRQ1211 (Type), skull, lateral view. (**B**) Oblique view of the preorbital fossa region. (**C**) Right occlusal view of cheek teeth.

"Hipparion" uzunagizli

Holotype. AS94/203 (fig. 11.17A,B).

Referred Specimens. AS 89/54, AS92/609 (Loc. 8B, 9.918 Ma).

Type Locality. Sinap Loc. 114, 9.967 Ma.

Etymology. From the Turkish "uzun agizli," meaning long-jawed.

Age. Late Miocene, early Vallesian (MN 9), 9.967–9.918 Ma.

Geographic Range. Western Anatolia, Turkey.

Diagnosis. A medium-sized to small hipparionine with a relatively long, wide snout; POB long with anterior limit of the lacrimal being placed more than one-half the distance from the anterior orbital rim to the posterior rim of the POF; orbital surface of the lacrimal bone with a prominent lacrimal foramen; POF reduced with subtriangular shape and anteroventral orientation giving way to an egg-shape and anteroposterior orientation, posterior pocketing sharply reduced but retaining substantial medial depth posteriorly, peripheral outline weak, anterior rim weakly present; nasal notch incised nearly to P2; dP1 persistent and functional; maxillary cheek teeth strongly curved mediolaterally, moderately high crowned, fossette ornamentation complex, posterior wall of postfossette distinct or confluent with posterior enamel wall, pli caballin varies from double to single, hypoglyph deeply incised (but not encircled), protocone lingually flattened-labially rounded and isolated until later wear frequently becoming connected to the protoloph, protoconal spur absent, premolar and molar protocone more lingually placed than hypocone, P2 anterostyle elongate; mandibular molars with rounded-to-elongate metaconids/metastylids; molar ectoflexid separates metaconid/metastylid; pli caballinid absent; protostylid absent on occlusal surface; ectostylids absent; linguaflexid shallow and U-shaped, preflexid and postflexid with simple margins.

Remarks. *"Hipparion" uzunagizli* is known to occur in Stratigraphic Unit 5, Locs. 114 (type) and 8B. Its most distinguishing features are its elongate, broad snout (at level

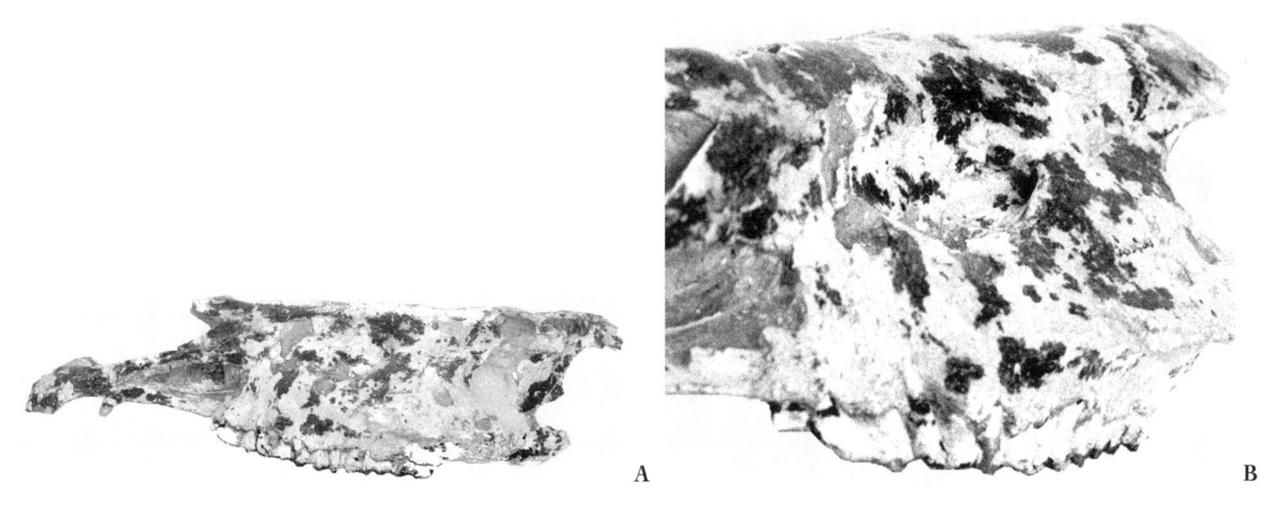

Figure 11.17. *"Hipparion" uzunagizli.* (**A**) AS94/203 (Type), skull, lateral view. (**B**) AS94/203 (Type), skull, oblique view of the preorbital fossa.

of the posterior I3), its further reduction of the POF, and evolutionary advances in its cheek tooth morphology.

Examination and comparison of skull character states (see table 11.4) reveal that *Hipparion uzunagizli* differs from *H. primigenium, Cormohipparion "occidentale"* and *C. sinapensis* in the transformation of the POF from a subtriangular shape with an anteroventral (C4 = D) orientation to an egg shape with an anteroposterior orientation (C4 = F). This stage of evolution anticipates just the transformation hypothesized by Bernor et al. (1980, 1989) for the Old World Groups 1–3 (= *Hipparion* s.s.). Preorbital fossa posterior pocketing (C5 = B/C, C) and peripheral border outline (C8 = C) are reduced compared to *H. primigenium, C. "occidentale,"* and *C. sinapensis,* whereas curvature of the maxillary cheek teeth remains primitive (C17 = A,B), as for *C. sinapensis* (see table 11.5).

Maxillary cheek teeth exhibit a mosaic of primitive and advanced characters (table 11.5): fossette ornamentation (C19) is as in *Cormohipparion sinapensis,* variably complex (A) or moderately complex (B); the posterior wall of the postfossette (C20) is consistently distinct from the posterior enamel wall of the cheek tooth (B; advanced over *C. sinapensis*); pli caballin (C21) varies from being double (A) to single (B), which is a condition like *C. sinapensis* and like early members of *Hipparion* s.s. (Bernor 1985; Bernor et al. 1989); hypoglyph (C22) is mostly deeply incised (B), as in *H. primigenium* and *C. "occidentale"*; protocone shape tends to have an oval modality (C,D,E,F) and is isolated (C24 = B) until late wear, when it connects to the protoloph (A); protoconal spur (C25) is consistently absent (C), in contrast with *Hippotherium primigenium,* which exhibits a significant incidence of reduced structures in earlier wear (Bernor and Franzen 1997); premolar (C26) and molar (C27) protocone/hypocone alignment is consistently like *H. primigenium, C. "occidentale,"* and *C. sinapensis* in the more lingual placement of the protocone. The P2 anterostyle is also consistently elongate (C28 = A), an advance over *C. sinapensis.*

Figure 11.2A shows that, although cheek tooth length (M9) of the AS94/203 (type) skull lies within the lower range of *Cormohipparion "occidentale"* from Xmas Quarry and Hans Johnson Quarry, it has a much longer snout (M1) than those populations. Figure 11.2C further shows that AS94/203 has a very wide snout at the level of the posterior I3; interestingly, there are individuals from MacAdams Quarry that fall close to or inside the lower border of the Xmas Quarry ellipse and others, clearly separated from these in their longer and wider snouts, more closely approximating AS94/203 in their dimensions (fig. 11.2C). The POB length (M32) of *Hipparion* sp. nov. 1 versus its POF length (M33) falls within the Xmas Quarry ellipse (fig. 11.2E,F) and is close to *C. sinapensis* and *"H."* sp. nov. 2 in this regard.

"Hipparion" kecigibi

Holotype. MNHNTRQ9001 (fig. 11.18A–C).

Referred Specimen. MNHNTRQ9002.

Type Locality. Sinap Loc. S01.

Etymology. After the Turkish word "keçigibi," meaning goatlike.

Age. Late Miocene, early Vallesian (MN 9), 9.683 Ma.

Geographic Range. Western Anatolia, Turkey.

Diagnosis. A medium-sized hipparionine with a short-wide snout; POF subtriangular shaped and anterposteriorly oriented, deeply pocketed posteriorly, medially deep, with moderately delineated peripheral border outline and no anterior rim; nasal notch retracted nearly to P2; maxillary cheek teeth strongly curved mediolaterally, moderately high crowned, having an estimated mesostyle height of ≤45 mm, fossette ornamentation moderately complex, posterior wall of postfossette distinct, pli caballin double, hypoglyph deeply incised, protocone oval and isolated from

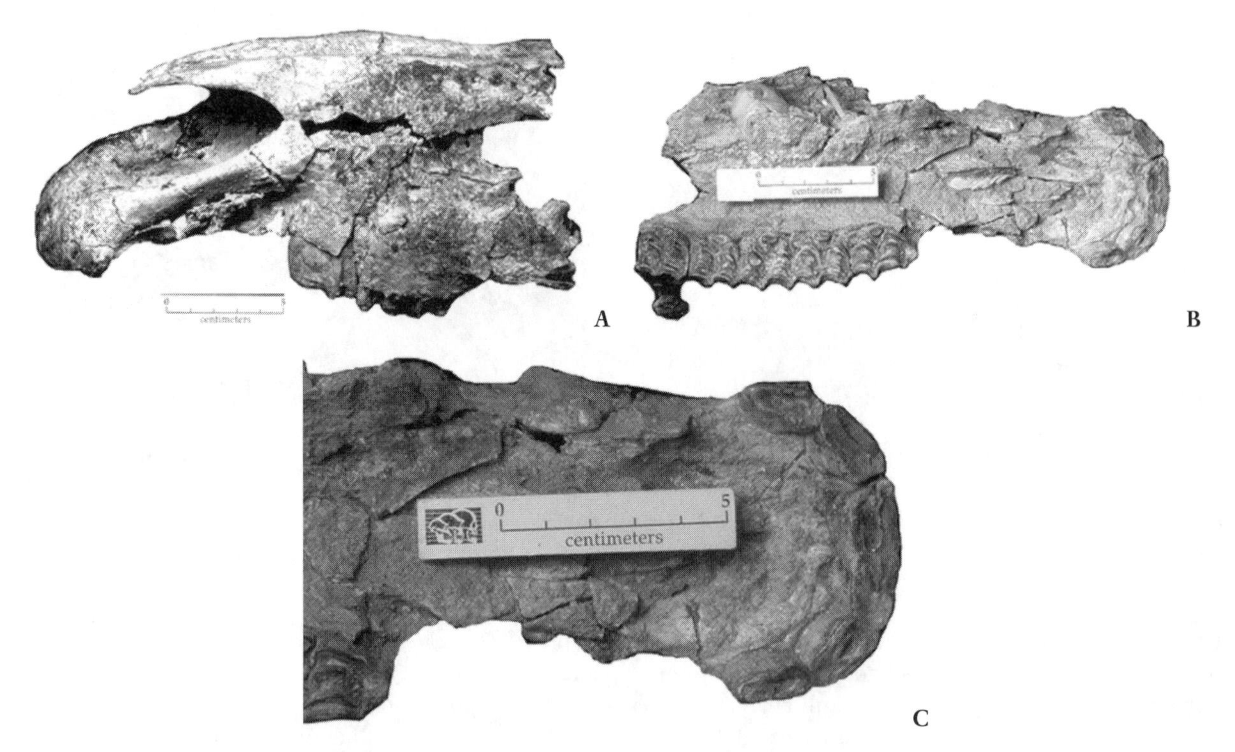

Figure 11.18. *"Hipparion" kecigibi.* (**A**) MNHNTRQ9001 (Type), skull, lateral view. (**B**) MNHNTRQ9001 (Type), skull, occlusal view of cheek teeth. (**C**) MNHNTRQ9001 (Type), skull, close-up of incisor region.

the protoloph, protoconal spur occurs as weakly formed on P2 but absent on other cheek teeth except early wear M3, premolar and molar protocone more lingually placed than hypocone except early wear M3.

Remarks. The type specimen, MNHNTRQ9001, is the best preserved of the three individuals from the standpoint of skull character state representation. This specimen is conservative and resembles both *Hippotherium primigenium* and *Cormohipparion "occidentale"* in most of its characters: C2, C5–C7, and C10–C14. It is like *C. sinapensis* in its moderately prominent peripheral border outline (C8 = B). It differs from *H. primigenium, C. "occidentale," C. sinapensis,* and *H.* sp. nov. 1 in its POF subtriangular shape with an anteroposterior orientation.

The maxillary cheek teeth of MNHNTRQ9001 and MNHNTRQ1064 are very similar. They are similar to *Cormohipparion "occidentale," C. sinapensis,* and *Hipparion uzunagizli* in their moderately complex fossettes (C19 = B), distinct posterior wall of the prefossette (C20 = B) and isolation of the protocone (C24 = B). They are similar to *H. primigenium* in their occasional occurrence of a reduced protoconal spur (C25 = B). Specimen MNHNTRQ1064 exhibits the primitive retention of anteroposteriorly aligned premolar protocone-hypocone (C26 = A). *"Hipparion" kecigibi* is markedly variable in pli caballin (C21 = A, B, C, D) and hypoglyph morphology (C22 = A, B, C, D), making comparisons difficult. Protocone shape is mostly oval (C23 = C).

"Hipparion" kecigibi is distinct from all Sinap taxa described above in its snout proportions. Figure 11.2A,B positions MNHNTRQ1064 in the middle of the Xmas Quarry ellipse for M1 (snout length) versus M9 (length P2–M3). Figure 11.2C,D shows that MNHNTRQ1064 has the same snout width (M15) as the type specimen for *"H." uzunagizli* but a much shorter snout length (M1); MNHNTRQ9001 (the type for this taxon) has nearly as great an M15 dimension as the other two specimens. Snout length (M1) versus width (M15) proportions are functionally significant, in that they reflect feeding adaptation. We believe that the differences seen here in M1 × M15 dimensions between *"H." uzunagizli* and *"H." kecigibi* are significant at the species level. Figure 11.2G shows that MNHNTRQ9001 has a POF depth dimension (M35) that is greater than those of the Xmas Quarry sample and two specimens belonging to *"H." uzunagizli* (AS92/609 and AS94/203 are within the ellipse).

Specimen MNHNTRQ9001 also has P2, M2, and M3 exposed in the maxilla, allowing mesostyle height measurements. Although this specimen is in a rather early wear stage, its crown heights are low: P2 = 29.8 mm, M2 = 32.5 mm, and M3 = 32.3 mm. In this primitive characteristic, *"Hipparion" kecigibi* is more like *Cormohipparion sinapensis* than *Hipparion* sp. nov. 1 and members of the *Hipparion* s.s. group.

"Hipparion" *sp. 1*

Referred Specimens. AS92/458 juvenile skull and mandible (fig. 11.19A,B); AS92/289 (MC III).

Locality. Sinap Loc. 91, Stratigraphic Unit 5, 9.977 Ma.

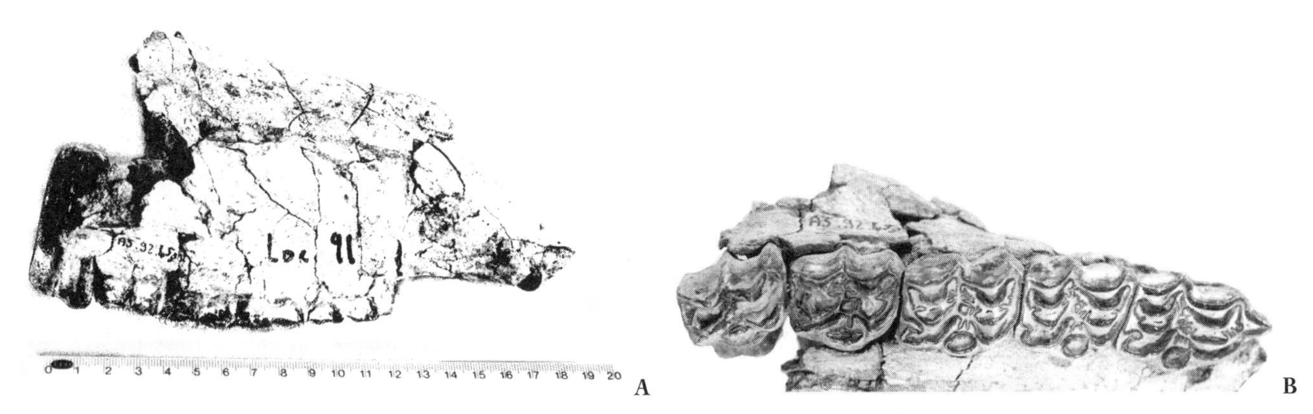

Figure 11.19. *"Hipparion"* sp. 1. **(A)** AS92/458, skull, lateral view. **(B)** AS92/458, skull, occlusal view of cheek teeth.

Remarks. Specimen AS92/458 is a juvenile skull with associated mandibular dentition from Loc. 91, ~9.977 Ma (fig. 11.19A,B). The facial region features a highly reduced POF, with M32 × M33 (fig. 11.2E,F) dimensions and accompanying reduction of the POF medial depth and posterior pocketing that would appear to be more advanced than holds for the rest of the hypodigm of this taxon. This may in part be explained as being an artifact of this individual's young age (M2 just coming into occlusion and M3 not yet erupted). However, we do not believe that this morphology can be considered a normal ontogenetic variant for the other three species of hipparion we recognized above; the POF is simply too reduced in it size and depth. Specimen AS92/458 also has greater crown heights (both measured and estimated) than other Sinap taxa within the Unit 3–5 stratigraphic interval (M1 = 54 mm; M2 = 57 mm), suggesting that there may be a third hipparion taxon here.

Because little deciduous character state data is available, only m1–2 are considered here. These two molar teeth are similar to *Hippotherium primigenium* in the following characters: metastylid lacks a spur (C37 = B); ectoflexid separates metaconid/metastylid (C39 = B); they lack ectostylids on the permanent dentition (C43 = B; character common to all Eurasian hipparions except late stage-of-wear specimens of *H. primigenium* from the Dinotheriensande, Germany). Although the predominant morphology for molar metastylid (C36) is round (A), Bernor and Franzen (1997) found that some Dorn Dürkheim individuals had a round-elongate morphology (A,B), as is found in this individual. Specimen AS92/458 is unlike *H. primigenium* in its complete absence of a pli caballinid (C41 = C), somewhat wider linguaflexids than most individuals of *H. primigenium* (C45 = C); simple preflexid and postflexid margins (C46 and C47 = B). In all these differing character states, AS92/458 compares closely with members of the Old World *Hipparion* s.s. clade.

From our earlier discussion of metapodial morphology, recall that Unit 3 has a very elongate MC III (AS 92/289) that is unique to the Unit 2–6 interval and may represent an additional taxon in itself. These two individuals exhibit morphological characters typical of advanced members of *Hipparion* s.s. and may be indicative of a third taxon from the Unit 3–5 interval.

"Hipparion" *sp. 2*

Referred Specimens. AS95/1054 associated complete metatarsal III and proximal metatarsal IV

Locality. Sinap Loc. 12, Stratigraphic Unit 6, 9.590 Ma.

Remarks. Specimen AS95/1054 is distinct from all other metapodials known from Sinap by virtue of its large size. The estimated body mass based of AS95/1054A (MT III) is 291 kg, a value well outside the range of body mass estimates for other metapodials known from Sinap. It appears that this specimen may belong to a rare large taxon not otherwise represented in the Sinap Formation. Specimen AS95/1054A also appears relatively short for its large size. The ratio diagram showing available measurements of the MT III AS95/1054A (fig. 11.12I) does not exclude the possibility that AS95/1054 represents a large individual of *Hippotherium primigenium,* although the M6 dimension for AS95/1054A is outside the range of the Höwenegg sample. This possibility cannot be evaluated definitively at this time. Another possibility would be to refer AS95/1054A to the same taxon as the MT III specimens assigned to Pattern 3 based on PCA analysis. This last option would have implications for the likelihood that various other mp III specimens could be grouped taxonomically.

Correspondence of Recognized Sinap Taxa to Additional Sinap Material

There is a considerable amount of material in addition to the specimens assigned to the taxa we have recognized here from Units 4–6, and some material from Units 12 and 13 as well. However, there are no direct associations between the specimens we have listed and postcrania, with the remarkable exception of the Loc. 108 specimen AS93/826. An interesting issue is whether skeletal elements other than those already included in our taxonomy reflect the diversity we report above and to what extent postcranial specimens can be matched with the taxa that we have described here based on dental and cranial characters. Of the postcranial elements we have analyzed, MC III and MT III are clearly the most useful for establishing and testing species distinctions because of their size diversity and proportions, which can be linked to functional and ecomorphologic attributes. Our use of bivariate log-ratio and principal components plots provides a consistent representation of the number and kind of "metapodial" taxa known from each locality and stratigraphic unit. Tables 11.11 and 11.12 summarize the plausible associations between metapodial III specimens and the taxa we have recognized principally on the basis of skulls.

We recognize four taxa of hipparion based on MC III from Stratigraphic Units 4–6, ~10.53–10.31 Ma (table 11.11):

1. *Cormohipparion sinapensis* from Locs. 108 (Unit 4) and 8B (Unit 5), which represents the primitive morphotype most closely related to the North American *C. occidentale* sample;
2. *"Hipparion"* sp. 1 from Locs. 72 (Unit 4) and 12 (Unit 6), an elongate MC III form that we interpret as representing the "cursorial morph."
3. A short, lightly built MC III from Unit 5 Loc. 8B and Unit 6 Locs. S01 / OZ01, OZ02, and 12, which we assign to aff. *"Hipparion" kecigibi* due to its indirect but close field association with the type skull, MNHN-TRQ9001, and the paratype skull, MNHNTRQ9002, at Loc. S01 / OZ01, and the plausible ecomorphologic correlation between the short, broad snout and short gracile goatlike morphology; and
4. A small, Pattern 4 MC III from Loc. OZ02, Unit 6, different from aff. *"H." kecigibi* from OZ02 in size and proportions, which we refer to cf. *"H." uzunagizli,* based mainly on the small apparent size of the *"H." uzunagizli* crania and the MC III relative to other Sinap specimens.

We recognize five hipparion taxa between Units 4 and 6 based on MT III specimens (table 11.12):

1. *Cormohipparion sinapensis,* the primitive morphology, from Locs. 108 (Unit 4), 8B (Unit 5), and 12 (Unit 6);
2. Aff. *"Hipparion" kecigibi,* the short, slender metapodial morphology from localities in upper Block Unit 5 and Block Unit 6: based on MC III specimens (see table 11.11) from Locs. 8B (Unit 5), 12, OZ01. OZ02 and S01; also based on MT III specimens (table 11.12) from Locs. 8B and OZ02. 8B (Unit 5) and OZ02 (Unit 6);
3. *"Hipparion"* sp. 1, the "cursorial morph" from Locs. 91 (Unit 5), S01 / OZ01 (Unit 6), 12 (Unit 6);
4. Cf. *"Hipparion" uzunagizli,* the smaller cursorial form from Locs. 114 (Unit 5) and 12 (Unit 6); and
5. A very large form, *"Hipparion"* sp. 2 from Loc. 12 (Unit 6).

The recognition of the above taxa is made possible by the general agreement between our bivariate, log-ratio diagram, and PCA analyses. Our first step was to propose a taxonomy based on metapodials, in which specimens were assigned to morphs (see tables 11.11, 11.12). The next step was to relate these morphs to taxa already recognized based on other elements. This resulted in the taxonomy described above and the assignments shown in the "Likely Taxon" columns in tables 11.11 and 11.12. The rationale for our taxonomic hypotheses shown in these tables is described here.

Associations among MC III and MT III specimens were made possible because the principal components for the third metacarpals and third metatarsals reflect similar morphological axes. It appears that metacarpal and metatarsal results replicate each other. For example, the Höwenegg sample and Loc. 108 specimens plot in the same space relative to each other for both the MC III and MT III specimens.

One way of comparing and suggesting associations between MC III and MT III specimens that is useful for taxonomic classifications is to plot MC III and MT III values together relative to the same axes. This was done by using the Höwenegg and Loc.108 specimens as a key for matching MC III and MT III values. Specifically, we scaled and rotated the MC III principal component scores such that the mean principal component scores for MC III from Höwenegg and Loc. 108 are equal to the equivalent scores for MT III from the same locations. We further shifted the origin of the resulting plot (see fig. 11.17) to correspond to the mean principal components for the Loc. 108 specimens. This plot provides a visual method for comparing metapodials. Thus, a species with relatively long and slender MC III and MT III should occupy the same space in figure 11.17. Furthermore, if the morphology of the Loc. 108 specimens is considered primitive, as their similarity to Xmas Quarry specimens suggests, this plot has the added advantage of showing more derived specimens farther from the origin. Inspection of figure 11.20 clearly indicates the two derived trends in metapodial morphology already noted—shortening and elongation.

The variation in metapodial III values shown in figure 11.20 was partitioned into hypothetical taxa, based on our previous categorizations by morphological pattern (see tables 11.8, 11.9) and the results of our bootstrap procedure (table 11.10). To recognize a taxon for every morphological pattern would likely result in excessive splitting. At the other extreme, confining recognized taxa to groups supported by significant bootstrap statistics is probably too conservative. Thus, groupings according to pattern and bootstrap groups were used in conjunction with body mass estimates (Scott et al., chapter 16, this volume) and reference to figure 11.20 to assign specimens to metapodial "morphs" (see tables 11.11, 11.12). Although morphological pattern assignments are not meant to be specific taxonomic hypotheses and were made independently for MC III and MT III, metapodial morphs are understood as taxonomic hypotheses and were defined with reference to both MC III and MT III.

Both the Pattern 1 MC III and MT III specimens show similarities to the Xmas Quarry specimens and include specimens from Loc. 108. On this basis, we argue that they likely belong to the same taxon. The Pattern 1 and Pattern 3 MT III specimens could not be split based on the results of our bootstrap procedure, suggesting that they may represent the same taxon. However, the bootstrap procedure rejected the hypothesis that the Pattern 1 and Pattern 3 MC

Table 11.11. Metacarpal III Specimens Sorted by Size, Morphology, and Likely Taxon

Specimen Number	Stratigraphic Unit	Locality	Age (Ma)	Body Mass Estimate (kg)[1]	Size Category[1]	Morphological Pattern (PCA)	Group (PCA with bootstrap test)	Morphology	Likely Taxon
AS92/228	5	8B	9.92	160.4	Moderately large	1	Primitive	Primitive	*Cormohipparion sinapensis*
AS93/604A	4	108	10.14	170.3	Moderately large	1	Primitive	Primitive	*Cormohipparion sinapensis*
AS92/238	5	8B	10.14	165.1	Moderately large	1	Primitive	Primitive	*Cormohipparion sinapensis*
AS92/289	4	72	10.08	226.9	Moderately large plus	1	Primitive	Primitive/Cursorial	"*Hipparion*" sp. 1 = Cursorial
AS91/420	6	12	9.59	161.9	Moderately large	2 (mc)	n/a	Cursorial	"*Hipparion*" sp. 1 = Cursorial
AS92/237	5	8B	9.92	197.0	Moderately large	3	Höwenegg-like	Ozansoy Short	aff. "*Hipparion*" *kecigibi*
AS93/840	6	12	9.59	151.9	Moderately large	3	Höwenegg-like	Ozansoy Short	aff. "*Hipparion*" *kecigibi*
MNHNTRQ1125	6	Oz02	9.59	177.4	Moderately large	3	Höwenegg-like	Ozansoy Short	aff. "*Hipparion*" *kecigibi*
MNHNTRQ1127	6	Oz02	9.59	188.2	Moderately large	3	Höwenegg-like	Ozansoy Short	aff. "*Hipparion*" *kecigibi*
MNHNTRQ1126	6	Oz02	9.59	138.9	Somewhat large	4 (mc)	n/a	Smaller	cf. "*Hipparion*" *uzunagizli*
MNHNTRQ1129	6	S01/Oz01	9.68	175.3	Moderately large	6	n/a	"Goatlike"	aff. "*Hipparion*" *kecigibi*

[1]Estimated body masses and size category assignments are after Scott et al. (this volume).

Table 11.12. Metatarsal III Specimens Sorted by Size, Morphology, and Likely Taxon

Specimen Number	Stratigrapic Unit	Locality	Age (Ma)	Estimated Body Mass (kg)[1]	Size Category[1]	Morphological Pattern (PCA)	Group (PCA with bootstrap test)	Morphology	Likely Taxon
AS93/332	4	108	10.14	222.1	Moderately Large	1	Primitive	Primitive	*Cormohipparion sinapensis*
AS93/827A	4	108	10.14	223.9	Moderately Large	1	Primitive	Primitive	*Cormohipparion sinapensis*
AS92/240	5	8B	9.92	204.7	Moderately Large	1	Primitive	Primitive	*Cormohipparion sinapensis*
AS93/1213a	6	12	9.59	218.8	Moderately Large	1	Primitive	Primitive	*Cormohipparion sinapensis*
MNHNTRQ1160	6	Oz02	9.59	216.6	Moderately Large	3	Primitive	Ozansoy short	aff. *"Hipparion" kecigibi*
MNHNTRQ1163	6	Oz02	9.59	207.9	Moderately Large	3	Primitive	Ozansoy short	aff. *"Hipparion" kecigibi*
MNHNTRQ1164	6	Oz02	9.59	206.6	Moderately Large	3	Primitive	Ozansoy short	aff. *"Hipparion" kecigibi*
MNHNTRQ1168	6	Oz02	9.59	211.0	Moderately Large	3	Primitive	Ozansoy short	aff. *"Hipparion" kecigibi*
AS92/251	5	8B	9.92	234.2	Moderately Large	3	Primitive	Ozansoy short	aff. *"Hipparion" kecigibi*
AS93/7	5	91	9.98	220.6	Moderately Large	2a	Cursorial	Cursorial	*"Hipparion"* sp. 1 = Cursorial Morph
MNHNTRQ1167	5	S01/Oz01	9.68	227.0	Moderately Large	2a	Cursorial	Cursorial	*"Hipparion"* sp. 1 = Cursorial Morph
AS91/780	6	12	9.59	205.5	Moderately Large	2b	Cursorial	Cursorial	*"Hipparion"* sp. 1 = Cursorial Morph
MNHNTRQ1161	6	S01/Oz01	9.68	220.5	Moderately Large	2b	Cursorial	Cursorial	*"Hipparion"* sp. 1 = Cursorial Morph
MNHNTRQ1166	6	S01/Oz01	9.68	224.4	Moderately Large	2b	Cursorial	Cursorial	*"Hipparion"* sp. 1 = Cursorial Morph
AS93/52	5	91	9.98	209.7	Moderately Large	2b	Cursorial	Cursorial	*"Hipparion"* sp. 1 = Cursorial Morph
AS91/373	6	12	9.59	160.5	Somewhat Large	4	Cursorial	Smaller	cf. *"Hipparion" uzunagizli*
AS93/1207a	6	12	9.59	153.5	Somewhat Large	4	Cursorial	Smaller	cf. *"Hipparion" uzunagizli*
AS95/1054a	6	12	9.59	291.4	Quite Large	5	Large	Larger	*"Hipparion"* sp. 2 = Larger Morph

[1]Estimated body masses and size category assignments are after Scott et al. (this volume).

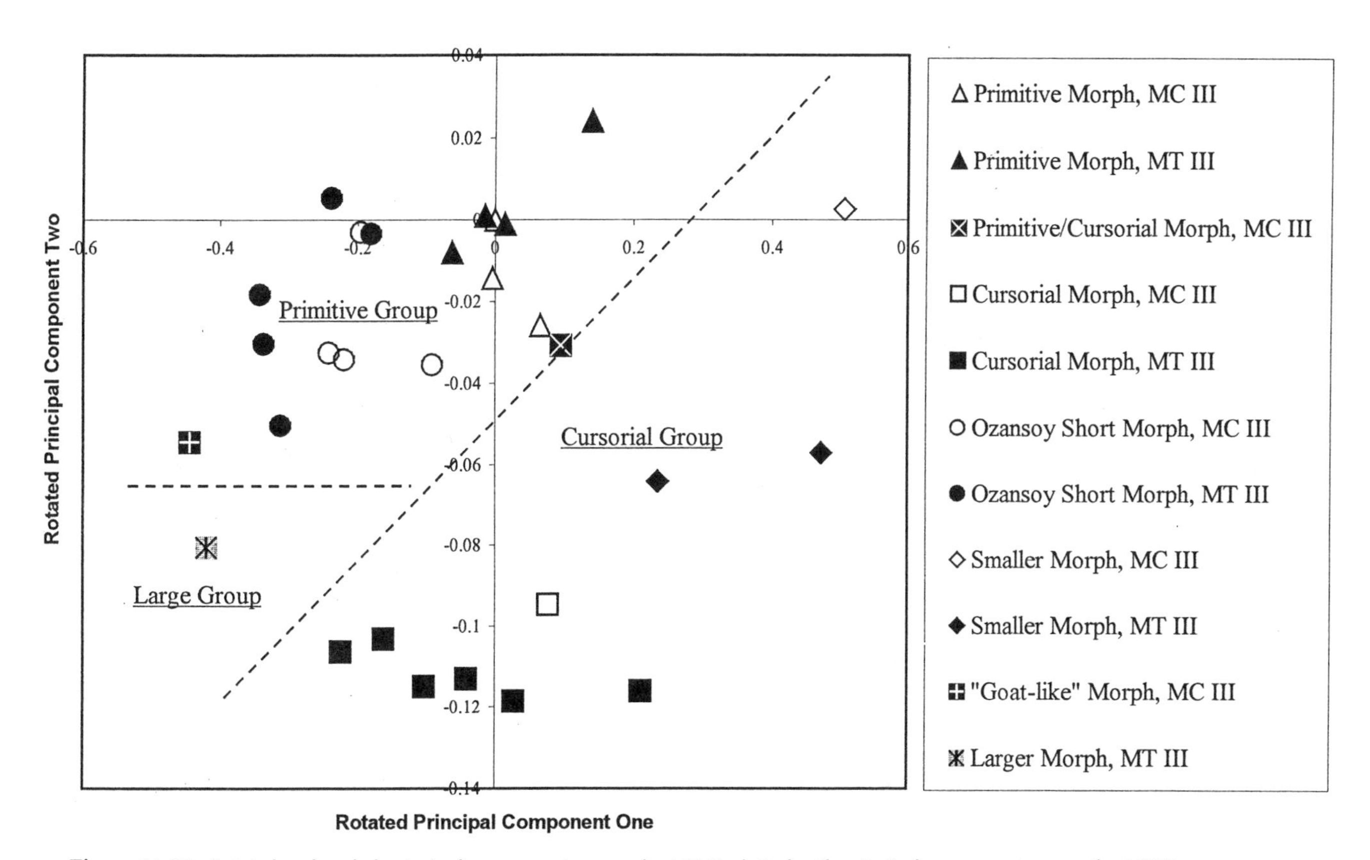

Figure 11.20. Rotated and scaled principal component scores for MC III plotted with principal component scores for MT III.

III specimens belong together. We conclude that the Pattern 3 morph has distinct MC III values (specifically, they are relatively and absolutely short) compared with both the Pattern 1 and the Xmas Quarry metapodials. Thus, the Pattern 3 specimens suggest a species with a different locomotor adaptation than either *Cormohipparion sinapensis* or *C. "occidentale."* With this in mind, we assign the Pattern 1 metapodials to the Primitive morph. Similarly, we place the Pattern 3 specimens in the Ozansoy Short morph. This morph is named for its relatively short length and Ozansoy specimens from his Loc. 02 (our Loc. OZ02), from which most of the included specimens originate.

The above assignments are confirmed by the bivariate plot of MC III maximum length (M1) versus distal articular width (M11) (fig. 11.8A,B) with one exception. Accordingly, one size grouping corresponding to the Primitive morph is depicted from Unit 4. The AS92/289 MC III that was assigned to Pattern 1 and hence the Primitive morph differs from all others known from Sinap by the M1 and M11 dimensions. This specimen is derived from Loc. 72 in Unit 4, and by virtue of its great length, it may represent a taxon distinct from the other Pattern 1 MC III specimens. Notably, it appears to plot on the border between the Primitive Group and the Cursorial Group in figure 11.20. Specimen AS92/289 could be interpreted as a transitional form between a more primitive taxon and a more cursorial taxon.

The grouping of the Pattern 3 MT III specimens is also confirmed by the bivariate plot of M1 versus M11 (fig. 11.11B) and the log-ratio diagram for Loc. OZ02 of Unit 6 (fig. 11.12H). The four Pattern 3 specimens from OZ02 (MNHNTRQ1160, MNHNTRQ1163, MNHNTRQ1164, and MNHNTRQ1168) all plot below the Höwenegg ellipse for the M1 × M11 dimensions. Similarly, the Pattern 3 MT III specimens can be recognized by a distinct ratio diagram profile for AS92/251, MNHNTRQ1160, MNHNTRQ1164, and MNHNTRQ1168. This profile includes low values for M10 and M11, and M3 and M4 dimensions that are not very different.

The Pattern 2a, 2b, and 4 MT III specimens could not be split according to our bootstrap results and accordingly are assigned to the Cursorial Group. The Pattern 2 (MC III) specimen plots with the Pattern 2a and b MT III specimens in figure 11.20. These specimens are all relatively slender and represent a species with a cursorial habitus. Thus, we assign them to the cursorial morph. The Pattern 4 MT III specimens are much smaller in terms of estimated body mass than the Pattern 2a and b MT III samples. On this basis, we recognize the smaller morph and also include in this morph the Pattern 4 (MC III) specimen, which is also small in estimated body mass.

The single goatlike specimen (Pattern 6, MNHNTRQ 1129) could belong to the same taxon as the Pattern 3 MT III specimens, but appears as an outlier when compared to the Pattern 3 MC III. This specimen is from a different locality (OZ01) than any of the Pattern 3 MC III or MT III specimens and it may be from a related species or part of the same lineage. We place this specimen in its own group, the goatlike morph, but recognize that it may be related to those specimens placed in the Ozansoy short morph.

A final morph must be recognized for the MT III AS95/1054A, which based on its very large estimated body mass appears to represent a different taxon than any of the other Sinap metapodials. Further justification of this specimen's unique identity can be drawn from figure 11.20, where AS95/1054A plots outside other specimen clusters. Thus, we assign this specimen to the larger morph.

Our next step was to interpret metapodial morphs with respect to named species. The "Likely Taxon" columns in tables 11.11 and 11.12 reflect a set of hypotheses regarding species identity and their correspondence to named taxa. We again used the Loc. 108 specimens as a key for our overall interpretations. As shown in table 11.12, our character state univariate, bivariate, log-ratio, and PCA analyses do provide a congruent estimate of one taxon from Unit 4, which includes Loc. 108. This suggests that all specimens from Loc. 108 may belong to a single species, *Cormohipparion sinapensis*. This particular hypothesis predicts that the metapodials from Loc. 108 should show affinities to *C. "occidentale"* from Xmas Quarry, and this is so. Therefore, we suggest the hypothesis that the Loc. 108 metapodials and all those assigned to the Primitive morph belong to *C. sinapensis*. Besides the Loc. 108 specimens, *C. sinapensis* is represented by AS92/228 and AS92/238 (both MC III) and AS92/240 (MT III) from Loc. 8B (Unit 5) and AS93/1213a (MT III) from Loc. 12 (Unit 6), providing evidence for its protracted presence in the Sinap sequence.

With one exception, all Pattern 1 metapodials are considered likely representatives of *Cormohipparion sinapensis*. The one exception to this conclusion is AS92/289, an MC III from Loc. 72. This specimen is relatively long compared with other members of the Primitive morph suggesting a more significant cursorial adaptation. Figure 11.20 shows this specimen on the edge of the Primitive Group. This fact, combined with the lower stratigraphic position of Loc. 72 in Unit 4, suggests the hypothesis that AS92/289 may represent a transitional form to the cursorial morph, represented by several specimens beginning with specimens from Locality 91. The juvenile skull and mandible from Loc. 91 assigned to *"Hipparion"* sp. 1 may represent the same species as AS92/289. Together, they describe a species with characters similar to advanced members of *Hipparion* s.s. The elongate metatarsals from Loc. 91 and other specimens assigned to the Cursorial morph may all belong to the *"Hipparion"* sp. 1 lineage, which would by our evidence occur in Units 4 through 6. Accordingly, AS92/289 and the Cursorial morph metapodials are assigned to *"Hipparion"* sp. 1 = Cursorial morph in tables 11.11 and 11.12. Note that table 11.11 suggests two competing hypotheses for the taxonomic status of AS92/289 (Unit 3, Loc. 72). One interpretation is supported by PCA analysis that emphasizes AS92/289's similarity to other Primitive morph specimens. In contrast, the ratio diagram analysis emphasizes AS92/289's exceptional length and suggests the competing hypothesis that AS92/289 belongs in the *"Hipparion"* sp. 1 = cursorial morph lineage.

The three specimens of the Smaller morph are notable for their morphological similarity and their small body mass estimates. Among named taxa recognized here, *"Hipparion" uzunagizli* appears to have been the smallest. Thus, we equate the Smaller morph with cf. *"H." uzunagizli*. The results of our bootstrap analysis do not exclude the possibility that the three specimens of the Smaller morph belong to the same species as those specimens assigned to the cursorial morph. However, we find this to be unlikely because the Smaller morph specimens are distinct in both size and shape (smaller and more elongate).

The goatlike morph is represented by MNHNTRQ1129, the shortest metapodial in the sample. Given that this specimen was collected at OZ01, stratigraphically and geographically very close to the S01 specimens of *"Hipparion" kecigibi*, we offer the hypothesis that this MC III morphology might belong to *"H." kecigibi*. The short, broad snout and short limbs both might be adaptations for exploiting short grass resources on uneven substrates. We pose the hypothesis here that this may have included hill slopes with rocky or unstable surfaces, hence goatlike. The MT III specimens assigned to the Ozansoy short morph could also be grouped with MNHNTRQ1129 as possible representatives of *"H." kecigibi*, based on their relatively short length and flat M3–M4 profile on the ratio diagrams. By extension, the Ozansoy Short morph metacarpals may also be related to *"H." kecigibi*. We would suggest two competing hypotheses for the Ozansoy Short morph metapodials. One hypothesis recognizes two taxa possibly corresponding to the Ozansoy short morph and goatlike morph (this hypothesis is represented in the morph column of tables 11.11 and 11.12). A second hypothesis would group the Ozansoy short morph MT III and MC III specimens with MNHNTRQ 1129. The resulting taxon might be *"H." kecigibi* and accordingly they are referred to aff. *"H." kecigibi* in the aforementioned tables. Under either hypothesis, it appears likely that the Ozansoy short morph and the goatlike morph were related.

As previously noted it is impossible to exclude the possibility that the single MT III from Loc. 12, AS95/1054A, assigned to the larger morph represents *Hippotherium primigenium*. One other possibilty would be that this specimen belongs to the same taxon as those specimens assigned to the Ozansoy Short morph. If this latter possibilty were the case, then it would be remiss to assign the goatlike MC III, MNHNTRQ 1129, to the same taxon. It would be extremely unlikely that the large morph, goatlike morph, and Ozansoy short morph together constitute a single species based on size alone. None of the other specimens discussed appear to have been as large as AS95/1054A and we suggest that the best designation for this specimen is *"Hipparion"* sp. 2 = larger morph.

The number of hipparionine taxa present in Units 4–6 appears most likely to have been five and at a minimum four. Different judgments are possible regarding how MP III specimens might be grouped into taxa, but these alternate hypotheses do not materially change the number of taxa represented as being either four or five in toto. Even when the entire cursorial group is lumped into a single taxon, our analysis here still derives four hipparion taxa in the Stratigraphic Block Units 4–6.

Discussion

The early Vallesian hipparion sequence in the Sinap Formation contrasts fundamentally with those of Central and Western Europe in both the species occurring at the "Hipparion Datum" and the lineages that are derived immediately succeeding this event. The first taxon in the Sinap sequence is referable to a new species of the North American genus *Cormohipparion, C. "sinapensis,"* and the local evolutionary sequence immediately succeeding the FAD are derivatives of *C. sinapensis* and not *Hippotherium primigenium.* Unknown until now is the strong provinciality of hipparion lineages that is evident very early on in the Vallesian, MN9: there is an apparent early divergence of lineages, with *Hippotherium primigenium* and its evolved species occurring in central and western Europe (Bernor et al. 1989, 1996, 1997) and *C. sinapensis* and its apparent derivatives occurring in Anatolia, and perhaps a broader southeast European-southwest Asian faunal province.

The various analyses made here show that the first occurring hipparion had relatively low crowned teeth. Both skull and postcranial morphology of the Unit 4 taxon *Cormohipparion sinapensis* sp. nov. reveal characteristics that suggest their direct derivation from North American *Cormohipparion,* not *Hippotherium primigenium* s.l.

Woodburne (1996) has recently argued that there are likely multiple species included in the *Cormohipparion "occidentale"* hypodigm of MacFadden (1984), and our own comparisons of European hipparions with populations of *C. "occidentale"* from Xmas Quarry, Hans Johnson Quarry, and MacAdams Quarry verify Woodburne's (and originally V. Eisenmann's, pers. comm., 1981, American Museum Hipparion Workshop) observations.

Beginning with Sinap Loc. 72 of Stratigraphic Unit 4, there is probably a second species of hipparion characterized by having more elongate limbs than *C. sinapensis.* There is some ambiguity about exactly how many taxa occur between Sinap Stratigraphic Units 4–6, but by Unit 6, there are likely four or five taxa represented. Unit 5 may include as many as four taxa. Among these are the type of *Hipparion uzunagizli* and *"Hipparion"* sp. 1. Metapodial III specimens found in Unit 5 also suggest that *"H." kecigibi,* known primarily by a type low in Unit 6, was also probably present in Unit 5. Both *H. uzunagizli* and *"H." kecigibi* are plausibly derived from *C. sinapensis* and differ most fundamentally from each other in their snout proportions: *H. uzunagizli* having a long, wide snout and *"H." kecigibi* having a short, wide snout. *"Hipparion"* sp. 1 has a more strongly reduced POF and apparently elongate-narrow metapodials; the skull-postcranial association posed here remains equivocal because of the lack of their direct association. Metapodial III evidence further suggests that *C. sinapensis* also occurred in Unit 5.

What is unequivocal is the striking contrast between the MN 9 evolutionary record in central and western Europe and that documented in Sinap. Between the ~10.69 Ma FAD at Sinap and the Unit 5, ~10 Ma, the Sinap Formation records an apparent punctuated speciation event in hipparion evolution, which is previously undocumented anywhere in Eurasia or Africa. At least some of the taxa derived from this evolutionary event are referable to *Hipparion* s.s., a group that was previously first documented from MN 10 of Spain (*Hipparion melendezi*) and Iran (*Hipparion gettyi*) (Bernor et al. 1996). The Sinap Unit 2–6, hipparions generally exhibit morphology adaptive for open country running in their postcranial morphology and like North American *Cormohipparion "occidentale,"* were likely grazers (Hayek et al. 1992; see also later in this discussion). Fundamentally then, their adaptation was different from *Hippotherium primigenium* and its Spanish Vallesian relatives *Hippotherium koenigswaldi* and *Hippotherium catalaunicum,* which inhabited warm temperate to subtropical forest environments and likely incorporated a large percentage of browse in their diet (Bernor et al. 1996, 1997).

The alpha taxonomy of North American *Cormohipparion "occidentale"* is clearly in need of revision; it is unequivocal that there are several species mixed within this nomen. The Sinap hipparions have been demonstrated to have an overall size and several characters and dimensions closest to *C. "occidentale"* s.l. The Sinap hipparions, as well as *Hippotherium primigenium,* do have some skull characteristics that are more primitive than *C. "occidentale,"* as characterized from Xmas, Hans Johnson, and MacAdams Quarries, thus suggesting that the sister taxon of the *C. sinapensis-H. primigenium* clade was likely a form somewhat intermediate in morphology between *C. quinni* and *C. "occidentale"* s.l.

Although the hipparion fossil record from later Sinap Stratigraphic Units 12 and 13 is not as extensive as the preceding units, it is clear that their evolutionary diversification continued apace. In the later Turolian levels, we find the elongate, very gracile metapodials typical of small hipparions known from southeast Europe and southwest Asia. From what we know of the later Turolian of this region, these taxa may be members of either (or both) the *Cremohipparion* and *Hipparion* s.s. lineages.

Conclusions

The Sinap Formation hipparion fauna is remarkable for its documentation of the entry of the North American genus *Cormohipparion,* after a discernible interval of stasis, there was a short interval speciation burst at the end of MN 9. According to our interpretation of the Sinap MN 9 hipparion record, *C. sinapensis* is the only hipparion known from first entry at 10.692 m.y. until the occurrence of the type partial skeleton from Loc. 108 at 10.135 m.y. Between 10.080 and 9.590 m.y., a matter of 0.49 m.y., the Sinap section records 3–4 additional taxa (fig. 11.21). A longer limbed cursorial form (Loc. 72, ~10.080 Ma) that we associate with

Location	Polarity	GPTS	Age	Block Unit	Begin	End	Hipparion Events
	Int						
84	R7	C4Ar.2r	9.367	7	9.50	9.25	
83	R7	C4Ar.2r	9.452				
11	R7	C4Ar.2r	9.483				
12, OZ02	N6	C4Ar.2n-C5n.1n	9.590	6	9.75	9.50	Hipparion: "g", lg, l-j, cur, Cormo
OZ01, S01	N6	C4Ar.2n-C5n.1n	9.683				H. "goat-like" & H. cursorial
8A	R6	C5n.1r	9.886	5	10.00	9.75	
8B	R6	C5n.1r	9.918				Hipparion "goat-like" FAD and Cormo
114	N5	C5n.2n-1n & -2n	9.967				Hipparion "long-jaw" FAD and Cormo
108/8	N5	C5n.2n-1n & -2n	9.970				
91	N5	C5n.2n-1n & -2n	9.977				*Hipparion* sp. 1 FAD
72	N5	C5n.2n-1n & -2n	10.080	4	10.25	10.00	*"Hipparion"* cursorial form FAD
108	N5	C5n.2n-1n & -2n	10.135				*Cormohipparion* n.sp. skeleton
106	N5	C5n.2n-1n & -2n	10.206				
113	N5	C5n.2n-1n & -2n	10.306	3	10.50	10.25	
89	N5	C5n.2n-1n & -2n	10.406				
87	R5	C5n.2n-3r	10.452				
93	N4	C5n.2n-3n	10.488				
121	N4	C5n.2n-3n	10.526	2	10.75	10.50	
94	N4	C5n.2n-3n	10.551				
122	N4	C5n.2n-3n	10.577				
107	N4	C5n.2n-3n	10.653				
4	N4	C5n.2n-3n	10.692				*Cormohipparion* FAD
88	N3	C5n.2n-4n	10.730				
64	N3	C5n.2n-4n	10.765	1	11.00	10.75	
104	N3	C5n.2n-4n	10.868				
65	N3	C5n.2n-4n	10.899				
24, 24A	—	—	16-15				

Figure 11.21. Sinap Vallesian hipparion species events.

a skull with a sharply diminished preorbital fossa (Loc. 91, ~9.977 Ma), *"Hipparion"* sp. 1. The sharp reduction of the preorbital fossa with the likely association of an elongate metapodial is suggestive of an adaptation for open country living and grazing. Within stratigraphic Unit 5 (9.977–9.886 Ma, or a 0.091 m.y. interval), we document the cooccurrence of *C. sinapensis* sp. nov., *"Hipparion"* sp. 1, *"H." uzunagizli* (Loc. 114, 9.967 Ma) and *"H." kecigibi* (9.918 Ma). *"Hipparion" uzunagizli* is characterized as having a skull with an elongate snout and broad gape and is associated with small, slender metapodials, whereas *"H." kecigibi* is characterized as having a short, broad muzzle with a broad incisor gape (lower Unit 6, Loc 8b, 9.918 Ma), and most plausibly very short, stocky metapodials (upper Unit 5, Loc 8b). We surmise that whereas *"H." kecigibi* probably exploited short grass (based on its broad and short snout; Bernor and Armour-Chelu 1999) possibly on rocky slopes, *"H." uzunagizli,* with its relatively longer metapodials and more elongate snout, plausibly favored tender shoots that rose above short grass (see Owen-Smith 1982, 1988; Bernor and Armour-Chelu 1999). The fifth taxon we recognize, *"Hipparion"* sp. 2, is represented by a very large, heavy-limbed form from Loc. 12 (Unit 6, 9.590 Ma), where specimens attributable to *C. sinapensis, "Hipparion"* sp. 1, *H. kecigibi,* and *"H." uzunagizli* are also known to occur. The occurrence of as many as four or five bona fide taxa from a single locality is, as far as we know, unprecedented in Old World Neogene hipparion faunas.

In all likelihood, the Sinap hipparions give us a limited insight into the initial Old World hipparion radiation. In the vast expanses of Eurasia and Africa, an immigrating species of *Cormohipparion* met environments ranging from tropical forest to open country woodland and was clearly capable of adapting to a great range of these environments. The Sinap horses were apparently adapting to the more open country spectrum of habitats available in the Old World, whereas in central and western Europe, hipparion populations adapted to the more equable subtropical to warm temperate forests. *Cormohipparion sinapensis* is a stem Old World hipparion that has no detectable apomorphy that could preclude its proximate relationship to primitive members of the *Hippotherium, Cremohipparion,* or *Hipparion* s.s. clades. More phylogenetic research is needed to further elucidate these relationships.

Acknowledgments

We thank Prof. Berna Alpagut for permission to study material under her care. We also thank Prof. Michael O. Woodburne for his comments on this manuscript. R.B. thanks the National Science Foundation (Grant EAR 125009), Alexander von Humboldt Stiftung and Prof. Dr. Siegfried Rietschel, director of the Staatliches Museum für Naturkunde, Karlsruhe, and the staff and curators at Karlsruhe for providing all research support during his study of the Sinap hipparion collection. M.F. thanks the Academy of Finland and J.K. thanks the National Science Foundation (NSF EAR 9304302) for supporting the research at Sinap.

Appendix: Definitions of Eigenvectors and Eigenvalues for Metacarpal III and Metatarsal III

Table 11.13. Eigenvalues for Metacarpal III

Principal Component	Eigenvalue	Percentage Variance Explained
1	0.1549	96.4
2	0.0027	1.7
3	0.0014	0.9
4	0.0009	0.6
5	0.0008	0.5
6	0	0.0

Table 11.14. Eigenvectors for Metacarpal III

	Eigenvector Principal Components					
Variable	1	2	3	4	5	6
M2/GEOMEAN	0.9955	0.0453	0.0101	0.0037	0.0632	0.0531
M3/GEOMEAN	−0.066	0.7605	−0.1343	0.4566	0.1425	0.413
M4/GEOMEAN	0.0139	−0.0154	0.2378	−0.2137	−0.735	0.5976
M5/GEOMEAN	−0.024	−0.4805	0.5148	0.5309	0.31	0.3544
M6/GEOMEAN	−0.0077	−0.4072	−0.771	−0.0174	0.1441	0.4675
M10/GEOMEAN	−0.0617	0.1508	0.2565	−0.681	0.5645	0.354

Table 11.15. Eigenvalues for Metatarsal III

Principal Component	Eigenvalue	Percentage Variance Explained
1	0.1599	96.4
2	0.0033	2.0
3	0.0015	0.9
4	0.0009	0.6
5	0.0003	0.2
6	0.0000	0.0

Table 11.16. Eigenvectors for Metatarsal III

Variable	Eigenvector for Principal Components					
	1	2	3	4	5	6
M2/GEOMEAN	0.9971	0.0431	0.0145	−0.0068	−0.0276	0.0543
M3/GEOMEAN	−0.0608	0.3995	−0.3052	−0.1486	−0.6714	0.5203
M4/GEOMEAN	−0.0033	0.0419	−0.538	−0.0443	0.6716	0.5058
M5/GEOMEAN	−0.0343	0.1159	0.6693	−0.5902	0.1984	0.387
M6/GEOMEAN	0.006	−0.8508	0.0722	0.1366	−0.22	0.4515
M10/GEOMEAN	−0.0301	0.3154	0.405	0.7803	0.0986	0.3419

Literature Cited

Bernor, R. L., 1985, Systematic and evolutionary relationships of the hipparionine horses from Maragheh, Iran (late Miocene, Turolian age): Palaeovertebrata, v. 15, no. 4, pp. 173–269.

Bernor, R. L., and M. Armour-Chelu, 1997, Later Neogene hipparions from the Manonga Valley, Tanzania, *in* T. Harrison, ed., Neogene paleontology of the Manonga Valley, Tanzania: New York, Plenum Press, pp. 219–264.

Bernor, R. L., and M. Armour-Chelu, 1999, Toward an evolutionary history of African hipparionine horses, *in* T. Brommage, and F. Schrenk, eds., African biogeography, climate change and early hominid evolution: Oxford, Oxford University Press, pp. 189–215.

Bernor, R. L., and J. Franzen, 1997, The equids (Mammalia, Perissodactyla) from the late Miocene (Early Turolian) of Dorn-Durkheim 1 (Germany, Rheinhessen): Courier Forschungs-Institut Senckenberg, v. 197, pp. 117–186.

Bernor, R. L., and J. Harris, 2003, Systematics and evolutionary biology of the late Miocene and early Pliocene hipparionine equids from Lothagam, Kenya, *in* M. Leakey, and J. Harris, eds., Lothagam: The dawn of humanity in eastern Africa: New York, Columbia University Press.

Bernor, R. L., and S. T. Hussain, 1985, An assessment of the systematic, phylogenetic, and biogeographic relationships of Siwalik hipparionine horses: Journal of Vertebrate Paleontology, v. 5, pp. 32–87.

Bernor, R. L., and D. Lipscomb, 1991, The systematic position of *Plesiohipparion* aff. *huangheense* (Equidae, Hipparionini) from Gulyazi, Turkey: Mitteilung der Bayerischen Staatslammlung für Palaontologie und Historische Geologie, v. 31, pp. 107–123.

Bernor, R. L., and D. Lipscomb, 1995, A consideration of Old World hipparionine horse phylogeny and global abiotic processes, *in* E. Vrba, et al., eds., Paleoclimate and evolution, with emphasis on human origins: New Haven, Connecticut, Yale University Press, pp. 164–177.

Bernor, R. L., and H. Tobien, 1989, Two small species of *Cremohipparion* (Equidae, Mammalia) from Samos, Greece: Mitteilungen der Bayerischen Staatssammlung für Palaontologie und Historisches Geologie, v. 29, pp. 207–226.

Bernor, R. L., M. O. Woodburne, and J. A. Van Couvering, 1980, A contribution to the chronology of some Old World Miocene faunas based on hipparionine horses: Géobios, v. 13, pp. 25–59.

Bernor, R. L., K. Heissig, and H. Tobien, 1987, Early Pliocene Perissodactyla from Sahabi, Libya, *in* N. T. Boaz, A. El-Arnauti, A. W. Gaziry, J. de Heinzelin, and D. D. Boaz, eds., Neogene paleontology and geology of Sahabi: New York, A. R. Liss, pp. 233–254.

Bernor, R. L., J. Kovar-Eder, D. Lipscomb, F. Rögl, S. Sen, and H. Tobien, 1988, Systematics, stratigraphic and paleoenvironmental contexts of first-appearing hipparion in the Vienna Basin, Austria: Journal of Vertebrate Paleontology, v. 8, pp. 427–452.

Bernor, R. L., H. Tobien, and M. O. Woodburne, 1989, Patterns of Old World hipparionine evolutionary diversification and biogeographic extension, *in* E. H. Lindsay, V. Fahlbusch, and P. Mein, eds., Topics on European mammalian chronology: New York, Plenum Press, pp. 263–319.

Bernor, R. L., J. Kovar-Eder, J.-P. Suc, and H. Tobien, 1990, A contribution to the evolutionary history of European late Miocene age hipparionines (Mammalia: Equidae): Paleobiologie Continentale, v. XVII, pp. 291–309.

Bernor, R. L., M. Kretzoi, H.-W. Mittmann, and H. Tobien, 1993a, A preliminary systematic assessment of the Rudabánya hipparions: Mitteilungen der Bayerischen Staatssammlung für Palaontologie und Historisches Geologie, v. 33, pp. 1–20.

Bernor, R. L., H.-W. Mittmann, and F. Rögl, 1993b. The Götzendorf hipparion: Annales Naturhistorishe Museum, Wien, v. 95A, pp. 101–120.

Bernor, R. L., G. Koufos, M. O. Woodburne, and M. Fortelius, 1996, The evolutionary history and biochronology of European and southwestern Asian late Miocene and Pliocene hipparionine horses, *in* R. L. Bernor, V. Fahlbusch, and H.-W. Mittmann, eds., The evolution of western Eurasian Neogene faunas: New York, Columbia University Press, pp. 307–338.

Bernor, R. L., H. Tobien, L.-A. Hayek, and H.-W. Mittmann, 1997, *Hippotherium primigenium* (Equidae, Mammalia) from the late Miocene of Höwenegg (Hegau, Germany): Andrias, v. 10, pp. 1– 230.

Bernor, R. L., T. Kaiser, L. Kordos, and R. Scott. 1999. Stratigraphic context, systematic position and paleoecology of *Hippotherium sumegense* Kretzoi, 1984 from MN10 (late Vallesian) of the Pannonian Basin: Mitteilungen Bayerische Staatslsammlung für Paläontologie und historische Geologie, v. 39. pp. 1–35.

Eisenmann, V., 1995, What metapodial morphometry has to say about some Miocene hipparions, *in* Paleoclimate and evolution—with emphasis on human origins: Vrba et al., eds., New Haven, Connecticut, Yale University Press, pp. 148–163.

Eisenmann, V., M.-T., Alberdi, C. de Giuli, and U. Staesche, 1988, Studying fossil horses, Volume I: Methodology, *in* M. O. Woodburne, and P. Y. Sondaar, eds., Collected papers after the "New York International Hipparion Conference, 1981": Leiden, Brill, pp. 1–71.

Gabunia, L. K, 1959, Histoire du genre Hipparion. USSR Academy of Sciences, Moscow [translated from Russian]: Paris, BRGM.

Getty, R. 1982. The anatomy of domestic animals: Philadelphia, Saunders, 1211 pp.

Gromova, V., 1952, Les Hipparion (d'après les matérieaux de Taraklia, Pavlodar et autres). Travaux de l'Institut paleontologique, Académie des Sciences de l'URSS, v. XXXVI, 473 pp.

Hayek, L., R. L. Bernor, N. Solounias, and P. Steigerwald, 1992, Methods in determining the dietary adaptations of extinct hipparionine equids, *in* A. Forsten, M. Fortelius, and L. Werdelin, eds., Bjorn Kurten—A Memorial Volume: Annales Zoologici Fennici, v. 28, pp. 187–200.

Hulbert, R. C., 1988, *Cormohipparion* and *Hipparion* (Mammalia, Perissodactyla, Equidae) from the late Neogene of Florida: Florida State Museum Biological Sciences Bulletin, v. 33, no. 5, pp. 229–338.

Hulbert, R., and B. J. MacFadden, 1991, Morphological transformation and cladogenesis at the base of the adaptive radiation of Miocene hypsodont horses: American Museum Novitates, v. 3000, pp. 1–61.

Jungers, W. L., A. B. Falsetti, and C. E. Wall, 1995, Shape, relative size, and size-adjustments in morphometrics: Yearbook of Physical Anthropology, v. 38, pp. 137–161.

Kappelman J, S. Sen, M. Fortelius, A. Duncan, B. Alpagut, J. Crabaugh, A. Gentry, J. P. Lunkka, F. McDowell, N. Solounias, S. Viranta, and L. Werdelin, 1996, Chronology and biostratigraphy of the Miocene Sinap Formation of central Turkey, *in* R. L. Bernor, V. Fahlbusch, and H.-W. Mittmann, eds., The evolution of western Eurasian Neogene mammal faunas: New York, Columbia University Press, pp. 78–95.

Lunkka, J.-P., Fortelius, M., Kappelman, J., and Sen, S., 1999, Chronology and mammal faunas of the Miocene Sinap Formation, Turkey, *in* J. Agusti, L. Rook, and P. Andrews, eds., Climate and environmental change in the Neogene of Europe: Cambridge, Cambridge University Press, pp. 238–264.

MacFadden, B. J., 1980, The Miocene horse Hipparion from North America and from the type locality in southern France: Palaeontology, v. 23, no. 3, pp. 617–635.

MacFadden, B. J., 1984, Systematics and phylogeny of *Hipparion, Neohipparion, Nannippus* and *Cormohipparion* (Mammalia, Equidae) from the Miocene and Pliocene of the New World: American Museum of Natural History Bulletin, v. 179, no. 1, pp. 1–196.

MacFadden, B. J., and M. F. Skinner, 1977, Earliest known Hipparion from Holarctica: Nature, v. 265, pp. 523–533.

Nickel, R., A. Schummer, and E. Seiferle, 1986, The anatomy of the domestic animals 1. The locomotor system of the domestic animals: Berlin, Springer Verlag, 499 pp.

Owen-Smith, N., 1982, Factors influencing the consumption of plant products by large herbivores, *in* B. J. Huntley, and B. H. Walker, eds., The ecology of tropical savannas: Berlin, Springer-Verlag, pp. 359–404.

Owen-Smith, N., 1988, Niche separation among African ungulates, *in* E. Vrba, ed., Species and speciation: Transvaal Museum Monograph No. 4, pp. 167–171.

Ozansoy F., 1957, Faunes de mammifères du Tertiaire de Turquie et leurs révisions stratigraphiques: Bulletin of the Mineral Research Exploration Institute Turkey (Foreign Edition), v. 49, pp. 29–48.

Ozansoy F., 1965, Étude des gisements continent aux et de mammifères du Cénozoïque de Turquie: Mémoires Societe Géologique de France, Nouvelle Série, v. 44, pp. 1–92.

Qiu, Z., W. Huang, and Z. Guo, 1988, Chinese hipparionines from the Yushe Basin: Palaeontologica Sinica, ser. C, v. 175, no. 25, pp. 1–250.

Scott, R. S., 1996, Relative faunal abundance and body size distribution of large mammals from the Sinap Formation of central Anatolia [M.A. thesis]: Austin, Texas, University of Texas, 63 pp.

Sen S., 1986, Contribution à la magnétostratigraphie et à la paléontologie des formations continentales Néogènes du pourtour Méditerranéen. Implications biochronologiques et paléobiologiques Thèse d'Etat University Paris 6, Mémoires Sciences Terre 86(19): Paris, University of Paris, 209 pp.

Sen, S., 1991, Stratigraphie, faunes de mammifères et magnétostratigraphie du Néogène de Sinap Tepe, Province d'Ankara, Turquie: Bulletin du Muséum National d'Histoire Naturelle, ser. 4e, v. 12, no. C3/4, pp. 243–277.

Swisher, C. C. III, 1996, New ^{40}Ar/^{39}Ar dates and their contribution toward a revised chronology for the late Miocene nonmarine of Europe and West Asia, *in* R. L. Bernor, V. Fahlbusch, and H.-W. Mittmann, eds. *The evolution of western Eurasian Neogene faunas:* New York, Columbia University Press, pp. 64–77.

Webb, D., and R. C. Hulbert, 1986, Systematics and evolution of *Pseudhipparion* (Mammalia, Equidae) from the late Neogene of the Gulf Coastal Plain and the Great Plains: Contributions, Geology, University Wyoming Special Papers, v. 3, pp. 237–272.

Woodburne, M. O., 1989, Hipparion horses: A pattern of endemic evolution and intercontinental dispersal, *in* D. R. Prothero, and R. M. Schoch, eds., The evolution of Perissodactyls: New York, Oxford University Press, pp. 197–233.

Woodburne, M. O., 1996, Reappraisal of the *Cormohipparion* from the Valentine Formation, Nebraska: American Museum Novitates, v. 3163, pp. 1–56.

Woodburne, M. O., and R. L. Bernor, 1980, On superspecific groups of some Old World hipparionine horses: Journal of Paleontology, v. 54, pp. 1319–1348.

Woodburne, M. O., B. J. MacFadden, and M. Skinner, 1981, The North American "Hipparion Datum," and implications for the Neogene of the Old World: Géobios, v. 14, pp. 1–32.

Woodburne, M. O., R. L. Bernor, and C. C. Swisher III, 1996, An appraisal of the stratigraphic and phylogeneteic bases for the "Hipparion Datum" in the Old World, *in* R. L. Bernor, V. Fahlbusch, and H.-W. Mittmann, eds., The evolution of western Eurasian Neogene faunas: New York, Columbia University Press, pp. 124–136.

12

Rhinocerotidae (Perissodactyla)

M. Fortelius, K. Heissig, G. Saraç, and S. Sen

For the Rhinocerotidae, the Miocene was a time of maximum species richness and ecological diversity. It was also a time of evolutionary change, driven partly by changes in the physical and biotic environment and partly by palaeogeographic changes (Bernor et al. 1996c; Fortelius et al. 1996b). The first hypsodont rhinoceroses appeared in the late early Miocene, and the late Miocene saw the radiation of forms adapted to increasingly open habitats, as evidenced, for example, by increasing body size and hypsodonty. This trend was particularly marked in Asia, and the Anatolian rhinoceros communities of the late Miocene represent some of the westernmost occurrences of typical Asian taxa, especially of the Aceratheriini of the *Chilotherium* clade.

The Neogene land mammal faunas of Eurasia are mostly from single localities, often without stratigraphic context. Only rarely is there anything like a sequence with successive localities in demonstrable stratigraphic superposition. The chronological framework for analyzing these faunas is vague and of low resolution. The best hope of improving this situation lies in the study of stratigraphically resolved sequences such as those of Sinap, so that they can be used as calibration standards for regional biozonations.

For the rhinoceroses, the problem is complicated because the material from the few other sequences that exist (especially Maragheh, Iran) has not been revised for decades. Furthermore, there are no comprehensive treatments of the Eurasian rhinoceroses except as part of rare family-level overviews, such as Osborn (1900) and Heissig (1973, 1989), and unresolved possible synonymies are common. For the Anatolian Neogene rhinoceroses, the highly condensed synopsis of Heissig (1975) has been the standard reference for a long time, but detailed descriptions and illustrations are needed to develop a stable interpretation and practice.

Because a review of the Neogene rhinoceroses of Eurasia is far beyond the scope of this chapter, we have little choice but follow what, to the best of our understanding, is common usage. We reluctantly agree with Cerdeño (1996) that more work is required before a stable taxonomy is feasible. With one exception, we have tried to retain "current usage" of names and we have refrained entirely from creating new taxa. Our suprageneric taxonomy follows Heissig (1989). The elasmothere taxonomy follows the recent revision of Antoine (2000).

The Sinap Formation (Ozansoy 1957, 1965; Öngür 1976; Sen 1991; Lunkka et al., chapter 1, this volume) has in the past yielded a fine collection of fossil rhinoceroses, curiously ignored by Ozansoy but partly documented in unpublished manuscripts by Sen (1970) and Saraç (1994). It is therefore somewhat surprising that the fossil collection brought together by the Sinap project (see Sen, Introduction, this volume) is relatively poor in rhinoceros material, particularly for the localities of the middle Sinap member. The situation is further complicated by the sudden termination of the Sinap project, as described by Sen (this volume, Introduction). The present treatment must therefore be regarded as preliminary and subject to uncertainties and inadequacies not usually acceptable in a description of this kind.

In an attempt to compensate for these difficulties, some rhinoceros material previously collected from the Sinap Formation has been included here, based on the manuscripts by Sen (1970) and Saraç (1994) and additional material supplied by these authors. Such additional material has been included and revised only to the extent that it adds taxon occurrence information at some Sinap locality. For further details, the reader is referred to the manuscripts themselves, available from the authors. We hope that this important material will be studied more thoroughly in the near future.

Materials and Methods

All dental measurements given here (see appendix tables 12.1–12.3) were taken by MF according to Fortelius (1990) and Fortelius et al. (1993). Measurements of postcrania were taken by MF and GS according to Guérin (1980). The photographs were taken under field conditions, except for figures 12.2, 12.3, 12.5, 12.13, and 12.16, which were reproduced from old photographic prints.

Information about faunal lists and age of fossil land mammal localities was obtained from the March 2000 version of the NOW (Neogene Old World) database (Bernor et al. 1996a,c; Fortelius et al. 1996b). The database is being continuously revised by the members of the NOW Advisory Board, and the latest public core dataset may be downloaded from the website http://www.helsinki.fi/science/now/. Other datasets may be requested from MF or from the NOW office (mikael.fortelius@helsinki.fi).

Fossil rhinoceros material from the institutions listed below was studied selectively, especially Anatolian material and material from the major Turolian localities of Samos and Pikermi (Greece) and Maragheh (Iran), as well as the Baodean localities of China.

Sinap Material

The fossil collection of the Sinap project was created during seven field seasons in the years 1989–1995. Most of it represents surface collection, but a substantial portion was also obtained by trenching. The rich collection from Loc. 49 was mostly obtained through excavation, and details are available on request from the senior author. Specimens were numbered sequentially in the order of cataloging, starting each year with number 1. Specimens are identified by the prefix AS (for Ankara Sinap), the year, and the catalog number, separated by periods (e.g., AS.95.123). Until 1993, the prefix used was plain S. The collection is stored in the Museum of Anatolian Civilizations in Ankara. For information on the geology and dating of localities, see Kappelman et al. (1996), Kappelman et al. (chapter 2, this volume), Lunkka et al. (1999). Measurements of Sinap specimens are given in appendix tables 12.1–12.3. The collection of the Maden Tetkik ve Arama Enstitüsü (MTA) in Ankara stems mostly from major trenching operations undertaken in the 1950s and the two following decades.

Abbreviations

General: C.V. = coefficient of variation, dex = right, DP = upper deciduous (pre)molar, dp = lower deciduous (pre)molar, Loc. = Locality, M = upper molar, m = lower molar, mc = metacarpal, mt = metatarsal, P = upper premolar, p = lower premolar, sin = left.

Measurements: AP = anteroposterior diameter, APD = distal anteroposterior diameter, APP = proximal anteroposterior diameter, APS = minimum anteroposterior diameter of shaft, JAPD = anteroposterior diameter of distal joint surface, JAPP = anteroposterior diameter of proximal joint surface, JWD = width of distal joint surface, JWP = anteroposterior diameter of distal joint surface, L = length, LB = buccal length, LL = lingual length, W = width, WD = distal width, WP = proximal (mesial) width, WS = minimum width of shaft.

Institutions: BMNH = Natural History Museum, London; BSPHGM = Bayerische Staatssammlung für Paläontologie und historische Geologie, München; MNHN = Muséum National d'Histoire Naturelle, Paris; MTA = Maden Tetkik ve Arama Enstitüsü, Ankara; NRM = Swedish Museum of Natural History, Stockholm; PDTFAU = Paleoantropoloji, Dil ve Tarih-Cografya Facültesi, Ankara Üniversitesi; PIU = Paleontological Institute, University of Uppsala; SMNS = Staatliches Museum für Naturkunde, Stuttgart.

Catalog of Fossil Material

Lower Sinap Member

Brachypotherium brachypus (Lartet in Laurillard 1848)

Taxonomy:

Aceratheriinae
Teleoceratini
Brachypotherium brachypus (Lartet in Laurillard 1848)

Restricted synonymy:

1981 *Brachypotherium brachypus* Gürbüz fig. 2
1994 *Brachypotherium brachypus* Saraç pl. 4, figs. 1–3

Sinap Material. Loc. 125: astragalus dex AS.95.454, cuboideum dex AS.95.453, mt II dex AS.94.143; MTA collection (İnönü I): mt III dex 06-INÖ-77/1607.

Age. Loc. 125 is stratigraphically older than Locs. 24 and 24A, but beyond this, the age is not known. Locs. 24 and 24A are found in redeposited ash from a volcanic event, possibly related to a basalt flow dated at 15–16 Ma (Kappelman et al., chapter 2, this volume). The localities are unfortunately situated in an isolated block in the middle of a fault zone, and the block's stratigraphic relationship to the surrounding strata remains elusive (Lunkka et al., this volume).

Remarks. The metatarsal from İnönü I (Locs. 24 and 24A of the Sinap project) is unmistakable; it is very similar to a specimen from Sofça figured by Heissig (1976, fig. 39). Only field identifications and a few measurements are available for the fossils from Loc. 125. Judging from the measurements (appendix table 12.3), the metatarsal at least seems more slender than is typical for the species, and the material may represent a more primitive brachypothere species than *Brachypotherium brachypus*.

Discussion. *Brachypotherium* is a conservative and long-lived genus with a wide geographic range in western Eurasia (Heissig 1996). It is one of several rhinoceros lineages to develop short legs and relatively high crowned teeth, but its paleoecology remains enigmatic. A hippopotamus-like lifestyle is possible, but this suggestion lacks direct support. Judging by dental wear, the animal seems to have been a mixed feeder (Fortelius 1990; Fortelius and Solounias 2000).

Hoploaceratherium tetradactylum (Lartet, 1837)

Taxonomy:

Aceratheriini
Hoploaceratherium tetradactylum (Lartet, 1837)

Restricted synonymy:

1994 *Hoploaceratherium tetradactylum* Saraç pl. 2, fig. 1

Sinap Material. MTA collection (İnönü I): juvenile maxilla dex with DP2-DP4 06-INÖ-77/1667.

Age. Close to but <15.2 ± 0.3 Ma (see above and Kappelman et al. 1996).

Remarks. The single specimen is very similar to the corresponding specimen from Paşalar described by Fortelius (1990) as *Aceratherium* sp. aff. *tetradactylum*. The DP2 is characteristically elongated, especially in its buccal part, and the ectoloph is strongly inflected at the metacone on all the teeth. The protocone of DP3 and DP4 shows moderate constriction both mesially and distally.

Discussion. This species is not represented in the collections of the Sinap project, nor did Gürbüz (1981) list it from the locality. It is, however, present at the Anatolian localities of Paşalar and Çandir which, like İnönü I, also have Begertherium and Brachypotherium (Heissig 1976; Fortelius 1990).

Hoploaceratherium is part of a plesion that has yet to be revised. Cerdeño (1996) synonymized *Hoploaceratherium* with *Acerorhinus*, but we have retained the genus here, partly because we feel that the complete loss of horns in *Acerorhinus* justifies separation at the generic level and partly to avoid premature changes. This species was a plesiomorphic rhinoceros, best regarded as a browser ecologically similar to the living small southeast Asian rhinoceroses.

Hispanotherium grimmi Heissig, 1974

Taxonomy:

Rhinocerotinae
Elasmotherini
Hispanotherium grimmi Heissig, 1974

Restricted synonymy:

1981 *Hispanotherium grimmi* Gürbüz fig. 2
1994 *Begertherium grimmi* Saraç pl. 5, figs. 1–3
1996 *Begertherium* cf. *B. grimmi* Kappelman et al. table 6.2

Sinap Material. Loc. 24A: M sin superior ectoloph part AS.89.111, astragalus sin AS.92.667, mt III sin proximal part AS.92.664, mt IV sin proximal part AS.91.400; MTA collection (İnönü I): maxilla sin with DP1–M3 06-INÖ-0802, mc III dex 06-INÖ-77/1773.

Age. Matrix is probably derived from volcanic activity at ~15–16 Ma (see above and Kappelman et al. 1996); the fauna indicates a late MN 5 or early MN 6 age, with co-occurrence of *Listriodon splendens* and *Bunolistriodon latidens* (Gürbüz 1981; Fortelius et al. 1996b).

Remarks. The material is similar to that described by Heissig (1974, 1976) and does not add anything critical to previous knowledge of the taxon. The complete upper toothrow shows hypsodont molars and strongly molarized premolars with thick cement coating (Saraç 1994, pl. 5, fig. 1).

Discussion. The tangled taxonomy and nomenclature of the taxon (Heissig 1976; Fortelius and Heissig 1989; Cerdeño 1995) has recently been clarified by Antoine (2000), whom we follow here.

These elasmotherines were the earliest hypsodont rhinoceroses in the Old World, and show grazerlike dental wear (Fortelius 1990). They were also relatively cursorial, as befits animals that first evolved in the open habitats that were beginning to appear in central Asia at this time (Bernor et al. 1996c).

Rhinocerotidae indet.

Sinap Material. Loc. 79: M dex inferior fragment (protoconid) AS.92.103, astragalus dex AS.92.97, calcaneum AS.92.96, proximal mt II sin AS.92.101; Loc. 80: proximal radius dex AS.92.109.

Age. Locs. 79 and 80 are situated north of the major fault in the Sinap-Delikayinçak area and are far outside any stragraphically measured section. Based on general lithostratigraphic relationships, they are thought to represent a stratigraphic position close to or lower than the lower Sinap member (J. Kappelman, pers. comm.).

Remarks. Judging from field notes by MF, at least some of the material from Locs. 79 and 80 may well represent *Alicornops simorrensis*, but without access to the specimens, it has not been possible to verify this. *Alicornops simorrensis* was present in Anatolia from MN 6 to MN 7 + 8 (Heissig 1996), so its presence in the lower Sinap member is to be expected. This record must be regarded as extremely tentative, however.

Middle Sinap Member

Acerorhinus zernowi (Borissiak 1905)

Taxonomy:

Aceratheriinae
Aceratheriini
Acerorhinus zernowi (Borissiak 1905)

Restricted synonymy:

1990 *Chilotherium* sp. Sen p. 250
1994 *Chilotherium (Acerorhinus) zernowi* Saraç pl. 8, fig. 3
1994 *Chilotherium (Chilotherium) samium* Saraç pl. 10, fig. 1a,b
1996 *Acerorhinus* cf. *A. zernowi* Kappelman et al. table 6.2

Sinap Material. Loc. 49: skull AS.95.747 + AS.95.24, partial skull with sin tooth row AS.93.823, sin maxilla and upper toothrow AS.94.554, maxillary fragment dex with M1–M3 AS.93.1074, sin M1 or M2 AS.94.500, male mandible AS.90.96 + AS.92.150, partial mandible AS.94.315-316 lacking anterior portion, i2 sin AS.95.72 (male), p2 dex AS.94.1466, p2 sin AS.94.1414, dp2 dex AS.91.188, tibia sin distal part AS.92.138, astragalus sin AS.91.731; MTA collection (Ozansoy's Loc. IB = Loc. 1): male skull (thought to have been subsequently lost) maxilla dex with DP1–M3 06-SIN-0136; MTA collection (Ozansoy's Loc. II = Loc. 12): maxilla sin with P2–M3 06-KAY-5, maxilla sin with P3–M3 06-KAY-11, mandibular ramus dex 06-KAY-12.

Age. The interpolated magnetochronologic ages of these localities are: Loc 12, 9.6 Ma; Loc. 1, 9.3 Ma; Loc. 49, 9.1 Ma (Kappelman et al., chapter 2, this volume).

Remarks. Unfortunately, no photographic documentation is available of the skull AS.95.747 from Loc. 49, discovered late in the 1995 season. Approximate measurements and a brief description are offered here based on preliminary field notes taken by MF. The skull is well preserved but lacks most of the face anterior to P4. The nasal bones were recovered from the surface at the beginning of the same season in which specimen AS.95.24 was recovered. The distance from the tip of the nasals to the nuchal crest is ~485 mm, the distance from the posterior rim of the orbit to the nuchal crest is ~320 mm, the total height of the skull at M1 ~135 mm, and the height of the occiput ~240 mm. The facial crista is confluent with the bulbous anterior rim of the orbit, which is placed above M2 and is not elevated, as it is in *Chilotherium*. The postglenoid process is stout and has a separate vertical semicylindrical joint surface for the mandible, in contrast with *Chilotherium*, where the process is weaker and the joint surface oblique and partly confluent with the glenoid joint surface. The zygomatic arch is quite deep and has a weakly sigmoid outline in lateral view (fig. 12.1A). The nasals are quite long and separated by a strongly developed median groove. They display a characteristic blunt beak separated from the posterior portion of the bone by a distinct shoulder. The upper cheek teeth of the skull are highly similar to those of specimen AS.93.823 (Fig. 1B), with moderately developed buccal folding and a

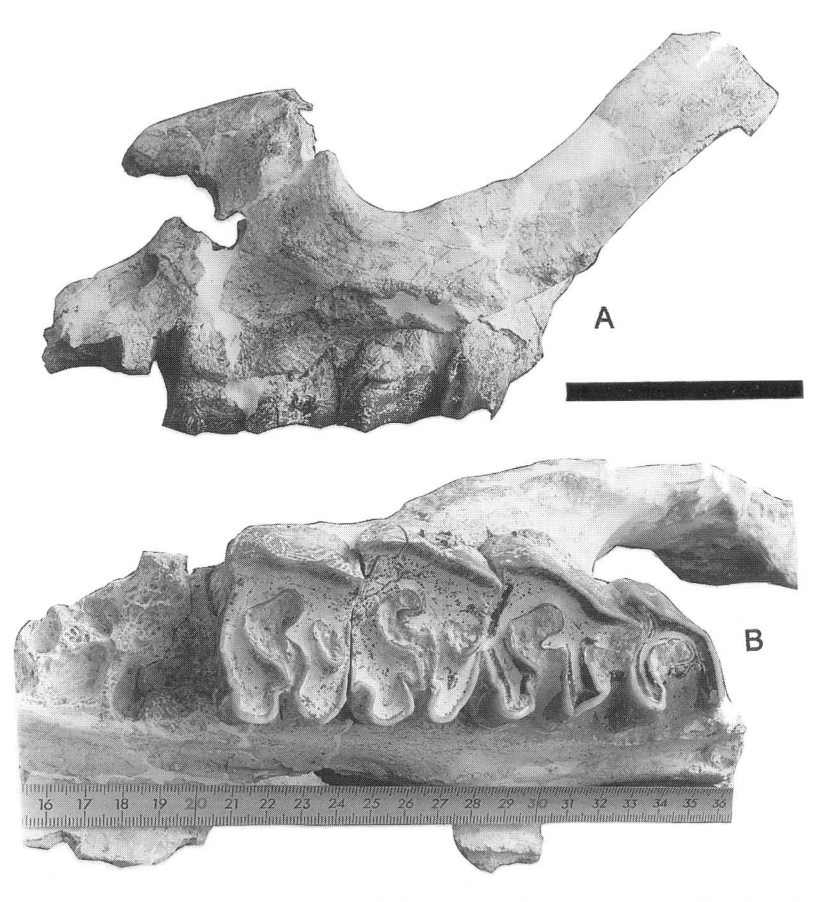

Figure 12.1. (A) Lateral view of partial skull AS.93.823 of *Acerorhinus zernowi* from Loc. 49. Scale bar = 10 cm. (B) Same specimen, occlusal view of left maxillary tooth row. Ruler in image.

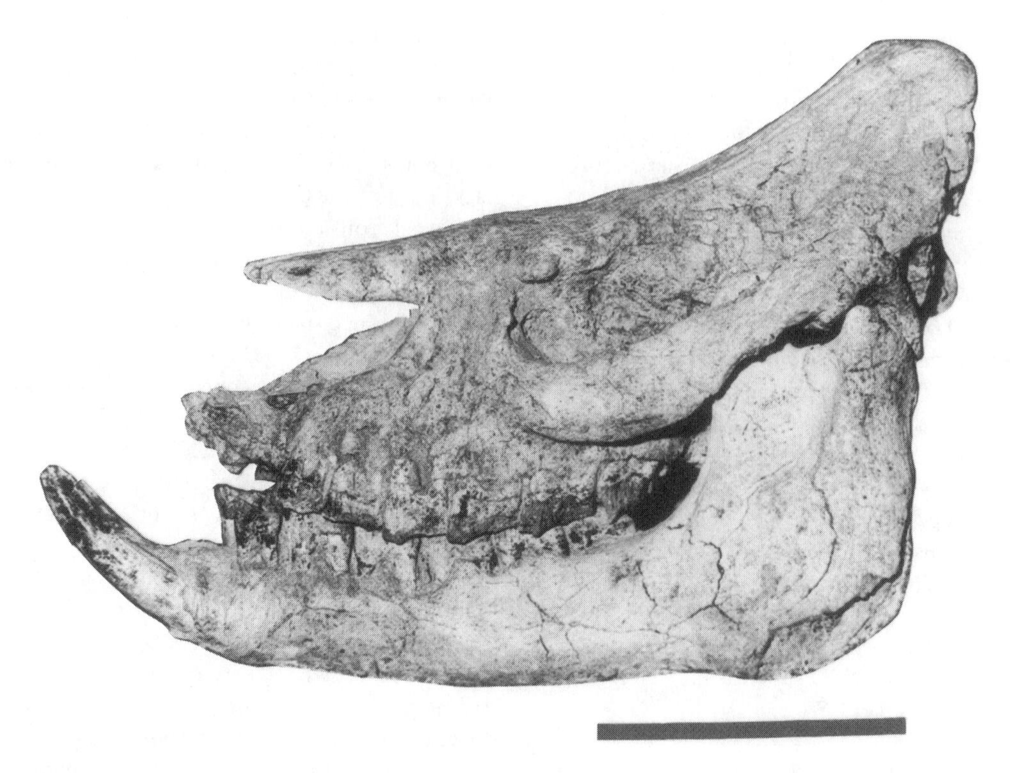

Figure 12.2. Male skull of *Acerorhinus zernowi* from Ozansoy's Loc. IB (=Loc. 1), MTA collection. Based on the only known photograph. The specimen is thought to have been lost. Scale bar ~20 cm.

moderately constricted molar protocone that is not flattened lingually. The M3 is relatively short.

Unfortunately, the unnumbered skull from Loc. 1 appears to have been lost and thus can only be described from the single photograph known to exist (fig. 12.2). At least in general characteristics, it appears to be very similar to the specimen described above. When last examined, the specimen was well preserved and complete, with an associated mandible in place. The occiput is elevated and the nasals curve gently to a blunt tip. There is a weak but distinct elevation of the frontals above the orbit, which, again, is not itself elevated. The narial incision is deep, reaching the anterior margin of M1. The preorbital bar is consequently narrow, with the orbit situated above M2. There is a small anteorbital apophysis and a larger supraorbital one.

The facial crista is present but appears to be quite weak. The zygomatic arch is gently curved and relatively slender. The upper cheek teeth have well-developed buccal folds, including a strong paracone rib and a distinct inflexion at the metacone. The ascending ramus of the mandible makes a slightly open angle with the body. M3 is relatively short and p2 relatively long. The tusk curves quite steeply upward and appears to be oriented almost directly forward, in contrast to the less curved, more horizontal, laterally flaring tusks typically seen in *Chilotherium*. The apparently rounded lateral face is also characteristic of *Acerorhinus* and unlike the angled lateral face of a *Chilotherium* tusk. Unfortunately, no measurements of this specimen are known.

The maxillary dentitions 06-SIN-0136 and 06-KAY-5 are very similar to each other. The teeth are mesodont and have well-developed buccal folds and a moderate cement covering (fig. 12.3). The protocones of M1 and M2 are moderately constricted by mesial and distal folds, and their lingual side is very slightly flattened, with rounded corners to the cusp. The M3 is relatively short (the lingual side is compressed). The maxillary dentition 06-KAY-11 is more worn but shows the rounded lingual cusps that distinguish it from *Chilotherium*.

The mandible (AS.90.96 + AS.92.150) has a long, relatively narrow symphysis and a large, upturned tusk with a rounded cross-section of the lateral side (fig. 12.4 A–B). The mandibular ramus is of even depth and begins to taper toward the symphysis only above p2. The angle of the mandible is somewhat expanded but not turned out toward lateral. The cheek teeth are low crowned and have a rounded metalophid without a distinct ectoflexid (except

Figure 12.3. Right maxillary toothrow 06-SIN-0136 of *Acerorhinus zernowi* from Ozansoy's Loc. IB (=Loc. 1). MTA collection. Occlusal view. Scale bar ~10 cm.

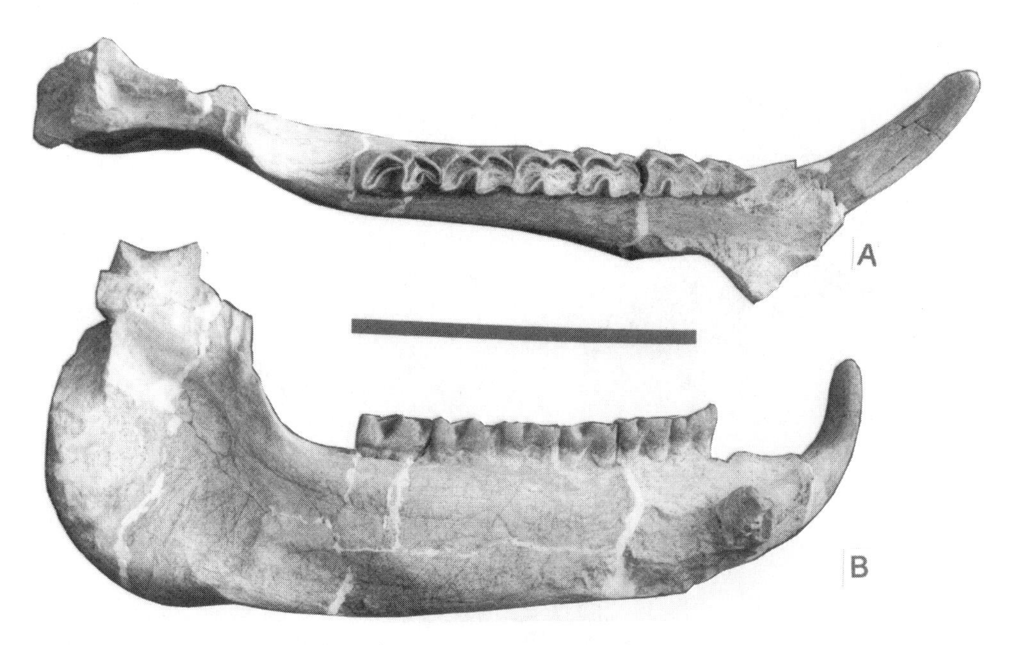

Figure 12.4. Left male hemimandible AS.92.150 of *Acerorhinus zernowi* from Loc. 49. (**A**) Lingual view; (**B**) occlusal view. Scale bar = 20 cm.

on the large p2), and with trigonid and talonid basins that have a V-shaped cross-section. Other mandibles all essentially correspond to this description. The glenoid joint is preserved on AS.94.315 and shows the typical rhinocerotid double arrangement, with a clearly separate, posteromedial cylindrical joint surface that embraces the stout postglenoid process of the skull. In *Chilotherium* from the same locality (e.g., specimen AS.94.316), this arrangement is modified, so that a semicontinuous curved joint surface articulates both with the temporal and the postglenoid process, which is shorter and more tapered.

Of the postcranial remains from Loc. 49, only a distal tibia (AS.92.138) and an astragalus (AS.91.731) can be confidently referred to this species, based on their narrow joint surfaces and pronounced trochlear relief, quite unlike the shallow and broad ankle joint of *Chilotherium*.

Discussion. Heissig (1975) assigned all Anatolian *Acerorhinus* remains to *A. zernowi* (Borissiak 1914), described from the MN 9 locality Sebastopol in the Crimea. This remains the best match for the Sinap material, which spans MN 10 (Locs. 12, 1) to MN 11 (Loc. 49). However, *A. tsaidamense* from Qaidam (Bohlin 1937) is also in several respects similar to the Loc. 1 skull: the occiput leans backward rather than forward, the facial crista is weak, the zygomatic arch is slender, and the angle of the mandible is slightly open (Bohlin 1937, fig. 164). The flattened lingual cusps of the upper molars of *A. tsaidamense* do, however, appear derived in comparison with the more plesiomorphic, rounded cusps of *A. zernowi* (cf. Borissiak 1915, pl. II; Bohlin 1937, pl. VIII, fig. 1). The skulls and mandibles from Tung-gur assigned to *A. zernowi* by Cerdeño (1996, figs. 2, 3) are similar to the Loc. 1 skull in occipital morphology but differ in having a more massive zygomatic arch, a shallower narial incision, longer nasals, and a more vertically oriented ascending ramus on the mandible, all characters in common with the type material of *A. zernowi*. Both *A. zernowi* from Tung-gur and *A. tsaidamensis* were long-limbed forms compared with short-limbed *Chilotherium*, although both were more robust than middle Miocene members of this plesion, such as *Hoploaceratherium tetradactylum* and "*Aceratherium incisivum*" from the Jilancik Beds in the Turgai (Cerdeño 1996). The postcranial *Acerorhinus* material from the Sinap Formation is too incomplete for meaningful comparison of limb proportions, however.

Acerorhinus zernowi was a plesiomorphic rhinoceros, not far removed in terms of ecology from *Hoploaceratherium tetradactylum*. Judging from its dental mesowear pattern (Fortelius and Solounias 2000) it was a browser or a browser with a limited mixed-feeding capability.

Acerorhinus sp. nov.

Restricted synonymy:

1996 *Acerorhinus* cf. *A. zernowi* Kappelman et al. table 6.2 (in part)

Sinap Material. Loc. 26: mc III dex proximal part AS.91.229, mc IV dex AS.90.241 (same individual?); Loc. 33: p2 dex AS.89.279; Loc. 58: p2 dex AS.90.184; MTA collection (level of Loc. 26): Unnumbered adult and juvenile skulls, maxilla dex with upper dentition P4–M3 06-AKK-011.

Age. All these localities are in the upper fossiliferous level of Upper Kavakdere, with a magnetostratigraphic age of 8.1 Ma (Kappelman et al., chapter 2, this volume).

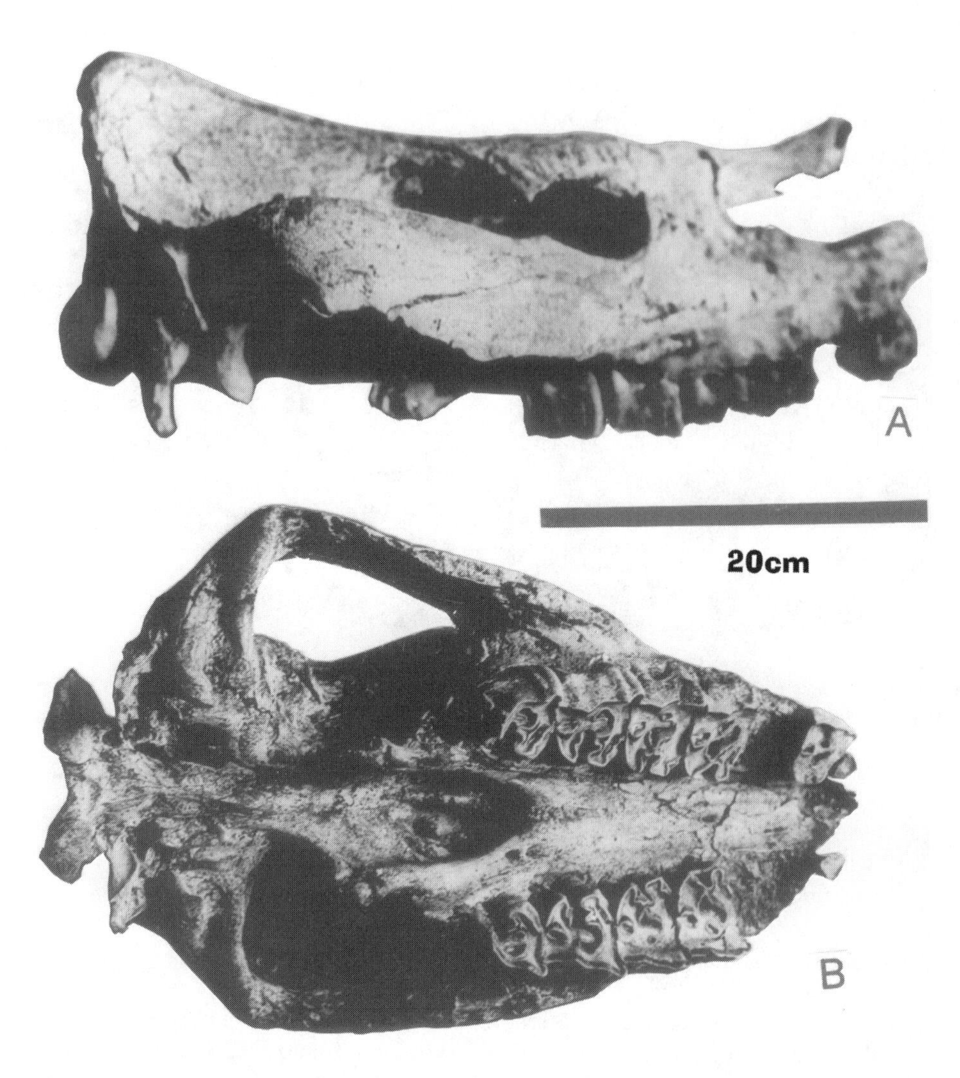

Figure 12.5. Unnumbered skull of *Acerorhinus* sp. nov. from Upper Kavakdere, MTA collection. (**A**) Lateral view; (**B**) ventral view. Scale bar ~20 cm.

Remarks. The adult MTA skull (fig. 12.5A–B) has a highly derived and suggestive combination of characters. It is long and has a concave profile in lateral view, with parietals and nasals both distinctly elevated above the plane of the frontals. The face is short and the orbit high, with a strong postorbital process above M3. The facial crest is strong and the skull tapers abruptly from frontals to nasals, in contrast with the gradual tapering invariably seen in *Chilotherium*. The zygomatic arch is nearly horizontal and very deep, about one-half of the height of the skull itself. The upper molars are hypsodont, with advanced folding of the enamel, and the premolars are relatively very broad and generally large (fig. 12.5B). Features typical of *Acerorhinus* include the strong ribs of the upper molars, the inflexion of the ectoloph at the metacone, the lingually pointed hypocones, and especially, the large and broad premolars. The juvenile skull shows essentially the same set of characters at an earlier ontogenetic stage.

The two isolated p2 specimens are unmistakable owing to their large size and characteristic wear profile descending from an acute tip at the paraconid; they unambiguously record the presence of the genus there. The metacarpals are shortened (fig. 12.6), and might be expected to belong to *Chilotherium kowalevskii*, a species with strongly shortened podials. They are, however, a much better match for *Acerorhinus palaeosinensis* in the Lagrelius collection than for any *Chilotherium* with which we have been able to compare them, and we prefer to associate them with *Acerorhinus*. Compared with mc III UMP M3831a, AS.91.229 has a proximal articular surface only somewhat more extended toward the posterior and has the same strongly developed lip below the articular surface on the plantar side. The mc IV UMP M3831c is also very similar to AS. 90.241, which again has a more anteroposteriorly extended proximal articular surface. The facets between mc III and mc IV are also relatively larger in the Kavakdere form. The small figures of Pavlow (1915, pl. V) permit only the most approximate comparison, but as far as can be judged, the *Acerorhinus* from Tchobrouchi also has metacarpals of about the same proportions as the Kavakdere form.

The main differences relative to derived *Chilotherium* (*C. anderssoni* and *C. persiae*) include larger and less vertical articular surfaces between mc III and the accessory metacarpals, more strongly curved mc IV with relatively broader

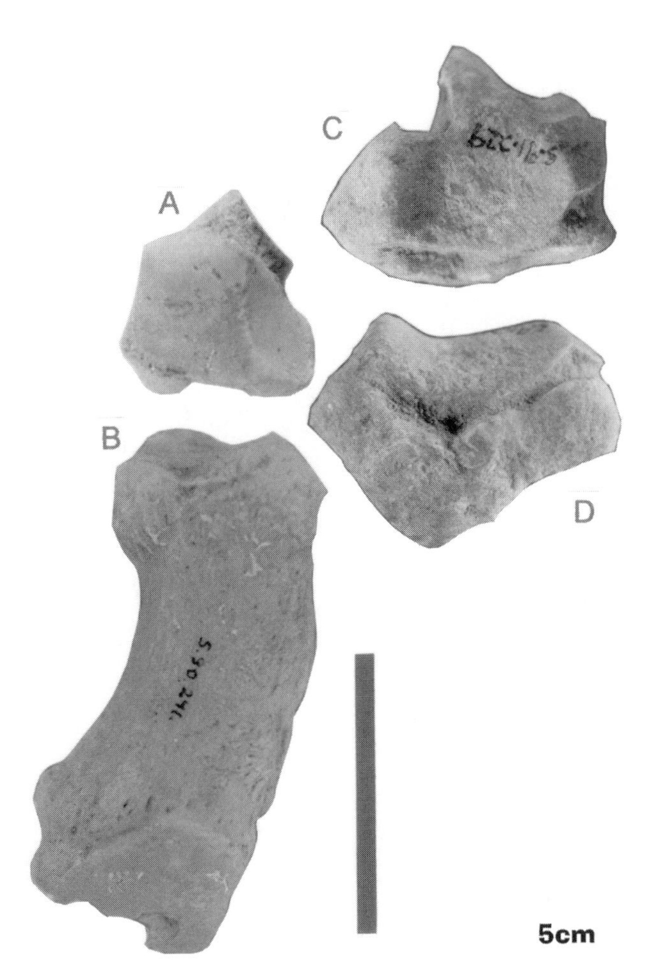

Figure 12.6. Right metacarpals IV and III (same individual?) of *Acerorhinus* sp. nov. from Loc. 26. **(A)** Proximal view of mc IV AS.90.241. **(B)** Plantar view of same specimen. **(C)** Proximal view of mc III (proximal part) AS.91.229. **(D)** Plantar view of same specimen. Scale bar = 5 cm.

articular surfaces both proximally and distally, a less developed ridge on the distal trochlea, and presence of an articular facet for mc V.

Discussion. It seems that two successive species of *Acerorhinus* are recorded in the Sinap Formation, the later one close to the roughly contemporaneous species described from Tchobroutchi by Pavlow (1915) as "*Aceratherium incisivum*" and from Udabno by Tsiskarishvili (1987) as "*Aceratherium* sp." These are derived forms, similar in several aspects to the Chinese *Acerorhinus palaeosinensis,* with a flat skull roof; a high orbit; and shortened, strongly splayed metapodials. The skull is much more elongated in the Kavakdere form than *Acerorhinus palaeosinensis* or in the material from Tchobroutchi, however, and the nasals have a different shape, with a peculiar dorsad twist at the tip. A long-skulled form virtually identical to the Kavakdere species, with the same peculiar nasal morphology, is, however, known from the late Miocene (?Turolian) locality Marmar in Tajikistan (S. Sharapov, pers. comm.). It seems likely that these Turolian forms represent at least one and perhaps two hitherto unrecognized species of *Acerorhinus,* perhaps representing a west Asian clade, but without detailed study of the material this cannot now be determined (cf. Cerdeño 1996, p.17).

The evolutionary history of the group is treated briefly under General Discussion later in this chapter. We note here in passing that Cerdeño's (1996) suggestion to transfer the derived members of the *Acerorhinus* to *Chilotherium* implies either multiple detailed homoplasy (at the least, the shape of mandibular symphysis, tusk position, morphology and proportions of premolars and molars, and construction of temporomandibular joint), or a very late origin of *Chilotherium* from *Acerorhinus.* The former alternative appears inherently unlikely, whereas the latter is contradicted by extensive stratigraphic evidence.

The trends seen in the evolution of *Acerorhinus* seem to indicate a parallel evolution with *Chilotherium,* but the precise nature of the adaptation of these highly successful open-habitat rhinoceroses remains somewhat enigmatic.

Chilotherium kiliasi (Geraads and Koufos 1990)

Restricted synonymy:

1996 *Chilotherium* cf. *C. samium* Kappelman et al. table 6.2 (in part)

Sinap Material. Loc. 49: upper toothrow sin AS.93.963, P4 dex AS.90.98, M2 dex AS.91.695, mandible AS.93.810 (female), mandible AS.93.809 (male), mandible AS.93.815 (female), partial mandibular ramus sin AS.91.701, mandibular ramus sin AS.94.566, partial mandibular ramus dex AS.94.537, m3 dex AS.91.690, m3 sin AS.90.100, m3 dex AS.90.97, juvenile mandible sin AS.93.1193.

Age. The magnetostratigraphic age estimate for Loc. 49 is 9.1 Ma (Kappelman et al., chapter 2, this volume).

Remarks. This is a mesodont form with moderately reduced premolars and a moderately short M3 compared with most more derived species of *Chilotherium*. The upper teeth have weak paracone styles and relatively flat buccal walls with a weak inflexion at the metacone (fig. 12.7). The protocone is constricted from mesial and distal and distinctly flattened lingually—a good distinguishing character from *Acerorhinus* from the same locality, in which M3 is also clearly shorter. The buccal walls of the lower teeth are rounded rather than angled (as in cf. *Chilotherium* from the middle Sinap) and the hypolophids of the premolars have strikingly strong transverse portions that are somewhat recurved. The mandibular ramus of the female mandible AS.93.810 tapers gradually toward the anterior (fig. 12.8), whereas in male specimens (e.g., AS.93.809), it remains equally deep almost to the symphysis, presumably because a deep root for the large tusk is present. The morphology of the lower molars differs from that of *Acerorhinus* cf. *A. zernowi* from the same locality in the following characters: the tooth is somewhat higher crowned, has a less expanded base, more vertical walls, and stronger ectoflexids.

Figure 12.7. Left upper toothrow AS.93.963 of *Chilotherium kiliasi* from Loc. 49. Occlusal view. Scale bar = 10 cm.

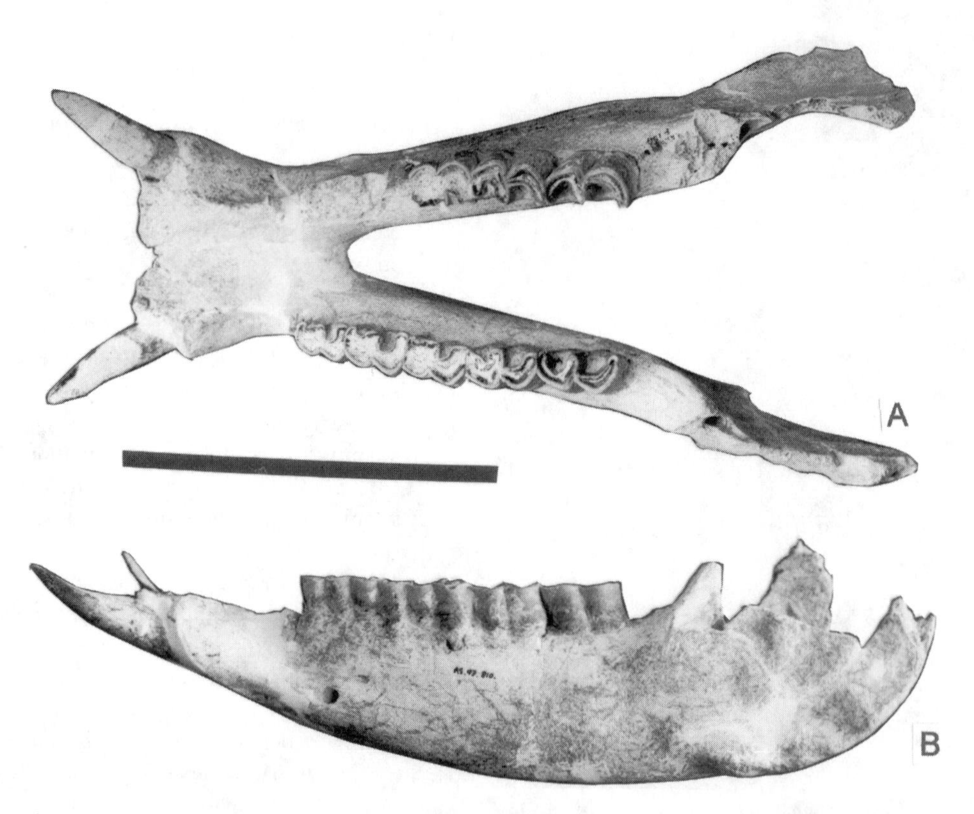

Figure 12.8. Female mandible AS.93.810 of *Chilotherium kiliasi* from Loc. 49. (**A**) Mandible in dorsal view. (**B**) Mandible in lateral view. Scale bar = 20 cm.

The trigonid and talonid basins are more open, with U- to V-shaped cross sections progressively opening up along the tooth row from mesial toward distal. The relative proportions of the cheek toothrow differs dramatically from *C. kiliasi* and *A. zernowi* from the same locality (fig. 12.9). The distinction from the more hypsodont *Chilotherium* from the same locality is described below.

Discussion. The taxonomy of plesiomorphic *Chilotherium* is highly problematic, not least owing to the nature of the type material of *C. samium* (Weber 1905) (an old individual with very worn teeth from an unknown horizon at Samos). *Chilotherium wimani* Ringström, 1924 is a taxon of approximately the same grade of evolution as *C. samium,* but no direct comparison has been undertaken. No skulls of *C. wimani* have been figured, and unfortunately for us, both the lower dentitions figured by Ringström (1924, pl. VIII, figs. 1,2) have worn teeth, making comparison with the Sinap material difficult.

Ironically, the recently described *"Aceratherium"* (=*Chilotherium*) *kiliasi* from Pentalophos I (Geraads and Koufos 1990) is also based on an old individual with worn teeth. Furthermore, the hypodigm of *"Aceratherium" kiliasi* includes a female mandible that clearly belongs to *Acerorhinus*

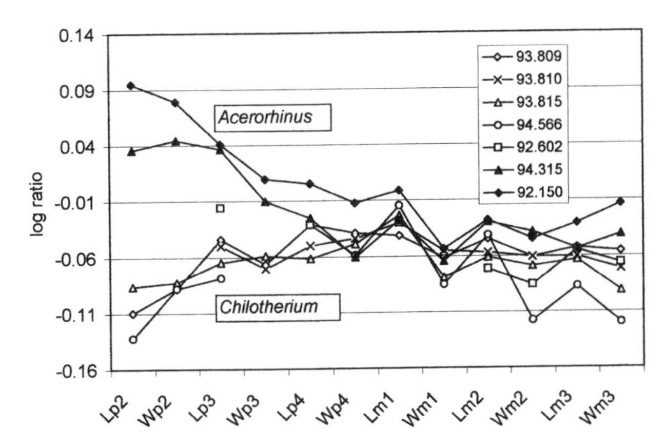

Figure 12.9. Log-ratio diagram of proportions of the lower cheek tooth row of *Acerorhinus zernowi* versus *Chilotherium kiliasi*, both from Loc. 49 only. Standard = *Chilotherium* sample from Maragheh. *Acerorhinus* has relatively much larger premolars, although the difference in molar size and proportions is minimal. As usual, m1 is the tooth showing the least difference. The legend gives Sinap specimen numbers without the prefix "AS."

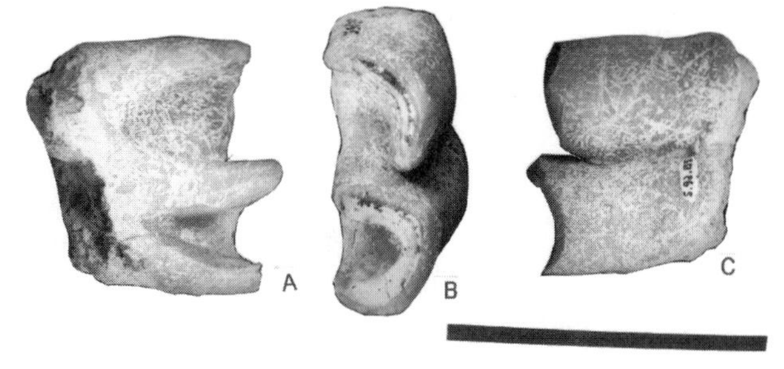

Figure 12.10. Left m3 AS.92.155 of *Chilotherium* cf. *C. habereri* from Loc. 49. **(A)** Lingual view; **(B)** occlusal view; **(C)** buccal view. Scale bar = 5 cm.

(Geraads and Koufos 1990, pl. 3, figs. 2,3,5), as testified by its narrow symphysis, large premolars, obliquely worn p2, and weak ectoflexids on all the characteristically broad cheek teeth, whereas the *Chilotherium* mandible figured (Geraads and Koufos 1990, pl. 2, figs. 3,4) is damaged and lacks the posterior molars.

The lower tooth morphology described above matches that of *Chilotherium kiliasi* well, especially in the strongly developed and recurved hypolophids of the premolars (Geraads and Koufos 1990, pl. 2, fig. 4), a trait apparently missing in *C. wimani* (Ringström 1924, pl. VIII, figs. 1,2), which is furthermore distinctly larger overall. The relatively large upper premolars (fig. 12.7) are also similar to those of the type skull of *C. kiliasi* (Geraads and Koufos 1990, pl. 3, fig. 4). Given the uncertain taxonomy of primitive *Chilotherium,* we tentatively assign the Sinap material to the nomen to which a specific morphological tie can be demonstrated, without implying any statement regarding synonymy. For the purpose of this chapter, we provisionally restrict the name *C. samium* to the type material.

This is a very primitive *Chilotherium,* probably not far removed from the basic, medium-sized aceratherine, with a browser-to-mixed-feeder lifestyle.

Chilotherium cf. *C. habereri*

Restricted synonymy:

1996 *Chilotherium* cf. *C. samium* Kappelman et al. table 6.2 (in part)

Sinap Material. Loc 49: m3 sin AS.92.155, juvenile mandible AS.90.313.

Age. The magnetostratigraphic age estimate for Loc. 49 is 9.1 Ma (Kappelman et al., chapter 2, this volume).

Remarks. The m3 differs from that of *Chilotherium kiliasi* in being distinctly more hypsodont, having more open trigonid and talonid basins, conspicuously thinner enamel lining of the basins, and a less recurved hypolophid outline in occlusal view (fig. 12.10). The lower milk molars (fig. 12.11) show a relatively small and slender dp2, a clear difference from the unidentified Kavakdere *Chilotherium* (see below). Originally a small peglike dp1 was present on the right side, but this was lost during later preparation. The milk teeth have the characteristic hypoplastic band near the base of the crown commonly observed in forms that have recently evolved or are in the process of evolving higher tooth crowns (e.g., very common in *C. persiae* from Maragheh). The m3 differs from the Kavakdere *Chilotherium* in having lingual cusps with short lingual cusps with rounded lingual walls.

Discussion. The high-crowned material from Loc. 49 is too hypsodont to belong to any of the plesiomorphic *Chilotherium* species discussed so far; it is close to the intermediate grade of evolution (especially crown height) represented by the Chinese species *C. habereri.* This specimen probably represents the Anatolian form that Heissig (1975, 1996) referred to *C. habereri.* Comparison with Ringström's (1924) plates and original specimens in the Uppsala collection shows that this tooth differs from the more derived *C. anderssoni* in being somewhat smaller, having a relatively longer metalophid with a slightly flattened buccal wall, and the hypolophid showing a slight flexion toward distal at the distal end, which gives the lophid a slightly sigmoid outline in occlusal view instead of the selenoid recurved profile seen in the Chinese species. To a lesser degree, all these differences also separate the Loc. 49 material from the Chinese *C. habereri,* and instead unite it with the large *Chilotherium* from Maragheh (e.g., MNHN MAR 1905.10).

There is a difficulty in that at least two species of *Chilotherium* are found at Maragheh, and pending revision of the material, one must apply the nomenclature with caution. The larger and more derived Maragheh species, which is similar to but larger than the hypsodont *Chilotherium* from Loc. 49, is the more common in the BMNH and MNHN collections and appears to correspond to the type material of *C. persiae* (de Mequenem 1924). Specimen

Figure 12.11. Juvenile mandible AS.90.313 of *Chilotherium* cf. *C. habereri* from Loc. 49. Scale bar = 10 cm.

AS.92.155 is about 15% smaller in linear dimensions than this form, but morphologically indistinguishable from it. A connection with *C. kowalevskii* appears unlikely, as that species has unreduced or perhaps secondarily enlarged premolars and anterior milk molars. The *Chilotherium* cf. *C. habereri* of Anatolia might represent an early stage of the evolution of the *C. persiae* lineage, but this cannot now be more than a speculation.

Chilotherium indet.

Sinap Material. Loc. 49: DP1 sin AS.94.509, DP1 sin AS.94.1382, i2 sin AS.94.572 (male), i2 sin AS.94.1451 (male), i2 sin AS.91.699 (female), p2 dex AS.94.312, p2 dex AS.94.1412, radioulna sin part AS.94.582; Loc. 50: DP1 sin AS.90.23, Loc. 34: Male mandible with ramus and symphysis sin AS.92.602; Loc. 26: astragalus sin AS.89.286; Loc. 33: radius sin proximal part AS.89.171, distal tibia dex AS.90.52, astragalus dex AS.89.215; Loc. 42: mt II sin proximal part AS.90.78, mt IV sin proximal part AS.89.422; MTA collection (level of Loc. 26): subadult mandible 06-AKK-013.

Age. The magnetostratigraphic age estimate for Loc. 49 is 9.1 Ma (Kappelman et al., chapter 2, this volume). The corresponding age for Loc. 34 is 8.4 Ma and 8.1 Ma for Locs. 26 and 33. Locs. 50 and 42 lack geochronologic age estimates. Loc. 42 (=Çobanpinar) is now placed in MN 13 (Kappelman et al., chapter 2, this volume; Van der Made, chapter 13, this volume).

Remarks. The indeterminate *Chilotherium* material from Loc. 49 consists mostly of worn teeth that are difficult to identify with confidence. There is no indication that an additional taxon is present at the locality. The indeterminate *Chilotherium* material from the Upper Kavakdere Locs. 34 and 26 and from Loc. 42 (Çobanpinar) all appears to represent one or more hypsodont species, smaller than *C. persiae,* a species that has previously been reported from these levels (Saraç 1994). The male mandible AS.92.602 shows lower molars with elongated and strongly flattened lingual walls on the lingual cusps, enclosing rather narrow trigonid and talonid basins (fig. 12.12). The tendency for the elongated entoconid to form an occlusal high point, almost as in brachydont hippomorph perissodactyls, is also a similarity with *C. persiae.* The subadult mandible 06-AKK-013 shows the lingually flattened molar morphology and a row of deciduous teeth, of which dp2 and dp3 are strikingly large (about the size of dp4). A smaller form similar to *C. persiae* but with relatively larger anterior cheek teeth might represent *C. kowalevskii* (cf. Pavlow 1913, pl. IV,

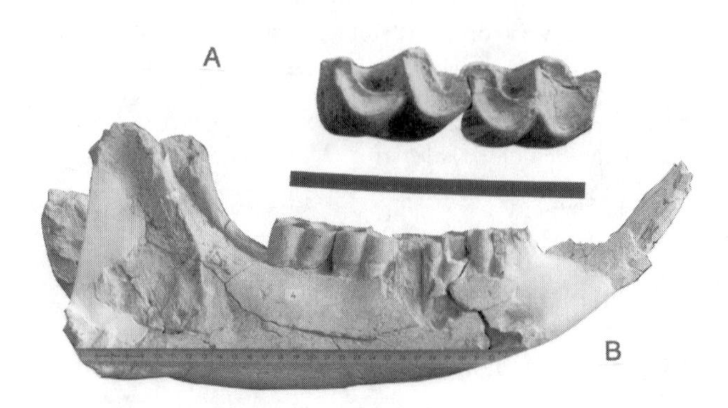

Figure 12.12. Left male hemimandible AS.92.602 of *Chilotherium* indet. from Loc. 34. **(A)** Detail of molars in occlusal view. Scale bar = 10 cm. **(B)** Buccal view of mandible. Ruler in image.

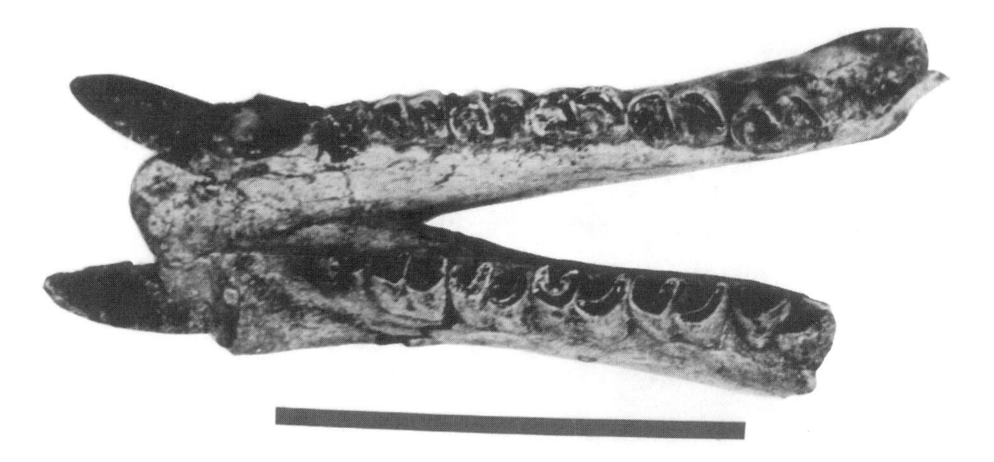

Figure 12.13. Male mandible 06-SIN-0135 of cf. *Chilotherium* sp. (primitive) from Ozansoy's Loc. IB (=Loc. 1). MTA collection. Scale bar = 20 cm.

fig. 8; de Mequenem 1924, p. 145; and Saraç 1994, pl. 12, fig. 3a), but the evidence is hardly conclusive. The tibia 06-AKK-017 is short and similar to a specimen figured by Pavlow (1913, pl. IV, fig. 23). The small astragali have broad trochleas with low relief and thus probably belong to *Chilotherium* rather than *Acerorhinus.*

Discussion. A characteristic that *C. kowalevskii* shares with *C. persiae* (but not with the more derived of the east Asian species of the genus) is the tendency for the lingual walls of the lower teeth to become elongated and strongly flattened, frequently to the extent of closing off the sinuses partly or completely. This presumably corresponds to the high degree of flattening of the lingual cusps of the upper teeth also seen in these west Asian forms, and may indicate that the west and east Asian species belong to separate clades.

As are other derived species of *Chilotherium,* these animals were most probably mixed feeders, judging from dental wear showing moderate rounding of the cusp tips. Grazing, even on fresh grass, leaves a considerably more rounded wear signal, at least in living ungulates (Fortelius and Solounias 2000).

cf. *Chilotherium* sp. (primitive)

Restricted synonymy:

1970 *Chilotherium* Sen Plate IX:2
1996 *Chilotherium* cf. *C. samium* Kappelman et al. table 6.2 (in part)

Sinap Material. Loc. 72: radius sin distal fragment AS.92.217, mt III dex AS.91.312; Loc. 12: associated forelimb dex AS.93.1210 (humerus distal fragment, radius proximal part, complete tetradactyl manus), calcaneum dex fragment AS.95.423; Loc. 51: mandibular rami dex and sin AS.90.132, male i2 dex part AS 90.131, humerus dex and sin AS.90.134, partial ulna AS.90.160, astragalus dex AS.91.387; MTA collection (Ozansoy's Loc. IB = Loc. 1): male mandible 06-SIN-0135; PDTFAU collection: mt II dex unnumbered; Şenyürek's Loc. F2 ("Aşağı yoncalık") in the middle Sinap member.

Age. The magnetostratigraphic age estimates are 10.1 Ma for Loc. 72, 9.6 Ma for Loc. 12, and 9.3 Ma for Loc. 1. Loc. 51 is probably close to Loc. 1 in age, based on general lithostratigraphic relationships and biochronology (Lunkka et al. 1999; Kappelman et al., chapter 2, this volume).

Remarks. The adult male mandible 06-SIN-0135 (fig. 12.13) has a long, broad symphysis, broader than in *Acerorhinus* (fig. 12.4) and *Subchilotherium* (Heissig 1972, pl. 8, fig. 2; Tsiskarishvili 1987, p. 53), and large tusks directed almost directly forward. The symphysis is hollowed-out on the ventral side, as in *Chilotherium* or *Acerorhinus* and unlike *Aceratherium.* The cheek teeth are plesiomorphic, mesodont, with v-shaped sinuses, long paralophids, and a distinct protoconid angle (ectoflexid) to the metalophid profile, which is characteristically "square" in occlusal view, especially on the premolars. The premolars are relatively large for *Chilotherium* and the planar buccal walls with distinct ectoflexids are unlike any other *Chilotherium.* The less complete specimen from Loc. 51 (AS.90.131-132) is similar in all particulars, including the relatively large premolars and the angled lophids.

The humeri AS.90.134 from Loc. 51 are relatively broad, short bones (fig. 12.14), with strong deltoid crests extending relatively further distally than in longer-limbed forms, such as the *Acerorhinus* from Tung-gur (Cerdeño 1996, fig. 6A). The distal end is relatively narrow, and the fossa olecrani narrow and high. The bone as a whole is somewhat shorter and broader than in *Aceratherium incisivum* (Hünermann 1989, fig. 10, table 5), but not nearly as shortened as the humeri of typical *Chilotherium* (e.g., Ringström 1924, pl. VIII, figs. 3, 4).

The right forelimb AS.93.1210 was collected during an undocumented excavation after the 1993 field season and was unfortunately not properly studied. The humerus is very similar to the bones described above. The metacarpals

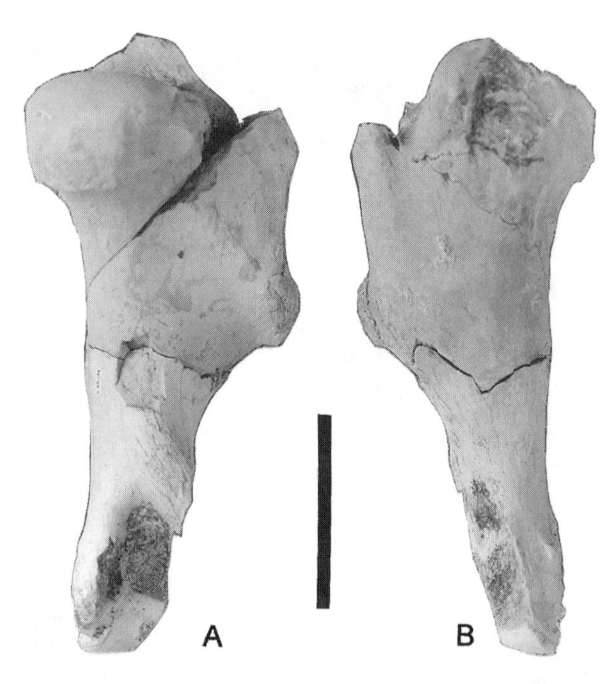

Figure 12.14. Humerus AS.90.134 of cf. *Chilotherium* sp. (primitive) from Loc. 51. (**A**) Left humerus; (**B**) right humerus. Scale bar = 10 cm.

(fig. 12.15) are, generally speaking, small and slender, not far from the proportions seen in *Aceratherium incisivum* (Hünermann 1989, figs. 31, 32), a species that also has a well-formed mc V. Compared with the measurements given by Hünermann (1989, table 15) all the metapodials are somewhat shorter and less flattened than in *A. incisivum,* however. The metacarpals are much smaller and perhaps relatively shorter than those of *A. tsaidamensis* (Bohlin 1937) and much smaller and stouter than the specimens from Tung-gur assigned to *A. zernowi* by Cerdeño (1996). They are still distinctly less shortened than in typical *Chilotherium.*

The astragalus is shorter and has a broad trochlea with low articular relief compared with *Acerorhinus* (Cerdeño 1996, pl. 7B,D), more like that of *Chilotherium* (Ringström 1924, pl. IX, fig. 3).

The mt III from Loc. 72 and the mt II from Şenyürek's Loc. F2 are both small and slender, similar to *Aceratherium incisivum* but again, somewhat shorter and less flattened (Hünermann 1989, fig. 62, table 15). The mt III is much shorter than a specimen from Tung-gur referred to *Acerorhinus zernowi* by Cerdeño (1996, fig. 9C). Like the metacarpals, the metatarsals are less shortened than in typical *Chilotherium.*

Discussion. The ventrally hollowed-out mandibular symphysis and the flattened tusks shows that this plesiomorphic aceratherine taxon belongs in the *Chilotherium-Acerorhinus* group. The shape of the lower cheek teeth, especially the long paralophids, excludes *Acerorhinus* but might fit an early *Chilotherium,* less derived than *C. kiliasi,* in which the metalophids are already rounded as in later *Chilotherium.*

The postcranial bones could fit an early *Chilotherium* well, being close to the primitive state (as represented by

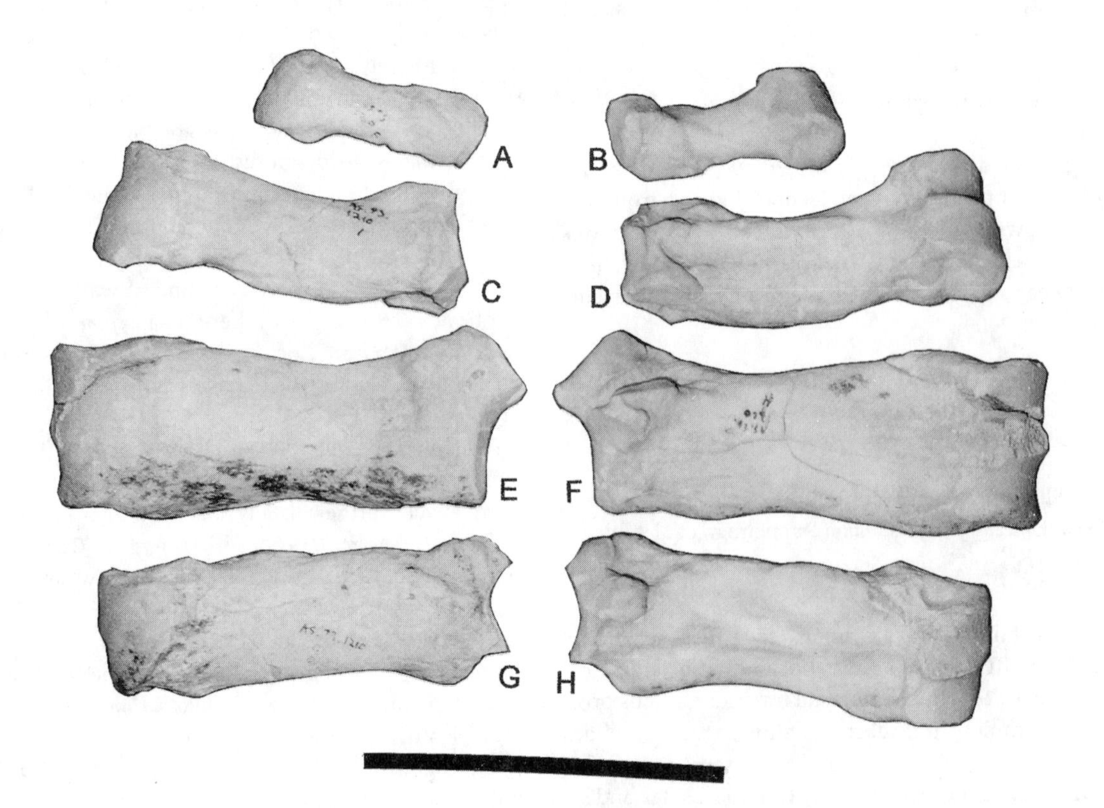

Figure 12.15. Metacarpals of right forelimb AS.93.1210 of *Chilotherium* sp. (primitive) from Loc. 12. Left row: plantar view. (**B**). Right row: palmar view. (**A,B**) mc V; (**C,D**) mc IV; (**E,F**) mc III; (**G,H**) mc II. Scale bar = 10 cm.

Aceratherium) but somewhat shortened, as would be expected. There is no direct association between the mandibles and the postcranial elements, but the shared "primitive *Chilotherium*" characteristics strongly suggest that they belong together. The material may, of course, represent more than one species.

It seems that this taxon is more primitive than any of the *Chilotherium* species so far described (see discussion under *C. kiliasi* above), but we have refrained from creating a new name, at least until the status and relationship of the existing nomina *C. samium, C. kiliasi,* and *C. wimani* are resolved. It differs from all known *Chilotherium* and from *Subchilotherium* in its angular lower cheek tooth morphology and its relatively large premolars. The earliest records of undisputable *Chilotherium* from the eastern Mediterranean reported by Heissig (1996) are from MN 10, a result in accordance with their first occurrence near the MN 10-11 boundary at Loc. 49 in the Sinap Formation. It is conceivable that cf. *Chilotherium* from the Vallesian Sinap localities could be close to the origin of *Chilotherium* s. str.

Stephanorhinus pikermiensis (Toula 1906)

Taxonomy:

Rhinocerotinae
Rhinocerotini
Stephanorhinus pikermiensis (Toula 1906)

Restricted synonymy:

1996 *Stephanorhinus* sp. (*pikermiensis*-group) Kappelman et al. table 6.2

Sinap Material. Loc. 33: isolated i1 AS.89.353, partial astragalus sin AS.90.50; MTA collection (Kavakdere): juvenile maxilla and toothrow dex with DP1–M1 06-AKK-0084, astragalus dex 06-AKK-038.

Age. Loc. 33 and probably all MTA localities are in the upper fossiliferous level of Upper Kavakdere, with a magnetostratigraphic age of 8.1 Ma (Kappelman et al. chapter 2, this volume).

Remarks. The incisor is large and stout for an i1, capped by a bean-shaped enamel crown very similar to the *Dicerorhinus orientalis* specimen figured by Ringström (1924, pl. 1, fig. 4). The DP1 is small relative to the other teeth and DP2 is elongated as is typical of the genus but is too worn to show much morphology (fig. 12.16). The DP3 and DP4 show a distinct paracone style and fold close to the buccomesial corner of the tooth, as figured by Weber (1904, pl. 16, fig. 1) and Ringström (1924, pl. 1, figs. 1, 2). The M1 shows the same feature, and, like DP4, a strong inflection of the ectoloph over the metacone (cf. Ringström 1924, pl. 1, fig. 1). The astragalus has an asymmetrical trochlea with relatively deep relief and a relatively short neck. The distal articular surface is relatively broader than in *Ceratotherium*.

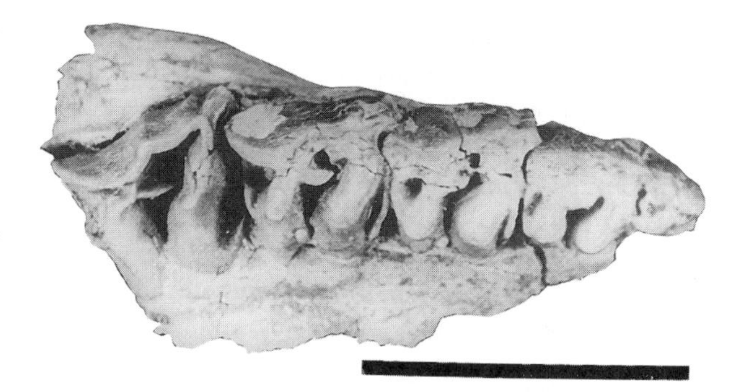

Figure 12.16. Left upper toothrow with DP1–M1 06-AKK-0084 of a subadult *Stephanorhinus pikermiensis* from Upper Kavakdere. MTA collection. Scale bar = 10 cm.

Discussion. The incisor AS.89.353 is unequivocal evidence for *Stephanorhinus*, as *Ceratotherium* lacks incisors entirely and the i1 of the aceratherines, if present at all, is much smaller. With this confirmation, the identification of the other specimens described above may also be regarded as secure. *Stephanorhinus* was evidently present as a rare taxon alongside the somewhat larger and more graviportal *Ceratotherium*.

The large two-horned Rhinocerotini of the later Neogene have long been placed in the wastebasket taxon *Dicerorhinus*, but gradually, over the past several decades, they have been split between *Stephanorhinus* Kretzoi 1942 (extended down from the early Pleistocene type species *S. etruscus*) and *Lartetotherium* Ginsburg 1974 (extended up from the middle Miocene type species *L. sansaniense*). This process has happened by diffusion rather than taxonomic revision, and the result is confused. We have provisionally used *Stephanorhinus* here to emphasize the similarity of the *S. pikerminensis* with the Pliocene *S. megarhinus* and the Pleistocene *S. kirchbergensis,* all part of a well-defined and close-knit clade or lineage (Fortelius et al. 1993), but acknowledge that use of Toula's (1906) name *Dihoplus* is also possible for the late Miocene species. The status of the east Asian *S. orientalis* relative to the roughly coeval *S. pikermiensis* (which has priority) remains unresolved, but the former appears to be of distinctly larger body size throughout its stratigraphic range (unpublished research by M. Fortelius).

We must mention here that Cerdeño's (1995) cladistic analysis proposes a major reinterpretation of rhinoceros taxonomy involving *Stephanorhinus*. Briefly, her preferred cladogram unites *Stephanorhinus*, restricted to dolichocephalic forms, with the dolichocephalic elasmotherines *Ninxiatherium* and *Elasmotherium,* based only on two of the most homoplastic characters imaginable for the Rhinocerotidae: skull length and loss of upper incisors (Osborn 1903; Heissig 1981). It does not help that Cerdeño splits skull lengthening into "normal zygomatic width," "dolichocephaly," "long nasal length," and "backward inclination of ascending ramus [of the mandible]," because all these categories record the same basic (secondary) lengthening of the skull. Neither does it help that the character state for loss of I2 is missing for *Lartetotherium* in her matrix. For this particular aspect, Cerdeño's analysis effectively substitutes superficial and homoplastic similarity for the detailed evidence from cranial and dental morphology that supports the conventional interpretation (Fortelius and Heissig 1989; Fortelius et al. 1993). We find the proposed reclassification difficult to accept, but a full refutation is beyond the scope of this chapter. For the elasmotheres, Cerdeño's classification was decisively rejected by Antoine (2000).

Ceratotherium neumayri (Osborn 1900)

Taxonomy:

Dicerotini
Ceratotherium neumayri (Osborn 1900)

Restricted synonymy:

1975 *Diceros neumayri* Heissig table 8
1991 *Diceros neumayri* Sen p. 260
1996 *Ceratotherium* cf. *C. neumayri* Kappelman et al. table 6.2
1996 Rhinocerotidae indet. (large) Kappelman et al. table 6.2

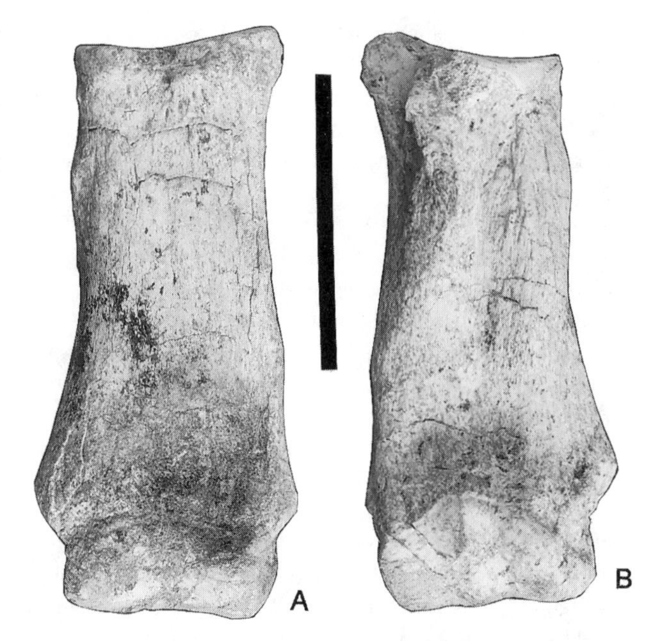

Figure 12.17. Left metatarsal III AS.94.1286 of *Ceratotherium neumayri* from Loc. 42 (Çobanpinar). (**A**) Plantar view; (**B**) palmar view. Scale bar = 10 cm.

Sinap Material. Loc. 12: humerus dex distal part AS.95.319, radioulna dex proximal part AS.95.348, magnum dex AS.95.333, tibia dex AS.95.339, astragalus dex AS.95.422; Loc. 49: p4 or p3 dex AS.91.187, associated astragalus and calcaneum AS.94.1362; Loc. 42: mt III sin AS.94.1286; MTA collection (Sinap): juvenile palate with milk molars 06-SIN-0138, juvenile maxilladex 06-SIN-0149, juvenile maxilla sin 06-SIN-0272, juvenile maxilla dex 06-KAY-21, juvenile mandibular ramus sin with dp1–m1 06-SIN-0134, juvenile mandibular ramus sin with dp1–m1 06-SIN-0273, mc III sin 06-KAY-10; MTA collection (Kavakdere): DP2–DP3 dex 06-AKK-031, humerus sin 06-AKK-032, mc II sin 06-AKK-034, mc III sin 06-AKK-035, mt III sin 06-AKK-036, mt IV sin 06-AKK-037; MTA collection (Çobanpinar): juvenile right maxilla with DP1–DP3, figured by Sen (1970, pl. II, fig. 1), since apparently lost; PDTFAU collection: unnumbered tibia from Şenyürek's Loc. F2 ("Aşağı yoncalık"); MNHN collection (Yassiören): dp3 sin TRQ 1048.

Age. The magnetostratigraphic age estimates are 9.6 Ma for Loc. 12 and 9.1 Ma for Loc. 49. Loc. 42 (Çobanpinar) is here placed in MN 13 (Van der Made, chapter 13, this volume). The other Sinap localities are almost certainly from the middle Sinap member, with an age span of 10.7–9.3 Ma (Kappelman et al., chapter 2, this volume). The taxon, or more properly lineage, thus has a range in the Sinap Formation of ~11–6 m. y.

Remarks. The milk upper dentitions from the middle Sinap localities feature a large DP1; a distally displaced paracone rib not only on DP2 but on the posterior milk molars as well; and a mesiolingually projecting, unconstricted protocone base, all characteristics of the Dicerotini. The specimen from Çobanpinar also has a large DP1 but differs from the earlier form in having strongly developed metacone styli on DP2 and DP3 (cf. Heissig 1975, p. 148). The mandibles show equally distinct characters: deep rami that taper strongly toward anterior, large dp1, shallow buccal folding, and relatively large hypolophid of the lower cheek teeth.

The humerus 06-AKK-032 is very short and stocky, easily distinguished from the more slender humerus of *Stephanorhinus*. The astragalus AS.95.348 from Loc. 12 has a trochlea with quasi-equal lateral and medial ridges, unlike the markedly asymmetrical trochlea of *Stephanorhinus* and corresponding to the distal articulation of tibia AS.95.339 from the same locality. The postcrania from Loc. 12 are assigned to *Ceratotherium* primarily by size, as only one large species is known to occur there. The astragalus AS.94.1362 from Loc. 49 is similar to the one from Loc. 12 but is larger. The metapodials are short and plantopalmarly flattened, as is typical of the genus, including the living *C. simum*. They differ from *Stephanorhinus* especially in that the distal trochlea has low relief and a broad, shallow keel (fig. 12.17).

Discussion. It appears that several related taxa of Dicerotini are found in the eastern Mediterranean late Miocene (cf. Heissig 1975), but pending a review of the group, we have conservatively placed them all in the conventional taxon, following Kaya (1994). The material from Loc. 12 and Loc. 1 (Sinap project as well as the MTA collection) seems to be consistently of a smaller size than the material from Loc. 49 and later. It is not possible here to judge whether a single evolving lineage underwent size increase at the beginning of the Turolian, or whether a second, replacing taxon immigrated at this time. Kaya (1994) interpreted the Anatolian material to represent a single lineage that increased in size from MN 9 to MN 12. Tshiskarishvili (1987) described the Vallesian form from Eldari-2 in the Caucasus as a separate species, *Diceros gabuniai*.

General Discussion

Chronology and Correlation

A range chart of the Sinap rhinoceroses is given in table 12.1. The rhinoceros assemblage from İnönü I, with *Brachypotherium brachypus, Hoploaceratherium tetradactylum,* and *Hispanotherium grimmi,* is similar to that of Paşalar (Fortelius 1990) and identical to that of Çandir (Heissig 1976). It thus seems to be representative for the early-middle Miocene transition of Anatolia as a whole, and perhaps western Asia generally. The European immigrant *Alicornops simorrensis* appears in the latest middle Miocene, apparently replacing *Hoploaceratherium* (Heissig 1996), but is at most ambiguously recorded at Sinap.

These Anatolian localities have usually been placed in MN 6 (Mein 1989; Steininger et al. 1996). If Locs. 24 and 24A (İnönü I) are indeed closely related to the volcanic events of ~15–16 Ma (Kappelman et al. 1996), these localities would be close to the MN 5/6 boundary in the correlations of Steininger et al. (1996). If the calibration of Krijgsman et al. (1996) is used, İnönü I is placed deep within MN 5. This raises the question of provincial diachrony in the appearance and disappearance of taxa, as one of the most characteristic MN 6 species, *Listriodon splendens,* is present at the locality. *Listriodon splendens* is known to disappear from the eastern Europe and the eastern Mediterranean one MN unit before it becomes extinct in western and central Europe, and it seems quite likely that it also arrived there earlier, given the probable south Asian origin of the group (Fortelius et al. 1996a). For the rhinoceroses, the situation is ambiguous in that *Hispanotherium* disappears from western Europe before MN 6, *Brachypotherium brachypus* has a long range spanning MN 5–6 in a large area, whereas *Hoploaceratherium tetradactylum* is only recorded in western Europe from Sansan, France, the type locality of MN 6 (Heissig 1996; NOW database March 2000). The question of diachrony and the MN system is too complicated to be discussed further here, but it should be mentioned that Alroy et al. (1998) found no evidence of directional diachrony in the MN system as a whole, as represented in the NOW database. A conservative interpretation of the evidence for İnönü I seems to be that it should be placed close to the MN 5/6 transition, always keeping in mind that the MN units strictly speaking do not have boundaries that can be expressed in units of time (de Bruijn et al. 1992).

The range of *Acerorhinus zernowi* at Sinap extends the MN 9–10 given by Heissig (1996) trivially if at all. The derived *Acerorhinus* from the Upper Kavakdere localities (MN 11/12 boundary or MN 11) extends the range of the genus in Anatolia and points to a continuity between the west and east Asian populations.

The primitive cf. *Chilotherium* represents both a stage of evolution and an age interval preceding the appearance of true *Chilotherium* in the eastern Mediterranean in MN 10 (cf. Heissig 1996). True *Chilotherium* first appears at Sinap at Loc. 49, and appears to have entered during the hiatus between Loc. 1 at 9.3 Ma and Loc. 49 at 9.1 Ma, probably within the temporal equivalent of later MN 10.

The name *Chilotherium kiliasi* (Geraads and Koufos 1990) is used here for the first time for material other than the hypodigm from Pentalophos I. As discussed above, the name may be a synonym of *C. samium* (Weber 1905), based on problematical specimens from an unknown horizon of Samos. Heissig (1996) recognized *C. samium* from the eastern Mediterranean from MN 10/11 to MN 11, a short interval that matches the occurrence of *C. kiliasi* at Pentalophos I

Table 12.1. Range Chart of Sinap Rhinoceros Taxa Identified at the Species Level

Sinap Locality	24	72	12	51	1	49	34	26	42
Age (Ma)	15–16?	10.1	9.6	?	9.3	9.1	8.4	8.1	~6
Brachypotherium brachypus	X								
Hoploaceratherium tetradactylum	O								
Begertherium grimmi	X								
Acerorhinus zernowi			O		O	X			
Acerorhinus sp. nov.								X	
cf. *Chilotherium* sp. (primitive)		X	XXXX	X	X				
Chilotherium kiliasi						X			
Ch. cf. *C. habereri*						X			
Chilotherium indet. (derived)							X	X	?
Stephanorhinus pikermiensis								X	
Ceratotherium neumayri			X		O	X		X	X

Notes: X, occurrences documented during the Sinap project; O, occurrences known only from previous collections; ?, questionable occurrences. Loc. 51 was interpolated to minimize range extensions of the occurring taxa.

Source: Lunkka et al. (1999).

and Loc. 49 perfectly. The mammal fauna of Pentalophos I, the type locality of *C. kiliasi,* is quite similar to that of Loc. 49: of 15 large mammal genera at Pentalophos, seven or eight are also found at Loc. 49, at least four of them represented by the same species (NOW database, March 2000). The faunal dating of Pentalophos I is problematic, but an age somewhat earlier than that of the MN 10 localities of the Axios valley has been proposed (de Bonis and Koufos 1999). The presence of *Dinocrocuta gigantea* at Pentalophos I (de Bonis and Koufos 1999), cited in favor of an earlier Vallesian age, does not necessarily constitute a difference from Loc. 49, which has an indeterminate percrocutid (Viranta and Werdelin, chapter 8, this volume). The magnetostratigraphic age of Loc. 49 at 9.1 Ma is just after the temporal equivalent of the MN 10/11 transition, according to the correlation of Steininger et al. (1996) or within late MN 10, according to Krijgsman et al. (1996). The presence at Loc. 49 of a second, rare, and more derived species of *Chilotherium* and the apparently greater degree of reduction of p2 in the Loc. 49 specimens of *Chilotherium kiliasi* are consistent with (but certainly do not prove) the interpretation that Loc. 49 is somewhat younger than Pentalophos I. The presences at both localities of *Chilotherium kiliasi* together with *Acerorhinus zernowi* and *Ceratotherium neumayri* might be taken as an indication that the difference in age cannot be a major one, however.

Chilotherium kowalevskii is one of the more distinct species of *Chilotherium* and a likely identification for at least some of the Upper Kavakdere *Chilotherium* material. The type locality is Grebeniki, with an MN 11–12 correlation (NOW database March 2000), and Heissig (1996) gives the range in the eastern Mediterranean as MN 10–11 to MN 11–12. The magnetostratigraphic correlation of the Upper Kavakdere localities is 8.4–8.1 Ma, spanning the MN 11/MN 12 boundary, according to Steininger et al. (1996) or within MN 11, according to Krijgsman et al. (1996).

The hypsodont *Chilotherium* of Loc. 49 is based on only two specimens, but as already discussed, it is clearly distinct from the Kavakdere form and represents the form referred to *C. habereri* by Heissig (1975, 1996). The range given by Heissig (1996) for *C. habereri* in the eastern Mediterranean is MN 10–11, which comfortably includes the magnetostratigraphic age estimate of Loc. 49 (9.1 Ma, which is within MN 10) (Steininger et al. 1996; Krijgsman et al. 1996). We use the name *Chilotherium* cf. *C. habereri* for this form, in acknowledgment of differences from the Chinese *C. habereri,* and suggest that it may be related to *C. persiae.*

Heissig (1996) lists *Stephanorhinus pikermiensis* only from Samos and Pikermi. The taxon is listed as '"*Dicerorhinus*" *schleiermacheri*' by Bernor et al. (1996b) from Pikermi (8.3–8.2 Ma, correlated with MN 11/12) and from Samos Main Bone Beds (≥7.1 Ma, correlated with MN 12). The occurrence of the species in the upper (8.1 Ma) level of Upper Kavakdere matches this range well.

Ceratotherium neumayri has a very long range in the eastern Mediterranean, MN 9 to MN 12–13, according to Heissig (1996). More or less the same range (Loc. 12 to Loc. 42) is represented in the Sinap Formation, but the material is unfortunately too incomplete to allow study of the evolution of this clade. All that can be said is that the early material seems smaller than the late material, as already noted by Heissig (1975) and recently confirmed by Kaya (1994), and that the material from Loc. 42 shows dental change in the direction of the Pliocene and Recent plagiolophodont representatives of the genus. The changes seen are of a magnitude that by common large mammal standards would justify recognition of separate morphospecies and possibly genera, and it seems that closer study of this group might be rewarding.

Paleoecology

The rhinoceroses of the Sinap Formation record part of the substantial faunal changes that took place from the beginning of the middle Miocene to the end of the late Miocene. The general trend is the same as for other large herbivores: a shift toward larger and more hypsodont species, evidently better adapted to cope with increasingly seasonal environments and their tougher and more abrasive forage.

It is unfortunate that there is a major gap in the record between the early middle Miocene Loc. 24 and 24A (İnönü I) and the MN 9 localities of the middle Sinap member, because a complete turnover of the rhinoceros fauna occurred during this missing interval. Before the gap, there is an ecologically diverse assemblage of species at Sinap, including the fairly generalized and brachydont *Hoploaceratherium,* the large and short-legged mesodont *Brachypotherium,* and the cursorial and hypsodont *Hispanotherium.*

When we first pick up the rhinoceros record after the gap, we see an entirely different assemblage at Sinap, dominated by aceratherine species, which are probably descendants, in a broad sense, of *Hoploaceratherium* and augmented by a recent immigrant from Africa, the large and hypsodont *Ceratotherium.*

The stratigraphic and zoogeographic evidence summarized in the taxonomic discussion (above) suggests that an *Acerorhinus zernowi*-like form evolved in the late middle Miocene of Central Asia (Cerdeño 1996), being essentially a somewhat more robust version of earlier forms such as *Hoploaceratherium tetradactylum* and Borissiak's (1927) middle Miocene Turgai aceratherine (both of which Cerdeño refers to *Acerorhinus*). This form gave rise to a lineage of forms increasingly convergent on *Chilotherium,* from *A. tsaidamensis* and *A. hezhengensis* to *A. palaeosinensis* and, perhaps, specialized forms like *A. cornutus* and *Sinorhinus brancoi* (Ringström 1924; Qiu et al. 1988; Heissig 1989, 1996). The main trends include increased folding of the dental enamel, shortening of the nasals, a shifting of the eye to a more elevated location, and, probably as part of the same complex, a flattening of the skull roof. The postcranial skeleton was eventually reduced to *Chilotherium*-like proportions at the stage of *A. palaeosinensis* (Ringström

1924), but the manus seems to have remained pentadactyl and the metapodials seem to have been both more flattened and more splayed than in *Chilotherium,* as discussed above. *Acerorhinus fuguensis,* described from the late Miocene of Fugu by Deng (2000), appears to be a large form of *A. palaeosinense,* with allometric development of a sagittal crest.

It appears that the earlier form of *Acerorhinus* represented at Sinap is somewhat derived relative to *A. zernowi,* whereas the later form shows a mixture of unique characters and similarities with *A. palaeosinense* and the probably conspecific *"Aceratherium incisivum"* from Tchobrouchi (Pavlow 1915). This suggests a continuously evolving population over much of Asia in the earlier part of the late Miocene. The reason for the greater diversification of this clade in east Asia may be related to the absence of the African Dicerotini there.

The most primitive forms of *Chilotherium* appear to have been little if at all removed from the primitive aceratherine lifestyle, at least as far as diet goes. The weak postcranial skeleton and the flat-topped skull with a high orbit found in the later forms could be interpreted to indicate a hippopotamus-like lifestyle, but, as in the case of *Brachypotherium,* no direct evidence seems to exist. A problem with this reasoning is that it makes it difficult to exclude the more derived *Acerorhinus* from the "hippo guild," particularly in view of its flattened and strongly splayed metapodials. Sympatry of several hippo-like species is difficult to envision, even if the dental evidence points to dietary differences between the species. The lack of any modification of the choanae in these forms may also argue against an aquatic life style.

The material of *Ceratotherium* is too fragmentary to allow assessment of evolutionary change at Sinap, but Heissig (1975) and Kaya (1994) reported size increase in this lineage within Anatolia as a whole. The youngest material from the Sinap Formation comes from Loc. 42 (Çobanpinar) and shows dental changes in the direction of living *Ceratotherium simum,* a true grazer.

A large species of the Eurasian genus *Stephanorhinus,* more cursorial and less hypsodont than *Ceratotherium* but only slightly smaller, is rare but present in the MN 12 assemblage from Kavakdere.

The number of sympatric rhinoceros species is three at most levels, reaching four in the MN 10/11 and MN 12 assemblages. The middle Miocene assemblage is composed of a smallish, plesiomorphic browsing form, a large, short-legged mixed feeder, and a cursorial form, apparently a grazer or a mixed feeder leaning heavily toward the grazing end of the spectrum (Fortelius 1990). In the Vallesian assemblage from the middle Sinap member, only forms in the browser or browser-mixed feeder range are documented. There is still a spread of sizes, but most species are small to medium-sized, and only *Ceratotherium* is large. Loc. 49 at ~9.1 Ma ago has the same structure, with three smallish to medium-sized aceratherines and *Ceratotherium.* Two of the three aceratherines represent *Chilotherium,* which is already at this time beginning to evolve hypsodont cheek teeth. The main change from this level to Kavakdere at ~8.4–8.1 Ma is that only one *Chilotherium* is now present, and instead a second large form appears: *Stephanorhinus pikermiensis.* Compared with *Ceratotherium, Stephanorhinus* appears to be somewhat smaller, more cursorial, and less adapted to feeding on abrasive foods, but the material from Sinap is not sufficiently complete to address this question locally.

Perhaps the most interesting overall feature of the rhinoceros assemblages from the Sinap Formation is the absence in the middle Sinap member of forms clearly adapted to feeding on tough or abrasive vegetation. This is remarkable, as such forms were present both before and after that interval, and might be taken as evidence that unusually mesic (forested?) conditions prevailed during the early part of the late Miocene. This is also the interval during which the mammal taxonomic richness peaks and hominoid primates are recorded from the region.

Acknowledgments

We thank Prof. Dr. Berna Alpagut, Prof. Louis de Bonis, Dr. Elmar P. J. Heizmann, Dr. Jeremy J. Hooker, Dr. George Koufos, Dr. David Lordkipanidze, Dr. S. Sharapov, Dr. Solveig Stuenes, and Dr. Ilhan Temizsoy for permission to study material under their care and for discussions, and fellow members of the Sinap project for all the fun. Special thanks to Celâl Metin of Kazan, King among Taxi Drivers.

Appendix

Appendix Table 12.1. Measurements of Sinap Dental Specimens

Taxon/ Specimen	Tooth	Measurement (mm)			
		LB	LL	WD	WP
cf. *Chilotherium* indet. (primitive)					
AS.90.132	p4	38.8	37.9	26.6	24.2
	m1	39.4	40.3	26.8	25.4
	m2	42.1	42.9	26.1	25.1
Chilotherium kiliasi					
AS.93.963	P3	—[1]	31.5	44.8	46.6
	P4	—	31.0	47.0	48.6
	M1	—	35.0	48.3	49.6
	M2	45.5	37.4	46.9	53.3
	M3	50.9	41	—	46.9
AS.93.1074	M1	—	34.0	—	—
	M2	46.3	38.7	51.0	57.6
	M3	50.8	37.0	—	47.7
AS.93.809	p2	21.5	21.6	15.1	13.9
	p3	30.0	28.0	22.4	17.4
	p4	36.0	34.5	26.7	23.8
	m1	38.0	—	27.7	24.7
	m2	43.4	41.6	26.5	26.7
	m3	44.0	45.1	25.4	25.5
AS.94.566	p2	20.5	20.0	15.1	13.0
	p3	27.7	—	—	
	p4	36.0	34.0	25.5	21.6
	m1	40.4	40.1	26.2	25.2
	m2	43.7	42.2	23.3	23.4
AS.93.810	p3	29.6	27.2	22.2	19.1
	p4	34.5	34.0	26.4	23.0
	m1	39.0	36.0	28.1	24.4
	m2	42.2	41.2	26.7	24.9
	m3	42.6	44.3	24.6	24.4
AS.93.815	p2	22.8	21.7	15.3	14.0
	p3	28.6	28.0	22.8	20.1
	p4	33.6	33.6	26.1	22.5
	m1	39.6	36.9	26.6	25.8
	m2	41.8	40.2	24.7	26.2
	m3	41.0	44.0	23.4	23.5
AS.91.701	m2	42.6	41.2	25.6	26.3
	m3	44.5	46.1	23.1	24.6
AS.94.537	m2	43.1	44.3	23.8	23.7
	m3	41.6	41.1	22.0	22.0
AS.94.316	m3	45.0	42.9	26.4	25.5
	m4	—	38.6	19.0	22.7
AS.93.1193	dp2	31.2	30.6	15.2	14.3
	dp3	39.5	37.9	18.5	18.6
	dp4	42.7	38.7	22.0	22.2
Chilotherium cf. *C. habereri*					
AS.90.313	dp2	26.9	27.2	14.0	12.8
	dp3	35.8	37.6	18.0	17.2
	dp4	40.1	41.8	21.5	19.2
	m1	41.0	—	24.0	—
AS.92.155	m3	—	48.5	23.2	24.2
Chilotherium indet. Upper Kavakdere					
AS.92.602	p3	32.0	—	—	19.6
	p4	36.0	—	—	22.1
	m1	39.0	—	—	—
	m2	40.2	—	24.4	25.2
	m3	44.5	44.1	23.0	24.9
Acerorhinus zernowi					
AS.93.823	P4	36.0	33.2	51.6	51.7
	M1	43.0	31.4	48.8	54.2
	M2	49.0	35.9	44.9	52.5
	M3	48.0	—	—	48.1
AS.95.747	P4	39.8	35.6	47.1	49.0
	M1	42.9	34.4	47.8	49.6
	M2	46.1	38.8	43.2	48.7
	M3	47.1	41.1	—	46.1
AS.94.554	M1	41.0	32.6	45.5	50.4
	M2	45.0	35.6	43.7	51.9
	M3	52.0	34.4	—	47.0
AS.92.150	p2	34.6	34.3	22.2	19.6
	p3	35.7	36.5	26.7	24.0
	p4	39.2	37.9	28.4	26.2
	m1	41.7	39.6	27.8	28.2
	m2	43.9	45.1	27.7	27.6
	m3	45.3	47.4	25.9	28.1
AS.94.315	p3	36.1	34.2	25.5	24.4
	p4	36.5	34.7	25.4	24.4
	m1	38.7	39.2	25.8	27.5
	m2	44.9	41.2	27.5	28.1
	m3	45.0	44.1	25.2	25.9
Acerorhinus sp. nov.					
AS.89.279	p2	—	32.9	22.3	18.2
AS.90.184	p2	36.7	34.4	23.1	19.7
Ceratotherium neymayri					
AS.91.187	p4/3	44.0	—	33.2	29.1

[1]—, Cannot be measured.

Appendix Table 12.2A. Statistics of Dental Measurements of *Chilotherium kiliasi* and *Acerorhinus zernowi* from Sinap, Upper Teeth

Variable/ Taxon	P4				M1				M2				M3		
	BL	LL	WP	WD	BL	LL	WP	WD	BL	LL	WP	WD	BL	LL	WP
hilotherium kiliasi															
N of cases	1	2	2	2	0	2	1	1	2	2	3	3	2	2	2
Minimum	37	31	48.6	47	—[1]	34	49.6	48.3	45.5	37.4	53.3	46.9	50.8	37	46.9
Maximum	37	32.7	53.3	51.1	—	35	49.6	48.3	46.3	38.7	57.6	51	50.9	41	47.7
Median	37	31.9	51	49.1	—	34.5	49.6	48.3	45.9	38.1	56.6	50.3	50.9	39	47.3
Mean	37	31.9	51	49.1	—	34.5	49.6	48.3	45.9	38.1	55.8	49.4	50.9	39	47.3
C.V.	1	0.04	0.07	0.06	—	0.02	1	1	0.01	0.02	0.04	0.04	0	0.07	0.01
Acerorhinus zernowi															
N of cases	2	2	2	2	3	3	3	3	3	3	3	3	4	3	4
Minimum	36	33.2	49	47.1	41	31.4	49.6	45.5	45	35.6	48.7	43.2	47.1	34.4	46.1
Maximum	39.8	35.6	51.7	51.6	43	34.4	54.2	48.8	49	38.8	52.5	44.9	52	42.6	48.1
Median	37.9	34.4	50.4	49.4	42.9	32.6	50.4	47.8	46.1	35.9	51.9	43.7	49.5	41.1	47.2
Mean	37.9	34.4	50.4	49.4	42.3	32.8	51.4	47.4	46.7	36.8	51	43.9	49.5	39.4	47.1
C.V.	0.07	0.05	0.04	0.06	0.03	0.05	0.05	0.04	0.04	0.05	0.04	0.02	0.05	0.11	0.02

Appendix Table 12.2B. Statistics of Dental Measurements of *Chilotherium kiliasi* and *Acerorhinus zernowi* from

Variable/ Taxon	p2				p3				p4			
	BL	LL	WP	WD	BL	LL	WP	WD	BL	LL	WP	WD
Chilotherium kiliasi												
N of cases	3	3	3	3	4	3	3	3	4	4	4	4
Minimum	20.5	20	13	15.1	27.7	27.2	17.4	22.2	33.6	33.6	21.6	25.5
Maximum	22.8	21.7	14	15.3	30	28	20.1	22.8	36	34.5	23.8	26.7
Median	21.5	21.6	13.9	15.1	29.1	28	19.1	22.4	35.3	34	22.8	26.3
Mean	21.6	21.1	13.6	15.2	29	27.7	18.9	22.5	35	34	22.7	26.2
C.V.	0.05	0.05	0.04	0.01	0.04	0.02	0.07	0.01	0.03	0.01	0.04	0.02
Acerorhinus zernowi												
N of cases	4	4	4	4	2	2	2	2	3	3	3	3
Minimum	29.6	28.4	18.6	20.5	35.7	34.2	24	25.5	36.5	34.7	24.3	25.4
Maximum	34.6	34.3	21	22.2	36.1	36.5	24.4	26.7	39.2	37.9	26.2	28.4
Median	32.1	30.1	19.6	21.3	35.9	35.4	24.2	26.1	38.5	36.4	24.4	27.1
Mean	32.1	30.7	19.7	21.3	35.9	35.4	24.2	26.1	38.1	36.3	25	27
C.V.	0.08	0.08	0.05	0.04	0.01	0.05	0.01	0.03	0.04	0.04	0.04	0.06

Sinap, Lower Teeth

m1				m2				m3			
BL	LL	WP	WD	BL	LL	WP	WD	BL	LL	WP	WD
4	3	4	4	6	6	6	6	9	9	9	9
38	36	24.4	26.2	41.8	40.2	23.4	23.3	41	41.1	22	22
40.4	40.1	25.8	28.1	43.7	44.3	26.7	26.7	45	46.1	25.5	26.4
39.3	36.9	25	27.2	42.9	41.4	25.6	25.2	43.4	44.3	24.4	23.4
39.3	37.7	25	27.2	42.8	41.8	25.2	25.1	43	43.9	24.1	23.9
0.03	0.06	0.02	0.03	0.02	0.03	0.06	0.06	0.03	0.04	0.05	0.06
3	2	2	2	3	2	2	2	2	2	3	2
38.7	39.2	27.5	25.8	42	41.2	27.6	27.5	45	44.1	25.9	25.2
42	39.6	28.2	27.8	44.9	45.1	28.1	27.7	45.3	47.4	28.1	25.9
41.7	39.4	27.9	26.8	43.9	43.2	27.9	27.6	45.2	45.8	27.8	25.6
40.8	39.4	27.9	26.8	43.6	43.2	27.9	27.6	45.2	45.8	27.3	25.6
0.04	0.01	0.02	0.05	0.03	0.06	0.01	0.01	0	0.05	0.04	0.02

Appendix Table 12.3. Measurements of Postcranial Material from Sinap

Bone/ Taxon	Locality	Number	Side	Measurement (mm)								
				L	WP	APP	WS	APS	WD	APD	JWD	JAPD
Humerus												
cf. *Chilotherium* sp. primitive	12	AS.93.1210	rt.	—	—	—	—	—	—	—	—	—
	51	AS.90.134	rt.	350.0	145.0	118.0	53.0	58.0	128.0	89.0	92.0	—
Ceratotherium neumayri	12	AS.95.319	rt.	—	—	—	66	65	152	—	99	—
Radius												
cf. *Chilotherium* sp. primitive	12	AS.93.1210	rt.	—	77.0	40.0	—	—	—	—	—	—
Acerorhinus zernowi	12	06-KAY-14	rt.	290.4	83.1	55.9	42.7	30.8	88.0	55.2	—	—
Ceratotherium neumayri	12	06-KAY-22	rt.	—	107.4	67.4	—	—	—	—	—	—
	12	06-KAY-8	rt.	363.0	110.0	71.0	58.4	44.3	118.7	73.0	—	—
	12	AS.95.348	rt.	387.0	102.0	84.0	56.0	41.0	100.0	76.0	89.0	48.0
Tibia												
Chilotherium indet. Upper Kavakdere	Kavakdere	06-AKK-017	lt.	276.0	123.7	103.3	42.8	43.4	83.2	58.6	—	—
	33	AS.90.52	rt.	—	—	—	—	—	85.0	57.0	66.0	—
Acerorhinus zernowi	49	AS.92.138	lt.	—	—	—	—	—	93.3	62.1	67.6	—
Ceratotherium neumayri	Oz F2	None	.	400.0	105.0	—	59.0	53.0	103.0	80.0	—	—
	12	AS.95.339	lt.	382.0	—	—	61.0	50.0	105.0	80.0	77.0	63.0
mc II												
cf. *Chilotherium* sp. primitive	12	AS.93.1210	rt.	105.0	39.0	34.0	30.0	15.0	34.0	33.0	31.0	33.0
Ceratotherium neumayri	Kavakdere	06-AKK-034	lt.	108.2	32.0	47.2	40.4	21.2	48.7	41.4	41.6	—
mc III												
cf. *Chilotherium* sp. primitive	12	AS.93.1210	rt.	122.0	44.0	41.0	34.0	15.0	44.0	36.0	40.0	36.0
Chilotherium indet. Upper Kavakdere	Kavakdere	06-AKK-015	rt.	—	—	38.9	36.5	16.6	45.9	33.6	40.2	—
	Kavakdere	06-AKK-029	rt.	135.0	56.6	48.7	41.4	18.5	51.5	40.4	45.3	—
Acerorhinus sp. nov.	26	AS.91.229	rt.	—	55.0	46.0	—	—	—	—	—	—
Ceratotherium neumayri	12	06-KAY-10	lt.	—	60.3	49.2	—	22.3	—	—	—	—
	Kavakdere	06-AKK-035	lt.	—	65.0	51.7	45.7	25.4	—	—	—	—
Begertherium grimmi	Inönü I	06-inö-77/1773	rt.	169.9	54.4	41.0	42.7	19.4	53.7	38.4	47.0	—
	24A	AS.92.664	lt.	—	57.0	—	40.0	21.0	—	—	—	—
mc IV												
cf. *Chilotherium* sp. primitive	12	AS.93.1210	rt.	97.0	31.0	33.0	25.0	14.0	31.0	33.0	29.0	33.0
Chilotherium indet. Upper Kavakdere	Kavakdere	06-AKK-016	rt.	92.5	32.9	35.7	25.1	13.7	31.1	28.8	28.6	—
	Kavakdere	06-AKK-030	rt.	89.4	32.8	39.7	27.5	21.0	31.8	36.4	29.7	—
Acerorhinus sp. nov.	26	AS.90.241	rt.	95.5	36.3	37.1	29.0	14.0	36.0	26.6	32.7	26.6

mc V												
cf. *Chilotherium* sp. primitive	12	AS.93.1210	rt.	61.0	14.5	19.5	13.5	9.8	23.0	19.1	17.9	19.1
mt II												
cf. *Chilotherium* sp. primitive	12	nonr	rt.	100.7	21.4	31.1	18.9	16.9	28.4	31.9	27.5	31.9
Chilotherium indet. Upper Kavakdere	Kavakdere	06-AKK-020	rt.	85.2	26.0	33.2	25.4	19.0	32.8	30.6	28.8	—
mt III												
cf. *Chilotherium* sp. primitive	72	AS.91.312	rt.	106.0	35.2	—	30.1	12.6	40.7	29.4	33.8	—
Chilotherium indet. Upper Kavakdere	Kavakdere	06-AKK-021	rt.	101.5	40.6	—	34.5	18.1	46.1	33.7	37.8	—
Ceratotherium neumayri	Kavakdere	06-AKK-036	lt.	175.3	57.9	52.2	49.7	26.5	68.8	47.3	56.1	—
	42	AS.94.1286	lt.	170.0	63.0	47.0	57.0	24.0	80.0	42.0	58.0	42.0
Brachypotherium brachypus	Inönü I	06-INÖ-77/1607	rt.	116.1	53.3	49.4	44.2	19.8	63.1	41.4	51.7	—
Brachypotherium sp.?	125	AS.94.143	.	86.0	—	38.0	39.0	23.0	49.0	38.0	44.0	38.0
mt IV												
Chilotherium indet. Upper Kavakdere	Kavakdere	06-AKK-022	rt.	83.4	36.3	33.5	26.9	15.8	33.3	33.5	28.1	—
Ceratotherium neumayri	Kavakdere	06-AKK-037	lt.	160.0	45.5	49.8	34.4	29.1	48.0	48.8	41.3	—
Begertherium grimmi	24A	AS.91.400	lt.	—	41.6	43.7	—	—	—	—	—	—
Astragalus				**L**	**W**	**AP**	**WP**	**APP**	**JWP**	**JAPP**		
Chilotherium indet. Upper Kavakdere	Kavakdere	06-AKK-018	lt.	67.1	77.0	43.9	—	—	—	—		
	33	AS.89.215	rt.	67.0	79.0	—	65.0	55.0	69.0	37.0		
	26	AS.89.286	lt.	64.0	76.0	—	61.0	51.0	66.0	33.0		
Acerorhinus zernowi	49	AS.91.731	lt.	61.5	78.5	51.2	65.9	50.0	69.3	39.5		
Ceratotherium neumayri	12	AS.95.423	rt.	77	93	—	82	68	74	52		
	49	AS.94.1362a	—	95.0	100.0	—	93.0	—	84.0	58.0		
Brachypotherium sp.?	125	AS.95.454	rt.	—	—	—	—	—	—	—		
Begertherium grimmi	24A	AS.92.667	lt.	82.0	88.0	55.0	78.0	—	77.0	44.0		

Literature Cited

Antoine, P.-O., 2002, Phylogénie et evolution des Elasmotheriina (Mammalia, Rhinocerotidae): Mémoires du Muséum National d'Histoire Naturelle, v. 188, pp. 1–249.

Alroy, J., R. L. Bernor, M. Fortelius, and L. Werdelin, 1998, The MN system—regional or continental?: Mitteilungen der Bayerischen Staatssammlung von Paläontologie und Historische Geologie, v. 38, pp. 243–258.

Bernor, R. L., F. Fahlbusch, H.-W. Mittmann, and S. Rietschel, 1996a, The evolution of western Eurasian Neogene land mammal faunas: The 1992 Schloss Reisensburg workshop concept, *in* R. L. Bernor, V. Fahlbusch, and H.-W. Mittmann, eds., The evolution of western Eurasian Neogene mammal faunas: New York, Columbia University Press, pp. 1–4.

Bernor, R. L., N. Solounias, C. C. Swisher III, and J. Van Couvering, 1996b, The correlation of three classical "Pikermian" mammal faunas—Maragheh, Samos, and Pikermi—with the European MN unit system, *in* R. L. Bernor, V. Fahlbusch, and H.-W. Mittmann, eds., The evolution of western Eurasian Neogene mammal faunas: New York, Columbia University Press, pp. 137–154.

Bernor, R. L., V. Fahlbusch, P. Andrews, H. de Bruijn, M. Fortelius, F. Rögl, F. F. Steininger, and L. Werdelin, 1996c, The evolution of western Eurasian Neogene mammal faunas: A chronologic, systematic, biogeographic and paleoenvironmental synthesis, *in* R. L. Bernor, V. Fahlbusch, and H.-W. Mittmann, eds., The evolution of western Eurasian Neogene mammal faunas: New York, Columbia University Press, pp. 449–471.

Bohlin, B., 1937, Eine Tertiäre säugetier-faunaaus Tsaidam: Palaeontologia Sinica, v. 14, no. 1, pp. 1–109.

Bonis, L. de, and G. Koufos, 1999, The Miocene large mammal succession in Greece, *in* J. Agustí, L. Rook, and A. Andrews, eds., Evolution of Neogene terrestrial ecosystems in Europe: Cambridge, Cambridge University Press, pp. 205–237.

Borissiak, A., 1914, Mammifères fossiles de Sebastopol. I.: Trudy Geologicheskago Komiteta Novaja Seria, v. 87, pp. 1–154.

Borissiak, A., 1915, Mammifères fossiles de Sebastopol. II.: Trudy Geologicheskago Komiteta Novaja Seria, v. 137, pp. 1–47.

Borissiak, A., 1927, *Aceratherium depereti* n. sp. from the Jilancik beds: Isvestia Akademii Nauk SSSR, v. 21, pp. 769–786.

Bruijn, H. de, R. Daams, G. Daxner-Höck, V. Fahlbusch, L. Ginsburg, P. Mein, and J. Morales, 1992, Report of the RCMNS working group on fossil mammals, Reisensburg 1990: Newsletters in Stratigraphy, v. 26, pp. 65–118.

Cerdeño, E., 1995, Cladistic analysis of the family Rhinocerotidae (Perissodactyla): American Museum Novitates, v. 3143, pp. 1–25.

Cerdeño, E., 1996, Rhinocerotidae from the Middle Miocene of the Tung-Gur Formation, Inner Mongolia, China: American Museum Novitates, v. 3184, pp. 1–43.

Deng, T., 2000, A new species of *Acerorhinus* (Perissodactyla, Rhinocerotidae) from the Late Miocene in Fugu, Shaanxi, China: Vertebrata Palasiatica, v. 38, no. 3, pp. 203–217.

Fortelius, M., 1990, Rhinocerotidae from Pasalar, middle Miocene of Anatolia (Turkey): Journal of Human Evolution, v. 19, pp. 489–508.

Fortelius, M., and K. Heissig, 1989, The phylogenetic relationships of the Elasmotherini: Mitteilungen der Bayerischen Staatssammlung von Paläontologie und Historische Geologie, v. 29, pp. 227–233.

Fortelius, M., and N. Solounias, 2000, Functional characterization of ungulate molars using the abrasion-attrition wear gradient: American Museum Novitates, v. 3301, pp. 1–38.

Fortelius, M., P. Mazza, and B. Sala, 1993, *Stephanorhinus* (Mammalia, Rhinocerotidae) of the European Pleistocene, with a revision of *S. etruscus* (Falconer, 1868): Palaeontographica Italica, v. 80, pp. 63–155.

Fortelius, M., J. van der Made, and R. L. Bernor, 1996a, Middle and Late Miocene Suoidea of central Europe and the eastern Mediterranean: Evolution, biogeography and palaeoecology, *in* R. L. Bernor, V. Fahlbusch, and H.-W. Mittmann, eds., The evolution of western Eurasian Neogene mammal faunas: New York, Columbia University Press, pp. 348–377.

Fortelius, M., J. van der Made, and R. L. Bernor, 1996b, A new listriodont suid, *Bunolistriodon meidamon* sp. nov., from the Middle Miocene of Anatolia: Journal of Vertebrate Paleontology, v. 16, pp. 149–164.

Geraads, D., and G. Koufos, 1990, Upper Miocene Rhinocerotidae from Pentalophos-1, Macedonia, Greece: Palaeontographica, v. A 210, pp. 151–168.

Guérin, C., 1980, Les rhinocéros (Mammalia, Perissodactyla) du Miocène terminal au Pleistocène supérieur en Europe Occidentale. Comparaison avec les espèces actuelles: Documents des Laboratoires de Geologie Lyon, v. 79, no. 1/2/3, pp. 1–1184.

Gürbüz, M., 1981, Inönü (KB Ankara) Orta Miyosenindeki *Hemicyon sansaniensis* (Ursidae) türünüm tanimlanmasi ve stratigrafik yayilimi: Türkiye Jeoloji Kurumu Bülteni C, v. 24, pp. 85–90.

Heissig, K., 1972, Paläontologische und geologische Untersuchungen im Tertiär von Pakistan. 5. Rhinocerotidae (Mamm.) aus den unteren und mittleren Siwalik-Schichten: Abhandlungen der Bayerischen Akademie der Wissenshaften, Mathematisch-Naturwissenschaftliche Klasse, NF, v. 152, pp. 1–112.

Hessig, K., 1973. Die Unterfamilien und Tribus der rezenten und fossilen Rhinocerotidae (Mammalia): Säugetierkundliche Mitteilungen v. 21, pp. 25–30.

Heissig, K., 1974, Neue Elasmotherini (Rhinocerotidae, Mammalia) aus dem Obermiozän Anatoliens: Mitteilungen der Bayerischen Staatssammlung von Paläontologie und Historische Geologie, v. 14, pp. 21–35.

Heissig, K., 1975, Rhinocerotidae aus dem Jungtertiär Anatoliens: Geologisches Jahrbuch B, v. 15, pp. 145–151.

Heissig, K., 1976, Rhinocerotidae (Mammalia) aus der Anchitherium-fauna Anatoliens: Geologisches Jahrbuch B, v. 19, pp. 3–121.

Heissig, K., 1981, Probleme bei der cladistischen Analyse einer Gruppe mit wenigen eindeutigen Apomorphien: Rhinocerotidae: Paläontologische Zeitschrift, v. 55, pp. 117–123.

Heissig, K., 1989, The Rhinocerotidae, *in* D. R. Prothero, and R. M. Schoch, eds., The evolution of Perissodactyls: New York, Oxford University Press, pp. 399–417.

Heissig, K., 1996, The stratigraphical range of fossil rhinoceroses in the late Neogene of Europe and the eastern Mediterranean, *in* R. L. Bernor, V. Fahlbusch, and H.-W. Mittmann, eds., The evolution of western Eurasian Neogene mammal faunas: New York, Columbia University Press, pp. 339–347.

Hünermann, K. A., 1989, Reconstruction des *Aceratherium* (Mammalia, Perissodactyla, Rhinocerotidae) aus dem Jungtertiär vom Höwenegg/Hegau (Baden-Württemberg, BRD): Zeitschrift der Geologischen Wissenschaften Berlin, v. 10, pp. 929–942.

Kappelman, J., S. Sen, M. Fortelius, A. Duncan, B. Alpagut, J. Crabaugh, A. Gentry, J. P. Lunkka, F. McDowell, N. Solounias, S. Viranta, and L. Werdelin, 1996, Chronology and biostratigraphy of the Miocene Sinap Formation of central Turkey, *in* R. L. Bernor, V. Fahlbusch, and H.-W. Mittmann, eds., The evolution of western Eurasian Neogene mammal faunas: New York, Columbia University Press, pp. 78–95.

Kaya, T., 1994, *Ceratotherium neumayri* (Rhinocerotidae, Mammalia) in the Upper Miocene of western Anatolia: Turkish Journal of Earth Sciences, v. 3, pp. 13–22.

Krijgsman, W., M. Garcés, C. G. Langereis, R. Daams, J. van Dam, A. J. van der Meulen, J. Agusti, and L. Cabrera, 1996, A new chronology for the middle to late Miocene continental record in Spain: Earth and Planetary Science Letters, v. 142, pp. 367–380.

Lunkka, J.-P., M. Fortelius, J. W. Kappelman, and S. Sen, 1999, Chronology and mammal faunas of the Miocene Sinap Formation, Turkey, *in* J. Agusti, P. Andrews, and L. Rook, eds., The evolution of Neogene terrestrial ecosystems in Europe. Hominoid evolution and climatic change in Europe, Volume 1: New York, Cambridge University Press, pp. 238–264.

Mein, P., 1989, Updating of MN zones, *in* E. H. Lindsay, V. Fahlbusch, and P. Mein, eds., European Neogene mammal chronology: New York, Plenum Press, pp. 73–90.

Mequenem, R. de, 1924, Contribution à l'etude des fossiles de Maragha: Annales de Paléontologie (Paris), v. 13, pp. 133–160, v. 14, pp. 1–36.

NOW, 2000, NOW (Neogene Old World) database: Can be accessed at http://www.Helsinki.fi/science/now.

Osborn, H. F., 1900, Phylogeny of the rhinoceroses of Europe: Bulletin of the American Museum of Natural History, v. 13, pp. 229–267.

Osborn, H. F., 1903, The extinct rhinoceroses: Memoirs of the American Museum of Natural History, v. 1, no. 3, pp. 75–164.

Öngür, T., 1976, Kizilcahamam, Camlidere, Celtikci ve Kazan dolayinin jeoloji durumu ve jeotermal enerji olanaklari, Unpublished Report, Maden Tetkik ve Arama Enstitüsü, Ankara.

Ozansoy, F., 1957, Faunes de mammifères du Tertiaire du Turquie et leurs révisions stratigraphiques: Bulletin of the Mineral Resource Exploration Institute of Turkey (Foreign Edition), v. 49, pp. 29–48.

Ozansoy, F., 1965 Études des gisements continentaux et de mammifères du Cénozoïque du Turquie: Mémoires de la Societé Géologique de France, Nouvelle Série, v. 44, pp. 1–92.

Qiu, Z., J. Xie, and D. Yan, 1988, A new chilothere skull from Hezheng, Gansu, China: Scientia Sinica, v. 31, pp. 493–502.

Pavlow, M., 1913, Mammifères tertiaires de la Nouvelle Russie. Avec un article géologique du Prof. A.P. Pavlow. 1-re Partie. Artiodactyla, Perissodactyla (*Aceratherium kowalevskii* n.s.): Nouveaux Mémoires de la Societé Impériale des Naturalistes de Moscou, v. 17, no. 3, pp. 1–68.

Pavlow, M., 1915, Mammifères tertiaires de la Nouvelle Russie, 2-e Partie: Nouveaux Mémoires de la Societé Impériale des Naturalistes de Moscou, v. 17, no. 4, pp. 1–78.

Ringström, T., 1924, Nashörner der hipparion-fauna nord-Chinas: Palaeontologica Sinica, v. 1, no. C4, pp. 1–156.

Saraç, G., 1994, The biostratigraphy and palaeontology of the Rhinocerotidae (Mammalia Perissodactyla) of the continental Neogene sediments in the Ankara region (Turkish, with an English abstract) [Ph.D. thesis]: Ankara, Turkey, Ankara University, 214 pp.

Sen, S., 1970, Türkiye Miosen ve Pliosen rhinoseros'larinin odontolojik özellekleri [M.Sc. Thesis]: Ankara, Turkey, Ankara University, 53 pp.

Sen, S., 1991, Stratigraphie, faunes de mammifères et magnétostratigraphie du Néogène de Sinap Tepe, province d'Ankara, Turquie: Bulletin de la Museum National d'Histoire Naturelle ser 4e, v. 12, no. C3/4, pp. 243–277.

Steininger, F. F., W. A. Berggren, D. V. Kent, R. L. Bernor, S. Sen, and J. Agusti, 1996, Circum-Mediterranean Neogene (Miocene-Pliocene) marine-continental chronologic correlations of European mammal units, *in* R. L. Bernor, V. Fahlbusch, and H.-W. Mittmann, eds., The evolution of western Eurasian Neogene mammal faunas, New York, Columbia University Press, pp. 7–46.

Tsiskarishvili, G. V., 1987, Pozdnetvetichnye nosorogi (Rhinocerotidae) Kavkaza (The late Tertiary rhinoceroses of the Caucasus): Tbilisi, Republic of Georgia, Metsniereba, 141 pp.

Weber, M. C., 1904, Über tertiäre Rhinocerotiden von der Insel Samos: Bulletin de la Societé des Naturalistes de Moscou, Nouvelle Série, v. 17, pp. 477—501.

Weber, M. C., 1905, Über tertiäre Rhinocerotiden von der Insel Samos II: Bulletin de la Societé des Naturalistes de Moscou, Nouvelle Série, v. 18, pp. 344–363.

13

Suoidea (Artiodactyla)

J. van der Made

Suoids from the lower and middle Sinap Formation were first reported by Ozansoy (1957b, 1965). The first revision was undertaken by Pickford and Ertürk (1979). Suoid material from Sinap has also been mentioned by Hünermann (1975), Sickenberg et al. (1975), and Fortelius et al. (1996b), and recently, material was described by Van der Made and Han (1994) and Van der Made (1996b, 1997a). Nevertheless, major parts of the various collections remained unstudied, even as recent collecting increased the available material. Not all identifications have been sufficiently argued, doubts have been published on specific and generic assignations, and suoid taxonomy has recently been a subject of some debate (e.g., Pickford 1978, 1988, 1995; Pickford and Ertürk 1979; Van der Made 1990a,b, 1994, 1995, 1996a,b, 1997a,b,c 1998; Pickford and Moyà Solà 1994, 1995). Here I describe the new material, discuss the identification of the older material, update the taxonomy, and place the result in a stratigraphic context.

Ozansoy (1957b, 1965) named the Sinap Series. The stratigraphy has subsequently been developed by Öngür (1976), Sen (1991), and Kappelman et al. (1996), who considered the unit to be a single formation. The lower and middle Sinap Formation are now considered to span MN 7+8 to MN 13, ranging in time from ~12 Ma to ~6 Ma, with a densely sampled interval between 10.9 and 9.3 Ma (Kappelman et al., chapter 2, this volume; contra Lunkka et al. 1999). This densely sampled interval corresponds to the early and middle Sinap members of Ozansoy, their boundary roughly corresponding to the middle/late Miocene boundary, marked by the local appearance of hipparionine horses at ~10.7 Ma (Kappelman et al., chapter 2, this volume; Bernor et al., chapter 11, this volume). No suoids are known from Ozansoy's Pliocene-Pleistocene upper Sinap member. For an up-to-date summary of the geology and chronology of the Sinap Formation, see Lunkka et al. (chapter 1, this volume) and Kappelman et al. (chapter 2, this volume).

Abbreviations

Measurements are in mm, unless otherwise indicated. All measurements are taken according to Van der Made (1996b). DAP = maximum anteroposterior diameter, DAPd = distal DAP, DLL = maximum labiolingual diameter in incisors, DMD = maximum mesiodistal diameter in incisors, DT = transverse diameter, DTa = anterior DT, DTm = DT of the middle lobe in dp4, DTp = posterior DT, DTpp = DT of the third lobe in m3, Li = width of the lingual side of a cm, La = width of the labial side of the lower male tusk, Po = width of the posterior side of the same.

Other abbreviations: AS = Ankara Sinap, Cm, cm, Cf, cf = Male and female upper and lower canines.

The fossils studied are kept in the following institutions: BMNH = Natural History Museum, London (formerly British Museum [Natural History]); FISF = Forschungsinstitut Senckenberg, Frankfurt; GSP = Geological Survey of Pakistan, Islamabad; HGSB = Hungarian Geological Survey, Budapest; IM = Indian Museum, Calcutta; IPS = Instituto de Paleontología, Sabadell; IPUW = Institut für Paläontologie der Universität, Wien; KME = Krahuletz Museum, Eggenburg; LPUM = Laboratoire de Paléontologie, Université de Montpellier II, Montpellier; MGL = Museum Guimet, Lyon; MHMN = Museu Històric Municipal de Novelda, Spain; MNCN = Museo Nacional de Ciencias Naturales, Madrid; MNHN = Muséum National d'Histoire Naturelle, Paris; MTA = Maden Tetkik ve Arama Enstitüsü, Ankara; NMB = Naturhistorisches Museum, Basel; NMW = Naturhistorisches Museum, Wien; PIMUZ = Paläontologisches Institut und Museum der Universität, Zürich; PDTFAU = Paleoantropoloji, Dil ve Tarih-Cografya Facültesi, Ankara Üniversitesi; UCBL = Université Claude Bernard, Lyon.

Systematic Paleontology

Synonymies in this section are restricted to papers that describe or discuss Anatolian suoid fossils. The material of the Sinap project and of earlier collections is listed in the tables with measurements. Previously described material is not redescribed, but the appropriate references are given.

Order Artiodactyla
Family Paleochoeridae Matthew, 1924
Subfamily Schizochoerinae Golpe-Posse, 1972
Genus *Taucanamo* Simpson, 1945
Taucanamo inonuensis Pickford and Ertürk, 1979

Restricted synonymy:

1979 *Taucanamo inönüensis* nov.; Pickford and Ertürk: 147–148, Fig. 5, Pl. 2, figs. 5–7.
1991 *Taucanamo inönüensis* Pickford ve Ertürk; Gürbüz: 87.
1990 *Taucanamo inonuensis* Pickford and Ertürk, 1979; Fortelius and Bernor: 521–524, Figs. 7–8.
1994 *Taucanamo inonuensis;* Van der Made and Han: 41–42.
1996 *Taucanamo inonuensis* (Pickford and Ertürk, 1979); Fortelius, Van der Made and Bernor: 352, 374.
1997 *Taucanamo inonuensis* Pickford and Ertürk, 1979; Van der Made: 129–130, 134.
1998 *Taucanamo inonuensis;* Van der Made: 234, 236, 260–261.

Sinap Material. İnönü I, site 3 (MTA collection): AKI3/4 holotype, skull with left canine and P1 and right P1, P3–M3; AKI3/582—left P3; AKI3/581 left M1; AKI3/584 right upper canine; AKI3/583 left upper canine; AKI3/453 right upper canine; unnumbered left M1; unnumbered right M1, left P4, P3.

Remarks. This is the type material of *Taucanamo inonuensis,* described in detail by Pickford and Ertürk (1979). Additional material from Paşalar was described and figured by Fortelius and Bernor (1990). The main characteristics are a larger size than *T. sansaniense* and the relatively large M3/m3 with complex third lobes.

İnönü I corresponds to Locs. 24 and 24A of the Sinap project, possibly associated with a lava flow dated by whole rock K-Ar analysis at 15.2 Ma (Lunkka et al. 1999, chapter 1, this volume; Kappelman et al., chapter 2, this volume). The precise relationship between the fossiliferous sediment and the lava flow is not clear, however, and others consider the fauna to be younger than the lava and anterior to the entry of *Hipparion* (Daams et al. 1999).

Discussion. The systematics of *Taucanamo* and related suoids have changed much in recent years. Until recently, these suoids were placed in the Doliochoerinae, Tayassuidae (=Dicotylidae), but they were moved to the Schizochoerinae, Palaeochoeridae by Van der Made (1994).

Pickford and Ertürk (1979) described the material from İnönü I as a new species because of its large size. Van der Made (1993) noted an overall size increase in time in the *Taucanamo sansaniense-inonuensis* group but also noted that some localities, such as Münzenberg, do not fit the trend; the possibility of two contemporaneous species was left open. Fortelius et al. (1996b) mentioned the possibility that *T. inonuensis* and *T. sansaniense* are distinct, geographically separated lineages. Van der Made (1997b) considered *T. sansaniense* and *T. inonuensis* to be members of a lineage that also includes the early Aragonian form *T. primum.* Van der Made (1998) argued that large material from MN 5 (including fossils from Münzenberg and Sandelzhausen) represents a species of a different lineage that he named *T. muenzenbergensis.*

Taucanamo primum, T. sansaniense, and *T. inonuensis* form a sequence in which the cheek teeth become larger and more elongate, M3/m3 and the premolars tend to become relatively larger, and the third molars also more complex. The first species is known from MN 4 and early MN 5 (Artenay, Els Casots, Bézian, La Romieu), the second from late MN 5 and early MN 6 (Göriach, Sansan), and the third from MN6 (İnönü I, Paşalar, Stätzling, Castelnau).

Genus *Schizochoerus* Crusafont and Lavocat, 1954
Schizochoerus sinapensis Van der Made, 1997

Restricted synonymy:

1978 *Schizochoerus* cf. *gandakasensis;* Pickford, 32–33 (skull from Sinap), Text fig. 4.
1979 *Schizochoerus* cf. *gandakasensis* (Pickford, 1976); Pickford and Ertürk, 150–152, Figs. 8–11, Pl. 3, figs. 6–7.
1994 *"Schizochoerus* cf. *gandakasensis"*; Van der Made and Han, 42–43, Fig. 6.
1996 *Schizochoerus* sp.; Fortelius, Van der Made and Bernor, 352–353, 374.
1998 *Schizochoerus sinapensis* n. sp.; Van der Made: 134.

Sinap Material. Unknown locality, lower or middle Sinap member (MTA collection): MTA1953 skull with left I1–M3 and right I2–P4 and M2–3.

Remarks. The species is known only from a skull from Sinap described by Pickford (1978) and Pickford and Ertürk (1979), of unknown exact provenance (M. Pickford, pers. comm.). Van der Made and Han (1994) made morphometric comparisons of the dentition with other schizochoerine species. They found the main characteristics to include sublophodont molars, wide upper premolars, relatively small M3, and an overall size inferior to that of *S. vallesiensis.*

Fortelius et al. (1996b) and Van der Made (1998) supposed the specimen might come from the lower Sinap Formation, because it is intermediate between *Taucanamo* (the possible ancestor of *Schizochoerus*) and *S. vallesiensis.* According to Kappelman et al. (chapter 2, this volume), the lower Sinap member in the central Sinap area spans only a short interval (~10.9–10.8 Ma) immediately preceding the middle Sinap member (~10.7–9.3 Ma).

Discussion. When Crusafont and Lavocat (1954) described *Schizochoerus,* they placed it in the subfamily Suinae, but noted similarities with Listriodontinae. Ozansoy (1965) followed this classification. Nikolov and Thenius (1967) placed *Schizochoerus* in the subfamily Listriodontinae and believed it to be a descendant of *Bunolistriodon*. Golpe-Posse (1972) named the tribe Schizochoerini, within the Suinae. Pickford (1978) and Pickford and Ertürk (1979) described the skull from Sinap and concluded that the genus is not a suid, but belongs to the Doliochoerinae, Tayassuidae, and probably evolved from *Taucanamo*. The "Doliochoerinae" were recognized as a family, the Palaeochoeridae, with the subfamilies Palaeochoerinae and Schizochoerinae, the latter including *Schizochoerus* (Van der Made 1996a, 1997a).

This skull from Sinap was first described as *Schizochoerus* cf. *gandakasensis* (Pickford 1978; Pickford and Ertürk 1979). However, the elongate premolars of the species "*gandakasensis*" suggest that it does not belong to *Schizochoerus* but to the genus *Yunnanochoerus*, whereas the short and wide premolars of the skull from Sinap show that it is a true *Schizochoerus* (Van der Made and Han 1994). The skull from Sinap indicates a smaller species than *S. vallesiensis*, which was named *S. sinapensis* (Van der Made 1997b).

Schizochoerus may have evolved from *Taucanamo? muenzenbergensis* (Van der Made 1998). The genus shows a rapid evolution with tendencies toward lophodont dentition, larger body size, and relatively larger last molars. Three species are recognized: *S. anatoliensis* (Çandir), *S. sinapensis* (lower or middle Sinap), and *S. vallesiensis* (middle Sinap, Nesebar, Can Purull, and La Tarumba) (Van der Made 1998). The latter species seems to have dispersed from the eastern to the western Mediterranean during the course of the Vallesian.

Schizochoerus vallesiensis Crusafont and Lavocat, 1954

Restricted synonomy:

1957 *Schizochoerus arambourgi* n. sp.; Ozansoy b: 33, 43, Pl. 1, fig. 3.
1957 *Schizochoerus arambourgi;* Ozansoy: 21, 23.
?1957 *Schizochoerus pachecoi;* Ozansoy: 23.
1965 *Schizochoerus arambourgi* nov. sp.; Ozansoy: 61–62, Pl. 6, fig. 1.
1965 *Schizochoerus* cf. *arambourgi* nov. sp.; Ozansoy: 16–17, 62–64, Pl. 6, fig. 8.
1975 *Schizochoerus;* Becker-Platen, Sickenberg and Tobien: 91.
1978 *Schizochoerus vallesiensis* Crusafont and Lavocat; Pickford: 35–36.
1979 *Schizochoerus vallesiensis* Crusafont and Lavocat 1954; Pickford and Ertürk: 150.
1990 *Schizochoerus vallesiensis;* Van der Made: 100, 104.
1991 *Schizochoerus vallesiensis* Crusafont and Lavocat, 1954; Sen: 250.
1994 *Schizochoerus vallesiensis;* Van der Made and Han: 35–43, Pl. 3, figs. 1, 3, 6.
1996 *Schizochoerus vallesiensis* (Crusafont and Lavocat 1954); Fortelius, Van der Made and Bernor, 352–353, 374.
1998 *Schizochoerus vallesiensis* Crusafont and Lavovat, 1954; Van der Made: 131, 134.

Sinap Material. Unknown localities, middle Sinap (MNHN collection): Yas 27 mandible with left i2, canine, p4 (in alveolus) and right canine, dp2–4 and m1–2; Yas 28 right maxilla with P1–2; 3337 palate with left P3–M3 and right P3 and M1–3; Yas 30 p3.

Remarks. Ozansoy (1965) provided detailed descriptions and figures of the material from Sinap. Important characters include the lophodont P4, the nearly lophodont molars, the large general size, and relatively large M3. Ozansoy (p. 62) also mentioned but did not describe an incisor from the "couche supérieure du Sinap moyen [upper bed of the middle Sinap]," which he attributed to *Schizochoerus* cf. *arambourgi*.

Ozansoy (1957b) figured a palate from the middle Sinap Formation (or series, in his terminology). This is specimen MNHN 3337, and there also seems to be an incisor and a canine. The level corresponds to "couche 20 (marne compacte brune) [bed 20 (compact brown marls)]" of Ozansoy (1965). In addition, Ozansoy (1957b, p. 43) cited *Schizochoerus* from the "calcaires marneux blancs [white marly limestones]," the lowest unit he recognized in the middle Sinap "Formation," most probably equivalent to the "double white" marker paleosol at ~10.7 Ma in the terminology of the Sinap project (Lunkka et al., chapter 1, this volume). This specimen seems to be mandible Yas 27 in the MNHN, where the Ozansoy collection is stored.

Discussion. Ozansoy (1957b) figured a palate from the Middle Sinap as *Schizochoerus arambourgi* n. sp., but did not give a description, diagnosis, or definition. Later, Ozansoy (1965) did give a detailed description of the specimen, but assigned it to *Schizochoerus* cf. *arambourgi* and described a mandible as *Schizochoerus arambourgi* nov. sp. The species is considered a junior synonym of *S. vallesiensis* (Nikolov and Thenius 1967; Pickford 1978; Pickford and Ertürk 1979; Van der Made 1997a).

Family Suidae Gray, 1821
Subfamily Listriodontinae Gervais, 1859
Tribe Kubanochoerini Gabunia, 1958
Genus *Kubanochoerus* Gabunia, 1955
Kubanochoerus cf. *mancharensis*

Restricted synonymy:

1979 *Libycochoerus khinzikebirus* (Wilkinson, 1976); Pickford and Ertürk, page 145, figure 4.
1989 *Libycochoerus* cf. *khinzikebirus* (Vilkinson); Tekkaya, 157.
1991 *Libycochoerus* cf. *khinzikebirus* (Vilkinson); Gürbüz: 87.
1996 *Kubanochoerus khinzikebirus* (Wilkinson, 1976); Fortelius, Van der Made and Bernor, 355, 375.
1996 *Kubanochoerus mancharensis* n. sp.; Van der Made b, 55–58, Pl. 11, figs. 7, 9, 10, Pl. 12, fig. 12.

Sinap Material. İnönü I (MTA collection): AKI3/779 left M1 or M2.

Remarks. The single specimen is a huge bunodont upper molar. It is wide compared with molars of *Kubanochoerus massai* and *K. robustus*. It was described and figured by Pickford and Ertürk (1979), who interpreted it as a second molar of the large species *"Libycochoerus" khinzikebirus*. Subsequently it was interpreted as a first molar of the still larger species *K. mancharensis* (Van der Made 1996b).

Tribe Listriodontini Gervais, 1859
Genus *Bunolistriodon* Arambourg, 1963
Bunolistriodon latidens (Biedermann, 1873)

Restricted synonymy:

1979 *Listriodon cf. lockharti* (Pomel, 1848); Pickford and Ertürk, 144–145 (the material from İnönü), Pl. 1, fig. 9, Plate II, fig. 1 and 2.
1989 *Listriodon* cf. *lockharti* (Pomel); Tekkaya, 157.
1990 *Listriodon lockharti;* Fortelius and Bernor, table 1 (partially).
1991 *Listriodon* cf. *lockharti* (Pomel); Gürbüz: 87.
1996 *Bunolistriodon latidens* (Biedermann, 1873); Fortelius, Van der Made and Bernor, 150–162, Figs. 1, 4, 6.
1996 *Bunolistriodon latidens* (Biedermann, 1873); Van der Made b, 77–78, Pl. 4, fig. 4, Pl. 18, figs. 1–9.
1996 *Bunolistriodon latidens* (Biedermann, 1873); Fortelius, Van der Made and Bernor, 353–354, 374.
1996 *Listriodon* cf. *L. latidens;* Kappelman et al.: 91.

Sinap Material. İnönü I, site 3 (MTA collection): AKI3/438 left mandible with dp4–m1; AKI3/586 left mandible with dp4–m2; AKI3/7 symphysis with left i1–3 and right i1–2; AKI3/6 left mandible with p4–m3; AKI3/11 right mandible with m1–3; AKI3/573–574 right i1–2; AKI3/166 right M2–3; AKI3/780 skull with left P3–M3 and right P2–M2; AKI3/9 left maxilla with P4–M3; AKI3/577–576 right P3–4; AKI3/162 m3; AKI3/575 left p2; AKI3/588 left upper male canine; AKI3/772 right upper male canine; AKI3/327 right upper male canine; AKI3/332 right upper male canine; AKI3/325 right upper male canine; AKI3/572 left i2; AKI3/571 left I1; AKI3/162 right I2; AKI3/50 right I2; AKI3/49 left I2.

Remarks. The material was briefly described and figured by Pickford and Ertürk (1979) and in more detail by Van der Made (1996b). The main characteristics of the species are sublophodont dentition, canines with a large section, and a long radius of curvature and wide (DMD) and flat (DLL, index I) incisors.

Discussion. The material was initially believed to belong to a form close to *Bunolistriodon lockharti*. Since the recognition that *B. latidens* is a valid species rather than a synonym of *B. lockharti* (Van der Made and Alférez 1988), the material from İnönü I was referred to *B. latidens* and material from Paşalar was named *B. meidamon* (Fortelius et al. 1996a,b). A lineage consisting of three species is recognized: *B. adelli, B. latidens,* and *B. meidamon* (Van der Made 1996b), characterized by progressive increase in incisor width (DMD) and index, decrease in incisor thickness (DLL), increase in canine size, and increase in the index of some cheek teeth. This trend is interpreted as indicating the following sequence from old to young: İnönü I, Paşalar, Çandir (Fortelius et al. 1996a; Van der Made 1996b).

Genus *Listriodon* von Meyer, 1846
Listriodon splendens von Meyer, 1846

Restricted synonymy:

1957 *Listriodon* n. sp.; Ozansoy b: 43.
1965 *Listriodon piveteaui* nov. sp.; Ozansoy: 15–16.
1975 *Listriodon splendens* Meyer; Hünermann: 153–154, not the canine described on p. 154.
1975 *Listriodon splendens;* Becker-Platen, Sickenberg and Tobien: 23, 25, 28, 32, 49, 69, 97.
1979 *Listriodon splendens* H. von Meyer 1846; Pickford and Ertürk, 142 and 144 (the "second form with smaller canines and less elongated upper central incisors"), Figs. 1–3, Pl. 1, figs. 2–5 and 8, figs. 1, 6 and 7 partially.
1979 *Listriodon* cf. *lockharti* (Pomel, 1848); Pickford and Ertürk, 144–145 (material from Sariçay).
1989 *Listriodon* sp.; Tekkaya, 157.
1990 *Listriodon* cf. *splendens;* Fortelius and Bernor, 510–511, fig. 1, 2 e–h, 5 c–d.
1990 *Listriodon splendens;* Fortelius and Bernor, table 1.
1991 *Listriodon piveteaui* n. sp. (Ozansoy, 1965; 16, nomen nudum); Sen, 250.
1994 *Listriodon* aff. *L. splendens;* Hunter and Fortelius, 105–124.
1996 *Listriodon splendens* Von Meyer, 1846; Van der Made b , 98–115, Pl. 25, figs. 7–8, Pl. 35, figs. 1–18, Pl. 36, figs. 1–14, Pl. 37, figs. 1–9, Pl. 38, figs. 1–15, Pl. 39, figs. 1–10, Pl. 40, figs. 1–6, Pl. 41, figs. 1–14, Pl. 42, figs. 1–8, Pl. 43, figs. 1.5, Pl. 44, figs. 1–2.
1996 *Listriodon splendens* (Meyer, 1846); Fortelius, Van der Made and Bernor, 353–355, 374.

Sinap Material. Loc. 24A: upper male canine left AS.92.644; İnönü I, site 3 (MTA collection): AKI3/42 mandible with p4–m3; AKI3/41 right mandible with p3–m3; AKI3/333 symphysis with left and right i1–2; AKI3/777 symphysis with left i1 and right i1–3; AKI3/384 right p3; AKI3/357 right p3; AKI3/778 right m3; AKI3/514 left m2; AKI3/252 tip of upper male canine; AKI3/424 right upper male canine; AKI3/450 left M3; AKI3/451 right M3; AKI3/452 left P4; unnumbered right M1; unnumbered left P4; unnumbered right P3; AKI3/418 left di2; AKI3/450 right I2; AKI3/579 left I2; "Couche 46" of Ozansoy, lower Sinap member (MNHN collection): unnumbered skull with right P3–M3 and left P2–4.

Remarks. The material from İnönü I was included in a general description of the species and one of the specimens from İnönü I was figured (Van der Made 1996b). The molars have clear lophs. Wide incisors (with large mesiodistal diameter) are common for the species, but within this

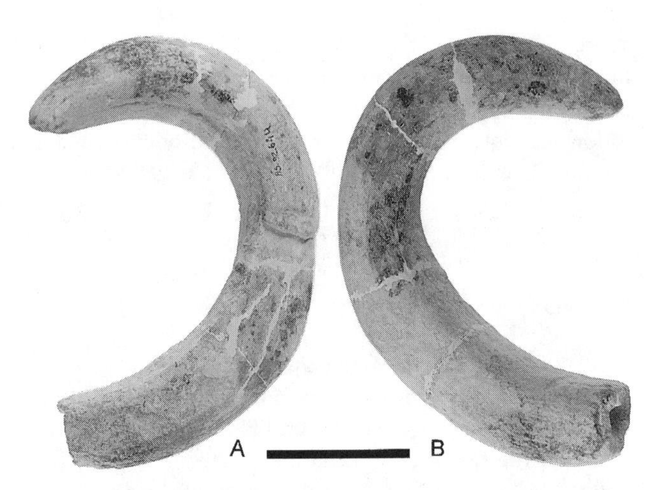

Figure 13.1. Right upper male canine AS.92644 of *Listriodon splendens* from Loc. 24A (level of İnönü I). (**A**) Lateral view; (**B**) medial view. Scale bar = 50 mm.

species the incisors from İnönü I are relatively narrow. There is one new specimen: AS.92.644, a right upper male canine (fig. 13.1). The tusk is large and complete but broken in four places. It measures 35.1 mm across, below, and parallel to the plane of the main wear facet. This value is high, but not outside the ranges for *Listriodon splendens,* not even for the MN 6 forms, which tend to have smaller canines. The radius of curvature, measured along the outside of the tooth (Ro) changes from ~8 cm near the tip to ~10 cm near the root and along the inner side (Ri), it is 3–6 cm. This radius (Ro and Ri) is ~7 and 8 cm in *B. latidens* from the type locality Veltheim and 12–13/16.5 cm in *B. meidamon* from Çandir (Van der Made 1996b, table 6). In *L. splendens,* the radius of curvature increased from MN 6 to MN 9; for MN 6, values of Ri vary between 15 and 85 mm and the Ro between 40 and 100 mm. The new material from Loc. 24a at İnönü I fits in these ranges and is assigned to *Listriodon splendens.*

A skull of *L. splendens* from the lower Sinap Formation, kept in the MNHN, was described and figured by Van der Made (1996b, pl. 43, figs. 1–3).

Discussion. Ozansoy (1965, p. 15–16) mentioned "*Listriodon pivetaui* nov. sp." from "couche 46 [bed 46]" of the lower Sinap Formation but did not describe the material, nor did he indicate the reasons for naming a new species. Therefore *L. piveteaui* is a nomen nudum. This citation is probably based on the skull that is kept in the MNHN, which was figured and discussed by Van der Made (1996). Like the material from İnönü I, the specimen from Sinap was assigned to *L. splendens.*

Listriodon splendens shows various evolutionary trends, one of them being the increase in width of the incisors (Van der Made 1996b; Fortelius et al. 1996a,b). Incisor morphometrics suggest that the İnönü I material represents an early stage of evolution of the lineage.

Subfamily Suinae Gray, 1821
Tribus Dicoryphochoerini Schmidt-Kittler, 1971

Genus *Propotamochoerus* Pilgrim, 1925
Cf. *Propotamochoerus provincialis* (Gervais 1859)

Sinap Material. Loc. 42 (Çobanpinar): AS.93.1124 left DP4; AS.94.1331 right dp4; Çobanpinar 1 (PDTFAU-collection): unnumbered right d2, d3 and m1; Çobanpinar (PDTFAU-collection): unnumbered left d4 and m1.

Remarks. There are an upper and a lower fourth milk molar (fig. 13.2A–F). In the older collections from Çobanpinar, there are two lower tooth rows with milk teeth and the first molar. The specimens have the common suine morphology.

The cheek teeth are small (fig. 13.3); the milk teeth are as small or smaller than the smallest specimens of *Propotamochoerus palaeochoerus* and are close in size to those of *P. provincialis* and *P. hyotherioides.* The permanent molars are close to the largest specimens of *P. palaeochoerus,* close to those of *P. provincialis,* and larger than those assigned to *P. hyotherioides.* Molars of *Microstonyx erymanthius* are larger.

This locality was placed in MN 12 by Mein (1977) but it was omitted from the updated chart of Mein (1990). It now appears that a better correlation might be MN 13 (see below and Van der Made et al., chapter 14, this volume).

Discussion. The common Suinae from the late Miocene are *Propotamochoerus, Microstonyx,* and *Hippopotamodon.* The first *Sus, S. arvernensis,* is recorded from MN 14 (Guérin and Faure 1985) and is cited from Dinar Akçaköy and Çalta in Turkey (Hünermann 1975; Guérin et al. 1998), localities that are currently placed in MN 15 (De Bruijn et al. 1992; Sen et al. 1998). *Hippopotamodon* and *Microstonyx* are much larger than the material from Çobanpinar.

The dp3s from Çobanpinar are small and the m1 large compared with the average of the *Propotamochoerus palaeochoerus* material, suggesting some reduction in size of the anterior cheek teeth. This phenomenon is not yet well studied in the genus *Propotamochoerus. Sus* tends to have relatively smaller premolars than does *Propotamochoerus,* but this reduction is hardly present in *S. arvernensis.* The m1s from Çobanpinar are much larger than the homologues of *Sus arvernensis,* the *Propotamochoerus* sp. from Europe (fig. 13.3), and *P. hysudricus* from the Indian subcontinent (measurements for this species are given by Pickford [1988]; see scatter diagrams by Van der Made and Han [1994]). A "Suinae sp." was cited from Karain (Fortelius et al. 1996b). There is an m1 from Karain that is so much wider than those from Çobanpinar that it seems unlikely that they belong to the same species. The size of the m1 seems to place the material from Çobanpinar close to *P. provincialis.* Because the evolution of *Propotamochoerus* is not well enough known and the material from Çobanpinar is poor, the material is assigned to cf. *Propotamochoerus provincialis.*

Propotamochoerus provincialis entered Europe in MN 13. Its possible presence in Çobanpinar adds to the evidence that the locality is not MN 12, but MN 13 (see the general discussion below and Van der Made et al., chapter 14, this volume).

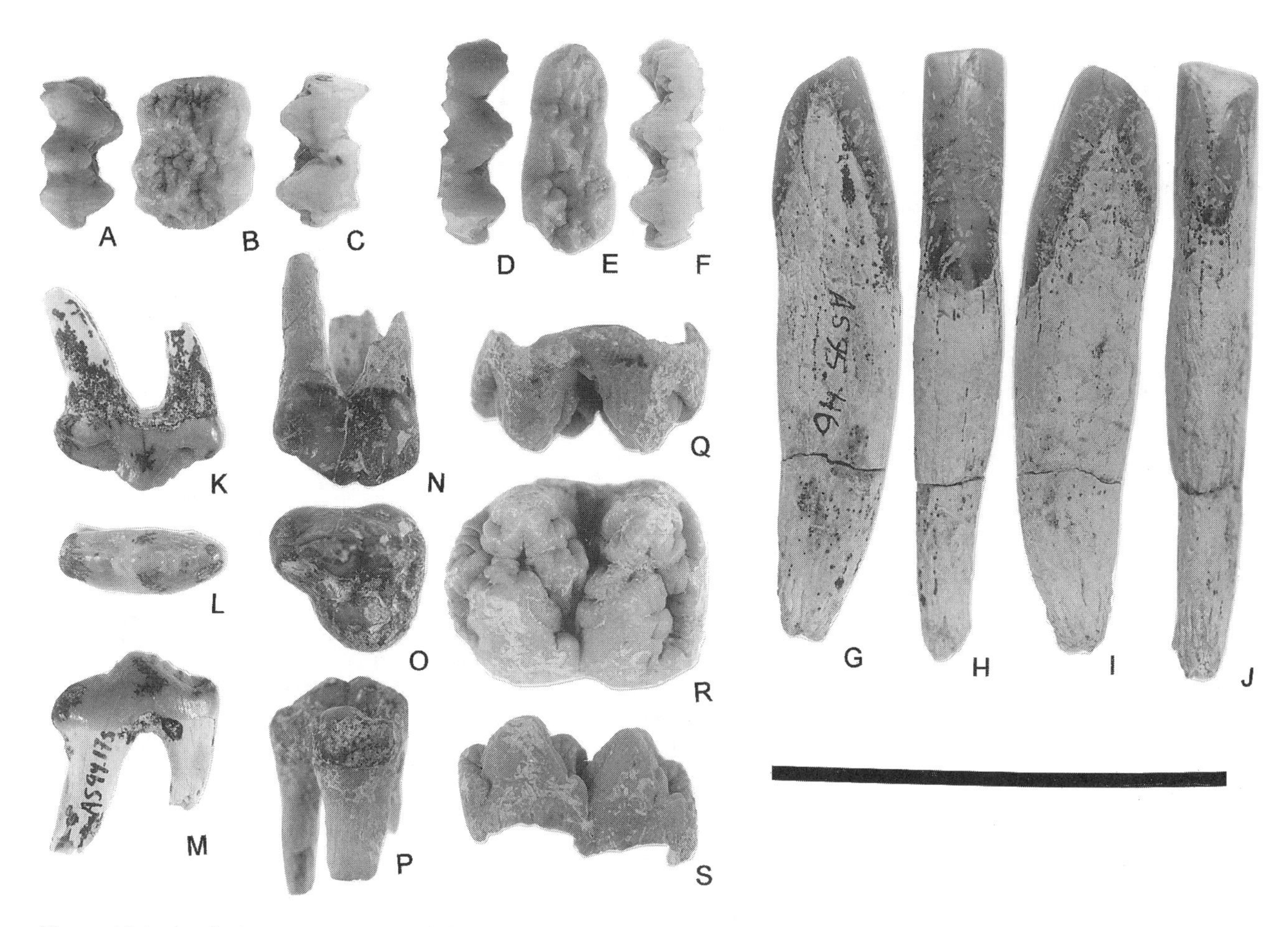

Figure 13.2. (A–C) The DP4 AS.93.1124 of cf. *Propotamochoerus provincialis* from Loc. 42 (Çobanpinar). (D–F) The dp4 AS.94.1331 of cf. *Propotamochoerus provincialis* from Loc 42. (G–J) Right i1 AS.95.46 of *Microstonyx major* from Loc. 49 (Igbek). (K–M) Left P1 AS.94.175 of *Hippopotamodon antiquus* from Loc. 114. (N–P) Left P3 AS.91.104 of *Microstonyx erymanthius* from Loc. 33 (upper Kavakdere). (Q–R) Right M2 AS.90.37 of *Microstonyx erymanthius* from Loc. 26 (upper Kavakdere). Scale bar = 100 mm.

Material from Kavakdere assigned to cf. *Korynochoerus* sp. (Hünermann 1975) is included here in *Microstonyx erymanthius* (see below).

Genus *Hippopotamodon* Lydekker, 1877
Hippopotamodon antiquus (Kaup, 1833)

Restricted synonomy:

1957 *Dicoriphochoerus meteai* n. sp. (type); Ozansoy b, 43.
1965 *Dicoryphochoerus meteai* nov. sp.; Ozansoy: 58–61, Pl. 7, figs. 2 and 4.
1975 *Sivachoerus giganteus;* Becker-Platen, Sickenberg and Tobien: 32.
1975 *Dicoryphochoerus-Microstonyx*-Formenkreis; Becker-Platen, Sickenberg and Tobien: 32.
1975 *Sivachoerus giganteus* (Falconer and Cautley); Hünermann: 154–156, Fig. 4.
1975 *Dicoryphochoerus-Microstonyx*-Formenkreis; Hünermann: 154.
1979 *Microstonyx major* (Gervais) 1848–52; Pickford and Ertürk: 143 (Esme Akçaköy).
1979 *Hippopotamodon meteai;* Pickford and Ertürk: 143, 147.
1990 *Microstonyx antiquus;* Van der Made: 101, 104.
1991 *Dicoryphochoerus meteai* Ozansoy, 1965; Sen: 250.
1996 *Hippopotamodon antiquus* (Kaup, 1833); Fortelius, Van der Made and Bernor, 358, 377.

Sinap Material. Loc. 114: AS.94.194 partial skull with palate and right P4–M3; AS.94.175 left P1; Loc. 37: AS.92.393 lateral metacarpal (distal part); "Couche 23" of Ozansoy (MNHN collection): unnumbered mandible with left i1–3, p2–m3, and right i1–2, female canine and p1–m3; unnumbered left mandible with p2 and p4–m1; middle Sinap (MTA-collection): MTA 1953 symphysis with left i1–3 and right i1–2, male canine and p1; MTA 1957 palate with left and right DP3–M1; MTA 1957 right M3.

Remarks. A mandible from middle Sinap was described and figured by Ozansoy (1965). The specimen has the morphology of the *Hippopotamodon-Microstonyx* group, and the remains of the canine indicate that it belonged to a female. The diastemata are short for such a large animal. The p1 is a relatively large tooth. The p4 has a large and well-separated metaconid. The third lobe of the m3 has a simple morphology. The incisors are relatively large, but with

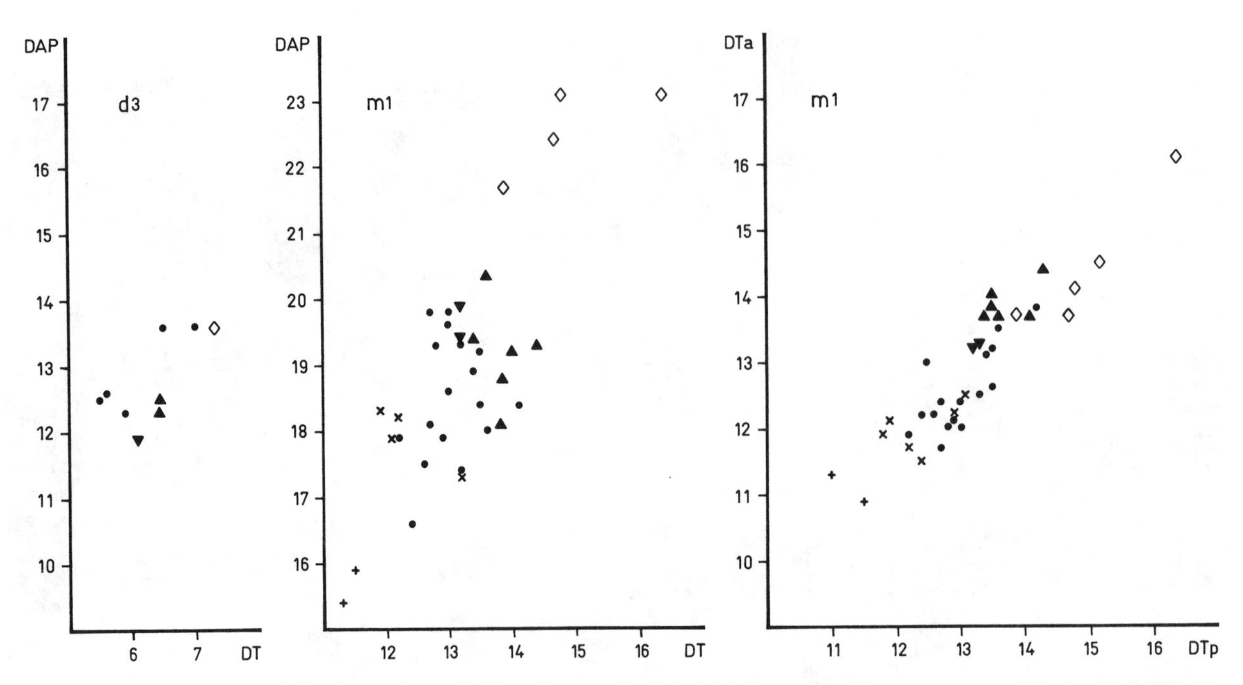

Figure 13.3. Bivariate plots of the dp3 and m1 of selected Dicoryphochoerini. Diamonds = *Microstonyx erymanthius brevidens* from Dorn Dürkheim (SMF; Van der Made 1997). Crosses = *Sus arvernensis* from Perpignan (NMB). Oblique crosses = *Propotamochoerus* sp. from Samos (NMW), Baccinello V3 (NMB), and Maramena (JGUM). Filled circles = *P. palaeochoerus* from Hainburg (NMW), Rudabánya (HGSB), Castell de Barberá (IPS), Montréjeau (MNHN), Eppelsheim (HLD), Hennersdorf (NMW), Magersdorf (IPUW), Wien III Belvedère (NMW), Wissberg (HLD) (only molars in association with premolars are plotted). Triangles = *P. provincialis* from Montpellier (LPUM). Inverted triangles = cf. *P. provincialis* from Çobanpinar.

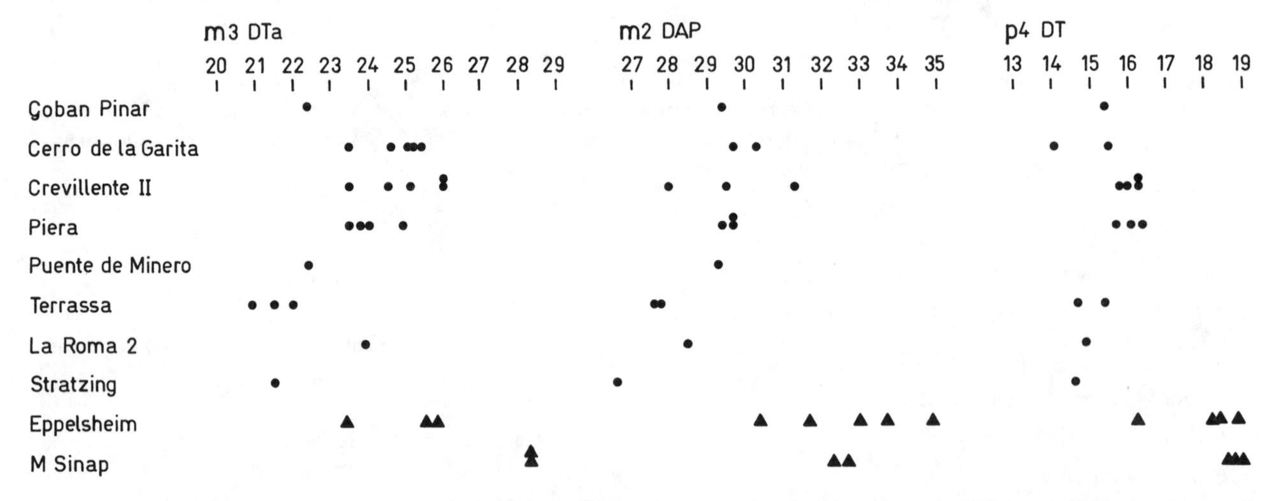

Figure 13.4. Size m3, m2, and p4 of *Hippopotamodon antiquus* (triangles) compared with *Microstonyx major* (filled circles). The localities are in approximate stratigraphical order. MN 9: middle Sinap, Eppelsheim (HLD). MN 10: Stratzing (NMW), La Roma 2 (Van der Made et al. 1992), Terrassa (IPS). MN 11: Puente Minero (MNCN), Piera (IPS), Crevillente II (MHNM). MN 12: Cerro de la Garita (MNCN). MN 13: Çobanpinar.

crowns that may not have been very high. The lower limit of the crowns is clearly marked; in later *Microstonyx major/ erymanthius*, the enamel of the crown may become thinner over a long distance, making the base of the crown less well defined. The specimens indicate individuals of huge size. There is additional material from Sinap in the MTA museum, sharing the main characteristics. A lower male canine with verrucosic section is rather large.

The palate from Loc. 114 also represents a *Hippopotamodon* of huge size (figs. 13.4, 13.5). The P4 has very well separated buccal cusps, a well-developed paraendoconule (the sagittal cusp of Pickford 1988) and metaendoconule (terminology of Van der Made 1996b).

A distal lateral metapodial was recovered from Loc. 37. The specimen has the typical suid morphology, is large and in particular it is wide.

Loc. 114 has an interpolated magnetochronologic age of 10.0 Ma, and Ozansoy's material is probably of similar age, and almost certainly from within the 10.7–9.3 Ma age span of the middle Sinap member (Kappelman et al., chapter 2, this volume). Loc. 37 is probably somewhat younger but control is lacking.

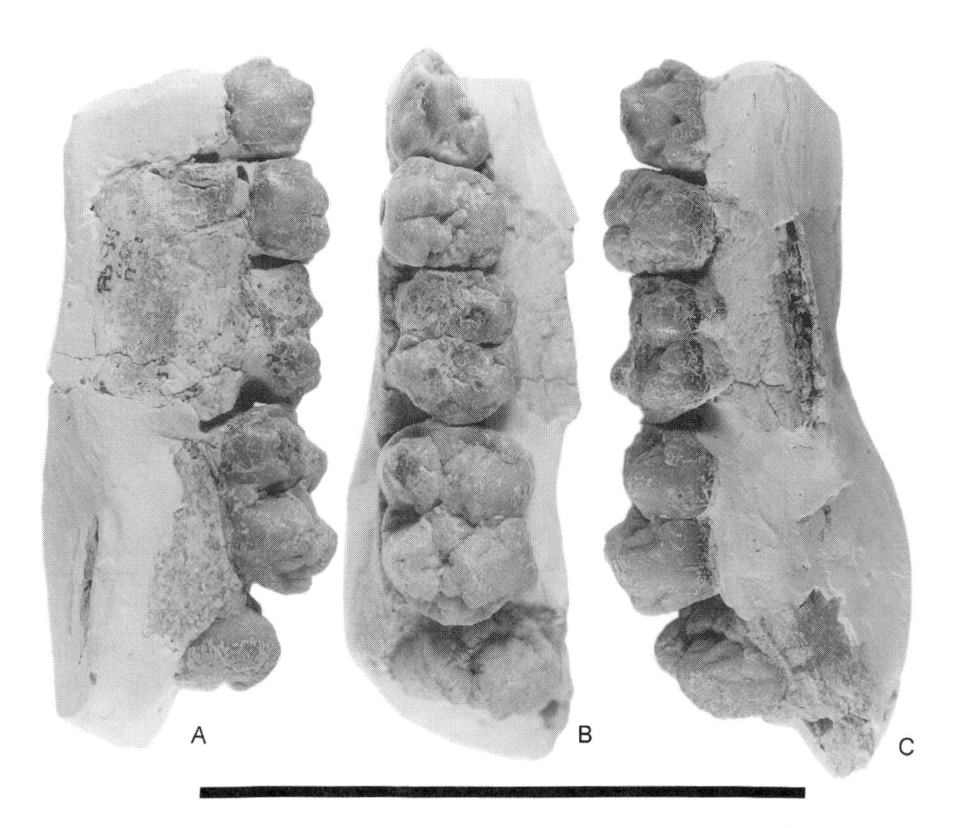

Figure 13.5. Left maxilla with P3–M3 AS.94.175 of *Hippopotamodon antiquus* from Loc. 114. (**A**) Buccal view. (**B**) Occlusal view. (**C**) Lingual view. Scale bar = 200 mm.

Discussion. Ozansoy (1957b) included "*Dicoriphochoerus meteai* n. sp." in a faunal list for middle Sinap. No description, diagnosis, or figure was given; the name is a nomen nudum. Ozansoy (1965) described a mandible from Sinap as the new species *Dicoryphochoerus meteai*. This is technically a valid description. However, the material is included here in *Hippopotamodon antiquus,* following Fortelius et al. (1996b).

The genus *Hippopotamodon* was introduced by Lydekker (1877) with type species *H. sivalense.* Pilgrim (1926) ignored this and introduced the genus *Dicoryphochoerus,* with type species *D. titan,* for large Suinae with large canines and the genus *Microstonyx* for similar suids, but with small canines. Later, Pickford and Ertürk (1979) and Pickford (1988) synonymized *D. titan* and *H. sivalense,* thus eliminating the genus *Dicoryphochoerus.* Pickford and Ertürk (1979) believed *H. meteai* to be close to *H. sivalense.*

Van der Made and Hussain (1989) suggested that *Microstonyx* might be a synonym of *Hippopotamodon,* because the difference in canine size does not seem to be enough to maintain them as different genera. Van der Made (1990) included the Turkish material in *Microstonyx antiquus.* Fortelius et al. (1996b) placed the material from middle Sinap and from a number of other Turkish localities in *Hippopotamodon antiquus* and speculated that *H. antiquus* and *H. sivalense* might be synonymous.

Mandible IM B 740 of *Hippopotamodon sivalense* has a huge male lower canine, which is much larger than that of *H. antiquus* (fig. 13.6). The cheek teeth have approximately the same size in both cases. Mandible IM B 539 of *Hippopotamodon sivalense* from Hasnot has huge incisors (fig. 13.6). The incisors have long labiolingual diameters (DLL) relative to their mesodistal diameters (DMD). The same can be observed in a series of isolated incisors attributed to the species. The long DLL is a common feature in Suidae with incisors that are not very hypsodont. The more hypsodont incisors of *Microstonyx major, M. erymanthius,* and *Sus* tend to have relatively shorter DLL. *Hippopotamodon antiquus* from Europe and Turkey (including the type of *D. meteai*) tend to have smaller incisors, with a relatively short DLL. The distance between the clusters of *H. sivalense* and *H. antiquus* is long compared to that between the smallest and largest specimens assigned to *M. erymanthius. Hippopotamodon sivalense* and *H. antiquus* differ in the size of their incisors and canines.

According to Van der Made and Han (1994), the *Hippopotamodon* canines are not extremely large for Suinae; it seems more likely that those of *H. antiquus* are reduced than that the canines of *H. sivalense* increased in size. The reduction in size in the anterior dentition is noted already in MN 9 in the European-Anatolian *Hippopotamodon,* whereas *H. sivalense* maintained large incisors and canines millions of years later. This might be taken as a derived character that *H. antiquus* shared with *M. major* and *M. erymanthius. Dicoryphochoerus meteai* is morphologically and metrically very close to *H. antiquus* and shares the derived character of reduction of the anterior dentition, justifying its inclusion in *H. antiquus.*

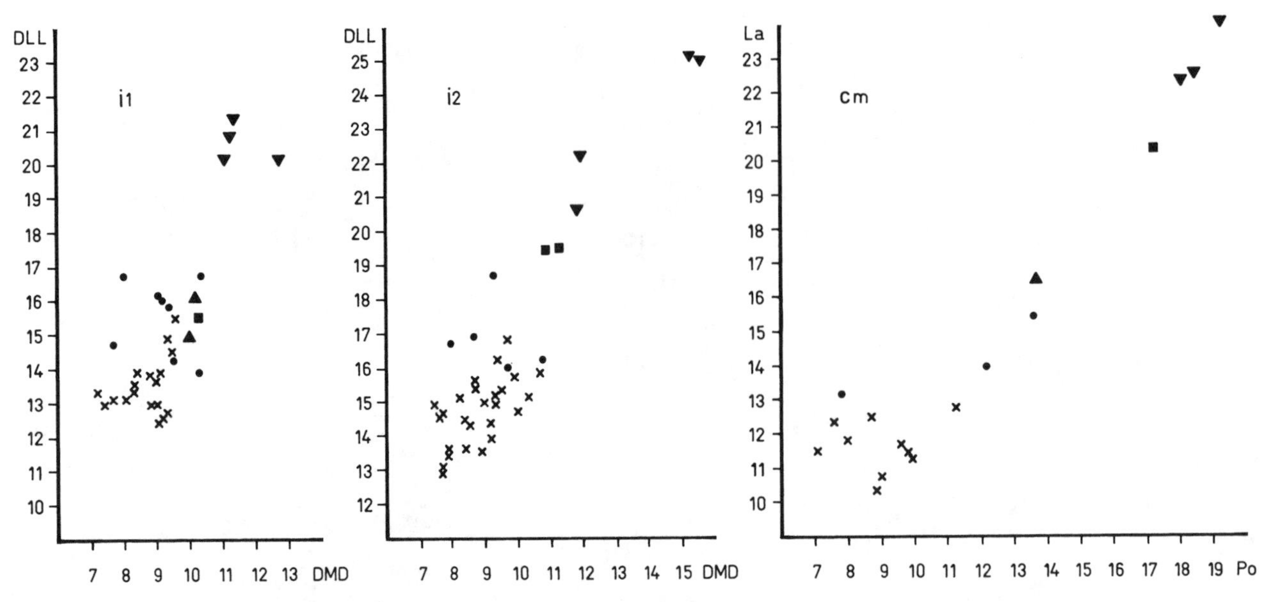

Figure 13.6. Bivariate plots of the i1, i2, and cm of *Hippopotamodon*. Oblique crosses = *Microstonyx erymanthius* from various localities (canines only from Dorn Dürkheim) (FISF; Van der Made 1997). Filled circles = *M. major* from Stratzing (NMW), Alisses (MGL), Las Pedrizas (MNCN), Crevillente II (MHNM), Terrassa (IPS). Triangles = *H. antiquus* from Eppelsheim (HLD). Squares = *H. antiquus* from M Sinap. Inverted triangles = *H. sivalense* from Pakistan (HGSP, IM).

Genus *Microstonyx* Pilgrim, 1926

Microstonyx major (Gervais, 1848–1852)

Restricted synonomy:

1957 *Microstonyx erymanthius* Roth et Wagner; Ozansoy b: 41.
1965 *Sus erymanthius* Roth et Wagner 1854; Ozansoy: 57.
1975 *Dicoryphochoerus-Microstonyx*-Formenkreis; Hünermann: 154, 156 (partially).
1979 *Microstonyx major* (Gervais) 1848–52; Pickford and Ertürk: 143 (partially).
1996 *Microstonyx major* (Gervais, 1848–1852); Fortelius, Van der Made and Bernor, 357–358, 377.
1996 *Microstonyx major/erymanthius;* Fortelius, Van der Made and Bernor, 358, 377 (partially).

Sinap Material. Loc. 49: AS.94.1799 right P3–M3; AS.95.46 right i1; Loc. 42 (Çobanpinar; MTA-collection): MTA AAEÇ1 right mandible with d4–m1; AAEÇ2 left P2–M1; AAEÇ8 left m1; AAEÇ1019 left m3; AAEÇ3 right M3; AAEÇ5 left M2; AAEÇ4 left M3; Çobanpinar (PDTFAU-collection): ÇBP40 right M2–3; ÇBP40 right M2–3; ÇBP621 left mandible with p2–m3; ÇBP718 right p4; ÇBP40 left M3.

Remarks. The specimens from Çobanpinar have the typical dicoryphochoerine morphology. The size of the cheek teeth is in the upper range of the overlap between *Microstonyx major* and *M. erymanthius.* This is also the case for the P2 and P3.

The skull fragment from Loc. 49 has a dentition of a similar size as that known from Çobanpinar. P3 and P4 have metacones that are not as well separated as in the Kavakdere specimens, being more like *Microstonyx major* than *M. erymanthius* in this character.

Pickford and Ertürk (1979) described *Microstonyx major* from Evciköy. This is the name of the village near Çobanpinar, and it seems likely that they meant this locality (Sen, pers. comm.).

The magnetochronologic age of Loc. 49 is 9.1 Ma (Kappelman et al., chapter 2, this volume), roughly time-correlative with the MN 10/11 boundary (Steininger et al. 1996) or late MN 10 (Krijgsman et al. 1996). As noted earlier, Loc. 42 (Çobanpinar) is probably best assigned to MN 13.

Discussion. Ozansoy (1957b, 1965) assigned the material from Çobanpinar to *Microstonyx erymanthius* and apparently did not consider *M. major.* Pickford and Ertürk (1979) assigned it to *M. major* and considered *M. erymanthius* to be a junior synonym, until recently the common opinion (Trofimov 1954; Hünermann 1968; Thenius 1972; Ginsburg 1974). Accepting *M. major* and *M. erymanthius* as two distinct species, the material from Loc. 49, probably should be assigned to *M. major.*

Microstonyx erymanthius (Roth and Wagner, 1854)

Restricted synonymy:

1957 *Microstonyx* sp.; Ozansoy b: 42.
1965 *Sus* sp.; Ozansoy: 57–58.
1975 *Sus* sp.; Becker-Platen, Sickenberg and Tobien: 81.
1975 *Dicoryphochoerus-Microstonyx*-Formenkreis; Hünermann: 154, 156 (partially).
1975 cf. *Korynochoerus* sp.; Hünermann: 154 (partially).
1979 *Microstonyx major* (Gervais) 1848–52; Pickford and Ertürk: 143 (partially).

1996 ?*Microstonyx erymanthius* (Roth and Wagner, 1854); Fortelius et al.: 377.
1996 *Microstonyx erymanthius* (Roth and Wagner, 1854); Fortelius, Van der Made and Bernor, 358, 377
1996 *Microstonyx major/erymanthius;* Fortelius, Van der Made and Bernor, 358, 377 (partially).
1996 *Microstonyx* sp.; Kappelman et al.: 91.

Sinap Material. Loc. 33: AS.91.104 left P3; Kavakdere (MTA-collection): MTA 1962 left mandible with dp2–m1; MTA 1963 left mandible with dp2–m1; MTA 1959 right maxilla with P4–M2; MTA 1950 skull with left I3 and P2–M3 and right I2 and P2–M3; Kavakdere (PIMUZ-collection): ASV left mandible with dp2 and dp4–m1.

Remarks. The material includes a skull fragment from Kavakdere. The skull has an elongated snout and long diastemata. The canine alveolae are small and suggest that it was a female. Female pigs tend to have much shorter diastemata than male pigs, although this is a highly variable character (Van der Made 1991a). On the left side, the distance P2–Cf is 57.0 mm and the distance P2–I3 is 85.0 mm. There is no P1. The zygomatic arches are not very massive, but project strongly laterally; they are only a short distance above the occlusal surface. The palate extends only a little more than 14 mm behind the M3. Most of the cheek teeth are comparable in size to those from Dorn Dürkheim, which were assigned to *Microstonyx erymanthius brevidens* (Van der Made 1997a). The P3 and particularly the P2 are much smaller and the I2 is much longer than in Dorn Dürkheim and resemble specimens from Samos in the NMB.

Some of the other specimens from Kavakdere have molars that are even smaller than those from Dorn Dürkheim. The P3 from Loc. 33 is very small. It has a separate metacone and a protocone that is separate and within the cingulum, and its cingula are in general well marked (fig. 13.2N–P). The P4 from Loc. 26 has the paracone and metacone well separated (fig. 13.2Q–S). This specimen and two molars from the same locality are small.

Locs. 26 and 33 of the Sinap project in the upper Kavakdere are correlated with chron C4.lr at ~8.1 Ma (Kappelman et al., chapter 2, this volume). The old collections from Kavakdere do not indicate exact locality or level, but virtually all rich fossil occurrences are at the same level as Locs. 26 and 33.

Discussion. Pickford and Ertürk (1979) did not distinguish *Microstonyx erymanthius* from *M. major,* the species they listed from Kavakdere. The small size of some of the specimens is possibly the cause for the citation of *Korynochoerus* sp. from Kavakdere (Hünermann 1975).

Van der Made and Moyà-Solà (1989) noted that there are important size differences within the material assigned to *Microstonyx major* and recognized *M. major erymanthius* as a subspecies. Two possible explanations were offered: geographical differences and size decrease with time. Van der Made et al. (1992) compared Spanish material with that from France and Greece (Pikermi and Samos, Greece): the material from Pikermi and Samos is smaller and has a number of progressive characters, notably longer I2–3, and more reduced canines. Again, two possible explanations were offered: the Greek material is younger, and the Spanish material maintained primitive characters. Fortelius et al. (1996b) treated *M. erymanthius* as a distinct species, with a stratigraphical distribution that overlaps with that of *M. major.*

The oldest material currently assigned to *Microstonyx erymanthius* is from MN 11 and is placed in the subspecies *M. e. brevidens* (Van der Made 1997a). It is relatively small, retains relatively short I2–3, and may retain the upper and lower first premolar. *Microstonyx erymanthius* samples from Pikermi, Samos, and Kavakdere have longer incisors and appear to have the second and third premolars reduced in size (fig. 13.7). The lower first premolar has disappeared and even the upper is absent in many individuals. There might be an increase in size in the latest populations in MN12. It is at present not clear whether the material from MN13 represents large *M. erymanthius* or *M. major.*

Discussion

The distribution in time of the suoid finds in the Sinap area is not even. The suoid-producing localities of İnönü I-level (AKI3 and Loc. 24A) are relatively rich in suoids, with four species recorded. The few finds thought to derive from the lower Sinap member represent two species. The lower Sinap member is thought to be roughly coeval with the later part of MN 7+8. There are altogether about eight suoid localities from the late Miocene middle Sinap member, with a total of five species spanning 10.9 to ~6 Ma (MN9–MN12/13). The maximum number of suoid taxa recorded from a single locality is two, except for the suoid-rich İnönü I.

The known (or at least published) distribution of the Sinap suoids suggests European (*Taucanamo, Bunolistriodon latidens, Listriodon splendens,* cf. *Propotamochoerus provincialis, Hippopotamodon antiquus, Microstonyx major, M. erymanthius*), north or east Asian (*L. splendens, Microstonyx*), Indian (*Kubanochoerus, M. major*) or African (*Kubanochoerus*) affinities. *Schizochoerus* is mainly known from the area around the Black Sea; it dispersed to Spain very late in its history for a brief period (Pickford 1988; Van der Made and Hussain 1989; Fortelius et al. 1996b; Van der Made 1996b). Although the north Asian record is insufficiently known, it seems likely that many of the European suoids came from or through north Asia, and the Sinap suoids might be part of a primarily Asian biogeographic complex, some elements of which reached as far west as France and Spain.

The ranges of the suoid taxa, the Sinap area localities and some correlated localities are indicated in figure 13.8. The localities of the İnönü I level (AKI 3 and Loc. 24A) yielded *Taucanamo inonuensis, Bunolistriodon latidens, Listriodon*

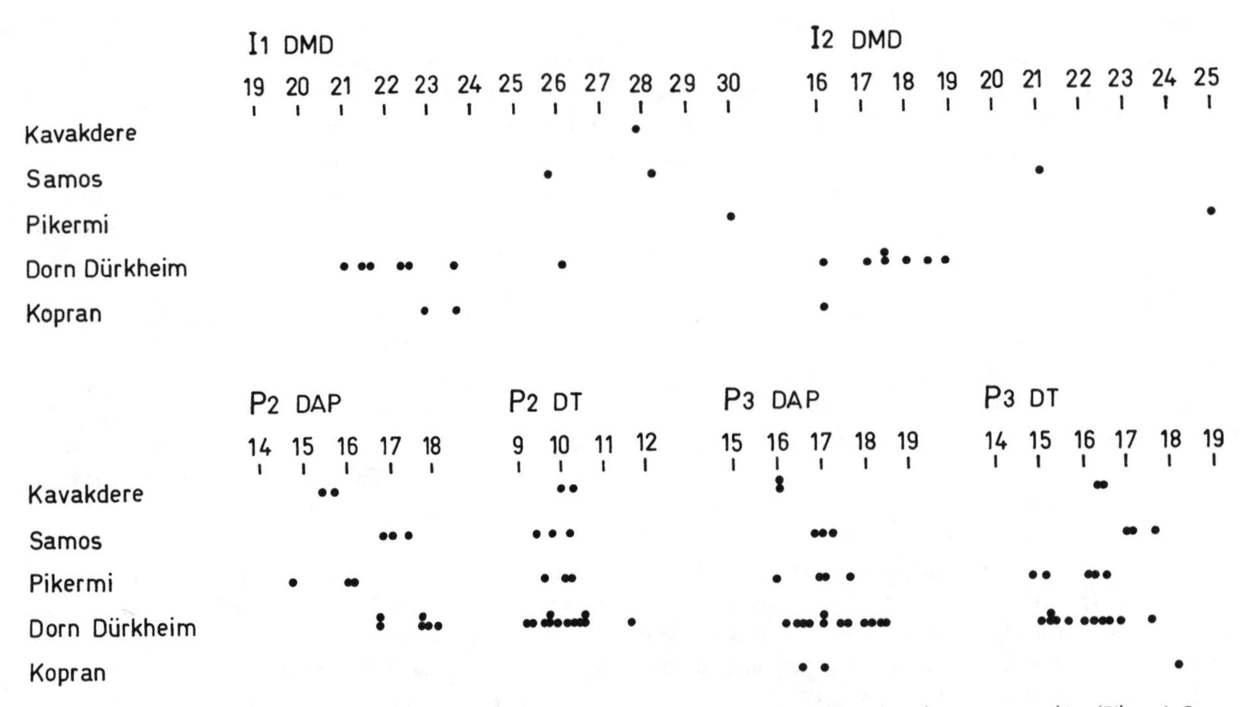

Figure 13.7. Size of selected teeth of *Microstonyx erymanthius brevidens* (Kopran, Dorn Dürkheim) and *M. e. erymanthius* (Pikermi, Samos, Kavakdere). The localities are in approximate stratigraphical order. MN 11: Lower Maragheh-Kopran (NMW), Dorn Dürkheim (FISF; Van der Made 1997). MN 12: Pikermi (NMW, casts Helsinki, MGL; incisors Gaudry, 1862–1867), Samos (NMB), Kavakdere and Loc. 33.

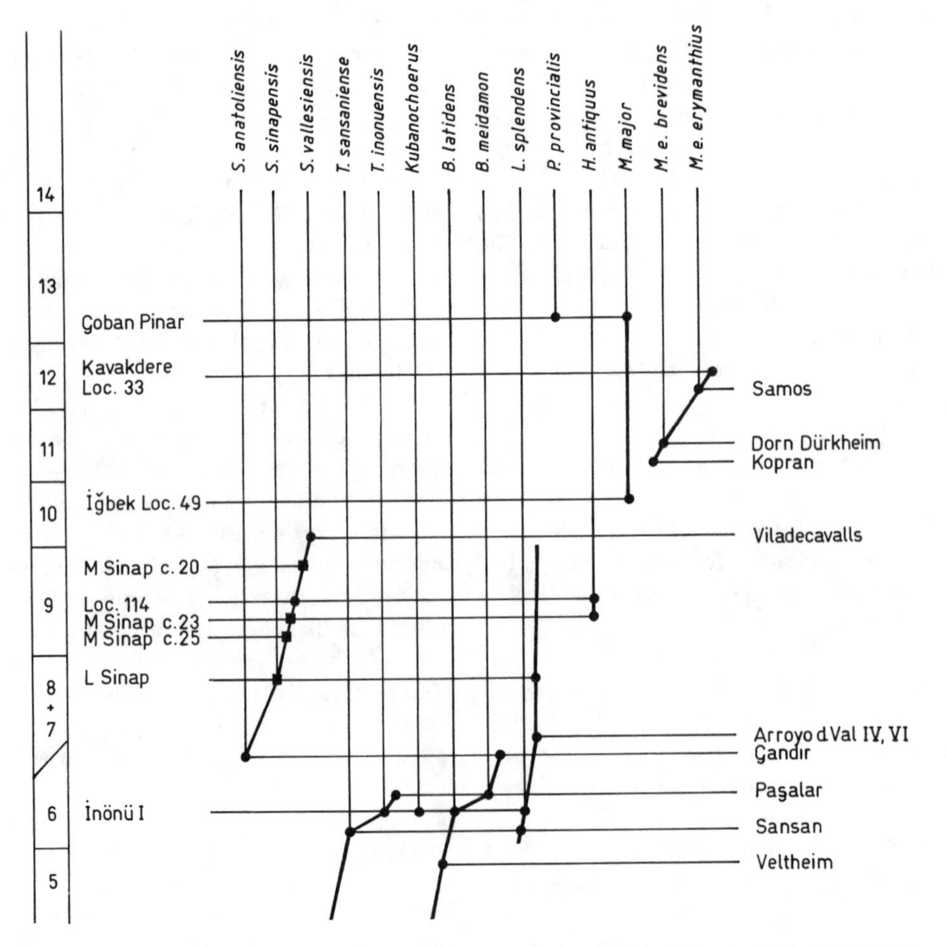

Figure 13.8. The suoid lineages from the Sinap Tepe area. On the left: MN units (Mein 1975, 1990; De Bruijn et al. 1992). Next column: localities in the Sinap Tepe area. Right column: localities outside the Sinap Tepe area. Center of the figure: suoid lineages. Thick lines indicate suoid lineages; evolution is indicated by the oblique lines.

splendens and *Kubanochoerus*. If *T. inonuensis* is interpreted to be a descendant of *T. sansaniense*, İnönü I should be younger than Sansan (Van der Made 1993, 1997a, 1998).

A lineage *Bunolistriodon latidens-B. meidamon* implies that İnönü I is close in age to Veltheim (probably MN 5, or else earliest MN 6) and Mala Miliva (MN 5) and older than Prebreza, Paşalar, and Çandir (all in MN 6) (Van der Made 1996b). The *Listriodon splendens* lineage evolved much slower than the previous lineage, but implies that İnönü I is close to Paşalar and Çandir and hence older than Arroyo del Val and Manchones.

A long and rapidly evolving *Schizochoerus anatoliensis-S. sinapensis-S. vallesiensis* lineage was assumed (Van der Made 1998) and implies that the specimen assumed to be from lower Sinap (type of *S. sinapensis*) is younger than the material from Çandir (type material of *S. anatoliensis*) and that the material from middle Sinap, Nesebar, and Viladecavalls is still younger. This interpretation is consistent with the conventional stratigraphic positions of these localities.

In Europe, *Hippopotamodon antiquus* is known from a few MN 9 localities. The species was supposed to be typical for MN 9, whereas in MN 10, it is replaced by *M. major* (Van der Made 1990b; Van der Made et al. 1992). This seems to be the case also in the Sinap Tepe area, where *H. antiquus* is found in the MN9 correlative, lower portion of the middle Sinap member, and the material from Loc. 49 (correlative with late MN 10) is tentatively assigned to *M. major.*

The oldest material assigned to *Microstonyx erymanthius* is from MN 11. A lineage *M. e. brevidens-M. e. erymanthius* has been proposed (Van der Made 1997a). The lineage shows increase in I2–3 length, a moderate increase in M3 length, and a reduction in relative premolar size. The type material of the first subspecies is from Dorn Dürkheim (MN 11) and of the second subspecies is from Pikermi (MN 12; inferred age 8.3–8.2 Ma) (Bernor et al. 1996). *Microstonyx e. brevidens* is also present in Kopran (lower Maragheh; MN 11; inferred age 8.24–9 Ma) (Steininger et al. 1996; Bernor et al. 1996) and *M. e. erymanthius* in Samos (MN 12/13; inferred age 7.66–7.1 or 6.2 Ma) (Steininger et al. 1996) and Kavakdere (MN 12; 8.1 Ma) (Kappelman et al. chapter 2, this volume). This lineage is well represented in the fossil record and may prove of great stratigraphic interest.

Çobanpinar yielded *Microstonyx major.* The species has an overlapping stratigraphical distribution with the *M. erymanthius* lineage. The locality was placed in MN 12 (Mein 1977), but was omitted by him from an updated correlation chart (Mein 1990). It appears to be lithostratigraphically higher than the Kavakdere localities, which have been correlated with magnetochron C4r.1r at ~8.1 Ma and with MN 12 (Kappelman et al., chapter 2, this volume). The possible presence of *Propotamochoerus provincialis* and the certain presence of a camel (Van der Made et al., chapter 14, this volume) in this locality, which both are known from MN 13 onward, suggest that the locality might be MN 13 (alternatively, these taxa might have entered Anatolia earlier than they entered Europe).

The suoid record in the Sinap area shows important changes in composition and diversity. *Taucanamo inonuensis, Kubanochoerus,* and *Bunolistriodon latidens* are found in İnönü I but not in younger localities. A sudden change in the suoid faunal composition in southwestern Asia as well as in Europe was found to have occurred late in MN 6 (Van der Made 1990a,b, 1993) or between MN 6 and MN 7+8 (Fortelius et al. [1996b]; this study did not consider the possibility of species ranging through parts of MN units). This turnover event has been interpreted as resulting from climatical changes involving a temperature decrease and increasing seasonality (Van der Made 1993; Fortelius et al. 1996b,c) and is probably ultimately related to a major climatic shift indicated by the isotopic sediment record of the deep ocean between ~15 and 12.5 Ma (Miller et al. 1995). The extinction of the lineages to which *T. inonuensis, K. mancharensis,* and *B. latidens* belong formed part of this process. İnönü I represents a time relatively long before the effects were seen in the Anatolian suoid record, as it is not *B. latidens* (present in İnönü I), but its descendant *B. meidamon* that went extinct.

In Europe, where the record is more abundant and apparently more complete than in Anatolia and in particular the Sinap area, species diversity remained high until the late Vallesian, with up to five species of suoid being found at some localities. A major extinction event occurred between MN 9 and 10 and/or early in MN 10 (~9.7 Ma) (Moyà Solà and Agustí 1990; Krijgsman et al. 1996; Steininger et al. 1996), after which no localities with more than two suid species are known from western Europe. Only the larger body size classes (typical of more open landscapes) survived (Fortelius et al. 1996b). In western Europe, the event is known as the "Mid-Vallesian Crisis" and had an effect on many mammal groups, as well as on flora and is believed to have been a global event, involving a decline in temperature (Van der Made 1990a,b,c; Fortelius et al. 1996). A recent study of European rodent associations suggests a cooling event at ~9.4 Ma and a gradual change from warm-cool to dry-wet seasonality between 9.4 and 8.2 Ma, related to global events (Van Dam 1997). A corresponding change is perhaps recorded in the Sinap Tepe area by the replacement of *Hippopotamodon antiquus* with *Microstonyx major* by 9.1 Ma and the extinction of the *Schizochoerus* lineage, which had a very long record in this area.

Acknowledgments

I thank Mikael Fortelius for inviting me to describe the Sinap suoids and for all the energy he put in this chapter, although he declined to be an author. I thank the following persons for access to material in their care, or help of another type: J. Agustí, L. Alcalá, B. Alpagut, M. Arif, Basu, B. Engesser, Ç. Ertürk, J. Franzen, A. Gentry, L. Ginsburg, E. Güleç, E. P. J. Heizmann, J. Hooker, G. Höck, K. A. Hünermann, T. S. Hussain, L. Kordos, P. Montoya, J. Morales, S. Moyà Solà, M. Philippe, K. Rauscher, G. Saraç, S. Sen, E. Ünay, W. Vasicek, and F. Wiedenmayer, and the referees R. Bernor, M. Pickford, and S. Sen, who provided useful comments on the contents as well as on the English. Support was received from projects PB-93-0114 and PB96-1026-C03-02 and the "Unidades Asociadas" program of the Consejo Superior de Investigaciones Científicas, as well as from the Sinap project.

Appendix

Appendix Table 13.1. Measurements of Lower Deciduous and Associated Permanent Dentition of the Suoidea from the Sinap Tepe Area

Species, Locality, and Specimen Number	Side	Measurement (mm) i2 DMD DLL	cm Li La Po	d2 DAP DTa DTp	d3 DAP DTa DTp	d4 DAP DTa DTm DTp	p4 DAP DTa DTp	m1 DAP DTa Dtp	m2 DAP DTa DTp
Schizochoerus sinapensis									
Yassiören, Middle Sinap MNHN Yas 27	l	4.7	9.5 9.5 7.7				12.2 6.7 6.7		
	r		9.5 9.3 7.5	±7.5 —[1] —	±8.4 — —	16.5 5.2 6.1 7.4		15.6 9.0 10.3	— 11.6 —
Bunolistriodon latidens									
Inönü I, MTA AKI3/438	l					— — 11.7 11.4		— — >13.0	
Inönü I, MTA AKI3/586	l					20.6 8.9 11.3 11.0		16.4 13.1 12.1	20.3 16.6 15.9
Cf. *Propotamochoerus provincialis*									
Çobanpinar 1, PDTFAU —	l					— — — 9.7		19.8 13.2 13.2	
Çobanpinar 1, PDTFAU ÇBP 12	r			— — 4.5	11.9 4.4 6.1			19.4 13.2 13.2	
Microstonyx major									
Çobanpinar, MTA AAEÇ1	r					— — 11.2 12.1		23.4 15.4 15.6	
Microstonyx erymanthius									
Kavakdere, PIMUZ ASV—	l			13.5 5.4 6.6		24.4 8.3 9.9 10.6		21.9 13.2 13.9	
Kavakdere, MTA 1963	l			13.2 4.3 5.4	13.9 5.4 6.6	25.9 8.9 10.8 11.3		25.2 15.8 15.8	
Kavakdere, MTA 1963	l			12.4 4.4 5.1	13.1 5.3 6.7	24.9 8.5 10.3 10.9		22.3 — 14.7	

[1]—, measurement could not be taken because of damage of the fossil.

Appendix Table 13.2. Measurements of the Lower Permanent Dentition of the Suoidea from the Sinap Tepe Area

Species, Locality, and Specimen Number	Side	I1 DMD DLL	I2 DMD DLL	I3 DMD DLL	Cm/Cf Li/DAP La/DT Po	P1 DAP DTa DTp	P2 DAP DTa DTp	P3 DAP DTa DTp	P4 DAP DTa DTp	M1 DAP DTa DTp	M2 DAP DTa DTp	M3 DAP DTa DTp DTpp
							Measurement (mm)					
Bunolistriodon latidens												
Inönü I, MTA AKI3/7	l	11.8	13.6	13.5								
		8.3	9.1	7.7								
	r	11.4	13.6									
		8.1	11.8									
Inönü I, MTA AKI3/6	l								15.4	—[1]	18.5	30.7
									9.7	12.5	14.8	17.7
									..[2]	—	14.9	15.8
												12.5
Inönü I, MTA AKI3/11	r									16.0	19.9	29.9
										12.1	15.8	17.6
										—	15.1	16.1
												12.8
Inönü I, MTA AKI3/573-574	r	10.8	14.8									
		7.9	9.1									
Listriodon splendens												
Inönü I, MTA AKI3/42	l								16.8	—	21.9	35.6
									12.6	—	18.6	21.2
									13.8	±14.9	20.3	19.6
												14.0
Inönü I, MTA AKI3/41	r							17.2	17.3	—	22.3	35.4
								9.6	12.8	—	18.6	21.4
								10.1	13.7	—	20.0	19.7
												14.0
Inönü I, MTA AKI3/333	l	11.4	15.0									
		9.5	11.6									
	r	11.9	9.7									
		15.4	11.9									
Inönü I, MTA AKI3/777	l	11.4										
		10.0										
	r	11.5	13.4	±11.8								
		10.0	11.1	±9.5								
Hippopotamodon antiquus												
Middle Sinap, MNHN —	l	10.4	11.3	11.0			22.9	24.2	23.2	22.4	32.3	54.0
		>14.8	19.5	13.9			9.2	13.7	17.3	19.2	24.9	28.3
							10.2	—	18.8	—	24.7	25.5
												20.2
	r	10.3	10.9		Cf	20.2	22.7	25.8	23.8	24.2	32.6	52.9
		15.5	19.4		16.6	6.9	9.2	13.0	16.9	18.4	25.1	28.3
					10.7	7.8	10.2	14.2	18.9	19.5	25.5	24.5
												19.7
Middle Sinap, MNHN —	l						..		25.8	±26.5		
							8.6		16.3	19.2		
		18.7	—									
Middle Sinap, MTA 1953	l	10.9	10.9	12.1								
		≥16.7	20.2	15.8								
	r	11.1	10.7		22.4	18.0						
		≥16.8	21.1		20.3	7.5						
					17.3	7.7						
Microstonyx												
Çobanpinar, PDTFAU ÇBP621	l						16.8	—	22.6	23.5	29.4	46.3
							6.4	—	—	16.1	20.5	22.4
							7.9	—	15.4	16.8	20.5	21.5
												16.7

[1]—, measurement could not be taken because of damage of the fossil

[2].., measurement was not taken for some other reason. (For instance, part of the bone was still covered by sediment.)

Appendix Table 13.3. Measurements of the Upper Deciduous and Permanent Dentition of the Suoidea from the Sinap Tepe Area

Species, Locality, and Specimen Number	Side	Measurement (mm)						
		P1 DAP DTa DTp	P2 DAP DTa DTp	D3 DAP DTa Dtp	D4 DAP DTa DTp	M1 DAP DTa DTp	M2 DAP DTa DTp	M3 DAP DTa DTp DTpp
Schizochoerus vallesiensis								
Yassiören, MNHN Yas28	r	9.6 5.1 5.4	10.3 5.1 6.1					
Bunolistriodon latidens								
Inönü I, MTA AK13/166	r						21.4 21.3 21.1	26.0 23.5 18.4
Hippopotamodon antiquus								
Middle Sinap, MTA 1957	l			22.0 10.9 15.6	24.3 20.6 19.3	27.6 25.4 24.7		
	r			21.6 10.6 15.6	23.9 20.9 29.9	27.9 24.9 24.7		
Microstonyx major								
Çobanpinar, PDTFAU ÇBP40	r						—[1] — 25.0	41.6 27.1 24.6 13.2
Çobanpinar, PDTFAU ÇBP40	r						29.4 25.2 24.3	43.6 27.6 26.1 13.6

[1]—, measurement could not be taken because of damage of the fossil.

Appendix Table 13.4. Measurements of the Upper Permanent Dentition of the Suoidea from the Sinap Tepe Area

Species Locality, and Specimen Number	Side	I1 DMD DLL	I2 DMD DLL	I3 DMD DLL	Cx DAP DT	P1 DAP DTa DTp	P2 DAP DTa DTp	P3 DAP DTa DTp	P4 DAP DTa	M1 DAP DTa DTp	M2 DAP DTa DTp	M3 DAP DTa DTpp
		Measurement (mm)										
Taucanamo inonuensis holotype												
Inönü I, AKI3/4	l				11.2 7.1	>6.5 3.2 3.2						
	r					—[1] 3.2 —		10.5 5.7 6.9	9.7 9.4	10.9 8.9 9.4	12.2 10.6 10.6	14.3 10.4 9.5
Schizochoerus sinapensis												
Lower Sinap, MTA1953	l	≥9.3 ≥8.5	7.0 5.6	5.5 4.3	— 16.3	7.8 4.5 4.3	9.5 5.7 88.4	±12.6 — 10.7	9.9 11.2	13.6 11.7 12.1	17.0 14.3 14.2	20.1 14.9 14.5
	r		7.5 5.7	5.1 4.2	±19.2 15.7	8.1 4.4 4.5	9.8 6.0 7.9	— — 10.9	9.9 11.6		16.5 14.6 14.4	19.4 14.9 14.7
Schizochoerus vallesiensis												
Yassiören, Middle Sinap, MNHN 3337	l							14.1 7.6 10.8	11.7 13.8	18.8 15.9 16.3	22.2 19.4 18.3	27.2 20.4 17.6
	r							13.9 7.6 10.6		18.3 16.1 16.4	22.1 19.8 18.9	27.8 — —
Bunolistriodon latidens												
Inönü I, MTA AKI3/780	l							15.7 9.0 14.8	13.6 14.4	16.9 16.3 16.3	— — —	26.6 20.9 18.9
	r						11.4 6.5 9.1	15.7 9.0 14.6	13.3 13.5	17.0 16.2 16.5	— 20.9 —	10.8
Inönü I, MTA AKI3/9	l								13.2 15.0	— — —	21.4 19.6 21.3	±26.1 <23.7 —
Inönü I, MTA AKI3/577-576	r							14.8 9.4 14.3	13.2 15.1			
Hippopotamodon antiquus												
Loc. 114, AS94-194	r								21.6 25.8	26.9 24.0 23.1	34.0 29.6 30.5	50.0 ..[2] .. ..
Microstonyx major?												
Loc. 49, AS94-1799	r							17.5 12.4 16.8	16.5 20.1 20.7	23.7 — 25.7	30.7 26.0 —	— 28.7 —
M. major												
Çobanpinar, MTA AAEÄ2	l						17.3 8.9 11.4	18.5 11.1 16.6	17.6 21.2	23.8 20.1 20.4		
Kavakdere, MTA 1959	r								16.4 18.4	19.9 19.8 19.8	28.4 24.0 22.1	
Kavakdere, MTA 1950	l			— 6.5			15.7 7.8 10.3	16.0 12.2 16.4	16.0 19.1	23.6 20.2 19.6	31.1 24.8 23.6	39.4 26.5 21.9 12.9
	r		≥27.9 8.4				15.4 7.9 10.0	16.0 11.1 16.3	16.0 19.0	23.6 20.1 19.6	29.7 24.9 24.4	40.1 26.5 22.9 13.2

[1]—, measurement could not be taken because of damage of the fossil.

[2].., measurement was not taken for some other reason. (For instance part of the bone was still covered by sediment.)

Appendix Table 13.5. Measurements of the Isolated Teeth of the Suoidea from the Sinap Tepe Area

Species	Locality	Specimen Number	Tooth	Side	Measurement (mm)			
					DAP	DTa	DTp	DTpp
Taucanamo inonuensis	Inönü I	MTA AKI3/582	P3	l	10.7	5.9	7.1	
		MTA AKI3/581	M1	l	10.6	9.3	9.2	
		MTA AKI3/584	Cx	r	14.6	9.6		
		MTA AKI3/583	Cx	l	±14.6	±9.3		
		MTA AKI3/453	Cx	r	14.6	9.1		
		MTA—	M1	r	12.1	9.7	9.7	
		MTA—	M1	l	11.9	9.4	9.5	
		MTA—	P4	l	9.6	8.1	8.2	
		MTA—	P3		8.2		8.5	
Schizochoerus vallesiensis	M Sinap	MNHN Yas 30	P3		14.0	8.3	10.3	
Kubanochoerus mancharensis	Inönü I	MTA AKI3/779	M1	l	44.4	44.9	42.9	
Bunolistriodon latidens	Inönü I	MTA AKI3/162	M3		33.6	19.3	17.3	12.5
		MTA AKI3/575	p2	l	14.7	7.0	7.9	
		MTA AKI3/588	Cm	l	31.2	28.5		
		MTA AKI3/772	Cm	r	tip			
		MTA AKI3/327	Cm	r	31.9	36.0		
		MTA AKI3/332	Cm	r	27.7	34.0		
		MTA AKI3/325	Cm	r	32.6	41.4		
Listriodon splendens	Inönü I	MTA AKI3/384	p3	r	17.7	10.1	11.5	
		MTA AKI3/357	p3	r	18.3	10.8	12.4	
		MTA AKI3/778	m3	r	35.3	19.7	18.6	13.6
		MTA AKI3/514	m2	l	24.2	19.1	18.7	
		MTA AKI3/252	cm	?	tip			
		MTA AKI3/424	cm	r	tip			
		MTA AKI3/450	M3	l	27.9	23.8	22.2	
		MTA AKI3/451	M3	r	27.5	24.0	22.5	
		MTA AKI3/452	P4	l	15.5	17.9		
		MTA —	M1	r	—[1]	18.2	—	
		MTA —	P4	l	—	≥17.7		
		MTA —	P3	r	17.1	10.9	18.7	
	Loc. 24A Inönü I	AS92-644	Cm		39.6	35.1		
Cf. *Propotamochoerus provincialis*	Locality 42	AS94-1331	d4	r	22.0	8.1	8.6	10.0
	Çobanpinar	AS94-1124	D4	l	15.9	12.3	13.2	
Hippopotamodon antiquus	M Sinap	MTA 1957	M3	r	54.6	37.6	28.9	20.8
	Loc. 114	AS94-175	P1	l	18.3	6.6	6.4	
Microstonyx major	Çobanpinar I	MTA AAEÇ8	M1	l	25.0	15.2	15.4	
		MTA AAEÇ1019	M3	l	48.9	23.6	22.6	17.8
		PDTFAU ÇBP718	p4	r	21.9	12.3	15.0	
		MTA AAEÇ3	M3	r	—	—	24.0	16.6
		MTA AAEÇ5	M2	l	31.6	25.2	25.5	
		MTA AAEÇ4	M3	l	46.7	28.1	24.5	16.4
		PDTFAU ÇBP40	M3	l	—	28.6	—	—
Microstonyx erymanthius	Loc. 33 Kavakdere	AS91-104	P3	l	16.8	9.1	15.7	

Species	Locality	Specimen Number	Tooth	Side	DMD/DAP	DLL/DTa
Bunolistriodon latidens	Inönü I	MTA AKI3/572	i2	l	14.6	9.0
		MTA AKI3/571	I1	l	28.9	11.6
		MTA AKI3/162	I2	r	15.8	7.6
		MTA AKI3/50	I2	r	18.3	8.1
		MTA AKI3/49	I2	l	18.5	9.2
Listriodon splendens	Inönü I	MTA AKI3/418	di2	r	8.9	5.8
		MTA AKI3/450	I2	r	13.3	7.1
		MTA AKI3/579	I2	r	12.2	6.1
Microsotnyx major?	Loc. 49 Igbek	AS95-46	i1	r	9.2	14.5

[1]—, measurement could not be taken because of damage of the fossil.

Appendix Table 13.6. Measurements of the Postcranial Skeleton of the Suoidea from the Sinap Tepe Area

Species	Locality	Specimen Number	Bone	Side	Measurement (mm)	
					DAPd	DTd
Bunolistriodon latidens	Inönü I	MTA AKI3/122	Distal central metapodial	..	22.5	22.7
Hippopotamodon antiquus	Loc. 37 Inönü II	AS92-343	Distal lateral metapodial	r	21.7	17.2

Literature Cited

Agustí, J., 1981, Roedores miomorfos del Neógeno de Cataluña [Ph.D. thesis]: Barcelona, University of Barcelona, 208 pp.

Arambourg, C., 1963, Continental vertebrate faunas of the Tertiary of North Africa, *in* F. C. Howell, and F. Boulière, eds., African ecology and human evolution: Chicago, Aldine, pp. 55–64.

Becker-Platen, J. D., O. Sickenberg, and H. Tobien, 1975, Die Gliederung der känozoischen sedimente der Türkei nach vertebraten-faunenruppen: Geologisches Jahrbuch, Reihe B, v. 15, pp. 19–100.

Bernor, R. L., N. Solonias, C. C. Swisher III, and J. A. Van Couvering, 1996, The correlation of three classical "Pikermian" mammal faunas—Marageh, Samos, and Pikermi—with the European MN unit system, *in* R. L. Bernor, V. Fahlbusch, and H.-W. Mittmann, eds., The evolution of western Eurasian Neogene mammal faunas: New York, Columbia University Press, pp. 137–154.

Biedermann, W.G.A., 1873, Petrefacten aus der Umgegend von Winterthur 4. Reste aus Veltheim: Winterthur, Switzerland, J. Westfehling, 16 pp. + 9 plates.

Bruijn, H. de, R. Daams, G. Daxner-Höck, V. Fahlbusch, L. Ginsburg, P. Mein, J. Morales, E. Heizmann, D. F. Mayhew, A. J. van der Meulen, N. Schmidt-Kittler, and M. Telles Antunes, 1992, Report of the RCMNS working group on fossil mammals, Reisensburg 1990: Newsletters on Stratigraphy, v. 26, no 2/3, pp. 65–118.

Colbert, E. H., 1934, An upper Miocene suid from the Gobi Desert: American Museum Novitates, v. 690, pp. 1–7.

Crusafont, M., and R. Lavocat, 1954, *Schizochoerus* un nuevo género de suidos del Pontienese inferior (Vallesiense) del Valles Penedes: Notas y Comunicaciones del Instituto Geologico y Minero de España, v. 36, pp. 79–90.

Daams, R., A. J. van der Meulen, M. A. Alvarez Sierra, P. Peláez-Campomanes, and W. Krijgsman, 1999, Aragonian stratigraphy reconsidered, and a re-evaluation of the middle Miocene mammal biochronology in Europe: Earth and Planetary Sciences Letters, v. 165, pp. 287–294.

Dam, J. A. van, 1997, The small mammals from the Upper Miocene of the Teruel-Alfambra region (Spain): Paleobiology and paleoclimatic reconstructions: Geologica Ultraiectina, v. 156, p. 204.

Fortelius, M., and R. L. Bernor, 1990, A provisional systematic assessment of the Miocene Suoidea from Paşalar, Turkey: Journal of Human Evolution, v. 19, pp. 509–528.

Fortelius, M., J. van der Made, and R. L. Bernor, 1996a, A new listriodont suid, *Bunolistriodon meidamon* sp. nov., from the Middle Miocene of Anatolia: Journal of Vertebrate Paleontology, v. 16, pp. 149–164.

Fortelius, M., J. van der Made, and R. L. Bernor, 1996b, Middle and Late Miocene Suoidea of central Europe and the eastern Mediterranean: Evolution, biogeography, and paleoecology, *in* R. L. Bernor, V. Fahlbusch, and H.-W. Mittmann, eds., The evolution of western Eurasian Neogene mammal faunas: New York, Columbia University Press, pp. 348–377.

Fortelius, M., L. Werdelin, P. Andrews, R. L. Bernor, A. Gentry, L. Humphrey, H.-W. Mittmann, and S. Viranta, 1996c, Provinciality, diversity, turnover and paleoecology in land mammal faunas of the later Miocene of western Eurasia, *in* R. L. Bernor, V. Fahlbusch, and H-W. Mittmann, eds., The evolution of western Eurasian Neogene mammal faunas: New York, Columbia University Press, pp. 414–448.

Gabunia, L. K., 1955, New representative of the Suidae from the middle Miocene of Belometschetskaia (Northern Caucasus): Doklady Akademia Nauk USSR, v. 102, no. 6, pp. 1203–1206.

Gabunia, L. K., 1958, On a skull of a horned pig from the middle Miocene of the Caucasus: Doklady Akademia Nauk Azerbaijan SSR, v. 18, no. 6, pp. 1187–1190.

Gaudry, A., 1862–1867, Animaux fossiles et géologie de l'Attique: Paris, F. Savy, 75 plates.

Gervais, P., 1848–1852, Zoologie et paléontologie françaises (animaux vertébrés) ou nouvelles recherches sur les animaux vivants et fossiles de la France: Paris, Bertrand, v. 1–3.

Gervais, P., 1859, Zoologie et paléontologie françaises, 2nd edition: Paris, Bertrand, 544 pp.

Ginsburg, L., 1974, Les Tayassuidés des phosphorites du Quercy: Palaeovertebrata, v. 6, pp. 55–85.

Golpe Posse, J. Ma., 1972, Suiformes del Terciario Español y sus yacimientos: Paleontolgía y Evolución, v. 2, pp. 1–197, 7 plates.

Gray, J. E., 1821, On the natural arrangement of vertebrose animals: London Medical Repository, v. 15, no. 1, pp. 296–310.

Guérin, C., and M. Faure, 1985, Les Suidae (Mammalia, Artiodactyla) du Pliocène de la Formation de Perpignan (Roussillon). Hommage à Charles Depéret: Paléontologie et Géologie Néogènes en Roussillon, p. 22.

Guérin, C., M. Faure, and S. Sen, 1998, Le gisement de vertébrés Pliocènes de Çalta, Ankara, Turquie. 8. Suidae: Geodiversitas, v. 20, no. 3, pp. 441–453.

Gürbüz, M., 1981, İnönü (KB Ankara) Orta Miyosenindeki *Hemicyon sansaniensis* (Ursidae) türünün tanimlanmasi ve stratigrafik yayilimi: Bulletin of the Geological Survey of Turkey, v. 24, pp. 85–90.

Hünermann, K. A., 1968, Die Suidae (Mammalia, Artiodactyla) aus den Dinotheriensanden (Unterpliozän + Pont) Rheinhessens (Südwestdeutschland): Schweizerische Paläontologische Abhandlungen, v. 86, pp. 1–96 + 1 plate.

Hünermann, K. A., 1975, Die Suidae aus dem Türkischen Neogen, *in* O. Sickenberg ed., Die Gliederung des höheren Jungtertiärs und Altquartärs in der Türkei nach Vertebraten und ihre Bedeutung für die internationale Neogen-Stratigraphie: Geologisches Jahrbuch, Reihe B, v. 15, pp. 153–156.

Hunter, J. P., and M. Fortelius, 1994, Comparitive dental occlusal morphology, facet develpment, and microwear in two sympatric species *Listriodon* (Mammalia, Suidae) from the Middle Miocene of western Anatolia (Turkey): Journal of Vertebrate Paleontology, v. 14, no. 1, pp. 105–126.

Kappelman, J., S. Sen, M. Fortelius, A. Duncan, B. Alpagut, J. Crabaugh, A. Gentry, J.-P. Lunkka, F. McDowell, N. Solounias, S. Viranta, and L. Werdelin, 1996, Chronology and biostratigraphy of the Miocene Sinap Formation of central Turkey, *in* R. L. Bernor, V. Fahlbusch, and H.-W. Mittmann, eds., The evolution of western Eurasian Neogene mammal faunas: New York, Columbia University Press, pp. 78–95.

Kaup, J. J., 1833, Description d'ossements fossiles de mammifères 2: Darmstadt, Germany, J. G. Heyer, 31 pp.

Krijgsman, W., M. Garcés, C.G. Langereis, R. Daams, J. van Dam, A.J. van der Meulen, J. Agustì, and L. Cabrera, 1996, A new chronology for the middle to late Miocene continental record in Spain: Earth and Planetary Science Letters, v. 142, pp. 367–380.

Lunkka, J.-P., M. Fortelius, J. Kappelman, and S. Sen, 1999, Chronology and faunas of the Miocene Sinap Formation, Turkey, *in* J. Agusti, L. Rook, and P. Andrews, eds., The evolution of Neogene terrestrial ecosystems in Europe: New York, Cambridge University Press, pp. 238–264.

Lydekker, R., 1878, Notices of Siwalik mammals, Records of the Geological Survey of India, v. 11, pp. 64–104.

Made, J. van der, 1988, Iberian Suoidea (Pigs and Peccaries). Coloquio Homenaje a Rafael Adrover: Bioeventos y Sucesiones faunísticas en el Terciario Continental Iberico, Sabadell, December 14–16, pp. 20–21.

Made, J. van der, 1990a, Iberian Suoidea: Paleontologia i Evoluciò, v. 23, pp. 83–97.

Made, J. van der, 1990b, A range chart for European Suidae and Tayassuidae: Paleontologia i Evoluciò, v. 23, pp. 99–104.

Made, J. van der, 1990c, Paleobiogeography of *Hippopotamodon* and *Microstonyx* in relation to climate, *in* IX RCMNS Congress, Barcelona 1990, Abstracts, p. 223.

Made, J. van der, 1991a, Sexual bimodality in some recent pig populations and application of the findings to the study of fossils: Zeitschrift für Säugetierkunde, v. 56, pp. 81–87.

Made, J. van der, 1991b, Climatical changes and species diversity in Suoidea. XIII INQUA Congress, Beijing, August 2–9, 1991, p. 365.

Made, J. van der, 1993, Artiodactyla and the timing of a Middle Miocene climatical change: Premier Congrès Européen de Paléontologie, Lyon, France, July 1994, p. 128.

Made, J. van der, 1995, When hobby horses are pigs and when opinions converge . . . : Paleontologia i Evoluciò, v. 28–29, pp. 275–277.

Made, J. van der, 1994, Suoidea from the Lower Miocene of Cetina de Aragón, Spain: Revista Española de Paleontología, v. 9, no. 1, pp. 1–23.

Made, J. van der, 1996a, *Albanohyus,* a small pig (Suidae) of the Middle Miocene: Acta Zoologica Cracoviense, v. 39, no. 1, pp. 293–303.

Made, J. van der, 1996b, Listriodontinae (Suidae, Mammalia): Their evolution, systematics and distribution in time and space: Contributions to Tertiary and Quaternary Geology, v. 33, no. 1–4, pp. 3–254.

Made, J. van der, 1997a, The fossil pig from the Late Miocene of Dorn-Dürkheim 1 in Germany: Courier Forschungs-Institut Senckenberg, v. 197, pp. 205–230.

Made, J. van der, 1997b, Systematics and stratigraphy of the genera *Taucanamo* and *Schizochoerus* and a classification of the Palaeochoeridae (Suoidea, Mammalia): Proceedings of the Koninklijke Nederlandse Akademie van Wetenschappen, v. 100, no. 1/2, pp. 127–139.

Made, J. van der, 1997c, On *Bunolistriodon* (=*Eurolistriodon*) and kubanochoeres. Proceedings of the Koninklijke Nederlandse Akademie van Wetenschappen, v. 100, no. 1–2, pp. 141–160.

Made, J. van der, 1998, *Aureliachoerus* from Oberdorf and other Aragonian pigs from Styria: Annalen des Naturhistorischen Museum Wien, v. 99A, pp. 225–277.

Made, J. van der, and F. Alférez, 1988, Dos suidos (Listriodontinae) del Mioceno inferior de Córcoles (Guadalajara, España), Coloquio homenaje a Rafael Adrover: Bioeventos y sucesiones faunisticas en el Terciario continental Iberico, Sabadell, December 14–16.

Made, J. van der, and D. Han, 1994, The Suoidea from the hominoid locality Lufeng (Yunnan, China): Proceedings of the Koninklijke Nederlandse Akademie van Wetenschappen, v. 97, no. 1, pp. 27–82.

Made, J. van der, and S. T. Hussain, 1989, *"Microstonyx" major* (Suidae, Artiodactyla) from Nagri: Estudios Geológicos, v. 45, pp. 409–416.

Made, J. van der, P. Montoya, and L. Alcalá, 1992, *Microstonyx* (Suidae, Mammalia) from the Upper Miocene of Spain: Géobios, v. 25, no. 3, pp. 395–413.

Made, J. van der, and S. Moyà-Solà, 1989, European Suinae (Artiodactyla) from the Late Miocene onwards: Bolletino de la Società Paleontologica Italiana, v. 28, no. 2/3, pp. 329–339.

Matthew, W. D., 1924, Third contribution to the Snake Creek fauna: Bulletin of the American Museum of Natural History, v. 50, pp. 59–210.

Mein, P., 1975, Proposition de biozonation du Néogène Méditerranéen à partir des mammifères: Trabajos Sobre Neógeno-Cuaternario, v. 4, p. 112.

Mein, P., 1977, Tables 1–3, *in* M. T. Alberdi, and E. Aguirre, eds., Round table on mastostratigraphy of the W. Mediterranean Neogene: Trabajos Sobre Neógeno Cuaternario, v. 7.

Mein, P., 1990, Updating of MN zones, *in* E. H. Lindsay, V. Fahlbusch, and P. Mein, eds., European Neogene mammal chronology: New York, Plenum Press, pp. 73–90.

Meyer, H. von, 1846, Mitteilungen an Prof. Bronn gerichtet (Brief): Neues Jahrbuch für Mineralogie, Geologie und Paläontologie, pp. 462–476.

Miller, K. G., G. S. Mountain, The Leg 150 Shipboard Party, and Members of the New Jersey Coastal Plain Drilling Project, 1996, Drilling and dating New Jersey Oligocene-Miocene sequences: Ice volume, global sea level, and Exxon records: Science, v. 271, pp. 1092–1095.

Moyà-Solà, S., and J. Agustí, 1990, Bioevents and mammal successions in the Spanish Miocene, *in* E. H. Lindsay, V. Fahlbusch, and P. Mein, eds., European Neogene mammal chronology: New York, Plenum Press, pp. 257–373.

Nikolov, I., and E. Thenius, 1967, *Schizochoerus* (Suidae, Mammalia) aus dem Pleistozän von Bulgarien: Annalen des Naturhistorischen Museums, v. 71, pp. 329–340.

Öngür, T., 1976, Kisilcahamam, Camlidere, Celtikci ve Kazan dolayinin jeoloji durumy ve jeotermal enerji olanaklari, unpublished report: Ankara, Maden Tetkik ve Arama.

Ozansoy, F., 1957a, Postions stratigraphiques des fromations continentales du Tertiare de l'Eurasie au point de vue de la chronologie Nord-Americaine: Bulletin of the Mineral Research and Exploration Institute of Turkey, v. 49, pp. 11–28.

Ozansoy, F., 1957b, Faunes de mammifères de Tertiarie de Turquie et leurs revisions stratigraphiques: Bulletin of the Mineral Research and Exploration Institute of Turkey, v. 49, pp. 29–48.

Ozansoy, F., 1965, Études des gisements continentaux et de mammifères du Cénozoïque du Turquie: Mémoires de la Societé Géologique de France, Nouvelle Série, v. 44, pp. 1–92.

Pickford, M., 1976, A new species of *Taucanamo* (Tayassuidae, Mammalia) from the Siwaliks of the Potwar Plateau, Pakistan: Pakistan Journal of Zoology, v. 8, no. 1, pp. 13–20.

Pickford, M., 1978, The taxonomic status and distribution of *Schizochoerus* (Mammalia, Tayassuidae): Tertiary Research, v. 2, no. 1, pp. 29–38.

Pickford, M., 1988, Revison of the Miocene Suidae of the Indian subcontinent: Münchener Geowissenschaftliche Abhandlungen, Reihe A, Geologie und Paläontologie, v. 12, pp. 1–91.

Pickford, M., 1995, Old World suoid systematics, phylogeny, biogeography and biostratigraphy: Paleontologia i Evolució, v. 26/27, pp. 237–269.

Pickford, M., and Ç. Ertürk, 1979, Suidae and Tayassuidae from Turkey: Bulletin of the Geological Survey of Turkey, v. 22, pp. 141–154.

Pickford, M., and S. Moyà Solà, 1994, *Taucanamo* (Suoidea, Tayassuidae) from Els Casots, early middle Miocene, Spain. Comptes Rendus de l'Académie des Sciences, Paris, Série II, v. 319, pp. 1569–1575.

Pilgrim, G. E., 1925, Presidential address to the geological section of the 12th Indian Science Congress, *in* Proceeedings of the 12th Indian Science Congress: pp. 200–218.

Pilgrim, G. E., 1926, The fossil Suidae of India: Palaeontologica Indica, New Series, v. 8, no. 4, pp. 1–65, plates 1–20.

Pomel, A., 1848, Observations paléontologiques sur les hippopotames et les cochons: Archives des Sciences Physisques et Naturelle Genève, v. 8, pp. 155–162.

Roth, J., and A. Wagner, 1854, Die fossilen Knochenüberreste von Pikermi in Griechenland: Abhandlungen der Bayerische Akademie der Wissenschaften, München, v. 7, pp. 371–464.

Schmidt-Kittler, N., 1971, Die obermiozäne Fossillagerstätte Sandelzhausen 3. Suidae, Artiodactyla, Mammalia: Mitteilungen der Bayerischen Staatssammlung für Paläontologie und Historische Geologie, v. 11, pp. 129–170.

Sen, S., 1991, Stratigraphie, faunes de mammifères et magnétostratigraphie du Néogène de Sinap Tepe, Province d'Ankara, Turquie: Bulletin du Muséum National d'Histoire Naturelle, ser. 4e, v. 12, no. C3/4, pp. 243–277.

Sen, S., G. Bouvrain, and D. Geraads, 1998, Pliocene vertebrate locality of Çalta, Ankara, Turkey. 12. Palaeocecology, biogeography and biochronology: Geodiversitas, v. 20, no. 3, pp. 497–510.

Sickenberg, O., J. D. Becker-Platen, L. Benda, D. Berg, B. Engesser, W. Gaziry, K. Heissig, K. A. Hünermann, P. Y. Sondaar, N. Schmidt-Kittler, K. Staesche, U. Staesche, P. Steffens, and H. Tobien, 1975, Die Gliederung des höheren Jungtetiärs und altquartärs in der Türkei nach Vertebraten und ihre Bedeutung für die internationale Neogen-stratigraphie (Känozoikum und Braunkohlen der Türkei, 17): Geologisches Jahrbuch, Reihe B, v. 15, pp. 1–167.

Simpson, G. G., 1945, The principles of classification and a classification of mammals: Bulletin of the American Museum of Natural History, v. 85, pp. 1–350.

Steininger, F. F., W. A. Berggren, D. V. Kent, R. L. Bernor, S. Sen, and J. Agustí, 1996, Circum-Mediterranean Neogene (Miocene and Pliocene) marine-continental chronologic correlations of European mammal units, *in* R. L. Bernor, V. Fahlbusch, and H.-W. Mittmann, eds., The evolution of western Eurasian Neogene mammal faunas: New York, Columbia University Press, pp. 7–46.

Tekkaya, I., 1989, The vertebrate fossil fauna of Middle Miocene (Astaracian) of Sarilar village (İnönü-I) at Ayaş in Ankara. Aksay nitesi, Bilimsel toplanti bildirileri-I ODT (Middle East Technical University): Türkiye Bilimsel ve Teknik Arastirma Kurumu, pp. 156–162.

Thenius, E., 1972, *Microstonyx antiquus* aus dem Alt-Pleistozän Mittel-Europas. Zur taxonomie und evolution der Suidae (Mammalia): Annalen des Naturhistorischen Museum, Wien, v. 76, pp. 539–586.

Trofimov, B. A., 1954, The fossil suids of the genus *Microstonyx, in* Tertiary mammals, Part 2. On the Mammalia of the southern SSSR and Mongolia: Doklady Akademia Nauk SSR, v. 47, pp. 61–99.

14

Camelidae (Artiodactyla)

J. van der Made, J. Morales, S. Sen, and F. Aslan

Camelidae originated during the Eocene in North America. They began to diversify during the latter part of the Tertiary and continued to do so until the Pleistocene, when their diversity waned. The Camelidae reached South America during the Pliocene (Menégaz and Ortiz Jaureguizar 1995). The living forms are native to South America, Africa, and Asia and include *Lama guanicoe* (guanaco), the domestic llama and alpaca (both derived from the wild guanaco), and *Lama vicugna* (vicuña). The Camelidae reached the Old World during the Late Miocene The earliest Old World camels, believed to be descendants of *Procamelus* or *Megacamelus* (Pickford et al. 1995), are placed in various species of the genus *Paracamelus*. Later forms are assigned to the genus *Camelus* and include *Camelus bactrianus* (bactrian camel) and *Camelus dromedarus* (dromedary, known only as a domesticated form).

Camels tend to be rare in fossil collections, and therefore the date of their arrival in the Old World was initially underestimated: the date moved back as collections grew. At present, the earliest record is from MN 13 (Morales et al. 1980; MN = Neogene Mammal Units—Mein 1977; de Bruijn et al. 1995). However, the number of localities of this age that have good biostratigraphical or geochronological control is still small. The find from Çobanpinar is one of the oldest Old World camels and increases our knowledge of this form.

Age of the Locality of Çobanpinar

Mein (1977) placed the locality of Çobanpinar in MN 12 without explanation, which would make the camel from Çobanpinar the oldest known specimen in the Old World. As a result, an accurate estimate of the age of the locality and thus of the camel is important. We used rodents to assess the age of the locality.

During the 1995 summer field work at Çobanpinar, one of us (S. Sen) took a pilot sample of <100 kg of sediment from the level that contains the large mammal fossils. Screenwashing of the sample also yielded 24 isolated rodent teeth, some of them fragmentary, representing eight species (number of specimens is given in parentheses):

Byzantinia sp. I small (3)
Byzantinia sp. II large (3)
Parapodemus sp. (2)
Occitanomys cf. *provocator* de Bruijn, 1976 (5)
Cf. *Paraethomys* sp. (1)
Pseudomeriones cf. *rhodius* Sen, 1977 (8)
Tamias sp. (1)
Hystrix primigenia (Wagner, 1848) (1)

Byzantinia is a common genus in late Miocene localities of Turkey, where it is represented by two species that are well differentiated by size and morphology. In Greece several Turolian localities also yielded *Byzantinia*. It appears that this genus does not cross the Miocene/Pliocene boundary. *Occitanomys* cf. *provocator* was previously described from Pikermi (Chomateri), a locality dated late Turolian (de Bruijn 1976) or middle Turolian (de Bruijn et al. 1992). *Pseudomeriones rhodius* was found at Maritsa (Rhodes, Greece) and Ano Metochi 3 (northern Greece). These localities were attributed to late Turolian or, in the case of Maritsa, to early Ruscinian (Van der Meulen and Van Kolfschoten 1986). The Çobanpinar *Pseudomeriones* is similar in size to the Greek material, but the specimens from Maritsa are more derived, having more elongated M1 and M2 and shallower protosinusid on the m2. Taking into account the rodent association from Çobanpinar and the stage of evolution of some key taxa, a late Turolian age (MN 13) should be attributed to Çobanpinar.

Propotamochoerus provincialis first appeared in MN 13 and may be present in Çobanpinar (Van der Made, chapter 13, this volume), corroborating the age suggested by the rodents.

Measurements

All measurements are given in mm.

DAP Anteroposterior diameter, either occlusal or maximal in cheek teeth
DAPb Basal anteroposterior diameter in cheek teeth
DLL Labiolingual diameter in incisors
DMD Mesiodistal diameter in incisors
DMDb Basal mesiodistal diameter in incisors
DTa Transverse diameter of the anterior lobe in cheek teeth
DTm Transverse diameter of the middle lobe in the dp4
DTp Transverse diameter of the posterior lobe in cheek teeth

Description

Genus *Paracamelus* Schlosser, 1903
Type species: *Paracamelus gigas* Schlosser, 1903

Schlosser (1903) provided both species and genus names for material from China. Later other species were named and included in this genus: *P. alexejeevi, P. praebactrianus, P. bessarabiensis, P. alutensis,* and *P. aguirrei.*

Paracamelus cf. *aguirrei* Morales, 1984

Synonymy (for *P. aguirrei* and *P.* cf. *aguirrei*):

1902 Camelidae; Stromer: 110–111, fig. 1.
1973 *Paracamelus spec.;* Raufi and Sickenberg: 84–90, figs. 10c, 10f.
1980 *Paracamelus* sp.; Morales, Soria and Aguirre: 139–142, fig. 1a.
1984 *Paracamelus aguirrei* nova sp.; Morales: 135–161, figs. 16–19.
1993 *Paracamelus;* Pickford, Morales and Soria: 701, fig. 1.
1995 *Paracamelus;* Pickford, Morales and Soria: 641–648 (material from Venta del Moro), plates 79–81.

Sinap Material: Çobanpinar (Geological Survey of Turkey [MTA] collection): left mandible with dp3–m1, right mandible with dp2–m1, symphysis with right di1–dc and left i1 and root of i2, isolated left canine, all probably from the same individual. The specimens were collected during excavations at the site in 1977 by the MTA and are stored in the Museum of the MTA in Ankara, Turkey.

Description: The deciduous incisors are flat (small DLL) and become more asymmetrical from di1 to di3 (fig. 14.1, table 14.1). The crowns are high, crown bases being well marked. The base of the crown of the third incisor is much lower distally than it is mesially. The tips of the crowns are curved mesially and not distally. There is very little relief on the lingual sides of the incisors.

The deciduous canine has a low crown. The crown does not have two lobes as in giraffids. There is no additional lingual cusp or cingulum as in the i3 of *Paracamelus* from Venta del Moro (pl. 79, fig. 2 in Pickford et al. 1995), and the

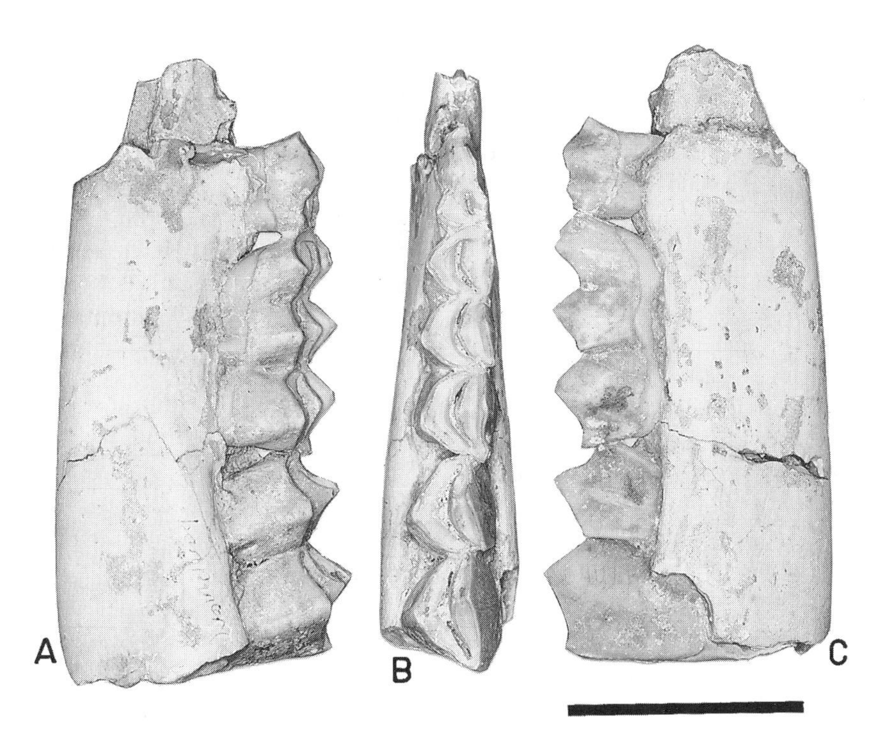

Figure 14.1. Right mandible of *Paracamelus* cf. *aguirrei* from Çobanpinar. **(A)** Buccal, **(B)** occlusal, and **(C)** lingual views. Scale bar = 5 cm.

Table 14.1. Measurements of the Teeth of *Paracamelus* cf. *aguirrei* from Çobanpinar

Side	Measurement (mm)	Tooth diI	di2	di3	Measurement (mm)	Tooth dc	d2	d3	d4	m1
L	DMD	12.5			DAP	16.5		20.0	46.7	43.6
	DMDb	10.4			DAPb				42.7	≤39.5
	DLL	9.3			DTa	6.7		8.9	14.2	≥19.1
					DTm				15.7	
					DTp			10.1	17.6	≥22.2
R	DMD	13.1	13.5	12.3	DAP	14.8	13.0	19.2	47.4	42.6
	DMDb	10.9	11.4		DAPb				43.3	≤38.1
	DLL	8.9	9.3	6.0	DTa	6.3	5.8	8.5	14.4	≥18.6
					DTm				15.6	
					DTp		5.4	10.4	17.6	±22.8

specimens are also much smaller and have lower crowns, as in the Venta del Moro specimen. The difference between this specimen and that from Venta del Moro and the low crown of the canine lead us to believe that the entire anterior dentition might be decidual. The cutting edge of the incisors is very curved, whereas in ruminants it tends to be much less curved.

The dp2 is a small tooth with a single cusp, from which an anterior and a posterior crest depart; there are two divergent roots. The dp3 is also relatively small. Three crests depart from the protoconid in anterior, postero-lingual, and postero-buccal directions. The postero-buccal crest leads to the hypoconid and then continues in a postero-lingual direction. The posterior lobe is much wider than the anterior lobe. The dp4 has three lobes. The tooth has a selenodont structure. Each lobe has a fossid that does not communicate with the adjacent fossid(s). There are no buccal pillars. The lingual wall is fairly flat. The crown is high. The m1 resembles the dp4, except that it lacks the anterior lobe and the crown is higher. The height of the entoconid is >37 mm.

Remarks: As is evident from the description, the combination of characters fits those of a camel but not a ruminant. The occlusal (42.6, 43.4 mm) and basal (≤38.1 mm, ≤39.5 mm) lengths of the m1 indicate a very large animal; for comparison, the length of the m1 of *P. gigas* is 34.7 mm (Zdansky 1926). The length of the M1 is more or less comparable to the m1 length. M1 lengths for *P. aguirrei* are 39 and 39.5 mm and for *P. alexejevi*, 30, 30, 32, 32, 32, and 32 mm (Morales 1984). The species of the genus *Camelus* are smaller than those of the genus *Paracamelus. Paracamelus gigas* is the type species of the genus *Paracamelus.* Schlosser (1903) named the genus and species by describing two teeth from China. He believed that the teeth represented a first and a second upper molar. Their measurements are given as 47 mm × 38 mm and 50 mm × 41 mm, respectively. The M2 was still in the maxilla. The supposed M1 is an isolated tooth and seems to be too large for an M1, but more likely represents an M2. This specimen was figured and should be considered as the type (lectotype, because Schlosser did not indicate the type). These specimens are even larger than their homologues from Venta del Moro. The material from Çobanpinar is referable to either *P. aguirrei* or *P. gigas.*

Discussion

Paracamelus from Çobanpinar is one of the oldest camels of the Old World. The genus has been cited or described from the latest Miocene (MN 13) at Venta del Moro (Morales et al. 1980; Morales 1984; Pickford et al. 1993; Pickford et al. 1995), Librilla (Alberdi et al. 1981) and in the Odessa limestone (Gabunia 1981). Venta del Moro is calibrated as 5.8 Ma, on the basis of biostratigraphy and paleomagnetism (Opdyke et al. 1996). The locality of Librilla is in beds overlying radiometrically dated rocks with ages between 6.2 ± 0.3 and 7.00 ± 0.03 Ma (Montenat et al. 1975). Webb (1965) cited camels from Eldar and the Ischim River. Eldar is correlated with MN 11 (Mein 1977) or MN 10 (de Bruijn et al. 1992), but a faunal list given by Gabunia (1981) does not include camels. We do not know the age of the Ischim River remains, nor can we confirm the presence of camels at this locality. The oldest record of *Paracamelus* in China is in the Yushe area, with an estimated age of ~5.5 Ma (Flynn 1997). A phalanx of a camel from Jalalabad (Raufi and Sickenberg 1973) might be of about the same age, but the accompanying fauna is scanty, and the age of the locality cannot be independently determined. The MN 13 *Paracamelus* specimens tend to be larger than the more recent camels (Morales 1984). The large size of the fossil from Jalalabad is similar to that of the specimen from Venta del Moro, suggesting that the locality is MN 13 or only slightly younger. Stromer (1902) described a camel bone from Wadi Natrun, a locality that is also correlated with MN 13 (Mein 1990).

There are thus six or seven localities in Europe, Asia, and Africa that are placed in MN 13 and/or have estimated ages between 6.2 and 5.5 Ma. The presence of camel remains at MN 10 or other pre–MN 13 localities cannot be confirmed: the accumulating evidence suggests that Camelidae dispersed during MN 13 from North America to the Old World.

Acknowledgments

We thank R. Bernor, M. Pickford, and an anonymous referee for comments and corrections to the English. J. van der Made thanks E. Güleç, who made it possible for him to visit Turkey, and E. Ünay and G. Saraç for help during his stay with the MTA. He also acknowledges support from the Ministerio de Cultura y Ciencia, the Dirección General de Investigación Científica y Tecnica (projects PB96-1026-C03-02 and PB98-0513), and the "Unidades Asociadas" program of the Consejo Superior de Investigaciones Científicas.

Literature Cited

Alberdi, M. T., J. Morales, S. Moyá, and B. Sánchiz, 1981, Macrovertebrados (Reptilia y Mammalia) del yacimiento finimioceno de Librilla (Murcia): Estudios Geológicos, v. 37, pp. 307–312.

Bruijn, H. de, 1976, Vallesian and Turolian rodents from Biotia, Attica and Rhodes (Greece): Proceedings of the Koninklijke Nederlandse Akademie van Wetenschappen, ser. B, v. 79, pp. 361–384.

Bruijn, H. de, R. Daams, G. Daxner-Höck, V. Fahlbusch, L. Ginsburg, P. Mein, J. Morales, E. Heizmann, D. F. Mayhew, A. J. van der Meulen, N. Schmidt-Kittler, and M. Telles Antunes, 1992, Report of the RCMNS working group on fossil mammals, Reisensburg 1990: Newsletters on Stratigraphy, v. 26, no. 2/3, pp. 65–118.

Flynn, L. J., 1997, Late Neogene mammalian events in North China: Mémoires et Travaux de l'École Practique des Hautes Études de l'Institut de Montpellier, v. 21, pp. 183–192.

Gabunia, L., 1981, Traits essentiels de l'évolution des faunes de Mammifères de la région mer Noire-Caspienne: Bulletin du Muséum National d'Histoire Naturelle, ser. 4, v. 3, no. C2, pp. 195–204.

Mein, P., 1977, Tables 1–3 *in* M. T. Alberdi and E. Aguirre, eds., Round table on mastostratigraphy of the W. Mediterranean Neogene. Trabajos sobre el Neogeno y Cuaternario, v. 7: Madrid, Museo Nacional de Ciencias Naturales.

Mein, P., 1990, Updating of MN zones, *in* E. H. Lindsay, V. Fahlbusch, and P. Mein, European Neogene mammal chronology: New York and London, Plenum Press, pp. 73–90.

Menégaz, A. N., and E. Ortiz Jaureguizar, 1995, Los Artiodáctilos, *in* M. T. Alberdi et al., eds., Evolución biológica y climática de la región pampeana durante los últimos cinco millones de años: Madrid, Museo Nacional de Ciencias Naturales, Consejo Superior de Investigaciones Científicas, pp. 311–337.

Meulen, A. J. van der, and T. van Kolfschoten, 1986, Review of the late Turolian to early Biharian mammal faunas from Greece and Turkey: Memorie de la Società Geologica Italiana, v. 31, pp. 201–211.

Montenat, Ch., L. Thaler, and J. Van Couvering, 1975, La faune de rongeurs de Librilla. Correlations avec les formations marines du Miocène terminal et les datations radiométriques du volcanisme de Barqueros (Murcia, Espagne meridionale): Comptes Rendus de l'Académie des Sciences, Paris, v. 281, pp. 519–522.

Morales, J., 1984, Venta del Moro: su macrofauna de mamíferos, y biostratigrafia continental del Mioceno terminal Mediterraneo: Madrid, Editorial de la Universidad Complutense de Madrid, 340 pp.

Morales, J., D. Soria, and E. Aguirre, 1980, Camelido finimioceno en Venta del Moro. Primera cita para Europa occidental: Estudios Geológicos, v. 36, pp. 139–142.

Opdyke, N., P. Mein, E. Lindsay, A. Pérez-González, E. Moissenet, and V. L. Norton, 1996, Continental deposits, magnetostratigraphy and vertebrate paleontology, late Neogene of eastern Spain: Palaeogeography, Palaeoclimatology, Palaeoecology, v. 133, no. 3–4, pp. 129–148.

Pickford, M., J. Morales, and D. Soria, 1993, First fossil camels from Europe: Nature, v. 365, p. 701.

Pickford, M., J. Morales, and D. Soria, 1995, Fossil camels from the Upper Miocene of Europe: Implications for biogeography and faunal change: Geobios, v. 28 no. 5, pp. 641–650.

Raufi, F., and O. Sickenberg, 1973, Zur Geologie und Paläontologie der Becken von Lagman und Jalalabad: Geologische Jahrbuch, v. 3, pp. 63–99.

Schlosser, M., 1903, Die fossilen Säugethiere Chinas nebst einer Odontographie der recenten Antilopen: Abhandlungen der II Classe der kaiserlich-königlichen Akademie der Wissenschaften, v. 22, no. 1, pp. 1–221, 14 pl.

Stromer, E., 1902, Wirbeltierreste aus dem mittleren Pliocän des Natrontales und einige Subfossile und recente Säugetierreste aus Aegypten: Zeitschrift der Deutschen Geologischen Gesellschaft, v. 54, pp. 108–115.

Webb, S. D., 1965, The osteology of *Camelops:* Bulletin of the Los Angeles County Museum, v. 1, pp. 1–54.

Zdansky, O., 1926, *Paracamelus gigas* Schlosser: Palaeontologia Sinica, v. 2, pp. 1–44.

15

Ruminantia (Artiodactyla)

A. W. Gentry

Sediments of the Sinap Formation outcrop in the vicinity of Yassiören, ~55 km northwest of Ankara, Turkey. Ozansoy (1955, 1957, 1965) named the formation and later modified its biostratigraphy and described a number of mammal fossils. Sen (1990; in Bouvrain et al. 1994) reviewed the faunas and concluded that the lower Sinap member is Astaracian-equivalent in age, the middle Sinap member Vallesian, near the MN 9/10 mammal zone boundary, and the upper Sinap member is probably lower Pleistocene. Ozansoy's Loc. Yassiören has often been placed in MN 9 (Steininger et al. 1996). Kappelman et al. (chapter 2, this volume) give the most recent interpretation of the formation.

This chapter examines the ruminants from the recent collections up to and including many of those of the 1994 field season. Successive temporal levels of Sinap localities with ruminants are shown in table 15.1, based on Kappelman et al. (chapter 2, this volume) and on correlations of ruminants. For other localities, the level is still unknown.

Size is important in allocating to species the teeth and postcranial bones at the various localities. In simple cases, such as at Loc. 12, limb bones slightly bigger than in gazelles were put into *Palaeoreas* sp., because a horn core shows that this species is the only one there that might have had limb bones of this size. Overlapping ranges of size are sure to have occurred between different species. Where one or several specimens are found to be close in osteological position and field numbers to specimens apparently of a slightly different size, I have chosen not to separate them at the species level.

Measurements are given in the specimen lists for each species. Occlusal lengths of teeth are summarized in tabular form for each species in tables at the end of the chapter. Tooth measurements are at the occlusal surface. Hypsodonty = m3 metastylid height as a percentage of length. Horn core compression = lateromedial basal diameter as a percentage of anteroposterior basal diameter.

Abbreviations in specimen lists: ap = anteroposterior, bp = basal pillar (entostyle of upper molars, ectostylid of lowers), hc = horn core.

BSPHG = Bayerischen Staatssammlung für Paläontologie und historische Geologie, Munich; MAFI = Hungarian Geological Survey, Budapest; NHMW = Naturhistorische Museum, Vienna; BMNH = Natural History Museum, London; MNHNP = Muséum National d'Histoire naturelle, Paris.

Localities other than Sinap mentioned in the text are as follows; where the MN zone is unknown, MM = middle Miocene, UM = upper Miocene. See Bernor et al. (1996a).

Al Jadidah, Saudi Arabia, MM (Thomas 1983); Arroyo de Val-Barranca, Spain, MN 7/8; Baringo deposits, Kenya, MM and UM and including the Ngorora Formation (Hill et al. 1985; Hill 1995); Belometscheskaya, Georgia, MN 6 (Gabuniya 1973); Bled Douarah, Tunisia, MN 7–9 (Robinson 1972; Geraads 1989b); Candir, Turkey, MN 6 (Köhler 1987); Catakbagyaka, Turkey, MN 7 (Köhler 1987); Chios, Greece, MN 5 (de Bonis et al. 1997); Corak Yerler, Turkey, MN 11 (Köhler 1987); Dhok Pathan, Siwaliks Group Pakistan, UM (Barry and Flynn 1989; Barry 1995); Ditiko 1, 2, and 3, Greece, MN 13 (Bouvrain 1988); Esme Akçakoy, Turkey, MN 9 (Köhler 1987); Fort Ternan, Kenya, MM and dated to 14.0 Ma (Gentry 1970); Garkin, Turkey, MN 11 (Köhler 1987); Grebeniki, Ukraine, MN 11 (Pavlow 1913); Hasnot, Siwaliks Group, Pakistan, UM (Barry and Flynn 1989); Inönü, Turkey, MN 6 (Geraads et al. 1995); Jebel Hamrin, Iraq, UM (Thomas et al. 1980); Kayadibi, Turkey, MN 11 (Köhler 1987); Kemiklitepe D, Turkey, MN 11 (Bouvrain 1994b); Kettlasbrunn, Austria, ?MN 9; Maragheh, Iran (lower Maragheh = MN 11, middle and upper Maragheh = MN 12) (Bernor et al. 1996b); Maramena, Greece, MN 13 (Köhler et al. 1995); Mont Lubéron, France, MN 12 (Heintz 1971); Namurungule Formation, Kenya, UM (Nakaya et al. 1984); Nikiti 1, Greece, MN 11 (Kostopoulos et al. 1996); Paşalar, Turkey, MN 6 (Andrews and

Table 15.1. Stratigraphic Attributions of Sinap Localities[1]

		Locality	
	MN	Kappelman et al.	Possible Attributions Based on Faunal Indications
Middle Sinap member	12		42, 78
		26	27, 28, 33
	11	34	
	10	49	63
		1	
		12	40, ?51
	9	8B	
		91	(13, 45, 96, and 101 around here
		72	based on *Prostrepsiceros* sp.)
		108	
			71
			14, 66
		89	
		94	
		4	
Lower Sinap member	8	64	
		104	
		65	
	5 or 6	24A	

[1]Hipparions have not been found at the localities placed before MN9.

Alpagut 1990); Pentalophos, Greece, MN 9/10 (Bouvrain 1997); Pikermi, Greece, MN 12 or late MN 11 (Solounias 1981b); Piram (formerly Perim) Island, India, UM (Prasad 1974); Prebreza, Serbia, MN 6 (Pavlovic 1969); Ravin de la Pluie, Greece, MN 10; Ravin des Zouaves no. 5, Greece, MN 11 (Bouvrain 1982); Rudabanya, Hungary, MN 9; Sahabi, Libya, UM with some Pliocene (Boaz et al. 1987; Geraads 1989b); Samos, Greece, MN 12 (+ some MN 11) (Solounias 1981a); Sansan, France, MN 6 (Filhol 1891); Sebastopol, Ukraine, MN 9 (Borissiak 1914); Siwaliks Group, Pakistan and India, Miocene-Pliocene (Barry and Flynn 1989); Sofça, Turkey, MN 7 (Köhler 1987); Steinheim, Germany, MN 7 (Heizmann 1976); Vathylakkos, Greece, MN 11/12 (Bouvrain 1982); Zelten, Libya, early MM (Thomas 1979).

Ruminants of the Lower Sinap Member

Ruminants in this member come from Loc. 24A of MN 5 or 6 and Locs. 64 and 65 of MN 8 (Steininger et al. 1996, pp. 31, 32; Kappelman et al., chapter 2, this volume).

Pecora sp., Family indet.

Sinap Material. Loc. 24A: 92.705 dex astragalus, height 14.7, width 7.9, ap 7.2.

Remarks. This pecoran astragalus is not well preserved and belongs to the smallest species to be described in this chapter. It could be moschid, cervid, or bovid, and is perhaps from a subadult individual. It is the size of a smaller neotragine (e.g., *Madoqua saltiana* among extant antelopes), which makes it smaller than the Miocene *Elachistoceras* from the Siwaliks and *?Homoiodorcas* sp. from Al Jadidah (Thomas 1977, 1981) and slightly smaller than six *Micromeryx* astragali at Rudabanya. *Micromeryx* is a moschid or cervid in the Orleanian to Vallesian of Europe, which has also been recorded in China (Qiu et al. 1981). In the Rudabanya astragali, however, width is 65% of the height and the anteroposterior dimension is 55% of the height, whereas in specimen 92.705, these ratios are 54% and 49%, respectively, making it slimmer all round.

Gentry (1990) described a tiny pecoran of unknown family from Paşalar as "*?Micromeryx* sp.," despite noting that its teeth resembled Bovidae more than Cervidae. Only one Paşalar astragalus (1985 S96) was small enough to fit this species; it is incomplete ventrally and has a height and width of ~18.0 × 11.0 mm. Thus its size and relative width (61% of height) again exceed those of the Loc. 24A astragalus.

Family Giraffidae Gray, 1821

Apparent giraffoids are found in the early Miocene of Spain (Morales et al. 1993). Giraffoids of the family

Climacoceratidae are found in Africa in the middle Miocene and probably in the early Miocene. Giraffidae appear early in the middle Miocene in Africa and the Indian subcontinent (Gentry 1994) and also in Europe (de Bonis et al. 1997).

Georgiomeryx georgalasi Paraskevaidis (1940) from Chios has teeth very like the primitive Zelten *Canthumeryx* Hamilton, 1973, but p4 is more advanced (de Bonis et al. 1997). *Giraffokeryx* Pilgrim, 1911, a name frequently applied to middle Miocene giraffids, is based on the Chinji *G. punjabiensis* Pilgrim, 1911, which has two pairs of horns, the main posterior pair being inserted widely apart, behind the orbits, and showing strong divergence (Colbert 1935). On p4 specimens, the transverse crest from protoconid to metaconid is truncated and its separated lingual end (the metaconid proper) forms a closed anterolingual wall with the paraconid.

The very similar *Injana-therium* Heintz, Brunet and Sen, 1981 of Arabia, first described from the late Miocene (see Brunet and Heintz 1983, p. 288), is also claimed for the middle Miocene of Al Jadidah as *I. arabicum* Morales et al., 1987. The African middle Miocene *"Palaeotragus" primaevus* of Fort Ternan (Churcher 1970) would almost certainly have had horns like *Giraffokeryx* and *Injana-therium*, but has slightly more advanced p4s. Its metapodials are long, as is also the metatarsal of *I. arabicum*.

The following is the only giraffid species known from the lower Sinap member.

Giraffidae sp.

Sinap Material. Loc. 24A: 92.623 two pieces of horns, diameters ~18.2 × 19.8; 91.409 sin dP4, mid wear, 27.0 × 23.1 (fig. 15.1A); 91.402 distal part of sin astragalus, width 41.3; 91.810 sin metatarsal lacking most of distal end. (fig. 15.2A), length ~484, minimum shaft width 31.0, ap width proximal articular surface 49.5, transverse width proximal articular surface 48.2; 91.811 dex metacarpal lacking medioposterior part of magnumtrapezoid facet, length 483, minimum shaft width 41.0, ap width of proximal articular surface ~39.2, transverse width of proximal articular surface 64.3, ap width across distal condyles 41.8, transverse width across distal condyles 64.7; 92.616 (in part) badly preserved long bone, perhaps a tibia; 92.631 partial dex proximal metacarpal, width ~54.0; 92.637 dex distal radius, width 61.2.

Loc. 64A: 92.4,16 two tooth fragments, one with rugose enamel.

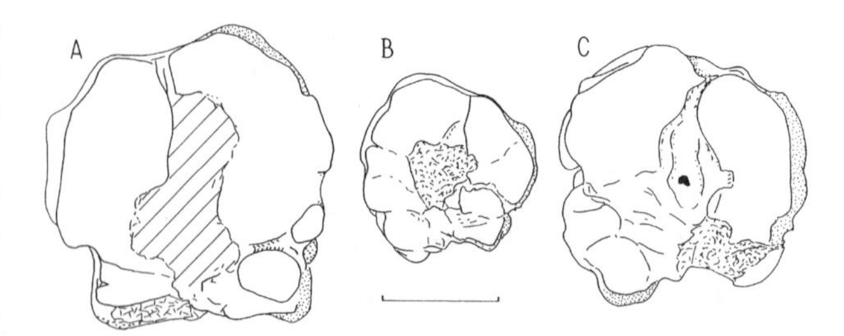

Figure 15.2. Giraffidae, proximal views of metatarsals. **(A)** Giraffidae sp. Loc. 24A 91.810, left. **(B)** *Palaeotragus coelophrys* Loc. 8 92.232, right. **(C)** *Helladotherium* sp. Loc. 33 91.178, right. Anterior toward the top. Diagonal lines = damaged bone. Scale bar = 20 mm for A, 40 mm for B and C.

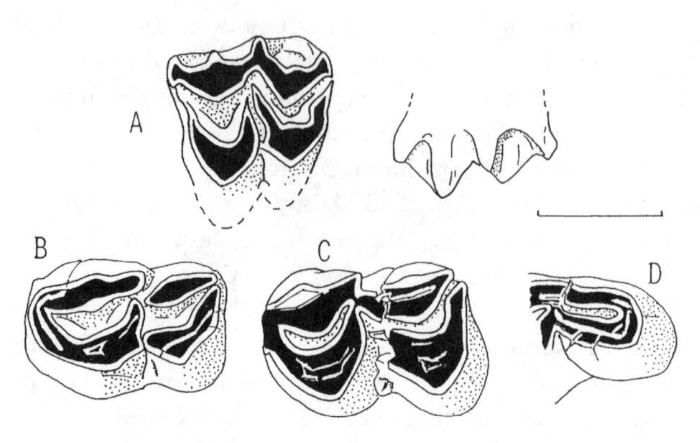

Figure 15.1. **(A)** Giraffidae sp. sin dP4 Loc. 24A 91.409 in occlusal and labial view, drawn from cast. **(B–D)** *Decennatherium macedoniae* Loc. 4 92.381, in occlusal view. **(B)** Sin p4, **(C)** Sin m2 **(D)** back of sin m3. Anterior toward the left. Black = dentine; dots = shading; broken lines = cracks. Scale bar = 20 mm.

Remarks. The two pieces of horns, 92.623, are certainly of small diameter, but the presence of dense bone internally is appropriate for a giraffid.

The dP4 91.409 (fig. 15.1A) is large, but a number of characters show that it is not a molar. First, it is very brachyodont and has a non-misshapen pattern of crescentic cusps and fossettes at an occlusal surface level low above the roots. Second, a strong labial cingulum passes as a wide and flat-topped ridge from the base of the mesostyle to the metastyle. Other characters are a strong, flangelike mesostyle, a small basal pillar, an anterolingual cingulum, and the anterolingual wall of the metaconule being interrupted near its lingual end by a small fold at the occlusal surface. If 91.409 is really a dP4, then an upper molar of the same species would have a length of ~32.0 mm, about the size of a large Turolian giraffid.

The Loc. 24A metapodials are long and slender. On the metatarsal, the anterior groove is pronounced down the entire length of the shaft and the longitudinal hollowing of the posterior surface is deep. There is no foramen at the top of the posterior surface. On the proximal articular surface of the metatarsal (fig. 15.2A), the anterior end of the division between ectocuneiform and naviculocuboid facets seems more nearly in a median position than in late Miocene giraffids. The posterolateral groove for the vestige of metatarsal V is prominent. The sliver of a probable metatarsal II is visible at the top of the medial surface.

The metacarpal has a prominent anteromedial angle on the magnumtrapezoid facet. Such an angle often occurs in ruminants, for example, the Fort Ternan giraffid (Churcher 1970, fig. 43).

Discussion. The morphology of the dP4 91.409 matches *Giraffokeryx* dP4 specimens from Prebreza (Pavlovic 1969, pl. 14, figs. 3–6). My own measurements of the Prebreza left dP4 and M1 lengths were 22.7 mm and 26.9 mm, so the Sinap dP4 is as large as the Prebreza M1. It is also larger than two Paşalar dP4 specimens of *Giraffokeryx* with lengths of 20.5 mm and 20.7 mm. A spur into the rear central fossette from the posterior wall of the metaconule is less obvious than it can be in some *Giraffokeryx*. Late Miocene giraffids have dP4 teeth (Rodler and Weithofer 1890 pl. 4, fig. 1; Arambourg 1959, pl. 14, fig. 2, pl. 15, fig. 6) that are more like permanent molars; for example, in the diminution of the labial basal cingulum on the metacone. They are also larger (like 91.409), although, interestingly, a *Bohlinia* dP4 (Geraads 1979, pl. 1 fig. 1) is not so large. (Dental material of *B. attica* from Pikermi seems very small in relation to the large size of the limb bones.) A *Palaeotragus coelophrys* dP4 from Sebastopol (Borissiak 1914, pl. 1, fig. 1) with a prominent lingual cingulum looks very different from 91.409.

If 91.409 were in fact a molar, it could belong to *Georgiomeryx,* but even the somewhat smaller M3 of *G. georgalasi* (de Bonis et al. 1997, fig. 3) fails to show the strong labial ridge across the base of the metacone and also looks less brachyodont. Specimen 91.409 is not a *Palaeomeryx,* because of the sharpness of the styles and the more fully crescentic shapes of the four crests.

The primitive features of the proximal surface of the metatarsal are more appropriate for a middle Miocene than a late Miocene giraffid, and the large size and length of the metapodials (fig. 15.3) are notable. (*"Palaeotragus" primaevus* from Fort Ternan and *Injana-therium arabicum* from Al Jadidah are both of middle Miocene age, and have smaller metapodials.) Giraffid postcranials from Candir (Köhler 1993, fig. 35) are, however, at least as large as Turolian *Samotherium boissieri* at Samos. The illustrated distal metacarpal from Candir appears to be ~70 mm wide across its condyles, slightly greater than the Loc. 24A specimen. The condyles are higher than in *S. boissieri,* and their outside edges less slanted in anterior view.

It is likely that African giraffids developed very long metapodials and eventually gave rise to *Giraffa.* The Loc. 24A metapodials seem to have advanced along this route and to be moving away from Eurasian *Palaeotragus coelophrys, Samotherium boissieri,* and sivatheres. This may suggest a relationship to the late Miocene *Bohlinia* and perhaps a recent arrival of the Loc. 24A species from Africa. Hence the frequently asserted relationship of *Bohlinia* to *Giraffa* (Gentry and Heizmann 1996, p. 380) might lie rather far back in time.

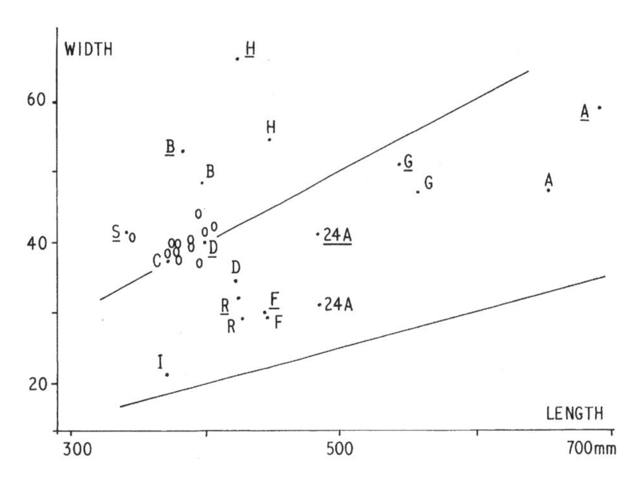

Figure 15.3. Transverse shaft width and total length of metapodials of Giraffidae sp. from Loc. 24A compared with those of other giraffids. The upper diagonal line is that along which width is 10% of length, the lower line is for 5%. Underlined abbreviations indicate metacarpals, others are metatarsals. Symbols: o = 12 left metatarsals of *Samotherium boissieri* from Samos; other readings denote means or single individuals. A = *Bohlinia attica*, B = *Birgerbohlinia schaubi*, C = *Palaeotragus coelophrys*, D = *Decennatherium pachecoi*, F = *"Palaeotragus" primaevus*, G = *P. germaini*, H = *Helladotherium duvernoyi*, I = *Injana-therium arabicum*, R = *P. roueni*, S = *S. boissieri*. Note that species to the right have longer metapodials and those to the top have thicker metapodials, but a small species like *I. arabicum* may still have relatively longer metapodials than a larger one like *P. germaini* because they are less thick. Sources: Author's measurements, Gaudry (1862), Arambourg (1959), Churcher (1970), Morales and Soria (1981), Morales et al. (1987), Montoya and Morales (1991).

Family Bovidae Gray, 1821

Subfamily Hypsodontinae Köhler, 1987

The once widespread Hypsodontinae did not outlast the middle Miocene and may be diphyletic to Bovidae if other bovids are all descended from forms close to *Eotragus, Homoiodorcas,* and *Tethytragus.* The horn cores have rather upright insertions and torsion that is clockwise on the right side, the molars are hypsodont, and the premolar rows are short. The torsion led in the past to discussion of their possible relationship to late Miocene *Oioceros* and to Caprinae (Pilgrim 1934; Gentry 1970). Their nomenclature is still unsettled (Azanza and Morales 1994, p. 275; de Bonis et al. 1998).

The type species of *Hypsodontus, H. miocenicus,* was founded on a conjoined m2 + m3 from Belometscheskaya, and Gabuniya (1973, fig. 33) later allocated to it a horn core from the same locality. This horn core showed slight twisting of its axis but its overall course was little curved. Gabuniya (1973, fig. 34, pl. 8, fig. 6) referred a similar but smaller and slightly compressed horn core to a new genus and species *Kubanotragus sokolovi.* Gentry (1970, p. 273) thought the shallow mandibular ramus of the holotype of *H. miocenicus* might make it a boselaphine. Thomas (1984) concurred and therefore used *Kubanotragus* for some hypsodontines of restricted temporal span (Barry 1995, p. 134) in the Siwaliks. Köhler (1987, p. 155) initiated studies of hypsodontines in Turkey and chose to use the name *Hypsodontus.* She founded *Turcocerus* for hypsodontines with shorter and more grooved horn cores than *Hypsodontus* and more curved horn cores than *Kubanotragus.*

Hypsodontus Sokolov, 1949

Type Species. *Hypsodontus miocenicus* Sokolov, 1949.

Remarks. *Hypsodontus* has somewhat curved horn cores without deep longitudinal grooving, hypsodont teeth, a notably short premolar row, and lower molars with flat lingual walls. The hypsodonty and short premolars can, in some species, reach remarkably advanced states for a middle Miocene pecoran.

Hypsodontus pronaticornis Köhler, 1987

Holotype. Right horn core with part of frontal, from Candir (Köhler 1987, fig. 14).

Sinap Material. Loc. 24A: 92.633 hc piece, possibly this species; 92.650 dex hc base, 28.7 × 24.1; 89.108 (in part) sin upper molar; 91.404 dex dP4–M2, M2 23.7 × 13.0, M1 19.9 × —; 92.616 (in part) dex P4, mid wear, 12.0 × 11.2; 92.634 sin P4+M1, mid wear, P4 11.4 × 10.0, M1 19.4 × 16.3; 92.635 fragments sin upper molar, early mid wear, ~22.5; 92.682 sin maxilla, M2+M3, mid wear, sin P4, P4 12.9 × 10.8, M2 23.6 × 19.1, M3 24.7 × 15.3; 92.737 upper molar fragment; 92.913 two sin upper molars, later mid wear; front one 22.5 × —, rear one 26.3 × 17.7; 92.574 dex mandible (composite specimen); 91.397 sin m3, mid middle wear, 32.2 × 10.9; 91.398 sin m3 early middle wear, 32.0 × 11.9; 92.632 (in part) sin lower molar; 92.679 sin mandible, broken m1, m2 + m3, middle wear, m1–m3 ~68.0, m2 21.2 × 12.7, m3 31.1 × 11.5, mandible depth below m1 36.3; 92.693 sin lower molar early middle wear, anterolingual wall missing, 18.9 × —; 92.734 sin dp3 unworn, 10.8 × 4.8; 91.410 distal sin tibia, width across astragalus facets 28.8; 92.732 dex astragalus; medial height 37.2, width 23.9, ap 23.6; 92.629 pieces of sin naviculocuboid and ectocuneiform, width of former ~29.0; 92.738 fragment metatarsal shaft; 92.655 sin scapula, articular end, ap facet for humerus ~37.0; 91.407 sin distal humerus, lat side; 91.406 dex cuneiform; 92.698 sin metacarpal, probably immature, without distal end, length ~250, minimum shaft width 19.1; 92.624 metapodial distal condyle, ap 20.9; 92.686 first phalanx.

Remarks. The large hypsodont teeth at Loc. 24A, in a late Miocene setting, could have been a small *Criotherium argalioides,* but the Hypsodontinae horn cores suggest otherwise. No molar teeth in early wear have come to light to enable the hypsodonty to be measured, it being usual for Hypsodontinae teeth to be found in a well-worn condition. The teeth and postcranial bones are larger than *Hypsodontus serbicus* Pavlovic (1969) from Prebreza and almost as large as *H. pronaticornis* from Paşalar and Candir. They are assigned to the latter species.

The poorly informative horn core 92.650 is the size of the smallest *Hypsodontus pronaticornis* from Candir. On what is preserved of it, the torsion is less strong than in the species described next. The P4 92.616 has a fairly flat labial wall. The dp3 92.734 is like Paşalar dp3 specimens of *H. pronaticornis* in the labial projection of the hypoconid and the split between parastylid and paraconid. The partial scapula 92.655 shows a rounded facet for articulation with the humerus. The metacarpal 92.698 seems to be longer than in the Tung Gur *Turcocerus grangeri,* which has lengths of 211 mm and 215 mm (Gentry 1970, p. 271) but is shorter than a Candir metacarpal of *H. pronaticornis* (Köhler 1993, fig. 35).

Turcocerus Köhler, 1987

Type Species. *Turcocerus grangeri* (Pilgrim 1934).

Remarks. The type species is from Tung Gur, Mongolia. *Turcocerus* was founded for *Hypsodontus*-like species with short curved horn cores having deep longitudinal grooves. The Turkish *T. gracilis* is the smallest hypsodontine, but Köhler's type species is much larger, and small size is evidently not a part of the generic concept. *Turcocerus* is a senior synonym of *Sinomioceros* Chen (1988).

Turcocerus gracilis Köhler, 1987

Holotype. Right horn core from Candir (Köhler 1987, fig. 23, pl. 5, fig. 1).

Sinap Material. Loc. 24A: 91.411 dex + sin hcs, 30.0 × 25.7 sin (see fig. 15.9A); 92.688 dex hc fragment; 92.641 sin dP2, 8.1 × 4.7; 92.710 sin upper molar early middle wear, 12.3 × 7.2. (fig. 15.4A); 92.706 dex mandible with m1 (broken)–m3, m2 14.1 × 6.6, m3 7.3 × 5.1, mandible depth below m3 24.0; 92.708 sin mandible fragment; 92.896 dex m3, middle wear (possibly this species), 14.8 × 6.5; 89.109 part of dex astragalus, width 15.3; 92.640 second phalanx; 92.654 distal sin immature radius, total width 20.2; 92.685 third phalanx, 20.3 × 10.9 high; 92.689 second phalanx; 92.694 sin proximal metacarpal; 92.707 dex astragalus, height 29.7, width 18.5, ap 16.7; 92.645 metapodial distal condyle.

Remarks. This species has hitherto been recorded only from Candir. The morphology and surface structure of the Loc. 24A horn cores are not suitable for immature horn cores of a larger hypsodontine species (cf. Gentry 1990, p. 544). They are larger than Candir or other Inönü *Turcocerus gracilis* (Geraads et al. 1995). They are mediolaterally compressed, with strong longitudinal grooves and ridges on the medial surface, the extremity ridges appearing as anterior and posterior keels (see fig. 15.9A), initially converging, without sinuses in pedicels, with a shallow postcornual fossa, raised mid-frontals' suture, and frontals probably level with the dorsal orbital rim.

The teeth of this species attain the size of smaller *Tethytragus* at Paşalar. Judged by the Candir specimens (Köhler 1987, fig. 25), the premolar rows might have been shorter than in *Tethytragus,* although longer than in other *Turcocerus* and *Hypsodontus* species. The Sinap upper molar 92.710 (fig. 15.4A) seems more hypsodont than an *Eotragus*

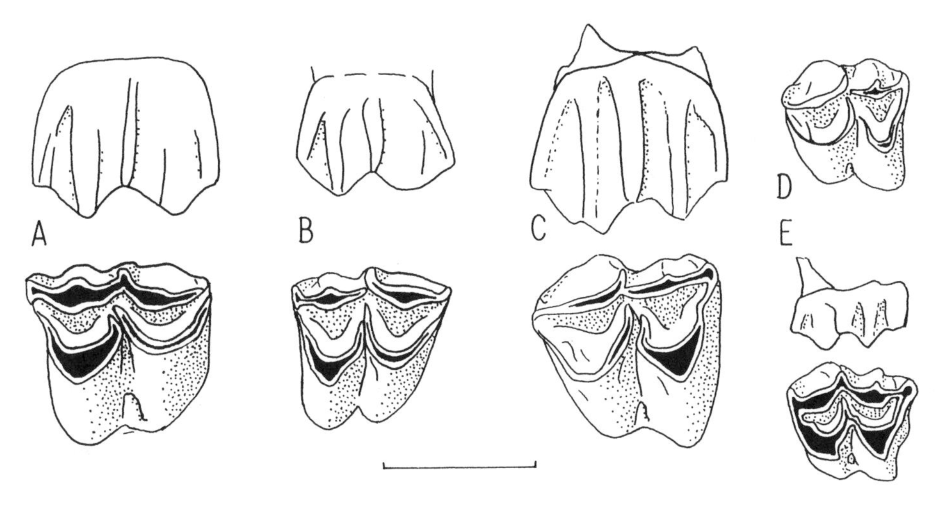

Figure 15.4. (**A**) *Turcocerus gracilis* Loc. 24A 92.710 sin upper molar in labial and occlusal views. (**B**) Bovidae sp., size of *Gazella* Loc. 64 92.601 sin M2/ in labial and occlusal views. (**C**) ?(Bovidae sp., size of *Gazella*) Loc. 64 92.14 dex upper molar in labial and occlusal views. (**D**) Bovidae sp. B Loc. 94 92.366 dex upper molar in occlusal view. (**E**) Bovidae, sp., size of *Gazella*, Loc. 64 92.10 dex dP4 in labial and occlusal views. Labial edge of occlusal views toward the top; **A** and **B** drawn from casts. Scale bar = 10 mm.

from Sansan (cast BMNH M44752), and the protocone is less narrowly pointed lingually. A right m3 in middle wear, 92.896, is wider overall than that of a gazelle, the enamel more rugose, the labial lobes more attenuated, the entostylid forms a more prominent corner, and the central fossettes are longer. The hypoconulid lobe is small and offset labially. It may be too small for this species.

Discussion. Systematic rearrangements of Hypsodontinae have not come to an end. Some brief observations can be made:

1. We should not overemphasize the scale of interspecific distinctions among hypsodontines; there are differences in horn core length, but the torsion is always weak and the degree of curvature is likely to be both individually variable and stronger in shorter horn cores.
2. The species *miocenicus* is distinguishable from *grangeri* (the oldest species name in the group) by its longer horn cores (unless they belong to one species showing an east/west cline) but by little else.
3. The species *serbicus* is very close or identical to *miocenicus.*
4. The species *pronaticornis* is close to *miocenicus-serbicus* but has larger teeth and relatively small horn cores (unless only female horn cores have been found).
5. Neither *sokolovi* nor the similar *gaopoensis* Chen (1990) are generically (or specifically?) distinct from *miocenicus.* Possible female horn cores of *gracilis* at Inönü somewhat resembled *sokolovi* (Geraads et al. 1995, p. 468).
6. The species *gracilis* may not be identical with *noverca,* as it appears to have less shortened premolar rows than any other hypsodontine.
7. The African *tanyceras* has very long and divergent horn cores in males, but females are hornless (Gentry 1970).
8. All the above forms could well go into one genus. In the present phase of studies, it is better to follow Köhler in accepting *Hypsodontus,* than to change it on the basis of unsubstantiated doubts about the tribal identity of the holotype of the type species. Hypsodontinae entered Europe in MN 5 and became extinct in Turkey by the end of MN 6 but later elsewhere (de Bonis et al. 1998; Köhler 1987, pp. 137, 221; Köhler 1993, p. 83).

Tribe ?Antilopini Gray, 1821
Bovidae sp., the size of *Gazella*

Sinap Material. Loc. 64: 92.10 dex P4 and dP4, unworn and middle wear respectively, 6.3 × 5.2, 8.1 × 6.2 (fig. 15.4E); 92.15 dex upper molar middle wear, 11.1 × 8.2; 92.22, 24–27 dex m3, dex M3 or M2, dex P3, sin upper molar all unworn, dex ?dP4 middle wear (the last one could be M1 or M2 of a smaller species), 22 m3 14.8 × 4.6 × ~10.0 high, 24 M3 or M2 11.9 × 7.6 × 9.8 high, 25 P3 7.4 × 6.3, 26 upper molar ~12.4 × 7.3, 27 upper molar or dP4 8.4 × 7.3. (fig. 15.5C,D); 92.32 (in part) dex upper molar middle wear 10.8 × 9.1; 92.34 sin dP3 late wear, 9.6 × 7.0; 92.68 (in part) sin upper molar early middle wear, 11.2 × 8.2; dex upper molar late middle wear, 9.4 × 9.2; 92.583 sin P4–M3 late wear, M1–M3 30.0 long, M1 8.4 × 10.3, M2 10.5 × 11.3, M3 11.3 × 10.2, P4 7.0 × 8.0; 92.601 sin dP4–M3, dP4 late wear, M1 middle, M2 early middle, M3 erupting, dP4 7.8 × 7.7, M1 8.9 × 7.2, M2 10.5 × 7.7 (fig. 15.4B); 92.820 dex P3 unworn, 8.0 × 5.5; 90.280 labial part sin lower molar, length ~12.4; 92.13 sin m3 unworn, 15.2 × 4.4 × 10.4 high; 92.17 dex m3 unworn (presumably the same individual as 92.13) 15.4 × 4.9 × ~10.1 high; 92.31 sin and dex mandible fragments (each with m1–2, the m3s still erupting or

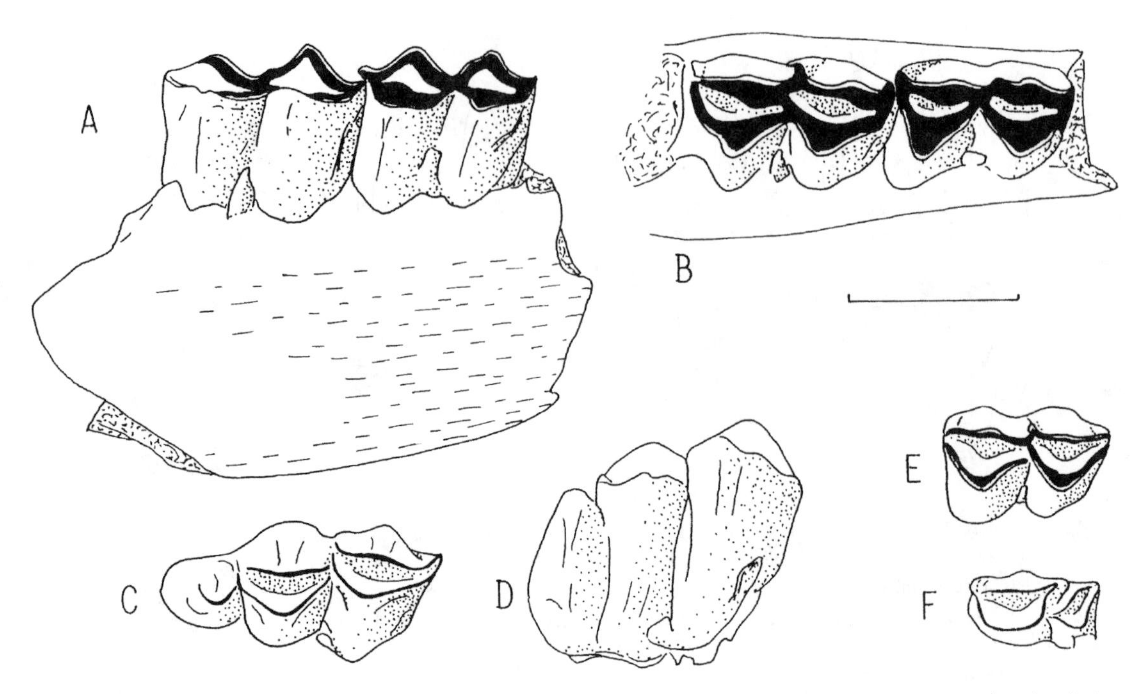

Figure 15.5. (A,B) Bovidae sp., size of *Gazella*, Loc. 64 92.31 dex mandible with m1 and m2 in lateral and occlusal views. (C,D) Bovidae sp., size of *Gazella*, Loc. 64 92.22 dex m3 in occlusal and labial views. (E) Bovidae, smaller sp. Loc. 64 92.30 dex m2 in occlusal view. (F) Bovidae, smaller sp. Loc. 64 92.817 sin p4 in occlusal view. Lingual edge of occlusal views toward the top. Scale bar = 10 mm.

unworn), dex m1 10.0 × 5.6, m2 12.1 × 5.6, mandible depth under m1 14.0 (fig. 15.5A,B); 92.35 fragment dex mandible with dp3 and part dp4 late wear; dp3 8.4 × 3.9; 92.67 dex ?dp3 late wear, 7.7 × 3.9; 92.816 sin lower molar early wear, 9.9 × 4.7 × 5.3 metastylid height; 92.821 fragments lower molar; 90.272 tip calcaneum; 91.481 fragment distal tibia; 91.482 fragment proximal metapodial (?dex metatarsal); 91.484 partial distal fibula; 91.497 most of patella (possibly this species); 92.6 dex immature tibia original length ~170, minimum shaft width 12.2; 92.7 sin astragalus, height 25.5, width 15.0, ap 13.9; 92.8 sin naviculocuboid (same individual as 92.7), width 17.7; 92.18 patella; 92.28 distal sin tibia (too small for 92.7), width 17.5; 93.389 partial proximal sin tibia; 90.270 dex unciform; 90.271 dex magnumtrapezoid; 92.21 partial dex scaphoid; 90.266 distal part of first phalanx; 90.267 proximal part of first phalanx; 90.268–269 metapodial immature distal condyle; 90.282 distal part of first phalanx; 91.277 second phalanx 17.6 × 9.1 high; 91.278 partial first phalanx; 91.279 third phalanx 16.4 × 9.2 high; 91.486 third phalanx 15.8 × 8.9 high; 91.487 first phalanx 32.6 × 11.2 high; 91.505 fragment first phalanx; 91.676 second phalanx 16.9 × 8.8 high; 92.1 distal immature metapodial condyle, ap 13.4; 92.11 (in part) metapodial distal condyle, ap 14.1; 92.19 partial first phalanx, probably not shorter than 25.0; 93.395 first phalanx 31.9 × 12.6 high; 93.397 second phalanx 19.2 × 11.5 high; 93.434 third phalanx 17.0 × 10.5 high.

Remarks. The three m3 specimens 92.13, 17, and 22 (fig. 15.5C,D) have height indexes as a proportion of length of 68%, 66%, and 68%, respectively—probably about the same as in Pikermi *Gazella capricornis,* the third lobe is offset labialwards (primitive), and they have an anterolabial cingular fold. There is a tiny basal pillar. Estimates involving the two known upper premolars suggest that the unknown lower premolar row may have had a length about 60% of that of the molar row. The ramus of the mandible fragments (fig. 15.5A,B) feels thicker than in Pikermi or later Sinap gazelles. There are few front limb elements of this species compared with hind limb bones or phalanges. Three upper molars from Loc. 64 may be too large for this species. They are 92.14 (fig. 15.4C) dex, early wear, 12.8 × 8.2; 92.32 (in part) dex, late wear, 11.5 × 11.4; and 92.68 (in part) sin, early wear, 12.8 × 8.2 × 9.5.

Discussion. The teeth in this species are too low crowned for *Turcocerus gracilis* and the premolar row may be relatively longer than in the larger middle Miocene species, *Tethytragus koehleri*. Gazelle horn cores have not been recorded in the Sinap succession below the level of Loc. 72, so this species from the lower Sinap member should not be assumed to belong to *Gazella*. It need not be an antilopine or a member of the Antilopini-Neotragini stem stock at all. It could be an early member of the Caprinae and/or a smaller relative of *Tethytragus koehleri*.

Bovidae ?species A.

Sinap Material. Loc. 64: 92.9 dex dP3 unworn, 7.7 × 4.6; 92.32 (in part) dex dP4 early wear, 7.6 × 6.0; 92.818 dex dP3 late wear, 8.0 × 5.1; 91.281 back dex m3 unworn, ~10.9 length × ~5.2 high; 92.30 dex mandible with dp2–m2 milk teeth late wear, molars early middle wear (rear central fossette on dp4, otherwise none on milk teeth; front of hypoconid not joined to rest of tooth on m1 or 2), dp2–4 19.6, dp2 4.3 × 2.4, dp3 5.4 × 3.0, dp4 10.1 × 4.9, m1 7.9 × 4.9, m2 9.8 × 4.8, mandible depth under dp2 11.0, under m1 13.2 (fig. 15.5E); 92.33 dex m1 or m2 late wear, 7.0 × 5.7; 92.67 dex dp2 4.9 × 2.9, sin dP2 or dP3 late wear, 6.1 × 3.7; 92.814 sin m3 (hypoconulid lobe missing) unworn, ~12.1 long; 92.815 sin lower molar early middle wear, 8.4 × 5.0; 92.817 sin p4 unworn, 7.7 × 4.0. (fig. 15.5F); 92.819 sin dp4 late wear 11.0 × 5.0; 92.823 dex incisor; 92.824 sin dp3 late wear, 6.0 × 3.1; 90.279 sin distal tibia, width 15.6; 91.266 fragments distal dex tibia; 91.270 parts of two calcanea; 91.271 dex ectocuneiform; 91.272 anterior part of sin naviculocuboid, width ~14.8; 91.273 partial proximal sin metatarsal; 91.500 dex astragalus height 19.0, width 11.0, ap 10.5; 91.504 dex naviculocuboid fragment, width 14.9; 91.499 sin calcaneum fragment; 91.501 distal fibula; 91.502 pisiform; 90.278 sin olecranon; 90.283 part of proximal sin metacarpal, transverse width × 14.9; 91.264 two fragments distal dex radii, width across articular surfaces of one 13.2; 91.265 fragment olecranon; 91.267 sin scaphoid height 9.2, width 5.9, ap 11.0; 91.269 sin unciform; 91.498 proximal radius fragment; 91.503 dex unciform; 90.273 second phalanx 14.9 × 4.5; 90.274 third phalanx 15.3 × 8.4 high; 90.275 third phalanx 15.3 × 8.0 high; 90.276 two sesamoids; 90.277 fragment shaft limb bone; 91.268 sesamoid; 91.274 two metapodial distal condyles, aps. 10.4, 10.0; 91.276 first phalanx 23.2 × 9.2 high; 92.3 third phalanx 16.0 × 8.7 high; 92.5 proximal first phalanx length ~21.7; 92.11 (in part) distal immature metapodial condyle, ap 9.1; 92.488 metapodial distal condyle, ap 7.9; 92.489 metapodial distal condyle, ap 10.0.

Loc. 65: 92.332 fragment skull; 92.326 fragment mandible with alveoli for p2–p4; 92.327 dex metatarsal, proximal fragment; 93.619 metapodial distal condyle, ap 11.2;

Sinap Material Possibly this Species. Loc. 64: 92.813 sin incisor; 91.275 sesamoid; 91.483 dex unciform; 92.812 lunate; 92.830 first phalanx distal fragment; 93.399 third phalanx ~15.2 × 8.1 high.

Remarks. In linear dimensions, these teeth seem to be about four-fifths the size of the preceding species (fig. 15.5E,B) so it is unlikely that all specimens have been correctly allocated. The m3 91.281 is lower crowned (height index ~48% instead of 67%) than in the preceding species. The p4 92.817 has complete fusion of paraconid with metaconid even in its unworn state (fig. 15.5F), but no p4 was assigned to the previous species. These fossils could be conspecific with Bovidae species A, described later.

Discussion. Fusion of paraconid with metaconid on p4 is known in some Miocene Cervidae (Alekseyeva 1915, pl. 2, figs. 26–28; Thenius 1950, fig. 4). It is also present in the probable bovid "*?Micromeryx*" sp. from Paşalar, Inönü, and Candir (Gentry 1990; Geraads et al. 1995), which is smaller than Loc. 64 92.817 (p4 length 6.2 at Paşalar). The Paşalar p4 is similar in that the lingual part of the entoconid is higher than the entostylid crest. The scaphoid 91.267 is about 15% higher than in extant *Gazella*.

Subfamily Caprinae Gray, 1821
Protoryx Major, 1891

Type Species. *Protoryx carolinae* Major 1891.

Remarks. *Protoryx*, along with *Palaeoryx*, *Pachytragus*, and other genera, were originally regarded as Hippotragini or as part of an extinct grouping Pseudotraginae. Gentry (2000) takes them as Caprinae. They are discussed later in the account of ruminants of the Middle Sinap Member (page 363).

?(*Protoryx solignaci*) (Robinson 1972)

Holotype. Cranium with part of the right horn core from the Beglia Formation at Bled Douarah (Robinson 1972, p. 75, fig. 1).

Sinap Material. Loc. 64: 90.281 fragments upper molar, length ~16.0; 91.495 fragments upper molar, length ~15.7; 92.69 sin upper molar (?M2) unworn, 17.5 × 7.7 × ~12.5 high (fig. 15.6B,C); 91.675 partial proximal dex radius; 93.385 partial distal sin radius; 92.581 dex distal tibia, total width 37.6, articular facets width 32.4, ap diameter 30.7; 93.393 fragment immature tibia; 91.479 patella; 93.440 fragment immature metatarsal; 90.264–265 first phalanges, length × least thickness 33.2 × 7.2, 33.3 × 7.0.

Loc. 65: 92.330 metapodial distal condyle, ap 20.3.

Remarks. The lower molar Loc. 64 92.69 (fig. 15.6B,C) is probably an m2 because the outlines of its front and back edges in side view converge little toward the base of the crown. It agrees with teeth of *Protoryx solignaci* in size, modest hypsodonty, nonflattened lingual wall, and presence of a basal pillar on m2. The postcranial bones and the metapodial condyle at Loc. 65 are of an appropriate size to be conspecific with the molar. A lateral half of an astragalus, 91.480 from Loc. 64 (height 32.9 mm) is too small for this species but larger than the gazelle-sized bovid.

Discussion. The lower molar is too low crowned to fit the *Hypsodontus pronaticornis* at Loc. 24A. In addition, the lingual wall of the entoconid is not flat enough and the basal pillar is too prominent. The same characters debar it from being one of the supposed Miocene "ovibovines." The Vallesian *Mesembriacerus melentisi* Bouvrain, 1975 is such a species, but its m1 is too small to match 92.69 and the m2 is too hypsodont. Specimen 92.69 is the same size as the *Tragoportax* species present at Loc. 49 in the upper Sinap member and could be the earliest boselaphine known in Turkey. However, it is more likely to belong to *Protoryx*

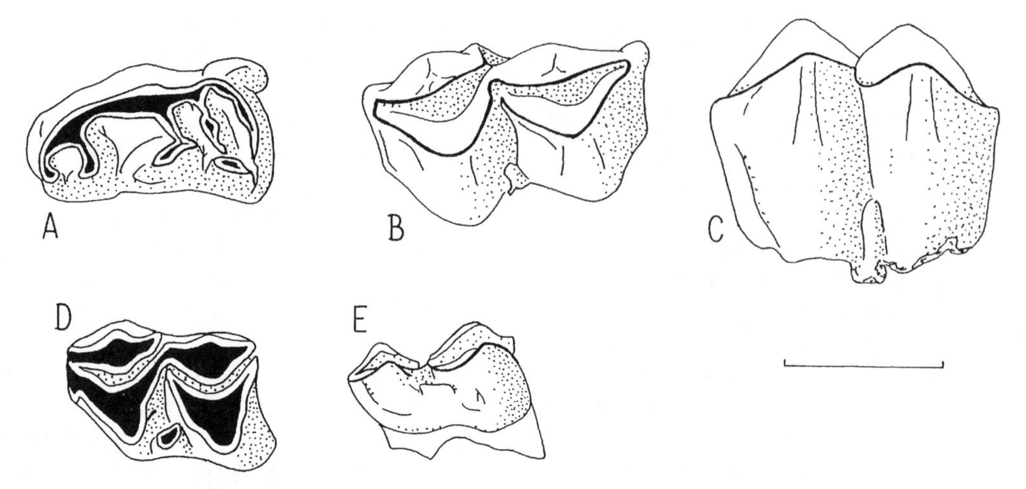

Figure 15.6. (A) *Protoryx solignaci,* dex p4 Loc. 4 94.14 in occlusal view. (B,C) ?(*Protoryx solignaci*), sin m2 Loc. 64 92.69 in occlusal and labial views. (D, E) Cervidae sp., sin lower molar Loc. 1 92.77 in occlusal and labial views. Anterior toward the left. Scale bar = 10 mm.

solignaci, which can be confused with boselaphines if the remains are too incomplete. In Turkey *P. solignaci* was thought to occur in MN 8 and 9 (Köhler 1987, p. 172).

Ruminants of the Middle Sinap Member

Family Cervidae Gray, 1821
Cervidae sp.

Sinap Material. Loc. 1: 92.77 sin m1 or m2 early middle wear, 12.3 × 8.3 × 3.8 high (fig. 15.6D,E).
Loc. 91: 93.41 sin upper canine.

Remarks. On the lower molar, the enamel is not rugose, the *Palaeomeryx* fold is weak, and the basal pillar is large. The front fossette opens to the rear, which looks primitive. No lingual projection of the metastylid remains. Its length is about that of m1 in a *Lucentia* species at Rudabanya, but it could be the m2 of a smaller species. The upper canine has the medial surface worn flat and a keel on its lateral surface. It is from a cervid and not *Dorcatherium.*

Discussion. The lower molar differs from the Paşalar ?*Stephanocemas* sp. by the weak indication of a *Palaeomeryx* fold and by its larger basal pillar. It is lower crowned and the crests less joined up than in the Pikermi *Pliocervus graecus.* It is thus more primitive than *P. graecus* and rather like the molars on a Steinheim mandible BMNH M5557, presumably of *Euprox* or *Heteroprox.* The *Palaeomeryx* fold is absent in *P. graecus* from the late Turolian of Maramena (Azanza 1995, p. 163). All this suggests a pre-Turolian date for Loc. 1.

The upper canine is from a smaller species than the Rudabanya *Lucentia* and is less reduced than in the late Miocene (mainly Turolian) *Cervavitus variabilis* (Alekseyeva 1913) of the eastern Paratethys region (Alekseyeva 1915, pl. 2, fig. 18). It is too large to belong to a *Micromeryx.* If it should be conspecific with the Loc. 1 molar, then the latter is likely to be an m2 and the single Sinap species is likely to be smaller than the Rudabanya *Lucentia.*

Family Giraffidae

By the late Miocene, it is possible to separate Giraffinae and Sivatheriinae within the Giraffidae. Sub-Paratethyan giraffines comprise mainly *Palaeotragus* and *Samotherium* Major, 1888, evidently different from the giraffes evolving in Africa. *Palaeotragus coelophrys* from its type locality of Maragheh (middle and upper Maragheh, according to Bernor et al. [1996b]) has advanced in p4 crest morphology to the extent of acquiring a closed lingual wall anteriorly and weakened fusion between the front of hypoconid, entoconid, and back of protoconid crests. *Palaeotragus roueni,* the Pikermi type species of the genus, is smaller and more advanced and has metapodials that are as long as or longer than in *P. coelophrys* but more slender (de Mecquenem 1924–1925, p. 160; also see fig. 15.3). The p4 of *P. roueni* has become molarized in that the entoconid is short and oriented anteroposteriorly along the lingual side, and the more crescentic hypoconid is labially opposite it. The well-known *Samotherium boissieri* from Samos is slightly bigger than *P. coelophrys,* and its teeth more hypsodont. Its p4 often shows the entoconid and protoconid united as one dominant crest, behind which is a reduced hypoconid.

The Algerian *Palaeotragus germaini* Arambourg 1959 of Vallesian-equivalent age has legs less lengthened than does the Fort Ternan *"P." primaevus,* so it may have more affinities with northern giraffids than with those later found in Africa. Hill et al. (1985, fig. 4) maintained that definite Giraffinae appeared in Africa at ~6.0 Ma.

Palaeotragus Gaudry, 1861a,b

Type Species. *Palaeotragus roueni* Gaudry, 1861a,b.

?(Palaeotragus roueni) Gaudry, 1861a,b

Holotype. Skull from Pikermi (Gaudry 1861a, pl. 7 figs. 1–3).

Sinap Material. Loc. 49: 90.122 stem of dex scapula (damaged posterolaterally), articular facet ap ~51.0, transverse 43.0.

Loc. 26: 91.243 second phalanx 32.2 × 22.5 high.

Remarks. These are the only two Sinap giraffid bones small enough to fit this species, but their identity is not definite.

Palaeotragus coelophrys (Rodler and Weithofer 1890)

Lectotype. Face with orbits and cheek dentitions and without horns in life (Rodler and Weithofer 1890, pl. 1, fig. 2), designated by Geraads (1978). It comes from Maragheh.

Sinap Material. Loc. 72: 92.221 most sin upper molar, early middle wear, ~34.6; 91.293 dex proximal metatarsal fragment; 91.300 ?sesamoid/pisiform; 91.323 dex proximal metatarsal fragment; 92.204 distal sin tibia, width 79.1; 91.328 proximal dex radius, width 98.0, width articular surface 95.7; 92.205 distal metacarpal, width 83.2, ap 52.9 (also piece of metapodial shaft); 92.50 second phalanx 46.6 × 38.5 high.

Loc. 91: 93.20 distal metapodial, width 71.0; 93.396 distal ?first phalanx; 93.410 proximal metapodial (probably medial dex metatarsal); 93.838 dex radius with parts of ulna, length 424, minimum shaft width 56.

Loc. 8: 92.232 proximal dex metatarsal, ap 62.7, transverse 59.5 (fig. 15.2B).

Loc. 12: 92.322 dex upper molar earliest wear, 34.7 × 21.7 × 18.5 high (mesostyle height 16.7); 91.379 sin calcaneum, total length 180, height 76.1; 91.383 fragment dex calcaneum; 91.650 sin ectocuneiform, ap 41.0, transverse 30.2; 93.1197 third phalanx 79.1 × 51.3 high (see fig. 15.8H).

Loc. 1: 92.56 dex P2 or P3 very worn, 18.3 × 13.8.

Loc. 49: 93.1008 two pieces of horns, diameter ~12.0, preserved length ~120; 91.702 sin maxilla P2–3, P4–M2 middle wear, M1–M3 ~92.0, M1 30.2 × 26.1, M2 32.6 × 25.0, P2–P4 ~65.2, P2 21.1 × 15.3, P3 21.2 × 18.3, P4 20.8 × 20.4 (fig. 15.7A); 91.703 dex maxilla with two molars, early middle wear, M(?M2), 30.0 × 25.7, M(?M3) 32.7 × 25.0; 92.885 dex upper molar, early mid wear, 32.3 × 22.0; 92.886 upper molar fragment; 91.704 dex mandible with p4–m3, bp small on m1, tiny on m3, intermediate on m2; early mid wear; p4 24.6 × 15.5, m1–3 97.0, m1 27.8 × 17.7, m2 30.8 × 18.5, m3 39.1 × 17.0 × 14.1 high (fig. 15.7B–D); 93.801 diastemal part of dex mandible i1–p3, p3 in early mid wear (?same bone as 91.704); p2 19.2 × 13.0, p3 23.8 × 11.1, mandible depth under p2 40.8 (fig. 15.7E); 93.693 sin dp2, early wear, 19.1 × 8.9 total width (fig. 15.7G); 93.699 sin mandible with dp3-erupting m1, lower deciduous premolars in early mid wear; dp3 24.2 × 10.0, dp4 39.3 × 12.6 (fig. 15.7F); 93.814 sin dp3 +dp4 in late mid wear and m(?m2) in early wear; dp3 19.6 × 7.7, dp4 29.8 × 10.7, lower molar (?m2) 25.6 × 13.2; 92.186 dex dp4, early wear; 32.6 × 9.0 × 10.6 height lingually between rear lobes and 10.2 height between front ones; 93.1157 sin lower molar, tiny basal pillar,

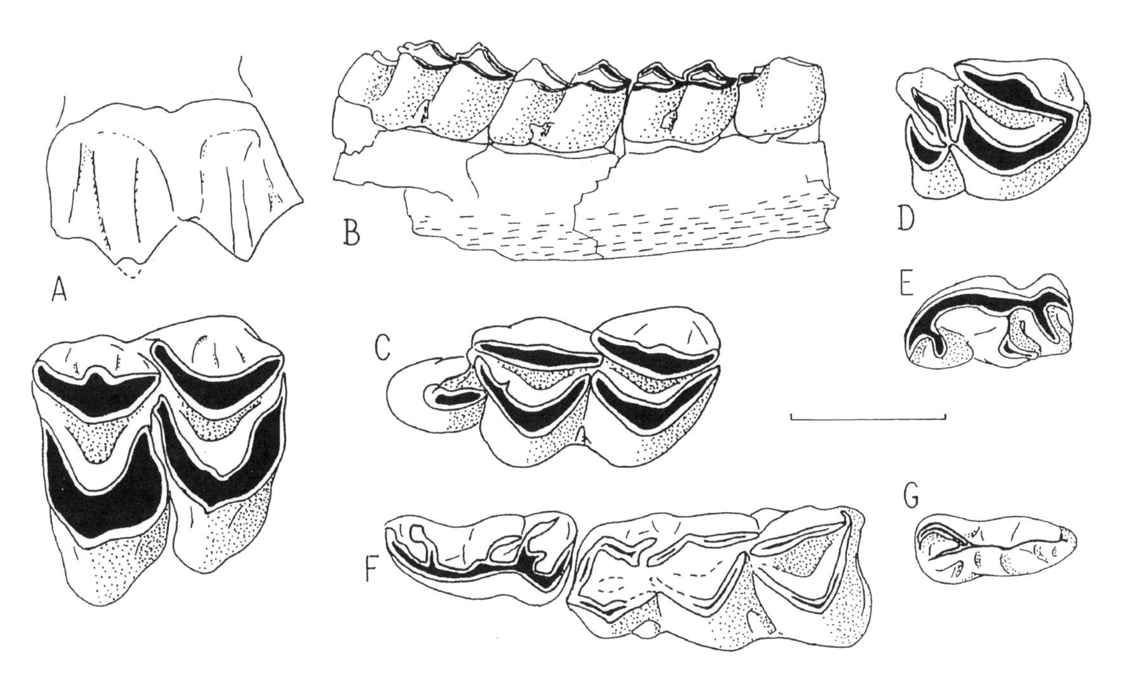

Figure 15.7. *Palaeotragus coelophrys* from Loc. 49. **(A)** Labial and occlusal views of sin M2 91.702. **(B)** Lateral view of dex mandible 91.704. **(C,D)** Occlusal views of dex m3 and p4 91.704. **(E)** Occlusal view of dex p3 93.801. **(F)** Occlusal views of sin dp3 and dp4 93.699. **(G)** Occlusal view of sin dp2 93.693. Anterior toward the left in **A, E,** and **F;** otherwise toward the right. Scale bar = 40 mm for **B,** otherwise 20 mm.

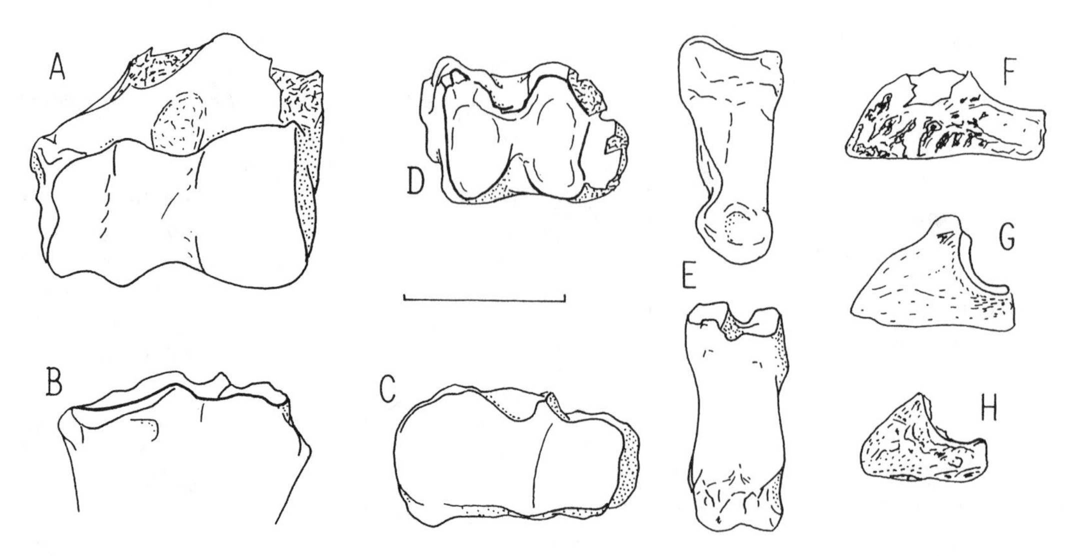

Figure 15.8. Giraffidae postcranial bones. (A) *Helladotherium* sp., distal dex humerus Loc. 26 89.134 in anterior view. (B,C) *Bohlinia attica*, proximal sin radius Loc. 51 90.154 in anterior and articular views. (D) *B. attica*, distal sin tibia Loc. 49 93.689 in articular view. (E) *Helladotherium* sp., first phalanx Loc. 27 89.492 in side and dorsal views. (F–H) third phalanges in side view. (F) *Helladotherium* sp., Loc. 33 89.351. (G,H) *Palaeotragus coelophrys* Loc. 4 93.478 and Loc. 12 93.1197. Anterior side of C toward bottom, of D toward top. Scale bar = 80 mm.

early mid wear, 26.7 × 15.7; 91.732 distal sin tibia, width 79.8; 92.135 immature distal dex tibia, width across astragalar facets 58.3; 92.141 sin femur, distal shaft; 93.1141 sin radius, length 473, proximal width 96.3, distal width 80.5; 92.152 sin scaphoid, height 40.8, width 25.9, ap 50.9.

The following remains from localities below Loc. 72 fit *Palaeotragus coelophrys* by size.

Loc. 4: 93.478 third phalanx, 62.6 × 41.4 high (fig. 15.8G).

Loc. 94: 93.542 shattered tooth.

Loc. 45: 92.671 fragment first phalanx.

Remarks. The horn pieces Loc. 49 93.1008 are referred to *Palaeotragus coelophrys* simply because giraffid teeth at Loc. 49 appear to belong to that species. They are notably narrow and have a rough exterior surface with inconstant planes and some angled corners. There are irregular sinuses at the base. However, postcranial bones at Loc. 49 show that three giraffid species are present.

The teeth at Loc. 49 can best be taken as *Palaeotragus coelophrys*. They come from at least one adult and two immature individuals. The crown height for the molars matches that shown for the Sebastopol *Achtiaria expectans* Borissiak (1914 pl. 1, figs. 9a, 10a) accepted here as *P. coelophrys*. On the p4, Loc. 49 91.704 (fig. 15.7D), the entoconid crest is still more or less transversely oriented and joins the crest immediately behind the protoconid. The protoconid also retains a transverse crest passing in front of the entoconid crest but terminating before it reaches the metaconid. This is more primitive than in a Maragheh specimen of *P. coelophrys* (de Mecquenem 1924–1925, pl. 2, fig. 8). As in most giraffid p4 specimens, the metaconid forms a closed wall along much of the lingual margin of the tooth. The p3 Loc. 49 93.801 (fig. 15.7E) has a transverse crest from the protoconid to the lingual cusp of the metaconid, but the metaconid does not form a lingual wall for the anterior part of the tooth. Such a lingual wall is also lacking in the dp3 of Loc. 49 93.699 (fig. 15.7F). The paraconid on 93.801 is less backwardly oriented, and therefore contributes less to closing the lingual side of the tooth than in the Maragheh *P. coelophrys* of de Mecquenem (1924–1925, pl. 20, fig. 8).

The crooked anterolingual front lobe of the dp4 93.699 (= pseudoparaconid of Hooker [1986 text fig. 54, after Hershkovitz] or paraconid plus prae- and postparacristids of Rössner [1996, fig. 154]) may be more like *Palaeotragus coelophrys* (de Mecquenem 1924–1925, pl. 20, fig. 6) than like a *Samotherium*. The backward bending of the lingual end of the metaconid crest on the dp3 of 93.699 is unlike *P. coelophrys* of either Borissiak (1914, pl. 2 fig. 5a) or de Mecquenem (1924–1925, pl. 20, fig. 6).

The morphology of giraffid lower premolars, especially p4, is important in taxonomy, yet shows an immense range of intraspecific variability (Morales and Soria 1981, fig. 7; Gentry 1990, fig. 5A–E).

The upper molar Loc. 12 92.322 is perhaps not too long to match the *Palaeotragus coelophrys* teeth at Loc. 49 and it is too high crowned to be *Bohlinia* (see below). The upper molar Loc. 72 92.221 is also accepted as *P. coelophrys*.

At Loc. 12, the calcaneum 91.379 is rather large, the third phalanx 93.1197 (fig. 15.8H) is about the size of *Samotherium boissieri* but relatively high, while the ectocuneiform 91.650 is rather small. The scaphoid Loc. 49 92.152 is about a fifth smaller than four *S. boissieri* in London, but its height actually exceeds theirs. Most of the limb bones at Loc. 72 are about the same size or slightly smaller than Samos *S. boissieri*, but the distal metacarpal 92.205 may be from a larger species. The radius Loc. 49 93.1141 is rather small and not very long. The third phalanx at Loc. 4 (fig. 15.8G) is anteroposteriorly shorter than in *S. boissieri,* and hence is taken as too small to go with the sivathere p4 at that locality.

Palaeotragus hoffstetteri Ozansoy (1965, p. 64) from Yassiören looks as if it belongs to the Sinap *P. coelophrys* described here. This name would be available, were future workers to disagree with my allocation of the material to the same species as at Maragheh.

Discussion. *Palaeotragus coelophrys* (and its synonyms) is the least advanced of Eurasian late Miocene giraffids. It seems to have existed from MN 9–12. The p4 specimens of figure 15.7D, de Mecquenem (1924–1925, pl. 20, fig. 8), and Bohlin (1926, fig. 56) could easily be the starting point for evolution toward both *P. roueni* and *Samotherium boissieri.* The species is, however, advanced on the middle Miocene giraffids at Paşalar and Chios by its larger size, metastylids on the lower molars no longer projecting lingually, relatively wider P2s and P3s, and more advanced lower premolars.

The teeth of *Bohlinia attica,* a long-legged giraffid in the late Miocene sub-Paratethyan faunas (see below), are about the same size as *Palaeotragus coelophrys* (table 15.2). The Sinap species differs from *Bohlinia* by higher crowned cheek teeth. On upper molars, the parastyles may be less prominent. On p4, the entoconid crest is still transversely oriented and contacts the protoconid, and the labial part of the transverse crest from protoconid to metaconid is still present unlike *Bohlinia* (Bohlin 1926, text fig. 203). The radius Loc. 49 93.1141 is too small and short to belong to *Bohlinia.*

The Sinap species differs from *Samotherium boissieri* by its slightly smaller size, lesser degree of hypsodonty, persistence of an incomplete metaconid transverse crest on p4, and no closed anterior part of the lingual wall on p3 or dp3. Basal pillars on dp4 can be less developed in *Samotherium* as seen in Rodler and Weithofer (1890, pl. 3, figs. 5–6). The upper premolar row is possibly slightly longer relative to the molar row than in *S. boissieri* (71% of molar row length as against a mean of 68% for six *S. boissieri*). It differs from *Palaeotragus roueni* by its larger size and the orientation of the entoconid crest on p4. The absence of closure of the anterior part of the lingual wall on p3 or dp3 is unlike some Pikermi examples of *P. roueni* (BMNH M8367 + 13063 and M13062) but like a Ditiko p3 of *P. roueni* (Geraads 1978, pl. 1, fig. 4).

The Sinap teeth are about the size of *Giraffa punjabiensis* (BMNH M13625; Colbert 1935, figs. 192–196), but the upper molars lack a fold in the rear central fossette arising from the metaconule.

Bohlinia Matthew, 1929

Type Species. *Bohlinia attica* (Gaudry and Lartet 1856).

Remarks. Also present in the sub-Paratethyan faunas from MN 10–13 with *Palaeotragus* and *Samotherium* is *Bohlinia,* with brachyodont teeth somewhat like those of *P. coelophrys.* It has greatly elongated legs like those of the Siwaliks-African *Giraffa,* but its ancestors may have already been present in Eurasia in the middle Miocene (see discussion of Loc. 24A "Giraffidae sp." above). Geraads (1979, p. 380) noted differences from *Giraffa* in the characters of the limb bones. *Bohlinia* limb bones are much longer and the articular ends are wider than in *S. boissieri,* despite the teeth being as small as in *P. coelophrys.* Because *Bohlinia* teeth are quite similar in size and morphology to those of *P. coelophrys,* it is important to note that the metapodials of *P. coelophrys* are much shorter than in *Bohlinia* (de Mecquenem 1924–1925, p. 160; also fig. 15.3). Geraads (1979) described *Bohlinia* from Ravin de la Pluie, but other reliably dated occurrences are Turolian.

Giraffa punjabiensis Pilgrim, 1911, which was present in the Siwaliks from about 7.0–7.5 Ma (Barry 1995) could be a close relative of *Bohlinia attica,* but not enough is known about either species to assess this relationship. It is not even certain that *punjabiensis* has the very long metapodials needed for it to be included in *Giraffa.* The premolars are broader and p2 relatively longer than in *Palaeotragus coelophrys* (compare Colbert 1935, fig. 193 with de Mecquenem 1924–1925, pl. 2, fig. 8).

Table 15.2. Length Measurements of Giraffid Cheek Teeth at Loc. 49 Compared with Those of *Palaeotragus coelophrys* and *Bohlinia attica*

	Tooth Measurement (mm)								
Taxon	P2	P3	P4	M1	M2	p4	m1	m2	m3
P. coelophrys	20.6 (2)[1]	21.9 (3)	18.8 (3)	25.6 (4)	29.4 (3)	24.0 (2)	27.4 (4)	30.0 (2)	38.0 (3)
B. attica	19.9 (2)	19.8 (3)	20.1 (4)	25.8 (3)	28.2 (4)	22.0 (1)	26.0 (1)	26.0 (1)	36.0 (1)
Loc. 49 giraffid	21.1 (1)	21.2 (1)	20.8 (1)	30.2 (1)	32.6 (1)	24.6 (1)	27.8 (1)	30.8 (1)	39.1 (1)

Sources: Author's measurements, Bohlin (1926), Borissiak (1914), Geraads (1979), Kostopoulos et al. (1996), Mecquenem (1924–1925).

[1]Numbers of individuals shown in brackets.

Bohlinia attica (Gaudry and Lartet 1856)

Lectotype. Geraads (1979, p. 378) accepted or nominated as lectotype a posterior set of limb bones from Pikermi (Gaudry 1865, pl. 40, figs. 1 [left],5,6,8). However, these bones were not collected until 1860 and could not have been part of an 1856 type series. If it should be impossible after a search to identify in any Pikermi collections material of *Bohlinia attica,* which was already in Paris by 1856, this outcome should be reported and the lectotype should become a neotype.

Sinap Material. Loc. 51: 90.154 proximal sin radius, width 125.0, width articular surface 110.3 (fig. 15.8B,C); 90.156 vertebral pieces including two immature thoracics, width centrum of first one ~48.0, dorsoventral centrum ~48.0, ap ~54.0, width centrum of second one 55.0, dorsoventral centrum ~49.0, ap ~48.0; 90.158 partial distal sin tibia, width ~82.0; 91.388 metapodial immature distal condyle, ap 47.0; 91.389 sin cuneiform.

Loc. 49: 93.822 + 181 dex radius, length 708, minimum shaft width ~66, proximal width 126.5; 93.689 distal sin tibia, width 101.7. (fig. 15.8D); 90.116 medial sin astragalus, height 83.8, ap 59.3; 90.117 proximal dex metatarsal, width ~64.7, ap 75.5; 93.1000 proximal dex metatarsal, width 78.2, ap 82.3; 93.803a + b part of shaft of metatarsal and an immature distal condyle, the latter with ap diameter 54.8; 93.803c first phalanx, length 102.1, height 51.6; 93.803d second phalanx, length 58.9, height 44.4; 93.938 first phalanx, length 102.3, height 51.4.

Loc. 34: 92.575 atlas vertebra, length 141.6, ventral centrum, length 64.3, width across anterior articulations 95.4.

Loc. 95: 92.747 partial sin astragalus, medial height 87.6.

Remarks. Horns or teeth attributable to *Bohlinia* have not been found in the Sinap collections. The proximal radius Loc. 49 93.822 was found to be the same bone as the distal end 93.181, and to be too long for assignment to any other giraffid than *Bohlinia.* Its length is less than those of 800, ?900, 850, and 788 mm recorded for *B. attica* (Gaudry 1865, p. 247; Geraads 1979, p. 382), but more than that of 680 mm recorded by Kostopoulos et al. (1996). It is much longer than the complete radius of *Palaeotragus coelophrys* also found at Loc. 49. The other Loc. 49 limb bones listed here are too large for *P. coelophrys* and usually too small to match Pikermi specimens of *Helladotherium duvernoyi.*

The giraffid remains at Locs. 34, 51, and 95 are not assigned to *Bohlinia* with much certainty. Loc. 51 is thought to be at an earlier stratigraphic level than Loc. 49, where *Bohlinia*'s presence is definite. The Loc. 34 atlas is larger than two middle Miocene giraffid atlases from Fort Ternan (Churcher 1970, figs. 14–18, 79–81). It is about the same size as a Samos atlas of *Samotherium boissieri* (BMNH unregistered, "207"), but its centrum is longer relative to the width across the facets for the occipital condyles of the skull. This size and relative length might fit *Bohlinia,* although they could also fit a large middle Miocene giraffid like that at Loc. 24A.

Decennatherium Crusafont Pairo, 1952

Type Species. *Decennatherium pachecoi* Crusafont Pairo, 1952.

Remarks. I take *Decennatherium* as a sivatheriine giraffid. There are two suprageneric groupings of sivatheres. The Asian *Bramatherium* Falconer, 1845, probably including *Hydaspitherium* Lydekker, 1878, is related to and perhaps congeneric with *Decennatherium* Crusafont Pairo, 1952, of Europe. These forms have enlarged anterior and smaller posterior ossicones (Lewis 1939, pl. 2), whereas the Spanish *Birgerbohlinia* Crusafont Pairo, 1952, has a large posterior pair and a smaller anterior pair, more reminiscent of later *Sivatherium.* The state of the cranial appendages is not known in the Pikermi *Helladotherium.*

Decennatherium macedoniae Geraads, 1989a

Holotype. Right mandible with complete dentition from Pentalophos (Geraads 1989a, fig. 1B, pl. 2, figs. 1,3).

Sinap Material. Loc. 4: 92.381 sin mandible with m2 + m3 in mid middle wear, separate p4 in mid middle wear, p4 29.7 × 16.0, m2 33.4 × 22.1, m3 48.2 × 19.4 (fig. 15.1B–D); 92.380 two dex lower incisors, sin lower canine, dex lower canines.

Possible occurrences:

Loc. 72: 93.886 labial wall of sin P, length 25.2.

Loc. 49: 90.101 dex lower canine.

Remarks. The mandible from Loc. 4 is from a bigger giraffid than the *Palaeotragus coelophrys* at Loc. 49. The canine of Loc. 4 92.380, probably the same individual as 92.381, is bigger than the *P. coelophrys* canine of Loc. 49 93.801. Linear dimensions of the teeth are about one-fifth greater. On p4, the entoconid is more anteroposteriorly oriented and the rear half of the tooth is anteroposteriorly longer, so that the whole tooth has more of the appearance of a molar. This is a similar morphology to that of *P. roueni.* Tiny basal pillars are still visible on the molars of 92.381. The identity of the teeth at Locs. 72 and 49 is not definite, but they are from bigger giraffids than *Palaeotragus coelophrys.*

Discussion. In size as well as morphology, the p4 (fig. 15.1B) agrees with those of *Decennatherium, Birgerbohlinia,* and *Helladotherium* (Morales and Soria 1981, fig. 7; Geraads 1989a, pl. 2, fig. 2; Montoya and Morales 1991, fig. 6; BMNH M4067). Turolian sivatheres may have slightly larger and wider teeth and their lower molars may lack basal pillars altogether, hence *Decennatherium* is the most likely genus for the Loc. 4 teeth. *Decennatherium macedoniae* was described from Pentalophos, and a larger *Decennatherium* occurs at Ravin de la Pluie (Geraads 1979). The Sinap Loc. 4 teeth are about the same size as at Ravin de la Pluie, and are here placed in *D. macedoniae.* They are smaller, perhaps by about one-tenth in linear dimensions, than Spanish Vallesian *D. pachecoi* (see Montoya and Morales 1991, table 2), but the p4 is slightly longer relative to the two molars. The difference between *D. macedoniae* and *D. pachecoi* lies in the p3 of the former having a stronger lingual wall behind the metaconid (Geraads 1989a, fig. 1; Morales and Soria 1981, fig. 7B). The p3 is not known at Sinap.

It is interesting that p3 specimens of both *Decennatherium* and *Birgerbohlinia* are more advanced than those of *Helladotherium* (de Mecquenem 1924–1925, pl. 20, fig. 1; BMNH M4067 [cast], M11397), in which the three transverse crests of metaconid, entoconid, and entostylid continue to be visible.

Geraads (1989a, p. 194) considered that *Samotherium pamiri* Ozansoy 1965, from Yassiören may be close to *Decennatherium macedoniae.* The P2–M2 (Ozansoy 1965, pl. 7, fig. 1) are larger than in the Loc. 49 *Palaeotragus coelophrys* by ~15% in linear dimensions.

Lower molars of late Miocene sivatheres differ from Pliocene-Pleistocene *Sivatherium giganteum* or *S. maurusium* in the more narrowly pointed labial lobes (protoconid and hypoconid), less outbowing of lingual ribs, and the presence of a basal pillar. The front central fossette is not transversely constricted across its center, as can sometimes be seen in *S. giganteum.*

Helladotherium Gaudry, 1860

Type Species. *Helladotherium duvernoyi* (Gaudry and Lartet 1856).

Helladotherium sp. indet.

Sinap Material. Loc. 63: 90.216 distal metacarpal, width ~89.6, ap 45.8.

Loc. 26: 89.294 base sin horn pedicel; 89.134 distal dex humerus, width 123, height medial condyle 75.7 (fig. 15.8A); 89.293 dex calcaneum, part of articular surface; 90.246 sin radius, distal lateral part.

Loc. 27: 90.39 sin naviculocuboid, width 106.5; 89.492 first phalanx, length 104.5, proximal width 49.5 (fig. 15.8E); 89.493 distal phalanx, width ~40.7.

Loc. 33: 91.126 partial distal dex tibia, width 108; 91.100 patella; 89.167 dex partial calcaneum; 89.244 partial sin calcaneum; 91.105 proximal dex metatarsal, ap 76.3; 91.178 proximal dex metatarsal, width 77.2, ap 83.0 (fig. 15.2C); 89.219 fragment distal dex radius; 91.122 distal sin radius, medial fragment; 89.199,200 proximal parts of sin metacarpal, approx length 335; 89.177,225 metapodial condyles, aps. 62.0, 60.1; 91.120 distal metapodial condyle ap 57.7; 89.351 partial third phalanx (fig. 15.8F).

Loc. 70: 93.1041 proximal dex metacarpal, width 105, ap 60.5.

Loc. 42: 91.801 upper molar central fossette; 89.413 much of dex calcaneum; 91.458 fragment calcaneum; 92.338 partial distal fibula; 92.340 sin metatarsal, partial proximal end; 93.1142 sin astragalus, anterolateral rim; 90.79 sin scaphoid, anterior part missing, height 55.8; 90.87 sin unciform, ap 42.0, maximum height 28.5.

Remarks. Like *Bohlinia,* this is another Sinap giraffid identified by postcranial bones. The distal humerus Loc. 26 89.134 (fig. 15.8A) is so large as to be either a *Helladotherium* or *B. attica.* Humeri of *B. attica* (BMNH M11399; also see Geraads 1979, pl. 2, fig. 2a) have vertically lower medial condyles and probably less prominent lateral condyles than in 89.134. Geraads (1989a, p. 192) records a distal humerus of *Decennatherium macedoniae* from Pentalophos with a transverse width of 111 mm, a little smaller than 89.134; he also notes the weakness of the lateral condyle. The horn base Loc. 26 89.294 is large and was evidently inserted widely apart from its partner and above the orbit. There is no reason to put it in a separate species from the humerus.

The partial large metacarpal Loc. 33 89.200 is definitely too short for *Bohlinia,* the third phalanx 89.351 (fig. 15.8F) is very large; hence other large giraffid material from Loc. 33 is also assigned to *Helladotherium* sp. The naviculocuboid at Loc. 27 is as large as examples of *H. duvernoyi* from Pikermi, and the large first phalanx at that locality (fig. 15.8E) is also assigned to the same species. The distal metacarpal at Loc. 63 is very wide, exceeding *Decennatherium pachecoi* (Montoya and Morales 1991, table 4), and hence more likely to be a sivathere than *Bohlinia.* The left scaphoid at Loc. 42 is definitely sivatheriine by size and the left unciform need not be taxonomically separated from it.

Discussion. The Sinap postcranial remains probably belong to a sivathere, but they are too robust for *Decennatherium macedoniae.* Geraads (1989a, p. 192) noted that Pentalophos limb bones assigned to *D. macedoniae* were longer and more gracile than in *Samotherium.* This is not true of the Sinap limb bones in question here, which are closer in size and proportions to the Turolian *Helladotherium,* and they are therefore named as *Helladotherium* sp. It is possible that *Helladotherium* (or *Decennatherium* itself, for that matter) is not generically separate from *Bramatherium,* the earliest bestowed sivathere name after *Sivatherium.*

Family Bovidae
Tribe Boselaphini Knottnerus-Meyer, 1907

Boselaphines are known sparsely from the middle Miocene and abundantly from the late Miocene. There are two suprageneric groups (Moya Sola 1983, p. 198, figs. 59, 60): (1) the middle Miocene *Austroportax latifrons* (Sickenberg 1929) of Europe and later allied forms such as *Pachyportax latidens* (Lydekker 1876) and *Selenoportax vexillarius* Pilgrim, 1937, which can be associated with modern *Boselaphus* and the bovine *Bubalus;* and (2) a group centering on *Miotragocerus* Stromer, 1928, and *Tragoportax* Pilgrim, 1937, which also appears in the middle Miocene, but is best known from the Turolian, wherein it is, along with *Gazella,* the most widespread of all bovids. This second group became extinct around the end of the Miocene. Moya Sola (1983) and Bouvrain (1988) accept a specialized concavity at the top of the lateral surface of the metatarsal as characterizing *Miotragocerus* in central Europe and use *Tragoportax* for the remaining late Miocene forms. I follow them.

Tragoportax Pilgrim, 1937

Type Species. *Tragoportax salmontanus* Pilgrim, 1937.

Tragoportax spp.

Sinap Material. Loc. 14: 89.71 sin dP3–dP4 (see fig. 15.21A,B), mid and early mid wear, 16.8 × 11.0, 18.2 × 12.9 [?Boselaphini].

Loc. 108: 93.833 sin mandible p3–m2 later mid wear, p3 14.8 × 6.8, p4 14.7 × 7.0, m1 13.8 × 9.3, m2 16.0 × 10.2; 94.213 sin upper molar early mid wear, 14.9 × 10.5.

Loc. 72: 91.317 sin p3 in mandible fragment, earliest wear, 14.7 × 6.0 × ~10.0 protoconid height.

Loc. 12: 91.339 dex upper molar (see figs. 15.21C, 15.22A) early mid wear, 18.7 × 11.8; 93.612 fragment dex calcaneum.

Loc. 49: 90.110 ?dex hc base, (?medial) surface missing, 36.1 × ~25.9; 91.744 part sin hc base, central sinus in pedicel, ~30.8 × ~22.4, anterior + posterolateral keels, both slight (fig. 15.9E); 93.1165 hc fragment, ~34.5 × ~20.0, other pieces; 94.442 distal hc; 90.91 dex M1 mid wear, dex M2 early mid, sin M?2 mid, dex P4 early mid, dex P3 mid, dex P2 late mid, sin P2 late mid, sin lower incisor, mandible with m1 + 2 late mid, front lobe sin lower molar mid wear, dex M1 15.4 × 15.2, dex M2 18.9 × 16.0, sin M?2 19.2 × 16.0, dex P4 12.6 × 10.2, dex P3 14.1 × 10.3, dex P2 14.7 × 10.2, sin P2 14.8 × 10.2, sin I width 9.0, sin mandible m1 14.5 × 9.8, m2 15.7 × 10.8 (see fig. 15.22C); 91.197 sin maxilla M2–3, early mid wear, M2 20.1 × 15.1, M3 19.8 × 13.8; 91.710–714 sin P3 and P4 late mid, dex and sin P2 late, sin P4 mid wear; 91.710 sin P3 13.2 × 11.3; 91.711 sin P4 11.7 × 10.3; 91.712 dex P2 12.8 × 10.0; 91.713 sin P2 13.7 × 10.1; 91.714 sin P4 13.2 × 10.7; 93.673 dex upper molar fragment, late mid wear; 93.688 sin P4–M2 (P4 early, M1 mid, M2 early mid wear), P4 11.6 × 9.1, M1 16.3 × 13.2, M2 18.4 × 13.2; 93.691 sin upper molar mid wear, 17.5 × 14.5; 93.720 sin upper molar, early mid wear, 18.9 × 13.0; 93.1176 sin upper molar, early mid wear 17.4 × 13.4; 90.113 mandibular diastema; 91.206 sin p3, early mid wear, 15.0 × 8.4; 91.707 sin mandible with m1–m3 + unerupted p4 and detached dp4, molars in early mid wear, p4 13.8 × 6.2, dp4 17.6 × 8.4, m1–m3 53.5, m1 15.0 × 8.4, m2 17.1 × 8.2, m3 22.6 × 8.0, mandibular depth under m1 22.2, under m3 27.0; 91.708 dex mandible p3–m1, mid wear; p3 15.8 × 5.7, p4 ~15.7 × 6.3, m1 15.7 × 10.0; 91.729 dex mandible m2–3, late mid wear; m2 17.3 × 10.9, m3 23.2 × 9.6; 91.730 dex mandible m2–3, mid wear, conjoined dex p3–4, early wear, dex lower molar late wear, p3 13.6 × 5.7, p4 16.2 × 6.1, lower molar 15.2 × 9.8, m2 15.4 × —, m3 22.4 × 10.0 (fig. 15.10C); 92.168 sin dp3, mid middle wear, 14.9 × 7.2; 92.201 dex mandible p2–m3, mid wear; p2–4 39.0, p2 11.2 × 5.6, p3 ~13.6 × 6.3, p4 14.4 × 7.3, m1–3 55.4, m1 15.8 × 9.2, m2 17.3 × 9.2, m3 23.0 × 9.2, mandible depth under p2 18.8, under m1 25.1, under m3 26.7 (fig. 15.10A,B); 93.659 dex p3 early wear, 15.3 × 6.3; 93.661 sin p3 late wear, 11.1 × 6.1 [?Boselaphini]; 93.684 sin m1 or 2, early mid wear, tiny bp hardly present, 18.5 × 9.4; 93.1036 dex mandible m2+3, early mid wear, m2 18.7 × 9.4, m3 25.2; 90.114 sin distal tibia, width 35.4; 93.986 dex distal tibia, width 36.5; 93.1080 dex distal tibia, width 36.4, and tarsus, calcaneum length 94.3, astragalus lateral height 43.1, width 26.0, ap23.5, naviculocuboid width 33.2; ectocuneiform ap 21.9; 93.674 dex partial calcaneum, stem length ~62.0; 93.1093 dex calcaneum fragment; 92.153 dex astragalus, lateral height 40.7, width 25.9, ap 22.3; 93.1154 dex naviculocuboid, width 33.2; 93.725 sin ectocuneiform, ap 19.2; 90.115 sin metatarsal, distal shaft; 92.179 distal metatarsal, width 32.9; 93.695 dex proximal radius, width ~37.7; 92.151 dex scaphoid, ap 23.4, width 13.3, height 20.0; 91.204 dex proximal metacarpal, width 33.0, ap 22.1; 92.178 distal metacarpal, width ~30.0; 93.682 distal metacarpal, width 30.2, ap 21.7; 92.171 immature distal metapodial, width ~29.0; 91.717 two first phalanges, two second phalanges, one-third phalanx; 93.642 proximal first phalanx; 93.929 distal first phalanx; 93.983 distal first phalanx; 93.1075 first and damaged second phalanges, length × height 47.1 × 21.6, 28.8 × 20.4; 91.748 third phalanx, 38.4 × 20.0 high; 93.1034 third phalanx, 35.5 × 20.7 high; 94.51 part axis vertebra.

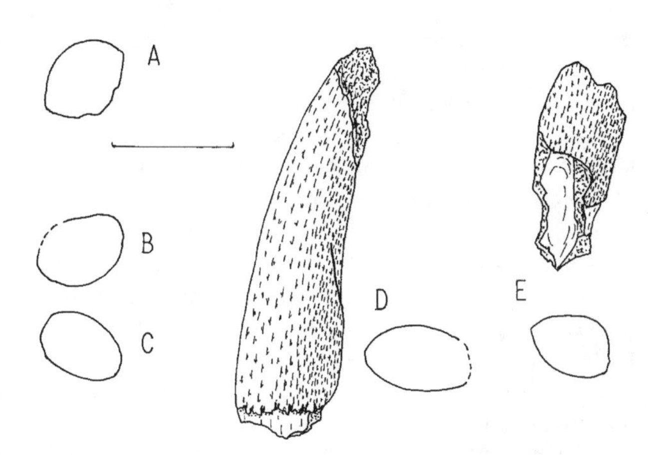

Figure 15.9. (A) *Turcocerus gracilis*, cross section dex horn core Loc. 24A 91.411. (B,C) *Tethytragus koehleri*, cross sections dex and sin horn cores Loc. 34 89.338 and 92.770. (D,E) *Tragoportax* sp., lateral view and cross section of sin horn cores Loc. 33 89.241 and Loc. 49 91.744. Anterior toward the left. Scale bar = 40 mm.

Loc. 26: 90.244 partial hc; 91.230 partial hc; 91.43 sin dP3–M1, early mid wear; dP3 c16.2, dP4 17.3 × 14.5, M1 18.9 × 12.0 × ~14.4 high; 91.61 sin P3, mid wear, 14.0 × 14.3.

Loc. 27: 89.325 back dex dp4, early mid wear.

Loc. 28: 89.116 dex mandible, damaged m1–3 ~71.0, p4 ~18.0.

Loc. 33: 89.186 dex hc base, 25.2 × 20.0; 89.239, 240 distal fragment and proximal (?base) dex hc, 33.8 × 20.4; 89.241 sin hc, ~36.3 × 23.7. (fig. 15.9D); 89.347 (?dex) hc, sinuses in pedicel, 29.2 × 19.7; 90.46 sin P4, late wear, 12.7 × 16.3; 91.76 incomplete dex upper molar, late wear; 91.127 sin upper molar, late mid wear, 18.0 × 18.9; 91.164 sin mandible dp3–m1, dPs in early mid wear, dp2–4 ~42.9,

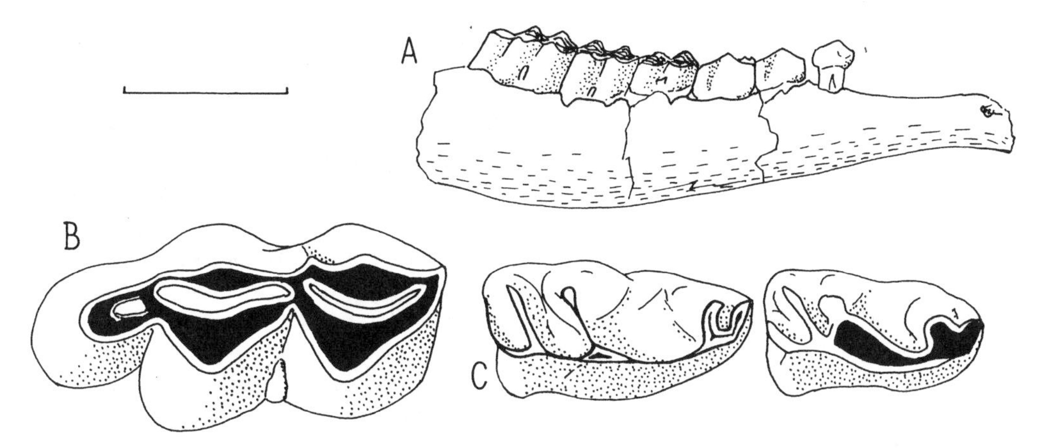

Figure 15.10. *Tragoportax* sp. from Loc. 49. (**A**) Lateral view of dex mandible 92.201. (**B**) dex m3 92.201 in occlusal view. (**C**) Dex p3 and p4 91.730 in occlusal view. Anterior toward the right. Scale bar = 40 mm for **A** and 10 mm for **B** and **C**.

dp3 13.0 × 6.4, dp4 22.6 × 7.0, m1 19.5 × 8.5, mandible depth below dp2 16.8, below m1 23.1.

Loc. 42: 91.796 (part) dex P2, late wear, 15.0 × 11.3; 93.1128 dex M1–3, damaged, late mid wear, small bps on M2 + 3, M1 18.5, M2 20.9 × 19.1, M3 22.0 × 17.5; 89.402–404 sin m3 unworn, most sin upper molar early wear, sin upper molar mid wear; 402: 28.0 × 10.1 × ~22.0 high; 404: 19.9 × 16.3; 91.454 dex p3, early wear, 14.7 × 5.1.

Loc. 99: 92.841 sin dp4–m1, early mid and early wear; 18.8 × 7.0, 14.3 × 7.8; 92.839 dex proximal metacarpal, ap 19.5, transverse 29.5.

Remarks. A meager amount of information on horn cores is available from fossils at Locs. 49 and 33. The horn core Loc. 49 91.744 (fig. 15.9E) is small and not very compressed. The horn cores Loc. 49 90.110 and Loc. 33 89.241 (fig. 15.9D) are larger, more compressed, and about the size of those of a cast (BMNH M15759) of the Siwaliks *?Miotragocerus gradiens* (Pilgrim 1937). The horn cores appear not to have a distal demarcation. They are variably compressed mediolaterally and approach having anterior and posterolateral keels. There is almost a posterior surface but no posteromedial keel. The horn cores curve backward but less than in *Protoryx* or *Tethytragus*. Divergence does not alter along their course.

For displaying tooth characters, Loc. 49 92.201 (figs. 15.10A,B) is the best available mandible. Boselaphine p3 teeth can be very long, as on the mandibular piece Loc. 49 91.708. No anterior lingual flange is present on p4 metaconids.

The dp4 Loc. 27 89.325 is quite large, with fairly prominent basal pillars, and with no stylid at the back of the middle lobe. The m3 Loc. 42 89.402 has a very small basal pillar. The upper molars 89.403 and 89.404 have parallel front and back edges in labial view. The P3 on the partial upper tooth row Loc. 26 91.61 has large styles and a prominent labial fold on the paracone. The posterior part is much widened posteriorly.

The dp4 and m1 Loc. 99 92.841 are low crowned, with rugose enamel and basal pillars. The selenodonty of the front lobe of the dp4 is too well developed for the specimen to be a cervid. The dp4 is slightly more than 10% longer than would have been expected in a Pasalar *Tethytragus koehleri* with an m1 of the same length as in 92.841. It is probable that the specimen is a boselaphine, and the proximal metacarpal 92.839 is assumed to be conspecific.

Three worn teeth at Loc. 33 (90.46, 91.76, and 91.127) are boselaphines at a locality that otherwise has *Pachytragus*. The teeth are large in comparison with either the *Pachytragus* or the boselaphine teeth at Loc. 49. Other signs of an increase in size of boselaphine teeth come from Loc. 42 89.402–404, 89.796 (in part), and 93.1128, and also from Loc. 27 89.325 and Loc. 28 89.116. There is no evidence from horn cores of increased size at Loc. 33 and above.

In attempting to assess the presence of Boselaphini at localities lower than Loc. 49, one has to differentiate the fossils concerned from *Protoryx solignaci* or related species. The upper molar Loc. 12 91.339 (see fig. 15.21C) looks satisfactory classed as a boselaphine. At Loc. 72, the p3 91.317 seems to be too long to fit a *Protoryx* or *Pachytragus* and too large for the smaller, related *Pseudotragus*. At Loc. 14, the dP3 of the maxilla 89.71 (see figs. 15.21A,B) is about 5–6% longer relative to the dP4 than in a *Pachytragus laticeps* Loc. 33 91.175 (see fig. 15.21D), so 89.71 is not entirely satisfactory classed as a boselaphine. At Loc. 108, the p3 on the mandible 93.833 is as long as the p4, which makes Boselaphini a more likely identification than various non-boselaphine genera. The upper molar 94.213 could be conspecific with the mandible and is the size of an M1 of the *Tragoportax* sp. of Loc. 49.

Thus it looks as if Boselaphini are present in Sinap from MN 9 onward. Köhler (1987, p. 139) noted an MN 9 boselaphine at Esme Akçakoy; her other records are of Turolian age.

Discussion. The cross section of the horn core Loc. 49 91.744 (fig. 15.9E) is more like *Tragoportax gaudryi* than *T. leskewitschi* or *T. amalthea* in that the level of maximum transverse thickness lies rather posteriorly and thereby accentuates the appearance of a posterior surface. The lateral surface is not flattened, unlike *T. amalthea*. The Dhok Pathan *?Miotragocerus vedicus* (Pilgrim 1939) is an example of a small-horned boselaphine species of late Miocene age, and it seems that such a species occurred also in the Sinap.

A small boselaphine from Kayadibi, called *Graecoryx recticornis* by Köhler 1987, has long horn cores, straight in profile, divergence lessening distally, and weak keels. It is probably different from the Sinap species. A frontlet of *Miotragocerus* from Kettlasbrunn (NHMW 1986/21) is about the size of the Sinap species, but the horn cores are more compressed and almost straight.

The Sinap horn cores look somewhat like the Maragheh *?Hispanodorcas rodleri* (Pilgrim and Hopwood 1928; Rodler and Weithofer, 1890, pl. 5, fig. 1, pl.6, fig. 1; de Mecquenem 1924–1925, pl. 3, fig. 3), except in having a stronger hint of keels and no increase in divergence distally. *?H. rodleri* is a species of uncertain relationships.

Boselaphine teeth at Loc. 49 come from at least four individuals, judged by right p3 specimens. These teeth are no larger than smaller dentitions of *"Graecoryx valenciennesi"* at Pikermi, but the P3 and P2 specimens are less enlarged or lengthened than in that species. This is probably a primitive, not an advanced condition. The narrow constrictions between front and back lobes of molars, narrow points of the labial lobes of the lower molars, and the poor paraconid-parastylid differentiation on p3 and p4 specimens also look primitive. The premolar row is shorter and the teeth are more robustly built, and perhaps more hypsodont, than in the Sebastopol *Tragoportax leskewitschi* (BMNH casts M15761–3, M15765). The teeth are about the size of those attributed to *T. gaudryi* at Kayadibi and smaller than later Turolian *Tragoportax* in Turkey (Köhler 1987, figs. 7, 10, 11). All this supports a Vallesian or early Turolian age for the Sinap localities where such Boselaphini occur. Quite probably more than one species is represented; the most likely identity would be *T. gaudryi*, unless the relatively short premolar row suggests a lineage evolving toward *T. rugosifrons*.

Tribe Antilopini

This tribe is allied to the probably paraphyletic Neotragini (Gentry 1992, 1994). The type genus, *Antilope* Pallas, 1766, has become nomenclaturally restricted among living species to *A. cervicapra* (Linnaeus, 1758), the spiral-horned blackbuck of India. The tribe also includes the abundant and widespread *Gazella* Blainville, 1816, and three related African genera. Both spiral-horned and nonspiral-horned Antilopini occur in the Sinap Formation. Spiral-horned forms will be considered first.

Spiral-Horned Antilopini

Some of the spiral-horned antelopes from the late Miocene sub-Paratethyan faunas have clockwise torsion on the right side—for example, *Oioceros* Gaillard, 1902 (including *Samotragus* Sickenberg, 1936; see Gentry and Heizmann 1996)—but most have anticlockwise torsion. Among these, three longstanding names still in use are *Palaeoreas* Gaudry, 1861, *Protragelaphus* Dames, 1883, and *Prostrepsiceros* Major, 1891. They were originally placed in the extant African tribe Tragelaphini but were later recognized as Antilopini or members of other tribes (Pilgrim and Hopwood 1928; Pilgrim 1939; Gentry 1971; Bouvrain 1992). Azanza et al. (1998) have raised the possibility that the *Oioceros* group with clockwise torsion may have evolved from an ancestry shared with *Tethytragus* rather than being related to those forms with anticlockwise torsion. More antelopes with anticlockwise spiraled horns, many of them relatively complete and belonging to new species, have been described since 1975 from Macedonia by Bouvrain and her colleagues, and more recently by Kostopoulos and Koufos in Thessaloniki. Thanks to the kind help of Geneviève Bouvrain in Paris, I have been able to examine much of the Macedonian material, which is so crucial to the study of the group.

Prostrepsiceros Major, 1891

Type Species. *Prostrepsiceros houtumschindleri* (Rodler and Weithofer 1890).

Remarks. *Prostrepsiceros* is the most prominent genus of antilopines with anticlockwise torsion on the right side. One may doubt that Major intended *P. houtumschindleri* to be the type species of *Prostrepsiceros*, but Pilgrim and Hopwood's (1928, pp. 84, 89) analysis shows how this is the consequence of his 1891 presentation of the matter. The species was first described from Maragheh (lower and middle Maragheh), and its horn cores have a strong posterolateral keel and often a weaker anterior one. Their spiraling is fairly open. I follow Bouvrain (1982) in not including *P. zitteli* (Schlosser 1904) in *P. houtumschindleri* (cf. Gentry 1971). The first *Prostrepsiceros* species to be discussed below, *P. elegans*, was originally placed in *Palaeoreas*.

Prostrepsiceros elegans (Ozansoy 1965)

Syntypes. This species has no holotype, but the type series was "un crâne avec des cornes et denture *in situ;* des cornes isolées; une mandibule" (Ozansoy 1965, p. 67) from Yassiören. The illustrated upper cheek tooth row (Ozansoy 1965, pl. 6, figs. 4, 4a) was presumably that of the skull. By comparison with an unpublished thesis photograph of a horn core (Ozansoy 1958, pl. 25, fig. 4), Köhler (1987, p. 202, fig. 93, pl. 10, fig. 2) felt able to use the name for a frontlet and other horn cores from Corak Yerler. Having seen the thesis photograph in Paris, I agree with her. Bouvrain (1994b, p. 180) noted that the fossils figured in Ozansoy's thesis were no longer to be found in Paris, but that five more horn cores of the type series were available. She used the name *?Palaeoreas* cf. *elegans* for horn cores from Kemiklitepe D. Neither Köhler nor Bouvrain nominated a lectotype for Ozansoy's species. It would be good to select one of the five syntype horn cores as lectotype and so preserve the concept of the species favored by Ozansoy, Bouvrain, Köhler, and myself. The relationship of *P. elegans* to *P. zitteli*, *P. woodwardi* and *Nisidorcas planicornis* needs assessment.

Sinap Material. Loc. 49: 91.738 sin hc base, 33.8 × 32.4; 91.739,740 pieces sin hc(s), perhaps the same one; 91.741 partial distal hc; 91.852–3 dex hc bases, 34.4 × 31.3, 35.3 × 31.0 (fig. 15.11C,D).

Remarks. There is very little compression of the horn cores (mean percentage of three examples is 92%) and no flattening of the lateral surface. The anterior keel is strong and the posterior one practically nonexistent, at least in its basal parts. The insertion is above the back of the orbits, the inclination is low in side view, and the divergence moderate. The spiraling is not very open.

In Paris, I was able to view the Yassiören horn cores of *Palaeoreas elegans* studied by Ozansoy. The index of the horn core illustrated here (fig. 15.11A,B) was 29.7 × 23.5, which gives it linear dimensions about one-fifth smaller and mediolateral compression (79%) exceeding that of the Loc. 49 horn cores. There is a well-marked anterior keel and distally a posterolateral and a more or less recognizable posteromedial keel. The spiraling is not very open. A narrow, deep supraorbital foramen is preserved. The postcornual fossa is, as a rule, moderate sized and quite deep, wider basally than at its top, and not elongated. A possible difference from the Loc. 49 horn cores is that the posterior part of the cross section seems to be extended posterolaterally rather than posteromedially. Other characters agree with the Loc. 49 horn cores.

The species also occurs at Corak Yerler (Köhler 1987, text fig. 93, pl. 10, fig. 2) and as ?*Palaeoreas* cf. *elegans* at Kemiklitepe D (Bouvrain 1994b, pl. 2). Thus it is known from MN 10–11. The Corak Yerler frontlet, of which I have seen a cast in Munich, has a horn core index of 34.1 × 26.5, giving it linear dimensions nearly one-tenth smaller and a degree of compression again exceeding that of the Loc. 49 horn cores. The cross sectional shape agrees with the Yassiören horn cores. The anterior keel is fairly blunt proximally and sharper distally, and there is a posterolateral keel distally. These keels are stronger than appears in Köhler's (1987) fig. 93, and the divergence, too, could have been slightly greater in life. The frontals are little raised between the horn bases.

Discussion. This species shows such close resemblance to *Prostrepsiceros zitteli* (Schlosser 1904) from Samos and other Turolian localities spanning MN 11–12 that I assign it to *Prostrepsiceros*. The horn cores of the lectotype *P. zitteli* (Schlosser 1904, pl. 6, fig. 5) and of another Samos specimen in Vienna, NHMW 1911 Samos V 130, have indexes at 25.3 × 27.9 and 25.8 × 27.7. Their linear dimensions are thus nearly one-fifth smaller (81%) than the Loc. 49 horn cores, and their compression is slightly anteroposterior rather than mediolateral (110% and 107%, respectively). It also looks as if in a progression from Loc. 49 and Yassiören

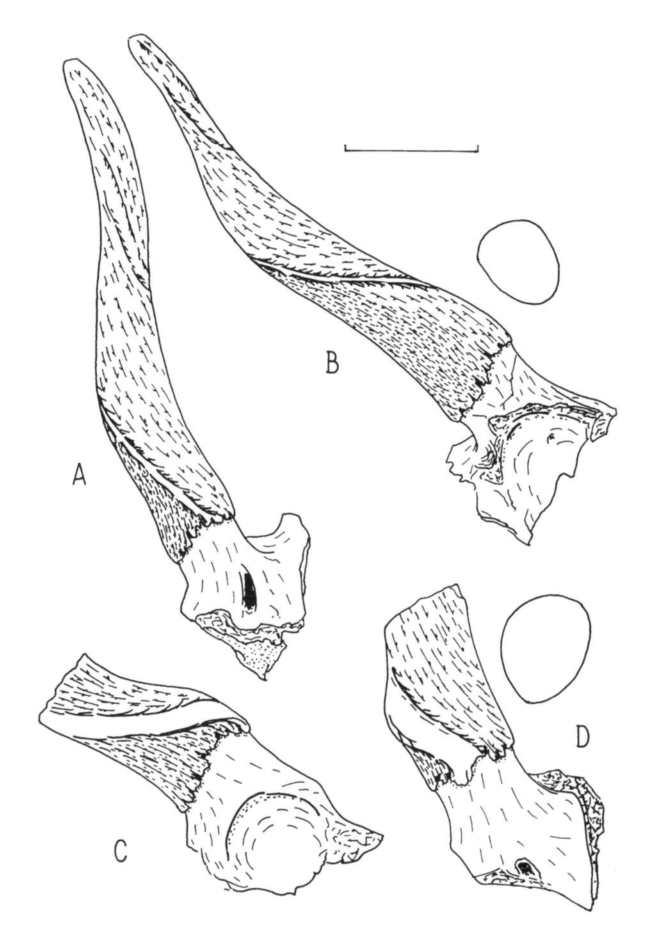

Figure 15.11. *Prostrepsiceros elegans.* **(A,B)** Anterior and lateral views of dex horn core from Yassiören (Ozansoy collection). **(C,D)** Lateral and anterior views of dex horn core Loc. 49 91.853. Anterior direction of cross sections toward bottom. Scale bar = 40 mm.

to Corak Yerler to Samos, the insertion position of the base of the anterior keel becomes anteromedial and then almost medial. Because of these slight differences, I continue to use both *elegans* and *zitteli* as names of separate species. Samos and Kemiklitepe horn cores have a more open spiral than Yassiören or Corak Yerler specimens. This is revealed in front view, where the lateral edge of the basal few centimeters of the horn core is more concave.

The holotype skull of *Prostrepsiceros woodwardi*, BMNH M4192 (Pilgrim and Hopwood 1928, pl. 7, fig. 1; Gentry 1971, pl. 5, fig. 3), also from Samos, is very like *P. zitteli* except that the horn core cross section is more flattened distally and, more importantly, the braincase roof looks steeper than it might have been in the Munich lectotype. Gentry (1971) regarded *P. woodwardi* as conspecific with *P. zitteli*, in which case some examples of *P. zitteli* show a steeply inclined cranial roof. The specimen M4192 was part of Major's 1889 collection and came from either Stefana or Vrysoula (Solounias 1981b, tables 3, 4, fig. 3). It is interesting that two conspecific *Prostrepsiceros* skulls from Nikiti 1 show differently inclined cranial roofs (Kostopoulos and Koufos 1996, pls. 2a, 3a). Fossilised bovid skulls are always liable to distortion and breakage across the orbital region between face and cranium.

Bouvrain (1982) used the name *Prostrepsiceros zitteli* for fossils from Ravin des Zouaves no. 5, in which the horn cores have a strong anterior keel and, distally, a weaker posterolateral one, and the slope of the cranial roof (Bouvrain 1982, fig. 5) is rather shallow. However, the anterior keel descends to a medial insertion, the spiraling is more open and the insertions more upright in side view, the horn cores look smaller in relation to overall size of the skull roofs (so that the dorsal orbital rims appear wide), and the frontals between the horn bases are very little raised. This animal could be a variant within *Prostrepsiceros elegans.*

In Paris, I noted the gradation between horn cores in the samples of *P. elegans* from Yassiören and Kemiklitepe D and *Nisidorcas planicornis* from Ravin des Zouaves no. 5 and Vathylakkos 2. A very close relationship must exist between them, and it may be significant that at Ravin des Zouaves no. 5, where both occur together, the *elegans-zitteli* lineage (see preceding paragraph) is unusually distinctive. *Nisidorcas* is discussed below.

Prostrepsiceros rotundicornis (Weithofer 1888)

Lectotype. Horn core from Pikermi (Weithofer 1888, pl. 18, figs. 1, 2), chosen by Pilgrim and Hopwood (1928, p. 21).

Sinap Material. Loc. 63: 90.213 dex hc (fig. 15.12A), 28.8 × 24.4.

Remarks. *Prostrepsiceros rotundicornis* was described from Pikermi and is the type species of *Helicotragus* Palmer, 1904, which Bouvrain (1982) regarded as a subgenus of *Prostrepsiceros.* It has horn cores almost devoid of keels; only a trace of an anterior keel can be detected or imagined. Gentry (1971) thought it was the most likely late Miocene species to be ancestral to *Antilope cervicapra. Prostrepsiceros fraasi* (Andree 1926) from Samos and Maragheh (and perhaps from Ravin des Zouaves no. 5) is a related, larger species with more openly spiraled horns. Gentry (1971) included *P. fraasi* in *P. rotundicornis,* but Bouvrain (1982) reseparated them and I again follow her. Both species are known from MN 11–12. *Prostrepsiceros libycus* Lehmann and Thomas, 1987 from Sahabi is another large species and shows open but weaker spiraling of its horn cores.

The Sinap horn core (fig. 15.12A) agrees well with Pikermi *Prostrepsiceros rotundicornis* but could be primitive in its small size and in being less divergent and more mediolaterally compressed. Compression at Pikermi is sometimes slightly in the anteroposterior plane (Gentry 1971, fig. 13). The Sinap horn core tapers fast and probably was not very long when complete. Other characters are that a slight keel descends to a posterior or posterolateral insertion and the inclination is low in side view. There is probably a postcornual fossa.

Two pieces of spiraled horn cores, Loc. 59 90.196, have a slight, blunt keel and diameters of about 22.6 × 19.8. They may be of Ruscinian age and are essentially indeterminate.

Prostrepsiceros aff. *vallesiensis* Bouvrain, 1982

Holotype. Skull from Ravin de la Pluie (Bouvrain 1982, figs. 2, 3).

Sinap Material. Loc. 40: 89.457 paired hcs, base of each side plus pieces, 33.0 × 23.7 (fig. 15.12B).

Loc. 77: 91.583 fragment hc; perhaps conspecific with Loc. 40 specimens.

Remarks. *Prostrepsiceros vallesiensis* is a small species hitherto known only from its MN 10 type locality, Ravin de la Pluie. It has a posterior keel and signs of an anterior keel.

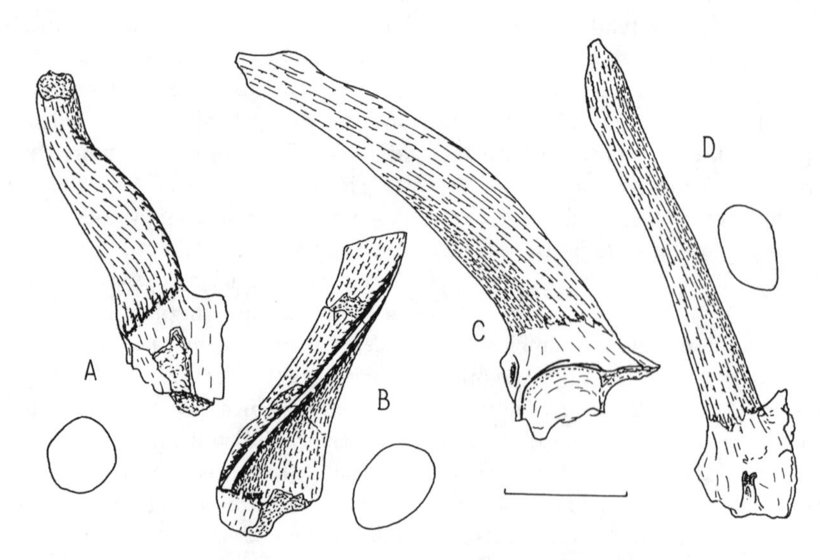

Figure 15.12. (**A**) *Prostrepsiceros rotundicornis,* dex horn core Loc. 63 90.213 in anterior view. (**B**) *P.* aff. *vallesiensis,* sin horn core Loc. 40 89.457 in anterior view. (**C,D**) *Prostrepsiceros* sp., dex horn core Loc. 91 92.455 in lateral and anterior views. Anterior direction of cross sections toward bottom. Scale bar = 40 mm.

The name was spelled *vallesienis* on page 118 of Bouvrain's (1982) paper, the page on which the species was formally founded, but appears almost everywhere else in the paper as *vallesiensis*.

The horn cores from Loc. 40 (fig. 15.12B) are smaller (by 10% or more in linear dimensions), more compressed (72%), and with weaker spiraling than in the *Prostrepsiceros elegans* of Loc. 49. An anterior keel is present and more of a trace of a second keel posterolaterally than posteromedially. Thus the lateral surface would originally have been flatter than the medial one. There are no sinuses in the pedicels. The horn cores are like *P. vallesiensis* except for being larger.

Discussion. *Prostrepsiceros vallesiensis* is similar to the later (MN 11–12) *P. houtumschindleri* of Maragheh. The so-called *P. zitteli* from Jebel Hamrin (Bouvrain and Thomas 1992) is close to *P. vallesiensis* and could well be evolving toward *P. houtumschindleri*. Kostopoulos and Koufos (1996) recorded a small-horned *P. houtumschindleri* from Nikiti 1 in Macedonia, and Gentry and Heizmann (1996) thought that *P. houtumschindleri* was also at Grebeniki (Pavlow 1913, pl. 1, fig. 19).

Prostrepsiceros sp.

A number of horn cores with a probably restricted stratigraphical range belong to this species. Teeth and postcranial bones of gazelle size within this same stratigraphical range are also assigned to this species.

Sinap Material. Loc. 14: 89.72 sin proximal metacarpal, width 18.9 ap 14.2; 89.96 sin distal radius, width 23.2.

Loc. 71: 91.459 dex mandible dp2–m2, milk teeth in late middle wear, lower molars in early wear, dp2–4 25.0, dp2 6.1 × 2.7, dp3 8.0 × 3.4, dp4 12.4 × 5.3, m1 10.8 × 5.3, m2 12.0 × 4.7, mandible depth under dp2 11.2, under m1 13.0; 91.460–461 base sin hc, fragment hc., 20.0 × 19.7; 91.462–4 sin upper molar late wear, 10.2 × 10.8; back sin m3 late mid wear; dex p4 mid wear, 9.5 × 4.6; 91.468–9 dex hc bases, 29.5 × 19.7, 29.0 × 21.3; 91.471 sin m2 mid wear, 11.5 × 7.0; 91.465 sin astragalus; height 21.1, width 12.7, ap 11.4; 91.466 first phalanx, distal, length ~29.8; 91.470 first phalanx, 35.4 × 12.6; 91.472 dex metatarsal, proximal posterior surface.

Loc. 91: 92.447 sin hc much damaged, 26.2 × 17.4; 92.454 frontlet with badly preserved hc bases; 92.455 paired hcs; 29.6 × 18.5, hcs apart ~61.5, width across supraorbital pits ~33.1 (fig. 15.12C,D); 93.8 sin hc, 18.3 × 14.6.

Loc. 8B: 92.244 sin hc base, 32.0 × 19.2; 92.245 dex hc base, ~31.6 × ~18.9; 92.268–270 three hc bases (dex, sin, dex) very damaged; 92.269 33.9 × 22.0; 92.229–230 dex and sin mandibles, p2–m3 late mid wear (same individual); 92.229, dex: m1–3 33.7, m1 7.5 × 6.4, m2 11.0 × 7.6, m3 15.7 × 7.4, p2–4 21.7, p2 5.8 × 3.0, p3 7.8 × 3.5, p4 8.5 × 4.2, mandible depth under p2 15.5, under m1 16.8, under m3 17.5; 92.231 sin mandible, p2–m3 early mid wear; m1–3 35.1, m1 10.0 × 5.6, m2 11.6 × 6.1, m3 14.2 × 5.2, p2–4 ~21.6, p2 5.8 × 3.0, p3 9.1 × 3.4, p4 9.4 × 4.0, mandible depth under p2 12.7, under m1 15.7, under m3 ~17.9; 92.233 sin p3, early mid wear, 7.6 × 5.2; 93.577 dex m3, early mid wear, 15.7 × 5.5 × 13.7 high; 93.580 sin m1 late wear, 8.1 × 8.9; 93.647 dex lower molar fragments; 89.61 sin metatarsal, incomplete distally, length ~152, minimum shaft width 10.9, proximal articular surface ap 16.9, transverse 16.0; 91.832 dex naviculocuboid, width 22.4; 93.576 fragment sin calcaneum; 89.91 dex radius with medial part of proximal end, incomplete distally, length ~142, minimum shaft width 15.0; 91.251 dex distal immature radius, width ~16.0; 91.831 metapodial, immature distal condyle, ap 14.8; 91.834 second phalanges, two distal fragments, widths 6.2, 5.3; 93.570 second phalanx, 19.6 × 12.1 high; 93.575 third phalanx, 18.7 × 11.4 high.

Loc. 13: 92.324 sin m2 early wear, 11.4 × 5.2.

Loc. 45: 92.762 sin hc 27.0 × 17.3; 92.407 fragment dex naviculocuboid; 92.408 proximal second phalanx; 92.670 three calcanea stems, two being bovid lefts, stem lengths 31.7, ~30.0; 92.755–756 dex + sin distal immature radii, width of 756 16.0; 92.757 sin astragalus, medial height 20.6, width 12.5, ap12.1; 92.758 probably partial sin magnumtrapezoid; 92.759 sin calcaneum, stem length 45.0.

Loc. 96: 92.369 dex hc, 28.0 × 19.1.

Loc. 101: 92.715 sin hc, 27.7 × 19.0.

Surface: 89.89 sin hc base (Loc. 13), 32.0 × 23.0; 89.528 sin hc (between Locs. 26 and 27), 28.2 × 20.0.

The following dental and postcranial remains at Loc. 91 could belong to *Prostrepsiceros* sp. or to either of the two other species of spiral-horned bovids at that locality.

Loc. 93.38,39 dex M2 or 3/, sin M1, mid and late wear, 11.8 × 10.1, 8.0 × 10.7; 93.45 dex P3 or P4 late mid wear, 7.0 × 7.7; 93.46 dex upper molar mid wear, 10.4 × 9.9; 93.47 dex upper molar late mid wear, 12.4 × 11.5; 93.48 sin M1 early mid wear, 9.9 × 7.9; 93.50 dex M1 late wear, 9.3 × 10.0; 93.58 dex M2 or M3 early mid wear, 10.9 × 8.2; 93.239 dex P4 early mid wear, 7.3 × 6.2; 93.414 dex P3 late wear, 6.5 × 7.7; 93.422 part dex upper premolar (?P2); 92.423 dex m3 mid wear, 16.2 × 7.7; 92.446 fragment dex mandible, broken m1+m2 mid wear, m2 11.6, mandible depth under m1 15.5; 92.596 sin mandible, dp2– incomplete m3, dp2–4 26.3, dp2 5.9 × 2.9, dp3 7.6 × 3.6, dp4 12.4 × 5.2, m1 9.4 × 5.6, m2 11.6 × 6.0, m3 ~14.0 × 4.0 × 11.2 high, mandible depth under dp2 12.3, under m1 14.5; 92.863 dex m3 early mid wear, 15.5 × 6.3; 92.864 dex m2 mid wear, 11.7 × 7.3; 93.15 sin p2–4 late mid wear, p2–4 21.9, p2 5.4 × 3.6, p3 8.9 × 4.2, p4 9.1 × 5.0; 93.37 dex m1 mid wear, 9.6 × 5.5; 93.40 dex p2, 6.0 × 3.2; 93.44 dex m1 early mid wear, 9.8 × 6.1; 93.49 sin m1 late wear, 8.7 × 6.8; 93.412 sin m1 early mid wear, 9.5 × 5.8; 93.10 dex distal tibia, tarsus, and proximal metatarsal, tibia width 23.1, calcaneum length 62.5, astragalus medial height 26.0, width 17.3, ap 16.2, naviculocuboid width 22.1, metatarsal 21.5 × 19.5; 93.16 much of sin calcaneum; 92.428 dex proximal metacarpal, transverse 19.0, ap 14.6; 93.13 distal metacarpal, sesamoids and set of phalanges; metacarpal width 20.9, first phalanx 36.8 × 14.1, second phalanx 21.9 × 13.4, third phalanx 24.1 × 13.9 high; 93.42 sin distal humerus (width 24.8, medial condyle height 18.9), proximal radius (25.2 wide), olecranon; 92.448 metapodial distal condyle, ap 12.0.

Remarks. The best preserved horn cores—for example, Loc. 91 92.455 (fig. 15.12C) and Loc. 101 92.715—are moderately long, mediolaterally compressed (mean for listed horn cores is 70%), without flattening of the lateral surface, and the anterior part of the lateral surface curves around to meet the medial surface at the anteriormost point of the horn core. These horn cores also exhibit an approach to an anterior keel; they are curved backward in side view, inserted close together, with divergence increasing from the base and decreasing distally. They have a very slight torsion that is anticlockwise on the right and are inserted above the orbits. Pedicels are sometimes very short or poorly demarcated. There is a small-to-moderate postcornual fossa, the dorsal orbital rims are quite narrow, frontals between horn bases are at a level no higher than orbital rims, and the small supraorbital pits are quite narrow.

At Loc. 71, both horn cores 92.468 and 92.469 are mediolaterally compressed and the latter has torsion.

The frontlet Loc. 91 92.454 shows a slight ascent of the bone surface in the midfrontals area immediately behind the horn bases, and thereafter the cranium roof probably turns down. It shows no sinuses in the pedicels. Pedicels at Loc. 8B are shorter than some other specimens (e.g., Loc. 96 92.369), and the horn cores perhaps more inclined backward. The horn core Loc. 8B 92.244 has a narrow supraorbital pit on the front of the pedicel like that of the surface horn core 89.528. Both these specimens also have an anteromedial ridge on the pedicel. The horn cores Loc. 8B 92.268–270 cannot be seen to possess differentiated pedicels at all; however, they are poorly preserved.

Teeth and postcranial bones that could match these horn cores are the size of *Gazella*. They differ from *Gazella* in that the m3 specimens Loc. 8B 93.577 and Loc. 91 92.596 are much higher (87%, ~80%, respectvely) than the *Gazella* m3 Loc. 33 91.94 (66%). The premolar rows are as long as in *G. capricornis,* which is longer than in *G. deperdita* (table 15.3). The p2 specimens are longer relative to p4 (64%) than is found in Turolian *Gazella* (table 15.3), and the two dp2 specimens at Locs. 71 and 91 are also rather long. It is interesting that two *Prostrepsiceros vallesiensis* m3 spec-

Table 15.3. Mean Occlusal Lengths for Some Lower Teeth of *Gazella dorcas* and Small Miocene Bovids

Taxon (Location and Age)	m1–3	m3	p2–4	p4	p2	A	B
Gazella dorcas (Extant)	35.8 (10)	16.0 (10)	18.7 (10)	7.9 (10)	5.0 (10)	52	63
G. deperdita (Mont Lubéron MN12)	36.4 (15)	15.9 (44)	20.7 (9)	8.7 (27)	4.9 (9)	57	56
G. capricornis (Pikermi MN12)	34.3 (12)	14.8 (5)	21.6 (6)	8.6 (11)	5.0 (6)	63	58
Hispanodorcas orientalis (Ditiko 3 MN13)	32.7 (3)	13.4 (3)	20.3 (2)	7.8 (4)	4.1 (2)	62	53
Nisidorcas planicornis (Vathylakkos 2 MN11)	37.6 (5)	15.7 (7)	21.0 (2)	8.6 (8)	4.8 (2)	56	56
Prostrepsiceros h. syridisi (Nikiti 1 MN11)	46.5 (15)	18.8 (16)	27.1 (10)	11.0 (18)	7.1 (7)	58	65
P. zitteli (Ravin de Zouaves 5 MN 11)	40.4 (1)	17.9 (1)	21.4 (1)	9.3 (1)	5.5 (1)	53	59
P. vallesiensis (Ravin de la Pluie MN 10)	35.5 (10)	14.9 (12)	18.2 (8)	7.4 (11)	4.1 (9)	51	55
Oioceros praecursor (Ravin de la Pluie MN 10)	34.5 (16)	14.6 (18)	19.8 (10)	8.3 (24)	4.2 (14)	57	51
Tethytragus langai (Arroyo de Val-Barranca MN 7/8)	43.8 (3)	19.2 (10)	29.8 (2)	12.0 (7)	7.8 (3)	68	65
Tethytragus koehleri (Pasalar and Candir MN 6)	41 (2)	17.4 (25)	26 (2)	10.8 (22)	7.7 (9)	63	71
Turcocerus gracilis (Candir MN 6)	38.1 (8)	16.7 (16)	22.8 (3)	8.7 (12)	5.5 (1)	60	63

Sources: Author, Heintz (1971), Bouvrain (1979), Bouvrain and Bonis (1985), Köhler (1987), Bouvrain and Bonis (1988), Bouvrain (1992, fig.11), Azanza and Morales (1994), Kostopoulos and Koufos (1996). The molar and premolar rows of *Tethytragus koehleri* were measured from Köhler (1987, figs. 39, 40); the m3, p4 and p2 of *Turcocerus gracilis* were measured from Köhler (1987, figs. 25).

Notes: Number of specimens is shown in brackets below each measurement. Top box: *Gazella* species; middle box: spiral-horned species; bottom box: Middle Miocene species. A = p2–4 as a percentage of m1–3 B = p2 as a percentage of p4.

imens from Ravin de la Pluie indicate a hypsodonty index ~76%, and an m3 of *Nisidorcas planicornis* from Vathylakkos 2 was as high as 88%.

The radius at Loc. 8B is longer than in *Gazella deperdita* from Mont Lubéron (Heintz 1971, p. 21), but less gracile than in extant *Gazella*. The naviculocuboid at the same locality is a little larger and perhaps too big for this species.

Discussion. This is an interesting species, presumably a member of the Antilopini and overlapping in the Sinap sequence the advent of *Gazella*. It is found in localities from the level of Loc. 14 to that of Loc. 8B inclusive. From the occurrences of horn cores, it looks as if Locs. 13, 45, 96, and 101 can be placed stratigraphically within this short span. The surface horn core 89.528, although originating topographically between Locs. 26 and 27, should not be as stratigraphically high as the level of these locations.

There are few differences between *Prostrepsiceros* sp. and *P. vallesiensis,* among them that in the former species, the horn core spiraling is weaker and the dorsal orbital rims are narrow. Presumably *Prostrepsiceros* sp. is a geologically earlier species.

Nisidorcas Bouvrain, 1979

Type Species. *Nisidorcas planicornis* (Pilgrim 1939)

Remarks. In the Sinap collection is a left horn core 91.260, with basal index 26.5 × 21.2, a surface find at the base of the north face of Delikayinçak, between Locs. 12 and 72 and topographically below both of them (fig. 15.13A,B).

This horn core is the only one like *Nisidorcas planicornis* except that Geraads and Güleç (1999, fig. 1D–E) recorded *N. planicornis* from Çobanpinar, which is Loc. 42 of the present chapter. The type locality for the species is Piram Island, India, and it was recognized in Europe by Bouvrain (1979). The species is found in Macedonia, Turkey, and Maragheh (de Mecquenem 1924, pl. 8, fig. 1; Köhler 1987; Kostopoulos and Koufos 1999) and lasted from MN 11 to MN 12.

The horn core 91.260 is very like the Indian holotype but slightly smaller and without sign of backward curvature in side view. Originally it was probably long. It has a mediolateral compression of 80%, is without a flattened lateral surface or definite keels, is inserted at a low inclination, and has moderate divergence diminishing distally, with torsion which would be anticlockwise on the right side. It is inserted above the back of the orbits, with slight raising of frontals between the horn bases, and no sinuses within the frontals.

Another (right) horn core, Loc. 82 92.838 with a basal index of 27.2 × 21.1, is too small for a *Pseudotragus* and its basal cross section is similar in size and shape (medial and lateral sides nearly parallel) to an example of *Nisidorcas planicornis* from Kayadibi (BSPHG 1985.XXVII.53, cast). However Loc. 82 is apparently the Pliocene (MN 17) locality Sarikol Tepe of Kostopoulos and Sen (1999), so my original assignment is probably wrong.

Nisidorcas planicornis is very like *Prostrepsiceros elegans,* as already noted. Its horn cores show variation in surface rugosity, degree of compression, and development of keels (Köhler 1987, p. 200). A skull from Vathylakkos 2 (Bouvrain 1979, figs. a,b), for example, has quite a sharp posterior keel on its horn cores, but frontlets and horn cores from Ravin des Zouaves no. 5 do not show this. Sometimes there is a blunt anterior keel or the posterior keel is accompanied by prominent grooves. A posterior keel is more likely to be evident on the distal part of the horn cores. The spiraling of the horn cores is no stronger than in *P. elegans.* Horn cores with stronger anterior keels look more readily identifiable as *P. elegans.* Given the variation in horn core characters within *N. planicornis,* it seems that *P. elegans* could be simply one extreme of the range in characters for the former species. Compared with the Loc. 49 horn cores of *P. elegans,* the horn core 91.260 is ~70% as large, more compressed mediolaterally (80% versus 92%), and lacks an anterior keel.

Nisidorcas planicornis is also not very different from the earlier species *Prostrepsiceros vallesiensis.* Keels are present

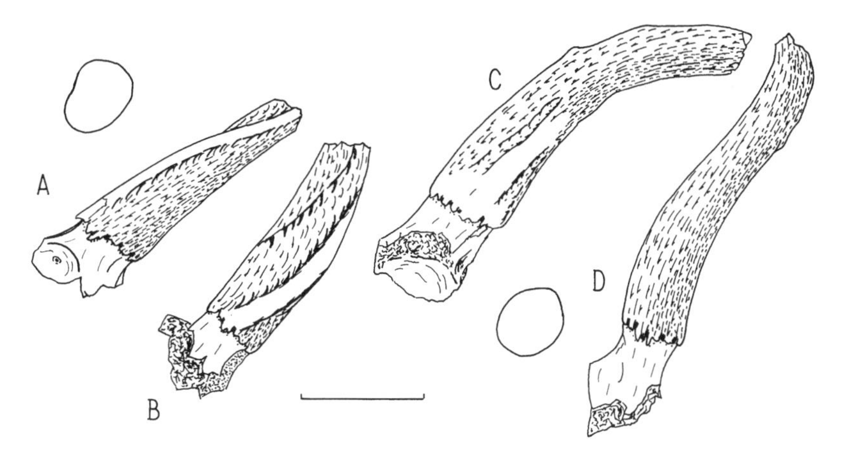

Figure 15.13. **(A,B)** *Nisidorcas planicornis,* lateral and anterior views of sin horn core 91.260 from Sinap surface. **(C,D)** *Sinapodorcas incarinatus* lateral and anterior views of sin horn core Loc. 91 93.26. Anterior direction of cross sections toward bottom. Scale bar = 40 mm.

but not accentuated on the horn cores of both, the degree of torsion is the same, and both show reduction of the entoconid transverse crest on p3 and p4. The differences from *P. vallesiensis* appear to be that the horn cores are less compressed, the occipital surface faces more directly backward, basisphenoid rises less abruptly in front of the basioccipital, the basioccipital is more triangular (perhaps because the posterior tuberosities are wider), the bulla more inflated, the molars probably more hypsodont, the lingual walls of the lower molars less outbowed, the premolar rows perhaps longer, and the entoconid reduction on p3 and p4 carried further.

Finally in connection with *Nisidorcas planicornis,* it may be noted that an immature female cranium BMNH M3683 from Piram Island, holotype of *Cambayella watsoni* Pilgrim (1939, p. 122, pl. 3, figs. 6–8), is very like the Vathylakkos skull. Its linear dimensions are about four-fifths the size of the latter, the braincase has little inclination of its roof except where it bends downwards posteriorly, the braincase probably widened posteriorly, the basioccipital is narrow anteriorly with a poor central longitudinal groove and anterior tuberosities without longitudinal ridges behind, and the auditory bulla is inflated and moderately sized. If M3683 should be conspecific with the holotype of *N. planicornis,* also from Piram Island, *Cambayella* would become an available senior synonym of *Nisidorcas* and in need of formal suppression.

Sinapodorcas Bouvrain, Sen and Thomas, 1994

Type Species. *Sinapodorcas incarinatus* (Ozansoy 1965).

Sinapdorcas incarinatus

Lectotype. A left horn core from the middle member of the Sinap Formation at Yassiören (Ozansoy 1965, pl. 8, fig. 2; Bouvrain et al. 1994, fig. 1).

Sinap Material. Loc. 91: 93.26 sin hc (fig. 15.13C,D), 24.3 × 20.9.

Remarks. This is a small species with openly spiraled and poorly compressed horn cores without keels. The horn core 93.26 (fig. 15.13C,D) is extremely similar to the lectotype. It is of small size, long and narrow, without a flattened lateral surface, with up to two strong persistent longitudinal grooves posterolaterally, curving backward in side view, curving increasingly outward above its base and then with decreasing divergence distally, inserted above the orbits, with a moderately deep postcornual fossa of a large area, without sinuses in the frontals, and probably with a small supraorbital pit.

Discussion. The open spiraling and absence of keels are like the later and much larger *Prostrepsiceros fraasi.* The openness of the spiraling exceeds that of *P. rotundicornis* of Pikermi. *Helicotragus major* Ozansoy (1965, p. 71, pl. 8, fig. 4) was also placed in *Sinapodorcas incarinatus* by Bouvrain et al. (1994). The specimen differs from *Prostrepsiceros* sp. by less compression (86% versus 70%), no approach to an anterior keel, slightly stronger spiraling, and a larger postcornual fossa.

Palaeoreas Gaudry, 1861a

Type Species. *Palaeoreas lindermayeri* (Wagner 1848).

Remarks. Some of the more obvious characters of *Palaeoreas,* as at present constituted, are horn cores rather large in relation to skull size, a strong posterior keel, strong longitudinal grooving and ridging sometimes giving rise to other recognizable keels, horn insertions close together, not very divergent, and with torsion but without open spiraling.

Palaeoreas lindermayeri itself from Pikermi has massive horn cores mounted on elevated frontals. Gentry (1971, 2000) related it to *Criotherium argalioides* Schlosser, 1904, which he had accepted in 1971 as an ovibovine, following Schlosser (1904, p. 27) and Bohlin (1935). *Palaeoreas asiaticus* from Garkin is at least nearly identical with *P. lindermayeri* (see below), but *P. zouavei* Bouvrain, 1980 from Ravin des Zouaves no. 5 is a separate and larger species. *Palaeoreas brachyceras* Ozansoy, 1965 is a nomen dubium.

Palaeoreas asiaticus Köhler, 1987

Holotype. Frontlet with horn cores from Garkin, Turkey (Köhler 1987, p. 204, fig. 94, pl. 10, fig. 1).

Sinap Material. Loc. 49: 92.159–160 dex and sin hcs, 35.0 × 26.6, 35.1 × 26.3 (fig. 15.14C,D); 92.187–189 sin p3 or 4 mid wear 10.4 × 5.2, sin m1 + front m2 late mid wear 10.2 × 7.0; sin m3 18.8 long; 93.685 two dex lower molars late mid wear 10.0 × 6.7, 12.7 × 8.1.

Remarks. The horn cores are presumably from one individual. They are mediolaterally compressed, without flattening of the lateral surface, with a strong posterior keel and a blunt anterior keel, inserted above the back of the orbits, with quite a low inclination in side view, inserted close together and not strongly divergent, with torsion that is anticlockwise on the right side.

The teeth Loc. 49 92.187–189 and 93.685 are marginally larger than *Gazella* and could belong to *Palaeoreas* or *Prostrepsiceros.* They are also only marginally smaller than teeth to be assigned below to *Pseudotragus,* so identifications are tentative. The front and back walls of the lower premolar are not set so transversely as in some of the Pikermi specimens attributed by Gentry (1971) to *Palaeoreas lindermayeri* (e.g., BMNH M15828). Paraconid-metaconid fusion is beginning at a low level and would be more apparent in later wear.

Discussion. *Palaeoreas asiaticus* was said to differ from *P. lindermayeri* by less massive horn cores, with a more compressed cross section and less divergence. The holotype cast in Munich has horn cores about one-eighth smaller than

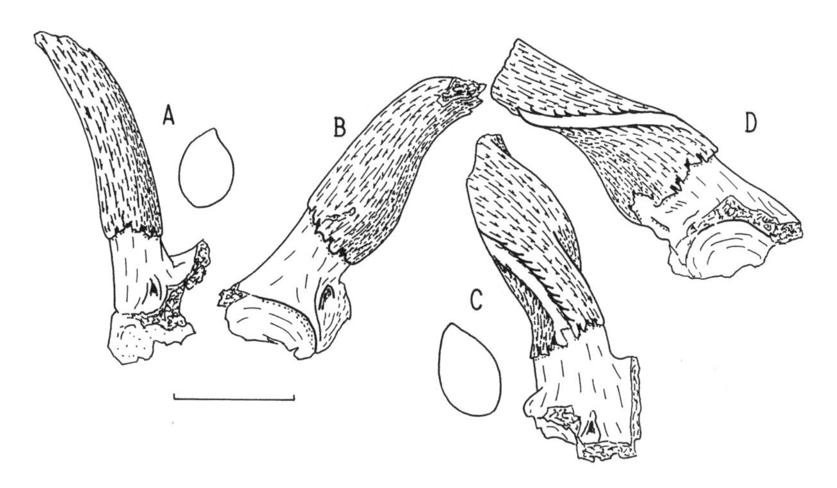

Figure 15.14. (A,B) *Palaeoreas* sp., anterior view of dex horn core and lateral view of sin horn core, both Loc. 91 92.453. (C,D) *P. asiaticus,* anterior and lateral views of dex horn core Loc. 49 92.159. Anterior direction of cross sections toward bottom. Scale bar = 40 mm.

P. lindermayeri in linear dimensions, and ~5% more compressed. The angle at which the left horn core is oriented on the frontlet is unreliable because of postmortem damage, and the divergence in life was probably no smaller than in *P. lindermayeri.* The Samos frontlet of *P. lindermayeri* illustrated by Gentry (1971, pl. 6, fig. 2) is the closest example of that species to *P. asiaticus.* It comes from Stefana and also has less tight torsion than Pikermi *P. lindermayeri.*

The Loc. 49 horn cores are in their turn about one-eighth smaller than the holotype of *Palaeoreas asiaticus* and their compression is about 76% (versus 84% in the holotype). The supraorbital pits may be smaller and are certainly smaller than in *P. lindermayeri.* The strong, sharp posterior keel and the blunt anterior keel are the same in *P. asiaticus* and the Loc. 49 horn cores.

Initially I would have preferred to sink *Palaeoreas asiaticus* in *P. lindermayeri,* but I shall continue to use the former name here, more easily to accommodate within it the somewhat smaller Loc. 49 fossils. Their size is more definitely too small for inclusion within *P. lindermayeri.*

It is problematic whether the Loc. 49 horn cores fall within the definition of *Palaeoreas brachyceras* Ozansoy, 1965. This species has no type specimen or illustration, but information was divulged in Ozansoy (1965) that it differed from *P. lindermayeri* by its horn cores being smaller, less robust, and with a less feeble second (i.e., posterior) keel, and from *P. elegans* by its horn cores being more robust, shorter, and more twisted. Tekkaya (1975) commented on the species and published six photographs of two horn cores, one said to belong to *P. elegans* and the other to *P. brachyceras.* His plate 1, figures 2, 4, and 6 show a horn core which is most probably *P. elegans,* and plate 1, figures 1, 3, 5 another more poorly preserved horn core. The index of the latter would be around 35 × 25 (71% compression), close to values for the Loc. 49 horn cores. However Tekkaya's captions refer figures 1, 2, and 3 to *P. brachyceras* and figures 4, 5, and 6 to *P. elegans,* so the situation continues to be confused. Köhler (1987, p. 206) regarded *P. brachyceras* as a nomen nudum. Bouvrain (1994b, p. 180) did not claim that any of Ozansoy's material still in Paris can be assigned to this name. I am just as unable as Köhler to use the name, and I take it as a nomen dubium.

Palaeoreas sp.

Sinap Material. Loc. 72: 91.330 dex p4 early mid wear 9.1 × 3.4; 91.324 sin calcaneum, stem height 41.5; 92.51 dex unciform; 92.284 second phalanx, proximal end; 92.286 two distal ends second phalanges.

Loc. 91: 92.453 paired hc bases, 27.3 × 18.4. (fig. 15.14A,B); 92.599 damaged dex hc, 18.5 × 16.2; 93.14 dex hc, 27.0 × 20.0.

Loc. 12: 91.365 dex hc base, 28.1 × 22.0; 92.117 second phalanx, 20.6 × 11.7 high; 92.118 part distal sin radius; 92.132 fragment distal second phalanx; 92.587 distal sin humerus, 25.7 × 18.4 high; 92.589 two first phalanges, 36.0 × 13.9 high; 92.590 two second phalanges, 19.7 × 11.8 high, 19.3 × 11.8 high; 92.591 three third phalanges, ~22.0 × ~12.4 high, ~23.0 × 12.9 high; 92.592 second phalanx, 23.6 × 15.0 high; 92.593 third phalanx, 24.3 × ~14.0 high; 92.594 metapodial distal condyle, ap 19.2; 92.595 first phalanx, 40.3 × 16.0 high; 93.855 second phalanx, 23.5 × 14.0 high; 93.862 dex astragalus, medial height 31.0, width 20.6, ap 19.0; 93.903 sin distal tibia, width 29.8, and tarsus, calcaneum length 69.5; astragalus, lateral height 33.4, width 20.6, ap18.8, naviculocuboid width ~25.8, ectocuneiform ap 21.9; 93.913 sin metatarsal, length ~237, proximal ap <26.4, transverse >24.2; 93.1181 distal metatarsal and complete phalanges; metatarsal 20.8 × 15.8; first phalanx, 35.3 × 13.6 high; second phalanx, 19.5 × 12.0 high; third phalanx, 21.0 × 12.4 high; 93.1202 dex proximal and distal tibia, part sin calcaneum; tibia proximal width 44.3, distal width 29.5.

Surface: 90.310 piece hc, spiraled and distal, from upper Kavakdere.

Remarks. The horn cores (fig. 15.14A,B) are small, moderately long, without a flattened lateral surface, the posterior keel basally strong and thereby conferring strong mediolateral compression, with a demarcation of medial and lateral surfaces meeting anteriorly rather than a keel as such, rather uprightly inserted, little divergence, with torsion (which would be anticlockwise on the right side), inserted above the orbits, dorsal part of orbital rims sloping downward rather than extending sideways, very large postcornual fossa, slight raising of frontals between the horn bases, no sinuses within the frontals, and narrow supraorbital pits.

The p4 Loc. 72 91.330 is rather large to be a *Gazella* p3. It is similar in size to p4 specimens of *Palaeoreas lindermayeri* except that the anterior and posterior edges of the tooth are not turned so strongly transversely. The metaconid crest is simple and positioned transversely across the tooth.

Teeth and postcranial remains attributed to this species are a little larger than those of *Prostrepsiceros* sp., which precedes it stratigraphically and coexists with it at Loc. 91. These remains are a little larger than in *Gazella* sp., which also overlaps the Sinap span of *Palaeoreas* sp. The identity of fossils other than horn cores at Loc. 91 is uncertain. They were listed above under *Prostrepsiceros* sp; another possible identity for them would be *Pseudotragus* aff. *capricornis*.

Discussion. *Palaeoreas* sp. is a smaller and less distinctive species than *P. lindermayeri* or the *P. asiaticus* of Loc. 49. According to linear dimensions, the Loc. 91 horn cores are about 70% of the size of those from Loc. 49, and their degree of mediolateral compression is the same. The Loc. 12 horn core is slightly bigger. These horn cores are placed in *Palaeoreas* because of the prominence of the posterior keel, but they could be close to the ancestry of *Helladodorcas*[1] *geraadsi* from Pentalophos (Bouvrain 1997).

It can be seen from figure 15.15 that the horn cores of various species of *Palaeoreas* form a series increasing in size from *Palaeoreas* sp. at Locs. 12 and 91 to the advanced *P. lindermayeri* at Pikermi. *Palaeoreas zouavei* has small horn cores relative to its skull size; hence it does not appear on this graph to be any larger than *P. lindermayeri*.

A strong posterior keel is also evident in a late Miocene horn core from the Namurungule Formation, Kenya, first designated *Palaeoreas* sp. and later assigned to *Ouzocerus* (Nakaya et al. 1984, p. 109, pl. 9, fig. 5; Nakaya 1994, p. 20). This is a rare example of a late Miocene bovid apparently identical in Eurasia and subsaharan Africa.

Antilopini without Spiraled Horns

Gazella Blainville, 1816

Type Species. *Gazella dorcas* (Linnaeus, 1758).

[1]*Helladodorcas* was founded as *Helladorcas*, but the stem of *Hellas* [Greece] is "*Hellado-*" as used in *Helladotherium*.

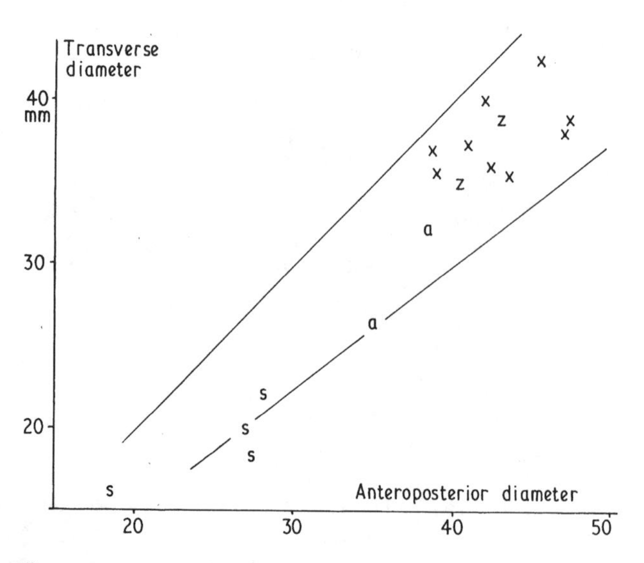

Figure 15.15. Graph of basal horn core diameters in *Palaeoreas*. The upper diagonal line is that along which the transverse diameter is 100% of the anteroposterior diameter, the lower line is 75%. Symbols: x = *Palaeoreas lindermayeri*, z = *P. zouavei*, a = *P. asiaticus*, s = *Palaeoreas* sp.

Remarks. *Gazella* is a widespread genus in the Old World. The 14 or more post-Pleistocene species graze and browse in semiarid and arid, open regions and are well adapted cursorially (Estes 1991). Their skulls, horn cores, and dentitions, however, are not notably distinctive, and the differentiation of Miocene species is difficult.

Gazella sp.

Sinap Material. Loc. 72: 91.320 base sin hc, 22.8 × 20.8; 92.207 dex mandible p2–m3 early mid wear, p2–4 22.4, p2 5.6 × 3.8, p3 8.0 × 4.4, p4 8.4 × 4.7, m1–m3 35.6, m1 9.0 × 5.8, m2 11.4 × 6.0, m3 15.3 × 4.9, mandible depth under p2 12.4, under m1 15.0; 92.213 dex upper molar early mid wear, 12.6 × 9.8; 92.218 distal second phalanx; 92.281 medial part dex distal tibia; 92.283 metapodial immature distal condyle, ap 10.8.

Loc. 12: 91.375 sin P2 early wear, 7.8 × 6.0; 92.321 dex mandible, m1–3, late mid wear, m1–3 36.0, m1 10.0 × 7.00, m2 12.0 × 7.4, m3 14.5 × 7.3; 92.323 part dex upper molar, late wear; 93.909 sin mandible, p3–m3 (fig. 15.16G) mid wear, m1–3 35.7, m1 10.3 × 6.2, m2 11.9 × 6.9, m3 14.4 × 5.8, p3 9.1 × 4.6, p4 9.5 × 3.8, mandible depth under m1 16.5, under m3 18.4; 93.843 distal sin tibia and most of astragalus; tibia distal width ~16.8; astragalus medial height 19.3, width 13.0, ap 12.0; 91.370 dex astragalus lateral height 19.0, width 11.8, ap 10.7; 93.843 distal sin tibia and most of astragalus; tibia distal width ~16.8, astragalus medial height 19.3, width 13.0, ap 12.0; 89.290 distal sin humerus, width 25.7, medial condyle height 18.4, minimum shaft width 14.3; 89.291 proximal dex radius; 93.1191 sin scapula, glenoid facet ap 20.5, transverse 14.0; 93.1200 distal dex radius, width 15.6; 93.1201 first phalanx, ~24.4 × 10.1 high, fragment third phalanx; 91.347 second phalanx,

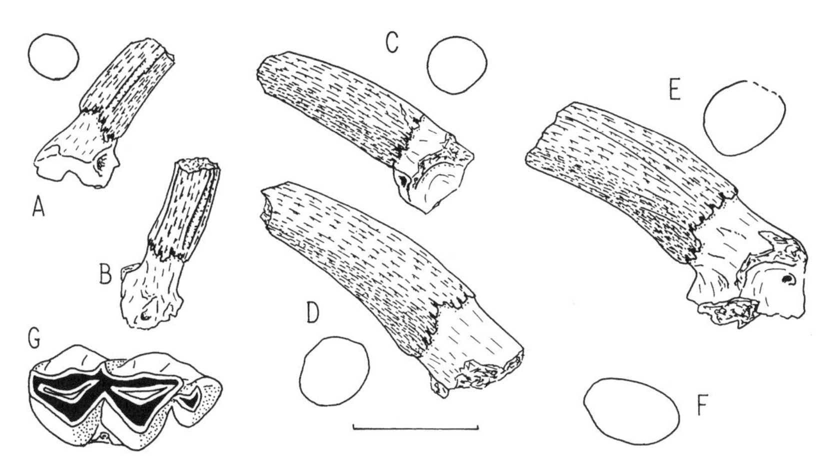

Figure 15.16. (**A**,**B**) Bovidae sp. B, sin horn core 94.164 of unknown locality in lateral and anterior views. (**C**–**G**) *Gazella* sp. (**C**) Dex horn core Loc. 49 90.112 in lateral view. (**D**) Dex horn core Loc. 49 90.111 in lateral view. (**E**) Dex horn core Loc. 33 89.268 in lateral view. (**F**) Cross section of dex horn core Loc. 33 89.221. (**G**) Occlusal view of sin m3 Loc. 12 93.909. Anterior direction of cross sections toward left for **A** and right for **C**–**F**. Scale bar = 40 mm for **A**–**F**, 10 mm for **G**.

16.3 × 10.2 high; 91.348 third phalanx, 17.5 × 10.9 high; 91.351 most of dex astragalus; 91.363 most of first phalanx, ~27.0.

Loc. 1: 92.319 dex P3–M1, late wear, M1 9.6 × 10.4, P3 8.8 × 8.1, P4 8.3 × 8.9; 92.57 dex mandible m1–3, late wear, m1–3 36.0, m1 9.5 × 6.0, m2 11.2 × 6.3, m3 15.3 × 6.3, mandible depth under m3 20.0; 93.529 sin mandible p3–m3, late wear, m1–3 35.3, m1 9.0 × 6.1, m2 10.5 × 6.0, m3 15.3 × 6.1, p3 8.9 × 4.6, p4 8.9 × 5.0, mandible depth under m1 17.4; 92.59 dex scapula, glenoid facet ap 21.5, transverse 17.6; 93.530 most of first phalanx. (All Loc. 1 bones may be from one individual.)

Loc. 49: 90.111–112 dex hcs, 27.6 × 22.4, 20.9 × 19.7 (fig. 15.16C,D); 91.743 dex hc, 27.0 × 22.6; 92.185 sin hc 23.1 × 19.6; 93.664 sin hc 23.9 × ~21.2; 92.195 dex P3 early mid wear, 8.4 × 6.1; 91.709 dex mandible p2–4 late mid wear, p2–4 20.6, p2 5.4 × 2.8, p3 7.2 × 3.7, p4 9.0 × 4.3, mandible depth under p2 ~10.9; 93.942 dex metatarsal, total length 149, minimum shaft width 10.5, proximal articular surface ap 18.0, transverse 16.9; 93.650 sin calcaneum stem 36.3; 93.732 distal dex tibia, width ~20.8, width across articular facets 18.5; 93.1010 dex tibia, part calcaneum, astragalus, naviculocuboid, ectocuneiform, length tibia 188, minimum shaft width 12.3, total distal width 20.5, astragalus lateral height 24.2, width 14.9, ap 13.6; naviculocuboid width 18.9; 92.183 metapodial immature distal condyle, ap 11.4; 93.979 distal first phalanx; 93.1077 second phalanx, 19.6 × 11.0; 93.1078 first and second phalanges, 31.9 × 12.5 high, 19.2 × 11.0; 93.1079 first and second phalanges, 31.7 × 13.1 high, 18.9 × 10.4.

Loc. 26: 89.318 dex mandible with m1 or 2 early mid wear, 11.1 × 4.7, mandible depth below M: 14.0.

Loc. 33: 89.150 part hc; 89.221 sin hc base, 33.0 × 24.2 (fig. 15.16F); 89.242 base dex hc, 26.7 × 21.6; 89.268 dex hc, ~29.6 × 22.2 (fig. 15.16E); 89.348 sin hc, 29.3 × 24.0; 91.4 distal sin hc (?too large for *Gazella*); 91.136 sin hc, 24.0 × 20.6; 91.29 part dex P4 late mid wear; 91.87 sin M1 later mid wear, 9.7 × 8.0; 91.96 dex upper molar front half, mid wear, length ~10.1; 91.94 sin m3 early wear, 15.6 × 5.4 × 10.3 high; 91.115 dex lower incisor, probably i1; 94.130 dex lower molar late wear, 11.5 × 7.0; 89.211 dex distal tibia, width across articular surfaces 21.5; 89.232 much of sin calcaneum; 91.84 medial partial proximal radius; 91.90 sin proximal metacarpal; 91.8 much of third phalanx; 91.153 proximal second phalanx.

Loc. 42: 89.417 base sin hc, 25.9 × 20.8; 89.438 base sin hc, 18.7 × 13.9; 91.428 sin hc, 29.8 × 23.8; 92.337 base sin hc, 22.7 × 17.0; 91.455 dex dP2, 7.2 × 5.0; 89.406–409 most dex P4, sin lower incisor, fragment mandible, fragment lower incisor, width 5.6; 91.796 (in part) sin upper molar mid wear 12.7 × 9.5; 91.456 part sin lower molar and part sin m3 mid wear, m3 ~13.9 long; 93.1125 paired mandibles, late mid wear, p2–4 22.0, p2 5.5 × 3.0, p3 8.1 × 3.9, p4 8.6 × 4.7, m1–3 34.3, m1 9.2 × 5.8, m2 10.9 × 5.8, m3 14.3 × 6.2, mandible depth under p2 14.9, under m1 16.3, under m3 ~19.8; 92.334 part sin calcaneum, stem height 41.0; 92.335 dex astragalus, lateral height 26.9, width 16.9, ap 15.3; 92.339 two metapodial distal condyles, ap 15.1; 91.440 second phalanx, 17.3 × 10.0 high; 91.441 first phalanx, proximal fragment; 93.1129 first phalanx, 33.8 × ~14.3 high; 93.1144 metapodial distal condyle ap 14.5.

Loc. 78: 91.606 part dex hc, 31.8 × 24.3; 91.613 dex hc, 29.1 × 23.2; 91.614 base sin hc, 27.4 × 22.5; 91.616 sin proximal radius, width 20.6.

Loc. 88: 92.804 dex hc, 22.0 × 17.4, likely length along front edge 80.

Remarks. *Gazella* horn cores have been found from the level of Loc. 72 upward. The bigger cores are the size of those of smaller *G. capricornis* of Pikermi, but the presence of smaller specimens—for example, Loc. 49 90.112 (fig. 15.16C)—suggests a pre–MN 12 age for the localities concerned. It is possible that the mediolateral compression of gazelle horn cores changes from just over 90% at Loc. 72 to just under 80% at the level of Locs. 42 and 78, but the sample is small.

The horn core Loc. 42 91.428 is too small for a *Pseudotragus* or small *Protoryx* species, but is larger and straighter than stratigraphically lower gazelle horn cores. Possibly a temporal change of species has taken place, as with the Boselaphini. If this is true, the horn core Loc. 42 89.438, which is the smallest gazelle in the Sinap collection, would have to be immature or female. Loc. 42 92.337 is as small as the Loc. 72 gazelle horn core, but the pedicel is much shorter; it, too, could be from a female. Potentially gazelline postcranial remains from Loc. 42 are large.

Gazella teeth are identified on little other than size. The upper molar Loc. 72 92.213 is slightly larger than in Pikermi *G. capricornis*. The lower cheek dentitions in Locs. 72, 49, and 42 are close in size and in relative length of the premolar row to *G. capricornis*. The p2 specimens, however, are slightly longer in comparison with p4 (67% at Loc. 72, 60% at Loc. 49 and 64% at the later Loc. 42). Compared with the good sample of *G. deperdita* teeth from Mont Lubéron (table 15.3), individual molar teeth of the Sinap *Gazella* are about 4% shorter and premolars about 6% longer. The m3 Loc. 33 91.94 has a height that is 66% of the occlusal length.

The p4 metaconid crest on Loc. 72 92.207 is turned forward at its lingual end. On the mandible Loc. 12 93.909, the p3 has a paraconid, the metaconid ridge is directed partly backward, the p4 metaconid has neither forward nor backward extensions at its lingual end, the molars have moderate-to-large basal pillars, and the lingual walls are little flattened. All are primitive characters. On Loc. 1 93.529, the p4 metaconid has a rear flange and forms a lingual wall with the entoconid, the p3 metaconid is directed obliquely backward, and basal pillars pass from moderate on m1 to small on m3. The p4 on Loc. 42 93.1125 again has its metaconid turned forward lingually, the paraconid on p3 and p4 is well marked, and a basal pillar is present on m2 and m3.

The P3 specimens of late Miocene gazelles are less reduced relative to P4 than in extant gazelles, but small samples and few associations make this character difficult to use. It is also very likely that the ratio is not identical in all living species.

Such a recognizable modern specialization as the enlarged hypoconulid lobe on m3, known in *Gazella* and some other Antilopini, is often not present in late Miocene gazelles. The enlargement involves the lingual wall of the hypoconulid being in line with the lingual wall of the more anterior part of the tooth and not offset labially. The m3 specimens of Loc. 72 92.207 and Loc. 12 93.909 are little advanced in this feature. The m3 Loc. 1 92.57 does have a large hypoconulid lobe, but the tooth is in late wear, by which stage a larger occlusal area of the lobe is to be expected in all bovids. Loc. 33 91.94 shows the advanced condition. The m3 hypoconulid of Loc. 42 93.1125 is also advanced and the lingual walls of the lower molars are probably flatter.

Three scapulae (Loc. 12 93.1191, Loc. 1 92.59, Loc. 33 91.10) have humeral facets that are transversely narrower than in extant gazelles. The left proximal metacarpal Loc. 33 91.90 has deep ligamentous insertion pits in its articular surfaces and is strongly drawn out anteriorly in its medial half. The tibia Loc. 49 93.1010 and the metatarsal Loc. 49 93.942 are shorter than in extant *Gazella dorcas* and slightly more robust. The ectocuneiform associated with the Loc. 49 tibia is partly fused with the naviculocuboid.

Discussion. The horn core Loc. 33 89.268 (fig. 15.16E) is of a fairly large late Miocene gazelle. Another Loc. 33 horn core, 89.221, looks like the Villafranchian *Gazella borbonica*, but is less compressed by about 5% (de Giuli and Heintz 1974). It is unlike other Sinap *Gazella* horn cores and fits awkwardly in MN 10–11. It differs from ?*Prostrepsiceros* sp. in that the level of widest cross section lies centrally (fig. 15.16F) rather than toward its back. In addition, the longitudinal grooving is deeper.

Gazelle teeth are smaller than *Tethytragus koehleri* from Paşalar or Candir. Basal pillars (ectostylids) on the lower molars are larger than in *T. koehleri*. The m3 in early wear at Loc. 33 is lower crowned than in ?*Nisidorcas* sp.

Metrical data on teeth of some Miocene *Gazella* and other bovids of similar size are listed in table 15.3. The lower cheek dentitions in Locs. 72, 49, and 42 are similar in size, in length of the premolar row relative to the molar row, and in length of p2 relative to p4 to ?*Prostrepsiceros* sp. at Loc. 8B. It can be seen that the premolar row of the late Turolian *Hispanodorcas orientalis* may be almost as long as in a gazelle, but otherwise neither the premolar row as a whole nor p2 of the Sinap mandibles are as much shortened as in the spiral-horned antilopines as far back as the Vallesian. This increases the likelihood that they are indeed attributable to *Gazella*.

It is difficult to use morphology to separate antilopine species in a collection of fossils from many localities spanning several million years, as at Sinap. However, Bouvrain and de Bonis (1985, p. 263) were able to separate two equally-sized species of spiral-horned antilopines by morphological characters in a sample of ~100 mandible pieces at Ravin de la Pluie.

?Tribe Antilopini
Genus indet.
Bovidae sp. A

Sinap Material. Loc. 108: 94.231 sin pedicel with base of hc, ~29.1 × ~21.8. 93.247 sin P3, early mid wear, 7.9 × 5.8; 93.278 partial upper molar; 93.606 palate with dex P3–M2, sin P2–M3, late mid wear; P2–4 21.9, P2 7.7 × 5.2, P3 7.6 × ~6.5, P4 7.2 × 7.6, M1–3 31.1, M1 9.5 × —, M2 11.0 × —, M3 11.3 × —; 94.243 sin P2, early mid wear, 6.3 × 4.3; 94.246 dex upper molar, probably M3, earliest wear, 11.0 × 7.6 × 6.0 mesostyle height; 94.254 dex upper molar, early wear, 10.6 × 9.0 × ~6.0 high (fig. 15.17A); 93.248 sin m3, early wear, 13.8 × 5.1 × 6.0 high; 93.833 (in part) sin mandible m2 and m3 mid wear, m2 10.8 × 7.3, m3 14.2 ×

6.6, mandible depth under m3 15.3; 94.204 sin mandible p4–m3 early wear, p4 8.5 × 3.7, m1–3 32.0, m1 8.0 × 4.2, m2 10.5 × 5.3, m3 14.0 × 4.1 × 6.4 high, mandible depth under m1 11.6; 94.211 dex mandible p4–m3, early wear, p4 7.4 × 3.1, m1 8.5 × 5.2, m2 10.2 × 4.8, m3 13.5 × 4.4 × 6.4 high (fig. 15.17B); 94.212 sin lower molar, mid mid wear, 10.6 × 6.5; 94.215 sin mandible with p3–m1, later mid wear for m1, p3 7.5 × 3.3, p4 7.0 × 3.8, m1 8.6 × 6.4, mandible depth under p2 14.8; 94.220 dex mandible p3–m3 late mid wear, p3 7.2 × 3.4, p4 8.0 × 3.9, m1–3 32.9, m1 9.1 × 6.5, m2 10.5 × 6.3, m3 13.9 × 5.8; 94.221 sin mandible, teeth in matrix block; 94.228 dex p2–4, late mid wear, p2–4 18.8, p2 5.4 × 2.6, p3 6.7 × 3.3, p4 7.7 × 4.3 (figs. 15.17C, 15.18C); 94.233 sin m3, early or middle mid wear, 13.9 × 5.9; 94.234 dex mandible, p3–m3 late mid wear, p2–4 ~17.1, p3 5.6 × ~3.6, p4 7.1 × ~4.7, m1–3 31.9, m1 7.6 × 5.0, m2 10.0 × 6.4, m3 14.2 × 6.2, mandible depth under p2 13.0, under m1 13.8 (fig. 15.18B); 94.245 dex lower molar, earliest wear, 11.0 × 5.1 × 6.4 high; 93.608 sin lower hind limb from distal tibia to third phalanx, in matrix; length metatarsal 135, minimum shaft width ~9.8, first phalanx 30.9 × 11.8 high, second phalanx 18.2 × 10.6 high, third phalanx 19.3 × 12.1 high.

Remarks. The completeness of preservation of a few mandibles of a small antelope at Loc. 108 reveals the most striking character of this species to be a lengthened diastema (fig. 15.18).

The tooth dimensions show a species smaller than a gazelle. The lower premolar row is about 56% of the length of the molar row, which is less than either *Gazella* sp. or ?*Prostrepsiceros* sp. at other middle Sinap localities. Three unworn m3 specimens (fig. 15.17B) show heights of 43%, 46% and 49% relative to length, very like Bovidae, ?sp. A in the lower Sinap member. On p4, the metaconid grows forward on the lingual side and contacts the paraconid (fig. 15.17C); again, resembling Bovidae, ?sp. A in the lower Sinap member. The p4 hypoconid projects sharply. The p3 metaconid is oriented almost transversely. The p2 94.228 has a metaconid crest growing backward and a blocklike hypoconid behind. The labial lobes of the lower molars become rather narrow as wear proceeds, the rear central fossettes may remain open until after earliest wear, incipient goat folds are present, a strong metastylid may be present, and basal pillars are small to moderate on m1 and m2. On most m3 specimens, but not on 94.233, the hypoconulid lobe is large with a central fossette and the lingual wall not displaced labially. This character, coupled with small size, make it reasonable to accept the species as being in the Antilopini or in the ancestral subfamilial group, from which Antilopini was later to appear.

The very incomplete horn core 94.231 is about the size of that from a gazelle. Detectable characters are a flat medial surface and the presence of a posteromedial, as well as a posterolateral corner. There are narrow supraorbital pits. The metatarsal 93.608 is shorter than the *Gazella* at Loc. 49 and more robust than living *Gazella*.

Discussion. The proportions of the long and shallow diastema can be assessed from table 15.4 and figure 15.18. A complete diastema is frequently not preserved in fossils, but its lengthened condition is obviously a specialization in this species. Long diastemas are characteristic of the extant browsing African antilopines *Litocranius walleri* and *Ammodorcas clarkei*. The better-known *L. walleri* feeds daintily on small green shoots amidst woody thorns (Kingdon 1982, p. 435; Estes 1991, p. 85) and has a longer premolar row and more brachyodont teeth than present-day gazelles. Both it and *A. clarkei* are arid country inhabitants. Highly selective feeding must be shared in common between *L. walleri* and Bovidae sp. A. The relative lengths of the premolar row are the same (56%), but whereas the premolar row of *L. walleri* is longer than in present-day gazelles, that of Bovidae sp. A is shorter than in the Sinap *Gazella* sp.

Bovidae sp. A shares some characters both with Bovidae ?sp. A of the lower Sinap member and with Bovidae sp. B to be described below, but the diastema length is not known for either of these species.

Gazella ancyrensis Tekkaya 1973, was founded for bovid remains from Yassiören of gazelle size or slightly smaller. The original material was meticulously described, compared, and properly illustrated, and the species is almost certainly represented in the present Sinap collection. An

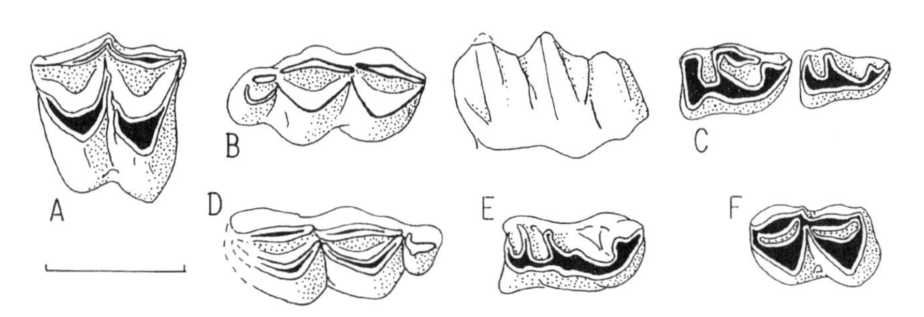

Figure 15.17. Loc. 108 Bovidae, all except E drawn from casts. (**A–C**) Bovidae sp. A, (**A**) Dex upper molar 94.254 in occlusal view. (**B**) Dex m3 94.211 in occlusal and lingual views. (**C**) Dex p4 and p3 94.228 in occlusal views. (**D,E**) *Pseudotragus* aff. *capricornis*, sin m3 93.296 and dex p3 94.224 in occlusal views. (**F**) ?(Bovidae sp. A), smaller sin lower molar 94.232 in occlusal view. Labial edge of **A** and lingual edges of **B–F** toward top. Scale bar = 10 mm.

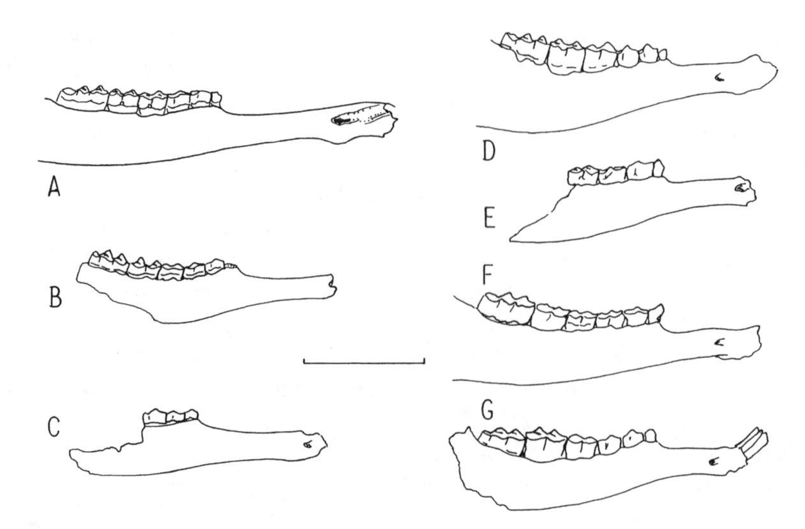

Figure 15.18. Lateral outlines of right mandibles to show diastema lengths. (**A**) Extant *Litocranius walleri*, BMNH 1932.12.17.1. (**B,C**) Bovidae sp. A, Loc. 108 94.234 and 94.228. (**D**) Extant *Gazella dorcas*, BMNH 1939.4064. (**E**) *G. capricornis* BMNH M13015. (**F**) *Nisidorcas planicornis*, Ravin des Zouaves no. 5 RZO 313. (**G**) *Prostrepsiceros vallesiensis*, Ravin de la Pluie RPL 655 (sin mandible, reversed). Scale bar = 40 mm.

Table 15.4. Measurements and Ratios for Individual Specimens or Illustrations of Length (= Front of P/2 Alveolus to Back of Mental Foramen) and Minimum Depth of Diastema

Taxon/Specimen Number	Measurement (mm)	Ratio (%)
?Homoiodorcas n. sp., Al Jadidah (Thomas, 1983, pl. 2 fig. 7b)	16.0 × 10.0	62
Palaeoreas lindermayeri, Pikermi BMNH M11448	~21.0 × 10.8	51
Prostrepsiceros vallesiensis, Ravin de la Pluie	17.0 × 8.0	47
Samotragus praecursor, Ravin de la Pluie	18.0 × 8.0	44
Nisidorcas planicornis, Ravin des Zouaves no. 5	18.5 × 8.0	43
Gazella ?deperdita, Samos BMNH M4177	16.2 × 7.0	43
Gazella capricornis, Pikermi BMNH M11504	20.0 × 9.3	47
Gazella capricornis, Pikermi BMNH M13017	23.5 × 8.0	34
Loc. 108 AS94-228	35.0 × 8.0	23
Loc. 108 AS94-234	30.0 × 7.0	23
Gazella dorcas, extant, mean of 10	14.3 × 8.0	56
Litocranius walleri, extant, mean 10	30.1 × 8.4	28

Notes: Readings for samples of two extant antilopines are also given. Percentage ranges for the last two species are 42–78% and 22–36%. Smaller percentages in the table indicate longer and/or shallower diastemata. Measurement is length by minimum depth.

illustrated mandible (Tekkaya 1973, pl. 3, figs. 2, 3, pl. 4) resembles Bovidae sp. A in size and premolar/molar row proportions, but no mandible with a diastema is illustrated, m3 lacks a central fossette on its rear lobe, and Tekkaya's (1973) measurements of M_3 suggest that the teeth are higher crowned. A horn core of *G. ancyrensis* (Tekkaya 1973, fig. 1, pl. 1, pl. 2, fig. 1) has basal diameters of ~19.0 × ~16.5 and is smaller than the single horn core pedicel of Bovidae sp. A. *Gazella ancyrensis* will be discussed again under Bovidae sp. B.

?(Bovidae sp. A)

Sinap Material. Loc. 108: 94.232 sin lower molar, early mid wear, 8.3 × 5.0 × ~5.0 high (fig. 15.17F); 93.254 second phalanx, 14.2 × 8.6 high; 93.603 third phalanx, 15.3 × 9.2 high; 93.605 first phalanx, 24.4 × 9.6 high.

Remarks. This left lower molar and three phalanges are small for Bovidae sp. A and may be from a still smaller species.

Bovidae sp. B

Sinap Material. Loc. 4: 93.329 part sin h/c base; 93.757,760 fragment sin h/c base, part another; 93.336 fragment labial wall upper molar; 93.762 dex m2, unworn, 10.1 × 3.8 × 8.0 high; 93.473 dex p4 late mid wear, 8.3 × 4.7; 94.116 dex p3, early wear, 7.5 × 3.8; 92.386 top tip of calcaneum; 93.754 fragment sin calcaneum; 93.850 sin astragalus fragment; 93.333 dex scaphoid, ap 12.9, height 10.1; 93.337 sin radius, partial proximal end; 93.338 sin scaphoid, ap 10.5, height 8.3; 93.355 fragment proximal sin radius; 93.365 fragment distal ulna; 93.425 large part of distal sin humerus, width ~22.6, medial height 16.2; 93.441 sin scapula, proximal ap 23.6, facet only 18.8; 93.467 sin cuneiform; 92.868 third phalanx, 16.4 × 9.8 high; 93.320 part first phalanx, probably immature; 93.339 part third phalanx; 93.340 immature second phalanx, probable length 17.0; 93.350 second phalanx, 15.7 × 9.7 high; 93.448 first phalanx, distal fragment; 93.455 metapodial, part distal condyle; 94.1 metapodial immature distal condyle, ap 12.7; 94.25 metapodial distal condyle, ap 11.4; 94.67 third phalanx fragment.

Loc. 94: 92.366 two sin, one dex upper molars unworn, sin upper molar (paracone cusp missing) 8.7 × 6.4 × 4.0 high, sin upper molar (protocone glued) 8.8 × ~6.9 × 4.7 high, dex upper molar 8.7 × 5.9 × 4.5 high (fig. 15.4D); 92.368 sin dP3 mid wear, 8.0 × 4.2; 92.548 sin upper molar, early mid wear, 8.0 × 6.3; 92.572 sin dp3 unworn, 7.5 × 3.3; 92.367 dex m1 or 2 unworn, 9.7 × 4.1 × 5.7 high; 92.556 dex mandible fragment, p3 early wear, 6.6 × 2.9; 92.567 dex dp3 early wear, 6.2 × 2.9; 93.543 sin incisor; 92.546 dex distal tibia, fragment; 92.561 patella; 92.796 femoral head; 93.118–9 sin astragali; 93.120 sin astragalus, lateral height 20.0, width 12.4, ap 11.1; 92.558 sin metacarpal, proximal fragment; 92.562 sin unciform; 92.571 dex cuneiform; 92.620 dex unciform; 92.541 second phalanx, 15.2 × 9.3 high; 92.542 first phalanges, proximal fragments; 92.543 second phalanges, distal fragments; 92.544 second phalanx proximal fragment; 92.545 second phalanges, distal fragments; 92.547 fragment metapodial condyle ap ~10.5; 92.560 second phalanx, 13.8 × 7.9 high; 92.563 two sesamoids; 92.565 distal second phalanx; 93.124 metapodial immature distal condyle, ap 10.8; 93.125–126 metapodial distal condyles, one being immature.

Loc. 66: 90.288 dex M1 ?almost unworn, 9.0 × 7.4 × 5.1, root divergence suggests a dP4; 90.289 sin lower molar early wear, 8.9 × 5.0 × 5.0; 90.290 fragment sin dp4 mid wear, total length ~8.0; 90.292 metapodial distal condyle ap 11.0; 90.293 first phalanx fragment.

No locality: 94.164 sin hc, 17.6 × 16.5 (fig. 15.17A,B).

Remarks. Gazelle horn cores have not been recorded below the level of Loc. 72. Gazelle-sized, rather high-crowned teeth in Locs. 8B, 13, 71, and 91 were grouped with the horn cores of *Prostrepsiceros* sp. Teeth and other remains in still lower stratigraphic levels near the base of the middle Sinap member, of slightly smaller size than a gazelle, are here referred to Bovidae sp. B. The material may include more than one species, because the two dp3 specimens at Loc. 94 are very different in size.

The basal diameters of the horn core Loc. 4 93.329 would have been about 20 mm, making it larger than *Elachistoceras* from the Siwaliks and smaller than ?*Homoiodorcas* sp. from Al Jadidah (Thomas 1977, 1981). There is a large shallow postcornual fossa and a small supraorbital pit, and the horn core is probably inclined backward. Another fragment of a small horn core base, Loc. 4 93.757, shows a large and shallow postcornual fossa. The horn core 94.164 ("Radio tower, south below, small gully, just above a field Kayincah side, Surface finding. Sevket [Sen]" on label; fig. 15.16A,B) is from a bovid smaller than a gazelle. It would originally have been of medium length, without much compression, probably irregular short keels, inserted above the orbits and moderately wide apart, inclined backward, with slight and constant divergence, no backward curvature, pedicel higher anteriorly than posteriorly, a deep postcornual fossa, probably no sinuses in the pedicels, frontals between the horn cores slightly above the level of the dorsal orbital rims, and large supraorbital foramina. The two scaphoids Loc. 4 93.333 and 93.338 are relatively higher (by about 10%) than in extant *Gazella*.

Discussion. The diastema length in Bovidae sp. B is not known, but its tooth characters are not distinguishable from Bovidae sp. A at Loc. 108 or Bovidae ?sp. A in the lower Sinap member. It appears to have about the same level of hypsodonty for example.

The illustrated right horn core of *Gazella ancyrensis* from Yassiören (Tekkaya 1973, fig. 1, pl. 1, pl. 2, fig. 1), already mentioned under Bovidae sp. A, is a good match for the horn core 94.164 of Bovidae sp. B. Unfortunately 94.164 was a surface find and the stratified Loc. 4 horn cores are more fragmentary. The *G. ancyrensis* horn core is small and has no backward curvature, so it cannot belong to a male gazelle. It could be from a female if the Sinap *Gazella* sp. had horned females, or it could be from an ancestor or earlier relative of *Gazella*. The second alternative is the more likely if the age of the *G. ancyrensis* type locality is MN 9, as suggested by Steininger et al. (1996, p. 34). A lectotype is needed for *G. ancyrensis*, but at present it is uncertain whether the horn core or the illustrated mandible would be the better choice.

Bovidae sp. C

Sinap Material. Loc. 8B: 91.250 second phalanx, 14.6 × 9.4 high.

Loc. 12: 93.906 dex m2 + 3 in matrix, mid wear, m2 8.2.

Loc. 49: 94.41 dex upper molar, late wear, 9.5 × 12.3; 94.44 sin P2, late wear, 6.4 × 4.8; 94.45 large part of dex upper molar, late wear, ~8.5 × 11.1; 94.47 dex P3, late wear, 7.1 × 7.5; 94.40 dex m3, late wear, 13.6 × 5.9; 94.43 dex m1–2, late wear; m1 8.5 × 5.7, m2 9.7 × 6.4; 94.48 fragment lower molar, late wear; 94.49 first phalanx, two immature fragments; 94.50 first phalanx, two immature fragments; 94.53 first phalanx, parts; 94.57 first phalanx, parts; 94.58 second phalanx, parts; 94.59 metapodial, damaged immature distal condyle.

Loc. 34: 92.787 very damaged dex astragalus, height ~20.0, width 12.6, ap 11.0.

Remarks. Upward from Loc. 14, wherein the gazelle sized *Prostrepsiceros* sp. is first known, there continue to be finds of bovids of smaller size than a gazelle. These are listed above, but little of interest can be deduced about them. The Loc. 34 astragalus is too small to fit *Gazella,* but may not date from before the appearance of that genus in the Sinap deposits.

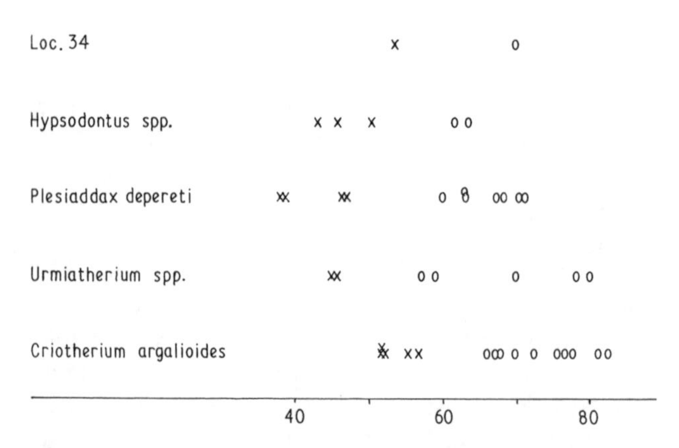

Figure 15.20. Histogram showing p4 length as a percentage of lengths of m3 (x) and of m2 (o) in Loc. 34 89.341 and in some other fossil bovids. Sources: Author's measurements, Bohlin (1935b).

Criotherium Major, 1891

Type Species. *Criotherium argalioides* Major, 1891.

Remarks. Previously, *Criotherium* has been accepted as an ovibovine. Gentry's (1971, p. 291) opinion of its relationship with *Palaeoreas* was strengthened (cf. Köhler 1987, p. 207) by acquaintance with the fossils of *P. zouavei*. In *P. zouavei,* the back surface of the occipital condyles lies in nearly the same plane as the occipital surface itself, and the posterior tuberosities of the basioccipital have some sign of a backward facing posterior surface. *Palaeoreas* now seems to be satisfactory as a member of the Antilopini alongside the other extinct genera considered here, so it is conceivable that *Criotherium* is also a notably large antilopine, and must have evolved characters in parallel with *Urmiatherium* Rodler, 1889, and other late Miocene supposed "ovibovines."

?*Criotherium* sp.

Sinap Material. Loc. 12: 91.777 piece of very large hc.

Loc. 49: 91.742 piece hc, one surface less convex than the other, 45.0 × 31.3; 91.202 first phalanx; length 48.3, proximal width 17.0.

Loc. 63: 90.214–215 parts of hc bases, one diameter of 90.214 ~55.0, two diameters of 90.215 50.4 × 47.2.

Loc. 34: 89.341 sin mandible with p4–m3, m1–3 77.9, p4 17.4 × 8.7, m1 22.1 × 10.8, m2 25.0 × 11.2, m3 32.7 × 10.3 (fig. 15.19A); 89.340 distal sin humerus, lateral part only; 89.339 distal sin tibia; 92.768 dex metatarsal, both ends missing, length ~230, minimum shaft width 17.0.

Loc. 26: 89.306 dex m3, mid wear, 29.3 × 12.0 × 14.3 high (fig. 15.19B).

Loc. 50: 90.24 sin maxilla with partial (?)dP4 + M1, early mid wear, dP4 ~19.5, M1 22.3 × ~14.9 × 14.7 high.

Loc. 88: 92.802 sin lower molar, mid wear, 19.6 × 12.8.

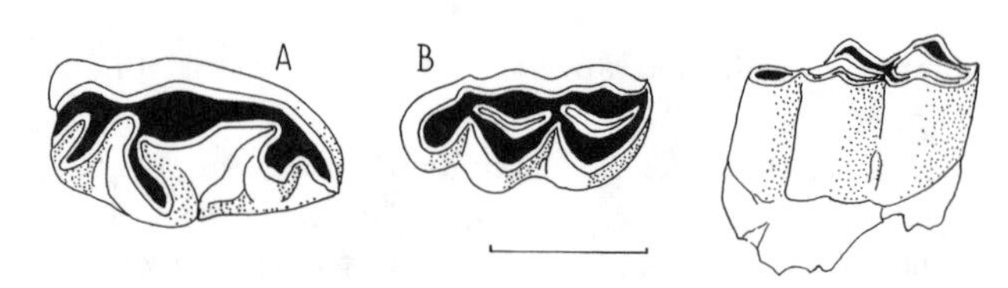

Figure 15.19. ?*Criotherium* sp. **(A)** Sin p4 Loc. 34 89.341 in occlusal view. **(B)** Dex m3 Loc. 26 89.306 in occlusal and labial views. Anterior to the right. Scale bar = 10 mm for **A**, 20 mm for **B**.

Remarks. The horn core Loc. 12 91.777 could be conspecific with Loc. 49 91.742. Neither are likely to be *Protoryx* because of too close an approach to keels at their front and back edges, they taper too fast distally, and they diverge too much distally. The horn core pieces Loc. 63 90.214–215 are large with surface indication of spiraling, and *Criotherium* is a possible identification for them.

The m3 Loc. 26 89.306 (fig. 15.19B) and probable dP4 + M1 Loc. 50 90.24 can be classed as ?*Criotherium* sp. The m3 is about the size of *C. argalioides* but is less distinctive and probably less high crowned. It has a tiny basal pillar. The teeth of the maxilla Loc. 50 90.24 are not very hypsodont.

The mandible Loc. 34 89.341 was identified as *Criotherium*. The molars have no basal pillars, and the metaconid crest of the p4 (fig. 15.19A) turns slightly forward at its lingual end and has no posterior flange. However, horn cores of the middle Miocene *Tethytragus* have been found at Loc. 34 and one needs to know if there is any reason why this mandible could not belong to the middle Miocene *Hypsodontus pronaticornis*. The relatively long p4 in relation to m2 and m3 (fig. 15.20), however, favors *Criotherium* as a more likely identity than either *Hypsodontus pronaticornis* or late Miocene bovids such as *Urmiatherium* and *Plesiaddax*. By size, the Loc. 34 postcranial bones would be conspecific with 89.341. There is a slight hollowing in the middle part of the posterior surface of the metatarsal 92.768.

Subfamily Caprinae Gray, 1821

Protoryx Major, 1891

Type Species. *Protoryx carolinae* Major, 1891.

Remarks. The late Miocene bovids *Palaeoryx, Protoryx,* and *Pachytragus* have been most fully described from Pikermi and Samos. Their tribal and subfamily attribution is still not agreed upon. Traditionally all three, together with some relatives, were regarded as Hippotragini or as in an extinct subfamily Pseudotraginae. Gentry (1971; 1992, fig. 6; 2000) proposed that they were Caprinae—the first an ovibovine and the last two Caprini; other authors have reacted variously to this idea (Solounias 1981a; Bouvrain and de Bonis 1984; Köhler 1987, 1993; Erdbrink 1988). *Norbertia hellenica* Köhler et al., 1995, is a related species from MN 13, more advanced toward modern Caprini.

Protoryx solignaci (Robinson 1972)

Holotype. Cranium with part of the right horn core from the Beglia Formation at Bled Douarah (Robinson 1972, p. 75, fig. 1).

Sinap Material. Loc. 4: 93.325 fragment labial wall upper molar; 94.32 large part of sin upper molar, length 20.2; 94.98 sin upper molar, fragment lingual wall; 92.377 dex lower molar, labial fragment; 93.315 dex p3 or 4, lingual fragment; 93.470 sin p3 unworn, 11.2 × 4.6; 94.14(+104) dex p4, early middle wear, 14.2 × 6.5 × 11.0 protoconid height (fig. 15.6A); 94.26 back dex m3; 93.475 metapodial immature distal condyle, ap 18.8; 94.97 second phalanx, distal fragment; 94.108 second phalanx, distal fragment; 94.111 first phalanx, proximal fragment.

Loc. 94: 92.554 dex upper premolar, labial wall only, length 13.3; 93.116 sin distal tibia, width 34.4; 92.795 sin astragalus, medial height 39.2, width 24.4, ap 23.3.

Loc. 104: 93.846 frontlet with hcs., ~53.7 × 28.7.

Remarks. The p4 Loc. 4 94.14 (fig. 15.6A) could be boselaphine, but the weakness of the transverse ridge of the entoconid and the foreshadowing of a lingual connection from metaconid to entoconid makes the present species an acceptable identification (Köhler 1987, fig. 51).

The p3 93.470 is too small to be from the same individual as p4 94.14. Its metaconid crest is aligned more diagonally backward than transversely, which is primitive and might allow it to be a small boselaphine. Otherwise (and if indeed it is not *Protoryx solignaci*), it may belong to a smaller species such as *Pseudotragus* aff. *capricornis.*

Discussion. The range of *Protoryx solignaci* was extended to Turkey by Köhler (1987, p. 172, figs. 52–54, pl. 3, figs. 3–4, pl. 4, fig. 1), who found it in MN 8 and 9 localities. It is larger than *Protoryx enanus,* a Turkish species of MN 7 age, and the horn cores are more compressed. It is not easy to distinguish from *Tragoportax* or *Miotragocerus* if the remains are insufficiently complete, but it does not have an abrupt diminution of horn core diameter distally, and the cranial roof is more inclined. The teeth are primitive in most characters, but on p4, the transverse crest of the entoconid is severed and the lingual end of the metaconid crest links with the lingual end of the entoconid, whereas on p3, the lingual end of the metaconid is sometimes isolated (BMNH M29228; Thomas 1983, fig. 1), as in some *Pachytragus.* The premolar rows are relatively shorter than in Boselaphini.

Pachytragus Schlosser, 1904

Type Species. *Pachytragus crassicornis* Schlosser, 1904.

Pachytragus laticeps (Andree 1926)

Holotype. Cranium with horn cores from Samos (Andree 1926, pl. 12, figs. 5, 9).

Sinap Material. Loc. 26: 89.317 sin p4, late mid wear, 15.0 × 9.2; 91.228 sin dp4, early mid wear, posterolingual lobe missing, ~21.3 × 8.9 × 6.5 high (see fig. 15.23A,B); 91.233 sin p2 early mid wear, 9.2 × 3.8.

Loc. 33: 91.86 sin P2, late mid wear, 13.1 × 9.3 (see fig. 15.22D,E); 91.168 dex dP2–M1, milk teeth in mid wear, M1 unworn, dP2 10.3 × 8.0, dP3 15.4 × 10.5, dP4 18.0 × 12.1, M1 20.9 × 11.5 × 20.0 high; 91.175 sin dP3–M1, wear as in 168 (?same individual as 91.174), dP3 15.3 × 13.9, dP4 17.4 × 11.9, M1 19.2 × 10.9 × 18.0 high (figs. 15.21D, 15.22B); 94.126 dex P3, late wear, 13.0 × 12.0; 90.14 dex m3 early mid wear, 25.2 × 8.7 × 22.5 high; 91.174 sin mandible dp3–m1, milk teeth in mid, m1 in early wear, dp3 12.6 × 5.9, dp4 22.4 × 8.0, m1 18.5 × 8.3, mandible depth under dp2 19.4, under m1 ~25.0 (fig. 15.23C,D).

The following fossils are either this species or a boselaphine.

Loc. 26: 89.495 femoral condyle; 89.496 thoracic vertebra; 91.57 distal immature second phalanx; 91.232 dex ectocuneiform.

Loc. 33: 91.16 lingual part dex upper molar mid wear; 91.75 rear half sin upper molar; 94.131 fragment upper molar labial wall; 91.27 dex incisor; 89.218 acetabular articulation of pelvis; 89.184 dex distal tibia, width across

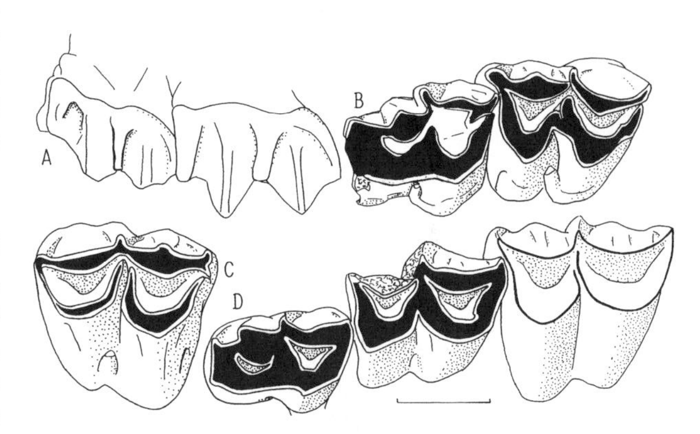

Figure 15.21. (A,B) *Tragoportax* sp., sin dP3 and dP4 Loc. 14 89.71 in labial and occlusal views. (C) *Tragoportax* sp., dex upper molar Loc. 12 91.339 in occlusal view. (D) *Pachytragus laticeps,* sin dP3–M1 Loc. 33 91.175 in occlusal view. Labial edge of occlusal views toward the top. Scale bar = 10 mm.

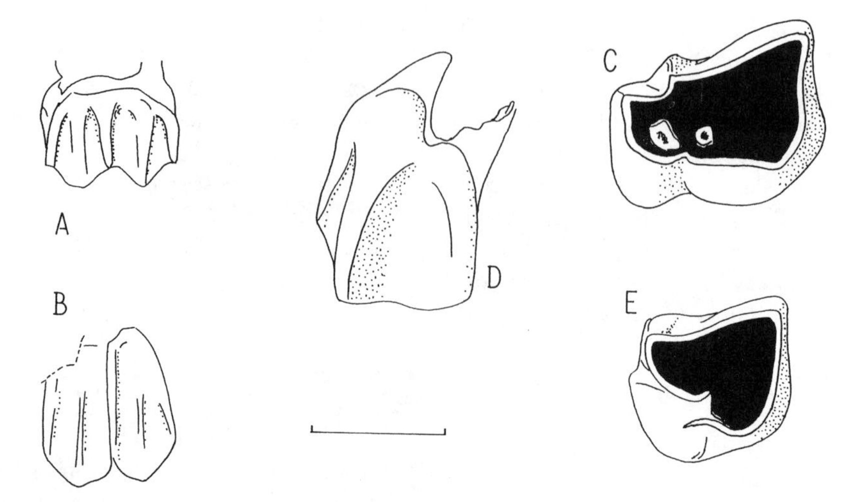

Figure 15.22. (**A,B**) Labial views of *Tragoportax* sp., dex upper molar Loc. 12 91.339 and *Pachytragus laticeps* sin M1 Loc. 33 91.175. (**C**) *Tragoportax* sp., sin P2/ Loc. 49 90.91 in occlusal view. (**D,E**) *P. laticeps* sin P2 Loc. 33 91.86 in labial and occlusal views. Anterior to the left, except for **A.** Scale bar = 20 mm for **A** and **B,** 10 mm for **C–E.**

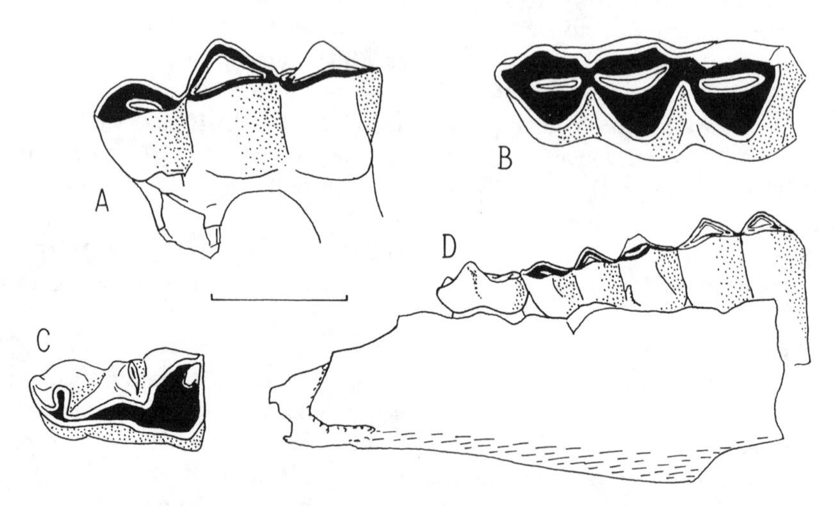

Figure 15.23. *Pachytragus laticeps.* (**A,B**) Sin dp4 Loc. 26 91.228 in labial and occlusal view. (**C**) Sin dp3 Loc. 33 91.174 in occlusal view. (**D**) Sin mandible with dp3–m1, Loc. 33 91.174, in lateral view. Anterior to the left. Scale bar = 10 mm for **A–C,** 20 mm for **D.**

articular surfaces 37.0; 91.102 dex distal tibia, width across articular surfaces 34.3; 89.216 sin calcaneum stem; 91.130 large part of sin calcaneum, stem length 72.5; 91.155 partial ectocuneiform; 91.121 sin scapula; 89.187 dex distal humerus, width ~54.0; 89.233 sin distal humerus, width ~48.0; 89.250 sin distal humerus, width 41.4, height medial condyle 29.5; 91.19 sin distal humerus, width 42.1, height medial condyle 31.2; 89.261 sin proximal radius, width across articulation 52.2; 91.3 sin scaphoid, ap 26.4, transverse 15.7, height 23.0; 91.160 axis vertebra.

Loc. 42: 91.442 two tooth fragments, one being the labial wall of a bovid upper molar; 91.436 dex unciform; 93.1143 metapodial distal condyle, ap 22.2.

Loc. 78: 91.626 dex proximal radius, width across articulation 53.5.

Remarks. The advanced teeth of *Pachytragus crassicornis* may be told from those of the boselaphine *Tragoportax* (figs. 15.21, 15.22) by some or all of the following characters: greater hypsodonty, molars with smaller basal pillars, selenodont crests of molars fusing earlier in wear to enclose the central fossettes, upper molars with stronger mesostyles and a flat or even concave labial wall of the metacone, lingual walls of lower molars less outbowed between the stylids, the premolar row shorter and anterior premolars smaller, paraconid and parastylid crests at the front of the lower premolars often turned more sharply transversely, the hypoconid on p4 more pointed labially, p4 metaconid more bulbous and without lingual flanges anteriorly or posteriorly, p4 metaconid with a more posterior connection to the protoconid, p4 paraconid smaller relative to parastylid, p3 metaconid crest to protoconid constricted or completely severed, and p3 metaconid often with a lingual flange passing posteriorly and joining the entoconid and entoconulid transverse crests to give a closed posterolingual wall of the tooth. Such character differences are less obvious when comparison is made between earlier or more primitive species.

By these criteria, some teeth at Locs. 26 and 33 are of *Protoryx* or *Pachytragus,* but not so obviously advanced as necessarily to be *Pachytragus crassicornis.* They are here referred to *Pachytragus laticeps.*

The dp3 on Loc. 33 91.174 (fig. 15.23C,D) is short for Turolian Boselaphini, but the dp4 is long relative to m1 for *Pachytragus.* The dp4 is about the same size as Loc. 26 91.228. The dP3 on Loc. 33 91.175 (fig. 15.21D) is shorter relative to dP4 than in the *Tragoportax* sp. of figure 15.21B and is without a basal pillar. The p4 Loc. 26 89.317 looks like *Pachytragus* except that the metaconid is not set posteriorly.

Discussion. The P2 Loc. 33 91.86 (fig. 15.22D,E) is too high crowned and anteroposteriorly short to be boselaphine and it is insufficiently squared off anterolingually and too small for *Criotherium argalioides.* It is also too large for the middle Miocene *Hypsodontus pronaticornis.* The dp4 Loc. 26 91.228 (fig. 15.23A,B) is smaller than *Criotherium argalioides* BMNH M4208 and has no basal pillars. The m1 of Loc. 33 91.174 is lower crowned and slightly longer than the ?(*Protoryx solignaci*) lower molar Loc. 64 92.69 (fig. 15.6B,C). The five teeth for which lengths can be compared with those of *Norbertia hellenica* from Maramena (p2, p4, P2, P3, M1) are, on average, one third larger.

A mandible with p3–m3 from Inönü of a size and morphology very similar to the Sinap species has been described by Geraads et al. (1995, figs. 7–9) under the name Genus et sp. indet, cf. *Protoryx.* Bouvrain (1997) described from Pentalophos an unnamed *Protoryx* of similar size to the Sinap species, which she believed was more advanced than *P. solignaci.*

Pseudotragus Schlosser, 1904

Type Species. *Pseudotragus capricornis* Schlosser, 1904.

Remarks. *Pseudotragus* is related to *Protoryx* and *Pachytragus* but is smaller. Its horn cores are not always larger than those of the MN 7 species *Protoryx enanus* Köhler (1987, p. 170). I regard *Tragoreas oryxoides* Schlosser (1904, p. 34, pl. 6, figs. 1, 9), type species of *Tragoreas,* as conspecific with *Pseudotragus capricornis. Sporadotragus* Kretzoi (1968, p. 164)—formerly the preoccupied *Microtragus* Andree (1926)—was synonymized with *Pseudotragus* by Köhler (1987, p. 178), whom I shall follow. It contained only one species, now to be called *Pseudotragus parvidens* (Gaudry 1861a,b). Erdbrink (1988, p. 135) and Bouvrain (1994b, p. 182) further synonymized *Pseudotragus* and *Sporadotragus* with *Protoryx,* but I shall not follow them.

Pseudotragus aff. *capricornis* Schlosser, 1904

Lectotype. Skull from Samos (Schlosser 1904, pl. 10, fig. 7, 7a, b), chosen by Pilgrim and Hopwood (1928, p. 39).

Sinap Material. Loc. 4: 92.831 dex lower molar, unworn, 15.6 × 5.6 × 11.9 high; 92.832 fragments dex lower molar and back m3; 92.833 dex dp4, late wear 16.0 × 7.5; 93.331 fragment sin p3 or 4, mid wear; 93.368 large part of dex m3, probably erupting; 93.370 dex lower premolar, labial fragment; 93.457 sin m3 early mid wear 19.4 × 8.3 × 10.0 high; 93.763 fragment dex lower molar; 94.7 central fossette ?lower molar; 94.21 dex lower incisor; 94.101 dex lower molar, part lingual wall; 93.376 dex astragalus, lateral height 37.1, width 22.7, ap 23.3; 89.511 condyles of distal sin humerus, width 29.7; 92.382 sin distal radius, width 27.9; 93.347 dex scaphoid, ap 18.7, width 10.3, height 14.2; 93.471 sin cuneiform; 93.451 sin unciform; 94.99 sin scaphoid; 89.512 third phalanx, ~30.0 × ~15.0; 93.323 third phalanx, ~28.0 × ~15.3; 93.335 proximal second phalanx, 13.4 high; 93.345 second phalanx, 25.8 × 16.7 high; 93.466 proximal immature second phalanx; 93.474 shattered distal metapodial.

Loc. 94: 93.115 sin distal humerus, width 32.9, medial condyle height 25.0; 92.619 sin naviculocuboid, width 24.2.

Loc. 89: 92.880 dex mandible, p3–m3 middle wear, m1–3 48.2, m1 12.6 × 8.6, m2 15.3 × 10.1, m3 21.2 × 9.2, p2–4 ~28.9, p3 12.0 × 5.0, p4 13.1 × 6.2, mandible depth under p2 19.5, under m1 23.9; 92.833 fragment dex proximal ulna.

Loc. 108: 93.260 dex P3 unworn, ~11.2 × 8.5; 93.264 dex p3 early middle wear, 10.8 × 5.3; 93.296 sin m3 ?unworn, 15.7 × 5.4 × 9.5 high (fig. 15.17D); 93.831 dex mandible dp3–m3, milk teeth in late wear, m1 in middle wear, m2 early wear, m3 unworn; bp small on m1, tiny on m3, dp3 7.6 × 4.0, dp4 10.9 × 5.0, m1 9.4 × 5.7, m2 11.5 × 5.0, m3 14.2 × 4.3 × ~11.0 high, mandible depth under dp2 11.6, under m1 15.2; 94.218 dex lower molar early middle wear, 12.7 × 6.6; 94.224 dex p3 early middle wear, 10.8 × 5.3 (fig. 15.17E); 94.226 sin fragmentary mandible with m1 and 2, latter in early middle wear and measuring 12.3 × 7.0.

Loc. 49: 93.824 cranium and palate, M3 in early mid, M1+M2 late mid wear, dex mandible, p4–m3, late mid wear, m1–m3 42.6, m1 11.5 × 7.9, m2 13.1 × 8.3, m3 18.9 × 8.4, mandible depth under m1 20.6, under m3 24.0; 94.39 fragment hc base, 42.4 × 26.6; 92.154 dex maxilla, P4–M2 early mid wear; M1 13.8 × 10.5, M2 15.0 × 9.8, P4 8.3 × 6.0; 93.1046 fragments upper molar, late wear; 92.194 dex lower molar mid wear, 13.8 × 7.6.

Loc. 51: 91.392 dex proximal ulna.

Loc. 26: 90.247 dex calcaneum, length 69.0.

Loc. 27: 90.9 sin mandibular ramus, two parts.

Loc. 33: 91.26 lingual fragment upper molar, early middle wear; 91.17 sin lower molar late wear, 13.8; 91.24 posterior partial dex dp4, late wear; 91.165 sin p2, 8.2 × 4.6; 89.157 partial distal metatarsal, width ~21.5, ap 15.6; 91.42 sin astragalus, medial height 26.8, width 17.0, ap 15.8; 91.124 distal metatarsal condyle ap 16.5; 89.212 sin distal humerus, lateral part; 89.213 third phalanx, 25.0 × 13.5; 89.214 partial first phalanx; 91.30 third phalanx, ~24.7 × ~12.4 high; 91.83 second phalanx, 21.0 × 12.1 high; 91.85 metapodial distal condyle, ap ~17.0; 91.91 third phalanx, 26.0 × 17.1; 91.118 third phalanx, 25.0 × 12.1 high; 91.133 two second phalanges, 20.0 × 12.2; 21.3 × 12.4 high; 94.124 second phalanx, 20.2 × 12.4 high.

Loc. 42: 89.443 dex calcaneum stem; 91.427 proximal first phalanx.

Remarks. The horn core Loc. 49 94.39 is very compressed (63%) and approaches having an anterior keel. It curves slightly backward. There is a sinus in the pedicel. It is not possible to say whether it is from the right or left side, but if it were a right, the lateral surface would be less convex than the medial one. The level of maximum transverse thickness is slightly behind the center of the anteroposterior span. Two very similar frontlets exist in the Şenyürek collection from Yassiören (?) F2. Their size may be too large for *Protoryx enanus,* and the horn cores look less compressed than in *P. solignaci.*

The cranium and palate Loc. 49 93.824 are preserved as two pieces, and horn cores are no longer associated with them. The back of the cranial roof slopes downward more than does the front part, temporal lines do not approach very closely posteriorly, the braincase sides are parallel in dorsal view and do not widen anteriorly or posteriorly, the back of M3 lies below the front edge of the orbit, there probably was a preorbital fossa, the median indentation at the back of the palate lies behind the level of the lateral ones, palatal foramina are level with the back half of the M2s, the nuchal crests are not especially prominent, each half of the occipital surface faces a little laterally as well as backward, the mastoids are moderately sized, the nuchal crests pass dorsally to most of the mastoids so that the latter's exposure is mostly posterior, the basioccipital is of even width with small anterior tuberosities and only slight longitudinal ridges behind them, a central longitudinal groove is little developed, auditory bullae are now absent but would have been moderately-to-well inflated, and tiny basal pillars are present on the upper molars.

Measurements taken on the skull are:

Width across braincase	56.3
Skull width across mastoids immediately behind external auditory meatus	67.7
Occipital height from top of foramen magnum to occipital crest	32.3
Minimum width across temporal lines on cranial roof	22.6
Width across anterior tuberosities of basioccipital	17.9
Width across posterior tuberosities of basioccipital	28.0
Width across palate between rear lobes of M2	22.0
Occlusal length M1–M3	39.3
Occlusal length and breadth of P4	9.4 × 11.0
Occlusal length and breadth of M1	12.1 × 12.7
Occlusal length and breadth of M2	14.9 × 13.4
Occlusal length and breadth of M3	14.9 × 12.0

Dental remains from Locs. 4, 49, 89, and 108 (fig. 15.17D, E) agree in size with *Pseudotragus* from other sites than Sinap and are slightly larger than Sinap teeth accepted as belonging to spiral-horned antelopes. The m3 Loc. 4 93.457 has a hypoconulid lobe with its lingual wall offset labially and notably flat; it also has a tiny basal pillar. The height is 52% of the occlusal length. The mandible of the skull Loc. 49 93.824 has teeth smaller than those of the *Tragoportax* at the same locality and its p4 is definitely unlike contemporary boselaphines. The maxilla Loc. 49 92.154 has a P4 of the same length as that of the gazelle Loc. 1 92.319, but the molars are longer, so it must be from a larger species with a relatively shorter premolar row. The lower molar and P2, 92.194 and 92.195, are probably conspecific if not the same individual. The lower molar has a goat fold anteriorly but no basal pillar. The p2 from Loc. 33 is robust enough to be this species but smaller than the contemporary *Pachytragus laticeps.*

The mandible Loc. 89 92.880 is only slightly larger than Loc. 49 93.824, but the p4 is notably long and the basal pillars are larger. It may be a different species.

The partial dp4 Loc. 33 91.24 has a prominent metastylid. The Loc. 108 teeth are from a larger and more hypsodont species than the abundant one there. The m3 Loc. 108 93.296 has its hypoconulid lobe offset labially and is without a central fossette. The anterolabial cingulum on the lower molar Loc. 108 94.218 is incipiently a goat fold.

The attribution of isolated teeth and postcranial bones to this species is not very certain. Those at Locs. 94, 4, and 108 occur at a time preceding the known appearance of similar sized spiral-horned Antilopini.

Discussion. *Pseudotragus* aff. *capricornis* is used here for certain Sinap fossils likely to be of aegodont and/or caprine bovids. It is not known whether only one species is included. The fossils are a little larger than *Tethytragus koehleri* and of a suitable size for *Pseudotragus capricornis* and *Pseudotragus parvidens,* both of which are smaller than *Pachytragus. Pseudotragus parvidens* has horn cores with little compression, a tendency to have an anterior surface, frontals very much raised between horn core bases, a strongly inclined cranial roof, the braincase narrowing posteriorly, quite a wide basioccipital, and a little-inflated (?secondarily reversed) auditory bulla. These characters are unlike the Loc. 49 fossils.

Pseudotragus capricornis has mediolaterally compressed horn cores, supraorbital pits rather wide apart, and wide dorsal orbital rims. The infraorbital foramen is situated low and forward on the lectotype; however, there is plaster of Paris in the vicinity of the tooth row. The face looks small in contrast to the massiveness of the horn cores. The premolar row may be relatively longer than in *P. parvidens.* The lectotype skull of *Tragoreas oryxoides,* as preserved, has a greater inclination of the cranial roof than shown in

Schlosser's (1904, pl. 6, fig. 9) picture. Furthermore, the inclination would have been greater in life than in the fossil as we have it now.

It is mainly the size, the compression of the horn cores, and the absence of the specializations of *Pseudotragus parvidens* that render *P. capricornis* similar to the Sinap species. However, the latter has a cranial roof that curves downward posteriorly, and the median indentation at the back of the palate is more posterior, so it is here taken as an allied but more primitive species. The cranial roof curvature looks more primitive than at Samos.

"Capra" bohlini Ozansoy, 1965, may be this species. No measurements were given for *C. bohlini*, but from the picture (Ozansoy 1965, pl. 8, fig. 1) it looks as though the distance across the lateral edges of the horn pedicels is too small to match *Protoryx solignaci*. Bouvrain et al. (1994, p. 376) implied that *"Capra" bohlini*, although definitely not a *Capra*, was also separate from another Sinap species akin to a small *Protoryx*.

By size and degree of mediolateral compression, the horn core Loc. 49 94.39 fits *Protoryx enanus* from Catakbagyaka and Sofça (Köhler 1987, figs. 47–50, pl. 3, fig. 2). However, the sinus system within the frontals extended up into the pedicel, and Köhler noted that this did not happen in *Protoryx enanus*. *Protoryx enanus* shows a further difference from *Pseudotragus* species in its supraorbital pits being very close to the horn pedicels. In addition, its frontals are not raised between the horn bases, which is unlike *Pseudotragus parvidens*.

Tethytragus at Loc. 34

Two bovid horn cores from Loc. 34 and a surface find were identified as *Tethytragus*, which is a middle Miocene genus and therefore anomalous at late Miocene localities. Other Loc. 34 species were dealt with above.

Tethytragus Azanza and Morales, 1994

Type Species. *Tethytragus langai* Azanza and Morales, 1994.

Tethytragus koehleri Azanza and Morales, 1994

Holotype. Left horn core from Candir (Köhler 1987, pl. 2 fig. 1).

Sinap Material. Loc. 34: 89.338 dex hc, 31.9 × 24.2 (fig. 15.9B); 92.770 sin hc. (fig. 15.9C).

Surface, near Loc. 33: 89.521 part sin hc base, 34.5 × 25.6.

Remarks. The left horn core at Loc. 34 is the same size as the right one but has less flattening of the lateral surface. The left core has a deep postcornual fossa and a narrow supraorbital pit at the base of the pedicel.

Discussion. I follow Köhler (1987) in regarding *Tethytragus* (formerly placed in *Caprotragoides*) as a characteristic middle Miocene genus and its Turkish representative as predecessor or ancestor of *Protoryx* near the MN 6/7 boundary (Köhler 1987, fig. 114). Thus the apparently late Miocene occurrence at Loc. 34 is highly problematic. The right horn core fits *Tethytragus* very well and I am reluctant to put it into *Protoryx* or any other genus.

The surface find 89.521 is probably a third *Tethytragus* horn core. It is rather fragmentary, with only about 60 mm of front edge and 20 mm of back edge preserved. It shows backward curvature and a flattened lateral surface.

Discussion

The Sinap ruminant list is rather long (tables 15.5, 15.6). The ruminants occur in a succession of faunas that precede and evolve toward the classical Turolian faunas of the sub-Paratethyan province. A high proportion of finds belong to species of gazelle size or smaller. This probably arises from modern excavating techniques, but it is also true that Sinap ruminants tend to be smaller than their later close relatives. The array of species in the middle Sinap member has a taxonomic spread like that of the classical Turolian localities, but they are less advanced. Sinap environments in the Vallesian must have been approaching the conditions of the sub-Paratethyan Turolian.

Locality 24A of the lower Sinap member (MN 5 or 6) contains a giraffid as large as at Candir and larger than others in the middle Miocene of Eurasia. Its metapodials suggest a possible relationship to *Bohlinia attica*. Evidently much remains to be discovered about middle Miocene giraffids. *Turcocerus* and *Hypsodontus*, both already known from the Turkish middle Miocene, are also present at Loc. 24A. Köhler (1987, 1993) believed that hypsodontines did not survive in Turkey as late as the end of the middle Miocene and that they were inhabitants of open country.

At Locs. 64 and 65 (MN 8), a different ruminant fauna is present, although the localities are thought to predate the advent of hipparionine horses; one bovid is probably *Protoryx solignaci*, another (Bovidae sp., size of *Gazella*) is about the size of *Turcocerus* but with lower-crowned teeth, and a third (Bovidae, ?sp. A) is smaller again and has still lower-crowned teeth and a p4 showing paraconid-metaconid fusion. An unknown giraffid is also present.

No cervids have been found in the lower Sinap member and only two cervid fossils, perhaps belonging to one species, have been found in the middle Sinap member. The morphology of the lower molar and the size of the canine support a pre-Turolian age for Locs. 1 and 91. Extreme rarity of cervids characterizes late Miocene faunas from Macedonia through Turkey to Maragheh and the Siwaliks, and contrasts with central and western Europe and sites north of the Black Sea.

In identifying the giraffes of the middle Sinap member, the long radius Loc. 49 93.822 + 181 was the basis for postulating *Bohlinia* at this and higher levels. Other postcranial bones showed that a second, somewhat smaller, species was also present; these postcranials were grouped with the Loc. 49 teeth here called *Palaeotragus coelophrys*. In several characters, the Loc. 49 lower premolars were more primitive

Table 15.5. Species List for Ruminants of the Lower and Middle Sinap Members with Locality Occurrences and MN Zone Spans

Species	Locality[1]	MN Zone Span
Lower Sinap member (Middle Miocene)		
Pecora sp., family indet.	24A	5 or 6
Giraffidae sp.	24A, ?64	5 or 6 (–8?)
Hypsodontus pronaticornis	24A	5 or 6
Turcocerus gracilis	24A	5 or 6
Bovidae sp., size of *Gazella*	64	8
Bovidae, ?species A	64, ?65	8
?(Protoryx solignaci)	64, 65	8
Middle Sinap member (Upper Miocene)		
Cervidae sp.	91, 1	9–10
Palaeotragus coelophrys	72, 12, 1, 49, ?(4, 94, 91, 8B, 45)	9–10
?(Palaeotragus roueni)	49, 26	10–12
Bohlinia attica	51, 49, 34; also 95	10–11
Decennatherium macedoniae	4, ?(72, 49)	9(–?10)
Helladotherium sp.	63, 26, 27, 33, 70, 42	10–12
Tragoportax spp.	108, 72, 12, 49, 26, 27, 33, 42, ?14; also 99	9–12
Tethytragus koehleri	34	?11
Prostrepsiceros sp.	14, 71, 91, 8B, 13, 45, 96, 101	9
Prostrepsiceros aff. *vallesiensis*	40, ?(77, 78)	10 (–?12)
Prostrepsiceros elegans	49	10
Prostrepsiceros rotundicornis	63	10
Nisidorcas planicornis	surface; ?82	
Sinapodorcas incarinatus	91	9
Palaeoreas sp.	72, 91, 12	9–10
Palaeoreas asiaticus	49	10
Gazella sp.	72, 12, 1, 49, 26, 33, 42, 78; also 88	9–12
Bovidae species A	108	9
Bovidae species B	4, 94, 66	9
Bovidae species C	8B, 12, 49, 34	9–12
?Criotherium sp.	12, 49, 63, 26, 34; also 50, 88	10–12
Protoryx solignaci	94, 104, ?4	9
Pachytragus laticeps	26, 33, 78	12
Pseudotragus aff. *capricornis*	4, 94, 89, 49, 51, 26, 33; ?108	9–12

[1]?, Uncertain occurrence; parentheses enclose two or more localities wherein occurrence is doubtful; localities following semicolon and "also" are not shown on table 15.1.

than *P. coelophrys* from Maragheh. Sinap *P. coelophrys* goes back to Loc. 4, which is also where a sivathere p4 is present, most probably of the Vallesian *Decennatherium*. Higher in the succession, above the Loc. 49 level, large postcranial bones are best attributed to *Helladotherium*. Interestingly, *Samotherium* was not found, but a possible *Palaeotragus roueni* occurs above the Loc. 49 level.

Beyond Sinap, *Bohlinia* is found from MN 10 (Ravin de la Pluie) to MN 13 (Ditiko 1, 2). *Decennatherium* is primarily Vallesian, whereas *Palaeotragus coelophrys* occurs from MN 9 to MN 12, with records from Sebastopol to Maragheh. *Decennatherium* was more of a woodland species than other giraffes (Köhler 1993, p. 74).

Sinap Boselaphini appear in MN 9. In MN 10, the *Tragoportax* horn cores at Loc. 49 are small and primitive, whereas anterior premolars are not noticeably elongated, all befitting a pre-Turolian age. In localities above Loc. 49, the boselaphine species is probably larger. Boselaphines have usually been taken as indicating woodland habitats, and Solounias et al. (1995) analyzed two *Tragoportax* species as mixed feeders or browsers. Bouvrain (1994a) noted, not least from limb bone proportions, that the Greek Turolian species *T. rugosifrons* inhabited more open environments than did other *Tragoportax*.

The Sinap spiral-horned Antilopini are discussed separately below.

Table 15.6. Stratigraphic Distribution of Ruminants in the Localities of Lower and Middle Sinap

	MN Zones (Localities)																	
	5/6	8	9	9	9	9	9	9	10	10	10	10	10	11	12	12	12	
	(24A)	(64, 104, 65)	(4, 94)	(89, 6, 13)	(71)	(108)	(72)	(91, 8B, 45)	(12)	(1)	(40)	(49)	(63, 51)	(34)	(26, 27, 33)	(42)	(78)	Other Localities
Lower Sinap member																		
Pecora sp., family indet.	X																	
Giraffidae sp.	X	-?																
Hypsodontus pronaticornis	X																	
Turcocerus gracilis	X																	
Bovidae sp., size of *Gazella*		X																
Bovidae, ?species A		X																
?Protoryx solignaci		X																
Middle Sinap member																		
Cervidae sp.								X	—	—X								
Palaeotragus coelophrys			?	-	-	-	X	—X	—X	—X	—	—X						
?(Palaeotragus roueni)												X	—	—	—X			
Bohlinia attica												X	—X	—X				95
Decennatherium macedoniae			X	-	-	-	?	-	-	-	-	?						
Helladotherium sp.													X	—	—X	—X		70
Tragoportax spp.						X	—X	—	—X	—	—	—X	—	—	—X	—X		
Tethytragus koehleri														X				
Prostrepsiceros sp.				X	—X		—	—X										
Prostrepsiceros aff. *vallesiensis*											X						?	?77
Prostrepsiceros elegans												X						
Prostrepsiceros rotundicornis													X					
Sinapodorcas incarinatus								X										
Palaeoreas sp.							X	—X	—X									
Palaeoreas asiaticus												X						
Gazella sp.							X	—	X	X	—	X	—	—	X	X	X	88
Bovidae species A						X												
Bovidae species B			X	X														
Bovidae species C								X	X	—	—	X	—	X				
?Criotherium sp.									X	—	—	X	X	X	X			50, 88
Protoryx solignaci			X															
Pachytragus laticeps															X	—	X	
Pseudotragus aff. *capricornis*			X	X	—	?	—	—	—	—	—	X	X	—	X			

Notes: Dark shading = definite occurrence; light shading = doubtful occurrences.

Gazella sp. occurs in the Sinap deposits from Loc. 72 upward. Its first definite appearance is thus just over halfway through the duration of MN 9, if MN 9 starts when hipparionine horses appeared. Some horn cores (e.g., Loc. 49 90.112) are smaller than most Turolian species, which fits a pre–MN 12 age for the localities involved.

Loc. 108, predating the appearance of known *Gazella* horn cores, contains Bovidae sp. A with lower-crowned teeth than in Miocene gazelles and a p4 with paraconid-metaconid contact. Both characters are reminiscent of what could be the same species at Loc. 64. At Loc. 108, the additional character of a long diastema can be seen. The antelope would have been a discriminating browser, but the nature of its habitat is unknown.

Teeth and other remains in low stratigraphic levels of the middle Sinap member, which are also of (or slightly smaller than) the size of a gazelle, are here attributed to Bovidae sp. B. Some may belong to the Loc. 108 Bovidae sp. A or they may be ancestors of later *Gazella* or *Prostrepsiceros.* Upward from Loc. 71, wherein *Prostrepsiceros* sp. is definitely known, there continue to exist antelopes smaller than gazelles, but little can be deduced about them. They are grouped under the label "Bovidae sp. C." The relationship of the small-horned *Gazella ancyrensis* Tekkaya 1973, to Bovidae spp. A or B or to other *Gazella* is uncertain. In this chapter, I describe the various small nonboselaphine antelopes as posible Antilopini, an attribution supported by the three central fossettes on the M_3 specimens in Bovidae sp. A. When more becomes known of such middle and early late Miocene species, it will be necessary to decide whether they are indeed close to or in the Antilopini (and therefore linked to the paraphyletic African Neotragini), or whether they have a nonantilopine origin within Eurasia, perhaps shared with ancestors of the Caprinae.

Within the Caprinae, *Protoryx solignaci* is present around the transition from lower to middle Sinap members, and *Pachytragus laticeps* appears much higher up (at the level of Loc. 26). Smaller tooth and skull material from Loc. 4 upward could belong to *Pseudotragus capricornis* or an earlier relative of that species. Köhler (1993, p. 74) saw this group as living in more or less moist open woodlands, but some species entering more open country "wherever there was sufficient bush shrub." On the evidence of masseteric morphology, Solounias et al. (1995) saw *Pachytragus laticeps* and *Pachytragus crassicornis* as mixed feeders.

The Sinap Spiral-Horned Antilopini

The Sinap spiral-horned Antilopini have to be compared with others of the sub-Paratethyan late Miocene, and a cladogram (table 15.7, fig. 15.24) is a convenient starting point. *Protragelaphus skouzesi* Dames 1883 on the cladogram is the type species of its genus and was first described from Pikermi. It is fairly large for a Miocene antilopine, and the horn cores are twisted around a straight axis rather than having open spiraling.

The cladogram shows the usual abundance of parallel advanced characters and few synapomorphies. It preserves three *Prostrepsiceros* species as a clade but joins *P. vallesiensis* with *Nisidorcas planicornis* as the primitive sister group to the rest. Two other cladograms with 43 instead of 41 steps

Table 15.7. Character Matrix for Nineteen Horn Core and Skull Characters for Species of Spiral Horned Antelopes

	Character																		
Taxon	1	2	3	4	5	6	7	8	9	10	11	12	13	14	15	16	17	18	19
Nisidorcas planicornis	0	0	1	0	0	1	0	0	0	0	0	1	0	0	1	1	0	1	0
Prostrepsiceros vallesiensis	1	0	1	0	1	1	0	0	0	1	0	0	0	1	0	1	1	1	0
Prostrepsiceros fraasi	0	1	0	0	0	0	1	0	0	1	?	1	1	1	1	0	1	0	1
Prostrepsiceros elegans + zitteli	0	1	0	1	0	1	0	0	1	1	1	?	?	?	?	0	?	0	1
Prostrepsiceros houtumschindleri	1	0	1	1	1	1	1	0	0	0	0	1	1	1	1	1	1	0	1
Palaeoreas lindermayeri	0	0	1	0	1	1	0	1	1	0	1	1	0	0	1	0	1	0	1
Protragelaphus skouzesi	0	0	1	0	0	0	0	1	0	1	1	1	0	0	0	1	1	0	0

Notes: 0 = primitive state, 1 = derived state. Some character states of *Protragelaphus skouzesi* were assessed from the more primitive *Ouzocerus gracilis*. The characters are listed below; for each, the cited condition or second alternative is considered derived. In character 1, an index of compression of < 80% was counted as advanced. Characters 1 and 2, and 7 and 8 are alternative advances from primitive intermediate conditions. Primitive states were taken as those of a Miocene gazelle such as *Gazella capricornis*.

1. Horn cores: not compressed/ mediolaterally compressed. 2. Horn cores: not compressed/ with slight anteroposterior compression. 3. Posterolateral keel: absent or moderate/ strong. 4. Anterior keel: absent or very weak/ moderate. 5. Horn cores inserted: above back of orbits (base of front edge above or anterior to back edge of orbit)/ more posteriorly. 6. Horn cores: slightly/ very inclined. 7. Horn cores: without/ with open spiraling. 8. Horn cores: without open spiraling/ with accentuated torsion of axis. 9. Cranial roof: slightly/ very inclined. 10. Braincase sides: widening posteriorly/ parallel or widening anteriorly. 11. Front of orbit above: front half of last molar/ back half of last molar. 12. Occipital surface faces: partly laterally/ mainly backward. 13. Longitudinal ridges behind anterior tuberosities of basioccipital: weak/ strong. 14. Basioccipital: without/ with central longitudinal groove. 15. Auditory bulla: moderately/ very inflated. 16. Length of p_2–p_4 in relation to m1–m3: longer/ shorter. 17. Paraconid and parastylid on p3+p4: indistinct/ separate. 18. Ectostylids on lower molars: larger/ smaller. 19. Lower molars without/ with incipient goat folds.

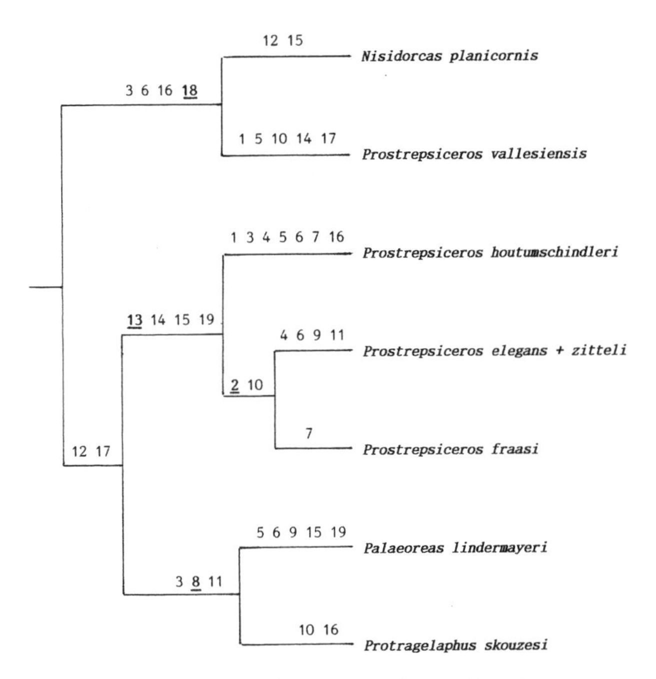

Figure 15.24. Cladogram for some spiral-horned antelopes. Number of steps is 41 and consistency index (number of steps minus one ÷ number of changes on tree) is 0.44. Character numbers from table 15.7; underlined characters = unique occurrences.

could be drawn, in one of which *N. planicornis* was the most primitive species isolated from a monophyletic *Prostrepsiceros,* and in the other of which *Prostrepsiceros fraasi* joins *Palaeoreas* and *Protragelaphus* whereas other *Prostrepsiceros* species group with *N. planicornis.* There is no great probability that the most economical cladogram is correct. It is a subtle and partly subjective exercise methodically to assign discrete alternative character states across a set of seven species. Variability in the infraspecific expression of keels has already been commented on, and it is only in the Turolian that characters become more distinctive.

I conclude that *Prostrepsiceros vallesiensis* is a primitive species, in which horn core torsion has not developed toward either the open spiraling of *Prostrepsiceros fraasi* or the twisting of the axis seen in *Palaeoreas* and *Protragelaphus* (fig. 15.25). Keels are present but not accentuated. *Prostrepsiceros* sp. at Sinap has only minimal torsion and could be still more primitive, but its assigned teeth are notably hypsodont, as also appears to be true of *Prostrepsiceros vallesiensis* and *Nisidorcas planicornis.* It will be interesting to see if the conclusion can be sustained that the teeth in this group are more hypsodont than in late Miocene gazelles.

Nisidorcas planicornis shows great variation in the strength of its keels. It differs little from the earlier *Prostrepsiceros*

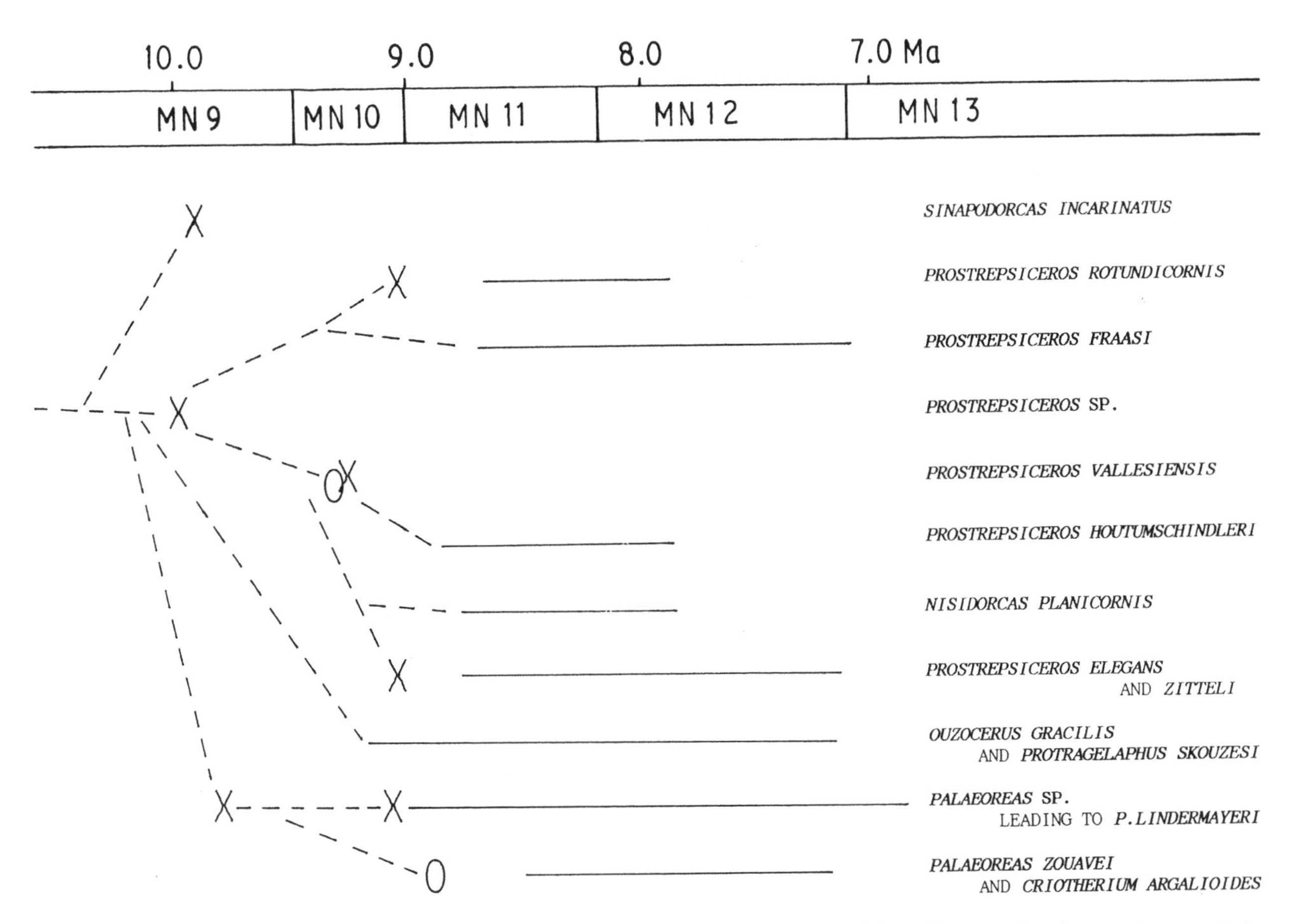

Figure 15.25. Time spans and a possible phylogeny for some spiral-horned antilopines of the Vallesian and earlier Turolian. Symbols: X = Sinap occurrences, O = single locality occurrences in Macedonia.

vallesiensis. It is also very close to *P. elegans* and could be ancestral, except that both occur together at Ravin des Zouaves no. 5, where, for some probably significant reason, the *P. elegans* is unusually distinctive. *Prostrepsiceros elegans* itself grades into *P. zitteli* in younger deposits.

Prostrepsiceros houtumschindleri is on a different line of descent from the *P. vallesiensis* stage than is *elegans-planicornis-zitteli.* It has a strong posterior keel like *Protragelaphus,* but retains fairly open spiraling. The Jebel Hamrin *Prostrepsiceros "zitteli"* of Bouvrain and Thomas (1992) may be evolving toward *P. houtumschindleri* from *P. vallesiensis.*

Prostrepsiceros rotundicornis and *P. fraasi* have vestigial keels and open spiraling linked with more upright insertions. The cranial roof remains little inclined. I retain them as a third lineage or stock within *Prostrepsiceros.* If the keels should be rudimentary instead of vestigial, then the two species could be more closely related to *Sinapodorcas incarinatus,* which may never have evolved keels to begin with.

Contrary to the cladogram, I prefer to see *Protragelaphus* and *Palaeoreas* as separate advances from a common ancestry with *Prostrepsiceros. Ouzocerus gracilis* Bouvrain and de Bonis, 1986 from Ravin des Zouaves no. 1 and Maragheh (Rodler and Weithofer 1890, pl. 5, fig. 2) is a suitable earlier stage leading or related to *Protragelaphus* and is more primitive than the latter in smaller size, horn insertions less far behind the orbits, frontals between horn bases less elevated, supraorbital pits less reduced, and cranial roof probably less inclined. The *O.* aff. *gracilis* at Maramena (Köhler et al. 1995, pl. 2) could disrupt this story if both its generic allocation and geological age are correct.

The Sinap *Palaeoreas* sp. is suitable as an early stage leading to *P. lindermayeri* and perhaps to *Helladodorcas geraadsi.*

The Sinap antilopine horn cores with anticlockwise torsion have been important in showing the similarity between early stages of species that subsequently became more clearly differentiated in the Turolian. Figure 15.25 shows how the Sinap horn cores predate their relatives. It is likely that infraspecific regional variation will become apparent in the group and that a reduction in the number of species and generic names will be possible. It would be good to place *Prostrepsiceros, Nisidorcas,* and *Sinapodorcas* in *Antilope.* Köhler (1993, pp. 64, 66) deduced "dry bush scrub and light wood with undergrowth" as a habitat for *P. vallesiensis* and more hilly country for the *Nisidorcas planicornis* at Kayadibi. Kostopoulos (2000) saw *N. planicornis* in broken cover in shrubby/mesophytic forests. Extant *Antilope cervicapra* is found in open or lightly wooded habitat but is less drought-resistant than present-day gazelles.

Teeth and horn core pieces suitable for *Criotherium* occur sparingly through the Sinap deposits. If the hypothesized relationship of *C. argalioides* to *Palaeoreas lindermayeri* were accepted, either *Criotherium* could become an antilopine or *Palaeoreas* could be removed from that tribe (also see below).

Oioceros, with opposite torsion in its horn cores to all forms discussed above, may not be an antilopine and has not yet been found in the Sinap formation.

Interpretation of Late Miocene Eurasian Ruminants

It was mentioned above that *Criotherium* looks like a large and specialized offshoot from an ancestry near the antilopine *Palaeoreas,* and that it is probably not related to late Miocene supposed "ovibovines" such as *Urmiatherium.* Other problems loom with the contents of the Ovibovini (Gentry 1996), not least the distant or nonrelationship of *Budorcas* with *Ovibos* (Bouvrain and de Bonis 1984). Now another ovibovine question can be raised. The small *Urmiatherium* (*"Parurmiatherium"*) *rugosifrons* (Sickenberg 1932) from Samos has clockwise torsion in its right horn cores, opposite to the anticlockwise torsion of *Criotherium.* It looks very much like a male (*"Samotragus crassicornis"*) of the Turolian *Oioceros rothi* (Wagner 1857), but with much shortened horn cores and elevated rugose surfaces on the frontals anterior to the horn insertions (Sickenberg 1933, pls. 4, 5; 1936, pl. 3). So, if horn cores with clockwise torsion, deep longitudinal grooving, and a tendency to proximal thickening were true synapomorphies, *Urmiatherium* could be a large specialized relative of *Oioceros,* just as *Criotherium* is of *Palaeoreas.* The concept of a group of specialized late Miocene ovibovines suddenly looks very weak. It may be that basicranial/cervical specializations simply evolved in parallel in the Vallesian *Mesembriacerus,* in *Urmiatherium,* and later in true ovibovines. As mentioned earlier, *Oioceros* is probably not an antilopine related to those with anticlockwise-spiraled horns but has an ancestry close to *Tethytragus,* first suggested as an option by Azanza et al. (1998).

The late Miocene sub-Paratethyan fauna was thought by its nineteenth-century discoverers to resemble modern African savanna faunas. This view has already been superseded, but the preceding paragraph takes another step in intensifying the taxonomic contrast with any later ruminant faunas at all. Cervids were rare in the sub-Paratethyan fauna, and the two common giraffids, *Palaeotragus* and *Samotherium,* were unrelated within the Giraffidae to living *Giraffa.* The sivathere is also probably unrelated to the later African and Siwaliks *Sivatherium.* The bovids now seem to comprise an extinct boselaphine group (*Tragoportax* spp.) unrelated to later Bovini or to the nilgai *Boselaphus tragocamelus;* two clusters of spiral-horned species embracing some large, specialized or even aberrant species; *Gazella;* and some Caprinae represented by *Protoryx* and its relatives.

The late Miocene ruminants must have been in a substantially different ecosystem from any later fauna. In the broadest terms, it looks as if the Old World northern hemisphere once contained a southern temperate zoogeographical province of lightly wooded or parkland country, suitable for this diverse and distinctive array of bovids and giraffids. Since the end of the Miocene, this Turolian fauna has been transformed. At some subsequent date or dates, the giraffids, *Tragoportax,* and the *Oioceros* group of spiral-horned bovids became extinct. The array of spiral-horned antilopines survive in Eurasia solely as the Indian blackbuck. *Gazella* remained widespread and

Table 15.8. Ruminant Mandibular Cheek Tooth Occlusal Length Measurements

Taxon (side)	Tooth Measurement (mm)									
	m3	m2	m2 or m1	m1	p4	p3	p2	dp4	dp3	dp2
Hypsodontus pronaticornis										
Left	32.2	21.2	18.9						10.8	
	32.0									
	31.1									
Bovidae sp., size of *Gazella*										
Left	15.2		~12.4							
			9.9							
Right	14.8	12.1		10.0					8.4	
									7.7	
Bovidae, ?sp. A										
Left	~12.1		8.4		7.7			11.0	6.0	
Right	~10.9	9.8	7.0	7.9				10.1	5.4	4.3
										4.9
?(Protoryx solignaci)										
Left		17.5								
Palaeotragus coelophrys										
Left			25.6					39.3	24.2	19.1
			26.7					29.8	19.6	
Right	39.1	30.8		27.8	24.6	23.8	19.2	32.6		
Decennatherium macedoniae										
Left	48.2	33.4			29.7					
Tragoportax spp.										
Left	22.6	16.0	18.5	13.8	14.7	14.8		17.6	14.9	
	28.0	15.7		14.5	13.8	14.7		22.6	13.0	
		17.1		15.0						
				19.5		15.0		18.8		
				14.3						
Right	23.2	17.3	15.2	15.7	~15.7	15.8	11.2			
	22.4	15.4		15.8	16.2	13.6				
	23.0	17.3			14.4	~13.6				
	25.2	18.7			~18.0	15.3				
						14.7				
Prostrepsiceros sp.										
Left	14.2	11.6		10.0	9.4	9.1	5.8	12.4	7.6	5.9
	~14.0	11.5		8.1	9.1	7.6	5.4			
		11.6		9.4		8.9				
				8.7						
				9.5						
Right	15.7	12.0		10.8	8.5	7.8	5.8	12.4	8.0	6.1
	15.7	11.0		7.5	9.5		6.0			
	16.2	11.4		9.6						
	15.5	11.6		9.8						
		11.7								
Gazella sp.										
Left	14.4	11.9		10.3	9.5	9.1				
	15.3	10.5		9.0	8.9	8.9				
	15.6									
	~13.9									
Right	14.5	12.0	11.1	10.0	8.4	8.0	5.6			
	15.3	11.4	11.5	9.0	9.0	7.2	5.4			
	15.3	11.2		9.5	8.6	8.1	5.5			
	14.3	10.9		9.2						
Bovidae sp. A										
Left	13.8	10.8	10.6	8.0	8.5	7.5				
	14.2	10.5		8.6	7.0					
	14.0									
	13.9									

(continued)

Table 15.8. Ruminant Mandibular Cheek Tooth Occlusal Length Measurements *(continued)*

	Tooth Measurement (mm)									
Taxon (side)	m3	m2	m2 or m1	m1	p4	p3	p2	dp4	dp3	dp2
Right	13.5	10.2	11.0	8.5	7.4	7.2	5.4			
	13.9	10.5		9.1	8.0	6.7				
	14.2	10.0		7.6	7.7	5.6				
					7.1					
Bovidae sp B										
Left			8.9					~8.0	7.5	
Right		10.1	9.7		8.3	7.5			6.2	
						6.6				
Protoryx solignaci										
Left						11.2				
Right					14.2					
Pachytragus laticeps										
Left				18.5	15.0		9.2	~21.3	12.6	
								22.4		
Right	25.2									
Pseudotragus aff. *capricornis*										
Left	15.7	11.5	13.8	9.4			8.2	10.9	7.6	
	14.2	12.3								
Right	21.2	15.3	15.6	12.6	13.1	12.0		16.0		
	18.9	13.1	12.7	11.5	10.8					
			13.8		10.8					

Table 15.9. Ruminant Maxillary Tooth Occlusal Length Measurements

	Tooth Measurement (mm)									
Taxon (side)	M3	M2	M/ indet.	M1	P4	P3	P2	dP4	dP3	dP2
Hypsodontus pronaticornis										
Left	24.7	23.6	~22.5	19.4	11.4					
			22.5		12.9					
			26.3							
Right		23.7		19.9	12.0					
Bovidae sp., size of *Gazella*										
Left	11.3	10.5	~12.4	8.4	7.0			7.8	9.6	
		10.5	11.2	8.9						
Right			11.1		6.3	8.0		8.1		
			11.9			7.4		8.4		
			10.8							
			9.4							
Bovidae, ?sp. A										
Left									8.0	6.1
Right								7.6	7.7	
									8.0	
?(*Protoryx solignaci*)										
Left			~16.0							
			~15.7							
Palaeotragus coelophrys										
Left		32.6	~34.6	30.2	20.8	21.2	21.1			
Right			34.7							
			30.0							
			32.7							
			32.3							

(continued)

Table 15.9. Ruminant Maxillary Tooth Occlusal Length Measurements *(continued)*

	Tooth Measurement (mm)									
Taxon (side)	M3	M2	M/ indet.	M1	P4	P3	P2	dP4	dP3	dP2
Tragoportax spp.										
Left	19.8	20.1	14.9	16.3	11.7	13.2	14.8	18.2	16.8	
		18.4	19.2	18.9	13.2	14.0	13.7	17.3~	16.2	
			17.5		11.6					
			18.9		12.7					
			17.4							
			18.0							
			19.9							
Right	22.0	18.9	18.7	15.4	12.6	14.1	14.7			
		20.9		18.5			12.8			
							15.0			
Prostrepsiceros sp.										
Left			10.2	8.0						
				9.9						
Right			11.8	9.3	7.3	7.0				
			10.4			6.5				
			12.4							
			10.9							
Gazella sp.										
Left			12.7	9.7			7.8			
Right			12.6	9.6	8.3	8.8				7.2
			~10.1			8.4				
Bovidae species A										
Left						7.9	6.3			
Right	11.3	11.0	10.6	9.5	7.2	7.6	7.7			
	11.0									
Bovidae sp. B										
Left			8.7						8.0	
			8.8							
			8.0							
Right			8.7	9.0						
Pachytragus laticeps										
Left				19.2		13.0	13.1	17.4	15.3	
Right				20.9				18.0	15.4	10.3
Pseudotragus aff. *capricornis*										
Left	14.9	14.9		12.1	9.4					
Right		15.0		13.8	8.3	~11.2				

successful by adapting to increasingly arid environments, and the Caprinae, also still widespread, became adapted or restricted to hilly or elevated terrain.

The Sinap ruminants show us an early stage in the evolution of the lost Turolian fauna.

Acknowledgments

Research on the Sinap Formation is supported by the General Directorate of Antiquities, T.C. Ministry of Culture and Tourism, and the University of Ankara, Faculty of Language, History and Geography, Department of Palaeoanthropology. I thank Berna Alpagut, John Kappelman, and Mikael Fortelius for inviting and allowing me to work on the Sinap ruminants. I thank Geneviève Bouvrain, Sevket Sen, and Denis Geraads for much help and hospitality in Paris, Dimitris Kostopoulos for information, and Mike Howarth for help with computer files.

Literature Cited

Alekseyeva, A., 1913, Nouvelle espèce de cerfs fossiles des environs du village Petroviérovka: Zapiski Novorossiiskago Obshchestva Estestvoispytatelei, v. 40, pp. 1–6.

Alekseyeva, A., 1915, Animaux fossiles du village Novo-Elisavetovka: Odessa, Tipografiya "Technik," pp. 1–453.

Andree, J., 1926, Neue Cavicornier aus dem Pliocän von Samos: Palaeontographica, v. 67, no. 6, pp. 135–175.

Andrews, P., and B. Alpagut, 1990, Description of the fossiliferous units at Pasalar, Turkey: Journal of Human Evolution, v. 19, pp. 343–361.

Arambourg, C., 1959, Vertébrés continentaux du Miocène supérieur de l'Afrique du Nord: Publications du Service de la Carte Géologique de l'Algérie, Nouvelle Série, Paléontologie, Mémoires, no. 4, pp. 1–159.

Azanza, B., 1995, The vertebrate locality Maramena (Macedonia, Greece) at the Turolian-Ruscinian boundary (Neogene). 14. Cervidae (Artiodactyla, Mammalia): Münchner Geowissenschaftliche Abhandlungen, v. A28, pp. 157–166.

Azanza, B., and J. Morales, 1994, *Tethytragus* nov. gen. et *Gentrytragus* nov. gen. Deux nouveaux Bovidés (Artiodactyla, Mammalia) du Miocène moyen: Proceedings Koninklijke Nederlandse Akademie van Wetenschappen, v. 97, pp. 249–282.

Azanza, B., M. Nieto, and J. Morales, 1998, *Samotragus pilgrimi* n. sp. a new species of Oiocerini (Bovidae, Mammalia) from the Middle Miocene of Spain: Comptes Rendus des Séances de l'Académie des Sciences, Sciences de la terre et des planètes, v. 326, pp. 377–382.

Barry, J. C., 1995, Faunal turnover and diversity in the terrestrial Neogene of Pakistan, *in* E. S. Vrba, G. H. Denton, T. C. Partridge, and L. H. Burckle, eds., Paleoclimate and evolution, with emphasis on human origins: New Haven, Connecticut, Yale University Press, pp. 115–134.

Barry, J. C., and L. J. Flynn, 1989, Key biostratigraphic events in the Siwaliks sequence, *in* E. H. Lindsay, V. Fahlbusch, and P. Mein, eds., European Neogene mammal chronology: New York, Plenum Press, pp. 557–571.

Bernor, R., V. Fahlbusch, and H.-W. Mittmann, 1996a, The evolution of western Eurasian Neogene mammal faunas: New York, Columbia University Press, pp. 1–487.

Bernor, R. L., N. Solounias, C. C. Swisher III, and J. A. Van Couvering, 1996b, The correlation of three classical "Pikermian" mammal faunas—Maragheh, Samos, and Pikermi—with the European MN unit system, *in* R. Bernor, V. Fahlbusch, and H.-W. Mittmann, eds., The evolution of western Eurasian Neogene mammal faunas: New York, Columbia University Press, pp. 137–154.

Boaz, N., A. El Arnauti, A. Gaziry, J. de Heinzelin, and D. Boaz, 1987, Neogene paleontology and geology of Sahabi: New York, Alan Liss, pp. 1–401.

Bohlin, B., 1926, Die familie Giraffidae: Palaeontologia Sinica, v. C4, no. 1, pp. 1–178.

Bohlin, B., 1935, Ueber die verwandtschaftlichen Beziehungen von *Criotherium argalioides* Forsyth Major: Bulletin of the Geological Institution of the University of Upsala, v. 25, pp. 1–12.

Bonis, L. de, G. D. Koufos, and S. Sen, 1997, A giraffid from the Middle Miocene of the island of Chios, Greece: Palaeontology, v. 40, pp. 121–133.

Bonis, L. de, G. D. Koufos, and S. Sen, 1998, Ruminants (Bovidae and Tragulidae) from the Middle Miocene (MN5) of the island of Chios, Aegean Sea (Greece): Neues Jahrbuch für Geologie und Paläontologie Abhandlungen, v. 210, pp. 399–420.

Borissiak, A., 1914, Mammifères fossiles de Sebastopol: Mémoires Comité Géologique St. Petersbourg, n.s., v. 87, pp. 1–154.

Bouvrain, G., 1975, Un nouveau bovidé du Vallésian de Macédoine (Grèce): Comptes Rendus des Séances de l'Académie des Sciences, v. D280, pp. 1357–1359.

Bouvrain, G., 1979, Un nouveau genre de Bovidé de la fin du Miocène: Bulletin de la Société Géologique de France, ser. 7, v. 21, pp. 507–511.

Bouvrain, G., 1980, Le genre *Palaeoreas* (Mammalia, Artiodactyla, Bovidae), systématique et extension géographique: Paläontologische Zeitschrift, v. 54, pp. 55–65.

Bouvrain, G., 1982, Révision du genre *Prostrepsiceros* Major 1891: Paläontologische Zeitschrift, v. 56, pp. 113–124.

Bouvrain, G., 1988, Les *Tragoportax* (Bovidae, Mammalia) des gisements du Miocène supérieur de Ditiko (Macédoine, Grèce): Annales de Paléontologie, v. 74, pp. 43–63.

Bouvrain, G., 1992, Antilopes à chevilles spiralées du Miocène supérieur de la province Gréco-Iranienne: Nouvelles diagnoses: Annales de Paléontologie, v. 78, pp. 49–65.

Bouvrain, G., 1994a, Un Bovidé du Turolien inférieur d'Europe orientale: *Tragoportax rugosifrons:* Annales de Paléontologie, v. 80, pp. 61–87.

Bouvrain, G., 1994b, Les gisements de mammifères du Miocène supérieur de Kemiklitepe, Turquie: 9. Bovidae: Bulletin du Muséum National d'Histoire Naturelle, ser. 4, v. 16, section C, no. 1, pp. 175–209.

Bouvrain, G., 1997, Les bovidès du Miocène supérieur de Pentalophos (Macédoine, Grèce): Münchner Geowissenschaftliche Abhandlungen, v. A34, pp. 5–22.

Bouvrain, G., and L. de Bonis, 1984, Le genre *Mesembriacerus* (Bovidae, Artiodactyla, Mammalia): Un ovibovinė primitif du Vallésian (Miocène supérieur) de Macédoine (Grèce): Palaeovertebrata, v. 14, pp. 201–223.

Bouvrain, G., and L. de Bonis, 1985, Le genre *Samotragus* (Artiodactyla, Bovidae), Une antilope du Miocène supérieur de Grèce: Annales de Paléontologie, v. 71, pp. 257–299.

Bouvrain, G., and L. de Bonis, 1986, *Ouzocerus gracilis* n.g., n. sp., Bovidae (Artiodactyla, Mammalia) du Vallésian (Miocène supérieur) de Macédoine (Grèce): Géobios, v. 19, pp. 661–667.

Bouvrain, G., and H. Thomas, 1992, Une antilope à chevilles spiralées: *Prostrepsiceros zitteli* (Bovidae). Miocène supérieur du Jebel Hamrin en Irak: Géobios, v. 25, pp. 525–533.

Bouvrain, G., S. Sen, and H. Thomas, 1994, Un nouveau genre d'antilope dans le Miocene supérieur de Sinap Tepe en Turquie: Revue de Paléobiologie, v. 13, pp. 375–380.

Brunet, M., and E. Heintz, 1983, Interprétation paléoecologique et relations biogéographiques de la faune de vertébrés du Miocène supérieur d'Injana, Irak: Palaeogeography, Palaeoclimatology, Palaeoecology, v. 44, pp. 283–293.

Chen, G. F., 1988, Remarks on the *Oioceros* species (Bovidae, Artiodactyla, Mammalia) from the Neogene of China: Vertebrata Palasiatica, v. 26, pp. 169–172.

Chen, G. F., 1990, Discovery of the genus *Kubanotragus* (Bovidae, Artiodactyla) from the Middle Miocene in Lantian District, Shanxi Province, China: Vertebrata Palasiatica, v. 28, pp. 1–8.

Churcher, C. S., 1970, Two new upper Miocene giraffids from FortTernan, Kenya, East Africa: *Palaeotragus primaevus* n. sp. and *Samotherium africanum* n. sp., *in* L. S. B. Leakey, and R. J. G. Savage, eds., Fossil vertebrates of Africa, v. 2: London, Academic Press, pp. 1–109.

Colbert, E. H., 1935, Siwalik mammals in the American Museum of Natural History: Transactions of the American Philosophical Society, n.s., v. 26, pp. 1–401.

Crusafont Pairo, M., 1952, Los jirafidos fosiles de Espana: Memorias y Comunicaciones del Instituto Geologico Barcelona, v. 8, pp. 1–239.

Dames, W., 1883, Hirsche und Mäuse von Pikermi in Attica: Zeitschrift der Deutschen Geologischen Gesellschaft, v. 35, pp. 92–100.

Erdbrink, D. P. B., 1988, *Protoryx* from three localities east of Marâgheh, N.W. Iran: Proceedings Koninklijke Nederlandse Akademie Wetenschappen, v. B91, pp. 101–159.

Estes, R. D., 1991, The behavior guide to African mammals: Berkeley, University of California Press, pp. 1–611.

Falconer, H., 1845, Description of some fossil remains of *Dinotherium,* giraffe and other Mammalia, from the Gulf of Cambay, western coast of India: Quarterly Journal of the Geological Society of London, v. 1, pp. 356–372.

Filhol, H., 1891, Etudes sur les mammifères fossiles de Sansan: Annales des Sciences Géologiques, v. 21, pp. 1–319. (Also issued as Bibliothèque de l'Ecole des Hautes Etudes: Section des Sciences Naturelles, v. 37, pp. 1–319.)

Gabuniya, L. K., 1973, Fossil vertebrates fauna of Belometscheskaya: Tbilisi, Republic of Georgia, Metsniereba, pp. 1–136 (in Russian).

Gaillard, C., 1902, Le bélier de Mendès ou le mouton domestique de l'ancienne Egypte: Bulletin de la Société d'Anthropologie et de Biologie de Lyon, v. 20, pp. 70–103.

Gaudry, A., 1860, Résultats des fouilles exécutées en Grèce sous les auspices de l'Académie: Comptes Rendus des Séances de l'Académie des Sciences Paris, v. 51, pp. 802–804.

Gaudry, A., 1861a, Notes sur les Antilopes trouvées à Pikermi (Grèce): Bulletin de la Société Géologique de France, sér 2, v. 18, pp. 388–400.

Gaudry, A., 1861b, Résultats des fouilles entreprises en Grèce sous les auspices de l'Académie: Comptes Rendus des Séances de l'Académie des Sciences Paris, v. 52, pp. 238–241.

Gaudry, A., 1865, Animaux fossiles et géologie de l'Attique: Paris, Libraire de la Société Géologique de France, pp. 1–476. (Whole work issued from 1862 to 1867. According to Pilgrim and Hopwood (1928, p. 97), section including bovids [pp. 241–323] was published in 1865.)

Gaudry, A., and E. Lartet, 1856, Résultats des recherches paléontologiques entreprises dans l'Attique sous les auspices de l'Académie: Comptes Rendus Hebdomadaire Séances de l'Académie des Sciences Paris, v. 43, pp. 271–274.

Gentry, A. W., 1970, The Bovidae (Mammalia) of the Fort Ternan fossil fauna, *in* L. S. B. Leakey, and R. J. G. Savage, eds., Fossil vertebrates of Africa, v. 2: London, Academic Press, pp. 243–324.

Gentry, A. W., 1971, The earliest goats and other antelopes from the Samos *Hipparion* fauna: Bulletin of the British Museum (Natural History) (Geology), v. 20, pp. 229–296.

Gentry, A. W., 1990, Ruminant artiodactyls of Pasalar, Turkey: Journal of Human Evolution, v. 19, pp. 529–550.

Gentry, A. W., 1992, The subfamilies and tribes of Bovidae: Mammal Review, v. 22, pp. 1–32.

Gentry, A. W., 1994, The Miocene differentiation of Old World Pecora (Mammalia): Historical Biology, v. 7, pp. 115–158.

Gentry, A. W., 1996, A fossil *Budorcas* (Mammalia, Bovidae) from Africa, *in* K. Stewart, and K. Seymour, eds., Paleoecology and paleoenvironments of late Cenozoic mammals: Tributes to the career of C. S. (Rufus) Churcher: Toronto, University Press, pp. 571–587.

Gentry, A. W., 2000, Caprinae and Hippotragini (Mammalia, Bovidae) in the Upper Miocene, *in* E. S. Vrba, and G. B. Schaller, eds., Antelopes, deer and relatives: Fossil record, behavioral ecology, systematics and conservation: New Haven, Connecticut, Yale University Press, pp. 65–83.

Gentry, A. W., and E. P. J. Heizmann, 1996, Miocene ruminants of central and eastern Tethys and Paratethys, *in* R. Bernor, V. Fahlbusch, and H.-W. Mittmann, eds., The evolution of western Eurasian Neogene mammal faunas: New York, Columbia University Press, pp. 378–391.

Geraads, D., 1978, Les Palaeotraginae (Giraffidae, Mammalia) du Miocène supérieur de la région de Thessalonique (Grèce): Géologie Méditerranéenne, v. 5, pp. 269–276.

Geraads, D., 1979, Les Giraffinae (Artiodactyla, Mammalia) du Miocène supérieur de la région de Thessalonique (Grèce): Bulletin du Muséum National d'Histoire Naturelle, sér. 4, v. 1, section C, no. 4, pp. 377–389.

Geraads, D., 1989a, Un nouveau Giraffidé du Miocène supérieur de Macédoine (Grèce): Bulletin du Muséum National d'Histoire Naturelle, ser. 4, v. 11, section C, no. 4, pp. 189–199.

Geraads, D., 1989b, Vertébrés fossiles du Miocène supérieur du Djebel Krechem el Artsouma (Tunisie centrale). Comparaisons biostratigraphiques: Géobios, v. 22, pp. 777–801.

Geraads, D., and E. Güleç, 1999, On some spiral-horned antelopes (Mammalia: Artiodactyla) from the Late Miocene of Turkey, with remarks on their distribution: Paläontologische Zeitschrift, v. 73, pp. 403–409.

Geraads, D., E. Güleç, and G. Saraç, 1995, Middle Miocene ruminants from Inönü, central Turkey: Neues Jahrbuch für Geologie und Paläontologie Monatshefte, v. 8, pp. 462–474.

Giuli, C. de, and E. Heintz, 1974, *Gazella borbonica* (Bovidae, Artiodactyla, Mammalia), nouvel élément de la faune villafranchienne de Montopoli, Valdarno inférieur, Pisa,Italia: Atti della Società Toscana di Scienze Naturali, Memorie, v. A81, pp. 227–237.

Hamilton, W. R., 1973, The lower Miocene ruminants of Gebel Zelten, Libya: Bulletin of the British Museum (Natural History) (Geology), v. 21, pp. 73–150.

Heintz, E., 1971, *Gazella deperdita* (Gervais) 1847 (Bovidae, Artiodactyla, Mammalia) du Pontien du Mont Lubéron Vaucluse, France: Annales de Paléontologie (Vertébrés), v. 57, pp. 209–229.

Heintz, E., M. Brunet, and S. Sen, 1981, Un nouveau Giraffidé du Miocène supérieur d'Irak: *Injanatherium hazimi* n. g., n. sp.: Comptes Rendus Séances de l'Académie des Sciences Paris, v. 292, pp. 423–426.

Heizmann, E. P. J., 1976, Die palaeontologische Erforschung des Steinheimer Beckens, *in* Meteorkrater Steinheimer Becken: Steinheim an Albuch, Bürgermeisteramt, pp. 29–45.

Hill, A., 1995, Faunal and environmental change in the Neogene of East Africa: Evidence from the Tugen Hills sequence, Baringo District, Kenya, *in* E. S. Vrba, G. H. Denton, T. C. Partridge, and L.H. Burckle, eds., Paleoclimate and evolution, with emphasis on human origins: New Haven, Connecticut, Yale University Press, pp. 178–193.

Hill, A., R. Drake, L. Tauxe, M. Monaghan, J. C. Barry, A. K. Behrensmeyer, G. Curtis, B. F. Jacobs, L. Jacobs, N. Johnson, and D. Pilbeam, 1985, Neogene palaeontology and geochronology of the Baringo Basin, Kenya: Journal of Human Evolution, v. 14, pp. 759–773.

Hooker, J. J., 1986, Mammals from the Bartonian (middle/late Eocene) of the Hampshire Basin, southern England: Bulletin of the British Museum (Natural History) (Geology), v. 39, pp. 191–478.

Kingdon, J., 1982, East African mammals, v. IIIC: London: Academic Press, pp. 1–393.

Knottnerus-Meyer, T., 1907, Uber das Tränenbein der Huftiere: Archiv für Naturgeschichte, v. 73, no. 1, pp. 1–152.

Köhler, M., 1987, Boviden des türkischen Miozäns (Känozoikum und Braunkohlen der Türkei 28): Paleontologia y Evolucio, v. 21, pp. 133–246.

Köhler, M., 1993, Skeleton and habitat of Recent and fossil ruminants: Münchner Geowissenschaftlichen Abhandlungen, v. A25, pp. 1–88.

Köhler, M., S. Moya Sola, and J. Morales, 1995, The vertebrate locality Maramena (Macedonia, Greece) at the Turolian-Ruscinian boundary (Neogene). 15. Bovidae and Giraffidae

(Artiodactyla, Mammalia): Münchner Geowissenschaftliche Abhandlungen, v. A28, pp. 167–180.

Kostopoulos, D. S., 2000, Functional morphology and paleoecological adaptations of *Nisidorcas planicornis* (Bovidae, Mammalia) from the late Miocene. Münchner Geowissenschaftliche Abhandlunger, v. A39, pp. 93–104.

Kostopoulos, D. S., and G. D. Koufos, 1996, Late Miocene bovids (Mammalia, Artiodactyla) from the locality "Nikiti-1" (NKT), Macedonia, Greece: Annales de Paléontologie, v. 81, pp. 251–300.

Kostopoulos, D. S., and G. D. Koufos, 1999, The Bovidae (Mammalia, Artiodactyla) of the "Nikiti-2" [NIK] faunal assemblage (Chalkidiki peninsula, N.Greece): Annales de Paléontologie, v. 85, pp. 193–218.

Kostopoulos, D. S., and S.Sen, 1999, Late Pliocene (Villafranchian) mammals from Sarikol Tepe, Ankara, Turkey: Mitteilungen der Bayerischen Akademie der Wissenschaften, v. 39, pp. 165–202.

Kostopoulos, D. S., K. K. Koliadimou, and G. D. Koufos, 1996, The giraffids (Mammalia, Artiodactyla) from the Late Miocene mammalian localities of Nikiti (Macedonia, Greece): Palaeontographica, v. A239, pp. 61–88.

Kretzoi, M., 1968, New generic names for homonyms: Vertebrata Hungarica, v. 10, no. 1/2, pp. 163–165.

Lehmann, U., and H. Thomas, 1987, Fossil Bovidae from the Mio-Pliocene of Sahabi, (Libya), *in* N. T. Boaz, A. El-Arnauti, A. W. Gaziry, J. de Heinzelin, and D. D. Boaz, eds., Neogene paleontology and geology of Sahabi: New York, Alan Liss, pp. 323–335.

Lewis, G. E., 1939, A new *Bramatherium* skull: American Journal of Science, v. 237, pp. 275–280.

Lydekker, R., 1876, Molar teeth and other remains of Mammalia: Memoirs of the Geological Survey of India, Palaeontologia Indica, ser. 10, v. 1, part 2, pp. 19–87.

Lydekker, R., 1878, Crania of ruminants from the Indian Tertiaries: Memoirs of the Geological Survey of India, Palaeontologia Indica, ser. 10, v. 1, part 3, pp. 88–171.

Major, C. I. F., 1891, Considérations nouvelles sur la Faune des Vertébrés du Miocène supérieur dans l'Ile de Samos: Comptes Rendus des Séances de l'Académie des Sciences Paris, v. 113, pp. 608–610.

Matthew, W. D., 1929, Critical observations upon Siwalik mammals: Bulletin of the American Museum of Natural History, v. 56, pp. 437–560.

Mecquenem, R. de, 1924–1925, Contribution à l'étude des fossiles de Maragha: Annales de Paléontologie, v. 13, pp. 135–160, v. 14, pp. 1–36.

Montoya, P., and J. Morales, 1991, *Birgerbohlinia schaubi* Crusafont 1952 (Giraffidae, Mammalia) del Turoliense inferior de Crevillente-2 (Alicante, Espana). Filogenia e historia biogeográfica de la subfamilia Sivatheriinae: Bulletin du Muséum National d'Histoire Naturelle, ser. 4, v. 13, sec. C, no. 3, pp. 177–200.

Morales, J., and D. Soria, 1981, Los artiodactilos de Los Valles de Fuentiduena (Segovia): Estudios Geologicos, v. 37, pp. 477–501.

Morales, J., D. Soria, and H. Thomas, 1987, Les Giraffidae (Artiodactyla, Mammalia) d'Al Jadidah du Miocène Moyen de la Formation Hofuf (Province du Hasa, Arabie Saoudite): Géobios, v. 20, pp. 441–467.

Morales, J., M. Pickford, and D. Soria, 1993, Pachyostosis in a lower Miocene giraffoid from Spain, *Lorancameryx pachyostoticus* nov. gen. nov. sp. and its bearing on the evolution of bony appendages in artiodactyls: Géobios, v. 26, pp. 207–230.

Moya Sola, S., 1983, Los Boselaphini (Bovidae Mammalia) del Neogeno de la peninsula Iberica: Publicaciones de Geologia, Universitat Autonoma de Barcelona, v. 18, pp. 1–236.

Nakaya, H., 1994, Faunal change of late Miocene Africa and Eurasia: Mammalian fauna from the Namurungule Formation, Samburu Hills, northern Kenya: African Study Monographs, supplementary issue, no. 20, pp. 1–112.

Nakaya, H., M. Pickford, Y. Nakano, and H. Ishida, 1984, The late Miocene large mammal fauna from the Namurungule Formation, Samburu Hills, northern Kenya: African Study Monographs, supplementary issue, no. 2, pp. 87–131.

Ozansoy, F., 1955, Sur les gisements continentaux et les mammifères du Néogene et du Villafranchien d'Ankara (Turquie): Comptes Rendus Séances de l'Académie des Sciences Paris, v. 240, no. 9, pp. 992–994.

Ozansoy, F., 1957, Faunes de mammifères du Tertiaire de Turquie et leurs révisions stratigraphiques: Bulletin of the Mineral Research and Exploration Institute of Turkey (Foreign Edition), v. 49, pp. 29–48.

Ozansoy, F., 1958, Etude des gisements continentaux et des mammifères du Cénozoique de Turquie [thesis]: Paris, Faculté des Sciences, Université de Paris, 158 pp.

Ozansoy, F., 1965, Etude des gisements continentaux et des mammifères du Cénozoique de Turquie: Mémoires de la Société Géologique de France, Nouvelle Série, v. 44, pp. 1–92.

Palmer, T. S., ed., 1904, Index generum mammalium: United States Department of Agriculture, Biological Survey Division, North American fauna, no. 23, pp. 1–984.

Paraskevaidis, I., 1940, Eine obermiocäne Fauna von Chios: Neues Jahrbuch für Mineralogie, Geologie und Paläontologie, Beilage-Bände, v. B83, pp. 363–442.

Pavlovic, M. B., 1969, Miozän-Säugetiere des Toplica-Beckens: Geoloski Anali Balkanskoga Poluostrva, v. 34, pp. 269–394.

Pavlow, M., 1913, Mammifères Tertiaires de la Nouvelle Russie. 1, Artiodactyles, Périssodactyles: Nouveaux Mémoires de la Société Impériale des Naturalistes, v. 17, pp. 1–67.

Pilgrim, G. E., 1911, The fossil Giraffidae of India: Memoirs of the Geological Survey of India, Palaeontologia Indica, n.s., v. 4, no. 1, pp 1–29.

Pilgrim, G. E., 1934, Two new species of sheep-like antelope from the Miocene of Mongolia: American Museum Novitates, no. 716, pp. 1–29.

Pilgrim, G. E., 1937, Siwalik antelopes and oxen in the American Museum of Natural History: Bulletin of the American Museum of Natural History, v. 72, pp. 729–874.

Pilgrim, G. E., 1939, The fossil Bovidae of India: Memoirs of the Geological Survey of India, Palaeontologia Indica, n.s., v. 26, pp. 1–356.

Pilgrim, G. E., and A. T. Hopwood, 1928, Catalogue of the Pontian Bovidae of Europe: London, British Museum (Natural History), pp. 1–106.

Prasad, K. N., 1974, The vertebrate fauna from Piram Island, Gujarat, India: Memoirs of the Geological Survey of India, Palaeontologia Indica, n.s., v. 41, pp. 1–23.

Qiu, Z., C. Li, and S. Wang, 1981, Miocene mammalian fossils from Xining Basin, Qinghai: Vertebrata Palasiatica, v. 19, pp. 156–173.

Robinson, P., 1972, *Pachytragus solignaci*, a new species of caprine bovid from the late Miocene Beglia Formation of Tunisia: Notes du Service Géologique de Tunisie, v. 37, pp. 73–94.

Rodler, A., 1889, Uber *Urmiatherium polaki* n. g., n. sp.: Denkschriften der Kaiserlichen Akademie der Wissenschaften, v. 56, pp. 315–322.

Rodler, A., and K. A. Weithofer,1890, Die Wiederkäuer der Fauna von Maragha: Denkschriften der Kaiserlichen Akademie der Wissenschaften, v. 57, pp. 753–772.

Rössner, G. E., 1996, Odontologische und schädelanatomische Untersuchungen an *Procervulus* (Cervidae, Mammalia): Münchner Geowissenschaftliche Abhandlungen, v. A29, pp. 1–127.

Schlosser, M., 1904, Die fossilen Cavicornier von Samos: Beiträge zur Paläontologie und Geologie Osterreich-Ungarns und des Orients, v. 17, pp. 28–118.

Sen, S., 1990, Stratigraphie, faunes de mammifères et magnétostratigraphie du Néogène de Sinap Tepe, province d'Ankara, Turquie: Bulletin du Muséum National d'Histoire Naturelle, ser. 4, v. 12, sec. C, no. 3–4, pp. 243–277.

Sickenberg, O., 1929, Eine neue Antilope und andere Säugetierreste aus dem Obermiozän Niederösterreichs: Palaeobiologica, v. 2, pp. 62–86.

Sickenberg, O., 1932, Eine neue Antilope, *Parurmiatherium rugosifrons* nov. gen. nov. sp., aus dem Unterpliozän von Samos: Anzeiger der Akademie der Wissenschaften in Wien, Mathematisch-Naturwissenschaftliche Klasse, v. 1, pp. 10–11.

Sickenberg, O., 1933, *Parurmiatherium rugosifrons,* ein neuer Bovide aus dem Unterpliozän von Samos: Palaeobiologica, v. 5, pp. 81–102.

Sickenberg, O., 1936, Uber *Samotragus crassicornis* nov.gen. et spec. aus dem Unterpliozän von Samos: Paläontologische Zeitschrift, v. 18, pp. 90–94.

Sokolov, J. J., 1949, On the remains of Cavicornia (Bovidae, Mammalia) from the middle Miocene of the north Caucasus: Doklady Akademii Nauk SSSR, v. 67, pp. 1101–1104.

Solounias, N., 1981a, The Turolian fauna from the island of Samos, Greece: Contributions to Vertebrate Evolution, v. 6, pp. 1–232.

Solounias, N., 1981b, Mammalian fossils of Samos and Pikermi. Part 2. Resurrection of a classic Turolian fauna: Annals of the Carnegie Museum, v. 50, pp. 231–270.

Solounias, N., S. M. C. Moelleken, and J. M. Plavkan, 1995, Predicting the diet of extinct bovids using masseteric morphology: Journal of Vertebrate Paleontology, v. 15, pp. 795–805.

Steininger, F. F., W. A. Berggren, D. V. Kent, R. L. Bernor, S. Sen, and J. Agusti, 1996, Circum-Mediterranean Neogene (Miocene and Pliocene) marine-continental chronologic correlations of European mammal units, *in* R. Bernor, V. Fahlbusch, and H.-W. Mittmann, eds., The evolution of western Eurasian Neogene mammal faunas: New York, Columbia University Press, pp. 7–46.

Stromer, E., 1928, Wirbeltiere in obermiocänen Flinz Münchens: Abhandlungen der Bayerischen Akademie der Wissenschaften, v. 32, pp. 1–71.

Tekkaya, I., 1973, Une nouvelle espèce de *Gazella* de Sinap moyen: Bulletin of the Mineral Research and Exploration Institute of Turkey, v. 80, pp. 118–143.

Tekkaya, I., 1975, Orta Sinap Bovinae Faunasi: Türkiye Jeoloji Kurumu Bülteni, v. 18, pp. 27–32.

Thenius, E., 1950, Die tertiären Lagomeryciden und Cerviden der Steiermark: Sitzungsberichte der Oesterreichischen Akademie der Wissenschaften, Mathematisch-Naturwissenschaftliche Klasse I, v. 159, pp. 219–254.

Thomas, H., 1977, Un nouveau bovidé du Nagri, plateau du Potwar, Pakistan: Bulletin de la Société Géologique de France, sér. 7, v. 19, pp. 375–383.

Thomas, H., 1979, Le rôle de barrière écologique de la ceinture Saharo-arabique: Arguments paléontologiques: Bulletin du Muséum National d'Histoire Naturelle, sér. 4, v. 1, sec. C, no. 2, pp. 127–135.

Thomas, H., 1981, Les Bovidés miocènes de la formation de Ngorora du Bassin de Baringo (Kenya): Proceedings Koninklijke Nederlandse Akademie van Wetenschappen, v. B84, pp. 335–409.

Thomas, H., 1983, Les Bovidae (Artiodactyla, Mammalia) du Miocène moyen de la formation Hofuf (Province du Hasa, Arabie Saoudite): Palaeovertebrata, v. 13, pp. 157–206.

Thomas, H., 1984, Les Bovidés anté-hipparions des Siwaliks inférieurs (plateau du Potwar, Pakistan): Mémoires de la Société Géologique de France, Nouvelle Série, v. 145, pp. 1–68.

Thomas, H., S. Sen, and G. Ligabue, 1980, La faune Miocène de la Formation Agha Jari du Jebel Hamrin (Irak): Proceedings Koninklijke Nederlandse Akademie van Wetenschappen, v. B83, pp. 269–287.

Wagner, A., 1848, Urweltliche Säugethier-Ueberreste aus Griechenland: Abhandlungen der Königlich Bayerischen Akademie der Wissenschaften, v. 5, pp. 333–378.

Wagner, A., 1857, Neue Beiträge zur Kenntnis der fossilen Säugethier—Ueberreste von Pikermi: Abhandlungen der Königlich Bayerischen Akademie der Wissenschaften, v. 8, pp. 111–158.

Weithofer, K. A., 1888, Beiträge zur Kenntnis der Fauna von Pikermi bei Athen: Beiträge zur Paläontologie Osterreich-Ungarns und des Orients, v. 6, pp. 225–292.

16

Abundance of "*Hipparion*"

R. S. Scott, M. Fortelius, K. Huttunen, and M. Armour-Chelu

The "*Hipparion*" Datum in the Sinap Formation represents the novel meeting of representatives from two distinct but parallel radiations. Large hipparionines belonging to the ancient radiation of the Equidae became a part of a basal herbivore fauna that includes members of the more recent adaptive radiation of the Bovidae. Hipparion immigration into the Old World has implications for environmental change, faunal turnover, and large mammal evolution scenarios. A complete understanding of the impact of this immigration requires quantification of the contribution of hipparionine immigrants to both species diversity and herbivore biomass. We address the second issue here and quantify the abundance of hipparionines relative to ruminants at several localities in the Sinap Formation. Changes in hipparionine abundance immediately following the "*Hipparion*" Datum can be evaluated in the context of the well-controlled chronology for the Sinap fossil localities (Kappelman et al., chapter 2, this volume) and can be matched with the apparent changes in hipparionine diversity recorded by Bernor et al. (chapter 11, this volume).

Accurate estimates of hipparionine abundance are obscured by the various processes that transform living organisms into fossil assemblages. Biomass is often calculated by multiplying the estimated mean body mass of a taxon by its density (number of individuals per area). Biomass is even more difficult to estimate when taking into account the imprecision associated with estimating body mass for extinct taxa. However, body size remains perhaps the single most important parameter in the ecology of a particular species (Damuth and MacFadden 1990) and should be considered whenever possible. Body size influences the basic metabolic requirements of an organism (Kleiber 1932; McNab 1963, 1990); influences factors such as longevity and generation time; and determines much of an organism's interaction with various aspects of its environment, including locomotion, feeding (Kay 1984), and predator avoidance (Scott 1979; Maiorana 1990; Scott et al. 1999). Body size is fundamental when assessing the ecological position of a taxon and its relationship with other taxa. Thus, we present estimates of hipparionine relative abundance in conjunction with body mass estimates when possible. These, in turn, are used together to model the limits of immigrant hipparionine contributions to herbivore biomass following the "*Hipparion*" Datum.

Quantitative estimates of abundance are often given as number of individual specimens (NISP) or minimum number of individuals (MNI). These methods can give varying estimates of relative abundance, and one method may be more accurate than the other, depending on the specifics of the sampling and taphonomy of the fossil assemblage in question. Thus, estimates of relative abundance should be presented in conjunction with relevant taphonomic data whenever possible. Similarly, both NISP and MNI should be calculated for each fossil assemblage. Here, we present relative abundance estimates based on both NISP and MNI in conjunction with taphonomic data concerning the numbers of fragmentary and associated specimens for each assemblage studied.

The goals of this study are to describe changes in hipparionine abundance relative to ruminants in the Sinap Formation at several representative sites and to report body mass estimates for hipparionines and ruminants from these sites whenever possible. We assess the varying estimates of hipparionine relative abundance with respect to the available taphonomic data and report a best estimate of relative abundance for each site studied. Finally, equid abundance was expressed relative to ruminant abundance and changes in equid abundance were modeled using an iterative logistic curve fit in SAS (version 8.02; Cary, North Carolina: SAS Institute) and a simple quantitative model to convert within-locality relative abundance of equids to between-locality absolute abundance and biomass estimates.

Materials and Methods

Fossils collected by the Sinap project were identified by project members under the supervision of the principal investigators and cataloged using a laptop computer and a relational database (Johnson et al. 1996). This relational database was queried both during and at the end of the summer 1995 field season for all specimens identified as being either artiodactyls or perissodactyls from seven fossil collecting localities in the Sinap Formation.

A locality is a stratigraphically and laterally discrete site with fossils in situ; each locality is thought to represent a single depositional unit. The seven localities, reported in stratigraphic order here, are: Locs. 64, 4, 122, 121, 72, 91, and 12. These localities were selected for their stratigraphic significance and the size of their fossil sample. Loc. 64 was selected because it yields the largest pre-"*Hipparion*" large mammal assemblage in the Sinap Formation, whereas Loc. 4 was chosen because it records the earliest occurrence of hipparionines in the Sinap Formation. During June and July of 1995, specimens identified as being equid or ruminant by database queries were examined by R.S.S. and K.H. in Kazan, Turkey, where the Sinap collections are housed. These specimens were then further identified to the lowest possible taxonomic level and all equids and ruminants were then further examined. All of the ruminant and equid specimens were identified according to side, element, part (proximal or distal when applicable), completeness, association (i.e., whether they were associated and/or articulated with other specimens), size category (for ruminant specimens only), age (juvenile or adult), and lower taxonomic group whenever possible. These identifications were sometimes supplemented by ad lib notes and descriptions.

Specimen Definition

Various indices of faunal abundance all rely on counting specimens that belong to a particular group, usually a particular taxon. Even the most simplistic index of faunal abundance, NISP, relies on the identifications of the investigators and their definition of what constitutes a specimen. Thus, a practical, operational definition of the term "specimen" is crucial to any discussion of relative abundance.

Holtzman (1979) devotes some discussion to the issue of what constitutes a specimen and, while invoking "traditional paleontological usage," defines a specimen as "all the remains that can be shown to derive from a single once living individual, provided that the remains include at least one *identifiable element*" (p. 78; emphasis added). This definition contrasts with that of Shotwell (1958, p. 272), who uses the terms "element" and "specimen" interchangeably to refer to Holtzman's "identifiable element" or "finite [number of parts] that can be identified when isolated but cannot be further subdivided without a significant loss of identifiability" (Holtzman 1979, p. 78).

Claims of "traditional paleontological usage" notwithstanding, both definitions ignore the essential practicality that a specimen in paleontology refers to whatever fossil or fossils receive a single catalog or museum accession number. Fossils found in different field seasons may later be shown to derive from a single individual, but operationally, they remain separate but potentially associated specimens. Similarly, multiple associated elements that are each easily identifiable on their own may be given a single catalog number. Thus, for the sake of clarity, we refer to all fossils with a single catalog number as a single cataloged specimen. Cataloged specimens are not to be confused with identifiable specimens, which we used to derive estimates of relative abundance.

Simply counting the number of cataloged specimens attributed to a different taxon would give an unsystematic estimate of their relative abundance: relative abundance would be confounded by the vagaries of catalog number assignment. The definitions of specimen given by Holtzman and Shotwell are more sensitive to what is desirable in an estimator of relative abundance.

If a specimen is considered to be all the remains that can be shown to derive from a single individual (Holtzman 1979), then, at its very best, the number of specimens would be equal to the number of once living individuals preserved in an assemblage. At its worst, the number of specimens would be the number of individuals that one investigator believes are preserved in an assemblage. Multiple fossils would be counted as only one specimen when they are thought to derive from a single individual. As a systematic measure of relative abundance, this technique leaves much to be desired.

Alternatively, Shotwell's definition of specimen would potentially count single fossils as multiple specimens. For example, a complete long bone would be counted twice; once for the proximal end and once for the distal end. Neither Shotwell's nor Holtzman's definition of specimen would yield estimates of relative abundance that are both systematic and sensitive to the processes by which a fossil assemblage is formed.

Rather than use cataloged specimens as the basic unit when estimating relative abundance, we define an identifiable specimen as a single fossil that is identified as belonging to a particular group relevant to the question under study (in this case, a ruminant or a hipparionine). The two criteria used to define a specimen are (1) identifiability and (2) physical boundaries. Multiple bones are counted as multiple identifiable specimens, and single bones, whose respective parts may be further identified but remain complete, are identified as single identifiable specimens. Thus, an identifiable specimen is a discrete, easily recognizable skeletal part. Identifiable specimens can be easily recognized and counted and will simply be referred to as "specimens."

Age and Size Categories

We followed the size categories for ruminants established by A. Gentry during the 1994 field season. Regression estimates of body weight derived from postcranial measurements

Table 16.1. Size Categories Used for Ruminant Specimens

Size Category	Description	Range of Estimated Body Weights (kg)
Very small	Smaller than *Gazella*	<15
Small	Size of *Gazella*	15–30
Medium	Larger than *Gazella*	30–75
Medium	*Prototoryx-Tragoportax* size	103
Large	Size of *"Hipparion" uzunagizli*	150
Large	Size of *Cormohipparion sinapensis*	200
Large	Size of *"Hipparion"* sp. 2	300
Very large	Giraffidae size	>500

Notes: Size categories were established by the authors during the 1995 field season based on notes by A. Gentry from the 1994 field season. Descriptions are based on A. Gentry's notes and body mass estimates reported here using the regression formulae of K. Scott (1990).

were later used to establish body mass means and ranges for each size category, and in a few cases, specimens were reclassified up or down one size category, based on regression estimates of body mass. In general, most initial subjective size category classifications were confirmed by body mass estimates indicating the robustness of the original framework. The size categories used in this chapter are shown in table 16.1.

Specimens were classified as belonging to a juvenile based on the presence of deciduous dentition, unfused epiphyses, or strongly visible lines of epiphyseal fusion (recently fused epiphyses). It is likely that specimens identified as juvenile encompassed a wide range of actual ages; hence these classifications are relative and probably not broadly applicable.

"Association" refers to whether a specimen was found in articulation with another in the field and cataloged as being associated or was subsequently determined to articulate with another specimen. Thus the definition of association used in the catalog is expanded here to include all specimens that are determined to articulate with each other (the definition used in the catalog refers only to specimens found in association during excavation). Each element or partial element (proximal or distal end) that articulates with another element or is cataloged as associated was scored as "associated." Elements could be associated with more than one other element but were scored as associated only once. The number of associated elements was divided by NISP and multiplied by 100 to define an index of degree of association.

Fragmentation and Association Indices

In general, specimens were ranked as complete or fragmentary. Complete specimens were all specimens that constituted a complete or nearly complete element such that any missing piece of the element would be unidentifiable on its own. Fragmentary specimens refer to specimens in which a large part of the identifiable portion of the element is missing. Examples range from proximal and distal ends of long bones to other, more severely fragmented specimens. Teeth represent a unique category in terms of fragmentation and isolated teeth are often well preserved. A complete isolated tooth is not strictly comparable to a complete long bone as an indicator of the degree of fragmentation in a fossil assemblage. Often isolated teeth are very common in even the most fragmented fossil assemblages. Thus, teeth were scored separately as either isolated, fragmentary, or part of an incomplete mandible or maxilla. An index of fragmentation was calculated by dividing the number of fragmentary specimens by the total number of complete and fragmentary specimens and multiplying the result by 100. Dental specimens were excluded from this calculation.

Relative Abundance Calculations

Relative abundance in this study is expressed as the abundance of one group relative to another. The two groups studied are the major basal herbivores—the ruminants and hipparionines. Thus, the relative abundance of the hipparionines is relative to the ruminants only and to not all of the specimens from the locality. The relative abundance of hipparionines is expressed as:

$$[\text{hipparionines} / (\text{hipparionines} + \text{ruminants})] \times 100$$

A number of different measures of abundance have been proposed, including NISP, MNI, and the weighted abundance of elements (WAE) (Shotwell 1955, 1958; Van Valen and Sloan 1965; Grayson 1978; Holtzman 1979; Gilbert et al. 1981; Badgley 1986; Marshall and Pilgram 1993). The abundance of hipparionines and ruminants was determined for this study using NISP, WAE, and two variants of MNI.

NISP is simply the number of specimens identifiable as belonging to a particular group (in this case, as hipparionines or ruminants). In some cases, and generally only in the case of associated specimens, more than one identifiable specimen was given the same catalog number. In these instances, an alphabetical designation was added to each individual identifiable specimen and all identifiable specimens were used to determine NISP.

WAE was determined by dividing the number of identifiable elements (e_i) by the number of identifiable elements per individual (m_i). The quantity m_i was introduced by Shotwell (1955) and refers to the number of skeletal parts in a single living individual that have a reasonable probability of being preserved and identified. Ribs and vertebrae are generally excluded from this number (Shotwell 1955), and determination of identifiable elements per individual relies on the judgment of the investigator (Shotwell 1955, 1958; Holtzman 1979; Badgley 1986). The number of identifiable specimens per hipparionine individual and ruminant individual is shown in table 16.2. As noted earlier, e_i is not the same as NISP. For example, a complete femur constitutes a single specimen and adds one to NISP. In contrast, a complete femur can be divided into two identifiable elements and so adds two to e_i. Thus, WAE is designed to compensate for the effects of fragmentation and variable numbers of elements per individual.

MNI is simply the number of the most abundant element. An element may occur multiple times in a single individual, in which case it is first divided by its representation in a single individual. For example, MNI might be given by the higher of either the number of right astragali or the number of calcanei divided by two. MNI-G is a variant of MNI in which the MNI of each size category of ruminant was first determined and these MNI values were then summed to determine MNI-G cumulatively for all size categories.

Various workers (Grayson 1978; Hill 1979; Hanson 1980; Brain 1981; Badgley 1986) have noted that many factors influence the formation of fossil assemblages, which in turn influence different measures of relative abundance. Perhaps the most significant of these factors are sample size, association, fragmentation, and differential preservation.

MNI and MNI-G always reduce the effective sample size and therefore increase random error (Van Valen and Sloan 1965). Furthermore, MNI always overestimates the abundance of rare taxa (Grayson 1978). These tendencies have been confirmed by computer simulations of faunal assemblages (Van Valen and Sloan 1965; Holtzman 1979; Gilbert et al. 1981), although the errors are reduced when the number of taxa is small (Gilbert et al. 1981). For these reasons, NISP may be preferable as a measure of abundance. However, for cases in which there are many associated specimens or there is differential preservation of taxa, NISP overestimates the abundance of well-preserved taxa and MNI is preferable as a measure of abundance. In general, as association of specimens increases, so does the accuracy of MNI (Badgley 1986). Similarly, the accuracy of NISP drops as differential preservation increases.

Fragmentation may influence estimates of relative abundance in unpredictable ways. MNI has often been considered less biased by fragmentation because fragments from the same element (and hence the same individual) are less likely to be incorporated into the estimate of relative abundance. Marshall and Pilgram (1993) found that MNI decreases with increasing fragmentation. In contrast, they found that NISP first increases with increasing fragmentation and then decreases at the highest levels of fragmentation. NISP may be the best measure of relative abundance at high levels of fragmentation because of biased undercounting (Marshall and Pilgram 1993). This conclusion generally assumes that highly fragmented assemblages always derive from separate bones and distinct individuals. One exception to this assumption is assemblages accumulated in conjunction with a high degree of trampling. Trampling will pulverize single elements into many fragments. When available, evidence of taphonomic factors such as trampling will influence the choice of the best estimator of relative abundance.

WAE has the advantage of usually being intermediate between NISP and MNI. WAE theoretically corrects for some differential preservation but involves a partly subjective correction factor, m_i. Effective sample sizes are higher with WAE estimates, which reduces random error, but the use of a subjective correction factor may be inappropriate (Badgley 1986). Other factors may already counter the effects of differential preservation, making WAE less useful as a measure of abundance. For instance, differential preservation in favor of larger taxa at a site may be countered by differential deposition in favor of smaller taxa with shorter generation times.

The approach we adopted here is to calculate all of these estimates of relative abundance and examine them in relation to the taphonomy of the sites under consideration. Thus, MNI is preferred at sites with greater association, whereas NISP is often preferred at sites with very high fragmentation. If estimates are similar, it is possible to bracket probable relative abundance within a very narrow range.

Body Mass Estimation

Body mass was estimated based on the nonlength measurements of limb bones and regression formulae of K. Scott (1990). The measurements described by Scott (1990) were taken by R.S.S. on all ruminant and equid limb bone specimens whenever possible. Estimates from each nonlength measurement on a specimen were averaged to determine estimated body mass. We used the regression formulae for bovids and ruminants to derive two body mass estimates for each ruminant specimen measured. Body mass estimates for the hipparionine specimens were based on the equid regression.

For purposes of estimating biomass, a parameter B was calculated for hipparionines and ruminants. In the case of hipparionines, $B_{\text{hipparionine}}$ was simply the mean of available body mass estimates for hipparionine specimens derived from all localities studied here. For ruminants, the mean of

Table 16.2. Identifiable Elements per Individual Used in the Computation of Weighted Abundance of Elements (WAE)

Identifiable Element	Number of Identifiable Elements per Ruminant Individual	Number of Identifiable Elements per Hipparionine Individual
Proximal humerus	2	2
Distal humerus	2	2
Proximal radius	2	2
Distal radius	2	2
Proximal ulna	2	2
Proximal femur	2	2
Distal femur	2	2
Proximal tibia	2	2
Distal tibia	2	2
Proximal metacarpal	2	2
Distal metacarpal	2	2
Proximal metatarsal	2	2
Distal metatarsal	2	2
Proximal accessory metapodial	0	8
Distal accessory metapodial	0	8
Phalanx 1	8	4
Phalanx 2	8	4
Phalanx 3	8	4
Accessory phalanx 1	0	8
Accessory phalanx 2	0	8
Accessory phalanx 3	0	8
Astragalus	2	2
Calcaneum	2	2
Cuboid	0	2
Cuneiform	2	2
Ectocuneiform	0	2
Entocuneiform	0	2
Lunar	2	2
Magnum	2	2
Navicular	0	2
Pisiform	0	2
Scaphoid	2	2
Trapezoid	0	2
Trapezoideum	0	2
Triquetrum	0	2
Unciform	2	2
Centrotarsale	2	0
Ectomesocuneiform	2	0
Malleolar	2	0
Molar	12	12
Premolar	12	14
Incisor	6	12
Canine	2	4
horn core	2	0
Total (m_{ij})	104	150

available body mass estimates was first determined for each size category across all seven localities. These values were then weighted according to the relative representation of each size category across all seven localities to yield a value for $B_{ruminant}$. The values of B were used in conjunction with a simplifying model of abundance to derive estimates of hipparionine biomass at each locality.

Abundance Model

Using the fossil record to model population changes first requires that there be a consistent relationship between the fossil assemblage and the life assemblage. Some fraction f of the life assemblage is preserved in the fossil assemblage. In practice, a model must assume that f is constant across the taxa preserved at each assemblage under consideration. Selecting the most appropriate estimator of abundance for a given assemblage's taphonomy is designed to increase the chances that this assumption is valid. The assumption of constant f is expressed below as Assumption 1. The variables referred to here are defined in table 16.3.

Assumption 1

The first assumption is:

$$f_q = \frac{N_{pq}}{n_{pq}} = \frac{\sum_p N_{pq}}{\sum_p n_{pq}}$$

When this assumption is valid, then:

$$n_{pq} = \frac{N_{pq}}{f_q}$$

The second assumption allows for the comparison of different fossil assemblages through time.

Assumption 2

Some component of the life assemblage under consideration must be constant through time. Thus, in the case of two taxa, at least one of the following three statements is assumed to be true:

$$n_{1q} \text{ is constant for all assemblages } q$$

or

$$n_{2q} \text{ is constant for all assemblages } q$$

or

$$(n_{1q} + n_{2q}) \text{ is constant for all assemblages } q$$

The first two cases assume that changes in the population of one taxon do not affect the population of the other (i.e., taxon populations vary independently of each other). The third case assumes that the two taxa are in competition and there is a one-to-one correspondence such that an increase in the population of one taxon is balanced by an equal and opposite decrease in the population of the other. These assumptions are special cases of a more general statement:

$$(i \cdot n_{2q} + n_{1q}) \text{ is constant through time and between localities}$$

(i.e., for all assemblages q).

When the parameter $i = 0$, then this assumption implies that n_{1q} must be constant; when $i = 1$, then $(n_{1q} + n_{2q})$ must be constant. Therefore, the parameter i expresses the degree to which a change in the abundance of one taxon must affect abundance of the other taxon: i can thus be referred to as a coefficient of infringement.

The hypothetical case $i = 1$ assumes that an individual of one taxon is competitively equivalent to an individual of

Table 16.3. Explanation of Variables

Variable	Explanation
N	Abundance in a fossil assemblage, arbitrary units
n	Abundance in a life assemblage, arbitrary units
f	Fraction of a life assemblage preserved in a fossil assemblage, unitless
p	Taxon, in this case p = hipparionines (h) or ruminants (r)
q	Fossil locality or assemblage, in this case q = Loc. 64, 4, 122, 121, 72, 91, or 12
i	Coefficient of infringement, unitless
C	Constant, expressed in $m$kg/km^2
m	Constant expressing units of abundance defined by a reference assemblage—in this case Loc. 64
B	Estimated body mass of a taxon, kg
A	Estimated density of a fossil taxon, m/100 km^2 or $m$kg/100 km^2
r	Intrinsic rate of population growth, m/1000 ky-km^2
K	Carrying capacity, m/100 km^2

the other taxon. This is quite clearly unlikely to be true and therefore a third assumption is actually implied in this case.

Assumption 3

The population density of both taxa under consideration is expressed in units that are competitively equivalent. A starting point here is to use units of biomass. Thus, N_{ij} and n_{ij} need to be expressed in biomass units.

Given the assumptions discussed above, we suggest the following:

$$n_{1q} = \frac{N_{1q}}{f_q}$$

$$f_q = \frac{N_{1q}(i) + N_{2q}}{n_{1q}(i) + n_{2q}}$$

$$n_{1q} = \frac{N_{1q}}{N_{1q}(i) + N_{2q}} (n_{1q}(i) + n_{2q})$$

$$C = (n_{1q}(i) + n_{2q})$$

$$n_{1q} = \frac{N_{1q}}{N_{1q}(i) + N_{2q}} C$$

As noted, C is a constant according to assumptions two and three. We may express C in terms of units of some multiplier of kilograms per square kilometer. Let this multiplier be m and let $n_{11} + n_{21}$ be equal to $100m$ kg/km^2. Thus, C is defined in terms of the first assemblage in a time series and is equal to $100m$ kg/km^2 when $i = 0$ or 1 or when $n_{11} = 0$.

We applied this simplifying model to the case of ruminants and hipparionines from the Sinap Formation as follows. The preferred abundance estimate for ruminants and hipparionines at each of the seven Sinap localities used in this study was multiplied by the body mass parameter (B) for ruminants and hipparionines and the results were saved as N_{hq} and N_{rq}, respectively. The constant C was set equal to $100m$ kg/km^2 at Loc. 64. Loc. 64 is a pre-"*Hipparion*" locality and therefore, $N_{h,64} = 0$. Estimates for $n_{h,64}$ and $n_{r,64}$ were therefore 0 and $100m$ kg/km^2, respectively. The model presented here was evaluated for the conditions $i = 0$ and $i = 1$ and the values of N_{hq} and N_{rq} based on the preferred abundance estimate at each locality. Thus, the absolute abundance of hipparionines at the other localities was estimated and expressed relative to the biomass of ruminants represented at Loc. 64, which is set equal to $100m$ kg/km^2. Dividing these biomass estimates by $B_{ruminant}$ and $B_{hipparionine}$, respectively, yields abundance estimates in units of $100m$/km^2.

These abundance estimates were in turn entered into a NLIN routine in SAS and paired with age estimates from Kappelman et al. (chapter 2, this volume) for the seven localities under consideration. The NLIN procedure was used to fit a sigmoid curve expressed by the Lotka-Volterra equation (Wilson and Bossert 1971) to the data. The results included asymptotic 95% confidence levels on r, the intrinsic rate of population growth, and K, the carrying capacity for hipparionines.

Results

Loc. 64

Loc. 64 is dated to ~10.77 Ma (Kappelman et al., chapter 2, this volume). The fossil assemblage is clearly dominated by bovids. Giraffids are also indicated and a smaller member of the hyaenid genus *Ictitherium*.

A total of 175 cataloged specimens were collected from Loc. 64. Of these, 21 were cataloged as indeterminate. One hundred thirty-two cataloged specimens were identifiable as ruminants and made up 150 identifiable specimens. Thirty-five identifiable specimens were placed in the category Very Small, 73 were Small, 16 were Medium, and 2 were Very Large (fig. 16.1A). Twelve specimens were identified as juvenile. The MNI of adult specimens is 6 and that of juveniles is 2. When MNI is calculated first for each size and age category and then summed, the total MNI-G is 11 (table 16.4). Teeth, phalanges, and podials, all of which are dense and have a high preservation potential, are the most common element types at Loc. 64 (fig. 16.2A). The Loc. 64 ruminant assemblage had a relatively high degree of fragmentation, with a fragmentation index of 57 (table 16.5). The degree of association was correspondingly low, with the possibility of only a few associated teeth (table 16.5).

Body mass estimates for Loc. 64 specimens were possible for 13 metapodials and tibiae (table 16.6). All but one of these estimates was <30 kg and most of these were <15 kg. This contrasts with the size category classifications of all Loc. 64 specimens, in which most specimens were classified as Small, corresponding to a body weight of ~15–30 kg. One specimen, AS92/581, a tibia, yielded a body mass estimate of 103 kg. Thus, larger bovids were present but probably rare or unpreserved due to a taphonomic bias.

Loc. 64 represents a small-bovid dominated assemblage that was probably accumulated after some degree of transport, leading to fragmentation and disassociation.

Loc. 4

Loc. 4 includes the earliest evidence for hipparionines in the Sinap Formation and dates to ~10.69 Ma (Kappelman et al., chapter 2, this volume). Like that of Loc. 64, the Loc. 4 assemblage is dominated by bovids. It also includes, in addition to hipparionines, giraffids, suids, rhinocerotids, felids, hyaenids, percrocutids, rodents, birds, and reptiles. There are 369 cataloged specimens from Loc. 4, and these yielded a total of 172 specimens that were identifiable as either ruminant or hipparionine. Of these, 169 were ruminants and three were hipparionines.

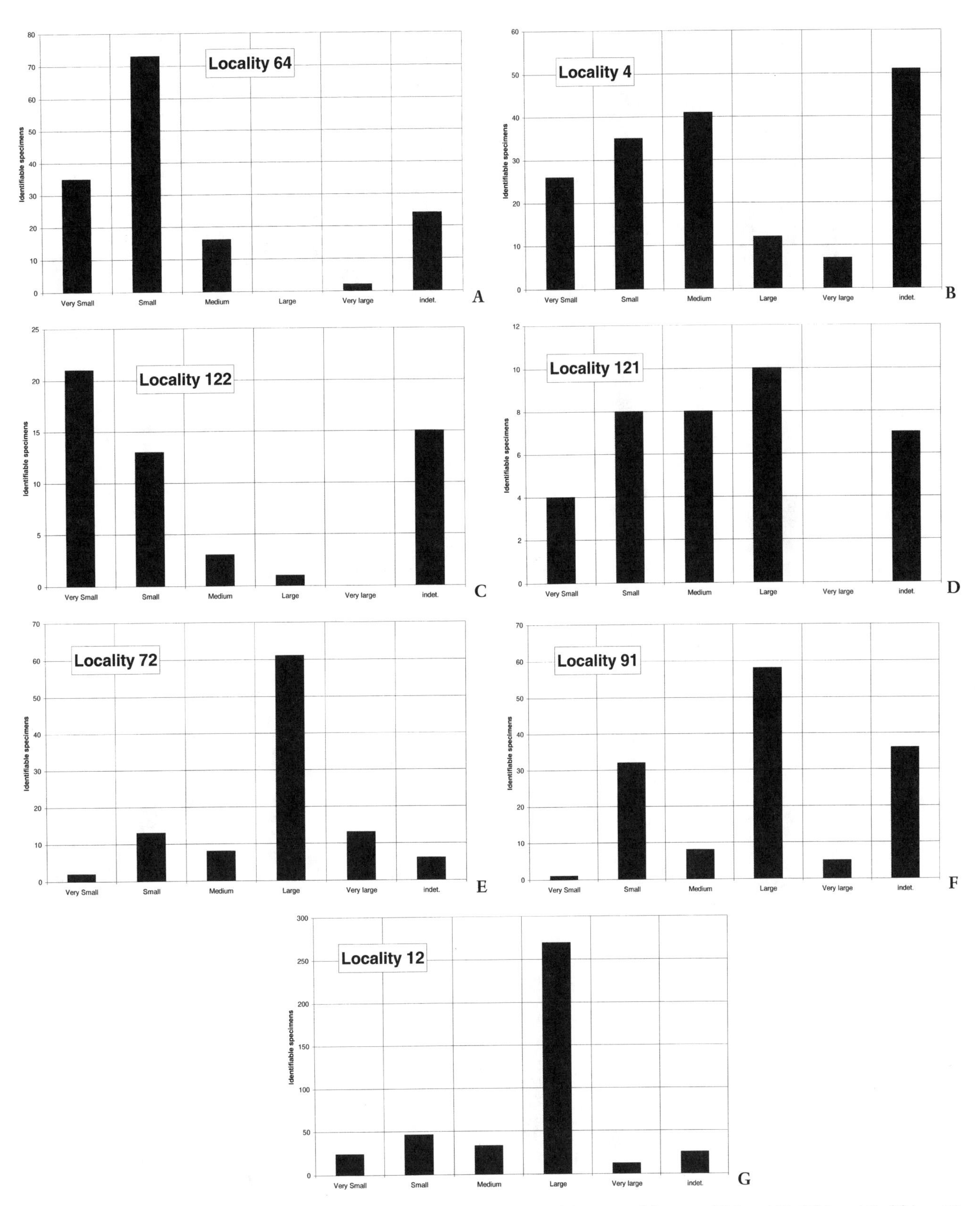

Figure 16.1. Frequency distribution of identifiable specimens by size category. (A) Loc. 64. (B) Loc. 4. (C) Loc. 122. (D) Loc. 121. (E) Loc. 72. (F) Loc. 91. (G) Loc. 12. Size categories are defined in table 16.1.

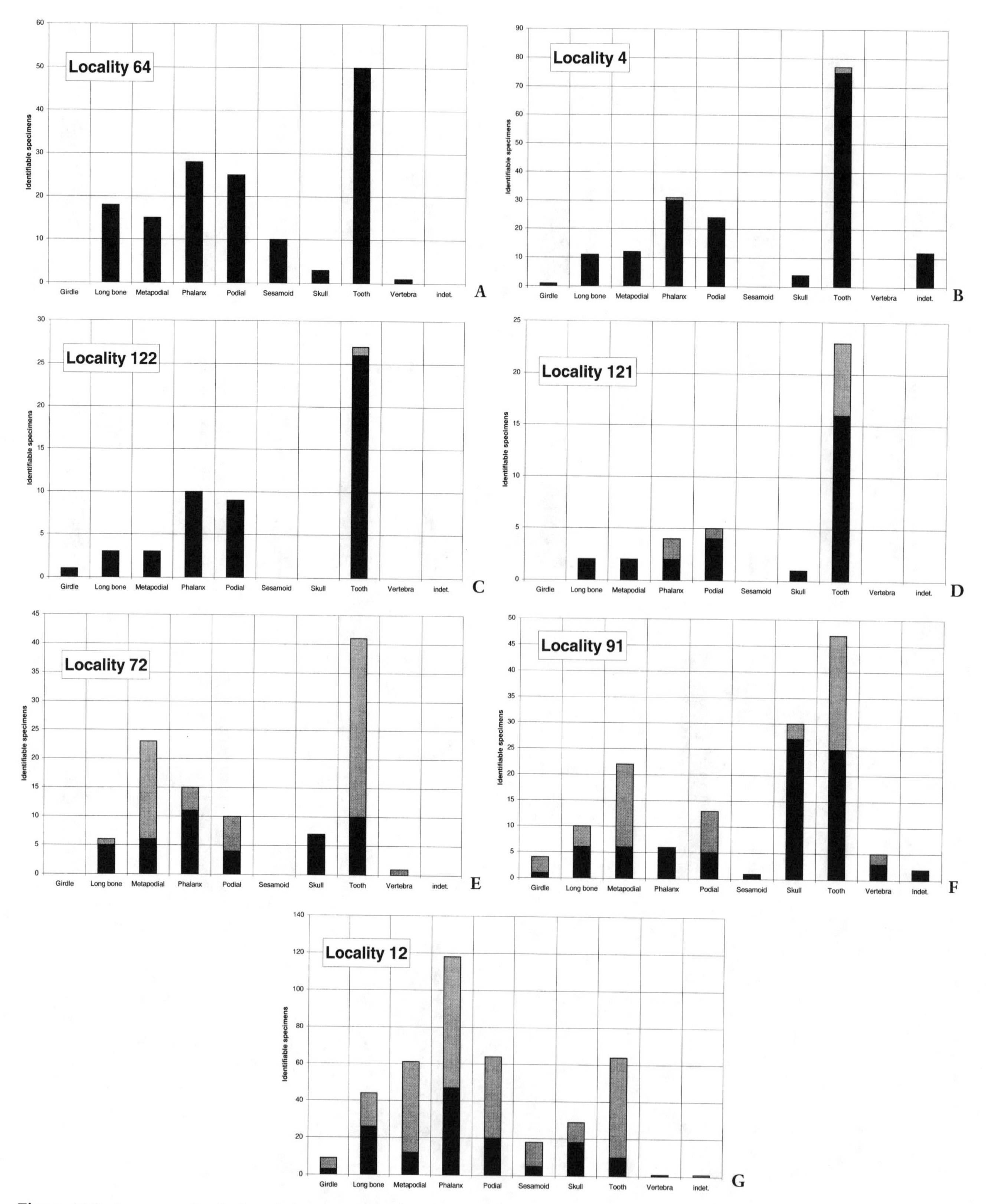

Figure 16.2. Frequency distribution of element types. **(A)** Loc. 64. **(B)** Loc. 4. **(C)** Loc. 122. **(D)** Loc. 121. **(E)** Loc. 72. **(F)** Loc. 91. **(G)** Loc. 12. Black denotes ruminants and grey denotes hipparionines.

Table 16.4. Raw Measures of Hipparionine and Ruminant Abundance by Locality

Locality	OTU	NISP	WAE	MNI	MNI-G
64	Ruminant	150	1.337	6	11
64	*Hipparionine*	*0*	*0.000*	*0*	*0*
4	Ruminant	169	1.404	4	9
4	*Hipparionine*	*3*	*0.013*	*1*	*1*
122	Ruminant	52	0.481	2	5
122	*Hipparionine*	*1*	*0.007*	*1*	*1*
121	Ruminant	27	0.260	1	4
121	*Hipparionine*	*10*	*0.067*	*1*	*2*
72	Ruminant	43	0.423	2	7
72	*Hipparionine*	*60*	*0.340*	*2*	*3*
91	Ruminant	82	0.760	11	6
91	*Hipparionine*	*58*	*0.493*	*3*	*4*
12	Ruminant	142	1.365	6	11
12	*Hipparionine*	*267*	*1.767*	*5*	*8*

Of the specimens identifiable as ruminants, 26 were categorized as Very Small, 35 as Small, 41 as Medium, nine as Large, and seven as Very Large (fig. 16.1B). Fourteen ruminant specimens were identified as juvenile. The MNI for adult ruminants is 4 and for juveniles is 1. The total MNI-G for ruminants is 9 (table 16.4).

The three specimens identifed as hipparionine are a deciduous premolar, a buccal tooth fragment (possibly of a milk tooth), and an accessory proximal first phalanx missing the epiphysis. These specimens suggest an MNI and an MNI-G of one juvenile.

The fragmentation and association indices for Loc. 4 were high and low, respectively (table 16.5), comparable to those for Loc. 64. Like Loc. 64, some degree of postmortem transport and breakage is probable for Loc. 4.

Body mass estimates for Loc. 4 specimens were possible for 12 ruminant specimens and range between 8 and 50 kg (table 16.6).

Locs. 122 and 121

Locs. 122 and 121 date to ~10.58 and ~10.53 Ma, respectively, and are separated by about 5 m. Both localities are just above the local "*Hipparion*" Datum and document an early increase in hipparionine abundance. The localities are situated on the same slope on the northeastern side of Sinap Tepe. *Hipparion* is present at both localities. However, in the case of the lower locality, Loc. 122, the only *Hipparion* specimen is AS95/684, a deciduous premolar and some associated tooth fragments. This specimen is a surface find and quite possibly may originally derive from Loc. 121, which lies directly upslope from Loc. 122. Thus, hipparionines appear to have been rare or absent at Loc. 122, but by the time of Loc. 121, they appear to have been more common. Indeed, Locs. 121 and 122 may record a period of rapid increase in hipparionine abundance.

The sample sizes at Locs. 122 and 121 are small, 123 and 57 cataloged specimens, respectively. Very Small, Small, and Medium bovids are the main groups represented in both assemblages (fig. 16.1C,D). Giraffids appear to be absent from both localities, possibly as a consequence of small sample size. Ten hipparionine specimens are identified from Loc. 121 compared with only one from the better-sampled Loc. 122 (table 16.4). As in Loc. 64, both localities are highly fragmented, showing a low degree of association,

Table 16.5. Fragmentation and Association Indices for Ruminant and Hipparionine Assemblages by Locality

Locality	OTU	Association Index	Fragmentation Index
64	Ruminant	Very low	57
64	*Hipparionine*	*Not present*	*Not present*
4	Ruminant	3	55
4	*Hipparionine*	*Three specimens only*	*Three specimens only*
122	Ruminant	4	73
122	*Hipparionine*	*One specimen only*	*One specimen only*
121	Ruminant	0	70
121	*Hipparionine*	*Very low*	*Mostly fragmentary*
72	Ruminant	0	79
72	*Hipparionine*	*0*	*52*
72	**Both**	**0**	**62**
91	Ruminant	26	54
91	*Hipparionine*	*9*	*62*
91	**Both**	**19**	**57**
12	Ruminant	27	47
12	*Hipparionine*	*45*	*37*
12	**Both**	**39**	**40**

Table 16.6. Estimated Body Weights of Ruminants from Selected Localities of the Sinap Formation

Specimen Number	Locality	Element	Estimated Body Mass (SD)
ST90-279	64	Tibia	8 (0.46)
ST90-268	64	Metapodial	13 (2.4)
ST90-269	64	Metapodial	20 (3.0)
ST91-274A	64	Metapodial	10 (2.0)
ST91-274B	64	Metapodial	10 (2.0)
ST91-488	64	Metapodial	7 (1.7)
ST91-489	64	Metapodial	8 (1.9)
ST92-1	64	Metapodial	22 (3.1)
ST92-11	64	Metapodial	25 (3.3)
ST92-20	64	Metapodial	12 (2.3)
ST92-28	64	Tibia	11 (0.21)
ST92-581	64	Tibia	103 (15)
ST93-429	64	Metapodial	13 (2.4)
ST94-753	4	Humerus	8 (2.4)
ST93-455	4	Metapodial	11 (2.2)
ST93-465	4	Metapodial	12 (2.3)
ST94-755	4	Radius	12
ST93-425	4	Humerus	13 (4.8)
ST94-768	4	Metapodial	15 (2.6)
ST94-25A	4	Metapodial	15 (2.6)
ST93-322	4	Metapodial	19 (2.9)
ST94-1	4	Metapodial	22 (3.1)
ST92-382	4	Radius	32
ST95-888	122	Tibia	8 (0.67)
ST95-649	121	Radius	35 (2.3)
ST95-650	121	Humerus	61 (6.6)
ST95-655A	121	Metatarsal	24 (15)
ST92-428	91	Metacarpal	24 (3.2)
ST92-448	91	Metapodial	19 (2.9)
ST93-10C	91	Tibia	27 (5.7)
ST93-10D	91	Metatarsal	26 (4.0)
ST93-13	91	Metapodial	34 (2.9)
ST93-2	91	Humerus	720[1] (1.5)
ST93-20	91	Metapodial	620[1] (146)
ST93-42A	91	Humerus	20 (5.2)
ST93-42B	91	Radius	22 (1.5)
ST93-838	91	Radius	649[1]
ST89-291	12	Radius	26 (2.6)
ST92-587	12	Humerus	22 (2.0)
ST93-913	12	Metatarsal	59 (26)
			144
ST95-1202	12	Tibia	44 (12)
ST89-290	12	Humerus	27 (8.1)
ST92-594	12	Metapodial	63 (3.4)
ST93-1181A	12	Metatarsal	29 (6.6)
ST93-1200	12	Radius	9
ST93-842	12	Humerus	217 (76)
ST93-903A	12	Tibia	48 (1.2)
ST95-163	12	Humerus	22 (3.5)

(continued)

Table 16.6. Estimated Body Weights of Ruminants from Selected Localities of the Sinap Formation *(continued)*

Specimen Number	Locality	Element	Estimated Body Mass (SD)
ST95-165	12	Metacarpal	32 (4.1)
			55
ST95-169/181	12	Radius	23 (2.2)
ST95-261	12	Metacarpal	15 (3.3)
			34
ST95-276A	12	Metapodial	12 (2.0)
ST95-311	12	Humerus	8 (1.9)
ST95-329	12	Metatarsal	40 (0.69)
ST95-337	12	Radius	42 (42)
			49
ST95-401	12	Radius	23 (0.33)
ST95-421	12	Metacarpal	19 (0.02)
ST95-574	12	Metatarsal	28 (2.3)
ST95-583	12	Radius	13
ST95-653	12	Radius	14 (0.66)
			17
ST95-663	12	Radius	748[1]

Notes: Estimated body mass is the mean of body mass predictions from nonlength measurements using the bovid regressions of Scott (1990). Standard deviations (SD) are shown when estimated body mass is based on more than one nonlength measurement. Body mass predicted by bone length is shown below the body mass estimate when available.

[1]Estimates based on the ruminant regressions of Scott (1990).

(table 16.5) and are dominated by denser element types (teeth, phalanges, and podials) (fig. 16.2C,D).

Some body mass estimates were possible for specimens from Locs. 122 and 121 (see table 16.6). Three specimens from Loc.121 yield estimates of 61, 35, and 21 kg. The only body size estimate available from Loc. 122 is 8 kg. These estimates confirm the importance of smaller body sizes at both localities (fig. 16.1D), although the increase in hipparionines at Loc. 121 adds a number of Large individuals to the size distribution at the site.

Loc. 72

Loc. 72 dates to ~10.08 Ma (Kappelman et al., chapter 2, this volume). The fossil assemblage includes hipparionines, bovids, giraffids, suids, hyaenids, rhinocerotids, orycteropodids, and chelonians. Hipparionines are more common here than in earlier localities, with relative abundance estimates ranging from 30.0% (MNI-G) to 58.2% (NISP) (table 16.7).

There are 183 cataloged specimens from Loc. 72, yielding a total of 43 and 60 identifiable specimens for ruminants and hipparionines, respectively. Ruminants were assigned to size categories as follows: Very Small (two), Small (13), Medium (eight), Large (one), and Very Large (13) (fig. 16.1E).

Six hipparionine specimens were juveniles, yielding an MNI of 1 for juveniles. The MNI for adult hipparionines is 2, and the MNI-G for all hipparionines is 3. Ruminants had an MNI of 2 for adults and 1 for juveniles and a total MNI-G of 7 (table 16.4).

Body mass estimates were possible for two hipparionine metapodials from Loc. 72: 172 kg and 234 kg (table 16.8).

Fragmentation was high for both ruminants and hipparionines at Loc. 72 and no associated specimens were identified among the ruminant and hipparionine specimens (table 16.5). The taphonomy of Loc. 72 appears similar to that for the earlier localities.

Loc. 91

Loc. 91, located on the east side of Sinap Tepe, is dated to ~9.98 Ma (Kappelman et al., chapter 2, this volume), and is only slightly older than Loc. 12. The Loc. 91 fossil assemblage

Table 16.7. Relative Abundance of Sinap Hipparionines by Locality

	Relative Abundance (%)			
Locality	NISP	WAE	MNI	MNI-G
64	0.0	0.0	0.0	0.0
4	1.7	0.9	20.0	10.0
122	1.9	1.4	33.3	16.7
121	27.0	20.4	50.0	33.3
72	58.2	44.6	50.0	30.0
91	41.4	39.4	21.4	40.0
12	65.3	56.4	45.5	42.1

Table 16.8. Estimated Body Weights of Hipparionines from Selected Localities of the Sinap Formation

Specimen Number	Locality	Element	Estimated Body Mass (SD)
ST93-289	72	Metacarpal	234 (47)
ST93-509	72	Metatarsal	172
ST92-419	91	Humerus	222 (24) 322
ST92-426	91	Radius	199 (36)
ST93-9	91	Metacarpal	182 (49) 283
ST93-11	91	Metatarsal	171
ST93-27	91	Femur	262
ST93-52	91	Metatarsal	210 (33) 262
ST93-56	91	Metacarpal	170 (35)
ST93-6	91	Metatarsal	183 (4)
ST93-7	91	Metatarsal	221 (79) 262
ST91-373	12	Metatarsal	160 (15) 240
ST95-422	12	Tibia	295 (46)
ST91-367	12	Metatarsal	180 (8)
ST91-419	12	Femur	227 (40)
ST91-420	12	Metacarpal	139 (9) 240
ST91-780	12	Metatarsal	206 (16) 276
ST93-1185A	12	Tibia	176 (19)
ST93-1193A	12	Metatarsal	164 (25) 269
ST93-1207A	12	Metatarsal	153 (15) 262
ST93-1213A	12	Metatarsal	219 (66) 279
ST93-1213A	12	Tibia	200 (27)
ST93-840A	12	Metacarpal	205 (49) 261
ST93-841A	12	Radius	184 (22) 186, 170[1]
ST93-860B,A	12	Humerus	176
ST93-860B,A	12	Radius	163 (31)
ST95-1054A	12	Metatarsal	291 (62) 322
ST95-131A	12	Metacarpal	154 268
ST95-149	12	Metacarpal	175 (36)
ST95-161	12	Metatarsal	175 (36)
ST95-418	12	Radius	160 (50)
ST95-513	12	Metacarpal	175 (28) 283
ST95-513	12	Radius	165
ST95-514-63	12	Metacarpal	187 272
ST95-603	12	Humerus	187 (6)
ST95-606	12	Humerus	194
ST95-615	12	Metatarsal	147 (20) 276

Notes: Estimated body mass is the mean of body mass predictions from nonlength measurements using the equid regressions of Scott (1990). Standard deviations (SD) are shown when estimated body mass is based on more than one nonlength measurement. Body mass predicted by bone length is shown below the body mass estimate when available.

[1]Estimate based on length including the ulna (measurement U1 of Scott 1990).

includes hipparionines, bovids, giraffids, hyaenids, rhinocerotids, and chelonians. Hipparionine relative abundance is similar to that determined for Loc. 72, with relative abundance estimates ranging from 21% (MNI) to 41% (NISP) (table 16.7).

There are 139 cataloged specimens from Loc. 91. These yielded 140 specimens that were identifiable as either ruminant or hipparionine. Of these, 82 were ruminants and 58 were hipparionines.

Ruminants belonging to the Very Small size category appear to have been rare; only one specimen classified as Very Small. The remainder of the ruminant specimens were either Small (32), Medium (eight), Very Large (five), or Size Indeterminate (36) (fig. 16.1F). The large number of Size Indeterminate specimens was probably due to the number of horn cores that were not possible to categorize according to relative size. Teeth and skull elements (mostly horn cores and partial mandibles) were the most commonly preserved ruminant elements (fig. 16.2F).

Of the 58 hipparionine specimens, one was classified as juvenile. Metapodials and teeth were the most common elements in the Loc. 91 hipparionine assemblage (fig. 16.2F). The abundance of hipparionine metapodials is unsurprising, given that the third metapodial is dense and large and hence more likely to be preserved than the smaller, more fragile bovid metapodials. The presence of accessory metapodials in hipparionines may also inflate their abundance relative to ruminants. Most surprising at Loc. 91 is the complete absence of hipparionine phalanges.

For the purposes of assessing degree of fragmentation and association (table 16.5), the Loc. 91 fossils were divided into ruminant and hipparionine assemblages. This was done for two reasons: (1) because ruminants and hipparionines have a different number of bones and elements with dissimilar physical attributes, the two groups will have different preservation potentials; and (2) ruminants and hipparionines may have exploited different habitats, possibly influencing their deposition after death. The first of these two factors is taphonomic whereas the second is paleoecological and taphonomic. Regarding the second factor, Shotwell (1955, 1958) speculated that multiple biotic communities might be preserved at a single site of deposition and be represented differentially based on their distance from the depositional setting. Thus, Shotwell divided fossil assemblages into distal and proximal communities. Fragmentation and association are relevant to this question, because differential fragmentation and association across taxa suggest degree of postmortem transport and hence membership in a proximal or distal community.

Fragmentation was greater and association less among the 58 hipparionine fossils identified from Loc. 91 than among the 82 identifiable specimens in the ruminant assemblage from Loc. 91 (table 16.5). The fragmentation index of hipparionines from Loc. 91 was 62, as opposed to an index of 54 for the ruminants. Even greater was the contrast in degree of association: the association index for the hipparionines was 9 compared with 26 for the ruminant specimens.

Figure 16.1F shows the distribution by body size category for Loc. 91. None of the Loc. 91 ruminant specimens yielded a body mass estimate <15 kg, and two Very Large specimens (presumed giraffids) exceeded 600 kg (table 16.6). The hipparionines from Loc. 91 range between 165 and 285 kg (table 16.8).

Loc. 12

Loc.12 (~9.6 Ma), located on the top of Delicayincak Tepe to the east of Sinap Tepe, is the best sampled and most diverse fauna of the localities we studied and includes hipparionines, bovids, giraffids, rhinocerotids, proboscideans, carnivorans, orycteropodids, and chelonians. The hipparionines, bovids, and rhinocerotids (in that order) dominate the assemblage in terms of number of specimens. Relative to the ruminants, hipparionine abundance at Loc. 12 ranges from 42% (MNI) to 65% (NISP) (table 16.7). Even by the lowest estimate (42%), hipparionines have increased to high levels of abundance in the approximately 1 m.y. between their first appearance and Loc. 12 times.

Most of the specimens from Loc. 12 were excavated during the 1995 field season and came from two distinct bone pockets. Trenches between these bone pockets revealed a low to zero bone density. Within these pockets, bone density was high and many associated elements were impressively preserved. For example, entire distal limbs of hipparionines were preserved intact and still articulated: even tiny sesamoids were preserved in articulation.

The taphonomy of Loc. 12 appears unique. The hipparionine assemblage has a high association index of 45 and a low fragmentation index of 37. The degree of association among the ruminant fossils is also high (association index of 27), comparable to that of the Loc. 91 ruminants. Similarly, fragmentation is also low (47) for the Loc. 12 ruminants, but is not as low as that for the Loc. 12 hipparionines (table 16.5).

Loc. 12 is an assemblage characterized by an abundance of hipparionines, a diversity of taxa, and by high levels of association and low levels of fragmentation, suggesting minimum postmortem transport.

Body mass estimates for a number of ruminants (table 16.6) and hipparionines (table 16.8) were derived for Loc. 12 specimens. Ruminants spanned all five size categories (fig. 16.1G) from Very Small (<15 kg) to one Very Large individual (presumably a giraffid) with an estimated body mass of 748 kg. One bovid specimen classified as Large yields a body mass estimate of 217 (standard deviation, 76) kg. This specimen establishes the presence of large bovids at Loc. 12; the large standard deviation suggests complex scaling relationships among different postcranial dimensions.

The body mass distribution of hipparionines from Loc. 12 ranges from <150 kg to nearly 300 kg (table 16.8). On the basis of body size alone, it is not unreasonable to argue that multiple hipparionine taxa were present at Loc. 12 (see Bernor et al., chapter 11, this volume).

Discussion

The oldest localities in the Sinap Formation are dominated mainly by small bovids, whereas subsequent localities record the first appearance and subsequent increase in abundance of the three-toed horse, *"Hipparion."* Estimates of relative abundance of hipparionines vary, depending on the specific measure used. For example, at Loc. 12, NISP gives a relative abundance estimate of 65% for hipparionines; in contrast, MNI-G yields an estimate of 42%. The WAE estimate suggests an intermediate abundance of 56%. Based on these varying estimates, multiple interpretations are possible for the Loc. 12 fauna. Hipparionines may be either the dominant basal herbivore (according to NISP) or a common and significant basal herbivore still outnumbered by ruminants (according to MNI and MNI-G). In either case, hipparionines are a critical element of the fauna, but the choice of relative abundance measure has implications for the precise role of hipparionines as well as for interpretations of faunal change.

This sensitivity of interpretation to the choice of abundance measure is particularly evident when Loc. 12 is compared to Loc. 91. According to NISP, hipparionine relative abundance increases from 41% to 65% over the short stratigraphic interval from Loc. 91 to Loc. 12. A very rapid increase in hipparionine numbers is implied by NISP. In contrast, according to MNI-G, hipparionine abundance remains stable from Loc. 91 to Loc. 12 (table 16.7). The choice of relative abundance measurement is essential to understanding faunal change and our study illustrates the importance of exercising caution when estimating relative abundance. We recommend the use of multiple measures of relative abundance.

In cases where different relative abundance measures are in significant disagreement, it is necessary to select the best measure of relative abundance based on sample size and taphonomic data. For example, in the cases of Locs. 121 and 122, the fossil sample size is small and the MNI and MNI-G estimates clearly appear to overestimate the abundance of hipparionines. NISP and WAE estimates are probably preferable. Association of specimens is low and there is no reason to argue for the higher hipparionine abundances indicated by MNI and MNI-G.

The situation is very different at Loc. 12, where high levels of association suggest that MNI or MNI-G may be the most appropriate estimate of relative abundance. Thus, based on taphonomic information such as the number of associated specimens and considerations of sample size, it is possible to suggest a probable best estimate of relative abundance drawn from among the four measurements of relative abundance calculated here.

We selected the best estimators of hipparionine and ruminant relative abundance from among NISP, WAE, MNI, and MNI-G for the localities included in our study. In the cases of Locs. 4, 122, and 121, NISP is likely the best estimate of relative abundance. No evidence of associated specimens at these localities argues against the use of either MNI or MNI-G. Instead, the small sample sizes at Locs. 121 and 122 suggest that NISP is likely the best estimate at these localities. Loc. 4 is probably best represented by either NISP or WAE, with NISP preferred for sake of consistency. At Loc. 72, the low rate of associated specimens disallows the use of MNI or MNI-G as a preferred estimate, but the very high estimate (58%) given by NISP compared with the other estimates suggests that WAE may be a better estimator for this locality. The larger number of hipparionine elements with a likelihood of being preserved could be responsible for an inflated estimate of abundance if NISP were used. The higher rates of associated specimens at Locs. 91 and 12 suggest that MNI-G may be the best estimate in these cases. Thus, the best estimators of relative abundance appear to be NISP at Locs. 4, 122, and 121; WAE at Loc. 72; and MNI-G at Locs. 91 and 12 (table 16.9).

Using the probable best estimates of relative hipparionine abundance, it is possible to assess the rate of increase in hipparionine abundance following the *"Hipparion"* Datum. We applied the abundance model described in the Methods section to the probable best estimates of relative abundance and evaluated for the conditions $i = 0$ and $i = 1$. These results are shown in table 16.9. The condition $i = 0$ corresponds to the hypothetical case in which ruminants and hipparionines do not compete. The results in table 16.9 suggest that biomass would have increased over 250% if hipparionine immigrants did not compete with established ruminant taxa. It would seem unlikely that such a change would occur in the absence of a major environmental change. Alternatively ($i >> 0$ or $i = 1$), ungulate herbivore biomass may have remained relatively constant, with the increase in hipparionines coming at the expense of ruminant taxa. The apparent increase in hipparionines reflected at the localities studied here would be due in part to a decrease in ruminant numbers. Immigration of a hipparionine species that was either competitively superior to Old World bovid or ruminant taxa or preadapted ("exapted" in Gould and Vrba 1982) to concurrent environmental change is then implied.

The density estimates for the case $i = 1$ (shown in shaded cells of table 16.9) were entered into a NLIN routine in SAS with age estimates for the seven localities under consideration from (Kappelman et al., chapter 2, this volume). As mentioned earlier, the NLIN procedure was used to fit a sigmoid curve expressed by the Lotka-Volterra equation (Wilson and Bossert 1971) to the data. The results included asymptotic 95% confidence levels on r, the intrinsic rate of population growth, and K, the carrying capacity for hipparionines. The results are shown in figure 16.3. The density estimates of abundance for each locality are plotted alongside their best fit Lotka-Volterra curve. The increase in hipparionine numbers recorded at the Sinap Formation clearly fits a sigmoidal pattern, indicating a good fit between the Lotka-Volterra model and the late Miocene *"Hipparion"* Datum data from the Sinap Formation. The theoretical predictions of the Lotka-Volterra model would thus appear applicable to the *"Hipparion"* Datum natural experiment.

A nonsigmoidal (and gradual) pattern of increase for hipparionines would imply a slow process of adaptation

Table 16.9. Estimated Absolute Abundances Calculated for Two Hypothetical Cases

Locality	Probable Best Estimate	$N_{hipparionine}$	$N_{ruminant}$	$N_{hipparionine}$ (kg)	$N_{ruminant}$ (kg)	Coefficient of Infringement (i)	$A_{hipparionine}$ ($m$kg/km)	$A_{ruminant}$ ($m$kg/km)	$N_{hipparionine}$ (m/100km)	$N_{ruminant}$ (m/100km)
64	NISP	0	150	0.0	12680.7	1	0.0	100.0	0.00	118.29
4	NISP	3	169	573.5	14286.9	1	3.9	96.1	2.02	113.72
122	NISP	1	52	191.2	4396.0	1	4.2	95.8	2.18	113.36
121	NISP	10	27	1911.8	2282.5	1	45.6	54.4	23.84	64.37
72	WAE	0.340	0.423	65.0	35.8	1	64.5	35.5	33.74	41.98
91	MNI-G	4	6	764.7	507.2	1	60.1	39.9	31.45	47.17
12	MNI-G	8	11	1529.4	929.9	1	62.2	37.8	32.53	44.73
64	NISP	0	150	0.0	12680.7	0	0.0	100	0.00	118.29
4	NISP	3	169	573.5	14286.9	0	4.0	100	2.10	118.29
122	NISP	1	52	191.2	4396.0	0	4.3	100	2.27	118.29
121	NISP	10	27	1911.8	2282.5	0	83.8	100	43.81	118.29
72	WAE	0.340	0.423	65.0	35.8	0	181.8	100	95.08	118.29
91	MNI-G	4	6	764.7	507.2	0	150.8	100	78.86	118.29
12	MNI-G	8	11	1529.4	929.9	0	164.5	100	86.03	118.29

Notes: Hypothetical cases are coefficient of infringement (i) = 1 or = 0. C = 100 *m*kg/km; body mass (*hipparionine*) = 191.2 kg; body mass (ruminant) = 84.5 kg. See table 16.3 for explanation of variables.

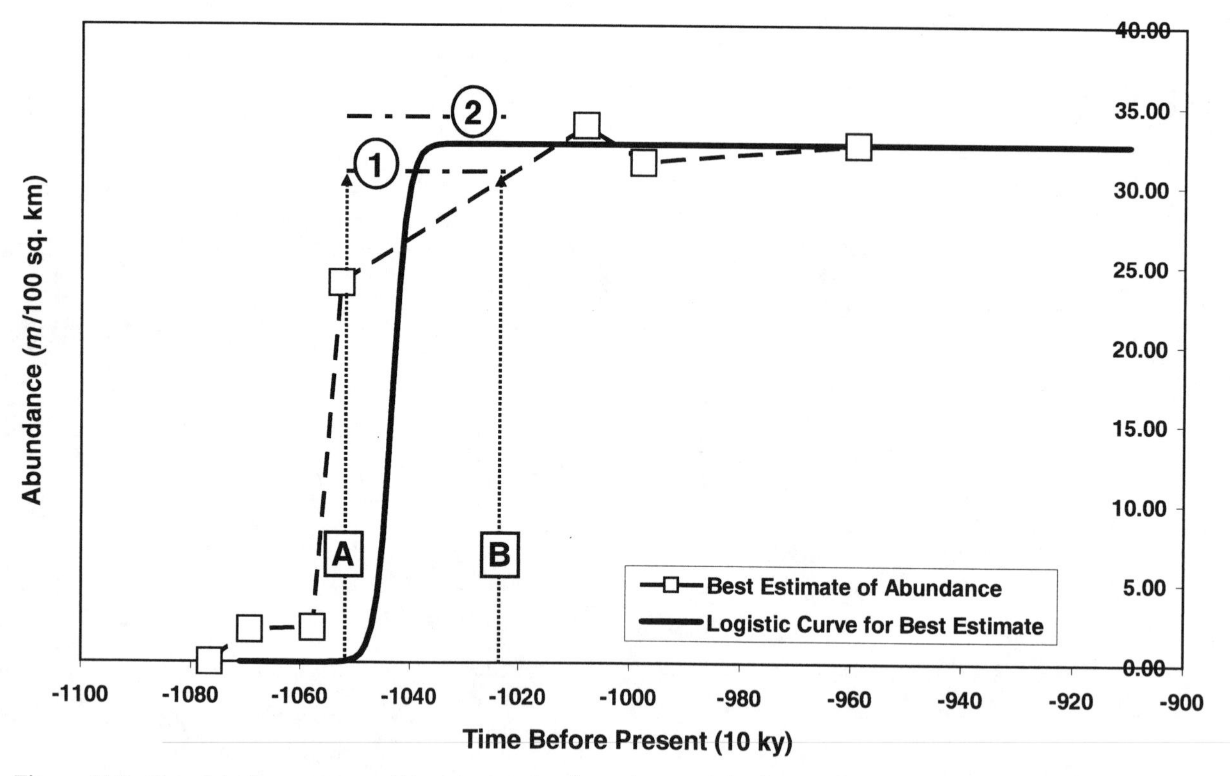

Figure 16.3. Plot of the best estimate of hipparionine abundance from each locality expressed in the units $m/100$ km² using the best-fit logistic growth curve. Symbols: [A] marks the time at which $N = K/2$ for the upper asymptotic 95% confidence limit for r and the best-fit estimate for b ($r = 1.0356$ $m/1000$ ky-km²; $t(K/2) = 10.52$ Ma); [B] marks the time at which $N = K/2$ for the lower asymptotic 95% confidence limit for r and the best-fit estimate for b ($r = 0.4195$ $m/1000$ ky-km²; $t(K/2) = 10.23$ Ma); (1) indicates the lower asymptotic 95% confidence limit for K ($K = 30.9$ $m/100$ km²); (2) indicates the upper asymptotic 95% confidence limit for K ($K = 34.3$ $m/100$ km²).

after immigration from the New World. Instead, the rapid and apparently sigmoidal increase of hipparionines after immigration suggests either open ecospace rapidly occupied by immigrant hipparionines and an accompanying explosive increase in basal herbivore biomass or the entrance of a competitively superior basal herbivore. It is worth noting that the peak of apparent hipparionine density is reached by ~10 Ma at Loc. 72. It is only after this time that the apparent diversification of *"Hipparion"* in Turkey ensues (see Bernor et al., chapter 11, this volume). Together, these events—the rapid increase to carrying capacity followed by a pulse of diversification—support a model of hipparionine diversification driven by competition and leading to niche separation and speciation.

Conclusions

An evaluation of the degrees of fragmentation and association of fossil specimens from seven localities in the Sinap Formation suggests preferred methods for estimating how taxonomic abundance varies between localities. An approach using a simplifying model allows estimates of hipparionine abundance relative to ruminants to be expressed in units of biomass and density. These estimates permit initial tests (using a nonlinear curve fitting procedure) of the patterns of increase in hipparionine numbers following their immigration from the New World. The results presented here suggest:

1. The increase of equids following the Old World *"Hipparion"* Datum fits a sigmoidal pattern (such as the Lotka-Volterra model);
2. The carrying capacity of immigrant equids depends on the degree to which *"Hipparion"* competed with the established ruminant dominated fauna at Sinap; and
3. If competition between ruminants and equids approached zero, large herbivore biomass at Sinap would have had to increase by >250% to accommodate the increase in equid abundance suggested by the data.

Thus we present for future study the hypothesis that hipparionine diversification was driven by competition leading to niche partitioning following a period of rapid population increase to a high carrying capacity.

Acknowledgments

We are grateful to Ilhan Temizsoy, director of the Anatolian Civilizations Museum, and Berna Alpagut, professor of paleoanthropology, Ankara University, for permissions to work with specimens from the Sinap Formation. Much appreciation goes to Trish Bennett from the Statistical Services Group at the University of Texas at Austin for her help fitting nonlinear curves using SAS Proc NLIN. We thank Derek Johnson for his work on and help with the Sinap catalog. We also express our gratitude to Ray

Bernor, John Kappelman, Ernie Lundelius, and Deborah Overdorff for comments on earlier drafts of this manuscript. This work was funded with grants from the University of Texas Fellowship (R.S.), the Academy of Finland to (M.F.), and the U.S. National Science Foundation (EAR 9304302) (J.K.).

Literature Cited

Badgley, C., 1986, Counting individuals in mammalian fossil assemblages from fluvial environments: Palaios, v. 1, pp. 328–338.

Brain, C. K., 1981, The hunters or the hunted? An introduction to African cave taphonomy: Chicago, University of Chicago Press, 365 pp.

Damuth, J., and B. J. MacFadden, 1990, Body size and its estimation, *in* J. Damuth, and B. J. MacFadden, eds., Body size in mammalian paleobiology: Estimation and biological implications: Cambridge, Cambridge University Press, pp. 1–10.

Gilbert, A. S., B. H. Singer, and D. Perkins, 1981, Quantification experiments on computer-simulated faunal collections: Ossa, v. 8, pp. 79–84.

Gould S. J., and E. S. Vrba, 1982, Exaptation—a missing term in the science of form: Paleobiology, v. 8, no. 1, pp. 4–15.

Grayson, D. K., 1978, Reconstructing mammalian communities: A discussion of Shotwell's method of paleoecological analysis: Paleobiology, v. 4, pp. 77–81.

Hanson, C. B., 1980, Fluvial taphonomic processes: Models and experiments, *in* A. K. Behrensmeyer, and A. Hill, eds., Fossils in the making: Chicago, University of Chicago Press, pp. 156–181.

Hill, A., 1979, Disarticulation and scattering of mammal skeletons: Paleobiology, v. 5, pp. 261–274.

Holtzman, R. C., 1979, Maximum likelihood estimation of fossil assemblage composition: Paleobiology, v. 5, pp. 77–89.

Johnson, D. D., J. Kappelman, and M. Fortelius, 1996, Advantages of microcomputer use for cataloging of fossil specimens: American Journal of Physical Anthropology Supplement Abstracts, v. 22, p. 132.

Kay, R. F., 1984, On the use of anatomical features to infer foraging behavior in extinct primates, *in* P. Rodman, and J. Cant, eds., Adaptations for foraging behavior in non-human primates: New York, Columbia University Press, pp. 21–53.

Kleiber, M., 1932, Body size and metabolism: Hilgardia, v. 6, pp. 315–353.

Maiorana, V. C., 1990, Evolutionary strategies and body size in a guild of mammals, *in* J. Damuth, and B. J. McFadden, eds., Body size in mammalian paleobiology: Estimation and biological implications: Cambridge, Cambridge University Press, pp. 69–102.

Marshall, F., and T. Pilgram, 1993, NISP vs. MNI in quantification of body-part representation: American Antiquity, v. 58, pp. 261–269.

McNab, B. K., 1963, Bioenergetics and the determination of home range size: American Naturalist, v. 97, pp. 133–140.

McNab, B. K., 1990, The physiological significance of body size, *in* J. Damuth, and B. J. MacFadden, eds., Body size in mammalian paleobiology: Estimation and biological implications: Cambridge, Cambridge University Press, pp. 11–24.

Scott, K. M., 1979, Adaptation and allometry in bovid postcranial proportions, Ph.D. dissertation, Yale University.

Scott, K. M., 1990, Postcranial dimensions of ungulates as predictors of body mass, *in* J. Damuth, and B. J. MacFadden, eds., Body size in mammalian paleobiology: Estimation and biological implications: Cambridge, Cambridge University Press, pp. 301–336.

Scott, R. S., J. Kappelman, and J. Kelley, 1999, The paleoenvironment of Sivapithecus parvada: Journal of Human Evolution, v. 36, pp. 245–274.

Shotwell, J. A., 1955, An approach to the paleoecology of mammals: Ecology, v. 36, pp. 327–337.

Shotwell, J. A., 1958, Inter-community relationships in Hemphillian Mid-Pliocene mammals: Ecology, v. 39, pp. 271–282.

Van Valen, L., and R. E. Sloan, 1965, The earliest primates: Science, v. 150, pp. 743–745.

Wilson, E. O., and W. H. Bossert, 1971, A primer of population biology: Sunderland, Massachusetts, Sinauer Associates, 192 pp.

Index

The suffix *t* on a page number indicates a table; *f* indicates a figure.

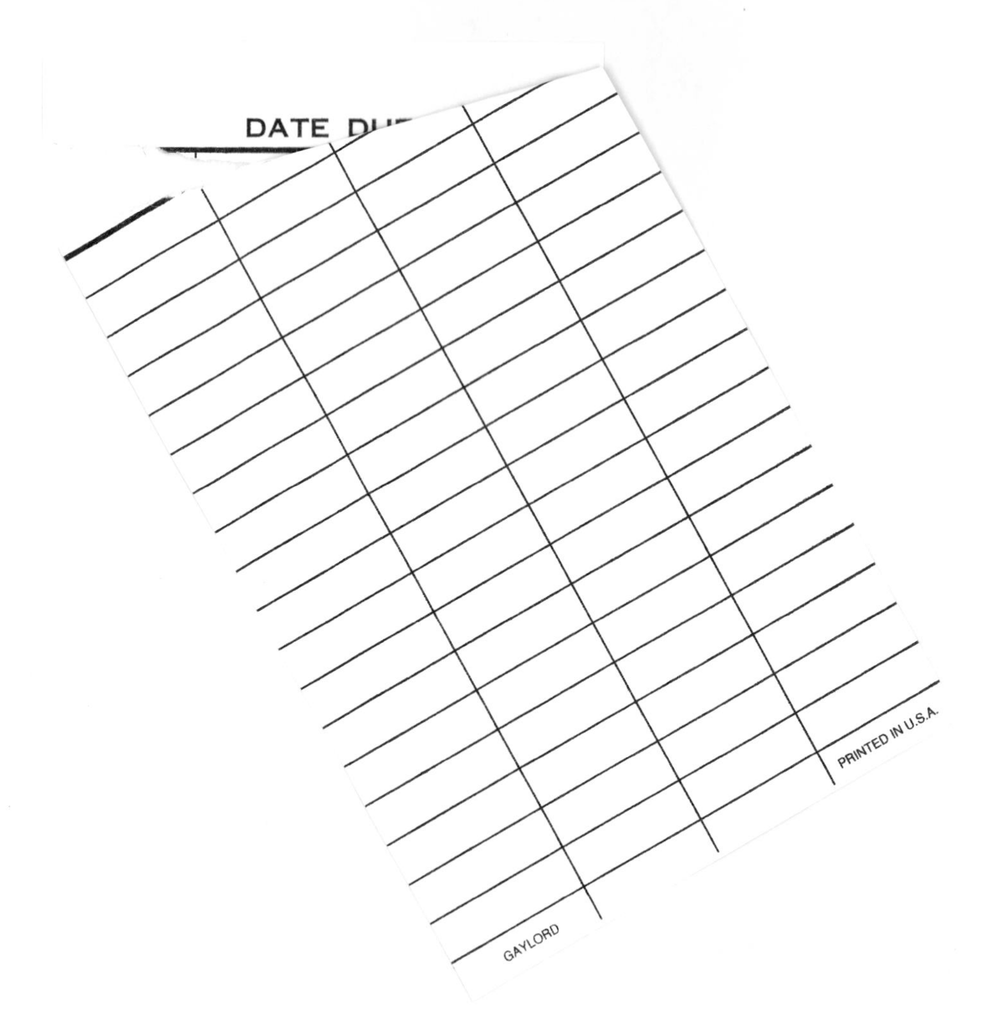
DATE DU
GAYLORD
PRINTED IN U.S.A.